Predictive soil mechanics

C. P. Wroth

Predictive soil mechanics

Proceedings of the Wroth Memorial Symposium held at St Catherine's College, Oxford, 27–29 July 1992

Edited by G. T. Houlsby and A. N. Schofield

Thomas Telford, London

Published by Thomas Telford Services Ltd, Thomas Telford House, 1 Heron Quay, London E14 4JD

Distributors for Thomas Telford books are
USA: American Society of Civil Engineers, Publications Sales Department, 345 East 47th Street, New York, NY 10017-2398
Japan: Maruzen Co. Ltd, Book Department, 3-10 Nihonbashi 2-chome, Chuo-ku, Tokyo 103
Australia: DA Books and Journals, 648 Whitehorse Road, Mitcham 3132, Victoria

First published 1993

A catalogue record for this book is available from the British Library

Classification
Availability: Unrestricted
Content: Collected papers
Status: Refereed
User: Geotechnical engineers and researchers

ISBN: 0 7277 1916 5

© Authors, 1992

All rights, including translation, reserved. Except for fair copying, no part of this publication may be reproduced, stored in a retrieval system or transmitted in any form or by any means, electronic, mechanical, photocopying or otherwise, without the prior written permission of the Publications Manager, Publications Division, Thomas Telford Services Ltd, Thomas Telford House, 1 Heron Quay, London E14 4JD.

Papers or other contributions and the statements made or opinions expressed therein are published on the understanding that the author of the contribution is solely responsible for the statements made and opinions expressed in it and that its publication does not necessarily imply that such statements and or opinions are or reflect the views or opinions of the organisers or publishers.

Typeset in Great Britain by Verse Graphics, Colchester
Printed in Great Britain by Redwood Books, Trowbridge, Wiltshire

Preface

This volume forms the proceedings of the Wroth Memorial Symposium, held at St Catherine's College, Oxford, from 27th to 29th July 1992. The purpose of the symposium and of this volume is to provide a lasting tribute to C. P. Wroth, our friend and colleague of many years, who had been Master of Emmanuel College, Cambridge for only one term at the time of his death on 3rd February 1991, after a short illness. Peter had spent most of his early career in the soil mechanics group at Cambridge. Prior to his return to Cambridge in 1990, he had been for ten years a most effective Head of the Department of Engineering Science at Oxford University, as well as having led the soil mechanics research group there. Peter Wroth's contributions to soil mechanics are recorded elsewhere, notably in the obituary in *Géotechnique* (volume 41, number 4, pages 631-635), which includes a full list of his publications.

The style of the symposium was a reflection of Peter's lively interest in all aspects of soil mechanics, and in particular his active encouragement of younger engineers. The topic of the symposium,"Predictive soil mechanics", was chosen to reflect Peter's own wide interests, and his appreciation that geotechnical engineers must develop techniques for predicting the performance of real structures.

The symposium included a number of special features. On the second day a lighthearted debate was held, and the proposal that "Modern in situ testing has made laboratory testing redundant" was firmly rejected, in spite of a spirited defence. Towards the end of the symposium a session was devoted entirely to predictions which had been made of the pile downdrag experiment at Bothkennar. We are grateful to Dr John Little for organising the prediction exercise and comparing the predictions of five teams with the measurements. His report appears in this volume after the main group of symposium papers.

This volume contains 49 papers, and it is a most fitting tribute to Peter's work that these were contributed by over 100 authors from many countries around the world. We are most grateful to all the authors for their efforts in the preparation of the papers.

This volume is organised as follows. The first four papers are introductions to the main themes of the symposium: soil properties and their measurement,

PREDICTIVE SOIL MECHANICS

prediction and performance, and design methods. The following 43 contributions form the main body of the symposium papers. They are arranged alphabetically by author, since many of the papers contribute to more than one of the themes of the symposium. The final two papers are the report on the Bothkennar prediction exercise, and a paper by the co-reporters summarizing their impressions, as younger engineers, of the symposium.

We wish to thank all those who contributed to the success of the symposium, and in particular our colleagues on the organising committee: Dr Harvey Burd, Professor John Burland, Dr Robert Mair, Professor David Muir Wood, Dr Dick Parry and Dr Gilliane Sills. A photograph and list of delegates is included at the end of this volume, and this indicates the names of the many people who acted as reporters, co-reporters, session chairmen, co-chairmen, discussion organisers and those who organised the debate: we wish to thank them all for their support. Finally we are most grateful to Mrs Rachel Wroth and other members of Peter's family, and also several friends, for their interest and participation in the symposium.

Guy Houlsby
University of Oxford,
Department of Engineering Science

Andrew Schofield
University of Cambridge,
Department of Engineering

June 1993

Contents

Theme report: The key material properties in geotechnical engineering need to be measured, instead of being specified.
J. GRAHAM ... 1

Theme report: Soil properties and their measurement.
C. SAGASETA ... 19

Theme report: Prediction and performance. H. J. BURD and G. T. HOULSBY ... 38

Theme report: Design methods. M. D. BOLTON ... 50

Field, in situ and laboratory consolidation parameters of a very soft clay. M. S. S. ALMEIDA and C. A. M. FERREIRA ... 73

Post-liquefaction settlement of sands. K. ARULANANDAN and J. SYBICO ... 94

A note on modelling small strain stiffness in Cam clay.
J. H. ATKINSON ... 111

Towards systematic CPT interpretation. K. BEEN and M. G. JEFFERIES ... 121

Determining lateral stress in soft clays. J. BENOIT and A. J. LUTENEGGER ... 135

The interpretation of self-boring pressuremeter tests to produce design parameters. B. G. CLARKE ... 156

An assessment of seismic tomography for determining ground geometry and stiffness. C. R. I. CLAYTON, M. J. GUNN and V. S. HOPE ... 173

The behaviour of granular soils at elevated stresses.
M. R. COOP and I. K. LEE ... 186

PREDICTIVE SOIL MECHANICS

Geotechnical characterisation of gravelly soils at Messina site: selected topics. R. CROVA, M. JAMIOLKOWSKI, R. LANCELLOTTA and D. C. F. LO PRESTI ... 199

A critical state constitutive model for anisotropic soil. M. C. R. DAVIES and T. A. NEWSON ... 219

The bearing capacity of conical footings on sand in relation to the behaviour of spudcan footings of jackups. E. T. R. DEAN, R. G. JAMES, A. N. SCHOFIELD, F. S. C. TAN and Y. TSUKAMOTO ... 230

Co-rotational solution in simple shear test. G. DE JOSSELIN DE JONG ... 254

Parameter selection for pile design in calcareous sediments. M. FAHEY, R. J. JEWELL, M. F. RANDOLPH and M. S. KHORSHID ... 261

Mobilisation of stresses in deep excavations: the use of earth pressure cells at Sheung Wan Crossover. R. A. FRASER ... 279

Consolidation of an accreting clay layer: solutions via the wave equation. R. E. GIBSON ... 293

The prediction of surface settlement profiles due to tunnelling. M. J. GUNN ... 304

Predicted and measured tunnel distortions associated with construction of Waterloo International Terminal. D. W. HIGHT, K. G. HIGGINS, R. J. JARDINE, D. M. POTTS, A. R. PICKLES, E. K. DE MOOR and Z. M. NYIRENDA ... 317

Modelling of the behaviour of foundations of jack-up units on clay. G. T. HOULSBY and C. M. MARTIN ... 339

Development of the cone pressuremeter. G. T. HOULSBY and N. R. F. NUTT ... 359

Predicting the effect of boundary forces on the behaviour of reinforced soil walls. R. A. JEWELL, H. J. BURD and G. W. E. MILLIGAN ... 378

Some thoughts on the evaluation of undrained shear strength for design. F. H. KULHAWY ... 394

CONTENTS

Attempts at centrifugal and numerical simulations of a large-scale in situ loading test on a granular material.
O. KUSAKABE, Y. MAEDA, M. OHUCHI and T. HAGIWARA 404

The behaviour of a displacement pile in Bothkennar clay.
B. LEHANE and R. J. JARDINE 421

Three-dimensional tests on reconstituted Bothkennar soil.
P. I. LEWIN and M. A. ALLMAN 436

Prediction of clay behaviour around tunnels using plasticity solutions. R. J. MAIR and R. N. TAYLOR 449

Settlement predictions for piled foundations from loading tests on single piles. A. MANDOLINI and C. VIGGIANI 464

In situ determination of clay stress history by piezocone model. P. W. MAYNE 483

Selection of parameters for numerical predictions.
D. MUIR WOOD, N. L. MACKENZIE and A. H. C. CHAN 496

Use of field vane test data in analysis of soft clay foundations.
H. OHTA, A. NISHIHARA, A. IIZUKA and Y. MORITA 513

Linear and nonlinear earthquake site response. M. J. PENDER 529

Observed and predicted response of a braced excavation in soft to medium clay. S. RAMPELLO, C. TAMAGNINI and
G. CALABRESI 544

Seismic and pressuremeter testing to determine soil modulus.
P. K. ROBERTSON and R. S. FERREIRA 562

Prediction and performance of ground response due to construction of a deep basement at 60 Victoria Embankment.
H. D. ST JOHN, D. M. POTTS, R. J. JARDINE and K. G. HIGGINS 581

An investigation of bearing capacity and settlements of soft clay deposits at Shellhaven. F. SCHNAID, W. R. WOOD,
A. K. C. SMITH and P. JUBB 609

Development and application of a new soil model for prediction of ground movements. B. SIMPSON 628

PREDICTIVE SOIL MECHANICS

Stability of shallow tunnels in soft ground. S. W. SLOAN and
A. ASSADI ... 644

Sliding resistance for foundations on clay till. J. S. STEENFELT ... 664

Sampling disturbance with particular reference to its effect
on stiff clays. P. R. VAUGHAN, R. J. CHANDLER, J. M. APTED,
W. M. MAGUIRE and S. S. SANDRONI ... 685

Development and application of a critical state model for
unsaturated soil. S. J. WHEELER and V. SIVAKUMAR ... 709

Predicting earthquake-caused permanent deformations of
earth structures. R. V. WHITMAN ... 729

The effects of installation disturbance on interpretation of in
situ tests in clay. A. J. WHITTLE and C. P. AUBENY ... 742

Shear modulus and strain excursion in the pressuremeter test.
R. W. WHITTLE, J. C. P. DALTON and P. G. HAWKINS ... 768

Analysis of the dilatometer test in undrained clay. H. S. YU,
J. P. CARTER and J. R. BOOKER ... 783

Predictions associated with the pile downdrag study at the
SERC soft clay site at Bothkennar, Scotland. J. A. LITTLE
and K. IBRAHIM ... 796

Recollections from the Wroth Memorial Symposium:
Predictive Soil Mechanics. S. E. STALLEBRASS,
S. M. SPRINGMAN and J. P. LOVE ... 819

Author index ... 835

Subject index ... 836

Symposium delegates ... 838

The key material properties in geotechnical engineering need to be measured instead of being specified

J. GRAHAM, University of Manitoba, Winnipeg, Canada

More than most branches of engineering, geotechnical engineering requires measurement of soil properties before design can be undertaken. The paper reviews five principal themes that are encountered in papers to the Symposium: laboratory testing, in situ testing, importance of stiffness and displacements, development of improved soil models, and effects of disturbance. It discusses these in a general framework of soil modelling, and suggests other areas where further work is needed.

Introduction
Though we sometimes forget how it distinguishes us from colleagues in other branches of engineering, the measurement of material properties is an essential feature of geotechnical engineering. Papers to the symposium dealing with the measurement of soil properties reflect a number of themes of active research activity, and suggest other topics where further work is needed.

I have taken the title for this theme report from the review paper by Wroth and Houlsby to the 11th International Conference of ISSMFE at San Francisco in 1985. In preparing the report I have been reminded of the clarity of Peter Wroth's writing, and the directness with which he expressed his ideas.

At the request of the symposium organizers, this is neither a General Report nor a State-of-Art Report. Instead, I have tried to relate the papers to my own interests and experience. This has suggested comments and questions which I hope will stimulate a closer examination of the papers themselves.

The need to measure soil properties
Why is the measurement of material properties so important in geotechnical engineering, and why does it have higher priority than in other branches of engineering?

Geological materials are inherently complex. They are particulate, and often highly variable. They are multi-phase. Their stress–strain behaviour is markedly non-linear, and usually inelastic. They have a strong tendency to change volume if drainage is permitted. The duration of the volume changing depends on the rate at which pore water pressures can equilibrate. Other time-dependent phenomena such as creep, relaxation and ageing are present.

In preparing advice for clients, geotechnical engineers must assess site variability, the material properties that are relevant for the problem being addressed, how these properties vary with position and with time, what analysis should be adopted, and what instrumentation should be installed to check that the expected levels of performance are being attained.

Wroth and Houlsby (1985) suggested that commonly used test programs are unlikely to provide sufficient data for highly complex models because

(a) the intrinsic accuracy of the tests may not be high due to sample disturbance
(b) considerable scatter may exist due to geological variations
(c) boundary conditions on the tests may be insufficiently controlled to allow unequivocal interpretation of the results.

Which properties?

The properties to be measured for a given project will fall into one of four categories which:

(a) categorize or classify the soils at the site, and at the same time define their variability
(b) identify compressibilities or stiffnesses
(c) define strength parameters
(d) describe rates at which water can move into or out of the soil.

If it is only necessary to categorize the material or log its variability, then non-quantitative empirical tests such as consistency limits are sufficient. However, most projects require quantitative, numerical data that can be used to synthesize an understanding of the behaviour of the whole deposit under envisaged loading, whether from construction, gravity, climate, or chemistry changes. The properties needed to permit this synthesis may be measured from samples recovered from the field and tested in the laboratory, or by using tools inserted directly into the deposit. Both procedures have inherent difficulties.

Laboratory testing applies known boundary stresses, displacements, and drainage conditions to elements of the soil. Such tests claim to produce direct measurements of the needed properties. Often, however, the stress states are indeterminate, and fixed in direction. The process may be criticized because of sampling disturbance; because the element

KEY MATERIAL PROPERTIES IN GEOTECHNICAL ENGINEERING

may not be fully representative of the deposit; and because the stress paths imposed in the test may not reflect those in the field in magnitude, direction or rotation.

As our knowledge of the behaviour of real soils increases, so our appreciation of the inadequacy of conventional laboratory testing grows. (Wroth, 1984)

In situ test probes are inserted into the ground either by direct pushing, or using self-boring technology. They then measure combinations of forces, displacements and pore water pressures. The volume of material tested by these probes is not well defined, nor are the boundary conditions of pressure, displacement or drainage. All in situ tests cause some measure of disturbance to neighbouring soil.

Any in situ test, when considered as a boundary value problem, is beset with difficulties. The boundaries of the problem are unknown and uncontrolled. (Wroth, 1984)

As a result, needed soil properties must be interpreted from analytical models of how the soil is being stressed. Values of the properties depend on the quality of the model.

Generally, in situ tests can be used for

(a) logging site variability
(b) determining parameters that describe the conditions in the deposit in terms of OCR, density index or, preferably, state parameter ψ
(c) measurements of strength and compressibility using parameters such as s_u, ϕ, E or G.

These three objectives are essentially independent, and not all in situ tools can be successfully used in all three. Both (b) and (c) require validation against field performance and also reliable mathematical models, if the tools are to be used confidently in all soils.

Two more features of Peter Wroth's work will be useful as we begin to look at themes that will arise in the symposium. First, is the value of solutions that are as simple as possible, while avoiding the simplistic. Second, is a rejection of raw empiricism. He suggested that successful relationships should be

(a) based on a physical appreciation of why the properties should be related
(b) set against a background of theory, even if idealized
(c) expressed in dimensionless variables so that advantage can be taken of the scaling laws of continuum mechanics.

Understanding the physics of soil behaviour leads to confidence in identifying proper (rather than spurious) relationships between variables. Theoretical modelling allows clearer understanding of confusing

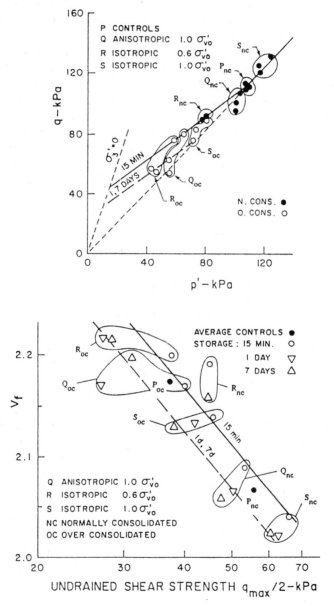

Fig. 1. Undrained shear strengths following storage and reconsolidation (Graham and Lau, 1988)

data, and the ability to extrapolate to other soils or problem types. Dimensionless variables allow insights and comparisons that are not possible when only dimensional results are used. Papers to the symposium use existing models to explain laboratory or field observations, and develop some interesting new models.

One of the most widely used models is Critical State Soil Mechanics (CSSM), to which Peter Wroth made significant contributions, both in its development and its application. Particularly important in understanding soil behaviour is an emphasis on combining q, p' plots with V, ln (p') plots in 'compression space'. Figure 1 shows some work with S.L.K. Lau on understanding disturbance due to stress release, and the reconsolidation procedures that are needed for best recovery of original undrained strengths. In this case, compression space has been represented by plotting V against log (s_u). Rationalization of the data was only possible when they were examined in compression space.

Elasto-plasticity and soil behaviour

Arising from CSSM is the elastic-plastic family of models which separate strains into elastic and plastic components. Models such as Modified Cam Clay (MCC) (Roscoe and Burland, 1968) and its subsequent developments, are inherently more general than earlier models which simply described non-linear behaviour. Modified Cam Clay was developed from tests on reconstituted soils.

Studies of the natural, soft clays that often cause problems with embankments and storage tanks have led to a realization that their behaviour is in some ways simpler than that of reconstituted clays (Bjerrum and Kenney, 1967; Mitchell, 1970; Crooks and Graham, 1976). Combinations of ageing and light overconsolidation produce microstructures that have relatively stiff responses to initial loading, but then become less stiff when some critical combination of stresses is reached. Gradually, the idea of the yield locus was realized, and its relationship with Cam clay modelling identified. The biggest difference between real clay and Cam clay may be that the elasticity is anisotropic, rather than isotropic (Muir Wood and Graham, 1990).

Modified Cam Clay predicts shear strains relatively poorly, especially when clay is heavily overconsolidated. This led to increased attention to the way in which shear stiffness varies with shear strain (Simpson et al., 1979; Burland, 1989). Non-linear models such as the hyperbolic model of Duncan and Chang (1970) typically require more testing and more parameters (Yin et al., 1990).

I have found the general framework of elastic-plastic soil mechanics useful for rationalizing initially confusing laboratory data. In addition to studies on yielding and stress release disturbance mentioned earlier, I have worked with J.-H. Yin on developing Bjerrum's 'time lines' into an elastic visco-plastic model that grows naturally out of Cam clay. My current work is on the dense 50/50 mixture of quartz sand and Na-bentonite (known as 'buffer') that has been proposed for the Canadian nuclear waste management program. The work is exploring extensions of elastic-plastic soil mechanics that include swelling and thermally induced straining. Further work will require an understanding of unsaturated behaviour, probably based on elastic-plastic modelling of unsaturated soils, such as that done by Alonso et al. (1990), and now by Wheeler and Sivakumar at this symposium. (References listed subsequently without dates are for papers to the symposium.)

My reason for mentioning this work is to emphasize our need for conceptual and qualitative models to assist our understanding of soil behaviour. Modelling will become a recurring theme in this report.

Principal themes

So far, we have thought about how the measurement of soil properties is important in geotechnical engineering. In what way is this reflected in papers to the symposium?

Almost two-thirds of the papers deal directly with laboratory or in situ measurement of soil properties, often in connection with design procedures or full-scale performance. The topics covered are wide-ranging, but allow identification of several themes which will be discussed in following sections. Sixteen papers address the important question of soil stiffness and displacements. There are papers on generalized stress paths, flow rules and plastic potentials, strain-based models, unsaturated soil mechanics, seismic tomography, centrifuge testing, and specially developed laboratory tests. The papers deal more with clays than sands, and rather more with laboratory testing than in situ testing. A small number of papers deal with localized soil types — residual granitic sand from Hong Kong, pyroclastic sands from Italy and Japan, and weakly cemented calcareous sand from Australia.

Topics which currently receive considerable attention in geotechnical engineering, but which are not represented here, include environmental and cold regions geotechnology, slopes and dams engineering, time dependency, constitutive modelling, and numerical analysis.

Papers to the symposium suggest five principal themes which we will examine in turn:

(*a*) laboratory testing
(*b*) in situ testing
(*c*) importance of stiffnesses and displacements
(*d*) development of improved soil models
(*e*) effects of disturbance on the measurement of soil properties using laboratory samples and in situ devices.

Laboratory testing

We saw earlier that understanding laboratory data requires conceptual models of how soils should behave under the conditions of testing. Several authors have contributed papers that assist with this. For example, de Josselin de Jong presents an interesting development of some earlier work he did with Wroth on interpreting direct shear tests; and Gibson discusses the consolidation behaviour of accreting clay layers.

Conceptual models must also be accompanied by detailed procedures which will produce consistent interpretations from measured data. It is good, therefore, to note a new approach by Muir Wood et al. who use data from a variety of test types to optimize parameter selection for MCC modelling. However, the procedure itself is general, and may also be applicable to the interpretation of other tests, including in situ tests such as CPTU and PMT.

Several of the papers on in situ testing deal directly with methods for interpreting desired information from measured data. In comparison, the papers on laboratory testing are generally less clear in their description of the technology involved, perhaps because of length restrictions.

It may be that laboratory testing is now seen as a mature technology. However, it is important to ensure that skills learned in the past are not neglected. Just recently, we surveyed current practice from 34 laboratories regarding the selection of material for drainage strips in triaxial testing. Materials being used included paper towels and blotting paper, as well as the hard filter papers that are normally recommended. The softer papers may simply not transfer water to the end filter stones under the pressures and durations of many tests.

Strength, stiffness, yielding and plastic flow all have to be examined using consistent, appropriate techniques that can provide information for the conceptual model being used. For example, in yielding studies, there is little point in examining q against ε_s plots for isotropic consolidation tests. Similarly, flow rules and plastic potentials should be examined using only the plastic component of straining in appropriate stress plots. It is not clear that this has always been done.

In situ testing

In situ testing is increasingly used to complement laboratory tests for design. However, we are probably not yet at a stage where models for interpreting in situ penetration tests have been definitively established. This means that the accuracy of derived soil properties is not completely assured.

Cone Penetration Test (CPT) and Piezocone Test (CPTU)

The cone measures site variability probably better than any other tool. It can generate information about the initial state of a soil, and a preliminary estimate of some design parameters for all soil types. Six papers show that the instrumentation and technology for CPT is currently achieving consensus in many areas. The importance of end area corrections, and the presentation of data in dimensionless forms are widely accepted.

While testing technology may be established, modelling of the test itself is still receiving attention. Two papers (by Been and Jefferies, and by Mayne) present techniques for evaluating OCR in clays. Both papers use critical state models but different combinations of the three measurements q_t, f, and u_{bt} that can be obtained from the piezocone.

New models of cone penetration testing continue to appear, but it is not yet clear that consensus has been reached on the modelling of undrained shear strength or overconsolidation ratios.

Pressuremeter (PMT)

Principal themes in the nine papers on pressuremeter testing (PMT) include the assessment of horizontal stresses, the interpretation of shear stiffness, and disturbance caused by insertion.

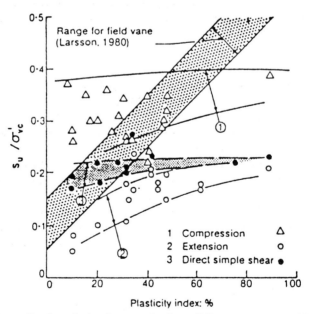

Fig. 2. Normalized undrained strengths from different test types (Graham et al., 1984)

KEY MATERIAL PROPERTIES IN GEOTECHNICAL ENGINEERING

The measurement of horizontal stress has been called by Jamiolkowski the 'black hole' of geotechnical engineering. The Self-Boring Pressuremeter (SBPT) appears to give lower bound values for horizontal stress, although experience and judgement are required (Benoît and Lutenegger). This clearly has important implications for the design of braced excavation systems. It may also influence design choices for embankments on soft ground, particularly if stage construction is needed.

It is fairly clear that field vane strengths can be used for embankment stability in soft lean clays, but not in plastic clays (Fig. 2). However, it is questionable whether there is yet an equivalent understanding of when pressuremeter strengths can be used reliably. Some earlier modelling began to approach this question, but further work may be required.

The important question of measuring shear stiffnesses by PMT and SBPT will be discussed later.

Two papers describe hybrid tools, the seismic cone (SCPT) which Robertson and Ferreira recommend to complement information from SBPM testing; and the cone pressuremeter (CPMT) which combines a cone and a displacement pressuremeter (Houlsby and Nutt).

These combination instruments widen the range of data collected and offset the weaknesses of one instrument against the strengths of another. They allow measurements of strength and stiffness in both sand and clay. If we now have seismic cones and cone pressuremeters, when do we get seismic cone pressuremeters? While it is exciting to see new tools being developed, care must continue to be taken to retain reliability and effectiveness.

Dilatometer Tests (DMT)

With three papers to the symposium, dilatometer testing has received less attention than other in situ tests. In the past, interpretation of DMTs has been mostly empirical. It is therefore valuable to have papers which use well-defined soil models in analyses using appropriate boundary conditions (Whittle and Aubeny; Yu, Carter and Booker). The papers suggest that the strength, rigidity index G/s_u and initial horizontal stress all influence lift-off pressures. Like others, I have questioned whether dilatometer testing can produce usable results from membrane movements that are small compared with the zone of disturbance. Papers to the symposium appear to suggest that the dilatometer will remain basically a logging tool, with interpretations based on local experience and empirical relationships.

Dynamic penetration testing

Standard Penetration Testing (SPT) is not represented in the papers. However, Crova et al. describe large-diameter penetration tests (LPT) in two different gravels from the Straits of Messina. The materials are broadly similar in grading and mineralogy, yet show differences in behaviour due to different geological ageing. Several forms of large diameter dynamic tests now exist in North America and in Asia, where they are used in gravelly deposits, tills, and many residual soils which cannot be readily explored using common in situ tests. It may now be appropriate to begin moving towards a standardized LPT.

Large scale testing
A number of papers describe the use of large scale models or elements of full scale structures to provide information for surface footings or pile foundations. These include plate loading tests (Kusakabe et al.) and full

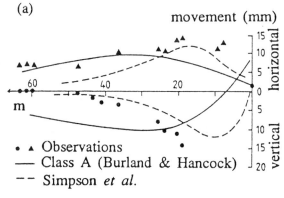

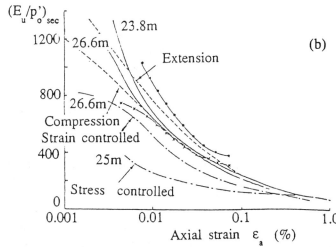

Fig. 3. (a) Observed and predicted ground surface displacements, New Palace Yard (Simpson et al., 1979); (b) Normalised secant stiffness for unweathered London clay (Hight et al., this volume)

scale load tests at Shellhaven (Schnaid et al.) for very large moving loads weighing up to 8 000 kN. Both of these projects were accompanied by numerical modelling so the mechanisms being observed could be well understood. A second group of large scale tests deals with piles. Examples include Constant Normal Stiffness (CNS) and Rod Shear Test (RST) modelling of pile–soil friction development in calcareous soils (Fahey, Jewell and Randolph); and instrumented pile testing in soft organic clay, where Lehane and Jardine report patterns of behaviour unlike those predicted by existing theories.

We can perhaps include here three projects that used centrifuge testing to model field-scale tests. They examine projects that are known to depend strongly on size and stress level – settlements and capacity of footings on sand (Kusakabe and Dean, with their respective coworkers); and liquefaction settlements on sand (Arulandam and Sybico).

Evaluating in situ soil structure still remains a difficult question which geotechnical engineering has traditionally approached by examining 'point' information through sampling, or 'line' information through logging probes such as CPT or down-hole resistivity. I believe I see additional effort being directed towards 'spatial' information, some of it using resistivity and tomographic techniques such as those outlined by Clayton, Gunn and Hope, and some using various forms of seismic probing (Clark and Guigné, 1988; Robertson and Ferreira). These approaches appear to be of increasing interest and importance as we move towards a greater involvement in geo-environmental problems.

Stiffness and displacements

Soil stiffness was a main theme at the European regional conference in Florence in 1991, and appears also in many of the papers to the symposium. The theme subdivides into three topic areas:

(a) a greater awareness of non-linear stiffness in both laboratory and in situ testing
(b) the development of hyperbolic non-linear finite elements, particularly for excavations
(c) the development of kinematic hardening models.

We will deal briefly with the first two topics here, and with the third in the following section. More detailed treatment of this topic can be found in the accompanying theme report by C. Sagaseta.

Simpson et al. (1979) showed that displacements round large excavations were poorly modelled using linear isotropic elasticity (Fig. 3a). They emphasized the importance of the non-linear response of soils to applied stresses, with much higher stiffnesses being associated with small strains, and lower stiffnesses with larger strains. The non-linearity of stress–strain response has significant influence on soil structure interaction, stress distributions in the soil mass, and displacement profiles around loaded areas and excavations (Jardine et al., 1986). Unless small-strain non-linearity is recognized, the interpretation of field measurements and in situ tests may be confusing and misleading (Burland, 1989).

This understanding led to an awareness that different testing procedures measure stiffnesses at different strain levels (Fig. 3b). Valuable results have been presented to the symposium that show agreement between shear moduli from laboratory and pressuremeter tests (Hight et al.). Additional comparisons of this type would be valuable. Stiffnesses at very low strains are best measured using seismic or resonant column testing. Thus, modelling the cyclic loading of offshore piles in calcareous sand appears to need high values of shear modulus that can be obtained only by dynamic methods (Fahey et al.). In triaxial testing, measurements of local strains directly on the specimen give higher stiffnesses at lower strains (Burland, 1989).

The pressuremeter can be used effectively to measure shear stiffness. However, since stiffnesses are strain dependent and decrease with disturbance, SBPT gives higher values of both strength and stiffness than PMT (Robertson and Ferreira; Whittle and Aubeny; Whittle and Dalton; and Houlsby and Nutt). Various techniques are now available for measuring the variation of shear stiffness with shear strain by making use of the reduced disturbance associated with unloading loops in pressuremeter tests.

Two important case studies in the symposium return to the topic with which we started this section; displacements round large excavations in London clay. Hight, St. John and their respective coworkers use moduli that are generally stiffer than those measured in triaxial tests (except when measured with local strain measuring devices), and closer to those derived from pressuremeter tests. It should be noted however that stress paths in the laboratory must follow the field paths if the correct stiffness strain relationships are to be observed.

Selected stiffness parameters depend on recent stress history and current stress state. They therefore appear to fall into the category of hypoelastic models like the hyperbolic modelling of Duncan and Chang (1970). Recent work with J.-H. Yin has shown that calibrations of this type are generally only valid for single values of OCR in clay, or state parameter ψ in sand.

These non-linear stiffness models appear to be used mostly in applications where the stresses decrease due to engineering activity, and the OCR is unchanged. However they have also been used with some success for embankments and foundations (R. Jardine, personal communication).

Most of the work that has been reported on strain-dependent stiffness is for stiff clays. In soft-to-firm lightly overconsolidated clays, plastic straining becomes relatively more important, and the benefits of elastic-plastic modelling are more obvious. In this case, the behaviour assumed by MCC and generically similar models may provide acceptable capabilities without going to the additional complexity of non-linear modelling.

Improved soil models

The variable stiffnesses that were examined in the previous section can also be modelled by inserting one or more regions of higher stiffness inside previously defined yield or failure envelopes (Al Tabaa and Wood, 1989; Burland, 1989; Velloso et al., 1989). The phenomenon is possibly one that is frequently present, even in medium-stiff clays. For example, a study of freeze-thaw effects at low stresses identified 'early yielding' in which initial straining was essentially linear, and was followed by non-linear response leading finally to failure (Fig. 4). Its significance was unclear at the time of writing.

Atkinson, in this symposium, suggests there is initially a region of small-strain elasticity. There is also a region of large, irrecoverable plastic strains outside a state boundary surface. Between these is a region representing the history of the soil, in which the stiffness decreases with

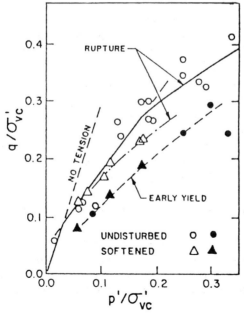

Fig. 4. 'Early yield' at low p'/σ'_{vc} (Graham and Au, 1985)

strain and the behaviour is highly non-linear. Methods for determining the different kinematic surfaces have not yet been finally established, nor is it clear what strain ranges define reversible and non-reversible behaviour. Calibration of these models would appear to be complex and time consuming, although they appear to offer important new insights in soil behaviour.

Non-linear modelling and kinematic bounding surfaces are alternative ways of solving the same problem. In principle, I suspect that solutions based on kinematic hardening of an elastic plastic framework will prove more useful because of their generally stronger basis in mechanics.

Soil deposits usually have $\sigma'_h \neq \sigma'_v$. These anisotropic in situ stresses must produce an anisotropic fabric and anisotropic behaviour.

Several of the symposium papers identify initial anisotropy of stress and fabric as being important in practice. Examples include papers by Fraser on lateral earth pressures on a deep braced excavation in Hong Kong, by Whittle and Aubeny in examining disturbance due to installation of in situ tools, and the two case studies on excavations in London clay described previously.

As mentioned earlier, anisotropy has been considered a reason for differences between the elliptical shapes of yield loci in MCC and the shapes of yield loci measured in natural clays. (The apparent symmetry of these latter loci about the K_0-line in s', t-plots may be an artefact of mapping procedures, Graham et al., 1988.) It would seem appropriate therefore to develop elastic-plastic models using rather different assumptions than MCC so that the behaviour of natural clays is more realistically included. Davies and Newson propose to model the anisotropy of natural soils using a yield locus that is symmetric about the K_0-line, and a non-associated plastic potential. The model, which is similar to that proposed by Mouratidis and Magnan (1983), requires one additional parameter to provide an anisotropic extension of MCC. It would be helpful to have additional information on how the yield locus, and especially the plastic potential, is derived from test data. Further clarification of the different capabilities of non-associated models and multiple-surface hardening models would be welcome. Triaxial testing can reasonably be criticized for the lack of generality of its stress states. The problem of generalizing from axisymmetrical to 3D stress space is not trivial in the laboratory or in the computer. The paper by Lewin and Allman on Bothkennar clay is therefore welcome. Their tests suggest that yield (failure) loci for this fairly representative soft clay can be modelled by the Lade–Duncan failure envelope. It will be interesting to see this work develop towards a modelling capability, and if it can be placed in an elastic-plastic framework. It would also be interesting to investigate yield surfaces for non-failing stress paths ($\eta_y < M$), that is, the surface of revolution of so-called 'cap' models.

Geotechnical engineers usually view stresses first, and then the resulting strain response. Yet the performance of our structures is often controlled by their displacement performance. It is therefore important that Simpson has put the telescope to the other eye, and noted that soils generally offer less resistance to continued straining in the direction they were previously following, than they do when the direction is changed. His model was originally derived for stiff clays like London clay, yet it has also been used in anisotropic soft clay in Singapore. Some commonly used assumptions, plus an interesting additional assumption that stress changes are only produced by elastic strains, allow prediction of clay behaviour in a wide range of states and stress paths. Credible values of K_0, s_u and E plotted against γ, have been modelled in both normally consolidated and overconsolidated states. The model is ideally suited for finite element analysis because it derives stress increments from strain increments.

Some time ago, I began to explore size effects in sands using critical state modelling in stress characteristics solutions. Until recently, there has been little evidence of normal consolidation lines in sands, and researchers such as Sladen and Oswell (1989) have chosen to work with the CSLs because they are easier to determine. It is therefore interesting to read that Coop has done high-pressure testing on sands and measured clearly defined normal consolidation lines, and to consider how this will influence critical state modelling in sands.

Disturbance

Considerable attention has been paid to soil disturbance in the symposium papers. Some projects used piston sampling, and others block samplers such as the Laval sampler and Sherbrooke sampler. However, it is important to remember that sampling is not the only process that causes disturbance, and care must also be taken during storage, extrusion, trimming and installation (Vaughan et al.). In my own experience, extra care during trimming and installation into test cells leads to improved results. When dealing with expansive clays such as Lake Agassiz clay from Winnipeg, or the dense sand-bentonites that I am now testing, we cannot allow early access to water at low stresses, otherwise non-typical stress–strain curves are recovered. It is also important not to use triaxial consolidation stresses that will damage the soil fabric, especially the fabric of soft or sensitive soils. Indeed, it was the Scandinavian and Canadian practice of only returning soft post-glacial clays to their in situ stresses that allowed development of the concept of yielding in natural clays.

Pursuing these ideas, we will see that Kulhawy and Mayne both use CIUC tests as the test of reference for undrained strengths. They then compare strengths from other tests with these. Strength depends quite strongly on the reconsolidation conditions (Fig. 1), and so it is necessary to define these conditions carefully if the tests are to be used for reference. Kulhawy presents data on the relationship between s_u and ϕ' from a variety of laboratory tests. As suggested earlier, the value of these data would be further improved if pressuremeter and cone penetrometer data could be added in the way field vane data were included in Fig. 2. Additional modelling of the relationships between different undrained shear strengths is still needed, and the results validated against field experience.

It is sometimes suggested that in situ tests are inherently better than laboratory tests because they examine 'undisturbed' soil. Yet, insertion of in situ tools must also cause some disturbance which can be reduced (but not eliminated) using self-boring technology. As we saw earlier, an important feature of the papers to the symposium is the attention directed to this question. Crova et al. report that tests which induce large straining in the surrounding soil can obliterate the effects of processes like ageing, cementation and other phenomena of early diagenesis which influence soil stiffness at small strains. This must pose doubts about the reliability of correlations between deformation moduli and penetration resistance in soils having well-developed structure. It also re-emphasizes the need for well-documented case studies like those presented to the symposium.

Concluding remarks

As we begin reading the symposium papers, we should perhaps remember our responsibility for providing reliable and cost-effective advice. Recently, we appear to have improved our ability to predict near-field and far-field displacements round large excavations. However, while we can often predict ultimate and indeed yielding conditions in foundations, embankments and slopes, it it not yet clear that we can generally predict settlements, durations of consolidation, or creep effects with equal confidence. Ohta and his co-authors give one of the few examples in the symposium where these factors have been considered.

The situation is even less clear in piled foundations, where it seems that analysis of load capacity is not yet fully established in all cases. Nor is it yet clear in the important question of earthquake engineering where three papers by Whitman, Pender, and Arulandam and Sybico, recent experiences in Mexico, China, the United States, and Australia, and concerns now being examined in Western Canada, suggest a need for a clearer understanding of the behaviour of soil deposits under earthquake loading. Will it be possible to build this understanding on work that has already been done on cyclic loading on offshore structures?

It used to be thought that calculations of displacement could be done assuming constant soil stiffness. Now, it is known that more care has to be taken. Are there other parameters that need closer attention? Hydraulic conductivity is one that comes to mind, because of its influence on consolidation rates, and the increasing importance of waste migration studies.

Much of the preceding discussion has involved conceptual and quantitative models, although we have avoided the details of how these can be developed or their parameters measured. Many soil models exist, and many numerical models, perhaps, are more than justified by the existing data base from instrumented case studies. The challenge now is how to develop ways of describing soil conditions that can be used effectively in analysis, and produce answers that can be used in practice. This will require mutual confidence between those who produce material properties, those who develop numerical models, those who provide advice to clients, and those who monitor field performance.

References

AL TABAA, A. AND WOOD, D.M. (1989) An experimentally based "bubble" model for clay. Proc. NUMOG III, Elsevier Applied Science, pp. 91–99.

ALONSO, A.A., GENS, A. AND JOSA, A. (1990). A constitutive model for partially saturated soils. Géotechnique Vol. 40, pp. 405–430.

BJERRUM, L. AND KENNEY, T.C. (1967). Effect of structure on the shear behaviour of normally consolidated quick clays. Proc. geotech. conf. on shear strength properties of natural soils and rocks, Oslo, Norway, Vol. 2, pp. 19–27.

BURLAND, J.B. (1989). 'Small is beautiful' — the stiffness of soils at small strains. Canadian Geotechnical Journal Vol. 26, pp. 499–516.

CLARK, J. I. AND GUIGNE, J.Y. (1988). Marine geotechnical engineering in Canada. Canadian Geotechnical Journal Vol. 25, pp. 179–198.

CROOKS, J.H.A. AND GRAHAM, J. (1976). Geotechnical properties of the Belfast estuarine deposits. Geotechnique Vol. 26, pp. 293–315.

DUNCAN, J.M. AND CHANG, C.-Y. (1970). Nonlinear analysis of stress and strain in soils. ASCE Journal of Soil Mechs. and Found. Division Vol. 96, pp. 1629–1653.

GRAHAM, J., CROOKS, J.H.A. AND LAU, S.L.K. (1988). Yield envelopes: identification and geometric properties. Geotechnique Vol. 38, pp. 279–300.

JARDINE, R.J., POTTS, D.M., FOURIE, A.B. AND BURLAND, J.B. (1986). Studies of the influence of non-linear stress–strain characteristics in soil-structure interaction. Géotechnique Vol. 36, pp. 377–396.

MOURATIDIS, A. AND MAGNAN, J.-P. (1983). Modèle élastoplastique anisotrope avec écrouissage pour le calcul des ouvrages sur sols compressibles. Laboratoire Central des Ponts et Chaussées, Rapport de recherche LPC 121.

MITCHELL, R.J. (1970). On the yielding and mechanical strength of Leda clays. Canadian Geotechnical Journal Vol. 7, No. 3, pp. 297–312.

MUIR WOOD, D. AND GRAHAM, J. (1990). Anisotropic elasticity and yielding of a natural plastic clay. Int. Journal of Plasticity Vol. 6, pp. 377–388.

ROSCOE, K.H. AND BURLAND, J.B. (1968). On the generalized stress–strain behaviour of 'wet' clay. In Engineering Plasticity (eds. J. Heyman and F.A. Leckie). Cambridge University Press, pp. 535–609.

SIMPSON, B., O'RIORDAN, N.J. AND CROFT, D.D. (1979). A computer model for the analysis of ground movements in London clay. Géotechnique Vol. 29, pp. 145–179.

SLADEN, J.A. AND OSWELL, J.M. (1989). The behaviour of very loose sand in the triaxial compression test. Canadian Geotechnical Journal Vol. 26, pp. 103–113.

VELLOSO, R.C., ASEVEDO, R.F. AND POOROOSHASB, H.B. (1989). Analysis of non-monotonic cubical triaxial tests with an elasto-plastic kinematic hardening constitutive model. Proc. NUMOG III, Elsevier Applied Science, pp. 248–255.

WROTH, C.P. (1984). The interpretation of in situ soil tests. Géotechnique Vol. 34, No. 4, pp. 447–492.

WROTH, C.P. AND HOULSBY, G.T. (1985). Soil mechanics—property characterization and analysis procedures. State-of-Art Report, 11th Int. Conf. Soil Mechs. and Founds. Eng., San Francisco, CA, Vol. 1, pp. 1–54.

YIN, J.-H., SAADAT, F. AND GRAHAM, J. (1990). Constitutive modelling of a compacted sand-bentonite mixture using three modulus hypoelasticity. Canadian Geotechnical Journal Vol. 27, pp. 365–372.

Soil properties and their measurement

C. SAGASETA, University of Cantabria, Santander, Spain

Introduction
The Symposium guidelines include under this Theme

(a) definition of appropriate soil properties
(b) their measurement by in situ or laboratory tests
(c) constitutive models.

This implies a broad scope, which, for comparison, in the last European Conference in Florence was covered by four General Reports (Sessions 1a, 1b, 2a and 2b). Fortunately, the guidelines add that '... not concentrating on fine details either of constitutive models or testing hardware'. On this basis, some significant aspects raised by the papers presented to the Symposium will be commented upon. The topics selected follow a more or less rational order, but do not cover all aspects of soil behaviour; nor do their relative extents express their relative importance. In the above-mentioned General Reports, a complete presentation can be found (Atkinson and Sallfors, 1991; Burghignoli et al., 1991; Sagaseta et al., 1991).

Of the 43 papers presented to the Symposium, 29 deal totally or partially with Theme 1. They are classified in Table 1 according to three main aspects: the feature of soil behaviour analysed, the method of measuring it (laboratory or field tests), and the main emphasis of the paper.

Most of the papers deal with soil behaviour at large strains and failure. However, a significant number of papers are concerned with the soil response at small strains. The precise ranges of strain will be commented upon later. The influence of past stress history continues to be an important point.

Half of the papers deal with laboratory tests. Most of them refer to the triaxial test, with special improvements for precise measurement and control of stress and strain. In two papers, the true triaxial test is used. Two papers address specifically the problem of sampling disturbance and its influence on the measured properties.

In field tests, attention has been paid to the effects of the insertion of the probe, which in some cases is the test itself (cone penetrometer), and in others presents a problem which should be minimized.

Table 1. Symposium papers dealing with soil properties and their measurement

Features of soil behaviour					Method of measurement													Main emphasis of the paper						Author(s)
					Laboratory					Field tests														
SH	SS	MS	LS	PC	S	T	TT	SS	O	I	SBP	CPT	CPM	DM	FV	SW	O	TI	SP	SM	ET	SS	PI	
				*					*														*	Almeida and Ferreira
				*					*										*					Arulanandan and Sybico
		*	*			*													*					Atkinson
*			*									*						*		*				Been and Jefferies
*		*				*				*	*			*					*				*	Benoit and Lutenegger
	*		*							*								*						Clarke
*		*	*			*										*		*						Clayton et al.
		*	*																*		*	*		Coop
*		*	*			*										*			*		*	*		Crova et al.
*			*			*														*	*			Davies and Newson
							*	*										*						De Josselin de Jong
		*	*			*				*									*	*		*		Fahey et al.
	*	*																		*	*			Gunn
*	*			*		*				*									*				*	Hight et al.
*			*														*	*						Houlsby and Nutt

			*																					*	Kulhawy
	*	*	*			*															*				Lewin and Allman
		*	*			*		*											*						Mayne
*		*	*			*						*						*	*						Muir Wood et al.
		*	*			*		*			*	*						*	*						Ohta et al.
*		*	*			*						*	*					*	*						Robertson and Ferreira
	*	*	*			*					*			*				*							Schnaid et al.
		*	*			*						*						*							Simpson
*		*	*			*						*					*					*			Steenfelt
*		*	*			*		*				*						*	*		*	*			Vaughan et al.
		*	*									*	*	*				*	*		*	*			Wheeler and Sivakumar
	*	*	*			*					*	*						*	*						Whittle and Aubeny
*		*	*			*						*	*					*	*						Whittle et al.
*		*	*			*						*	*					*	*						Yu et al.
14	8	13	20	2	2	11	2	4	2	3	8	6	2	4	2	3	2	13	12	9	7	3	4		29 papers

KEY:

Soil behaviour. SH: stress history SS/MS/LS: small/medium/large strains PC: permeability, consolidation
Laboratory tests. S: sampling T: triaxial TT: true triaxial SS: direct/simple shear O: others
Field tests. I: insertion SBP: self-boring pressuremeter CPT: cone penetrometer/piezocone CPM: cone pressuremeter DM: flat dilatometer FV: field vane SW: seismic waves O: others
Main emphasis. TI: test interpretation SP: soil properties SM: soil model ET: experimental technique SS: special soils PI: practical implications

A third aspect considered here is the main emphasis of the papers. About one third of them deal with test interpretation (usually of field tests), and a similar number refer to the analysis of some specific feature of soil behaviour. Soil modelling is treated in eight papers, either by presentation of new models or by application of already known models. Special soils are studied in a few cases.

Influence of stress history

The role of past stress history on soil behaviour has always been recognized. The parameters defining the stress history are the effective confining pressure, p'_0 (or the vertical stress σ'_{z0}), the overconsolidation ratio OCR, and the lateral stress coefficient K_0. Their influence will be examined in the following sections when describing the different features of soil behaviour.

Several papers deal with the measurement of these parameters: Benoit and Lutenegger, Clarke, Hight et al. and Whittle and Aubeny use or refer to the self-boring pressuremeter for determination of K_0. Although the SBP test is the most efficient tool for reducing soil disturbance, it seems that judgement is required for the determination of the lift-off pressure by inspection of the pressure–strain curves. Been and Jefferies and Mayne refer to indirect determination of the overconsolidation ratio from piezocone tests. Also indirect are the measurements of K_0 from CPT or CPM tests (Houlsby and Nutt, Whittle and Aubeny) or from the flat dilatometer (Whittle and Aubeny, Yu et al.). The interpretation of these tests is slowly moving away from empiricism.

Influence of strain level

Recognition of the importance of strain level in soil stress–strain behaviour is not new: non-linearity of soil response has been observed since the beginnings of modern soil mechanics. The existence of very stiff behaviour at small strains was identified in the seventies (Simpson et al., 1979) as one of the main reasons for discrepancies between predicted and observed soil deformations in foundations and excavations. It was also invoked to explain differences between static and dynamic moduli of soils. For cyclic loading, curves describing the variation of shear modulus and damping ratio with strain level have been used for years. However, only recently has the strain level been used as a reference parameter to describe general soil behaviour. In the Florence Conference, it was used as the key parameter in the three General Reports cited above. In this Symposium, the papers by Atkinson, Robertson and Ferreira and Simpson deal specifically with this point.

Three ranges of strain are identified:

(a) An early range in which soil behaviour is linear.
(b) A range in which non-linear behaviour is evident and soil stiffness decreases with shear strain. In cyclic loading, damping ratio increases in a similar way.
(c) A range in which irrecoverable strains are important and some type of failure takes place.

There are some differences in the nomenclature and threshold values considered in the Reports of the Florence Conference. The alternative names for the above ranges were 'small/medium/large' (Sagaseta et al., 1991) and 'very small/small/large' (Atkinson and Sallfors, 1991). The first one has the advantage of having been traditionally used for cyclic loading (Ishihara, 1982), and will be used in the following discussion. The upper limit for the small strain range is about 10^{-5}, although some experimental evidence (Georgiannou et al., 1991) suggests that for clays it could extend to 10^{-4} (Fig. 1). For large strains, the threshold of 10^{-2} (1%) is reasonable for monotonic loading, but for cyclic loading accumulation of irrecoverable strains and degradation with the number of cycles causes failure for lower levels of strain and hence the limit of 10^{-3} is adequate.

In field problems, large strains are present in excavations and embankments on soft soils. Medium strains are typical of foundations of structures and excavations in firm soils. Small strains are of importance in seismic problems. However, as noted by Burland (1989), small and medium strains always take place at some distance from the applied load or excavation. Hence, soil behaviour at small strains is important if

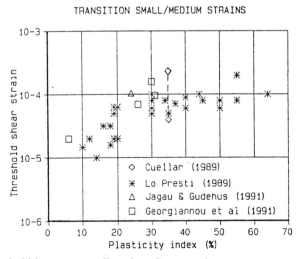

Fig. 1. Threshold between small and medium strains

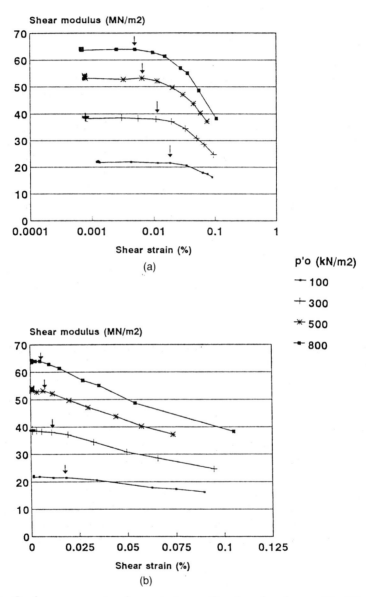

Fig. 2. G_0 from resonant column tests on Canales clay ($w_L = 60$, PI = 35, $w = 21\%$) (Cuéllar, 1989): (a) logarithmic scale; (b) natural scale

a knowledge of distant soil movements is needed. The analysis of surface settlements due to tunnelling is a remarkable example of this, as shown in the paper by Gunn et al.

Soil behaviour at small strains
Linear soil response

In this range, soil behaviour is linear, and hence the governing parameter is the shear modulus, G_0 (or G_{max}). This range is evident in G against γ curves as a horizontal plateau. However, these curves must be used with care, because the logarithmic scale commonly used for strains tends to overemphasize the constant modulus region. A simple exercise using for instance the hyperbolic model (which has no linear range) shows that an apparent plateau exists with a limit of about 0.01 times the 'reference strain' ($\gamma_r = \tau_f/G_0$). Rather surprisingly, if usual values are taken for τ_f and G_0, the apparent threshold of linear behaviour appears to be about 10^{-5} for sands and 10^{-4} for clays, in agreement with Fig. 1. A linear scale for the strains reveals more clearly the existence of a true linear range (Fig. 2).

The most important factor governing the value of G_0 is the effective confining pressure, p'_0. The dependence can be expressed as:

$$G_0/p_r = k \, (p'_0/p_r)^n \qquad (1)$$

where p_r is any reference pressure and the variable, k, is often referred to as 'modulus number'. The power n is about 0.5. It is interesting to note that the same formal expression was found using theoretical analyses by Wroth and Houlsby (1985).

The main factors influencing the value of the modulus number, k, are the past stress history (overconsolidation ratio) and the age of the soil (time at constant static pressure). Other factors, such as initial deviatoric stresses (K_0) or strain rate, are less important.

The variation of G_0 with OCR can be expressed either directly or through the void ratio, e (e, OCR and p'_0 are interrelated). In the first case, a linear variation with the logarithm of OCR is observed (Atkinson). The second option has been traditionally followed in cyclic loading, and there are a number of expressions of the type:

$$G_0/p_r = k_1 \frac{(e_1 - e)^2}{1 + e} (\sigma'_{z0}/p_r)^n \qquad (2)$$

in which the vertical stress, σ'_{z0}, is used instead of the mean pressure. The power n is in the range 0.40–0.60, the parameter e_1 about 2–3, and k_1 is about 8000–16,000 for granular soils and 100–3000 for clays (when p_r is taken as $1 \, kN/m^2$). The different expressions available do not always

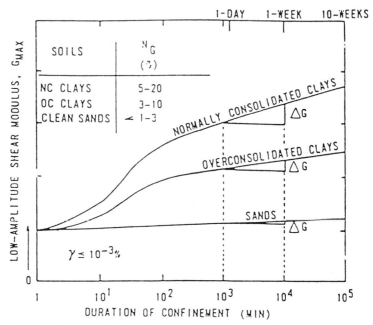

Fig. 3. Effect of age on G_0 (Anderson and Stokoe, 1978)

give coincident values, and discrepancies up to 50% can be found when they are applied to a given soil (Ishihara, 1982; Sagaseta et al., 1991).

The effect of soil age has been invoked as being responsible for differences between laboratory and field determinations of soil modulus. Anderson and Stokoe (1978) found a quasi-linear variation of G_0 with the logarithm of time of load application. The effect is small in sands, and maximum in normally consolidated clays (Fig. 3). Crova et al. present results in holocene and pleistocene sands, of similar nature and grading, with differences in soil modulus much greater than the ones predicted by Fig. 3. The authors attribute this to cementation in the pleistocene sand.

Measurement of G_0

Experimental evidence shows that moduli obtained from static and dynamic tests agree reasonably well when similar levels of strain are used (Georgiannou et al., 1991). Laboratory measurements of G_0 are usually carried out in torsion by means of the resonant column test (Richart et al., 1970), which can reach the required low strain levels. However, recent developments in load and strain measurement in triaxial and simple shear tests have allowed strain measurements as small as 10^{-5}, and this has produced an increasing use of these tests for this purpose (Tatsuoka, 1988; Atkinson and Sallfors, 1991; Atkinson; Hight et al.).

In the field, the most common method is seismic measurement of shear wave velocity v_s which is related with the modulus G_0 by:

$$G_0 = \rho v_s^2 \qquad (3)$$

The measurements are usually obtained using cross-hole or down-hole tests, or refraction from the surface (Crova et al., Robertson and Ferreira). Clayton et al. describe the use of seismic tomography, in which a number of sources and receivers are used. Spectral analysis of the propagation of surface waves (Abbiss, 1981) provides a means of soil profiling from surface measurements, even if soil modulus decreases with depth (Cuéllar et al., 1992) (Fig. 4).

Direct measurement of G_0 with the self-boring pressuremeter (Hight

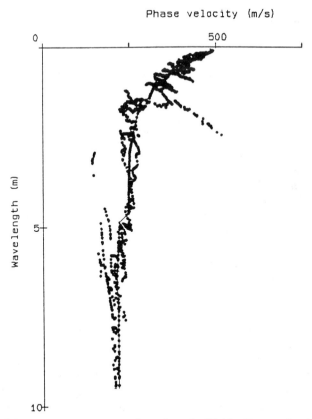

Fig. 4. Rayleigh waves measurement in a fly ash fill (Cuéllar et al., 1992): Los Barrios Power Plant

et al., Robertson and Ferreira and Whittle et al.) is influenced by the uncertainties in lift-off pressure mentioned earlier. Better results can be obtained by using unload–reload cycles, properly analysed with a non-linear model (Palmer, 1972). However, fitting the curve simultaneously for all strain levels is not always satisfactory, and moduli between 1/2 and 1/3 of the dynamic value are obtained. It should be noted that if unloading loops are used after significant strains have been generated, then insertion effects are not so important, and hence the cone pressuremeter can give useful results (Houlsby and Nutt).

Soil behaviour at medium and large strains
Modelling of non-linear soil response
For this range, non-linearity in the stress–strain behaviour becomes evident. Dilatancy also takes place, either in the form of volumetric strains in drained shear processes, or pore pressure variations in undrained cases. Irrecoverable strains occur when the loads are removed. The distinctive feature of the large strain range is the occurrence of some type of failure or ultimate state of the soil. The possibilities of modelling soil behaviour can be grouped in three main categories.

Equivalent linear elastic model. For medium strains, it is possible to assume linear elastic behaviour, using an equivalent secant modulus, selected according to the expected strain level from G against γ curves such as those depicted in Fig. 2. This method has been traditionally followed for problems of cyclic loading (in which case an equivalent damping ratio D is also needed). For sands, the curves of modulus ratio G/G_0 and damping ratio against shear strain depend on the value of the confining pressure, whereas for clays this dependence is very small or negligible. The influence of the overconsolidation ratio is small both for sands and clays. On the other hand, the curves are very sensitive to plasticity index (Vucetic and Dobry, 1991).

Non-linear elastic models. For this purpose, a number of models have been presented. The hyperbolic model (Kondner and Zelasko, 1963) is perhaps the most popular, although the Ramberg–Osgood (1943) model has been also used for many years in cyclic loading. Other models are possible, such as the one proposed by Jardine et al. (1986) used in the paper by Hight et al.; the power function proposed by Gunn et al., or the simple bilinear model of Simpson et al. (1979).

The usual ways of formulating these models are either a law of variation of secant modulus with stress or strain, or a function linking stress and strain, from which the tangent modulus G_t can be obtained by differentiation. The resulting tangent modulus is used incrementally, combined with Hooke's linear law. It should be noticed that the G_t against γ curves are not the same as those in the previous section (Fig. 2), in which secant and not tangent moduli were used. However, the

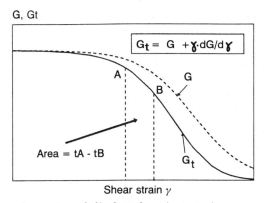

Fig. 5. Tangent and secant moduli plotted against strain

moduli are related as shown in Fig. 5. The use of the tangent modulus has the advantage of being easy to implement in finite element codes for incremental analysis. As the tangent modulus is the derivative of stress with respect to the strain, it is obvious that the area enclosed by the curve and the strain axis between two points is equal to the increment of stress between these two points (Fig. 5). This property is used by Simpson in the model presented to this Symposium.

Elasto-plastic models. The models of the two above groups are essentially elastic. Although the laws are non-linear, they keep some of the limitations of linear elasticity, namely, the same response in loading and unloading, and no dilatancy.

Elastoplastic models extend elasticity to take account of these two limitations, and have been increasingly used from the 'sixties in soil mechanics. Elastic unloading is assumed during stress reversals, and soil dilatancy is modelled by the use of appropriate expressions for the plastic potential.

At present, models based on the critical state concept are the most useful tools for modelling soil behaviour at medium and large strains. The original Cam clay and modified Cam clay models have been extended to include aspects such as non-linear normal and shear stiffnesses (Houlsby, 1985), or anisotropic K_0 consolidation (Davies and Newson, Whittle and Aubeny). Critical state models have provided a unified background in which the main features of soil behaviour can be included. In the paper by Coop, the behaviour of granular soils under high pressures is interpreted in these terms, introducing effects such as the breakage of particles into the general framework. The behaviour of partially saturated soils is now beginning to be understood in these terms (Alonso et al., 1990), and an experimental analysis is presented by Wheeler and Sivakumar.

Experimental measurement

With respect to the measurement of soil response in the laboratory, the triaxial test continues to be the most useful tool, specially with recent advances in stress and strain measurements and control, which have allowed application of different stress or strain paths (papers by Atkinson, Coop, Davies and Newson, Hight et al., and Wheeler and Sivakumar).

The true triaxial test permits a further step towards generalization, enabling virtually any stress or strain paths (Lewin and Allman). Muir Wood et al. also use results of this test to illustrate the determination of model parameters. This approach, within a probabilistic framework, has been successfully used for back-analysis of instrumented works (Ledesma et al., 1986).

Shear tests are used in three papers presented to the Symposium. De Josselin de Jong presents an extension of his previous work on interpreting the simple shear test to include co-rotational Jaumann stress derivatives in the finite strain formulation. His comments on his relationship with Peter Wroth's work (Wroth, 1984) are of special relevance for the aims of this Symposium. The simple shear test is extremely useful for reproducing several real problems such as seismic motions or soil-structure interactions. Fahey et al. make a special use of direct shear (constant normal stiffness, or CNS test) to analyse shaft friction of piles in calcareous sediments by imposing a variation of the normal stress as a function of the shear strain, deduced from cavity expansion conditions. It is worth noting that these conditions can be considered intermediate between drained (at constant normal effective stress), and undrained conditions (shearing at constant volume is equivalent to a CNS test with infinite normal stiffness). As a consequence, resulting cyclic stress-strain curves closely resemble the phenomenon of cyclic mobility of dense sands under undrained cyclic shear.

The direct simple shear test still keeps some of its capabilities, due to the easy possibility of performing large-scale or even field tests. Steenfelt presents the analysis of a contact problem using direct shear tests with sizes ranging from 10 cm × 10 cm to 1 m × 2 m on a glacial till.

With respect to field tests, the self-boring pressuremeter is the most useful for determining the soil stress–strain response. The papers by Clarke, Fahey et al., Hight et al., Robertson and Ferreira, Whittle and Aubeny and Whittle et al. present or analyse results of pressuremeter tests in this range. The increasing use of G against γ curves for interpretation is noted.

At this point, a comment is required about field tests, particularly about the cone penetrometer (CPT/CPTU). In other tests (SBP, DM and FV), a complete static test is performed after insertion of the probe, and the stress-strain behaviour of the soil can be determined, subject to limitations of dealing with complex boundary conditions. However, in CPT tests, the soil is brought to the limit state of steady flow, and then stress parameters (tip resistance, shaft friction, pore pressure) are measured, precisely in the region where failure has been reached. So, stiffness and initial stresses parameters are deduced indirectly from the failure stresses only (no strains are measured). For this reason, this test is treated in the next section.

On the other hand, the cone pressuremeter test (CPM) represents a hybrid case, and the pressuremeter phase of the test is really a stress–strain test. Its interpretation is presented in the paper by Houlsby and Nutt. In clays, the application of the cavity expansion theory (Houlsby and Withers, 1988) is the basis for determining soil parameters from the test results, and acceptable results are obtained for soil strength and stiffness. However, unrealistic high values for K_0 are obtained. In granular soils, soil dilatancy makes realistic theoretical interpretation very difficult.

Failure conditions

In this section, some comments are included about soil modelling or test techniques addressed to the definition of failure conditions, with no attempt being made to reproduce the complete stress–strain behaviour.

The papers by Kulhawy and Ohta et al., and partially the one by Whittle and Aubeny, are concerned with a classical problem in soil mechanics, namely the operating value of undrained shear strength of clays s_u to be used for design purposes. The authors all consider s_u not to be an intrinsic parameter. Following the proposals of Wroth (1984), they establish the correlations between values obtained from different tests through more fundamental properties, such as the angle of internal friction, ϕ, and appropriate soil models. Ohta et al. go a step further in the analysis, and propose to use this methodology inversely for deriving soil stress history and intrinsic parameters from the results of two different undrained tests. However, the use of any relationship (either theoretical or empirical) in a sense opposite to what it was derived for, must be done with care.

As expected, there are several papers dealing with the cone penetration test and related devices: piezocone and cone pressuremeter. Two papers refer to the interpretation of CPT and CPTU tests. The paper by Been and Jefferies presents a systematic process of interpretation, using a hybrid methodology which includes empirical methods and refined soil models. Using a similar approach, Mayne emphasizes the determination of OCR in clays from tip resistance and excess pore pressure measurements. Schnaid et al. present a real case in which s_u is determined from CPT, CPTU and FV. They show how pore pressure corrections reduce the scatter in cone factor estimations.

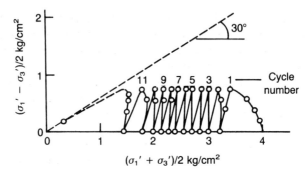

Fig. 6. Liquefaction of loose sand (Castro, 1969)

Influence of stress reversals

Stress reversals occur in cyclic loading twice per cycle. The effect is negligible in the small strain range (linear behaviour). For medium strains, an equivalent damping ratio is enough for most purposes.

For small strains, it was pointed out that no fundamental difference exists between soil response under static and cyclic loading. The same holds for failure conditions. For many years, liquefaction was regarded as a phenomenon specifically linked to cyclic loading, although some work was done on static liquefaction (Castro, 1969). Figure 6 shows the results of a classic liquefaction test on loose sand. In the effective stress path, three different phases can be easily identified:

(a) An undrained path (as in a conventional static test), which is stopped at about 50% of the maximum deviatoric stress.
(b) Repeated unloading and reloading to the same stress level. The overconsolidation ratio increases continuously and hence the stress path becomes more and more vertical. In the final cycles, its initial slope becomes positive.
(c) The last part is again an undrained path for a very loose soil, in which the deviatoric stress is not stopped, but is increased steadily until failure.

It is evident from the above description that no special phenomenon has taken place in the soil, which does not happen in static loading. So, 'liquefaction' and 'pore pressure build-up' are only terms to express the consequences of well-known features of soil behaviour.

From this point of view, soil modelling for cyclic loading is not intrinsically different from the monotonic loading case. The main difference is operational. A model which, for the test shown in Fig. 6, would have properly predicted the stress path until point 1 followed by a vertical unloading path, would have been considered as of good quality, with only a small relative error in the residual pore pressure. However, this small error would become unacceptable if a large number of cycles is applied (with the result of a 'frozen' behaviour with elastic closed loops down and back to point 1).

As a result, soil models of classical plasticity, even those including strain-hardening, are not suitable for cyclic loading, and special improvements have been found necessary in order to include irrecoverable strains for stress paths inside the yield surface. With this aim, a number of models have been proposed, whose description falls out of the scope of this Report: multi-surface models, bounding surface models, hypoplasticity and generalized plasticity, and endochronic theory. A general comparative description can be found in Sagaseta et al. (1991).

Besides these refined models, which at present are used only in research, the solution of engineering problems with stress reversals can be performed using the Masing (1926) rule. Based on thermodynamic considerations at the molecular level, the Masing rule states that if a virgin stress–strain law can be expressed as $\tau = f(\gamma)$, then unloading from a point (τ_a, γ_a) follows (Fig. 7):

$$\frac{\tau - \tau_a}{2} = f\left(\frac{\gamma - \gamma_a}{2}\right) \qquad (4)$$

This very simple rule allows any of the non-linear models to be applied to cyclic loading. Pyke (1979) and Vucetic (1990) have given

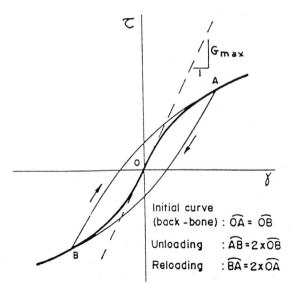

Fig. 7. *Masing behaviour*

complementary rules for adequately modelling any series of irregular cycles. One of the reasons for the popularity of the Masing rules is that any combination of elastic springs and Coulomb friction sliders, the so-called 'Iwan models' (Iwan, 1967), leads to Masing behaviour. Some authors have gone further with this approach. For instance, Kovári (1979) presents the potential capability of Iwan models for rock behaviour, including brittle points and elastic-frictional wedges, and these have been implemented in commercial finite element codes. Uriel (1991) shows how an assembly of these models in two- or three-dimensional cases can reproduce many features of soil response, such as dilatancy and stress reversal.

An extension of the problem of stress reversals is the case in which the stress path has a break, or an abrupt change in direction, measured by the angle θ (Fig. 8) ($\theta = 0$ means no break and $\theta = 180°$ is a stress reversal). Experimental data for this case have been presented by Atkinson et al. (1990) (Fig. 8). Soil stiffness is shown to increase with the angle of rotation θ of the stress path, with a Masing-type behaviour for $\theta = 180°$. From the foregoing comments, it follows that an extension of Iwan models to two dimensions can be expected to reproduce adequately this behaviour. The model presented by Simpson is an excellent example of this.

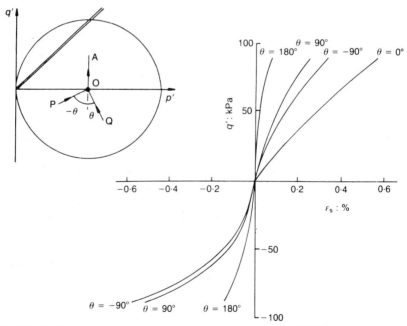

Fig. 8. Effect of corners in the stress path (Atkinson et al., 1990)

Sampling disturbance and insertion effects

These points are the main drawbacks of laboratory and field tests, respectively, and are considered specifically in several Symposium papers.

The first step in any laboratory or field test is to insert some kind of tool in the ground (either a sampler or a probe). This insertion produces some strains in the soil around it.

The strain path method (Baligh, 1985) has been the first successful approach to the analysis of this process in clays (undrained penetration). The soil is assumed to flow in a steady regime with respect to the tool. This is specially suitable for the cone penetration test, in which penetration goes on at constant rate. In other cases (FV, SBP, DM, samplers), insertion stops at some finite penetration, typically 200–500 mm., and then the test is started or the sample is withdrawn. In these cases, finite element simulations are also useful (Yu et al.).

The paper by Whittle and Aubeny includes the application of the strain path method to a number of insertion problems. The stresses are obtained from the strains by using refined models. For the simulation of the self-boring pressuremeter, the simultaneous action of a ring source (tube driving) and in inner sink (extracting device) are used. The results of the test are shown to depend on the extraction ratio f between the excavated and driven volumes (f = 1 means an ideal self-boring pressuremeter, f < 1 undercutting and f > 1 overcutting). This illustrates the potential use of theoretical methods of analysis in identifying parameters governing real problems.

In soil sampling, the strain path method suggests insertion strains of 1–2% in the sample axis (Baligh, 1985), depending on the sampler area ratio. These strains are clearly non-permissible. Moreover, they are not the only reason for sampling disturbance. Vaughan et al. present a thorough analysis of the causes of disturbance during soil sampling in clays, including stress release to zero total stresses, shear strains induced by the sampler insertion, and pore pressure changes due to cavitation or wetting in fissures or sand lenses. The effects of these are measured through the suction in the sample, which decreases with respect to the in situ value in normally consolidated clays, and increases in overconsolidated clays. This produces corresponding variations in undrained shear strengths measured without careful reconsolidation to field stresses. In stiff clays, recorded suctions can be up to 50–100% higher than the initial value. For soft clays, suction decreases make it necessary to apply correction factors to the shear strength measured without sample reconsolidation to in situ stresses (Ohta et al.).

References

ABBISS, C.P. (1981). Shear wave measurements of the elasticity of the ground. Géotechnique, Vol. 31, No. 1, pp. 91–104.

ALONSO, E., GENS, A. AND JOSA, A. (1990). A constitutive model for partially saturated soils. Géotechnique, Vol. 40, No. 3, pp. 405–430.

ANDERSON, D.G. AND STOKOE, K.H. (1978). Shear modulus: A time dependent soil property. Dynamic Geot. Testing, ASTM, STP 654, pp. 66–90.

ATKINSON, J.H., RICHARDSON, D. AND STALLEBRASS, S.E. (1990). Effect of recent stress history on the stiffness of overconsolidated soil. Géotechnique, Vol. 40, No. 4, pp. 531–540.

ATKINSON, J.H. AND SALLFORS, G. (1991). Experimental determination of stress-strain-time characteristics in laboratory and in-situ tests. Proc. Xth ECSMFE. Florence, Italy, Vol. 3, pp. 915–956.

BALIGH, M.M. (1985). Strain Path Method. J. Geot. Engg. ASCE, Vol. 111, No. GT9, pp. 1108–1136.

BURGHIGNOLI, A., PANE, V. AND CAVALERA, L. (1991). Modelling stress–strain-time behaviour of natural soils. Monotonic loading. Proc. Xth ECSMFE. Florence, Italy, Vol. 3, pp. 961–979.

BURLAND, J.B. (1989) Small is beautiful. The stiffness of soils at small strains. Proc. IXth L. Bjerrum Mem. Lecture. Can. Geot. J., Vol. 26, pp. 499–516.

CASTRO, G. (1969). Liquefaction of sands. PhD thesis, Harvard Univ. Soil Mech. Series No. 81.

CUÉLLAR, V. (1989). Uso geotécnico de los parámetros geofísicos. Técnicas Geofísicas Aplicadas al Reconocimiento Geotécnico. Gabinete de Formación y Documentación. CEDEX, Madrid.

CUÉLLAR, V., VALERIO, J. AND MUÑOZ, F. (1992). Spanish experiences on the determination of the dynamic properties of soils through the analysis of surface waves. Invited lecture. Proc. Xth World Conf. Earthq. Engg. Madrid, (in press).

GEORGIANNOU, V.N., RAMPELLO, S. AND SILVESTRI, F. (1991). Static and dynamic measurements of undrained stiffness on natural overconsolidated clays. Proc. Xth ECSMFE. Florence, Italy, Vol. 1, pp. 91–96.

HOULSBY, G.T. (1985). The use of a variable shear modulus in elastic-plastic models for clays. Computers and Geotechnics, Vol. 1, pp. 3–13.

HOULSBY, G.T. AND WITHERS, N.J. (1988). Analysis of the cone pressure-meter test in clays. Géotechnique, Vol. 38, No. 4, pp. 575–587.

ISHIHARA, K. (1982). Evaluation of soil properties for use in earthquake response analysis. Num. Models Geom., Zurich (Dungar, Pande and Studer, eds.), pp. 237–259.

IWAN, W.D. (1967). On a class of models for the yielding behaviour of continuous and composite systems. J. App. Mech., Vol. 34, E3, pp. 612–617.

JARDINE, R.J., POTTS, D.M., FOURIE, A.B. AND BURLAND, J.B. (1986). Studies of the influence of non-linear stress-strain characteristics in

soil-structure interaction. Géotechnique, Vol. 36, No. 3, pp. 377–396.
KONDNER, R.L. AND ZELASKO, J.S. (1963). A hyperbolic stress-strain formulation of sands. 2nd PACSMFE, pp. 289–324.
KOVÁRI, K.H. (1979). Models for the interpretation of plastic and brittle behaviour of rocks. Proc. IIIrd Int. Conf. Num. Meth. Geom. Aachen, Vol. 2, pp. 533–544.
LEDESMA, A., GENS, A. AND ALONSO, E. (1986) Identification of parameters in a tunnel excavation problem. 2nd NUMOG. Ghent, pp. 333–344.
MASING, G. (1926). Eigenspannungen und Verfestigung beim Messing. Proc. 2nd Int. C. App. Mech., pp. 332–335.
PALMER, A.C. (1972). Undrained plane strain expansion of a cylindrical cavity in clay: a simple interpretation of the pressuremeter test. Géotechnique, Vol. 22, No. 3, pp. 451–457.
PYKE, R.M. (1979). Soil models for irregular cyclic loading. J. Geot. Engg. Div., ASCE, Vol. 105: GT6, pp. 715–726.
RAMBERG, W. AND OSGOOD, W.R. (1943). Description of stress-strain curves by three parameters. Tech. Note 902, Nat. Adv. Comm. Aer., Washington, D.C.
RICHART, F.E., HALL, J.R. AND WOODS, R.D. (1970). Vibrations of soils and foundations. Prentice Hall, N.J.
SAGASETA, C., CUÉLLAR, V. AND PASTOR, M. (1991). Modelling stress-strain-time behaviour of natural soils. Cyclic loading. Proc. Xth ECSMFE. Florence, Italy, Vol. 3, pp. 981–999.
SIMPSON, B., O'RIORDAN, N.J. AND CROFT, D.D. (1979). A computer model for the analysis of ground movements in London clay. Géotechnique, Vol. 29, No. 2, pp. 149–175.
TATSUOKA, F. (1988). Some recent developments in triaxial testing systems for cohesionless soils. Advanced Triaxial Testing of Soil and Rock (Donaghe, Chaney and Silver, eds.). ASTM, STP 977, pp. 7–67.
URIEL, A.O. (1991). Behaviour of a physical and conceptual model based on an extension of Masing's model to 2-D. Proc. Xth ECSMFE. Florence, Italy. Disc. Sess. 2-b, Vol. 4 (in press).
VUCETIC, M. (1990). Normalized behaviour of clay under irregular cyclic loading. Can. Geot. J., Vol. 27, pp. 29–46.
VUCETIC, M. AND DOBRY, R. (1991). Effect of soil plasticity on cyclic response. J. Geot. Engg., ASCE, Vol. 117: GT1, pp. 89–107.
WROTH, C.P. (1984). Interpretation of in-situ soil tests. Proc. 7th Rankine Lect. Géotechnique, Vol. 34, No. 4, pp. 449–489.
WROTH, C.P. AND HOULSBY, G.T. (1985). Soil Mechanics. Property characterisation and analysis procedures. Proc. XIth ICSMFE, San Francisco, USA, Vol. 1, pp. 1–55.

Prediction and performance

H.J. BURD and G.T. HOULSBY, Oxford University

Introduction
Prediction and performance is a central theme of the Wroth Memorial Symposium. The purpose of this report is to highlight some of the predictive exercises described in the Symposium proceedings and to comment on the comparison between these predictions and the measured performance.

Prediction refers to the use of an analysis procedure, together with empirical correlation or some other method, to estimate the behaviour of a real structure. Prediction will usually relate to the response of a structure under a set of loads and conditions that would reasonably be expected to occur in practice; such predictions may in principle be compared with observed performance. In this sense, prediction is a quite different activity from design. *Design* involves the consideration of all of the possible factors that might affect the strength and serviceability of a construction, so that it is able to fulfil its function safely; a detailed discussion of the design process is given in the theme report to the Symposium by Bolton. In any design activity it is necessary to include all of the possible factors that might contribute to behaviour, including any extreme events and conditions which, although unlikely, are nevertheless possible. Design calculations therefore would not necessarily relate to measurable performance parameters. In a prediction exercise, however, it is necessary only to include conditions and loadings that are likely to occur in the prototype.

Performance is interpreted in this report as the measured response of any quantity relevant to structural behaviour. Performance may be measured at a range of physical scales. At the scale of a soil element, for example, the results of triaxial tests on clay are considered to be the performance, and the process of using a constitutive model to reproduce this behaviour is a prediction. The purpose of comparing performance and prediction in this case might be to assess the accuracy of a particular constitutive model for its use in numerical computations. This type of comparison between performance and prediction is useful for gaining an insight into the details of soil behaviour, but it does not give direct information about the appropriateness, or otherwise, of a given constitutive model for soil in the context of a particular engineering problem.

The performance of a model structure, usually at reduced scale, may be measured in the laboratory, in a centrifuge or as a field trial. These

PREDICTION AND PERFORMANCE

types of tests, if properly carried out, may be used to give valuable information, both about performance under working conditions and also about structural failure. This latter area of performance occurs rarely in working structures but can be simulated in a trial.

The prediction of the performance of full-scale structures is the ultimate goal of any predictive activity, and several papers on this topic appear in the Symposium proceedings. Performance measurements made on full-scale structures where failure has occurred (during a pile test, for example) are a particularly valuable source of data.

Many predictive activities exhibit two separate features; a soil model and an analysis procedure. The current state-of-the-art does not make it possible to choose a soil model that reflects accurately all aspects of soil behaviour; it is instead necessary to choose a constitutive model, and associated analysis procedure, that is most appropriate for the task in hand. In some cases these two features are necessarily linked; for example in a limit-equilibrium calculation a rigid-plastic soil model is invariably adopted. In a more detailed numerical calculation, however, there is considerably more scope for the choice of an appropriate constitutive model. It became clear during the discussion at the Symposium that the relationships between the soil constitutive model and the quality of the predictions are complex and not yet fully understood.

Prediction and performance of soil element behaviour

It is well established that the behaviour of clay is linked closely to its immediate stress history, and in particular may be affected by anisotropic consolidation. Two papers to the Symposium (by Davies and Newson and by Simpson) deal with the problem of modelling this behaviour, but using quite different approaches. Davies and Newson report a developed form of 'Modified Cam clay' in which the yield locus is rotated in stress space by a certain angle in order to reflect the effects of consolidation under K_0 conditions. The proposed model requires a single additional parameter to specify the degree of anisotropy. In this approach an existing model, 'Modified Cam clay', has been extended in order to achieve an improved comparison with measured performance. Such developments should in general be treated with caution as they can rapidly lead to models of extreme complexity. The authors show, however, that this model provides a good match with the results of K_0 consolidated triaxial tests described by Atkinson et al. (1977). A problem with the model is that it cannot describe the evolution of anisotropy.

An alternative approach is adopted by Simpson, who describes a model, developed in terms of a 'bricks on strings' analogy, in which the effects of anisotropic consolidation become automatic consequences of the structure of the model. This model is used to make predictions of K_0

consolidated plane strain tests on clay; these predictions are not compared with any experimental data, although the trends are similar to those given by Davies and Newson. In addition to providing a method for the prediction of stress paths in anisotropically consolidated clay, Simpson's model also implies a reduction of tangential stiffness with continuing straining in any one direction. This aspect of soil behaviour is increasingly being recognised as a feature which must be taken into account if realistic predictions of certain aspects of prototype behaviour, particularly settlements associated with tunnelling and excavations, are to be made.

Prediction and performance for tunnels

Five papers to the Symposium address the design of tunnels or the prediction of tunnel performance. These papers on deformation and failure analysis of tunnels form an interesting group since different analysis techniques are described, each relating to specific aspects of performance.

A numerical study of surface settlements caused by shallow tunnels is given by Gunn. A finite element approach is adopted, in which three different soil models are used. A linearly elastic model produces a rather flat settlement trough and a linearly elastic-perfectly plastic model is shown to generate slightly larger surface settlements with the displacements more localised above the tunnel crown. When a model in which soil non-linearity at small strain is included (derived from a simple power law relationship between deviator stress and deviator strain) the predicted response is significantly different; settlements are of much greater magnitude (this is presumably linked to the values of the material properties used in the analysis) and the settlement trough is much narrower (see Fig. 5 in Gunn). These data are replotted by Gunn in his Fig. 6, and compared with the error curve often used in practice for the estimation of surface settlements. This exercise illustrates the important influence that the form of the soil model can have on the quality of computed settlement predictions. It is unfortunate, however, that few data on measured performance are included in the paper. The one comparison between prediction and performance is given in Gunn's Fig. 7, in which ground loss measured in centrifuge tests (Davis et al., 1980) is shown to compare well with the finite element results.

This paper raises an interesting question about the form of soil model and the calculation procedure that should be adopted for the prediction of surface settlements associated with tunnelling. Several contributions were made during the discussion of this paper, from which it became clear that poor correlations between measured surface settlements and finite element predictions was a common experience.

An alternative approach to the analysis of displacements around

PREDICTION AND PERFORMANCE

tunnels is given by Mair and Taylor, who describe the use of cavity contraction analysis with an elastic-perfectly plastic constitutive model. A spherical cavity contraction model is shown to provide good predictions of displacements in front of an advancing tunnel heading; a cylindrical model is used to predict displacements above and to the side of a tunnel. For undrained materials the pattern of displacements implied by this analysis procedure does not depend on the details of the model used, and the general trend of displacement with distance from the tunnel can be explained simply by the assumptions of cavity contraction and constant volume. Differences between this simple model and the observed displacements from several sites are illustrated in Fig. 5 of Mair and Taylor and reproduced here as Fig. 1. These performance data indicate that the vertical dimension of the tunnel shortens more than the horizontal dimension. This feature of performance can be explained by lack of symmetry of the deformation, with the deformations at the tunnel wall being more elliptical than cylindrical. The line OA through the origin added to Fig. 1 represents predictions of movement assuming cylindrical cavity contraction and a ground loss of 2.5%.

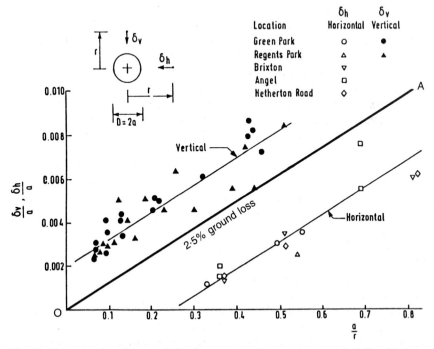

Fig. 1. Horizontal and vertical displacements adjacent to tunnels (after Mair and Taylor)

In a contribution to the discussion of this paper, Sagaseta suggested possible reasons for the asymmetrical deformations that are observed in practice. One of the suggested reasons was that the proximity of the ground surface to the tunnel has an important effect on the deformations. In order to test this hypothesis, Sagaseta described the use of a simple and elegant analytic model (Sagaseta, 1987) to predict the displacements for one of the case studies reported by Mair and Taylor (Attewell and Farmer, 1974) in which information is given about the tunnel depth. This calculation indicated that, when the position of the ground surface is included in the analysis, then the correct form of deformation asymmetry is predicted, although the magnitude of the difference between horizontal and vertical deformation was less than that observed in practice. It seems likely that anisotropy of in situ stresses is also an important effect although this is not readily included in the analysis.

It emerges from the data given by Mair and Taylor that simple soil models cannot be used satisfactorily to model the pore pressures generated around tunnels. A linear elastic–perfectly plastic soil model would predict that excess pore pressures fall to zero at the elastic–plastic boundary; in practice, however, excess pore pressures are generated over a substantially greater radius than this (Fig. 8 of Mair and Taylor). Mair and Taylor show that, when non-linear behaviour is assumed for the elastic component of the soil model, the cavity contraction analysis predicts pore pressure generation in the 'elastic' regime reasonably successfully. In the model adopted by Mair and Taylor, the shear modulus increases linearly with radius which implies that $G \propto 1/\varepsilon_r^{0.5}$. In fact any soil model in which the shear modulus increases with radial distance would predict non-zero pore pressures in the 'elastic' zone.

Predictions of tunnel collapse are made by Sloan and Assadi, who use a linear programming method to compute upper and lower bound solutions for the stability of shallow circular tunnels in soft ground. A series of predictions are made and one graph is included (Fig. 16 of Sloan and Assadi), in which upper and lower bound analyses are compared with some of the centrifuge results obtained by Mair (1979). A particular feature of this method is that the upper and lower bounds are claimed to be rigorous, which means that the exact solutions must lie within the bounds given. The upper and lower bounds presented are not in fact significantly closer than those obtained using analytical techniques by Davis et al. (1980); the value of Sloan and Assadi's work lies in the fact that their numerical method can much more readily be adapted to new types of problem. Their work accounts, for example, for increasing undrained strength with depth.

The three papers mentioned above describe specific analysis procedures for idealised tunnels. Predictions of displacements around the

Bakerloo Line tunnel caused by excavation for the Waterloo International Terminal are given by Hight et al. The predictions make use of a non-linear elastic–perfectly plastic model in which the elastic component requires the specification of ten material constants. In the analysis six distinct soil layers are modelled, leading to a large number of specified material properties. A substantial amount of in situ and laboratory tests were carried out in order to obtain these data. Predictions of deformations on the tunnel circumference are given in Figs. 9(a) and 9(b) of Hight et al. and these appear to be generally successful. In this analysis a sophisticated soil model is used; the extensive site investigation exercise carried out as part of this project is clearly necessary in order to specify the material properties needed for the analysis. No conclusions are reached by the authors as to whether the level of sophistication that is adopted is appropriate; it may be relevant to ask whether the modelling is too complex, about right or perhaps not detailed enough.

Predictions of performance of retaining structures and excavations

A group of papers on the analysis and performance of retaining structures are included in the Symposium proceedings. Three of these (Fraser, Rampello et al. and St. John et al.) refer to predictions carried out as part of a design process. Jewell et al. describe a general analysis technique for the prediction of deformations in reinforced soil walls. An interesting feature of these papers is the range of sophistication of analysis procedures, and the conclusions reached from the comparisons between prediction and performance.

Fraser describes a series of predictions for an excavation of 32 m depth, supported by a propped diaphragm wall. A series of predictions are made using a non-linear subgrade reaction model; further detail could be provided by a finite element analysis, although such a calculation is not reported in the paper. Two sets of predictions are compared with performance. In Fig. 4 of Fraser, predicted and observed wall displacements are shown to compare well. In Table 2 calculated strut loads are compared with those measured after construction. The agreement is again close. It is not entirely clear in both cases, however, whether these predictions were made before construction, or were in fact a back-analysis. This paper raises interesting questions about the complexity of calculation needed to make adequate predictions of performance. The use of a simple subgrade reaction model is more straightforward than a non-linear finite element analysis and appears to function well as a predictive tool. During the discussion of his paper, Fraser observed that significant settlements were caused during installation of the diaphragm wall. It is clear that construction processes of this sort lead to deformations that are difficult to include directly in a numerical model and yet they may have an important influence on wall performance.

Further indications of the relationship between model complexity and prediction quality are provided by Rampello et al. The construction described in this paper is an excavation in soft to medium clay supported by a braced sheet pile wall; the depth of the excavation is 11 m. A finite element analysis was carried out using the CRISP program in which a linearly elastic–perfectly plastic model was used to represent the undrained behaviour of the clay. Predictions were also carried out using various semi-empirical approaches. Finite element predictions of horizontal displacement are shown to agree well with measured performance (Fig. 5 of Rampello et al.), but agreement between performance and prediction for the surface settlements is poor. This type of finite element analysis predicts heave adjacent to the excavation, whereas the soil actually settled; the semi-empirical methods for prediction of these surface displacements, however, were more successful. The suggestion is given by Rampello et al. that the lack of success of the finite element predictions of surface settlements is due, at least in part, to the neglect of non-linear soil behaviour at small strain. Similar observations about the use of elastic analysis were made in relation to the New Palace Yard car park by Simpson et al. (1979). Rampello et al. report, however, that the horizontal movements of the sheet piles were predicted acceptably well. This leads to the interesting conclusion that certain aspects of performance may be represented well by a relatively simple model; to capture other features correctly it may be necessary to use a model of increased complexity.

A detailed set of predictions and comparisons with performance is given by St. John et al. using a similar numerical model to that used by Hight et al. These calculations refer to a deep excavation in London Clay (of depth about 19 m) supported by a secant pile wall. A series of comparisons between calculations of horizontal movement of the retaining wall and measured performance shows that the finite element procedure provides satisfactory predictions. An important feature of the predictions is that settlements adjacent to the excavation are also modelled fairly well (Fig. 15 of St. John et al.) although the predicted magnitude of the settlements tends to overestimate the range of measurements by a significant margin. During the discussion of this paper, St. John indicated that significant contributions to the measured movements around the excavation were caused by unforeseen events occurring during construction. Significant movements associated with ground loss during installation of the secant pile wall were recorded and unexpected local heave was measured during the grouting of mini-piles.

It would be difficult to include these two effects directly in the numerical model.

The contributions by St. John et al. and Rampello et al. combine to provide an interesting theme. Calculations of horizontal displacements adjacent to excavations seem to be relatively insensitive to the choice of small strain model adopted for the soil. Calculations of vertical displacement, however, require more sophisticated models in order to make accurate predictions.

Jewell et al. describe an analytical model that may be used to predict front panel displacements in reinforced soil walls. A simple analysis based on a limit-equilibrium approach is used to make predictions that compare well with measurements made during the test of a 3 m high reinforced soil wall. An interesting feature of this analysis is that facing panel movements are shown to be determined by the strength characteristics of the backfill soil and the stiffness of the reinforcement; backfill stiffness appears not to be an important parameter. Jewell et al. indicate the importance of using realistic values of soil friction angle (which may be significantly higher than that usually adopted for design) in order to make successful predictions.

Predictions of foundation performance

Several papers in the Symposium refer to prediction and performance of foundations. A selection of these are reviewed briefly below. Mandolini and Viggiani describe the prediction of the settlements of a piled raft foundation, using an interesting combination of settlement measurements made on a single pile, coupled with a linear interaction analysis. The basis of the analysis for the interaction between piles in a group is described by Caputo and Viggiani (1984) and illustrated using the results of a load test carried out on a two pile group of floating piles, in which a force Q is applied to one of the piles and the other is not loaded. The results of this pile test (reproduced in Fig. 2) indicate that the settlement of the loaded pile is a *non-linear* function of Q but the settlement of the unloaded pile is a *linear* function of Q. In order to capture this behaviour for a pile group, Caputo and Viggiani assume that the interaction between piles is a linear function of pile *loads* rather than the more usual method of assuming a linear interaction between pile *displacements* (Poulos, 1980). The authors characterise this interaction using interaction factors that depend only on pile spacing. It is possible in this way to carry out a pile group analysis in which the settlement of a given pile in a group is related to the load at the pile head by a suitable non-linear relationship (measured, perhaps, during a pile test) and to the loads at the heads of all of the other piles by a suitable linear relationship.

In the predictions described by Mandolini and Viggiani, a semi-

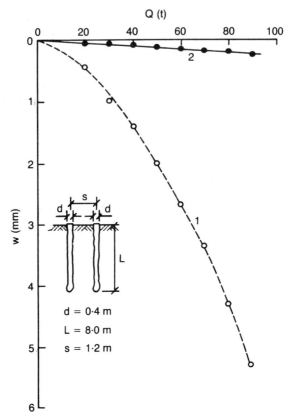

Fig. 2. Experimental load-settlement curves (after Caputo and Viggiani (1984))

empirical expression is used for the interaction factors; at spacings greater than 20 pile diameters interaction is assumed to be negligible. Predictions of the performance of two buildings (referred to as Tower A and Tower U) are presented. Tower A is founded on a total of 182 piles of diameter 1.5 m and Tower U is founded on a total of 314 piles of diameter 0.6 m. The foundation is a piled raft; the reinforced concrete raft slab varies in thickness from 1.2 to 3.5 m. In the approach adopted for these predictions it is necessary to assume either that the raft is rigid (in which case the settlement profile beneath the building would be linear with horizontal coordinate) or flexible (in which case the raft is assumed not to interact with the soil). Measured settlements indicate that the flexible raft solution provides better predictions.

Mandolini and Viggiani describe the use of a field test on a single pile to predict performance of a structure in which piles interact. A similar procedure in which a field test is used as part of a predictive exercise is

PREDICTION AND PERFORMANCE

described by Schnaid et al. This prediction was carried out in order to assess the suitability of an existing haul road for the transport of large structural modules on multi-wheeled transporters. A requirement on maximum surface deflection, and a factor of safety against failure were set for the haul road. A predictive exercise was carried out in which, at a particular test location, a set of in situ and laboratory tests were used to derive properties for an elastic–perfectly plastic soil model, and these were then used to predict the results of a foundation test carried out on the site. After the test was complete the load–displacement response was compared with the numerical predictions. The strength of the foundation was predicted well by the numerical analysis, but the calculated settlements were larger than those that were measured. This latter observation was assumed to indicate that an inappropriate value of stiffness was used for the numerical analysis. A reassessment of the test data was made to find a value of soil stiffness that would predict the load test more accurately. Once this had been achieved the expected performance of the road at other locations, where soil properties were different from those at the test site, could be predicted with some confidence. A plot of surface settlement obtained as the transporter passed a given station is given in by Schnaid et al. in their Fig. 15, confirming that the prediction gave a satisfactory indication of the performance. This procedure includes the comparison between prediction and performance at two levels. Performance measurements were first used to check the proposed predictive tools; further predictions of behaviour were then made and compared with the final measured performance.

Two papers address the problem of measurement of the performance of the foundations of jack-up units. The assessment of design procedures for these structures is made difficult by the lack of well-documented case histories, although the example described recently by McCarron and Broussard (1992) of measurements made on a jack-up unit during a storm in the North Sea provides an outstanding example of the sort of record that is required.

In the absence of well-documented field data, predictions must be tested against laboratory tests. Dean et al. report centrifuge tests of flat and conical footings in sand, whilst Houlsby and Martin report laboratory tests on spudcan footings in clay. Although the problem of 'spudcan fixity', which has a major impact on the certification of jack-up units, is essentially one of stiffness, this stiffness is affected significantly by the detailed size and shape of the yield locus at any penetration. Both papers compare the experimentally determined yield locus with simple theories, based principally on the widely used formulae of Hansen (1970). Simple theories are shown to compare well with the variation of vertical bearing capacity with depth of penetration for both clays and sands. Although these simple formulae also perform reasonably satisfactorily for cases where either vertical load and moment or vertical load and horizontal load are applied, they are much less satisfactory for the general case of vertical load, horizontal load and moment. A striking feature of the theoretical models currently being developed is the analogy which can be drawn with critical state models.

Conclusions

Prediction is the process by which the expected behaviour of a construction or soil element is estimated. This process is distinct from design which must allow for extreme and unlikely events. Predictions can sometimes be compared with performance, and their value is thereby enhanced enormously. These comparisons are vital in order to assess whether soil models and analysis procedures are appropriate for a particular problem.

Current work on prediction is rightly concentrating on the development of appropriate soil models and analysis procedures. Increasing the complexity of models would be expected to increase the cost of the prediction process but not necessarily the quality of the predictions. The quality also depends on the standard of the soil testing and interpretation methods used to derive the soil properties.

The importance of adopting a suitable soil model for satisfactory predictions is illustrated well in the Symposium proceedings. Horizontal movements of retaining structures are shown to be captured reasonably well by a simple elastic–perfectly plastic calculation. For other types of prediction, for example surface settlements associated with shallow tunnels and excavations, and pore pressures generated by tunnelling, more detailed models clearly need to be adopted for accurate predictions. A single calculation may provide a good fit to some features of performance but a poor comparison with others.

A recognition of the dramatic reduction of soil stiffness with increasing strain, even within the region for which a soil is conventionally treated as elastic, is essential if the quality of predictions for certain classes of problem is to be improved.

Several of the Symposium papers dealing with the general subject of deep excavations indicate that soil movements associated with construction procedures can be significant. These effects may have important consequences for performance, and yet are difficult to include directly in a numerical prediction.

It is clear that any predictive activity is enhanced by the use of a suitable field test to verify the analytical procedures. Back-analysis of the field test can be used to establish confidence in predictive methods. This type of approach is described by both Schnaid et al. and by Mandolini and Viggiani. Sensibly used, data obtained in this way is likely to enhance substantially the quality of any subsequent predictions.

References

ATKINSON, J.H., RICHARDSON, D. AND ROBINSON, P.J. (1987). Compression and Extension of K_0 Normally Consolidated Clay. ASCE J. Geo. Eng. Div., Vol. 113, No. 12, pp. 1468–1482.

ATTEWELL, P.B. AND FARMER, I.W. (1974). Ground Disturbance Caused by Shield Tunnelling in a Stiff Over-Consolidated Clay. Engineering Geology, Vol. 8, No. 4, pp. 361–368.

CAPUTO, V. AND VIGGIANI, C. (1984). Pile Foundation Analysis: A Simple Approach to Nonlinearity Effects. Estratto da Rivista Italiana di Geotechnica, pp. 34–51.

DAVIES, E.H., GUNN, M.J., MAIR, R.J. AND SENEVIRATNE, H.N. (1980). The Stability of Shallow Tunnels and Underground Openings in Cohesive Material. Geotechnique, Vol. 30, No. 4, pp. 397–416.

HANSEN, J. BRINCH (1970). A Revised and Extended Formula for Bearing Capacity. Bulletin No. 28, Danish Geotechnical Institute, Copenhagen, pp. 38–46.

MAIR, R.J. (1979). Centrifugal Modelling of Tunnel Construction in Soft Clay. PhD Thesis, Cambridge University.

MCCARRON, W.O. AND BROUSSARD, M.D. (1992). Measured Jack-up Response and Spudcan Seafloor Interaction for an Extreme Storm Event. Proc. 6th Int. Conf. on Behaviour of Offshore Structures, Vol. 1, London, July, pp. 349–361.

POULOS, H.G. AND DAVIS, E.H. (1980). Pile Foundations Analysis and Design. Wiley.

SAGASETA, C. (1987). Analysis of Undrained Soil Deformation Due to Ground Loss. Geotechnique, Vol. 37, No. 3, pp. 301–320.

SIMPSON, B., O'RIORDAN, N.J. AND CROFT, D.D. (1979). A Computer Model for the Analysis of Ground Movements in London Clay. Geotechnique, Vol. 29, No. 2, pp. 143–179.

Design methods

M.D. BOLTON, Cambridge University

Definition of the design process

We must begin, especially considering the man in whose memory this Symposium is held, by clarifying the question and defining our terms. To design may be to 'indicate, draw, form a plan of, contrive, or intend', but most civil engineers are acutely aware of the difference between intentions, plans, and achievements. Executable plans need to be robust enough to cope with uncertainties and deviations, but sufficiently detailed to permit resources to be allocated to specific construction activities. For the purpose of this report, 'design methods' will be taken to mean 'techniques contributing to the creation of executable plans'.

Four components can be recognized in the creation of a design. Firstly, there must be a clarification of the ultimate goal expressed in terms of specific performance requirements; secondly, an attempt to synthesize a complete solution; thirdly, an evaluation of whether the proposed solution would fully meet the requirements; and fourthly the revision and refinement of the proposal to obtain such compliance in an optimum fashion, such as at minimum cost. A possible fifth stage, which might be called 'design as you go' or 'the observational method', following Peck's Rankine Lecture, might involve the extension of the refinement phase to include construction control to ensure continuing compliance so that the goal is achieved in the face of all uncertainties and difficulties.

Not all ways of writing performance specifications, or of performing soil mechanics analyses, are equally useful in the creation of a geotechnical design. In particular, it will be suggested that the final phases of refinement and control have often been neglected in engineering—in teaching, research and practice—and that undue emphasis has been placed on analytical techniques which are too inflexible to be of much assistance in decision-making. Prediction (*pace* the organizing committee) is not enough.

Performance requirements
Many attributes of a constructed facility—size, function, appearance—are obvious or fall outside the focus of the geotechnical engineer. An objective statement of the geotechnical problem can be provided by a list of performance criteria. Such criteria can best be expressed in terms of the avoidance of any *critical event* comprising the activation of any *limit*

DESIGN METHODS

mode in any of a specified set of *design situations*; e.g. Bolton (1991). The limit modes must cover the various independent mechanisms by which the facility could come to grief, while a small number of design situations are intended to be sufficiently onerous to encompass all the combinations of loads, environmental agencies and soil-structure parameters which could occur during construction, normal working, and any foreseeable class of accident. Checks must be made of each limit mode in each design situation, unless some of these events can be seen to be inherently less critical than others. These dual definitions clarify the otherwise intractable problem of attempting to forestall every conceivable *limit state*, formed as a conjunction of every possible variable in some complex probability space.

Modern 'limit state design' methods, as exemplified by the Eurocodes currently being drafted by CEN, incorporate distinct criteria for safety and for distortion expressed respectively in terms of the prevention of 'ultimate' and 'serviceability' limit states. Ultimate (or, more descriptively, collapse) limit states endanger people; unserviceability implies excessive distortion which compromises the efficiency or economy of the facility. Bolton (1989) proposes that just five independent limit modes be recognized in geotechnical design:

(a) unserviceability arising through soil strain, e.g. differential settlement of a bridge abutment causing jamming of the deck bearings
(b) unserviceability arising through structural deformation, e.g. compaction of fill causing cracking in the concrete at the base of a retaining wall
(c) collapse arising through soil failure, e.g. catastrophic translation or rotation of a retaining wall as a monolith
(d) collapse arising through structural failure, e.g. collapse of a sheet-pile wall following progressive rupture of the anchors
(e) collapse of a structure arising without soil failure, e.g. differential settlement of a bridge abutment causing the deck to fall off its ledge.

Various design situations may need to be checked. Geometries, loads, water tables, and soil conditions, may each be a function of circumstances; consider over-dig during excavation in the construction phase of a retaining wall, the effects of abnormally heavy loads on the back-fill in service, or the raising of water levels due to the accidental bursting of a main. Clear statements of the standard combinations of loads and ground conditions have to be devised. Society may demand more severe situations to be checked for collapse than those used to evaluate serviceability. Designers may be asked to select 'worst credible values' of loads and resistances in collapse checks, but be permitted to employ less severe assumptions in serviceability checks if there is a safe and economic remedy, should the worst happen.

Pessimism over resistances should generally lead engineers to use critical state soil strengths rather than peak strengths in collapse analyses. Even smaller material resistances, mobilizable at small strains, may have to be used in serviceability checks. For example, lateral earth pressure coefficients used in structural serviceability checks of retaining walls must logically be greater than the active pressure coefficients used to check against soil collapse, since the strength of the soil will not yet have been fully mobilized. A medium–dense granular fill behind a retaining wall might be capable of mobilizing a peak angle of shearing of 42° ($K_a = 0.20$), a critical state angle at large strains of 35° ($K_a = 0.27$), and a value consistent with acceptable displacements in an adjacent structure of only $\phi_{mob} = 33°$ ($K_a = 0.29$). Many engineers checking for cracking of a reinforced concrete retaining wall would use the peak angle to determine coefficient $K_0 = 0.33$, even if they made no other allowance for locked-in stress due to compaction, which would be equivalent to mobilizing the even more conservative value $\phi_{mob} = 30°$. It is therefore likely that the critical calculations for the integrity of structures are those to limit deformation and cracking in serviceability limit states.

Synthesis
Before any evaluation can be made, there must be something to evaluate. At any level of detail, the designer should perceive options and be prepared to compare their positive and negative attributes. Taking a geotechnical perspective, a road in cutting could be created with in situ techniques (diaphragm or bored-pile walls, steel sheet-pile walls, soil nailing system), or with a variety of back-filled structures constructed in front of temporary earth supports, or simply by designing stable side slopes using drains as required. An experienced engineer would have developed prejudices regarding many of the alternatives, and might make a selection based on a qualitative assessment of the local circumstances. If the decision on structural form is effectively final, any subsequent calculations will serve mainly as a check on reliability. If the designer is inexperienced, or the task is unusual, the detailed evaluation of a number of alternatives may be thought necessary to find the optimum.

If new technology becomes available, such as soil nailing or reinforcement, some additional incentive may be necessary before designers are prepared to educate themselves in its use. The relative infrequency of novel solutions in the UK compared with other countries suggests that such incentives are absent. Fee competition can hardly assist, since it draws attention away from the cost-effectiveness of the completed facility. Perhaps the current enthusiasm for design-build contracts is

DESIGN METHODS

explained by the recognition by some clients that there needs to be an incentive to design efficiently.

The original synthesis of a complete proposal is clearly the sort of creative act classed by the dictionary as design, but any such conception then requires evaluation, revision and refinement if it is to attain specified performance targets. The writers of codes and standards tend to ignore the decision-making which leads to an original synthesis, and to describe as a 'design method' the system of check calculations to which the proposal is then subjected. This can, perhaps, be traced from the schism between architects who propose and structural engineers who evaluate. The geotechnical designer needs to do both, and must therefore be capable of imagining working models of reality rather than simply a balance sheet of calculations.

Evaluation by geo-structural analysis
The evaluation of the effects of forces on any deformable body is carried out with respect to three conditions: equilibrium, compatibility, and the material's stress–strain relations. Most soil mechanics evaluations are based on plasticity, where sufficient strain is invoked to permit every point within the zone of plastic deformation to develop its ultimate strength. This leaves the engineer to estimate an appropriate value for soil strength and imposed load, and then to solve for equilibrium alone. There are two fundamentally different methods of performing the calculations which offer upper and lower bounds to the theoretical collapse loads derived on the assumption that soil is ideally plastic and that the engineer's estimate of its strength is correct. Since they both rely on the structural engineers' approach of discretizing the continuum according to the function of the parts, and of assuming that the elements behave in certain simple ways, it may be appropriate to call this approach geo-structural analysis.

An upper bound to the collapse load is provided by the kinematic method in which a collapse mechanism is assumed. Either global equilibrium, or an equivalent balance between work, potential energy and plastic dissipation, then guarantees that the soil body will find a way of collapsing at or before the calculated value of applied load. A lower bound to the collapse load is provided by the statical method in which equilibrium is shown to be satisfied at every point within the soil body without having to mobilize more than the plastic strength available at that point. These two approaches can best be appreciated if their results are expressed in terms of the plastic strength which needs to be mobilized in the soil body to carry the weights and external forces demanded by the design situation. Kinematic and statical methods then provide lower and upper bounds to the required strength, respectively.

A great deal of importance is often attached to the question of whether some plastic calculation method invokes a kinematically admissible mechanism which satisfies the plastic compatibility rules at every point, or a statically admissible stress field which can be extended to infinity or beyond the region of any possible collapse mechanism. Failure to observe these strict conditions leads to the method being described as a limit equilibrium analysis, the bounding nature of which is unknown. Of course, if the method is based on observed mechanisms of soil behaviour, the scope for error should be small.

A much more significant question concerns the possibility of progressive failure, and the status of peak versus ultimate strength. Unless there is strong evidence to the contrary, the collapse design situation must be assumed to be one in which all dilatancy has been progressively eliminated within fully softened shear zones mobilizing critical state strengths. Although residual strengths on polished slip surfaces in pure clays are known to be even smaller than the critical state strengths of random aggregates, there is no evidence that intact soil bodies suffer deterioration which reduces their average angle of shearing below that of critical states, especially if additional steps are taken to limit strains to the pre-rupture regime.

It has recently been pointed out by Bolton and Sun (1991) that a combination of kinematic and statical methods can offer the designer a simple way of assuring against unacceptable deformations. The mobilization of soil strength is typified by plastic hardening along a loading curve, with much increased stiffness observed on unloading–reloading cycles. A similar situation obtains for annealed copper. If, in either case, the objective is to design structures which will be serviceable in that they distort with no more than about 1% shear strain (say) at any point, a plastic design can be carried out with the strength at the '1% proof shear strength' rather than at either the peak or ultimate shear strength. The usual plastic equilibrium equation will relate loads to required shear strengths (bearing capacity factors, stability numbers, etc.). The hardening curve will permit the mean shear strain to be deduced from the mobilized strength. The ratio between normalized surface displacement (settlement/width, translation/height, etc.) and mean mobilized shear strain can then be found from some appropriate plastic deformation mechanism which satisfies the kinematics; see Fig. 1.

Taking this view, it can be shown that mobilization factors—reduction factors on peak strength—of about 1.5 on the undrained shear strength, or about 1.2 on the drained shear strength, are necessary to reduce shear strains in moderate to strong soils to within about 1% and typical displacement ratios to less than 1%, which might generally be regarded as on the limits of unserviceability. The evaluation of the safety and serviceability of soil bodies can then be treated in a unified fashion using simple plastic analysis with the soil strength set at the lower of two values: the ultimate strength at large strains, and the strength mobilizable at permissible strains.

It must, of course, be recalled that the undrained shear phase of fine-grained soils will be followed by a transient flow phase in which the soil will consolidate (or swell) as positive (or negative) excess pore pressures dissipate as the ground comes into hydrostatic equilibrium or adopts a regime of steady seepage between the boundaries. It would be consistent with the simple plastic calculations described here to analyse the final, drained, soil condition in terms of the development ab initio of drained strength with shear strain, if such stress paths had been

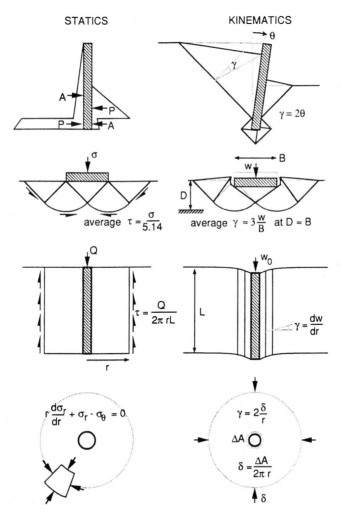

Fig. 1. Geo-structural mechanisms

investigated in element tests. Equally, an approximation of one-dimensional compression might be used for the transient flow phase, so that the increment of consolidation settlement could be calculated if the pore pressures at the end of the undrained phase could be estimated, and appropriate oedometer tests had been carried out. Centrifuge tests have proved valuable in testing simplified procedures of this sort, but more needs to be done.

Evaluation by continuum mechanics
It is usually better in principle to attempt a solution of the equilibrium, compatibility and constitutive equations at every point within the body, rather than rely on local mechanisms, of whatever sort. Algebraic solutions are possible to a restricted range of boundary value problems for ideal elastic material, and well documented in Poulos and Davis (1974), for example. The real problem is that soil is very non-linear even when it exhibits recoverable strains in executing a hysteresis loop on a stress–strain diagram. On taking the soil state beyond previous limits, irrecoverable plastic deformations occur. In performing such tests in other than a monotonic fashion some generalization of stress states is called for. The five-parameter Cam clay model, for example, is capable of representing isotropic hardening with elastic unloading and reloading inside a nest of yield surfaces. Even such a simple soil model can be used effectively only in the context of a numerical solution, probably using a finite element discretization of the continuum.

Failure by shear localization in heavily over-consolidated soil, the shaping of yield surfaces to account for inherent anisotropy, the introduction of hysteresis to replace elasticity, and the additional production of permanent plastic strains due to cyclic stress changes are essential extra components in some investigations. These greatly increase the numbers of parameters necessary to specify the pertinent model, and still leave some aspects of soil behaviour undescribed. Whittle and Aubeny, for example, in their paper 'The effects of installation disturbance on interpretation of in-situ tests in clays', describe the use of the 15-parameter MIT-E3 model which is capable of representing some, but not all, of these additional facets.

There is growing anxiety about the increasing number of parameters required to specify soil models. On philosophical grounds, Popper suggests that the best scientific theories generate the greatest number of testable propositions based on the smallest quantity of input data. On practical grounds, the designer will wish to limit the expenditure on soil testing to that which is truly necessary to guarantee economy and reliability. Note, however, that the number of model parameters is not the most appropriate measure of the complexity of input data. It is the diversity of soil tests demanded by the soil model, and the unambiguity of the means of fitting parameter values to this data, which is relevant. What should be sought is the soil model which makes maximum use of the information available from standard tests carried out on test paths which relate to the proposed facility.

The first essential feature of the Cam clay family of soil models is their capacity to scale automatically for different preconsolidation pressures, using data of plastic compression on a plot of v versus p'. Their additional capacity to relate gross plastic behaviour at different over-consolidation ratios—through boundary surfaces particular to the variant—can easily be calibrated by performing shear tests at a range of initial OCRs. The remaining feature of stiffness inside the boundary surface can be fixed in a number of ways, as long as sufficient basic data of unloading and reloading has been acquired. The inevitable hysteresis can either be modelled through piecemeal alterations of element stiffnesses in incremental analyses, taking cognizance of strain history, strain path, and stress level, or by a more thorough-going 'constitutive relation' fitted at the outset to the same data.

Refinement
If it is accepted, following Wroth and Houlsby (1985), that the efficient selection of continuum soil models is to be based not on their capacity to predict everything but their performance on the particular behaviour which is thought relevant, it might follow that the alternative geo-structural approach to evaluation will be even more efficient. If the mechanism is properly selected, the types of soil test (compression, extension, simple shear, etc.) which must be conducted on soils representative of different zones will immediately be apparent. The outcome will be in terms of sketches of soil movement patterns, simple equations between dimensionless groups, and corresponding charts. The consequences of parameter variations (i.e. uncertainties) and design modifications (i.e. revisions and refinements of the first guess) will be immediately obvious. To characterize the choice as between 'design charts' and 'doing the analysis properly' would severely mis-represent the situation, therefore.

Consider as an example the question of designing an embankment over a stratum of soft clay. The engineer should have in mind the following options:

(a) to spread the embankment with slopes and, if warranted, berms
(b) to use geotextiles (or other reinforcement) within the fill to retain vertical, or steep, faces
(c) to reinforce the base of the embankment to resist spreading
(d) to build the embankment in stages to capitalize on consolidation
(e) to use vertical drains in the clay to accelerate consolidation.

Whatever synthesis the designer first chooses will be arbitrary, and unlikely to reflect the optimum solution which will depend on relative land values, the urgency of completion, the tolerance of settlement in service, and aesthetic considerations, in addition to the more obvious matter of construction cost. What is wanted is a quick method of assessing the viability and effectiveness of each possible combination. Continuum analysis is not relevant at this stage: what is wanted is a structural engineer's mechanistic perception of the costs and benefits of the various components listed above. The author has been working with Dr Wing Sun to derive and validate plastic mechanisms applicable to this class of problem, using both statical and kinematic solutions based on the method of characteristics.

Sun (1990) showed that, whether plastic mechanisms or finite element solutions were applied to the analysis of these situations tested in centrifuge models, careful account had to be taken of strain history and future strain path in determining appropriate stiffnesses 'inside' the yield surface. Geo-structural calculations based on the same data gave answers that were just as reliable as those of the FE analysis. Extensions into the drained phase of behaviour based simply on Terzaghi's 1D consolidation theory were somewhat *more* successful than FE predictions. Modified Cam clay assumes isotropic elastic soil behaviour inside the yield surface which generates volumetric soil consolidation in the absence of shear strain: consolidation settlement vectors point to the region of greatest consolidation, inwards, beneath the embankment. The centrifuge models, and real embankments of course, continue to shear the subsoil as it consolidates, due to the influence of kinematic plasticity. The net effect can be that displacements during consolidation are vertical, or inclined outwards as with the earlier shear displacements. It is sometimes easier to guess an appropriate mechanism than the soil properties which are necessary to induce it.

Objective and subjective design methods

Within living memory geotechnical design was based on subjective judgements. A client would request a facility, and the consulting engineer would recall recent solutions for similar problems, and select a candidate solution on the basis of pattern recognition. A single bad experience with a particular material or technique would effectively banish it from use, irrespective of the reasons for its failure. A costly technique which was easy to design and which 'worked' in practice could soon be copied by other designers, become a standard solution written in to Codes of Practice, and thereby hold off any challenge from more efficient, novel solutions. Even when soil mechanics calculations came into use, the specified 'factors of safety' were ambiguously couched in terms of parameters which were not clearly defined, tending

DESIGN METHODS

to obscure the fact that any such factor could be shown, at whim, to be either satisfactory or unsatisfactory depending on the assumptions. The evaluation process has largely remained normative, therefore, rather than objective.

Structural design broke free of conventional thinking about a generation ago when the analysis of structures could account properly for the interconnection of members. Not only could the elastic analysis of frames be carried out, their eventual plastic collapse could also be predicted. Although stability and buckling problems can hasten flexural plastic collapse, and membrane or dome action can delay it, designers were able to devise geometrical rules based on simplified treatments of these effects, which were successful in enabling quick decisions to be taken during the elaboration of structural form. At the same time, finite element analyses had been developed which permitted an analysis to be performed of the proposed structure making far fewer assumptions regarding its behaviour and the interactions between its components. Large-scale physical tests, following the Merrison Report on box-girder bridge failures for example, were used to validate the methods of analysis.

It is now possible to design structures objectively to meet their performance targets. Such targets have to be specified, of course. Whilst 'collapse' ought to be unambiguous, the criterion of 'unserviceability' requires additional information. Burland and Wroth (1974) introduced simple mechanistic models of superstructure behaviour to define and set limits for the tolerable deformation of foundations. They treated whole buildings as being capable of characterization as beams deforming either in shear or flexure depending on their structural type, linked foundation distortion to tensile strain in the equivalent beam, and reported tests associating average tensile strain with the onset of visible cracking in panels, for example. This procedure for setting serviceability criteria for foundations embodies the scientific approach to design. Unserviceability is defined in terms of lack of fitness for purpose; unfitness is defined as including visible cracks in partitions; cracking is correlated with foundation distortion through a structural model. The procedure is objective because the terms are unambiguous, and it is scientific because every step of the process is open to challenge and reinterpretation.

It now seems that the techniques of Burland and Wroth can be carried below the superstructure into the foundations. The soil continuum itself can be declared to consist of geo-structural components. Experience shows that foundation distortion can be linked with soil strains in significant soil zones, and that the limitation of these strains can be achieved through the limitation of soil stresses to those mobilized at the limiting strain magnitude in appropriate stress–strain tests. Although much remains to be done, the technique of centrifuge modelling has been shown to provide—at moderate cost—the essential scientific feedback by which deformation mechanisms can be justified.

Figure 2 illustrates a procedure for geotechnical design which would permit creativity full rein while preserving an objectively scientific approach. The heart of the design method is the cyclic review of alternative schemes, and the successive modification of each scheme currently under consideration, leading to the proposal of a particular scheme with a particular configuration. This 'final' scheme can be subjected to a rigorous finite element analysis if desired, but it will be earnestly hoped at that stage that no flaws will be found, and that no fundamental changes will be made. Of course, the final construction will

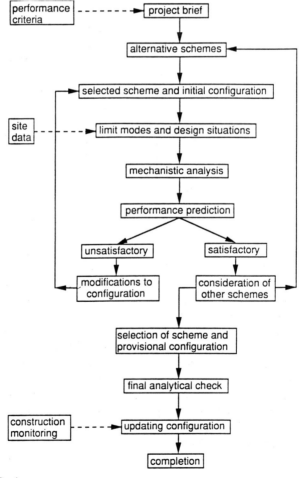

Fig. 2. Design process

DESIGN METHODS

not exactly follow the final design, and construction monitoring may be used in Peck's sense to 'design as you go'.

At the centre of the creative cycle in Fig. 2 is the geo-structural analysis of limit modes in design situations. It is suggested that this can best be fulfilled by geo-structural mechanisms based on plastic methods, as shown in Fig. 1. These can be framed to be sufficiently simple to enable comparisons of different techniques and geometries, descriptive enough to enable spatial problems of interaction to be recognized or avoided, and close enough to the truth to guarantee that any 'final' analysis will be unlikely to overturn their conclusions.

Examples from the symposium
Earth retention and soil reinforcement

Jewell, Burd and Milligan raise some important issues of design, analysis, and scientific evaluation in their paper 'Predicting the effect of boundary forces on the behaviour of reinforced soil walls'. Their contribution is based on their earlier work which established an understanding of the action of soil reinforcement through the use of limit equilibrium analyses. Clear mechanisms emerge from these studies, derived via the equilibrium of wedges, and enhanced by introducing consistent arguments related to deformations and kinematics. The pictorial presentation of reinforcement action, and the exposition of results in terms of dimensionless groups of parameters, fulfils exactly the conditions laid down above for the type of geo-structural mechanism which is capable of being used creatively in design, whilst remaining open to further scrutiny.

The prediction of wall displacements and reinforcement extensions is under review at a number of research centres. In order to test the success of such predictions, Bathurst conducted large-scale tests at the Royal Military College, Canada. The reinforcement tensions, and consequent wall displacements, were much smaller than had been anticipated. Subsequent investigations showed that this was due, in the main, to side-wall friction influencing the results of the 3 m high and 2.4 m wide wall. Jewell et al. produce a variant of their limit equilibrium analysis to take account of boundary forces, following an analysis by Bransby and Smith which was developed 17 years ago in identical circumstances, to salvage the data of reinforced earth models tested in a narrow chamber with insufficient side-wall lubrication. Jewell et al. confirm that their mechanism is capable of explaining the results of the RMC trials, by including the bearing thrust on the bottom wall panel, and friction on the sides of the test section.

Boundary friction will continue to challenge experimental workers, and 3D effects will continue to puzzle geotechnical designers. Jewell et al. point out, however, that it may well be inconsistent to include in design calculations each source of strength monitored in pilot tests. The design situation would have to assume minimal bearing capacity under the front panels unless particular trouble had been taken to assure its presence, and the effect of side-wall friction could not be relied upon unless the relative immobility of the side walls could be guaranteed. Even more uncertainty currently exists regarding the advisability of adopting the peak angle of shearing in well-compacted backfill (over 50°, probably, in plane strain), as opposed to fully dilated critical state state angles (perhaps 35° to 38°). The opportunity exists for research workers to demonstrate—if it is the case—that some reinforced fills do not strain enough to soften, and for engineers in the field to investigate whether a high degree of compaction in fills containing relatively extensible geotextiles is either possible or desirable. Until that is clarified, it is necessary for designers to take the pessimistic view—the worst credible scenario—consistent with the precepts of limit state design.

Excavations and tunnels

Sloan and Assadi describe an attractive numerical package for plastic analysis in their paper 'Stability of shallow tunnels in soft ground'. They have developed two finite element techniques which satisfy, separately, the conditions for statical and kinematic solutions to bound the equilibrium of perfectly plastic bodies—here taken to be undrained clay with any linear variation of strength with depth. In each case, the closest bound obtainable within a specified family of mechanisms is found through linear programming optimization. Statical solutions search through equilibrium solutions which can satisfy body forces and include any permissible stress discontinuities at the boundaries between elements. Kinematic solutions search through mechanisms which involve uniform strain within elements and compatible displacement discontinuities on their boundaries.

These solutions are therefore automated forms of the geo-structural analysis referred to earlier, and with sufficient degrees of freedom to be capable of approaching the 'correct' answer—as the authors demonstrate in the case of plane strain cavity collapse. Although tight bounds had already been produced manually by Davis et al. for the case of constant strength with depth, Sloan and Assadi show that efficient solutions can be generated for strength varying with depth. Bounds relative to their mean value were generally ±5% and not worse than ±10%, based on the ratio of the strength required in the soil at the depth of the cavity centre, and the vertical stress deficit at the cavity centre (i.e. the stress in the free field minus the cavity pressure). Although there are many dimensionless ways of presenting stability data, the key parameter is always the strength required for equilibrium divided by the stress

DESIGN METHODS

difference causing distortion. These values, written perhaps as a mobilization ratio $c_{mob}/\Delta\sigma$ and calculated at collapse using Sloan and Assadi, fell between 0.6 (for a cavity with cover/diameter ratio 1, in soil which lost half its strength between the crown and the ground surface) and 0.2 (for a cavity with cover/diameter ratio 5, in uniform soil), approximately.

Mair and Taylor in their paper 'Predictions of clay behaviour around tunnels' show that there are significant advantages in the medium strain range in modelling the soil movements around tunnels in terms of spherical cavity collapse into the approaching heading, followed by cylindrical movements onto the tunnel supports once they are in place. This geometrical understanding is enhanced by relating the magnitude of deformation to the stability number $\Delta\sigma/c_u$, using cavity expansion theory. The result is an admirable physical clarification, embodied in dimensionless charts of meaningful parameters, of the soil movements due to tunnelling. The authors use a linear–elastic perfectly plastic relation for the soil, but remark on its inability to characterize some significant elements of the data they present. This aspect of the problem is carried forward in another paper to the Symposium.

Mair et al. (1981) presented their results for 'ground loss' above shallow tunnels in terms of a 'load factor' which could be interpreted as c_{mob}/c_u. Gunn recalls Mair's centrifuge test data, and earlier finite element analyses based on Cam clay with simple κ-behaviour inside the yield surface, in his paper 'The prediction of surface settlement profiles due to tunnelling'. Gunn shows, as Bolton and Sun (1991) also remarked, that power curves fit the medium strain range of the shear stress-strain data of clays rather well. He uses an FE analysis with a soil model based on Tresca with a power law hardening curve to demonstrate that higher small-strain stiffness causes the soil deformations to localize, in better agreement with the physical evidence. The magnitude of 'ground loss'—proportional loss of cavity volume, expressed as a subsidence trough—corresponded with the earlier findings, namely that between about 1.5% and 3% ground loss might be expected at $c_{mob}/c_u = \frac{2}{3}$.

This finding conforms with the spirit of the geo-structural approach described earlier. The stress-strain function which Gunn used gave $c_{mob}/c_u = \frac{2}{3}$ at an 'axial strain' of about 0.75% which presumably relates to a shear strain in plane strain of 1.5%. The simplest geo-structural mechanism which may convey something of the kinematics, at least for deep cavities, is simple cylindrical contraction for which proportional volume change equates to shear strain mobilized at the cavity boundary. Crudely, therefore, a 1.5% shear strain in the soil near the cavity boundary might have been expected to relate to 1.5% ground loss. If it is generally essential to mobilize no more than $\frac{2}{3}c_u$, that is to have a

63

BOLTON

mobilization factor of at least 1.5, to limit ground deformations in undrained clay, Sloan's ±10% uncertainty regarding the bounding of the 'correct' value of c_{mob} is seen to be of negligible practical importance.

Hight et al. in their paper 'Predicted and measured tunnel distortions associated with construction of Waterloo International Terminal' were clearly commissioned, in the terminology of Fig. 2, to perform a 'final analytical check' rather than a recursive design routine. Nevertheless, it is interesting to compare their computations and observations with the estimates available through geo-structural thinking. The objective was to determine the changes in tunnel diameter on the Bakerloo line due to an average 180 kPa of unloading due to excavation above. This is a reversed bearing capacity mechanism with a bearing capacity factor of about 6, so the average change in mobilized shear strength must be about 30 kPa. The authors went to immense trouble to establish the in situ stresses ($K_0 \approx 1$ near the tunnels) and the small strain stiffnesses of the soils respecting their strain history – the essential input to any method of analysis.

The tunnels in question were close to the weathered/unweathered boundary for the London clay. Since the data was presented (inconveniently for these purposes) as normalized secant stiffness versus log strain, rather than normalized shear stress versus log strain, it was necessary to back-calculate a few points to find the strain required to mobilize 30 kPa. For an initial effective stress $p'_0 \approx 190$ kPa, the authors' data imply that a shear stress of 30 kPa can be mobilized at triaxial extensions of about 0.08% in the weathered clay and 0.22% in the unweathered clay: they remark on the strange drop of apparent stiffness with depth. However, a tunnel lying in the passive unloading zone will tend to take up the vertical extension of the soil surrounding it, so the increase in vertical diameter of a 5 m diameter tunnel might have been guessed, without further analysis, to lie in the range 4–11 mm, depending which value of extension was selected. The observed range of extensions was apparently 5–8 mm. This simple geo-structural analysis confirms and explains the overwhelming importance of stress–strain data in the first 0.25% of strain following some new excursion. In critical cases, especially of soil–structure interaction, a non-linear finite element solution respecting measured soil hysteresis will be indispensable.

This was demonstrated in the paper 'Prediction and performance of ground response due to construction of a deep basement at 60 Victoria Embankment' by St John et al. The authors were able to create a numerical simulation of the excavation process, including intermittent changes in the conditions of fixity of the secant-pile wall as floor slabs were cast. In cases such as this, where possible damage to neighbouring property was a concern, strain history effects in the soil had to be

DESIGN METHODS

accounted for in the stress–strain response anticipated in active and passive zones.

The authors conclude by showing that maximum wall displacements induced during the construction of various excavations in London clay: 30 mm due to 10 m of excavation was typical. Following the geostructural approach of Bolton et al. (1989, 1990) the lateral displacement should be divided by the effective depth of the wall in order to obtain an estimate of the principal strain in a plane element test on the surrounding soils. Although the depth of excavation, rather than the depth of the wall, was quoted, it seems likely that the induced principal strain is about 30 mm on 20 m, or about 0.15%. It is the soil strength which can be mobilized following construction of the wall, and after a further 0.15% principal strain, which could be entered into preliminary design calculations. The authors do not comment on whether this represents a carefully considered serviceability limit or whether, in some cases, more displacement could have been tolerated and a thinner wall or cheaper support system could have been provided to support soil which was mobilizing a greater proportion of its own strength.

Shallow foundations

Two papers on spud foundations for jack-up rigs reflect precisely this reporter's preference for self-contained approximate mechanisms of behaviour to clarify the task of design. Dean et al. present 'The bearing capacity of conical footings on sand in relation to the behaviour of spudcan footings of jack-ups', while Houlsby and Martin offer 'Modelling the behaviour of foundation jack-up units on clay'. As a conical footing penetrates a soil bed it continuously shears the soil through a succession of self-similar states scaled by the instantaneous depth of penetration which is used as a sort of hardening parameter for the soil-structure system. Both papers offer an interpretation of the plastic yielding of the soil bodies beneath the spuds in terms of an interaction between the vertical force, horizontal force, and moment applied by the structure. Both draw attention to the similarity of the emerging system model to the critical state material models, with the shapes of yield surface defined empirically rather than theoretically.

Houlsby and Martin aim to determine the penetration of a single spud in soft clay, first under surcharged vertical load on placement, then on recovery to working load, and then under storm conditions with shear and moment interactions. The self-consistency of interaction envelopes was demonstrated at 1 g in very soft clay models subject to strain-controlled loading, with control of the three degrees of freedom of the spud. They point to the greater uncertainty on the 'dry' side where, just as with material models, sliding and cracking rather than plastic deformation must be expected. Here, under a lifting leg, the resilience of the clay is significant and the authors may like to consider the use of power curves fitted to hysteresis loops rather than the identification of an equivalent-linear shear modulus.

Dean et al. consider the behaviour of a three-legged jack-up unit on sand. They develop plastic yield envelopes as described above, and also develop possible paths in load space through a simple structural analysis. This simple geo-structural idealization was validated through a sequence of ingenious model tests carried out in a drum centrifuge. The normalization of vertical force with penetration depth was achieved through an apparently constant N_γ value corresponding to quite low values of ϕ, at or below critical state values. Only 34° was mobilized in fine, medium dense Leighton Buzzard sand, for example. It is interesting to speculate on whether higher ϕ values would be mobilized in dense sand, or whether shear strains during placement are effective in causing dilation to critical states. The authors may need to consider the influence on ϕ of dilation or contraction at different levels of mean effective stress, especially since the stress under the 1.6 m diameter prototype spuds reported here are an order of magnitude smaller than those applicable in the much larger field-scale structures. The installation behaviour would, in any event, permit the re-evaluation of storm effects using the interaction diagrams presented here. The possibility of taking commercially valuable decisions regarding the moment-fixity of spuds, and therefore of the working envelope of particular rigs, is a good example of the virtue of simplified, wholistic, simulations for design and operation.

Schnaid et al. in their paper 'An investigation of bearing capacity and settlements of soft clay deposits at Shellhaven' provide an example of 'design as you go' in planning a haul route for an exceptional load. The success of the exercise demonstrates the value of careful investigation, testing and observation. A geo-structural analysis might have been performed with the following effect. The measured bearing capacity was 84 kPa from a load test, but this was reduced to 80 kPa as a correction for shape effect. However, the settlement at failure was of the order of 0.25 m which must have created a relative depression of about 0.3 m, giving a buoyancy effect of perhaps 4.5 kPa, so the true bearing capacity might have been about 75.5 kPa. The bearing pressure finally adopted was 37 kPa giving a mobilization factor of 2.04 against collapse. A triaxial sample with a compressive strength of 29.5 kPa mobilized a deviatoric stress of $29.5/2.04 = 14.5$ kPa at an axial strain of about 0.19% in a triaxial test, corresponding to a shear strain of $1.5 \times 0.19\%$ or $\gamma = 0.285\%$ which would have occurred in a plane test at a principal strain of 0.14%. Following Bolton and Sun (1991), the settlement of a 6 m wide strip on 5–7 m of the soft clay would have been predicted to be

DESIGN METHODS

simply 0.14% of 5–7 m, or 7–10 mm. The average settlement during passage was inferred to be 8–10 mm under the edge of the load.

Deep foundations
Mandolini and Viggiani present a paper 'Settlement predictions for piled foundations from loading tests on single piles'. Their objective is to demonstrate that linear elastic interaction coefficients can be used with an empirical correction to predict settlements of flexible piled rafts from single-pile tests. It becomes clear from their data, however, that the stress–strain data of the sands through which the piles penetrated must have been highly non-linear. Suppose, for example, that the vertical shear stress could be represented by a power law

$$\frac{\tau}{\tau_m} = \left(\frac{\gamma}{\gamma_m}\right)^b$$

Now let the shear strain in the soil be γ_0 at the pile interface radius r_0, reducing to γ at radius r. Vertical force balance means that $2\pi r \tau$ must remain constant at any radius, neglecting end effects. Equating the drag-down slope (dw/dr) to the shear strain γ, and integrating, it can then be shown that the settlement of a single friction pile is

$$w_0 = \gamma_0 r_0$$

while the drag-down at radius r can be written

$$\alpha = \frac{w}{w_0} = \left(\frac{r}{r_0}\right)^{1-1/b}$$

and the development of settlement of the pile itself mimics the original power law

$$\frac{Q}{Q_m} = \left(\frac{w}{w_m}\right)^b$$

where Q_m and w_m form a point on the load–settlement curve representing ultimate failure. Figure 3(a) shows the comparison between the expression for interaction factor α and the authors' results from dual-pile tests, and Fig. 3(b) shows the single-pile settlement curve, fitted for exponent b = 0.565. The fit is excellent. This shows that the authors' intention to extrapolate from single pile tests may be possible, but that the analysis should be non-linear. This in turn means that more thought should be given to integrating effects over a number of loaded piles, since superposition will not apply. The authors are, however, quite right to say that deformations are much more highly localized around piles

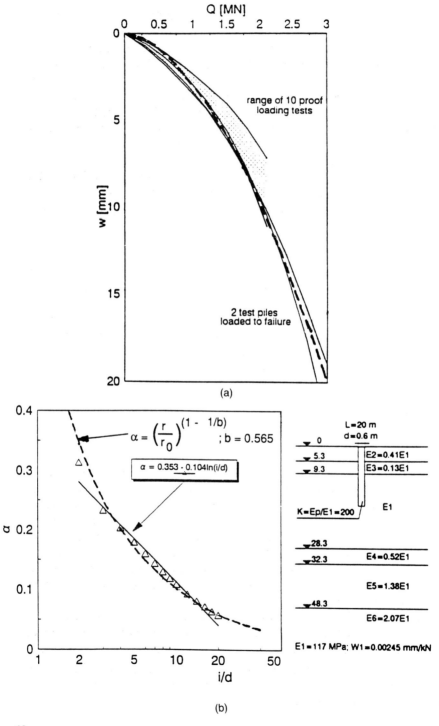

DESIGN METHODS

than linear–elastic theory suggests, so the integration need not involve piles at large separations.

Lehane and Jardine show data of pile installation by jacking, and later loading behaviour, in their paper 'The behaviour of a displacement pile in Bothkennar clay'. The skin friction was found to depend on residual friction within a superficial zone of disturbance within which anomalous pore pressures must have developed during jacking. Excellent data of transient radial stress and pore pressure at the pile, at various points along its length, presently defy simple interpretation. The effective radial stress is difficult to predict.

A similar observation is made by Fahey et al. in their paper 'Parameter selection for pile design in calcareous sediments'. They show a complex behaviour in terms of local dilation or contraction at the pile interface being offset by the response of the surrounding soil in cavity expansion or contraction. Constant normal stiffness tests in direct shear provided an analogous interaction, with the pile material sprung against the soil in the laboratory. Once again, the highly non-linear stress–strain relation of the soil at small shear strains proved essential in understanding the results. The higher the shear modulus of the soil, the faster the radial effective stress can be lost due to contraction. There is clearly some hope that extrapolations of pile field tests, and the use of CNS or other novel laboratory tests, will soon prove adequate for the more accurate and economic design of friction piles.

Conclusions

Far more than a method of prediction, the designer needs a descriptive mechanism which embodies the working of any geotechnical facility. The mechanism will need to represent the equilibrium and displacement of the various elements of the facility, especially at their boundaries. Overall function is much more important than local detail: in that sense, the mechanism may be like the structural engineer's idealization. Engineers' beam theory fails to deal with stress concentrations at joints, and ignores shear deformation in the axiom 'plane sections remain plane': it is invaluable, however, in the design of buildings. The symbolism of a geo-structural model should ideally have a closed algebraic form, or at least offer an unequivocal dimensionless scaling from measurable parameters to desired output.

Error is qualitatively different than approximation. What is desired is a soft-focus perspective of the workings of the system (with optional

---

Fig. 3(a) Results of loading tests on piles, and (b) computed values of the superposition factors and subsoil model (from Mandolini and Viggiani); – – – $Q/Q_{max} = (w/w_{max})^b$; $b = 0.565$

zoom!). In that sense, the full complexity of a finite element program or a centrifuge model test simply takes the engineer one step of the way: the next step must always be to attempt to produce a simplified model of the complex reality. Geo-structural models which have proved useful include: active and passive triangles, Prandtl's bearing capacity mechanism generalized for inclined loads, and the equilibrium of a cylindrical insert expanding, contracting, or translating in an infinite medium.

The non-linear relationship between stress and strain in soil is central to a correct understanding of soil deformations and ground displacements. Soil hardens through plastic strain in a fashion not unlike that of many metals or polymers. Whereas geotechnical engineers have been used to applying plastic theory to collapse at peak strength, structural engineers perform their plastic calculations at a yield stress far below ultimate tensile strength, because they wish to limit plastic deformations and avoid progressive failure. Geotechnical engineers should perform safe analyses of collapse with fully softened critical state strengths. They should, in this reporter's view, use elementary plastic calculations to limit deformations to the medium strain range (<1%, perhaps).

It is well known that significant non-linear behaviour is exhibited in all soils beyond about 0.01% strain, so linear elastic calculations are difficult to apply. What value of modulus should be taken? Evidence is growing of the success of an approach treating a stress-path test, on an element representative of some soil zone, as a curve of plastic mobilization. Plastic equilibrium relates applied stresses to mobilized soil strengths. Strains are deduced from the test. They are then entered in to a plastic deformation mechanism to predict boundary displacements. The approximation is good if the mechanisms are appropriate: hence the need for physical or numerical modelling. Various levels of sophistication are called for in fitting curves to stress–strain data: a simple power curve has proved very valuable.

We correctly use simple models to teach young engineers how the world works. We can extend this thinking now into most areas of geotechnical engineering. This must lead us to provoke more engineer-like behaviour in our students by setting exercises which go beyond prediction (what is the settlement?) to refinement (what width of foundation would limit settlement to 25 mm?), and then to synthesis (do I need piles here, or not?). Research projects often emphasize scientific exploration followed by numerical prediction, and fail to follow these up with the creation of simplified idealizations—'seeing the trees as a wood'. The responses of grant-giving bodies and academic referees often show a surprising ability to miss the point. One of the first referees to look over the reporter's plastic mechanism approach to predicting the settlement of foundations in two lines, taking non-linearity into account, wrote that 'all this could be done in an FE analysis!' Perhaps that is why, in the UK especially, practising engineers have undervalued research. We need to persuade industry to support our best post-doctoral workers, those people who have done the science and the numerical analysis, while they undertake the essential task of digesting and clarifying their understanding. This must be the best way of bridging the gap between 'research' and 'design' to the mutual advantage of both cultures.

References

BOLTON, M.D. (1989). The development of codes of practice for design, Proc. 12th ICSMFE, Vol. 3, pp. 2073–2076.

BOLTON, M.D., POWRIE, W. AND SYMONS, I.F. (1989, 1990). The design of in-situ walls retaining overconsolidated clay. Part I, Ground Engineering, Vol. 22, No. 8, pp. 44–48 (1989); Vol. 22, No. 9, pp. 34–40 (1990) Part II, Ground Engineering, Vol. 23, No. 2, pp. 22–28 (1990).

BOLTON, M.D. AND SUN, H.W. (1991). Designing foundations on clay to limit immediate movements, 4th Int Conf on Ground Movements and Structures, Cardiff.

BURLAND, J.B. AND WROTH, C.P. (1974). Settlement of buildings and associated damage, Conference on the Settlement of Structures, BGS, Cambridge.

MAIR, R.J., GUNN, M.J. AND O'REILLY, M.P. (1981). Ground movements around shallow tunnels in soft clay, 10th ICSMFE, Vol. 2, pp. 323–328.

PECK, R.B. (1969). Advantages and limitations of the observational method in applied soil mechanics, 9th Rankine Lecture, Geotechnique, Vol. 19, No. 2, pp. 171–187.

POULOS, H.G. AND DAVIS, E.H. (1974). Elastic Solutions for Soil and Rock Mechanics, Wiley, New York.

SUN, H.W. (1990). Ground deformation mechanisms for soil-structure interaction, PhD Thesis, University of Cambridge.

WROTH, C.P. AND HOULSBY, G.T. (1985). Soil mechanics—property characterization and analysis procedures, 11th ICSMFE, Vol. 1, pp. 1-55.

Field, in situ and laboratory consolidation parameters of a very soft clay

M.S.S. ALMEIDA and C.A.M. FERREIRA,
COPPE, Federal University of Rio de Janeiro

Vertical drains are used to accelerate the settlement of structures. Besides the traditional sand drains, a large number of prefabricated drains are available. This paper presents the behaviour of a test embankment on the very soft Rio de Janeiro clay with vertical drains. Horizontal and vertical coefficients of consolidation obtained by back-analysis of settlement and pore pressure data are compared with those obtained by means of laboratory and in situ tests.

Introduction

A comprehensive study on the behaviour of Rio de Janeiro very soft clay was initiated almost two decades ago. The first part of this research concentrated on failure behaviour and consisted of a thorough investigation of the reference site by means of in situ and laboratory tests followed by an excavation and a trial embankment I, both taken to failure. These studies have already been extensively reported (e.g. Lacerda et al., 1977; Werneck et al., 1977; Almeida, 1982; Ortigão et al., 1983).

The second part of this research concentrated on consolidation studies by means of complementary in situ tests (e.g. Soares et al., 1987; Danziger et al., 1992) and on the analysis of the stage constructed trial embankment II on vertical drains. This heavily instrumented test embankment, under consolidation since 1981, included two sections without drains, three sections with sand drains and two sections with prefabricated drains. Despite not being much used in developed countries, in Brazil sand drains are still used owing to the wide availability of sand borrowing areas, the narrower choice of prefabricated drain types and cheap labour. A preliminary settlement analysis of the central settlement plate of each section has already been reported (Almeida et al., 1989). Analysis of the settlement variation with depth and of the pore pressure data has been finalized recently (Almeida et al., 1990; Ferreira, 1991).

This paper analyses the behaviour of the test embankment. Emphasis is given to the pore pressure data, not yet reported, and to the reanalysis of settlement data, this time considering smear effects. The main purpose of the paper is to compare coefficients of consolidation obtained by back-analyses of field data with those obtained by laboratory and in situ tests.

Clay site and test embankment

Description of the soft clay site

The test site is situated in a swampy area at the back of Guanabara Bay, near the city of Rio de Janeiro. The soft clay deposit at the test site, located close to the Sarapuí River Bridge at the BR-040 motorway, is about 10 m thick and overlies sand layers. It is estimated that the soft clay deposit is about 6000 years old.

Soil properties of Sarapuí clay are shown in Fig. 1. Profiles of Atterberg limits and water content, Fig. 1(a), show that the water content is always greater than the liquid limit and both decrease with depth. The plastic limit is around 50% and shows much less change with depth than the liquid limit. The undrained strength profile (Ortigão and Collet, 1988) determined by the vane borer equipment is shown in Fig. 1(b). Undrained strength profiles determined by other procedures are also shown in Fig. 1(b), but the vane is the only one to show the higher crust strength. The average sensitivity determined by the vane test was 4.3. Pore pressures measured when driving a piezocone (Soares et al., 1987) are shown in Fig. 1(c) and illustrate the homogeneity of the clay deposit. For soft clays driving pore pressures are better parameters to deal with than point resistances (also shown in Fig. 1(c)), as these are of

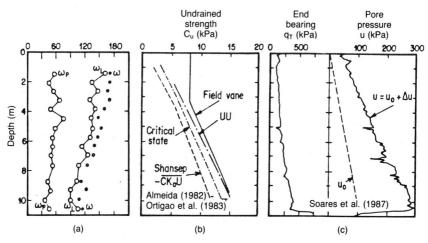

Fig. 1. Geotechnical properties of the Sarapuí clay site

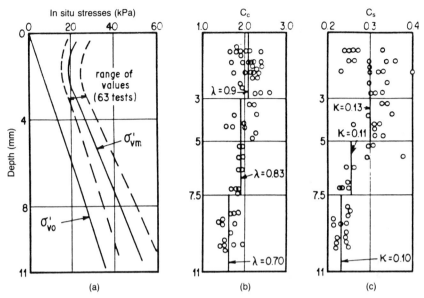

Fig. 2. Compressibility data (after Ortigão, 1980)

small magnitude. The stress history of the deposit is shown in Fig. 2(a). Yield stress (or preconsolidation stress) profiles σ'_{vm} have been obtained from high quality incremental and constant rate of strain consolidation tests (Lacerda et al., 1977; Martins et al., 1992). The in situ stress σ'_{vo} is also shown in Fig. 2(a). The water level is very close to the surface. For the crust ($z < 1.5$ m) the yield stress ratio (or OCR) is greater than 4, while below 3 m the yield stress ratio decreases with depth from 2.3 to 1.5 at the bottom of the layer. The main reasons for the overconsolidation are water level fluctuations, weathering and ageing (Almeida, 1982). Coefficients of compressibility (C_c) and swelling (C_s) are shown in Fig. 2(b,c) and illustrate the high compressibility of this clay. The fines content of the clay is 60% and the average organic content is 5%.

Determinations of c_v and c_h by laboratory and in situ tests

Laboratory studies of the consolidation behaviour of Sarapuí clay have been performed by a number of authors (e.g. Lacerda et al, 1977). Different types of radial consolidation tests were performed by Lacerda et al. (1977) in order to determine c_h values. These results are summarized in Fig. 3. Coefficients of consolidation are shown in Fig. 3(a). Values of c_v and c_h corresponding to the normally consolidated condition are about 1.2 and 2.4×10^{-8} m²/s, respectively. Figure 3(b) shows data of k_h and k_v for the normally consolidated condition, to be extensively used below. In situ constant head permeability tests carried

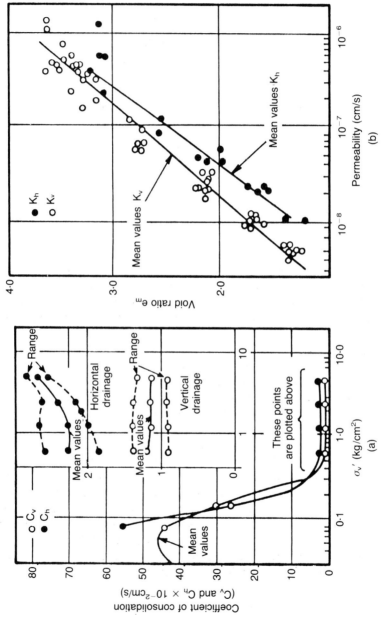

Fig. 3. Consolidation tests data (after Lacerda et al., 1977)

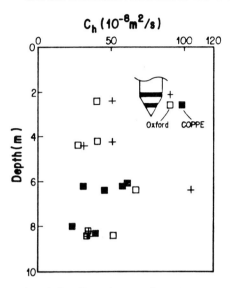

Fig. 4. Piezocone c_h data (after Danziger et al., 1992)

out in hydraulic piezometers installed at the test site (Werneck et al., 1977) yielded coefficients of permeability varying between 1.4 to 3.3×10^{-9} m/s, which are slightly lower than laboratory values. Pushed in place piezometers remould the soil and usually underestimate k values.

Values of c_h obtained from an earlier analysis (Sills et al., 1988) of piezocone dissipation tests with a four porous element probe were determined using Baligh and Levadoux's (1986) solution. However, the initial pore pressure distributions given by this solution are those of the Boston Blue clay which has a rigidity index almost ten times higher than that of Sarapuí clay.

Recent reanalysis (Danziger et al., 1992) of these tests used Houlsby and Teh's (1988) strain path solutions with a suitable rigidity index. These c_h values are summarized in Fig. 4. Excluding the single high c_h value obtained, all the others are in the range 24 to 67×10^{-8} m²/s. The above c_h data correspond to stress levels close to the in situ stress condition, i.e. to the overconsolidated state. Therefore, it is necessary to transform the c_h data from overconsolidated conditions to normally consolidated conditions. This can be done as proposed by Baligh and Levadoux (1986) using the following equation

$$c_h(nc) = \frac{C_s}{C_c} c_h(probe) \tag{1}$$

where $c_h(nc)$ and $c_h(probe)$ are the coefficients of consolidation for normally consolidated conditions and measured directly, respectively. Using the typical value $C_s/C_c = 0.13$ (Almeida, 1982) yields $c_h = 0.13 \times (24 \text{ to } 67) \times 10^{-8} \text{ m}^2/\text{s} = 3.1 \text{ to } 8.7 \times 10^{-8} \text{ m}^2/\text{s}$.

Past experience (e.g. Parry and Wroth, 1977) suggests that soil compressibility can be regarded as approximately isotropic ($m_v = m_h$), thus allowing estimation of c_v by

$$c_v = c_h \times \frac{k_v}{k_h} \tag{2}$$

Substituting the representative ratio $k_v/k_h = 0.5$ (see Fig. 3(b)) in equation (2) gives $c_v(\text{nc})$ within the range $1.6\text{--}4.4 \times 10^{-8} \text{ m}^2/\text{s}$.

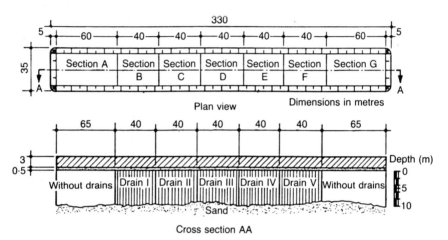

Fig. 5. Geometry of the test embankment

Table 1. Characteristics and geometry of the drains used

Section	Drains			
	Material	Type	Dimensions (mm)	Spacing† (m)
B	Sand	Closed mandrel	400	2.50
C	Sand	Open mandrel	40	2.50
D	Sand	Jetted	40	2.50
E	Polyetilen	Fibro-chemical	2.8 × 100	1.70
F	Polyester	Bidim OP-60	4.5 × 210	2.00

†Drains were installed in a square pattern

Test embankment and drain installation

The test embankment is about 300 m long and 30 m wide and consists of seven sections A to G, as shown in Fig. 5. Geotechnical studies have shown that the clay layer is quite homogeneous along the embankment length. Drain characteristics, dimensions and spacing are shown in Table 1. The fibro-chemical drain was pushed into the soil while the geotextile drain was driven.

Loading of the test embankment II was in two main stages and the final height reached was about h = 3.6 m (test embankment I, built quickly, failed at h = 2.8 m; Ortigão et al., 1983a). The first stage of loading was applied during 1981 in two steps: the first lasted about a month and reached h = 0.7 m; the second started 200 days later, lasted two to three months, and reached about h = 2.0 m. Section A was not further loaded. The second stage of loading, applied in early 1986 to sections B to G, was virtually instantaneous (about a week) and reached the final height. Embankment heights for each section and stage of loading are given in Table 2. The embankment material was a silty sand with average unit weights γ = 19.6 kN/m^3 and 15.8 kN/m^3, for the first and the second stages of loading, respectively.

The test embankment was heavily instrumented with settlement plates, magnetic settlement gauges, inclinometers and Casagrande and hydraulic piezometers.

The present paper concentrates on the analysis of the second stage of construction as the complicated loading history adopted for the first stage does not make analysis straightforward, particularly with regard to the pore pressure data. Preliminary settlement and pore pressure analyses of the first loading stage have been performed (Collet, 1985). Mid-depth pore pressures and final settlements at the end of the first

Table 2. Mid-depth excess pore pressure and surface settlements

Section	Stage I (last reading)			Stage II (last reading — 3300 days)		
	h (m)	u (kPa)	Δh (cm)	h (m)	u (kPa)	Δh (cm)
B	1.8	4.3–4.9	110	3.5	6.6–9.1	210
C	1.8	2.7–6.6	105	3.5	9.7–10.0	214
D	2.1	1.8–1.9	151	3.8	1.3–3.0	243
E	1.9	6.8–7.7	106	3.6	12.4–13.3	205
F	1.9	6.6–6.9	113	3.6	10.8–11.5	204
G	1.9	7.5–9.7	86	3.6	26.2–28.2	153

stage of loading are shown in Table 2. The first stage final applied stress is $\Delta\sigma_v \cong 30$ kPa (considering the embankment submersion). This stress increment is virtually constant with depth (as easily shown by theory of elasticity) due to the embankment width/clay depth ratio adopted (about three). Inspection of the stress history of the soil deposit (Fig. 3(a)) shows that the whole soft clay deposit becomes normally consolidated under $\Delta\sigma'_v \cong \Delta\sigma_v \cong 30$ kPa (the excess pore pressure is greatly dissipated, as shown in Table 2). Therefore, as far as stress conditions are concerned, the second loading stage was applied on a normally consolidated clay layer. Due to the embankment geometry and soil modulus, immediate settlements are negligible compared with consolidated settlements.

Radial drainage

The solution of the consolidation equation for pure radial drainage was given by Hansbo (1981) in which the delay in the process of consolidation caused by smear and well resistance was included. The well resistance, assessed according to the current approaches (Hansbo, 1981; Holtz et al., 1987) can be disregarded for the present vertical drain configurations. A key parameter in radial drainage is defined by

$$F_s(n) = F(n) + \left(\frac{k_h}{k'_h} - 1\right)\ln(s) \qquad (3)$$

where $F(n) = \ln(n) - 0.75$; n is the spacing ratio d_e/d_w; d_e is the diameter of the equivalent cylinder of soil influencing each drain — equal to 1.13S for square pattern; S is spacing; d_w is the diameter of the drain in the case of sand drains and equivalent diameter of the drain in the case of prefabricated drains (defined below); $s = d_s/d_w$; d_s is the diameter of the smeared zone; k_h is the horizontal coefficient of permeability of the undisturbed soil; k'_h is the correspondent coefficient of permeability in the smeared zone.

The equivalent diameter of the drain d_w in the case of prefabricated drains can be found by the equation of equal perimeter proposed by Hansbo (1979). Subsequent studies (Rixner et al., 1986) suggested simply to compute d_w by the average between width and thickness of the prefabricated drain. Other equations have also been proposed, such as that of the equal void area (Koerner, 1986). Hansbo's equation results in a greater value of d_w than the other proposals, so is more conservative. The value of d_w used here followed the manufacturers' recommendation of using Hansbo's equation multiplied by the factor 0.75, which turns out to be virtually the value given by Rixner's equation. In any case an erroneous evaluation of d_w has a rather limited influence on the consolidation process (Hansbo, 1987a). A misjudgement of d_w by 20%

Table 3. *Parameters for radial consolidation analysis*

Section	$d_w(m)$	$d_e(m)$	n	F(n)	s	$F_s(n)$ $k_h/k_h' = 3$	$F_s(n)$ $k_h/k_h' = 6$
B	0.40	2.83	7.1	1.25	2.0	2.64	4.71
C	0.40	2.83	7.1	1.25	2.0	2.64	4.71
D	0.40	2.83	7.1	1.25	1.0	1.25	1.25
E	5×10^{-2}	1.92	38.4	2.90	1.5	3.71	4.93
F	1×10^{-1}	2.26	22.6	2.40	1.5	3.21	4.43

will change the consolidation process by just 5%, which is small compared with other uncertainties of greater magnitude to be discussed below. The geometric factors used in the consolidation solutions are given in Table 3.

The ratio $s = d_s/d_w$ which appears in equation (3) was based on literature recommendations (Hansbo, 1981; Jamiolkowski et al., 1983, 1985; Holtz et al., 1987; Hansbo, 1987b). Values adopted were $s = 2$ for displacement type sand drains, $s = 1$ for jetted sand drains (also known as low displacement drains), and $s = 1.5$ for prefabricated drains.

The ratio k_h/k_h', also appearing in equation (5), is far more influential on the results than the choice of s value. Oedometer tests in good quality and on fully remoulded Sarapuí clay specimens gave average k_v/k_v' ratios between 5.2 and 6.9. These are upper bound values as the actual soil around the drain is not fully disturbed. Tests on Panigaglia and on Trieste soft Italian clays (Jamiolkowski et al., 1983) with plasticity indexes (I_p) not too different from that of Sarapuí clay yielded k_v/k_h' values between 1.5 and 3. For Porto Tolle clay, which has a smaller I_p, that ratio varied between 4 and 8. In the absence of data regarding k_v/k_h', Hansbo (1987b) has suggested substituting with k_h/k_v (i.e. to assume $k_h' = k_v$), which for normally consolidated Sarapuí clay varies between 1.6 and 2.1 (see Fig. 3). This suggestion was confirmed experimentally by Bergado et al. (1991) for the soft Bangkok clay by means of laboratory simulation of driving of prefabricated drains. Considering the experimental data available for the Sarapuí clay, as well as data in the literature, two permeability ratios will be used here: $k_h/k_h' = 3$, a reasonable and expected value; and $k_h/k_h' = 6$, the upper bound value. Values of $F_s(n)$ are given in Table 3.

Back-analysis of field data
Pore pressure analysis
The location of Casagrande and hydraulic piezometers in cross-section is shown in Fig. 6. Typical isocrones plotted with normalized depth are

shown in Fig. 7 for various times. These data suggest that pore pressure dissipated more quickly for the top clay layer, which would then have a higher coefficient of consolidation. This was confirmed by laboratory investigations (Costa-Filho et al., 1985) but not by in situ tests (Sills et al., 1988). A comparison of the mid-depth piezometer data for all

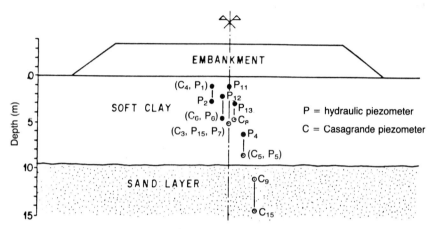

Fig. 6. *Instrumentation for pore pressure measurements*

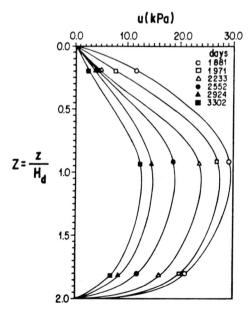

Fig. 7. *Pore pressure isocrones, section E, stage II*

sections is shown in Fig. 8. These diagrams show drain performance for each section.

Coefficients of vertical and horizontal consolidation c_v and c_h can be computed from pore pressure data using various procedures (Jamiolkowski et al., 1983, 1985). The method used here (Orleach, 1983; Jamiolkowski et al., 1985) allows these computations to be carried out independently of the generated pore pressure due to the applied loading. In the case of pure radial drainage the variation of pore pressure with time given by the equal strain solution can be rewritten (Orleach, 1983) as

$$\ln(u) = \alpha_0 - \alpha_1 t \qquad (4)$$

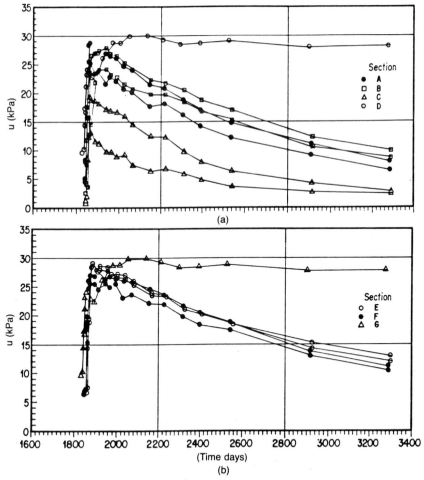

Fig. 8. Mid-depth pore pressure data

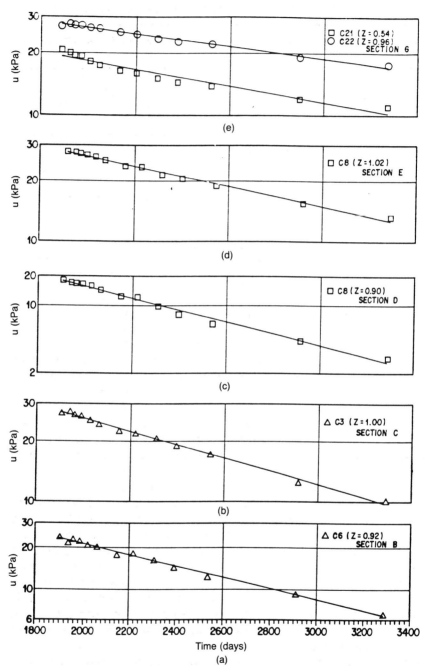

Fig. 9. Pore pressure dissipation data

Table 4. *Equations to compute c_h and c_v from pore pressure data*

Pure radial drainage ($Z > 0.65$)†	Pure vertical drainage ($T_v > 0.1$)
$c_h = \left((d_e)^2 \dfrac{F_s(n)}{8}\right)\alpha_1$	$c_v = 4\left(\dfrac{H_d}{\pi}\right)^2 \alpha_1$

†Range of validity for the present case (Ferreira, 1991)

Table 5. *Coefficients of consolidation ($\times 10^{-8}$ m^2/s) computed from pore pressure data*

Section	c_h ($k_h/k'_h = 3$)	c_h ($k_h/k'_h = 6$)	c_v	No. of piezometers
B	2.8	5.0	—	4
C	2.3	4.2	—	4
D	2.1	2.1	—	4
E	1.2	1.6	—	4
F	1.6	2.2	—	3
G	—	—	2.2–4.5	5

where α_0 and α_1 are constants which can be determined by linear regression between ln (u) and t. The coefficient of horizontal consolidation c_h can be obtained from the slope α_1 using the equations given in Table 4.

Pore pressure dissipation data plotted as log (u) against time is illustrated in Fig. 9 for most sections. All linear regressions to compute α_1 had correlation coefficients very close to unity. Pore pressure data corresponding to the time period of 50 days following the end of the second stage of load application were not included, as these presented some oscillation.

Average values of c_v and c_h (for two k_h/k'_h ratios) computed for each section, as outlined above, are shown in Table 5. For the sections with vertical drains all piezometers had normalized depths Z greater than 0.65 so that they would be subjected to pure radial flow. It is observed that c_h computed with $k_h/k'_h = 3$ results in c_v/c_h ratios greater than 1.0, which is unexpected. Computations with the upper bound $k_h/k'_h = 6$ result in more reasonable c_v/c_h ratios. An overall discussion on c_v and c_h values obtained by different procedures is presented below.

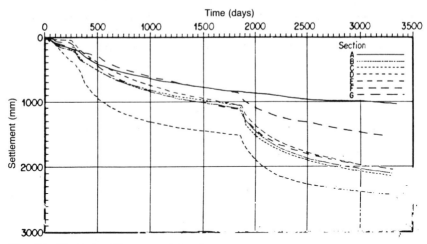

Fig. 10. *Central settlements plates data*

Table 6. *Coefficients of consolidation ($\times 10^{-8}$ m^2/s) computed from settlement data*

Section	c_h No smear	c_h ($k_h/k_h' = 3$)	c_h† ($k_h/k_h' = 6$)	c_v
		Central settlement plates		
B	4.2	8.9	15.8	—
C	7.5	15.8	24.9	—
D	5.7	5.7	5.7	—
E	6.0	7.7	10.2	—
F	6.8	9.1	12.6	—
G	—	—	—	29.7
		Magnetic settlement gauge‡		
B	3.9	8.1	14.5	—
D	4.2	4.2	4.2	—
E	4.6	5.9	7.8	—
F	4.9	5.4	7.5	—
G	—	—	—	22.6

†Upper bound values of c_h
‡The settlement gauge of section C was damaged

Settlement analysis
The central settlement data is shown in Fig. 10. A number of methods to compute final settlements and coefficients of consolidation from settlement data have been applied (Almeida et al., 1989) to test embankment II; it was concluded that Asaoka's method was the easiest and most reliable. Table 6 summarizes c_v and c_h values calculated from settlement data, these last ones with and without consideration of smear. Calculations considering combined drainage are not presented for two reasons: the first concerned doubts as to which c_v to adopt; the second is that calculations using $c_v = 0.5c_h$ (the computation was therefore iterative) resulted in c_h for combined drainage just 10% smaller than when it is not considered. Therefore the combined drainage has a rather small influence on c_h calculations if $c_v < c_h$ is used. However, if greater c_v values (obtained from settlement data) are used, the combined drainage has a much greater influence. Values of c_v for the second stage computed for section G were 29.7 and 22.6×10^{-8} m²/s, respectively, for the central settlement plate and for the top magnetic gauge. These values are quite high (about twenty times higher than the laboratory values presented above). The apparent reason for this discrepancy is the secondary consolidation (Almeida et al., 1992).

Discussion of results

A number of procedures have been used here to obtain c_v and c_h for the Sarapuí clay. Results of these analyses are summarized in Table 7. Despite resulting in reasonable c_v/c_h values, the upper bound values of c_h corresponding to $k_h/k'_h = 6$ are not included in Table 7 as they are not considered representative. The reasons for this are that they correspond to a fully remoulded clay condition around the drain, which is of little probability. Less homogeneous c_h values (see, for instance the wide range of settlement plate data in Table 6) were obtained with this hypothesis. The only way to clarify doubts regarding k_h/k'_h values is by performing laboratory tests simulating drain installations (Bergado et al., 1991). Inspection of Table 7 indicates reasonable differences among the

Table 7. Final comparison of c_v and c_h ($\times 10^{-8}$ m²/s)

Procedure/test	Depth interval (m)	c_v	c_h
Settlement plates	Whole layer	29.7	5.7–15.8
Magnetic gauges	Whole layer	22.6	4.2–8.1
Pore pressure data	3.3–8.3	2.2–4.5	1.2–2.8
Oedometer test	5.0–6.0	1.2	2.4
Piezocone test	2.2–8.2	1.6–4.4	3.1–8.7

various values shown. Differences in procedures and related boundary conditions are apparently the reason for those differences. Values of c_h, which show a better general agreement among themselves than c_v values, will be discussed first.

Coefficients of horizontal consolidation

Analysis of c_h values in Table 7 shows that the smallest values are those from the laboratory, $c_h(lab)$, and the greatest values are those from settlement data, $c_h(s)$. In the intermediate range are pore pressure data values, $c_h(u)$, and piezocone values $c_h(piez)$. The ratio $c_h(s)/c_h(lab)$ is around 2.5, which is expected and usually attributed to the macrostructure of the deposit (e.g. small lenses and draining layers). In the present case, however, the only apparent inhomogeneity is the stiffer and more permeable top crust. It is also possible that secondary consolidation had some influence on $c_h(s)$ computations.

Comparison between $c_h(s)$ and $c_h(u)$ indicates that the former is about three times greater than the latter. The non-linearity between effective stress and void ratios appears to be the reason for this discrepancy. Orleach (1983) performed similar analyses for North American clays and also found $c_h(s)/c_h(u)$ values of about three. Garassino et al. (1978) using conventional methods of analyses (different from those adopted here) for Porto Tolle clay, also obtained $c_h(s)/c_h(u)$ values of about three.

Comparison between $c_h(lab)$ and $c_h(piez)$ shows that the former is smaller than the latter for normally consolidated (nc) conditions. However for the in situ stress condition, i.e. the overconsolidated (oc) state, these two tests showed very good agreement (Danziger et al., 1992). Test differences and the way $c_h(oc)$ was transformed in $c_h(nc)$ for the piezocone data (see equation (1)) are the apparent reasons for the present differences. Robertson et al. (1992) have concluded that $c_h(piez)$ values were slightly higher than $c_h(lab)$ and in better agreement with $c_h(s)$, a finding also obtained here.

Coefficients of vertical consolidation

Analogous to c_h data, c_v (lab) and $c_v(s)$ are the smallest and the largest, respectively. The difference between $c_v(lab)$ and $c_v(s)$ is greater than usual. The two apparent reasons for the present c_v discrepancies appear to be the influence of secondary compression on $c_v(s)$ calculations (Almeida et al., 1992) and, of lesser importance, the top 2 m of crust which appears to have higher c_v.

Leroueil (1988) analysed data from 16 embankments and observed highly variable $c_v(s)/c_v(lab)$ ratios, with values ranging between 3 and 200. Neglecting two extremely high values, the average $c_v(s)/c_v(lab)$ ratio obtained by Leroueil was equal to 20, which is very close to that found here. Bergado et al. (1990) obtained about the same ratio for the Bangkok

Table 8. c_v values for the 24 m high embankment on Sarapuí clay (COPPETEC ET-15404, 1990)

Procedure	$c_v (\times 10^{-8}\,\text{m}^2/\text{s})$
Oedometer (IL and CRSC) tests	0.8–2.0
Piezocone tests	0.9–3.0
Asaoka's method ($\cong$100 surface marks)	1.0–2.0

clay, the properties of which are fairly similar to those of Rio de Janeiro clay.

The same order of magnitude of the three c_v determinations excluding $c_v(s)$, suggests that $c_v(u)$ is, unlike $c_v(s)$, much less influenced by secondary compression.

It is well known (Leonards and Girault, 1961) that secondary consolidation gets less influential with increasing applied stress ratio $\Delta\sigma_v'/\sigma_{vo}'$. There is limited evidence for Sarapuí clay to show that when secondary compression is negligible, good agreement can be obtained between c_v evaluated using different procedures. The evidence available was provided by a 24 m high embankment (COPPETEC ET-15404, 1990) built on a soft deposit not too far away from that in which the test embankment was built. This embankment has produced a $\Delta\sigma_v'/\sigma_{vo}'$ ratio of about 60, while the test embankment produced a $\Delta\sigma_v'/\sigma_{vo}'$ close to unity. The embankment should, ideally, fully displace the soft clay. However, owing to construction problems, an average of 3 m of thick soft clay remained under the embankment. Values of c_v for this underconsolidated and remoulded clay layer were carefully determined by laboratory tests, in situ tests, and Asaoka's method. These c_v values are given in Table 8 and the agreement obtained in this case is remarkable, particularly between c_v(lab) and c_v(s), though c_v(piez) values are also within the same range.

Conclusion

A test embankment has been constructed on a well studied soft clay site in which the rate of consolidation was accelerated using different types of vertical drains. Two sections without drains were used as reference. Field observations have been made for about ten years, and settlement and pore pressure data were back-analysed and coefficients of horizontal consolidation calculated considering smear effects. Well resistance was shown to be of no concern and combined drainage was duly taken into account whenever appropriate.

Back-calculated c_h values from settlement and pore pressure data have shown a reasonable agreement with laboratory and in situ values, the agreement being quite good when c_h from settlement data is not included in the comparison. Analogous to the above, c_v values from pore pressure data also compared quite well with c_v from laboratory and in situ tests. However the discrepancy between c_v computed from settlement data and the three other determinations is even greater than for c_h.

There is evidence, supported by numerical analyses, that secondary settlements are of great relevance here, thus making inappropriate back-analysis of settlement data on the hypothesis that the process is basically one of primary consolidation. Because vertical drains accelerate primary compression, the influence of secondary consolidation appears to be smaller on back-calculated c_h than on back-calculated c_v. One other factor that may have affected c_v calculations was the apparent greater c_v at the 2 m top crust. Good overall agreement of c_v values was obtained in a case history on Sarapuí clay in which (owing to the high loading ratio) the secondary compression was negligible. According to Leroueil and Jamiolkowski (1991) the most difficult parameter to define in soil mechanics is a representative coefficient of consolidation. Results of the analysis presented here appear to support this statement.

Acknowledgements

The authors are much indebted to the Brazilian Highway Research Institute (IPR–DNER) and to their colleagues at COPPE for the assistance during the whole research study. This paper was written while the first author was on sabbatical studies in Europe. Thanks are due to M. Jamiolkowski, G. Baldi and T. Lunne for the support during this period. S. Hansbo, J. Keaveny, S. Leroueil and Dick Parry, are greatly acknowledged for their comments and suggestions on the paper.

References

ALMEIDA, M.S.S. (1982). The undrained behaviour of the Rio de Janeiro clay in the light of critical state theories. Report TR 119, Cambridge University Engineering Department.

ALMEIDA, M.S.S.., LACERDA, W.A., FERREIRA, C.S. and TERRA, B.R. (1989). The efficiency of vertical drainage systems for ground improvement in a very soft clay. Proc. II Int. Conf. on Foundations and Tunnels, London, Vol. 1, pp. 393–398.

ALMEIDA, M.S.S., COLLET, H.B., CARVALHO, S.R.L. and FERREIRA, C.A.M. (1990). Complementary Studies on Trial Embankment II. Research Report (in Portuguese), Highway Research Institute, IPR/DNER.

ALMEIDA, M.S.S. et al. (1992). Influence of secondary settlements on the back calculation of the coefficient of consolidation. Paper in preparation.

ASAOKA, A. (1978). Observational procedure of settlement prediction. Soils and Foundations, Vol. 18, pp. 87–101.
BALIGH, M.M. and LEVADOUX, J.N. (1986). Consolidation after undrained piezocone penetration II: interpretation. ASCE, J. Geotech. Eng., Vol. 112, pp. 727–745.
BERGADO, D.T., AHMED, S., SAMPACO, C.L. and BALASUBRAMANIAN, A.S. (1990). Settlement of Bangna–Bangpakonk highway on soft Bangkok clay. ASCE, J. Geotech Eng., Vol. 116, pp. 136–155.
BERGADO, D.T., AKAMI, H., MAROLO, C.A. and BALASUBRAMANIAN, A.S. (1991). Smear effects of vertical drains on soft Bangkok clay. ASCE, J. Geotech Eng., Vol. 117, No. 10, pp. 1509–1531.
COLLET, H.B. (1985). The Trial Embankment II: conception and preliminary settlement analysis. Professorship thesis (in Portuguese), Dept of Civil Engineering, Fluminense Federal University.
COPPETEC-ET 150404 III (1990). Analysis of the behavior of S. Jose Powerplant Embankment by means of instrumentation, in situ and field tests. Report for Furnas Electricity Board (in Portuguese), COPPE, Fed. Univ. of Rio de Janeiro.
COSTA-FILHO, L.M., GERSCOVICH, D., BRESSANI, L.A. and THOMA, J.E. (1985). Discussion, Embankment failure on clay near Rio de Janeiro. ASCE, J. Geotech. Eng., Vol. 111, No. 2, pp. 259–262.
DANZIGER, F.A.B., ALMEIDA, M.S.S. and SILLS, G.C. (1992). The significance of the strain path analysis in the interpretation of piezocone dissipation data. Submitted to Géotechnique.
FERREIRA, C.A.M. (1991). Analysis of the piezometric data of an embankment on soft clay with vertical drains. MSc thesis (in Portuguese), COPPE, Fed. Univ. of Rio de Janeiro.
GARASSINO, A., JAMIOLKOWSKI, M., LANCELLOTTA, R. and TONGHINI, M. (1979). Behaviour of pre-loading embankments on different vertical drain with reference to soil consolidation characteristics. Proc. VII European Conf. on Soil Mechanics and Foundation Engineering, Brighton, Vol. 3, pp. 213–218.
HANSBO, S. (1979). Consolidation of clay by band-shaped prefabricated drains. Ground Engineering, Vol. 12, 5, pp. 16–25.
HANSBO, S. (1981). Consolidation of fine-grained soils by prefabricated drains. Proc. X Int. Conf. on Soil Mechanics and Foundation Engineering, Stockholm, Vol. 3, pp. 677–682.
HANSBO, S. (1987a). Fact and fiction in the field of vertical drainage. Int. Symp. on Prediction and Performance in Geotechnical Engineering, Calgary, pp. 61–72.
HANSBO, S. (1987b). Design aspects of vertical drains and lime column installation, Proc. 9th Southeast Asian Geotechnical Conf., Bangkok, Thailand, 8.1–8.12.

HANSBO, S., JAMIOLKOWSKI, M. and KOK, L. (1981). Consolidation by vertical drains. Geotechnique, Vol. 31, pp. 45–66.

HOLTZ, R.D., JAMIOLKOWSKI, M., LANCELLOTTA, R. and PEDRONI, S. (1987). Performance of prefabricated band-shaped drains. CIRIA – RPS 364, London.

HOULSBY, G.T. and TEH, C.I. (1988). Analysis of the piezocone in clay. ISOPT I, Orlando, Vol. 2, pp. 777–783.

JAMIOLKOWSKI, M., LADD, C.C., GERMAINE, J.T. and LANCELLOTTA, R. (1985). New developments in field and laboratory testing of soils. Proc. XI Int. Conf. on Soil Mechanics and Foundation Engineering, San Francisco, Vol. 1, pp. 57–153.

JAMIOLKOWSKI, M., LANCELLOTTA, R. and WOLSKI, W. (1983). Precompression and speeding-up consolidation. Proc. VIII European Conf. on Soil Mechanics and Foundation Engineering, Helsinki, Vol. 3, pp. 1201–1226.

KOERNER, R.M. (1986). Designing with Geosynthetics. Prentice-Hall, USA.

LACERDA, W.A., COSTA-FILHO, L.M. and DUARTE, A.E.R. (1977). Consolidation characteristics of Rio de Janeiro Soft Clay, Proc. Int. Symp. on Soft Clay, Bangkok, pp. 231–243.

LEONARDS, G.A. and GIRAULT, P. (1961). A study on the uni-dimensional consolidation test. Proc. V Int. Conf. on Soil Mechanics and Foundation Engineering, Vol. 1, pp. 213–218.

LEROUEIL, S. (1988). Recent developments in consolidation of natural clays. Canadian Geotech. J., Vol. 25, No. 1, pp. 85–107.

LEROUEIL, S. and JAMIOLKOWSKI, M. (1991). General Report, Session 1: Exploration of soft soils and determination of design parameters, Geo-coast, Japan, September 1991.

ORLEACH, P. (1983). Techniques to evaluate the field performance of vertical drains. MSc thesis, MIT, Cambridge, Mass.

ORTIGÃO, J.A.R., WERNECK, M.L.G. and LACERDA, W.A. (1983). Embankment failure on clay near Rio de Janeiro. ASCE, J. Geotech Eng., Vol. 109, No. 2, pp. 1460–1479.

ORTIGÃO, J.A.R. and COLLET, H.B. (1988). Errors caused by friction in field vane tests, Proc. Int. Symp. on Laboratory and Field Vane Shear Strength Testing, ASTM (STP 1054), Tampa, pp. 104–115.

PARRY, R.H.G. and WROTH, C.P. (1977). Shear properties of soft clays. Report presented at the Symp. on Soft Clay, Bangkok, Thailand.

RIXNER, J.J., KRAEMER, S.R. and SMITH, A.D. (1986). Prefabricated vertical drains, FWHA/RD-86/168, Federal Highway Administration, Washington, D.C.

ROBERTSON, P.K., SULLY, J.P., WOELLER, D.J., LUNNE, T., POWELL, J.J.M. and GILLESPIE, D.G. (1992). Estimating coefficients from piezocone tests. Canadian Geotechnical Journal, Vol. 29, No. 4, pp. 539-550.

SILLS, G.C., ALMEIDA, M.S.S. and DANZIGER, F.A.B. (1988). Coefficient of consolidation from piezocone dissipation tests in a very soft clay, ISOPT I, Orlando, Vol. 2, pp. 967–974.

SOARES, M.M., LUNNE, T., ALMEIDA, M.S.S. and DANZIGER, F.A.B. (1987). Piezocone and dilatometer tests in a very soft clay Rio de Janeiro clay, Proc. Int. Symposium on Geotechnical Engineering of Soft Soils, Mexico.

WERNECK, W.L.G., COSTA-FILHO, L.M. and FRANÇA, H. (1977). In situ permeability and hydraulic fracture tests in Guanabara bay clay. Proc. Int. Symp. on Soft Clay, Bangkok, pp. 399–416.

Post-liquefaction settlement of sands

K. ARULANANDAN and J. SYBICO, Jr., University of California, Davis

The mechanism causing the substantial settlement observed during and immediately after liquefaction in a centrifuge test is studied using resistivity measurements to characterize the soil structure. The permeability value at the point of initial liquefaction is shown to increase from the value at the initial state when the tortuosities and the pore shape factors at the appropriate states are used in the Kozeny-Carman equation. The mechanism causing the substantial increase in settlement is considered to be due to the increase in permeability. The higher the permeability, the higher is the rate of settlement and the larger is the incremental settlement in each time step. In addition, the increase in permeability causes the liquefaction front to propagate upwards faster and, therefore, the generation and dissipation processes affect a larger depth of the soil column during shaking, and hence, produces a larger settlement. The use of a modified permeability in a fully coupled finite element code, DYSAC2, provides a good measure of the observed settlement.

Introduction

The settlement behaviour of saturated sands under earthquake loading has led to several studies since the 1964 earthquakes in Alaska and Niigata. Historically, liquefaction has been associated with sand boils, excessive settlement and massive lateral deformation.

Field observations of earthquake induced settlements in saturated sands range from a fraction of an inch to 20 inches (Tokimatsu and Seed, 1987). In the case of pipeline facilities transitioning from firm ground to liquefiable soil deposits, ground settlements on the order of several inches may become significant with respect to damage potential. Hence, an understanding of deformation behaviour and methods of evaluation during and after liquefaction are necessary.

It has been postulated that as soon as sand liquefies, the settlement of particles or sand consolidation takes place followed by the flowing out of water from the voids (Florin and Ivanov, 1961). The boundary between settled and liquefied sand starts in the lower part of the liquefied stratum and moves up towards the surface. This process of settling, solidification and consolidation was represented by Scott (1986) by considering sedimentation first followed by consolidation in the growing solidified layer as a linear function of pore water pressure diffusion. Material properties such as permeability and coefficient of consolidation during pore water dissipation were obtained. It has been suggested that the process of settlement and pore water pressure dissipation occurs after the shaking stops (Hushmand et al., 1988). However, it has been observed from centrifuge model studies (Hushmand et al., 1988) and from studies conducted under the NSF sponsored VELACS Project (Arulanandan and Scott, 1992), that substantial settlement starts immediately after initial liquefaction. Settlement actually occurs when the shaking starts; however, it is relatively small (Lambe, 1981). Currently, numerical procedures predict only 20 to 30% of the settlement measured in the centrifuge.

The objectives of this paper are: (1) To study the mechanism causing the substantial settlement observed during and immediately after soil liquefaction in a centrifuge test using resistivity measurements to determine the change in soil structure and permeability during initial liquefaction; and, (2) To evaluate the observed settlement by using a modified permeability in a finite element computer code DYSAC2 (Muraleetharan et al., 1990).

Electrical characterization of transversely isotropic sands

An index which has been shown experimentally to depend on the porosity, particle shape and size distribution and the direction of measurement is the formation factor, F (Arulanandan and Kutter, 1978; De la Rue et al., 1959; Dafalias and Arulanandan, 1979; Fricke, 1924; Meredith et al., 1962; Wyllie, 1955). It is defined as the ratio of the conductivity of the electrolyte, which saturates a particulate medium consisting of non-conductive particles, to the conductivity of the mixture (Archie, 1942). Formation factor measurements have been used for determining volume changes during a pressuremeter test (Wroth and Windle, 1975), for evaluating the in situ porosity of non-cohesive sediments (Jackson, 1975), and for evaluation of in situ density and fabric of soils (Arulanandan, 1977).

In recent years, the formation factor measurements made in the horizontal and vertical directions, where the latter is the direction of placement of the sample, have been used to characterize the anisotropic nature of sand deposits due to particle orientation (Arulanandan and Kutter, 1978; Arumoli et al., 1985; Arulanandan and Muraleetharan, 1988; Dafalias and Arulanandan, 1979). These electrical parameters are

Vertical Formation Factor $\quad F_v = \dfrac{\sigma_s}{\sigma_v}$ (1)

Horizontal Formation Factor $\quad F_h = \dfrac{\sigma_s}{\sigma_h}$ (2)

where σ_s = conductivity of the solution, σ_v = conductivity of the soil in the vertical direction and σ_h = conductivity of the soil in the horizontal direction. It was also possible to develop theoretically, on the basis of electromagnetic theory, analytical relations between the average formation factor, $\bar{F}$, the porosity, n, and parameters associated with the shape and orientation of particles for transversely isotropic sands called the average shape factor, $\bar{f}$ (Dafalias and Arulanandan, 1979; Arulanandan and Muraleetharan, 1988; Arumoli, Arulanandan and Seed, 1985). This relationship is

$$\bar{F} = n^{-\bar{f}} \quad (3)$$

The average formation factor, $\bar{F}$, is calculated as

$$\bar{F} = \dfrac{F_v + 2F_h}{3} \quad (4)$$

Using the values of the shape and orientation parameters obtained independently (Dafalias and Arulanandan, 1979) from the results of thin section studies performed by Mitchell et al. (1976), the theoretically predicted values of F along the axis of transverse isotropy were shown to be in close agreement with the experimentally measured values

Fig. 1. Stacked ringed laminar box

(Dafalias and Arulanandan, 1979). This gave a sound theoretical confirmation of the experimentally known fact that F depends on the sand structure, especially on the orientation of particles, which cannot be accounted by other indices like relative density.

Centrifuge test
Centrifuge and laminar box
The tests were performed in a stacked rectangular ring apparatus as shown in Fig. 1. The inside dimensions of the rings are 7 inch × 14 inch and the thickness of each ring is 0.5 inch. A total of 20 rings comprise the box and adds up to a height of 10.24 inch with spacing between rings. The earthquake motions were simulated using an electro hydraulic shaker.

Electrical resistivity probe
Two 3 inch × 3 inch brass plates were glued on to 3/16 inch plexiglass and braced so that the plates are 3 inch apart as shown in Fig. 2 and the schematic drawing in Fig. 3. The plates were hung vertically into position inside the laminar box before the sand was poured in. The plates' faces were placed parallel to the direction of shaking to minimize disturbance to the sand during shaking. Since electrical resistance is affected by the area of plate–sand contact, and consequently the depth

Fig.2. Electrical resistivity probe inside the laminar box

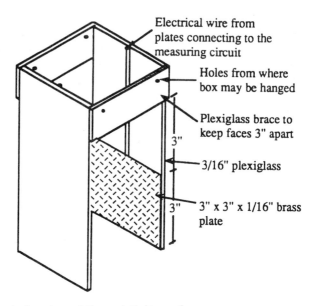

Fig. 3. Schematic drawing of the resistivity probe

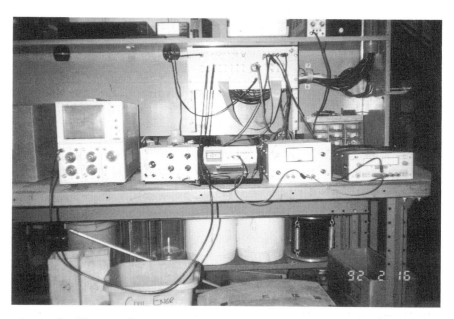

Fig. 4. Oscilloscope, function generator, multi-tester, negative and positive d.c. power supply

at which it is buried, the tops of the plates were positioned at least 2 inch below the surface. At a depth of 1.25 inch or more, the readings were found not to be sensitive to changes in height. The horizontal electrical resistance between the two plates is converted into volts by an electrical circuit and is sent through the existing data acquisition system. The circuit requires a function generator, a negative and positive DC power supply as shown in Fig. 4.

With the resistance measured and the dimensions of the electrical probe known, the conductivity can be calculated using the relationship

$$\sigma = \frac{1}{R}\frac{L}{A} \tag{5}$$

where σ is the conductivity, R is the electrical resistance in ohms, L is the length of the sample in cm and A is the cross-sectional area of the sample in cm^2. Since L and A are the same, the formation factor can alternatively be defined as the ratio of the 1/R readings. Thus, equations (1) and (2) can be also expressed as

$$F_v = \frac{R_{\text{soil in vert. dir.}}}{R_{\text{solution}}} \tag{6}$$

$$F_h = \frac{R_{\text{soil in hor. dir.}}}{R_{\text{solution}}} \tag{7}$$

Sample preparation

A thin plastic bag of slightly larger size than the inside dimensions of the laminar box was first placed inside to contain the sample. After the resistivity probe was hung in its desired position, an appropriate amount of water with known conductivity was poured into the box. Fine Nevada sand was then pluviated to the appropriate height in the laminar box. As the sample increased in height, pore pressure transducers and accelerometers were placed in positions at their desired locations. Two Linear Variable Differential Transformers (LVDT) were placed on top of thin balsa wood blocks to prevent them from sinking into the liquefied soil; balsa wood was found to prevent sinking better than aluminum square plates. This may be because the density of the wood is closer to the density of the liquefied soil than the aluminum. The weight of the LVDT's potentiometer on the balsa wood blocks prevented it from floating in the layer of water that developed at the surface after liquefaction. The method of using balsa wood to prevent sinking was first attempted by Fiegel (1992).

Test results

Nevada sand with gradation characteristics as shown in Fig. 5 was used in this study. The measured vertical and horizontal formation factors of the sand prepared at different porosities by the pluviation method of placement are shown in Fig. 6. It can be seen that an index of anisotropy, A, defined as

$$A = \sqrt{\frac{F_v}{F_h}} \qquad (8)$$

where F_v is the vertical formation factor in the direction of placement of the sample and F_h is the horizontal formation factor (Arulanandan and Kutter, 1978), is 1.04.

The centrifuge model at a porosity of 0.418 was accelerated to 50 g and subjected to a base acceleration as shown in Fig. 7. The initial permeability at this porosity was measured to be 6.56×10^{-5} m/s based on permeameter tests on this sand. An accelerometer, attached underneath the box, measured the horizontal acceleration. The formation factor plates were positioned in the middle of the box with the tops of the plates 2 inch beneath the surface. Six pore water pressure transducers (weight = 2.9 g, diameter = 0.3 inch, length = 0.4 inch) were placed inside the sample, in pairs, an inch to the left and right of the formation factor plates at depths of ¼ H, ½ H and ¾ H from the bottom. Two LVDTs, placed at 1.15 inch on each side of centre line, measured the settlement of the surface. The measured base accelerations, generated pore water pressure ratios, horizontal formation factor and surface settlement with respect to time are shown in Figs. 7 to 10, respectively. The prepared sample initially had a horizontal flat surface. After liquefaction, the surface formed a curvature with radius equal to the distance from the soil surface to the axis of the centrifuge rotation. The correction due to this curvature at the location of the LVDTs is equivalent to 9.92 inch at the prototype scale as shown in Fig. 11. The total measured settlement at the end of shaking has been corrected for the above curvature of the soil surface. The settlement time history was adjusted proportionally to be consistent with the corrected settlement at the end.

Prior to shaking, after the pore water pressures stabilized at 50 g, the centrifuge was stopped to observe the settlement due to curvature and it was found that the curvature did not form before shaking. It was difficult to determine at which point the curvature was formed during shaking based on the settlement time history. It is for this reason that the time history of settlement was adjusted proportionally to account for the curvature of the soil surface.

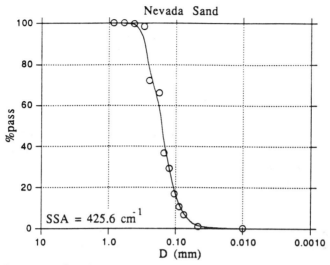

Fig. 5. *Grain size distribution curve*

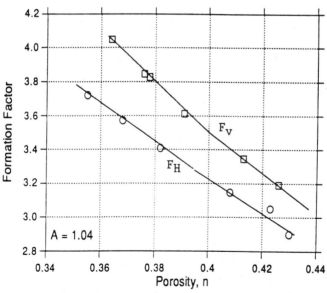

Fig. 6. *Vertical and horizontal formation factors of pluviated Nevada sand*

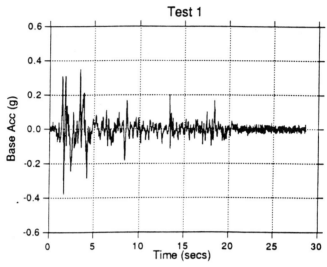

Fig. 7. Measured horizontal base acceleration

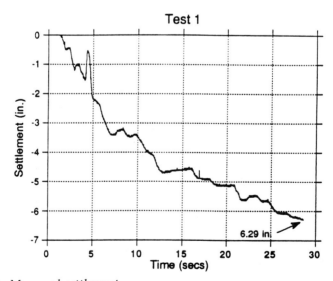

Fig. 8. Measured settlement

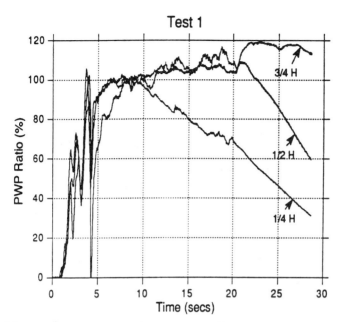

Fig. 9. Measured excess pore water pressure ratio

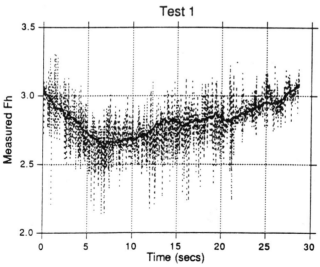

Fig. 10. Measured horizontal formation factor

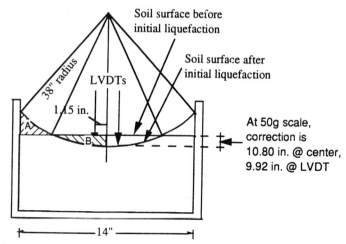

Fig. 11. *Correction due to surface curvature*

Analysis

The change in the value of F_h is related to the change in the value of permeability. The Kozeny–Carman equation,

$$k = \frac{\gamma_w}{\mu} \frac{1}{k_0 T^2 S_0^2} \frac{n^3}{(1-n)^2} \quad (9)$$

where k = permeability, γ_w = unit weight of water, μ = viscosity of water, k_0 = pore shape factor, T = tortuosity, S_0 = specific surface area and n = porosity, has been shown to be valid for the evaluation of the permeability of non-clay minerals.

The tortuosity, T, in eqn. (9) can be expressed in terms of the formation factor, F, and porosity, n, using the relationship (Wyllie, 1955).

$$n = cT^2 \quad (10)$$

where

$$c = \frac{1}{F} \quad (11)$$

and F is the formation factor measured in the same direction as the permeability measurement. Eqn. (9) becomes

$$k = \frac{\gamma_w}{\mu} \frac{1}{k_0 F S_0^2} \frac{n^2}{(1-n)^2} \quad (12)$$

relating permeability to the structure of the soil in terms of the formation factor, F, and the pore shape factor, k_0.

The pore shape factor, k_0, prior to shaking was estimated using the following properties:

(a) permeability
(b) the vertical formation factor calculated using the measured horizontal formation factor and the anisotropy index
(c) the specific surface area, $S_0 = 426$ cm^{-1}, evaluated using the grain size distribution as shown in Fig. 5, and
(d) the constant value μ/γ_w of 1.02×10^{-5} cm s. The estimated value of k_0 prior to shaking is shown in Table 1.

At the point of initial liquefaction, the soil particles lose full contact with each other; this creates an easier path for water flow. The creation of such flow paths reduces the tortuosity, T, and the pore shape factor, k_0. These two properties lead to an increase in the permeability at the point of initial liquefaction, where it is assumed there is negligible change in porosity. It is further assumed that the soil is homogeneous throughout the layer and the pore sizes become uniform as effective stresses approach zero. At this state, the pore shape factor, k_0 is assessed to be 2.5 and used in the Kozeny–Carman equation to evaluate permeability.

Equation (9) may also be expressed as

$$k = C_s \frac{\gamma_w}{\mu} \frac{1}{S_0^2} \frac{n^3}{(1-n)^2} \qquad (13)$$

where

$$C_s = \frac{1}{k_0 T^2} \qquad (14)$$

is called the shape coefficient. Since n has been assumed to be constant, the rest of the equation remains unchanged when comparing permeabilities at the initial and liquefied states. It is convenient to express the permeability value at a different state in terms of

$$k_{new} = C k_{initial} \qquad (15)$$

Table 1. Estimated values of k_0 prior to shaking

Test no.	Height (inches)	Porosity n	Permeability ×10^{-3} cm/s	F_h Measured	F_v A = 1.04	Estimated value of k_0
1	7.500	0.418	6.55	3.05	3.30	12.90

Table 2. Calculation of average factor

Test no.	(F_v) initial A = 1.04	(F_v) liquefied A = 1.00	k_0 initial Table 1	k_0 liquefied	C_{Liq}	C_{ave}
1	3.30	2.68	12.90	2.50	6.35	1.67

where k_{new} = permeability at a new state, $k_{initial}$ = initial permeability and C is a factor. (The values of C at the initial and liquefied states are 1.0 and C_{Liq}, respectively.)

The factor, C, is calculated using the values of C_s at the initial and the new states.

$$C = \frac{Cs_{new}}{Cs_{initial}} = \frac{(k_0 T^2)_{initial}}{(k_0 T^2)_{new}} = \frac{(k_0 F_v)_{initial}}{(k_0 F_v)_{new}} \qquad (16)$$

To calculate C_{Liq}, $(k_0 F_v)_{liquefied}$ was used in eqn. (16). Since the pore shape changes constantly as the particles settle down, an average factor C_{ave} representing the change in shape factor as a function of time was used to modify the permeability. The change in porosity after settlement contributes to the change in permeability as indicated in eqn. (9) and should be included in eqn. (16) if it is a post-liquefaction prediction. However, it is relatively small compared to the contributions due to changes in k_0 and F_v. The estimated values of C_{ave} are shown in Table 2. The analysis show that the permeability at the point of initial liquefaction increases 6–7 times due to structure change.

Evaluation of settlement

The computer code, DYSAC2 was used to predict the settlement of each test twice, using the initial permeability in the first prediction and the average or modified permeability in the second. The predicted settlements are compared with the measured settlements and are shown in Fig. 12. A reasonable agreement between the measured and evaluated settlements is seen to exist if the modified permeability is used in the calculations. A summary of the measured and the predicted settlements is shown in Table 3.

Mechanism causing settlement during earthquake shaking

The magnitude of the pore water pressure ratio of a layer for a given time interval during shaking is influenced by two major factors:

(a) the earthquake acceleration which tends to increase it, and
(b) the dissipation during consolidation which tends to reduce it.

Centrifuge model studies (Lambe, 1981; Arulanandan and Scott, 1992) have shown that the initial liquefaction occurs simultaneously in all layers. At a particular layer, once the initial sedimentation is completed after initial liquefaction, the pore water pressure decreases due to consolidation and an increase in pore water pressure pressure occurs due to the earthquake acceleration. The net effect of the two processes may be zero and the measured pore water pressure values may indicate that no dissipation/consolidation has occurred. A layer can therefore consolidate without visible change in the measured pore water pressure even when the pore water pressure is 100 percent. There will be no visible change in the measured pore water pressure until the shaking diminishes or until sufficient consolidation has taken place. It is due to this dynamic effect that an increase in permeability results in an increase in total settlement. The higher the permeability, the faster the rate of

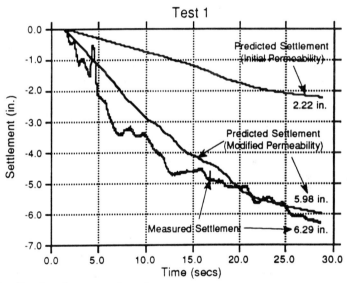

Fig. 12. Predicted and measured settlements

Table 3. Summary of measured and predicted settlements

Test no.	Measured settlement	Initial prediction of settlement	Modified prediction of settlement
1	6.29	2.22	5.98

settlement and the larger the incremental settlement in a given time interval.

The height of the soil column and the duration of shaking are important factors that influence the degree of the effect of a higher permeability on the total settlement. The larger the depth of the layers where initial sedimentation is completed after initial liquefaction, the larger the increase in settlement expected.

In the static case (no shaking during dissipation), an increase in permeability does not affect the total settlement. The soil column with a lower permeability has a lower rate of settlement and will simply require a longer period to dissipate the excess pore water pressure.

Conclusions

The fabric of the soil has been characterized using electrical properties. The influence of fabric on permeability during low effective stresses is estimated in terms of k_0, the pore shape factor, using the Kozeny–Carman equation for permeability. Permeability values at initial liquefaction and during the earthquake motion are necessary for the analysis of post-liquefaction settlement.

The application of the above approach to predict post-liquefaction vertical settlement in practice requires the evaluation of the increased permeability at the point of initial liquefaction. The porosity, the specific surface area and the initial permeability may be estimated using electrical methods. The assumed change in vertical formation factor and pore shape factor will provide an indication as to the permeability increase which is necessary for use in analytical predictions.

Three similar centrifuge tests were conducted using resistivity measurements to evaluate the post liquefaction settlements. The results of the settlements obtained in the three tests are in close agreement with the predicted values.

Acknowledgements

The support provided by the National Science Foundation Grant No. BCS89-12074 (The VELACS Project) is gratefully acknowledged. The results presented in this paper are from an M.Sc thesis on 'Post Liquefaction Settlement of Sands', submitted by the second author. The authors are grateful for the assistance given by Dr X. S. Li in the development of the electronic systems used in this study. They are also grateful to Prof. Ronald F. Scott for reviewing the paper and providing valuable suggestions.

References

ARCHIE, G.E. (1942). The Electrical Resistivity Log as an Aid in Determining Some Reservoir Characteristic. Trans. AIME, Vol. 146, pp. 54–61.

ARULANANDAN, K. (1977). Method and Apparatus for Measuring In-Situ Density and Fabric of Soils. Patented, Regents of the University of California, 1977.

ARULANANDAN, K. AND MURALEETHARAN, K. (1988). Level Ground Soil Liquefaction Analysis using In-Situ Properties Part I. J. Geotech. Eng'g. Div., ASCE, Vol. 114, No. 7, pp. 753–770.

ARULANANDAN, K. AND KUTTER, B. (1978). A Directional Structural Index Related to Sand Liquefaction. Proceedings of a Specialty Conference on Earthquake Engineering on Soil Dynamics, ASCE, Pasadena, CA, June 19–21, pp. 213–230.

ARULANANDAN, K. AND SCOTT, R.F. (1992). Project VELACS – Control Test Results. Submitted to ASCE.

ARULMOLI, K., ARULANADAN, K. and SEED, H.B. (1985). New Method for Evaluating Liquefaction Potential. J. Geotech. Eng'g. Div., ASCE, Vol. 111, No. 1, pp. 95–114.

CARMAN, P.C. (1956). Flow of Gases through Porous Media, Academia, New York.

DAFALIAS, Y. AND ARULANANDAN, K. (1979). Electrical Characterization of Transversely Isotropic Sands. Archives of Mechanics, Warsaw.

DE LA RUE, R.E. AND TOBIAS, C.W. (1959). On the Conductivity of Dispersions. J. Electrochem. Soc., Vol. 106, pp. 827–833.

FIEGEL, G.L. (1992). Centrifuge Modeling of Liquefaction in Layered Soils. M.Sc. Thesis, University of California, Davis.

FLORIN, V. AND IVANOV, P. (1961). Liquefaction of Saturated Sandy Soils. Proc. 5th International Conference, Soil Mechanics & Foundation Engineering, Paris.

HUSHMAND, B., CROUSE, C., MARTIN, G. AND SCOTT, F. (1988). Centrifuge Liquefaction Tests in a Laminar Box. Géotechnique Vol. 38, No. 2, pp. 253–262.

JACKSON, P.D. (1975). An Electrical Resistivity Method for Evaluating the In Situ Porosity of Non-cohesive Marine Sediments. Geophysics.

KOZENY, J. (1927). Ueber kapillare Leitung des Wassers im Boden, Wien, Akad, Wiss., Vol. 136, Pt. 2a, p. 271.

LAMBE, P. (1981). Dynamic Centrifuge Modelling of a Horizontal Sand Stratum, Sc.D Thesis, Dept. of Civil Engineering, Mass. Inst. of Technology, Cambridge, Mass.

MEREDITH, R.E. AND TOBIAS, C.W. (1962). Conduction in Heterogeneous systems. Adv. Electrochem., Electrochem. Engng., 2nd ed., J. Wiley and Sons Inc., New York.

MITCHELL, J.K., CHATOIAN, J.M. AND CARPENTER, G.C. (1976). The

Influence of Fabric in the Liquefaction Behaviour of Sand, Report to U.S. Army Engineering Waterways Experiment Station, Vicksburg, Berkeley.

MURALEETHARAN, K.K. AND ARULANANDAN, K. (1991). Dynamic Behavior of Earth Dams Containing Stratified Soils, Centrifuge 91, Proceedings of the International Conference Centrifuge 1991, Boulder, CO, June 13–14.

MURALEETHARAN, K., MISH, K., YOGOCHANDRAN, C. AND ARULANANDAN, K. (1991). Users Manual for DYSAC2.

SCOTT, R.F. (1986). Soil Properties from Centrifuge Liquefaction Tests. Mechanics of Materials, Vol. 5, pp. 199–205.

TOKIMATSU, K. AND SEED, H.B. (1987). Evaluation of Settlements in Sand due to Earthquake Shaking. J. Geotech. Eng'g. Div. ASCE, Vol. 113, No. 8, August, pp. 861–878.

WROTH, C.P. AND WINDLE, D. (1975). Analysis of the Pressuremeter Test Allowing for Volume Change. Géotechnique, Vol. 25, No. 3, pp. 598-603.

WYLLIE, M.R.J. (1955). Verification of Tortuosity Equations. Bulletin of the American Association of Petroleum Geology, Vol. 30, No. 2, pp. 266–267.

A note on modelling small strain stiffness in Cam clay

J.H. ATKINSON, City University, London

The highly non-linear and history dependent stress–strain behaviour of overconsolidated soil is inconsistent with the simple Cam clay models in which the behaviour for states inside the state boundary surface is taken to be elastic. Evidence from laboratory tests demonstrates that soil behaviour is essentially inelastic throughout loading except at very small strains.

Shear modulus in Cam clay

In the Cam clay family of constitutive equations the component of the compliance matrix which relates change in deviatoric stress δq to change in triaxial shear strain $\delta\varepsilon_s$ can be written as 1/3G, where G is thought of as equivalent to a shear modulus. For ordinary Cam clay this is

$$\frac{1}{3G} = \frac{1}{vp'}\left[\frac{\lambda - \kappa}{M(M - \eta')} + \frac{g}{3}\right] \quad (1)$$

where λ, κ and M have the usual meanings (Schofield and Wroth, 1968), $\eta' = q'/p'$ and the elastic shear parameter g, which is equivalent to the volume change parameter κ is given by $g/\kappa = K'/G'$. (In the original derivations of Cam clay elastic shear strains were neglected so that g did not appear; elastic shear strains were added for applications of Cam clay in finite element analyses so there is an inconsistency between the original derivation and the application of Cam clay.) Eqn. (1) applies also for undrained shearing so $G' = G_u$ and consequently it is not necessary to distinguish between total and effective stresses and shear moduli.

For states inside the state boundary surface the elastic shear modulus is given by

$$\frac{G_e}{p'} = \frac{v}{g} \quad (2)$$

while for states on the state boundary surface the shear modulus is given by eqn. (1). If elastic strains are neglected, as in the original derivation of Cam clay so that g = 0,

$$\frac{G}{p'} = \frac{M(M - \eta')v}{3(\lambda - \kappa)} \quad (3)$$

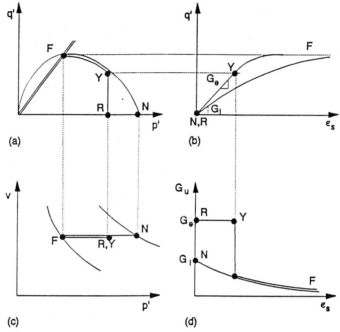

Fig. 1. Basic features of Cam clay

For isotropically normally consolidated soil $\eta' = 0$ at the start of shearing and

$$\frac{G_i}{p'} = \frac{M^2 v}{3(\lambda - \kappa)} \quad (4)$$

The essential features of Cam clay are illustrated in Fig. 1, which shows the behaviour of initially normally consolidated and lightly overconsolidated samples sheared undrained. Figs. 1(a) and (c) show the familiar state paths and Fig. 1(b) shows the stress–strain curves. The path RY is inside the state boundary surface and the behaviour is taken to be elastic. The stress–strain curve is linear with an elastic shear modulus G_e and the volumetric strains are zero for undrained shearing. At Y the state reaches the state boundary surface and for the path YF there are simultaneous elastic and plastic strains as the state moves on the state boundary surface. The path NF for the normally consolidated sample moves on the state boundary surface throughout the shearing; there is no purely elastic straining and the initial value of the shear modulus is G_i.

Fig. 1(d) illustrates the variations of G_u/p' (the tangent shear modulus normalised with respect to the current mean stress) with strain given by

Cam clay. For overconsolidated soil there is a yield point at Y where there is a very sudden reduction in shear modulus. Taking $G' = K'$ (corresponding to a value of Poisson's ratio $\nu' = 0.125$), $M = 1$ and $\lambda/\kappa = 4$ then, from eqns. (2) and (4), G_e will be about an order of magnitude greater than G_i. The general shapes of the two stiffness–strain curves in Fig. 1(d) are distinctly different at small strains.

Current knowledge of soil stiffness

Over the past 20 years or so there have been important developments in laboratory testing (principally hydraulic stress path triaxial cells, more accurate measurement of soil strains with internal gauges and dynamic measurements of soil stiffness at very small strains). These have led to new knowledge of soil stiffness summarised by Atkinson and Sallfors (1991).

Figure 2 illustrates the characteristic stiffness–strain curve for soil; this shows the variation of shear modulus G with shear strain on a logarithmic scale: a similar characteristic curve is obtained if G is plotted against the change of shear stress $\delta q'$. Curves for the bulk modulus K against volumetric strains or change of mean stress are similar except at large strains where the bulk modulus must increase (Atkinson, Richardson and Stallebrass, 1990). A comparison of Fig. 2 with Fig. 1(d) shows significant differences between the observed behaviour of soils and the qualitative predictions of Cam clay.

The essential features of soil stiffness shown in Fig. 2 are:

(a) There is a maximum stiffness G_0 at very small strains (usually less than about 0.001%). This can be measured reliably in dynamic tests

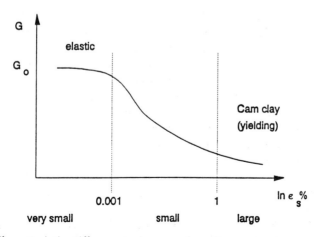

Fig. 2. *Characteristic stiffness–strain curve for soil*

using the velocity of shear waves. In these dynamic tests the damping is negligible at small amplitudes and the behaviour in the very small strain region can be considered as purely elastic. (If a typical value for G_0 is 100 MPa then a shear strain of 0.001% corresponds to an increment of shear stress $\delta q'$ of only 3 kPa.)

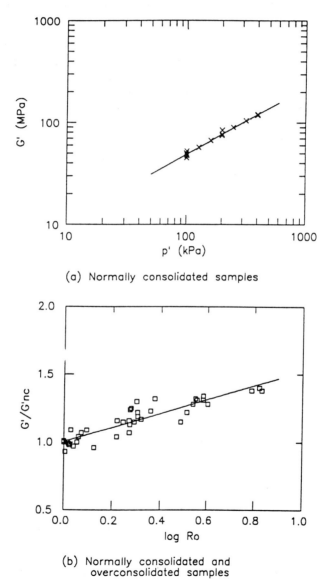

Fig. 3. *Stiffness of kaolin clay at very small strain (Viggiani, 1992)*

MODELLING STRAIN STIFFNESSES

(*b*) At large strains, greater than about 1%, the state of soil has reached the state boundary surface and it is clearly yielding. Unloading results in large irrecoverable plastic strains. In this large strain region a version of Cam clay with an appropriate yield surface is a suitable model.

(*c*) There is an intermediate, small strain region in which the stiffness decreases steadily with strain (or with change of shear stress) from G_0 to the value corresponding to yielding on the state boundary surface.

Within the framework of non-linear stiffness illustrated in Fig. 2 the stiffness at any particular strain depends on the current state, on the history (described by the overconsolidation ratio R_0) and on the recent history. In addition the stiffness depends on the recent history described by the rotation of the stress path (Atkinson, Richardson and Stallebrass, 1990).

In the very small strain region G_0 varies with some power n of the current stress and also with the overconsolidation ratio R_0. Wroth and Houlsby (1985) suggested

$$\frac{G}{p'_r} = \beta \left(\frac{p'}{p'_r}\right)^n \qquad (5)$$

where the reference pressure p'_r is included to make the relationship dimensionless. Recent work by Viggiani (1992) shows that, for clays, the very small strain shear modulus G_0 can be represented by

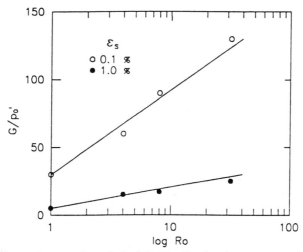

Fig. 4. *Stiffness of reconstituted glacial till soil related to overconsolidation ratio (Atkinson and Little, 1988)*

ATKINSON

$$\frac{G_0}{p'_r} = \beta \left(\frac{p'}{p'_r}\right)^n [1 + m \ln R_0] \qquad (6)$$

Fig. 3 shows results for kaolin clay for which $n = 0.65$ and $m = 0.22$. Eqn. (6) applies equally for overconsolidated samples and for normally consolidated samples for which $R_0 = 1$.

For the small strain region Wroth and Houlsby (1985) gave

$$\frac{G_e}{p'} = \left(\frac{G}{p'}\right)_{nc} [1 + c \ln R_0] \qquad (7)$$

where $(G/p')_{nc}$ is the value for a normally consolidated soil and c is a dimensionless constant. Fig. 4 shows results of undrained triaxial compression tests on a glacial till soil (Atkinson and Little, 1988). These results are in agreement with eqn. (7). In Fig. 4 there are two lines each corresponding to the stiffness at a different strain level and these recognise that the stiffness of soil is non-linear. (Note that eqn. (6) for stiffness at very small strain reduces to eqn. (7) for $n = 1$.) It is significant that in Figs. 3 and 4 the data for normally consolidated samples fall on the same lines as the data for overconsolidated samples indicating that overconsolidation has little effect on the basic nature of soil stiffness.

A further indication of the basic nature of soil stiffness can be found by comparing the variations of stiffness with strain for normally consolidated and overconsolidated samples of the same soil. Fig. 5

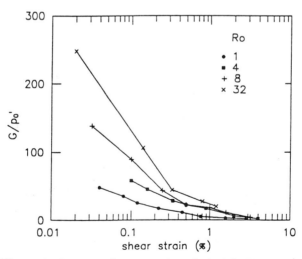

Fig. 5. Stiffness–strain curves for reconstituted glacial till soil (Atkinson and Little, 1988)

MODELLING STRAIN STIFFNESSES

shows a set of stiffness–strain curves from reconstituted samples of a glacial till soil (Atkinson and Little, 1988). Figs. 6(a) and 6(b) show similar sets of data for reconstituted samples of the Bothkennar soil (Allman and Atkinson, 1992). Figs. 5 and 6 show that there is no real

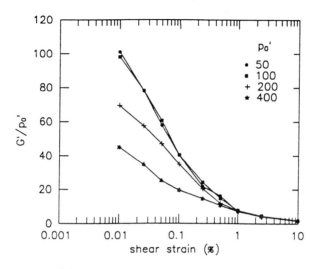

(a) Normally consolidated samples

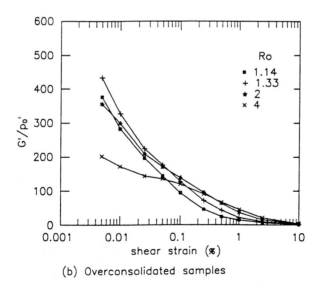

(b) Overconsolidated samples

Fig. 6. *Stiffness–strain curves for reconstituted Bothkennar soil (Allman and Atkinson, 1992)*

difference in the shapes of the stiffness–strain curves for normally consolidated and overconsolidated samples. It has also been noted earlier that the data in Figs. 3 and 4 show that the variations of the very small strain shear modulus G_0 and the small strain shear modulus G (at a given strain) with overconsolidation ratio both apply equally for normally consolidated and overconsolidated soils.

The state path for normally consolidated soil must move on the state boundary surface and there will be inelastic straining with continuous yielding and hardening (or softening on the dry side of critical) and irrecoverable plastic straining. The results shown in Figs. 5 and 6 indicate that the stress–strain behaviour of overconsolidated soil at states inside the state boundary surface should also be regarded as inelastic, with basic behaviour similar to that of normally consolidated soil rather than as essentially elastic, except at very small strains less than about 0.001%.

Modelling small strain stiffness

From the present evidence it is clear that soil is only linear and elastic within the range of very small strains corresponding to changes of stress of a few kPa. At large strains, when the state has reached the state boundary surface soil is yielding and a high proportion of the total strains is inelastic. The question then arises as to the nature of the straining within the intermediate, small strain region. One possibility is to regard this as essentially elastic, but non-linear and to use curve fitting techniques to obtain an empirical expression relating shear modulus G' to strain. This is the approach followed by Duncan and Chang (1970) and by Jardine et al. (1991); the method requires complex laboratory tests in which the stress paths mimic the in situ paths and numerical analyses which should stop and restart at each change in the direction of a stress path.

An alternative approach is to regard soil behaviour in the small strain region as inelastic with yielding and hardening with kinematic surfaces inside the state boundary surface. The simplest approach seems to be to adapt the Cam clay models by including additional surfaces (e.g. Mroz et al., 1979; Al-Tabbaa and Wood, 1989; Atkinson and Stallebrass, 1991). In these models the parameters remain the fundamental parameters required by Cam clay together with additional parameters which describe the relative sizes of the inner and outer surfaces.

In this case soil has a single yield curve representing the end of elastic behaviour between the small and very small strain regions in Fig. 2. Other surfaces, including the Cam clay state boundary surface, are plastic potential surfaces but they are not, strictly, yield surfaces as they do not separate elastic from inelastic behaviour.

Conclusion

Three characteristic regions of soil stiffness can be identified and these are shown in Fig. 2. There is a region where the behaviour is elastic but this corresponds to very small strains. There is a region of large strains where the behaviour is inelastic and there are large irrecoverable plastic strains. Between these there is a region where the stiffness decreases with strain and the behaviour is highly non-linear.

The basic nature of the stiffness–strain curves shown in Figs. 5 and 6 for normally consolidated and overconsolidated soils in the small and large strain regions, is similar and the variations of shear modulus (at a particular strain) with state and history shown in Figs. 3 and 4 are consistent for all overconsolidation ratios.

If the behaviour of normally consolidated samples in the small strain region is associated with inelastic behaviour involving yielding, hardening and plastic straining, then it is logical to conclude that the basic behaviour of overconsolidated soils should be regarded as inelastic too.

References

AL-TABBAA, A. AND WOOD, D.M. (1989). An experimentally based 'bubble' model for clay. Proc. NUMOG III, Elsevier Applied Science, pp. 91–99.

ALLMAN, M.A. AND ATKINSON, J.H. (1992). Mechanical properties of reconstituted Bothkennar soil. Geotechnique (to be published).

ATKINSON, J.H. AND LITTLE, J.A. (1988). Undrained triaxial strength and stress–strain characteristics of a glacial till soil. Canad. Geotech. J., Vol. 25, No. 8, pp. 428–439.

ATKINSON, J.H., RICHARDSON, D. AND STALLEBRASS, S.E. (1990). Effect of recent stress history on the stiffness of overconsolidated soil. Geotechnique, Vol. 40, No. 4, pp. 531–540.

ATKINSON, J.H. AND SALLFORS, G. (1991). Experimental determination of stress–strain–time characteristics in laboratory and in situ tests. Proc. 10th European Conf. on Soil Mechanics and Foundation Engineering, Florence, Vol. 3, pp. 915–956.

ATKINSON, J.H. AND STALLEBRASS, S.E. (1991). A model for recent history and non-linearity in the stress–strain behaviour of overconsolidated soil. Proc. 7th IACMAG '91, Cairns, pp. 555–560.

DUNCAN, J.M. AND CHANG, C.Y. (1970). Non-linear analysis of stress and strain in soils. ASCE, J. Soil Mech. Fdn. Eng. Div., Vol. 96, SM5, pp. 1629–1653.

JARDINE, R.J., POTTS, D.M., ST. JOHN, H.D. AND HIGHT, D.W. (1991). Some applications of a non-linear ground model. Proc. 10th European Conf. on Soil Mechanics and Foundation Engineering, Florence, Vol. 1, pp. 223–228.

Mroz, Z., Norris, V.A. and Zienkiewicz, O.C. (1979). Application of an anisotropic hardening model in the analysis of elasto-plastic deformation of soils. Geotechnique, Vol. 29, No. 1, pp. 1–34.

Schofield, A.N. and Wroth, C.P. (1968). Critical state soil mechanics. McGraw-Hill, London.

Viggiani, G. (1992). Private communications.

Wroth, C.P. and Houlsby, G.T. (1985). Soil mechanics – property characterisation and analysis procedures. Proc. 11th ICSMFE. San Francisco, pp. 1–55.

Towards systematic CPT interpretation

K. BEEN and M.G. JEFFERIES, Golder Associates

This paper describes a systematic interpretation of the cone penetration test with pore pressure measurement (CPTu) based on all three independent measurements from the CPTu, which is quasi-independent of site specific factors. The first step in the approach is to recognize the similarities of sands and clays within a critical state framework for soil behaviour, in particular the description of the initial state of the material. The CPTu can be used to determine the initial state, which largely determines subsequent material behaviour, in any type of soil. However, drainage conditions and the type of soil being penetrated by the cone affect the interpretation of initial state. The second important step in the interpretation is therefore to use the CPTu measurements to estimate the drainage conditions and mechanical behaviour (or type) of the soil being penetrated and to fold this understanding back into the assessment of initial state.

Introduction

The cone penetration test (CPT) is probably the most widely used in situ test in soil mechanics, with the possible exception of the SPT. Modern electronic instruments usually incorporate a pore pressure transducer, resulting in the so-called CPTu.

There appear to be three levels at which CPT data can be viewed and utilized. On the simplest level, the CPT is used by many engineers merely as a stratigraphic logging tool. It does an excellent job of this, especially with the additional pore pressure measurement facility of the CPTu, and although the CPT is not a downhole geologist it has demonstrated success over the past 30 years in providing a reliable method of soil classification. At the second level, some engineers use the instrument as an indexing measurement for the in situ state of the soil; methods of interpretation which result in OCR of clays and the relative density or state parameter of sands fall into this category. On the third level the instrument is used to determine strength parameters, such as c_u, or ϕ; this step requires calibration of the CPT results to independent strength measurements for any particular soil.

Each aspect of the CPTu interpretation (soil classification, in situ state or strength) is essentially carried out independently. There is only a loose link between the soil classification interpretation and the state or strength interpretation. Different methods of interpretation are also used for drained and undrained response of the soils to cone penetration. The first step in the interpretation is thus for the engineer to decide whether the soil is a 'sand' (drained) or a 'clay' (undrained). Partially drained soils, such as silts, are not adequately covered by existing interpretation methods.

A modern CPTu gives three separate readings, q_c, f_s and u, which are seldom used together to enhance the interpretation, except for those soil classification charts which use a normalized penetration resistance and a friction ratio on the axes. This paper provides a unified conceptual background, using all three CPTu measurements, that is applicable to all soil types including silts. The framework is quasi-empirical, but relies on critical state soil mechanics for its fundamentals. Wroth (1988) pointed out that an empirical method should be based on a physical appreciation of why the properties should be related; it should be set against a background of theory and expressed in terms of dimensionless variables. The method described satisfies these basic requirements.

It should be noted that the method applies only to right cylindrical electronic piezocones and procedures which conform to the proposed International Reference Test Procedure (de Beer et al., 1988). Standardization of the equipment and procedures is required for any such empirically calibrated methodology. In addition, the friction ratio is important and the penetrometers should therefore have an independent transducer to measure sleeve friction (as opposed to the so called 'subtraction cones').

Normalized CPT parameters

The normalized, or dimensionless variables, used for the CPTu interpretation in this paper are defined as follows:

$$Q = (q_c - \sigma_{v0})/\sigma'_{v0} \quad \text{or} \quad Q_p = (q_c - p_0)/p'_0 \tag{1}$$

$$F = f_s/(q_c - \sigma_{v0}) \tag{2}$$

$$B_q = (u - u_0)/(q_c - \sigma_{v0}) \tag{3}$$

where: q_c is the cone tip resistance corrected for end area effects (sometimes q_t), and therefore requires use of the CPTu, rather than the CPT

$\sigma_{v0}, \sigma'_{v0}$ are the vertical total and effective stresses before penetration

p_0, p'_0 are the mean normal total and effective stresses before penetration

f_s is the cone sleeve friction (corrected for end area effects)

u is the measured CPTu pore pressure, in this case 5 mm behind the conical tip

u_0 is the initial pore pressure in situ.

These dimensionless variable groups follow Wroth (1988). In this paper, the mean normal stress is sometimes preferred over the vertical stress. Mean normal stress allows more flexibility in the approach as soils with different K_0 values do not have to be treated separately. If measurements of K_0 are not available, values should be estimated (but preferably not from $K_0 = 1 - \sin\phi'$) and a simple linear relationship between p' and σ'_v exists. Many authors use σ_v in the above dimensionless groups and for the purposes of discussion in this paper the two can be treated as equivalent. However, when quantitative analyses are made, it is necessary to distinguish between the two definitions of Q (see eqn. (1)).

The parameter Q is normally used alone for CPT interpretation to determine in situ density, state or strength parameters. Q and F combined result in a good soil classification chart (see, for example, Robertson (1990) for a recent update). B_q has been used less frequently, but can be correlated successfully to OCR for a particular soil stratum (e.g. Wroth, 1988) although no universal relationship has been found. In addition, Houlsby (1988) suggests that $(q_c - u)/\sigma'_{v0} = Q(1 - B_q) + 1$ should be a simple function of OCR.

Relationships between parameter groups incorporated in the systematic approach

The approach described in this paper utilizes several relationships between the dimensionless parameter groupings noted above. These relationships have all been published previously and are briefly described below.

A systematic interpretation of the CPT should include the influence of drainage on penetration resistance. Piezocones (i.e. the CPTu) provide an indirect measurement of the influence of drainage by measuring the pore water pressure induced during penetration as well as the conventional tip resistance and sleeve friction. The additional pore pressure information can be included in the soil classification interpretation through the parameter grouping $Q(1 - B_q) + 1$ plotted against F, rather than simply Q against F (Jefferies and Davies, 1991). Figure 1 shows the usual CPT classification chart after Robertson (1990) and the extended version of Jefferies and Davies (1991).

For drained tests in sands, Been et al. (1987) use the relationship between Q_p and ψ (the state parameter, or offset from the critical state line on an $e - \log p'$ diagram, Been and Jefferies, 1985) shown on Fig. 2 to interpret the CPT. They point out that the correlations for different sand types appear to be related to the slope of the critical state line λ for each sand, but the relationships proposed are less than perfect. It is noted that the parameter k is the intercept of the Q versus ψ lines with

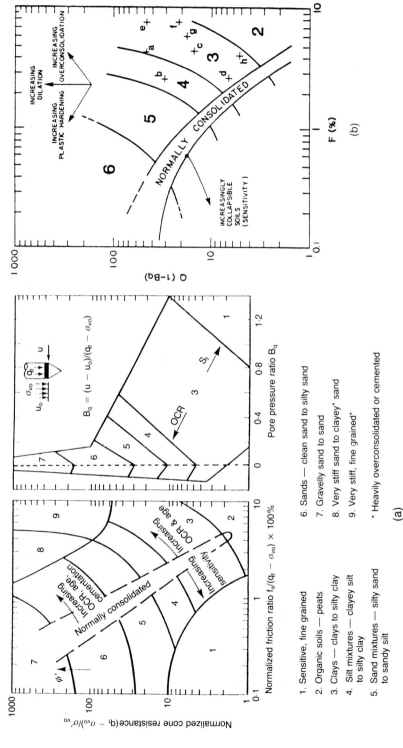

Fig. 1. CPTu soil classification systems (a) after Robertson (1990) and (b) proposed classification chart using $Q(1 - B_q)$ and F (after Jefferies and Davies, 1991)

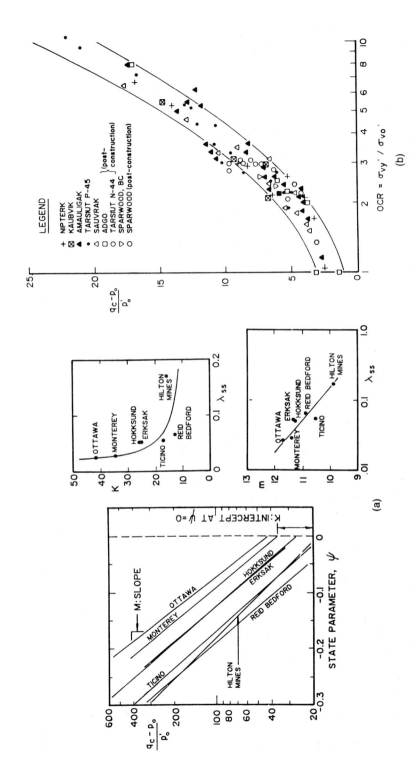

Fig. 2. Interpretation of in situ state from the CPT. (a) CPT interpretation for sands in terms of ψ and λ (after Been et al., 1987a), and (b) CPT interpretation for clays in terms of log OCR (after Crooks et al., 1988)

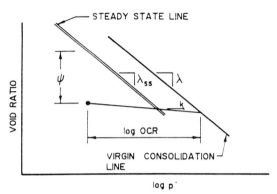

Fig. 3. *Definition of state for sands (ψ) and clays (log OCR)*

the $\psi = 0$ axis, corresponding to penetration of the cone into the material at its critical state. Penetration at the critical state is expected to be a strong function of the critical state friction angle (expressed as M in this paper), as well as the critical state volumetric hardening parameter λ. The relationship between k and critical state parameters can therefore be improved by incorporation of M into the interpretation. The parameter m, which is the slope of the Q versus ψ line, is also a function of soil type.

Q can readily be related to OCR in clays (Wroth, 1988; Sills et al., 1988; Crooks et al., 1988) and there are numerous relationships between OCR and other parameter groupings such as $(u - u_0)/\sigma'_{v0}$ in the literature (e.g. Mayne, 1988). Been et al. (1988) show how OCR is converted to a state parameter ψ value comparable to that for sands using the mod-Cam clay critical state model (Roscoe and Burland, 1968):

$$\log R = \log 2^\Lambda + \psi/(\kappa - \lambda) \tag{4}$$

where $\Lambda = 1 - \kappa/\lambda$ and R is OCR expressed in terms of p'_{max}/p'_0 rather than the conventional vertical stress definition.

Figure 3, from Been et al. (1988), shows the geometric relationship between these two parameters for clays. The mod-Cam clay model is necessary only to define the spacing ratio between the critical state line and the normal consolidation line. The critical state model therefore provides a basis for extending Q versus OCR relationships for clays into Q versus ψ relationships which are comparable to those for sands. For sands ψ and OCR are independent quantities and such a transformation is not possible, although in clays it appears that the transformation is reasonable as the influence of OCR clearly dominates that of ψ.

A difference between the Q–ψ relationships for sands and clays is drainage, or rather, pore water pressure. CPT penetration in sands is drained, while in clays it is essentially undrained and in silts it is

SYSTEMATIC CPT INTERPRETATION

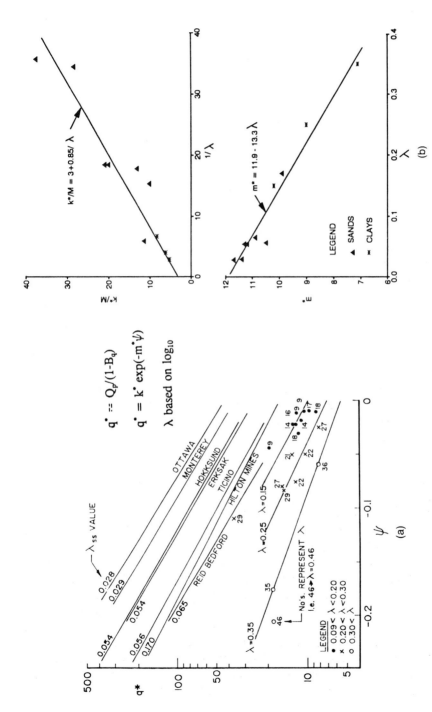

Fig. 4. Unified relationship of $Q_p(1 - B_q)$ to state parameter and critical state parameters M and λ

probably partially drained. Figure 4, adapted from Been et al. (1988), shows a unified relationship of $Q_p/(1 - B_q)$ versus ψ, giving:

$$q^* = Q_p/(1 - B_q) = k^* \exp(-m^*\psi) \tag{5}$$

The parameters k^* and m^* are given as functions of the critical state parameters M and λ. In algebraic form these are:

$$k^*/M = 3 + 0.85/\lambda \tag{6}$$

$$m^* = 11.9 - 13.3\lambda \tag{7}$$

It is noted that the relationship in eqn. (5) covers a range of material behaviours from fully drained sands to undrained clays. It is reasonable to expect that partially drained silts are encompassed by this relationship although this has not yet been demonstrated.

Basis for proposed interpretation

The proposed interpretation is intended to be a systematic, critical state based, interpretation of the CPTu. In this context, critical state soil mechanics is used in the sense of Wood (1990). The important factor is that a critical state model for soil behaviour must consider volume changes. The model itself should predict how strength (c_u or ϕ) changes with state (OCR or density and stress level). This feature distinguishes critical state behaviour from other models and is crucial for a unified framework, as it is amply evident that soil properties remain the same while a soil behaviour (such as dilation, c_u or ϕ) may change when stress or density changes occur. It is not sufficient to describe the same sand at two different states with different sets of parameters (e.g. ϕ).

Within the context of critical state models, the most important parameters describing the soil properties are: M, λ, κ and the yield function (for example, the mod-Cam clay function used to define the relationship between ψ and OCR in the previous section). In addition to *model parameters* for the soil such as M, λ and κ, it is necessary to know the initial state of the soil before a prediction of soil performance under loading can be made. The *initial state* is described by p'_0, K_0, and OCR or ψ. Permeability should also be known for problems which are neither perfectly drained nor undrained.

In a critical state framework, there are not many parameters which influence soil response during cone penetration, but they may be divided into model parameters and initial state parameters. The model parameters M, λ and κ describe the soil type and can readily be measured in the laboratory. Initial estimates of these parameters can also be made with reasonable accuracy by experienced engineers if the soil type is known. The initial state parameters OCR and ψ have a much larger influence on cone penetration and cannot be estimated a priori by

SYSTEMATIC CPT INTERPRETATION

engineers. The CPT should therefore be used to derive OCR and ψ. Permeability estimates can be made independently on the basis of dissipation tests during the CPTu.

The proposed systematic approach to CPT interpretation is thus developed as follows:

(a) Estimates or measurements of in situ stress conditions (σ'_v, K_0) are made.
(b) CPT is used in 'classification' mode to determine soil type, from which model parameters M and λ are estimated.
(c) CPT results are used to determine in situ state of soil.

The above is nevertheless inadequate for engineering design. The following are also required:

(d) CPTu dissipation tests to determine in situ permeability and pore pressures.
(e) Laboratory tests to determine soil model properties. In some cases laboratory tests will require the initial CPT data to estimate the in situ state to ensure that the boundary conditions in the laboratory test are correct.
(f) Reinterpretation of CPT results with the revised soil properties.
(g) Derivation of design parameters (such as undrained strength or ϕ) from the constitutive model as well as the CPT and laboratory data.

In practice, it is desirable that steps (b) and (c) are carried out automatically in the field during the CPT testing process. In addition, estimates of undrained strength or density can be made during step (c). However, step (c) requires knowledge of the soil type, as defined by the critical state parameter λ, and to a lesser extent M. These parameters may vary significantly from one soil layer to the next, so that it is not practical to use an engineer's estimate for the initial field interpretation. While it is reasonable to associate a single λ value with each soil (which is done in this paper), λ varies continuously during the soil profile as different soils are encountered. The only way to deal with variation in properties adequately is if the value of λ is estimated on a continuous basis from the penetration test data itself.

What is needed is a basis for an algorithm to transfer the soil classification data from the 'classification' part of the CPT interpretation directly into the second level, or state, part of the interpretation. As noted above, this must be done on a continuous basis through the profile.

Relationship between λ and soil type index

The CPT has proved itself a good tool for soil classification through several decades of application. It is a mechanical device that measures a

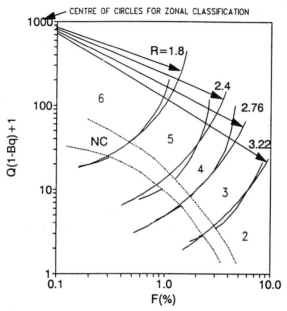

Fig. 5. Revised soil classification chart with concentric zonation (after Jefferies and Davies, 1992). Refer to Fig. 1 for description of zones

complex soil response to loading, and does not see the soil or estimate the grain size as an engineer or geologist might do to classify the soil. Given the CPT's success at soil classification, there must be a link between the measured mechanical behaviour of the soil and the soil type. This link should be expressed in a soil type index calculated from the CPTu measurements.

Figure 1 shows the existing CPT classification charts. Jefferies and Davies (1992) have shown that the boundaries on Fig. 1(b) can be approximated as concentric circles provided the vertical and horizontal scales are appropriately chosen (and unequal). Figure 5 shows the revised chart. With the concentric circle approximation, the soil type is indicated by the circle radius, which is thus used as a soil behaviour type index. A soil index I_c is defined as follows:

$$I_c = \sqrt{\{3 - \log[Q(1 - B_q) + 1]\}^2 + \{1.5 + 1.3(\log F)\}^2} \qquad (8)$$

In eqn. (8), the factor 1.3 is the mapping used to obtain a plot with concentric circles and the centre of the circles is at $\log[Q(1 - B_q) + 1] = 3$, $\log[F] = -1.5/1.3$. The logarithms are to the base 10. I_c is a material behaviour index, ranging from gravels to clays. Table 1, from Jefferies and Davies (1992), summarizes the classification index.

SYSTEMATIC CPT INTERPRETATION

Table 1. *Soil behaviour type from the classification index I_c*

CPT index I_c	Zone	Soil classification
$I_c < 1.80$	6	Sands—clean sand and gravel to silty sand
$1.8 < I_c < 2.4$	5	Sand mixtures—silty sand to sand silt
$2.4 < I_c < 2.76$	4	Silt mixtures—clayey silt to silty clay
$2.76 < I_c < 3.22$	3	Clays
$3.22 < I_c$	2	Organic soils

In the simple critical state model, the only material parameters which describe such differences in soil type are λ, M and to a lesser extent κ. The parameters describing the $Q_p/(1 - B_q)$ versus ψ relationship, k* and m* on Fig. 4, are also functions of λ and M. In order to proceed to the state interpretation (step (c)) of the CPT from the classification step (b) it is therefore only necessary to define a relationship between λ and M, and I_c.

In the absence of information, it is necessary as a first step to assume that M is a constant equal to 1.2, which will generally be sufficiently accurate for the first field interpretation of CPT logs. Figure 6 shows the authors' limited data currently available for a relationship between the index I_c and λ.

Points on Fig. 6 were generated from CPT data where there is also λ information from laboratory testing. For sands and silty sands, the λ

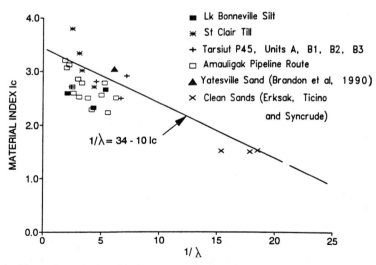

Fig. 6. *Tentative relationship between soil classification index I_c and λ*

values must be obtained from the critical (or steady) state line determined for the sand (Been and Jefferies, 1985; Been et al., 1991). These λ data are available for several sands that have been tested in large CPT calibration chambers (Been et al., 1987a, 1987b; Brandon et al., 1990). For clays and cohesive silts, λ is determined from oedometer consolidation tests on undisturbed samples, and is plotted against I_c derived from adjacent CPT data.

Based on the limited data of Fig. 6, it appears that a useful relationship could be developed between I_c and λ. Although further information is needed to fill out the relationship, a first order (linear) estimate of the relationship, with some subjective bias to the more reliable data sets (St Clair Till, Tarsiut P45 and Erksak sand), is defined by the equation:

$$1/\lambda = 34 - 10 I_c \qquad (9)$$

In Fig. 6 there are groups of data at each end of the line, with no data in the middle of the line. However, it is noted that there is a point representing Yatesville sand in amongst the data points for clays and silts which does suggest that this proposed unification is not unreasonable.

Summary and conclusions

In order to carry out a geotechnical analysis within a critical state framework, there are three types of parameters that must be known:

(a) The initial stress conditions (p'_0, K_0).
(b) The model parameters, related to the soil type (λ, M, κ, k, etc.).
(c) The initial state of the soil (OCR, ψ, etc.).

The engineer requires both laboratory and in situ test data to provide him with this information. Further design parameters may also be desirable, and can be derived from the constitutive model or particular types of laboratory tests.

In summary, the CPTu can provide the initial field estimate of these parameters prior to laboratory testing as follows:

(a) Estimate the in situ stress conditions (including K_0).
(b) Use the CPT measurements to determine $Q(1 - B_q) + 1$ and F, calculate the material index parameter (eqn. (8), Fig. 5), and make an initial estimate of the critical state volumetric hardening parameter λ (eqn. (9), Fig. 6).
(c) Derive the parameters k* and m* which describe the relationship between $q^* = Q_p/(1 - B_q)$ and ψ (eqns. (6) and (7), Fig. 4). In fact, λ is simply the linking parameter between k* and m* and I_c, and could therefore be eliminated from the equations.

(d) Calculate the in situ state parameter ψ or OCR (eqn. (5), Fig. 4), given the measured Q and B_q, calculated k^* and m^*, and the estimated in situ stresses.

The proposed approach provides the initial state of a soil and a preliminary estimate of some critical state parameters for all soil types. It is not necessary to know beforehand whether the material is drained or undrained during CPT penetration. The approach uses all 3 CPT measurements (q_c, f_s and u) which are currently common, with some generalized empirical relationships to come up with a soil classification index and a measure of in situ state. Additional relationships may be derived from the basic relationships.

The methodology also has limitations. The engineer is required to estimate the initial stress state K_0. A constant value of M has been assumed for all soils types, as M cannot really be determined without additional laboratory tests on each soil. The equations used are each empirically derived and are thus sensitive to the type and quality of data used in their derivation. However, the equations are straightforward and readily programmed into a CPT processing package, without a requirement for numerical solution methods.

It is the nature of the approach and its relationship to critical state soil mechanics that bears consideration. If it is accepted that critical state models provide a conceptual framework for engineers to gain insight into soil behaviour, and thus carry out better designs, then the approach provides the basis for incorporating the CPTu into this framework at the investigation phase of a design. In addition, variability of soil properties in the field is real, important and very difficult to incorporate into engineering. A continuous profiling tool such as the CPTu, interpreted in a systematic manner, is almost essential for improved engineering in the future.

References

BEEN, K., CROOKS, J.H.A. AND JEFFERIES, M.G. (1988). Interpretation of material state from the CPT in sands and clays. Proceedings, ICE Conference on Penetration Testing in the UK, Birmingham, UK, 89–92.

BEEN, K. AND JEFFERIES, M.G. (1985). A state parameter for sands. Geotechnique, Vol. 35, No. 2, 99–112.

BEEN, K., JEFFERIES, M.G., CROOKS, J.H.A. AND ROTHENBURG, L. (1987a). The cone penetration test in sands, Part 2: General inference of state. Geotechnique, Vol. 37, No. 3, 285–299.

BEEN, K., JEFFERIES, M.G. AND HACHEY, J.E. (1991). The critical state of sands. Geotechnique, Vol. 41, No. 3, 365–381.

BEEN, K., LINGNAU, B.E., CROOKS, J.H.A. AND LEACH, B. (1987b). Cone penetration test calibration for Erksak (Beaufort Sea) sand. Canadian Geotechnical Journal, Vol. 24, No. 4, 601–610.

BRANDON, T.L., CLOUGH, G.W., WASKIEL, A. AND MARSHALL, J.L. (1990). Liquefaction prediction based on calibration chamber testing of K_0 consolidated silty sands. Report to US Bureau of Reclamation No 9-PG-81-15890, Virginia Polytechnic Institute.

CROOKS, J.H.A., BEEN, K., BECKER, D.E. AND JEFFERIES, M.G. (1988). CPT interpretation in clays. First International Symposium on Penetration Testing, Orlando, Florida, Vol. 2, 715–722.

DE BEER, E.E., GOELEN, E., HEYNEN, W.J. AND JOUSTRA, K. (1988). Cone penetration test (CPT): International reference test procedure. First International Symposium on Penetration Testing, ISOPT-1, Orlando, Florida, Vol. 1, 27–51.

HOULSBY, G.T. (1988). Introduction to papers 14–19. Proceedings ICE Conference on Penetration Testing in the UK, Birmingham, UK, 141–146.

JEFFERIES, M.G. AND DAVIES, M.P. (1991). Discussion on Soil classification by the cone penetration test. Canadian Geotechnical Journal, Vol. 28, No. 1, 173–176.

JEFFERIES, M.G. AND DAVIES, M.P. (1992). Use of CPT data to estimate equivalent SPT $(N)_{60}$. Submitted to ASTM Geotechnical Testing Journal.

ROBERTSON, P.K. (1990). Soil classification using the cone penetration test. Canadian Geotechnical Journal, Vol. 27, No. 1, 151–158.

ROSCOE, K.H. AND BURLAND, J.B. (1968). On the generalized stress strain behaviour of wet clay. In Engineering Plasticity, eds. Heyman and Leckie, Cambridge University Press, 535–609.

SILLS, G.C., MAY, R.E., HENDERSON, T. AND NYIRENDA, Z. (1988). Piezocone measurements with four pressure positions. Proceedings ICE Conference on Penetration Testing in the UK, Birmingham, UK, 247–250.

WOOD, D.M. (1990). Soil behaviour and critical state soil mechanics. Cambridge University Press. 462 pp.

WROTH, C.P. (1988). Penetration testing – a more rigorous approach to interpretation. First International Symposium on Penetration Testing, Orlando, Florida, Vol. 1, 303–311.

Determining lateral stress in soft clays

J. BENOÎT, University of New Hampshire, and
A.J. LUTENEGGER, University of Massachusetts at Amherst

Determination of in situ lateral stress remains a difficult task despite two decades of significant innovation in field and laboratory measurement techniques. This paper summarizes and discusses the advantages and shortcomings of the various laboratory and field methods available from which direct or indirect measurements of lateral stress in soft clays can be obtained. Two case histories at soft clay research sites in the northern United States are used to illustrate the fairly wide scatter in the interpreted test data from the various tests and methods. The results suggest that the self-boring pressuremeter still remains the tool of reference and generally provides a lower bound value of in situ horizontal stress. However, other intrusive type methods are promising and can give relatively good estimates of the field lateral stress state.

Introduction

Accurate measurements of in situ lateral stress still elude the geotechnical profession despite the emergence of several innovative testing techniques which have been developed in the last twenty years. In his general report at the 1975 ASCE Conference on In Situ Measurement of Soil Properties held in Raleigh, North Carolina, Professor Wroth discussed the in situ methods available at the time for the measurement of initial stresses and suggested that a new device, called the self-boring pressuremeter, introduced in France (Baguelin et al., 1972) and in England (Wroth and Hughes, 1973) had the potential to measure lateral stresses in the ground with a significant degree of improvement over other existing methods. In the 1975 conference, most references were made to in situ methods involving the principle of hydraulic fracturing or the use of push-in total earth pressure cells. Comprehensive testing programs involving both tools were presented and discussed by Tavenas et al. (1975) and by Massarsch et al. (1975) indicating their advantages and shortcomings and pointing towards the self-boring pressuremeter as a potential for measuring in situ stresses with sufficient accuracy. Since then, several field devices have been introduced and have had varying degrees of success in evaluating the horizontal stress in the ground. These methods include the Marchetti flat plate dilatometer (DMT) and the K_0 stepped blade (ISB). Nearly two decades after Professor Wroth's review, the assessment of the in situ horizontal stress is still unresolved and is referred to by Jamiolkowski et al. (1991) as a 'black hole' in experimental soil engineering.

Several laboratory methods are also in existence for the determination of the lateral earth pressures. These methods often involve the use of the triaxial or the oedometer test apparatus to provide some direct measurement or, alternatively, involve relating the overconsolidation ratio (OCR) to the lateral stress ratio at-rest (K_0). Laboratory measurements are attractive in terms of parametric studies but are always hindered by sampling disturbance and often cannot reproduce effects from a number of mechanisms responsible for the geologic history of cohesive soils such as cementation, ageing, etc. Wroth, in his 1975 report, stated that there are no satisfactory laboratory methods of determining K_0 for a natural soil.

This paper presents and critiques various in situ and laboratory methods used for the determination of lateral stress in soft clays and discusses the advantages and shortcomings of the methods, especially in terms of their applicability to current geotechnical design methods. The discussion is about lateral stress evaluation with less emphasis on the coefficient of earth pressure at-rest, K_0, since the latter is not a measured value but is calculated, which can thus lead to significant errors unrelated to the in situ lateral stress (Tavenas et al., 1975; Massarsch et al., 1975). Two case histories involving the use of in situ and laboratory techniques to obtain values of lateral stress for design purposes at two natural clay sites are presented.

Current methods for estimating lateral stresses in clays

Methods for evaluating in situ lateral stresses can broadly be classified as either indirect or direct. Indirect methods involve first estimating an intermediate soil parameter such as friction angle (ϕ'), plasticity index (PI), stress history, etc., to arrive at either a lateral stress value or at a lateral stress ratio. Direct methods, on the other hand, implicitly involve the measurement of actual applied stresses either in the laboratory or in the field. In the case of at-rest conditions, the methods may be characterized as falling into one of the following areas:

(*a*) Indirect estimates of K_0 based on other soil properties.
(*b*) Indirect estimates of K_0 based on in situ penetration tests.
(*c*) Active or passive laboratory measurements of lateral stress on undisturbed or reconstituted samples.
(*d*) Intrusive field methods to measure lateral stress.
(*e*) Non-intrusive field methods to measure lateral stress.

While the first two areas are largely indirect methods, the other three are considered more direct but in many cases may involve an empirical approach not unlike an indirect estimate.

Indirect estimates of K_0 based on other soil properties

For a perfectly elastic isotropic material, the theory of elasticity gives the coefficient of earth pressure at-rest in terms of Poisson's ratio, ν as:

$$K_0 = \frac{\nu}{1-\nu} \qquad (1)$$

Soft saturated clays behaving in drained loading at very small strain levels may be expected to behave elastically and therefore eqn. (1) may be useful. However, these conditions are not easily satisfied and, furthermore, Poisson's ratio is difficult to evaluate.

The most frequently used empirical method to estimate the at-rest lateral earth pressure is the following simplification of the expression introduced by Jáky (1944) for normally consolidated soils:

$$K_0 = 1.0 - \sin \phi' \qquad (2)$$

where ϕ' = drained internal friction angle of the soil.

For typical friction angles of soft clay soils, 20–30°, this gives a narrow range of K_0 of 0.65–0.50. This expression is only valid for truly normally consolidated soils (OCR = 1) which exhibit zero cohesion during drained shear. The value of friction angle should ideally be evaluated from K_0 consolidated drained triaxial compression tests; however, this, of course, requires an initial estimate of K_0. Brooker and Ireland (1965) performed an experimental study to empirically relate K_0 to OCR and PI. Schmidt (1966), Massarsch (1979) and Mayne and Kulhawy (1982) developed similar expressions to evaluate K_0 for normally or overconsolidated clays also based on PI, friction angle and/or OCR. Mayne and Kulhawy (1982) showed that the simple expression from Jáky seems to hold true for a wide range of cohesive soils. Wroth (1984) reminded us that since, for normally consolidated clays, the drained friction angle is a function of PI, we may thus expect a secondary correlation to exist between K_0 and PI. A recent compilation by Kulhawy and Mayne (1990) developed an alternative expression based on a correlation between laboratory preconsolidation pressure and liquidity index (LI).

Indirect estimates of K_0 based on in situ penetration tests

In recent years, significant advances have been made in understanding the measured response of clays to static and dynamic penetration tests. Recognizing that the initial stress state, i.e. octahedral stress, may dominate measurements such as CPT cone tip resistance, CPTU porewater pressure, and SPT N-values, a number of empirical correlations

Table 1. Laboratory methods for evaluating K_0 or lateral stress in clays

Type of test	Method	Active or passive	References
Oedometer Rings	Directional Friction Ribbons	A	Terzaghi (1920)
	Null Pressure Ring	A	Brooker & Ireland (1965) Ofer (1982)
	Strain Gauge Instrumented Oedometer Ring	P	Sowers et al. (1957) Komornik & Zeitlen (1965) Som (1970) Silvestri & Morgavi (1982)
	Oedometer Ring with Flush Mount Pressure Transducer	P	Goodrich (1904) Thompson (1963) Ladd (1965) Holtz et al. (1986)
	Fluid Surrounded Oedometer Ring	P	Dyvik et al. (1985)
	Ratio of Directional Yield Stress	P	Zeevaert (1953) Simons (1965) Tavenas et al. (1975) Mahar and O'Neill (1983) Mesri & Castro (1987)
	Instrumented DSS Membrane	P	Dyvik et al. (1981)
Split Rings	Stress Meter	A	Sherif & Ishibashi (1981)
	3 Section Split Ring with LVDT	A	Senneset (1989)
Flexible Wall Triaxial Chambers	Stress Path Tests	A	Moore & Spencer (1972) Poulos & Davis (1972) Campanella & Vaid (1972) Chang et al. (1977) Berre (1982) Bauer & El-Hakim (1986) Lo & Chui (1991) Garga & Khan (1991)
	Double Cell Chamber	A	Yasuhara et al. (1988)
Slurry Consolidation Chambers	Flush Mounted Pressure Transducer	P	Abdelhamid & Krizek (1976)

have been suggested to obtain indirect 'first-order' approximations of K_0 (e.g. Kulhawy et al., 1989). These techniques may involve a number of substantial errors, deriving both from the quality of the in situ test results and the assumptions made in developing the data base. Consequently, the actual data usually show wide scatter, especially true for soft clays in which the major component of stress during the in situ test is likely to be excess porewater pressure. Also, in soft clay, the vertical effective stress contributes a dominant influence over the octahedral stress, making the lateral stress more difficult to assess. This approach should be used with extreme caution.

Active or passive laboratory measurements of lateral stress

Most laboratory methods for estimating K_0 involve performing an axial compression test on a soil sample while measuring the developed or applied lateral stress for the condition of no lateral strain. A summary of the methods which have been used for soft clays is presented in Table 1. The most common techniques revolve around the use of the triaxial test apparatus where lateral pressure is applied to the sample, or the oedometer test where a passive type pressure measurement is usually recorded. Unfortunately, these methods generally suffer from a number of deficiencies and do not produce true K_0 conditions, since either some finite amount of lateral strain is allowed to occur or shear stresses are developed during the loading as discussed by Bishop (1958). However, with the advent of closed-loop feedback schemes in triaxial testing, the chamber pressure can be adequately controlled to ensure a condition of zero lateral strain.

In addition to these likely errors, laboratory methods might be considered only to provide a lower bound estimate of K_0 for a particular soil since stresses are imposed by a sequence of simple loading and unloading to achieve the required stress history or to overcome inevitable sampling disturbance. This laboratory approach often implies de facto normalized soil behaviour, since the soil sample is loaded beyond its current oedometric yield stress to produce a normally consolidated condition. This implicitly involves a destructurization of the natural soil fabric and in effect, the creation of a new non-characteristic soil. Furthermore, these procedures do not adequately replicate effects of soil ageing and other factors and, therefore, do not reproduce typical field conditions of real natural soils even in young soil deposits. As described by Finno et al. (1990), a range of laboratory K_0 values is required to account for such simple factors such as sample disturbance effects and uncertainties associated with the true geologic history of the deposit.

An alternative technique for evaluating the current state of stress in a normally consolidated clay may be made by considering the loading history of a sample. For virgin compression in which K_0 is generally considered a constant and represents a lower bound value, the stresses defined at the preconsolidation stress also represent yield conditions; i.e. the current horizontal and vertical effective at-rest stresses are also the maximum past stresses, hence OCR = 1. The value of the vertical yield stress σ'_{pv}, is determined from a conventional oedometer test. An initial approximation of the value of the lateral yield stress σ'_{ph} may be obtained from an oedometer test conducted on a sample trimmed in the horizontal direction. Becker et al. (1987) have shown that this value is almost identical to the yield stress evaluated from the SBPM test. Therefore, one might reasonably estimate K_{onc} as the ratio $\sigma'_{ph}/\sigma'_{pv}$. This has previously been suggested by a number of authors as indicated in Table 1.

Additionally, the value of σ'_{ph} may be used to represent the upper

Table 2. In situ methods for measuring lateral stress in clays

Type of test	Method	Active or passive	References
Intrusive	Push-In Spade Cells	P	Massarsch (1975) Tedd and Charles (1981)
	Hydraulic Fracturing	A	Bjerrum and Anderson (1972) Lefebvre et al. (1991)
	K_0 Stepped Blade	A	Handy et al. (1982) Lutenegger and Timian (1986) Handy et al. (1990)
	Lateral Stress Cone	P	Huntsman et al. (1986) Jefferies et al. (1987) Campanella et al. (1990)
	Cone-Pressuremeter	P	Lutenegger (1990)
	Dilatometer	A	Marchetti (1980) Lutenegger (1988)
	Instrumented Model Piles	P	Kenney (1967) Azzouz and Morrison (1988) Coop and Wroth (1989) Bond and Jardine (1991)
Non-Intrusive	Self-Boring Pressuremeter	P A	Baguelin et al. (1972) Wroth and Hughes (1973)
	Self-Boring Load Cell	P	Dalton and Hawkins (1982) Huang and Haefele (1990)

bound value of K_0 as $K_{0\,max}$ which is equal to $\sigma'_{ph}/\sigma'_{vo}$. Thus it may also be shown that:

$$K_{0\,max} = K_{onc} \, OCR \qquad (3)$$

This eliminates the need to perform a horizontal oedometer test if another reasonable estimate of K_{onc} can be made by other field or laboratory techniques. This expression provides an upper bound of K_0 for simple unloading.

Field intrusive methods

A number of in situ intrusive or full-displacement techniques have been suggested for estimating the in situ at-rest lateral stress conditions in soft clays as summarized in Table 2. Other than the measurements using an instrumented pile which was first suggested by Kenney (1967), in situ techniques for evaluating K_0 started in the early 1970s with the use of the hydraulic fracturing test and push-in spade cell tests. In many cases, the tests listed in Table 2 may be used to provide an upper bound value of the true in situ value of at-rest lateral stress since significant soil displacement is required for insertion of the devices. All of these methods suffer from a lack of complete understanding of the complex mechanisms during installation. Consequently, these methods remain highly empirical in nature, like most in situ tests. Among the more common of these methods in use today are: push-in spade cells, hydraulic fracturing and Marchetti flat plate dilatometer. Other less common techniques include the K_0 stepped blade, the full displacement pressuremeter and the lateral stress cone. Instrumental model piles have also been used to measure post-insertion lateral stress.

The principle involved in these types of test is essentially the same; however, the operation differs somewhat with each of the tests. In soft clays, installation of any of these devices results in large positive excess pore water pressures. The methods either involve a measurement immediately following the installation of the in situ tool or a waiting period is allowed to account for dissipation of pore pressures and reduction of the effects resulting from the installation. The flat dilatometer, the K_0 stepped blade and the lateral stress cone are generally used to make measurements at the time of installation and thus have the ability to record a significant amount of information in a relatively short time. On the other hand, the push-in spade cell and the hydraulic fracturing test require a substantial period for pore pressure stabilization – days and often weeks – and thus generate potentially useful information but at a slower rate.

For both types of test, a total stress measurement is made which needs to be corrected in the case of tests where measurements are made immediately following installation; in the other case, the resulting value of stress may be taken as close to the actual in situ horizontal stress if sufficient time is allowed to elapse. However, of all these devices, regardless of the care involved in installing the probe, or the width to thickness ratio for the blade type devices, or the length of time between penetration and measurement even after more than a year, the at-rest conditions are violated simply as a result of installation. Furthermore, the devices create a destructured zone of soil in the immediate vicinity of the probe, i.e. where the measurement takes place. Tavenas et al. (1975) and Massarsch et al. (1975) clearly stated the repeatability and the resolution of their measurements, as well as the drawbacks from probe insertion, inevitably lead to a value only close to the in situ at-rest condition.

It should be remembered that within the normal test period none of these tests measure lateral stress under at-rest conditions. In a soft clay, given sufficient time, the assumption is usually made that the soil will return to an at-rest condition. Presumably, if the probe was left indefinitely, the stress gradient created by inserting the instrument would dissipate as a result of soil creep and stress equilibrium would be achieved.

One of these devices, the K_0 stepped blade, attempts to account for the disturbance by relating the test measurement to soil displacement (via blade thickness) and extrapolating to a condition of zero displacement. Unfortunately, in soft clays, all intrusive tests are considered large strain problems in terms of soil behaviour, and therefore the soil will essentially be in a state of plastic failure, even for a thin spade instrument. It should be expected then that in a soft clay deposit all of the intrusive methods should give essentially the same results with little dependence on geometry.

Field non-intrusive methods

All of the direct methods discussed thus far involve some level of disturbance either from sampling for laboratory testing or as a consequence of the insertion mechanisms of various in situ devices in the ground. In response to the problems associated with intrusive in situ testing methods, a device, called the self-boring pressuremeter, capable of insertion with minimum disturbance was introduced in the early 1970s by two independent research groups in France and in England. Although both ensuing devices are different in terms of geometry, operation and field installation procedures, the concept remains the same. In spite of several references made at the 1975 conference regarding the great potential of this promising field instrument, twenty years after its introduction there remains significant uncertainties in the lateral stress results obtained in soft clays and in other soils in general.

The French and the British probes are designed to measure the in situ lateral stress using two different approaches. The French probe, called the pressètre autoforeur (PAF), is a hollow cylinder covered by a rubber membrane which is inserted into the ground with its measuring cell slightly inflated such that the average diameter of the expandable section of the probe is equal to the diameter of the cutting shoe. Once at the test depth, the lateral stress is evaluated through a passive type measurement which consists of monitoring the internal pressure, during the stabilization period, necessary to maintain an average condition of zero lateral strain. The device uses the entire volume of the expandable probe section to evaluate the lateral stress and consequently the measured value is an average stress for the soil mass surrounding the probe. The British probe, originally called a Camkometer, is a similar device except that the membrane is not inflated during self-boring which, in theory, ensures a perfect cylindrical insertion. The probe measures the lateral stress by using a system of spring loaded feeler arms to detect movements of the expanding membrane and thus is capable, for a standard unit, of measuring the lateral stress in three independent directions on a single horizontal plane. In principle, this active measuring system is superior in accuracy but at the same time the additional measurements have had a tendency to seriously question the ability of the SBPM to measure the true in situ lateral stress. Dalton and Hawkins (1982) rationalized the different stresses measured by the British probe using the Mohr circle of stress representation to arrive at a major and minor direction and magnitude of the principal stress in the horizontal plane. What these measurements imply is that a condition of anisotropy exists in the horizontal plane. Although there should not be a reason for denying such state of stress, it is likely the magnitude difference between the major and minor stresses has been overestimated in the past due to a lack of understanding of the influence of several factors, both mechanical and behavioural, on the measurements.

To account for mechanical, electronic and data acquisition problems associated with early SBPM models, several interpretation methods, summarized by Lacasse and Lunne (1982) were introduced to help the determination of lateral stress in the ground. These methods, some of which are somewhat exotic, were often introduced to help bring the field measurements to a more credible level or at least to a level believed to be appropriate in soil mechanics. Consequently, these methods, with the exception of the lift-off by inspection method and the pore pressure method, transformed a direct measurement method into a semi-empirical method. Research into very soft to stiff clays with more recent SBPM models has shown that with improved electronic, mechanical and data acquisition systems (Benoît, 1983; Benoît and Clough, 1986; Findlay, 1991) the simple method involving a direct inspection of the pressuremeter curve is the best and should be the only interpretation method used to evaluate the horizontal stress in the ground from an SBPM test. The pore pressure method is too simplified given the complexity of the pore pressure field surrounding the probe during the early stages of the test.

Other factors are of major importance when using the SBPM in soft clays. Careful membrane and transducer calibrations are essential since any corrections and fluctuations constitute a significant portion of the measured stress. A complete revised calibration procedure for the SBPM membrane and transducers has been recently suggested (Findlay and Benoît, 1992). The revised procedure results from numerous calibrations under controlled conditions in the laboratory. In addition to calibrations for membrane stiffness and temperature effects, the excess pore pressures generated during insertion and during the initial stages of the expansion test must be adequately accounted for. Independently of the method of insertion, i.e. conventional cutting technique or jetting insertion, pore pressures will be generated in soft clays due to the inevitable imperfect insertions. The pore pressures can be positive or negative even in soft clays (Findlay, 1991). Other factors affecting the quality of the insertion and thus the results in soft clays have been investigated and show that cutting or jetting will essentially yield the same quality results if proper procedures are used to minimize disturbance (Atwood, 1990; Findlay, 1991).

Typically, SBPM expansions are non-uniform and thus, as some sections of the probe start to expand before other sections, the initial horizontal stress measured by those sections expanding late include a component of excess pore pressure from shear which must be accounted for. Finite elements analyses are currently in progress to evaluate the

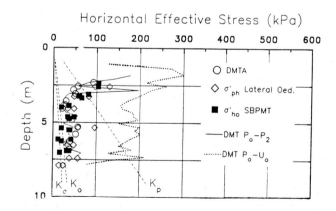

Fig. 1. *Horizontal effective stress profile — Pease*

field of pore pressures surrounding the probe in the usual case of non-uniform expansion.

A tool which operates under the same principle as the self-boring pressuremeter is the self-boring load cell pressuremeter (Dalton and Hawkins, 1982). The principle is similar to the SBPM with the exception that no expansion of the probe takes place during testing. An internal pressure must be regulated to maintain a condition of zero strain at the probe wall. This pressure is said to be equivalent to the horizontal in situ stress. However, experience with the SBPM expansion pressuremeter shows that the shape of the pressuremeter curve, especially at low strain, is critical in evaluating the quality of the test curve. A preliminary method introduced by Findlay (1991) shows that the use of the entire pressuremeter curve can help reject objectively tests of questionable quality.

Case histories of determining lateral stress

Results from laboratory and field techniques used in the evaluation of the in situ lateral stresses are presented using case histories from two soft clay sites located in the northeastern United States. The sites are the I-95 Portsmouth Pease Air Force Base site in southeast New Hampshire and the University of Massachusetts at Amherst Geotechnical Experimentation site in Amherst, Massachusetts. The results presented herein are part of an ongoing project where a wide range of field intrusive and non-intrusive techniques, as previously described, as well as laboratory tests are being used to assess the potential of these methods to evaluate the in situ lateral stress.

I-95 Portsmouth, New Hampshire site

The Pease Air Force Base site is located within the I-95 area previously investigated by Ladd (1972) during the construction of a highway interchange. As part of that study, a test section was constructed in stages and loaded to failure. The site profile for the current study consists of a deposit of very soft highly sensitive grey marine silty clay, with occasional silt and sand lenses, approximately 5.8 m thick with an overlying stiff mottled silty clay about 1.5 to 2.5 m in thickness. Below the very soft clay is a thin layer of glacial till deposit overlying bedrock. The groundwater condition at the site is hydrostatic as confirmed by a series of piezometers. The water table varies from the ground surface up to 1 m above ground due to seasonal fluctuations and frequent beaver damming. Geotechnical soil properties are described in detail in Findlay (1991) and NeJame (1991) but are briefly summarized in Table 3.

Figure 1 shows horizontal effective stress results from laboratory and in situ testing. The laboratory tests conducted to date consist of a series

Table 3. Typical soil properties at I-95 Portsmouth Pease AFB and University of Massachusetts at Amherst sites

Test location	Soil description	Water content (%)	Plastic limit (%)	Liquid limit (%)	Plasticity index (%)	Liquidity index (%)	Normalized NC strength* (S_u/σ'_{vo})	S_t
Pease I-95	Soft Marine grey silty clay with occasional silt and sand lenses	35–50	20–25	25–40	10–15	1.3–2.3	0.13–0.30	10–15
Amherst UMASS	Stiff to soft, brown to grey silty varved clay	40–69	26–34	37–55	9–26	1.2–2.9	0.18–0.24	5–19

*The normalized strengths are shown as a range from various laboratory tests using different stress paths.

of oedometer tests carried out on horizontally trimmed samples. Those results are presented as a range of effective preconsolidation pressures given the uncertainties associated with the Casagrande (1936) procedure. The in situ test results consist of lateral stresses from one of several SBPM soundings carried out at the site, using only the good quality tests as defined by Findlay (1991). The results shown are for a profile where the self-boring was accomplished using the conventional cutting technique. In addition, dilatometer test results are shown from a standard DMT profile using the P_0–u_0 and P_0–P_2 values, as well as DMTA dissipation test results also corrected for the porewater pressure existing at the site prior to probe insertion. To provide a basis for comparison of these results, the active, at-rest, and passive states of stress are presented assuming that the soil is all normally consolidated. The active and passive stress conditions provide approximate lower and upper bound conditions, respectively, for comparison with measurements by other techniques.

The results show that the DMT P_0–u_0 values seem to follow a trend very similar to that of the SBPM test results but, as expected, with a large magnitude difference. However, the P_0–P_2 values from the DMT tests follow a trend and magnitude near that of SBPM results in both the soft and stiffer part of the clay deposit. The DMTA dissipation P_0–u_0 values are also found to be in very good agreement with the SBPM data. In the stiffer part of the deposit, the DMTA P_0–u_0 values are slightly less than the SBPM data while in the soft clay the DMTA results are nearer the maximum SBPM effective horizontal stress values. The agreement is similar between the minimum values of effective preconsolidation stress from horizontally trimmed oedometer tests and the SBPM results.

University of Massachusetts at Amherst site
The test site at the University of Massachusetts is situated in the Connecticut River Valley of western Massachusetts and therefore the subsurface stratigraphy consists predominantly of lacustrine varved clay sediments overlying glacial outwash deposits resting on bedrock. Beneath a thin surficial layer of fill, the varved clay is overconsolidated but grades to a very soft near normal consolidated consistency below a depth of about 7.6 m. In this zone, which extends to a depth of approximately 19.8 m, the liquidity index as shown in Table 3 is at or greater than 1. However, below a depth of 4 m, the soil is soft enough that it will not maintain an open borehole. Based on measurements from a number of piezometers located at the site, the in situ pore water pressures are known to be hydrostatic. In general, the water table is located at a depth of about 1.5 m but fluctuates about 1 m throughout the year.

In situ and laboratory tests have also been conducted at this site throughout the profile to evaluate the in situ lateral stress conditions. The results of interpreted and measured lateral stresses are summarized in Fig. 2. The data for this site, as for the I-95 site, indicate that the values of lateral stress evaluated from laboratory (primarily oedometer-type tests) generally provide results which fall nearer the lower bound. This is to be expected since all of these tests suffer from unknown sample disturbance effects and essentially respond to simple loading/unloading sequences. This becomes more obvious in the upper part of the profile where other effects tend to produce conditions which deviate from a simple loading/unloading sequence to produce lateral stresses above those predicted for a normally consolidated condition even though the soil has a soft consistency.

Results of the intrusive field tests (spade cells, DMTA, cone pressuremeter) give values which generally fall nearer the estimated K_0 condition in the lower part of the profile but clearly give measured values which

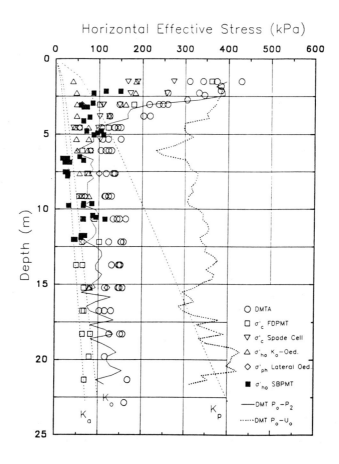

Fig. 2.

are much greater than those predicted by the normally consolidated passive condition in the upper part of the profile. In the soft clay, the post-insertion, post-consolidation measured stresses appear to be independent of the specific geometry of the instrument. This is also as expected since any disruption of the soil will initially produce a limiting condition. It also appears that the initial effective stresses from the DMT P_0–P_2 may be used to provide an initial estimate of at-rest conditions in these soils.

The results obtained with the SBPM suggest that at this site also the SBPM gives results which approximate a lower bound stress condition and compare well with laboratory values. It would be expected that the SBPM results in the upper part of the profile would be lower than results obtained from intrusive tests; this is indicated in Fig. 2.

Conclusions

From the results presented at the two research test sites described in this paper it is clear that the self-boring pressuremeter remains the tool of reference for the evaluation of lateral stress data. These results suggest that in soft clays, the SBPM generally provide a lower bound value for the in situ effective horizontal stress. Other methods such as intrusive in situ tests are promising in terms of providing a relatively good estimate of the field lateral stress state. Such techniques are simple, nearly operator independent and yield a mass of information which can be calibrated using more sophisticated methods such as the SBPM. In soft clays, the oedometer test carried out on horizontally trimmed samples can also provide reasonable estimates of in situ conditions.

The SBPM has theoretically the greatest potential, but its fulfillment may be achieved in clays only by a better understanding of the three-dimensional in situ cavity expansion problem, and the effects of non-uniform expansion with the resulting excess pore pressures at the time of lift-off. Still, to date, the SBPM is considered by most researchers familiar with in situ testing techniques to be the best tool to determine the lateral stress in the ground and, thus, should be used as the primary reference value for any correlations involving K_0. Furthermore, only quality results obtained in recent years should be used to formulate empirical relationships to evaluate K_0. Without the SBPM and perhaps some of the laboratory direct measurements, engineers would be lacking a reference scale from which other field or laboratory methods could be evaluated.

With the methods currently available, selection of a design value should require evaluating both ends of the range of results even though the scatter indicated in the interpreted test data may mostly stem from the differences in test technique. Simply choosing a mean value or a single interpreted stress profile could result in an unconservative approach depending on the design problem. It is expected that research in this area will narrow the range of interpreted values leading to both safe and economical designs. On large projects which warrant detailed investigations of lateral stress conditions, engineers should not rely on a single test method to evaluate in situ lateral stresses. For example, an error in the range of 10–25% could result in substantial additional cost to projects for which earth pressures are needed for design.

Acknowledgements

The authors are indebted to all the graduate students at both universities who have contributed through research in gathering the information and results presented in this paper.

References

ABDELHAMID, M.S. and KRIZEK, R.J. (1976). At-Rest Lateral Earth Pressure of a Consolidating Clay, Journal of the Geotechnical Engineering Division, ASCE, Vol. 102, GT7, pp. 721–732.

ATWOOD, M.J. (1990). Investigation of Jetting Insertion Procedures for Rapidly Deploying a Self-Boring Pressuremeter in Soft Clays. A Thesis Presented in Partial Fulfillment of the Requirements for the Masters Degree. University of New Hampshire, Durham, New Hampshire.

AZZOUZ, A.S. and MORRISON, M.J. (1988). Field Measurements on Model Pile in Two Clay Deposits, Journal of the Geotechnical Engineering Division, ASCE, Vol. 114, No. 1, pp. 104–121.

BAGUELIN, F., JÉZEQUEL, J.-F., LEMÉE, E. and LE MÉHAUTE, A. (1972). Expansion of Cylindrical Probes in Cohesive Soils, Journal of the Soil Mechanics and Foundations Division, ASCE, Vol. 98, No. SM11, pp. 1129–1142.

BAUER, G.E. and EL-HAKIM, A.Z. (1986). Consolidation Testing – A Comparative Study, ASTM STP 892, pp. 694–710.

BECKER, D.E., CROOKS, J.H.A., BEEN, K.A. and JEFFERIES, M.G. (1987). Work as a Criterion for Determining In Situ and Yield Stress in Clays. Canadian Geotechnical Journal, Vol. 24, No. 4, pp. 549–564.

BENOÎT, J. (1983). Analysis of Self-Boring Pressuremeter Tests in Soft Clay, Thesis Presented in Partial Fulfillment of the Ph.D. Degree, Stanford University, California.

BENOÎT, J. and CLOUGH, G.W. (1986). Principal Stresses Derived from Self-Boring Pressuremeter Tests in Soft Clay, The Pressuremeter and Its Marine Applications: Second International Symposium. ASTM STP 950, J.-L. Briaud and J.M.E. Audibert, Eds., American Society for Testing and Materials, pp. 137–149.

BERRE, T. (1982). Triaxial Testing at the Norwegian Geotechnical Institute, Geotechnical Testing Journal, ASTM, 5, pp. 3–17.

BISHOP, A.W. (1958). Test Requirements for Measuring the Coefficient of Earth Pressure at Rest, Proceedings of Conference on Earth Pressure Problems, Brussels, Vol. 1, pp. 2–14.

BJERRUM, L. and ANDERSON, K.H. (1972). In Situ Measurement of Lateral Earth Pressures in Clay, Proceedings 5th European Conference on Soil Mechanics and Foundations Engineering, 1, pp. 11–20.

BOND, A.J. and JARDINE, R.J. (1991). Effects of Installing Displacement Piles in a High OCR Clay, Geotechnique, Vol. 41, No. 3, pp. 341–363.

BROOKER, R.W. and IRELAND, H.O. (1965). Earth Pressures at Rest Related to Stress History, Canadian Geotechnical Journal, Vol. II, No. 1, pp. 1–15.

CAMPANELLA, R.G. and VAID, Y.P. (1972). A Simple K_0 Triaxial Cell, Canadian Geotechnical Journal, Vol. 9, No. 3, pp. 249–260.

CAMPANELLA, R.G., SULLY, J.P., GREIG, J.W. and JOLLY, G. (1990). Research and Development of a Lateral Stress Piezocone, Transportation Research Record 1278, pp. 215–224.

CASAGRANDE, A. (1936). The Determination of the Preconsolidation Load and Its Practical Significance, Proceedings of the 1st International Conference on Soil Mechanics and Foundation Engineering, Paper D-34.

CHANG, M.F., MOH, Z.C., LIU, H.H. and VIRANUVUT, S. (1977). A Method of Determining the In Situ K_0 Coefficient, Proceedings of the 9th International Conference on Soil Mechanics and Foundations Engineering, 1, pp. 61–64.

COOP, M.R. and WROTH, C.P. (1989). Field Studies of an Instrumented Model Pile in Clay, Geotechnique, Vol. 39, No. 4, pp. 679–696.

DALTON, J.C.P. and HAWKINS, P.G. (1982). Fields of Stress – Some Measurements of the In-Situ Stress in a Meadow in the Cambridgeshire Countryside, Ground Engineering, Vol. 15, No. 4, pp. 15–22.

DYVIK, R., LACASSE, S. and MARTIN, R. (1985). Coefficient of Lateral Stress from Oedometer Cell, Proceedings of the 11th International Conference on Soil Mechanics and Foundations Engineering, 2, pp. 1003–1006.

DYVIK, R., ZIMMIE, T.F. and FLOESS, C.H.L. (1981). Lateral Stress Measurements in Direct Simple Shear Device, ASTM STP 740, pp. 191–206.

FINDLAY, R.C. (1991). Use of the 9-Arm Self-Boring Pressuremeter to Measure Horizontal In Situ Stress, Stress Anisotropy, and Stress Strain Behavior in Soft Clays, A Thesis Presented in Partial Fulfillment of the Requirements for the Ph.D. Degree, University of New Hampshire, Durham, New Hampshire.

FINDLAY, R.C. and BENOÎT, J. (1992). Some Factors Affecting In Situ Measurement Using the Cambridge Self-Boring Pressuremeter, Submitted for Review to the Geotechnical Testing Journal, ASTM.

FINNO, R.J., BENOÎT, J. and CHUNG, C.K. (1990). Field and Laboratory Measured Values of K_0 in Chicago Clay, Proceedings of the Third International Symposium on Pressuremeters, ISP3, Oxford University Press, England.

GARGA, V.K. and KHAN, M.A. (1991). Laboratory Evaluation of K_0 for Overconsolidated Clays. Canadian Geotechnical Journal, Vol. 28, No. 5, pp. 650–659.

GOODRICH, E.P. (1904). Lateral Earth Pressures and Related Phenomena, Transactions of the American Society of Civil Engineers, Vol. 69, pp. 272–290.

HANDY, R.L., REMMES, B., MOLDT, S. and LUTENEGGER, A.J. (1982). In Situ Stress Determination by Iowa Stepped Blade, Journal of the Geotechnical Engineering Division, ASCE, Vol. 108, No. 11, pp. 1405–1422.

HANDY, R.L., MINGS, C., RETZ, D and EICHNER, D. (1990). Field Experience with the Back-Pressured K_0 Stepped Blade, Transportation Research Record 1278, pp. 125–134.

HOLTZ, R.D., JAMIOLKOWSKI, M.B., and LANCELLOTA, R. (1986). Lessons from Oedometer Tests on High Quality Samples, Journal of the Geotechnical Engineering Division, Vol. 112, No. 8, pp. 768–776.

HUANG, A.-B. and HAEFELE, K.C. (1990). Lateral Earth Pressure Measurements in a Marine Clay, Transportation Research Record 1278, pp. 156–163.

HUNTSMAN, S.R., MITCHELL, J.K., KLEJBUK, L. and SHINDE, S. (1986). Lateral Stress Measurement During Cone Penetration, Proceedings of the Specialty Conference on In Situ Testing, ASCE, pp. 617–634.

JÁKY, C. (1944). The Coefficient of Earth Pressure at Rest (in Hungarian), Magyau Mérnok és Epitéez Egylet Közdönye (Journal of the Society of Hungarian Architects and Engineers), Vol. 78, No. 22, pp. 355–358.

JAMIOLKOWSKI, M., LEROUEIL, S. and LO PRESTI, D.C.F. (1991). Theme Lecture: Design Parameters from Theory to Practice, Geo-Coast, Japan.

JEFFERIES, M.G., JONSSON, L. and BECH, K. (1987). Experience with Measurement of Horizontal Geostatic Stress in Sand During Cone Penetration Test Profiling, Geotechnique, Vol. 37, pp. 483–498.

KENNEY, T.C. (1967). Field Measurements of In Situ Stresses in Quick Clays, Proceedings of the Geotechnical Conference, Oslo, Vol. 1, pp. 49–55.

KOMORNIK, A. and ZEITLEN, J.G. (1965). An Apparatus for Measuring Lateral Soil Swelling Pressure in the Laboratory, Proceedings of the 6th International Conference on Soil Mechanics and Foundations Engineering, 1, pp. 278–281.

KULHAWY, F.H., JACKSON, C.S. and MAYNE, P.W. (1989). First Order

Estimation of K_0 in Sands and Clays, Foundations Engineering: Current Principles and Practices, ASCE, 1, pp. 121–134.

KULHAWY, F.H. and MAYNE, P.W. (1990). Manual on Estimating Soil Properties for Foundation Design, EPRI Research Report EL-6800.

LACASSE, S. and LUNNE, T. (1982). In Situ Horizontal Stress from Pressuremeter Tests, Proceedings of the Symposium on the Pressuremeter and Its Marine Applications, Paris.

LADD, C.C. (1972). Test Embankment on Sensitive Clay Specialty Conference on Performance of Earth and Earth-Supported Structures, ASCE, Purdue University, June.

LADD, R.S. (1965). Use of Electrical Pressure Transducers to Measure Soil Pressure, Report RG5-48, Department of Civil Engineering, MIT, 79 pp.

LEFEBVRE, G., BOZOZUK, M., PHILIBERT, A. and HORNYCH, P. (1991). Evaluating K_0 in Champlain Clays with Hydraulic Fracture Tests, Canadian Geotechnical Journal, Vol. 28, No. 3, pp. 365–377.

LO, S.-C.R. and CHUI, J. (1991). The Measurement of K_0 by Triaxial Strain Path Testing, Soils and Foundations, 31, pp. 181–187.

LUTENEGGER, A.J. and TIMIAN, D.A. (1986). In Situ Tests with K_0 Stepped Blade, Use of In Situ Tests in Geotechnical Engineeering, ASCE, pp. 730–751.

LUTENEGGER, A.J. (1988). Current Status of the Marchetti Dilatometer Test, Proceedings of the First International Symposium on Penetration Testing, Vol. I, pp. 137–155.

LUTENEGGER, A.J. (1990). Determination of In Situ Lateral Stresses in a Dense Glacial Till, Transportation Research Record 1278, pp. 194–203.

MAHAR, L.J. and O'NEILL, M.W. (1983). Geotechnical Characterization of Desiccated Clay, Journal of the Geotechnical Engineering Division, ASCE, Vol. 109, No. 1, pp. 56–71.

MARCHETTI, S. (1980). In Situ Tests by Flat Dilatometer, Journal of the Geotechnical Engineering Division, ASCE, Vol. 106, GT3, pp. 299–321.

MASSARSCH, K.R. (1975). New Method for Measurement of Lateral Earth Pressure in Cohesive Soils, Canadian Geotechnical Journal, Vol. 12, No. 1, pp. 142–146.

MASSARSCH, K.R., HOLTZ, R.D., HOLM, B.G. and FREDRIKSSON, A. (1975). Measurement of Horizontal In Situ Stresses, Proceedings of the Conference on In Situ Measurement of Soil Properties, ASCE, Raleigh, North Carolina, Vol. I, pp. 266–285.

MASSARSCH, K.R. (1979). Lateral Earth Pressure in Normally Consolidated Clay, Proceedings of the Seventh European Conference on Soil Mechanics and Foundations Engineering, Brighton, England, Vol. 2, pp. 245–250.

MAYNE, P.W. and KULHAWY, F.H. (1982). K_0–OCR Relationships in Soil,

Journal of the Geotechnical Engineering Division, ASCE, Vol. 108, GT6, pp. 851–872.

Mesri, G. and Castro, A. (1987). C_α/C_c Concept and K_0 During Secondary Compression, Journal of the Geotechnical Engineering Division, ASCE, Vol. 113, No. 3, pp. 230–247.

Moore, P.J. and Spencer, G.K. (1972). Lateral Pressures from Soft Clay, Journal of the Soil Mechanics and Foundations Division, ASCE, Vol. 98, pp. 1225–1244.

NeJame, L.A. (1991). Dilatometer Testing of the Marine Clay Deposit at Pease Air Force Base, New Hampshire, A Thesis Presented in Partial Fulfillment of the Requirements for the Masters Degree. University of New Hampshire, Durham, New Hampshire.

Ofer, Z. (1982). Lateral Pressure Developed During Compaction, Transportation Research Record 897, pp. 71–79.

Poulos, H.G. and Davis, E.H. (1972). Laboratory Determination of In Situ Horizontal Stress in Soil Masses, Geotechnique, Vol. 22, No. 1, pp. 177–182.

Schmidt, B. (1966). Discussion of Earth Pressures at Rest Related to Stress History, Canadian Geotechnical Journal, Vol. III, No. 4, pp. 229–242.

Senneset, K. (1989). A New Oedometer with Splitted Ring for the Measurement of Lateral Stress, Proceedings of the 12th International Conference on Soil Mechanics and Foundations Engineering, 1, pp. 115–118.

Sherif, M.A. and Ishibashi, I. (1981). Overconsolidation Effects on K_0 Values, Proceedings of the 10th International Conference on Soil Mechanics and Foundations Engineering, 1, pp. 785–788.

Silvestri, V. and Morgavi, R. (1982). Measurement of Lateral Stresses in One-Dimensional Consolidation Tests on a Sensitive Clay of Eastern Canada, Proceedings of the 19th Annual Engineering Geology and Soils Engineering Symposium, pp. 201–216.

Simons, N.E. (1965). Consolidation Investigation on Undisturbed Fornebu Clay, NGI Publ. No. 62.

Som, N.N. (1970). Lateral Stresses During One Dimensional Consolidation of an Overconsolidated Clay, Proceedings of the 2nd Southeast Asian Conference on Soil Engineering, pp. 295–307.

Sowers, G.F., Robb, A.D., Mullis, C.H. and Glenn, A.J. (1957). The Residual Lateral Pressures Produced by Compacting Soils, Proceedings of the 4th International Conference on Soil Mechanics and Foundations Engineering, 2, pp. 243–246.

Tavenas, P.A., Blanchette, G., Leroueil, S., Roy, M. and La Rochelle, P. (1975). Difficulties in the In Situ Determination of K_0 in Soft, Sensitive Clays, Proceedings of the Conference on In Situ Measurements of Soil Properties, ASCE, Raleigh, North Carolina, Vol. I, pp. 450–476.

TERZAGHI, K. (1920). Old Earth Pressure Theories and New Test Results, Engineering News Record, September 30, p. 633.

TEDD, P. and CHARLES, J.A. (1981). In Situ Measurement of Horizontal Stress in Overconsolidation Clay Using Push-In Spade-Shaped Pressure Cells, Geotechnique, Vol. 31, No. 4, pp. 554–558.

THOMPSON, W.J. (1963). Lateral Pressures in One-Dimensional Consolidation, Proceedings of the 2nd Asian Conference on Soil Mechanics and Foundations Engineering, pp. 26–31.

WROTH, C.P. and HUGHES, J.M.O. (1973). An Instrument for the In Situ Measurement of the Properties of the Soft Clays, Proceedings of the 8th International Conference on Soil Mechanics and Foundations Engineering Moscow, Vol. 1, Part 2, pp. 487–494.

WROTH, C.P. (1975). In Situ Measurement of Initial Stresses and Deformation Characteristics, Proceedings of the Conference on In Situ Measurements of Soil Properties, ASCE, Raleigh, North Carolina, Volume II, pp. 181–230.

WROTH, C.P. (1984). Interpretation of In Situ Tests, Geotechnique, Vol. 34, pp. 449–589.

YASUHARA, K., HIRAO, K. and UE, S. (1988). Effects of Long Term K_0-Consolidation on Undrained Strength of Clay, Proceedings of the International Conference on Rheology and Soil Mechanics, pp. 273–287.

ZEEVAERT, L. (1953). Discussion, 3rd International Conference on Soil Mechanics and Foundations Engineering, 3, pp. 113–114.

The interpretation of self-boring pressuremeter tests to produce design parameters

B.G. CLARKE, University of Newcastle upon Tyne

Pressuremeters have been in commercial use since 1957. Copies of the original instruments are still widely used to obtain parameters for design; these test dependent parameters are used directly in design rules based on observations of full scale structures. The rules can only be used if the instrument, installation and test procedures and interpretation conform with a standard specification. The development of self-boring instruments led potentially to the direct measurement of in situ stress, stiffness and strength. However, the interpretation is subjective and is dependent on the installation and testing techniques used. The parameters will differ from those used in other tests since the testing technique is different. Design parameters can be obtained from this test if a consistent framework of soil mechanics theories and experimental and field observations is used in the interpretation. It is now possible to determine with some confidence horizontal stress. Recent developments in the understanding of deformation behaviour of soils have led to the potential of converting the cavity stiffness measured in a pressuremeter test to a material stiffness, though the cavity stiffness is similar to the average stiffness backfigured from observations of full scale structures.

Introduction

Robertson (1986) proposed a list of applicability and usefulness of in situ tests. He stated that most in situ soil tests are used in sands, silts and clays. He concluded that the best in situ tests are electric piezocone tests and self-boring pressuremeter (SBP) tests both of which can only be used in soils. Prebored pressuremeters (PBP) and the weak rock self-boring pressuremeter (RSBP) can be used in rocks. Thus, the pressuremeter is seen as a versatile instrument that can be used to obtain quality data from tests in a variety of ground conditions.

In 1957, Menard developed a pressuremeter (MPM) to determine in situ soil parameters. The instrument was lowered down a borehole of slightly larger diameter than the instrument. The process of preboring changes the properties of the soil adjacent to the borehole which implies that a prebored pressuremeter test requires further interpretation in addition to basic cavity expansion theories if fundamental parameters are required. To overcome this Menard developed a series of design charts based on relationships between foundation performance and pressuremeter test results. This empirical approach is still widely used today and is continually being updated as case studies become available. This design approach, similar to that used with many penetration devices, is recognised internationally.

In 1968 a new type of pressuremeter, the self-boring pressuremeter, was developed (Jezequel et al., 1968) which, theoretically, would cause no disturbance to the surrounding soil since it could be drilled into the ground. The parameters obtained would be properties of the soil rather than of the test and installation procedure. Thus they could be used in any analyses that correctly model soil. The SBP allows tests to be carried out on minimally disturbed soil from which in situ stress, stiffness and strength can be derived. The results cannot be used in the design methods referred to above since the installation and testing techniques are different. This is no different from other tests since many soil parameters are a function of the sampling/installation technique and the testing technique.

In 1971 Wroth and Hughes (1973) developed the Cambridge SBP and since 1978 it has been used in site investigations to obtain design parameters. There are several types of self-boring instrument but those most commonly used are based on the Cambridge design. A detailed description of the equipment and installation and testing techniques is given by Windle and Wroth (1977) and Clarke and Smith (1992).

Factors affecting the analysis and interpretation of a test

SBP tests are assumed to be in undisturbed ground. Thus, an SBP test curve most closely represents the expansion of a cavity from the in situ stress conditions. The equipment, installation and testing techniques and interpretation of the tests have improved. However, until it is an established technique that has been validated in designs for structures which have been built and monitored then results from the tests have to be used with caution. It is common practice to compare results from one test with those from another test to justify the quality of data. This is not correct for the same reasons that SBP results cannot be used in MPM design rules. However, it does allow a statement on the quality of the data to be made and provide a framework within which the pressuremeter test results can be analysed. Examples of this technique are discussed below.

Any installation process will change the properties of the soil adjacent to the borehole cavity. Potentially, self-boring does not affect the ground but it can produce misleading results since there is no independent check that the instrument is drilled in correctly. For example, slight variations in drilling technique give similar-shaped test curves but the interpretation of those tests produces different results.

The process of self-boring imposes a shear stress between the soil and the probe which changes the in situ stress conditions though Clarke and Wroth (1984) have shown that the magnitude of the horizontal stress remains the same provided the soil is not at a state of failure prior to installation.

The membrane has a finite length and, therefore, the restriction imposed at the ends can affect the expansion of the membrane. It is often assumed that, if the total length of the expanding section is at least six times the diameter of the instrument, then the expansion, if monitored within the central third, is the same as that for an infinitely long cylinder. Yeung and Carter (1990) show that this may not be the case especially for tests at shallow depths.

The changes in dimensions of the borehole cavity are measured within the membrane. Corrections have to be applied to allow for the compression and thinning of the membrane. The correction for thinning is based on the assumption that the membrane material has a constant volume. This may not be the case.

The system compliance and fluctuation in data can cause difficulty in interpreting data.

It is unlikely that the pressuremeter will ever give truly undisturbed ground parameters. However, if careful installation follows a specified procedure then any changes in the ground prior to testing will be similar for different tests. The amount of change varies with the installation techniques and ground conditions. Self-boring has the least effect if correct drilling techniques are used.

Fundamental parameters for design
Horizontal stress
Only SBP tests can be used to give a direct assessment of in situ horizontal stress. The pressure at which the membrane begins to move during an expansion test is the horizontal stress. As the stiffness of the soil increases it becomes more difficult to determine the horizontal stress since the deformations due to the compliance of the instrument can be similar to the initial deformations in the soil.

It is often assumed that the SBP can be drilled into clays without changing the horizontal stress. This is unlikely because of the difficulty in ensuring that there is no displacement of clay around the instrument. The clay can be loaded if there is underdrilling (A in Fig. 1) and unloaded if there is overdrilling (B in Fig. 1). The clay is likely to relax onto the instrument for extreme cases of overdrilling. Thus, at the start of a test there is a horizontal stress acting on the instrument but this may not be the in situ stress.

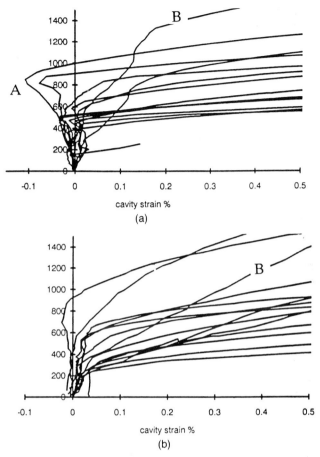

Fig. 1. The start of pressuremeter tests in a stiff clay. (a) Using spring loaded feeler arms, and (b) as (a) with 'Fahey' rollers

Figure 1 shows the start of several tests in London Clay. The main difference between the two sets of results is that improved transducers (Fahey and Jewell, 1990) were used at the second site. These reduce the errors due to the instrument behaviour but it is still difficult to interpret the tests because of installation problems such as overdrilling, underdrilling and deviation. Various methods have been proposed to assist in the assessment of horizontal stress (e.g. Lacasse and Lunne, 1982) but it is often found that it is best to establish the signature of each transducer and use an inspection method together with rules based on well-known soil mechanics theories.

The shape of the initial loading curve is often similar for each transducer in different tests and therefore by viewing several tests

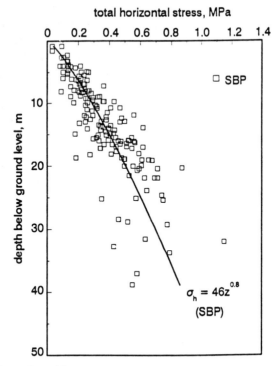

Fig. 2. *Values of total horizontal stress from fifteen sites in London Clay*

together it is possible to establish the signature of the transducer which reflects the system compliance. This signature is used to assist in the determination of a profile of total horizontal stress especially for tests in which there is a change in horizontal stress due to installation.

Figure 2 shows values of total horizontal stress for London Clay assessed using the inspection method. The data are taken from sites where the London Clay was within 3 m of the ground surface. The data are scattered though this scatter is apparent with other types of data from other tests. A fit to the data is

$$\sigma_h = 46\, z^{0.8} \tag{1}$$

This does not represent the variation in total horizontal stress with depth for London Clay because the in situ conditions at each site will differ as a result of different pore pressure regimes, geological conditions and construction activities. It does give a guide for preliminary interpretation of SBP tests.

Experience has shown that, in sands, there is inevitably some disturbance which can be small. It implies that the horizontal stress is reduced perhaps to the ambient pore pressure. This is expected since the voids ratio around the instrument must increase as sand grains are removed if the whole or part of a grain is in the path of the instrument.

The displacements in a clay caused by installation are small and can be neglected when interpreting a test for strength and stiffness. This is not necessarily the case for tests in sands since the cavity strain due to installation can exceed 0.5%. This affects the interpretation of a test. During installation the cavity wall is unloaded, while during a test the cavity wall is reloaded. There is evidence that the initial unloading stiffness and reloading stiffness are different and the reloading stiffness beyond the horizontal stress is different again. This hysteretic behaviour is seen in the unload–reload cycles used for stiffness evaluation. Thus, by inspection, a horizontal stress can be selected from a curve by noting a significant change in slope. This horizontal stress may not be the in situ stress because of the disturbance created during self-boring. Newman (1991) suggests that good quality tests in sands can be those in which the cavity strain at the horizontal stress is less than 0.2%

Installation techniques for drilling in rock use drill bits which create a cavity rather than rely on the instrument being pushed into the ground. The cavity created represents less than 0.25% cavity strain but it is

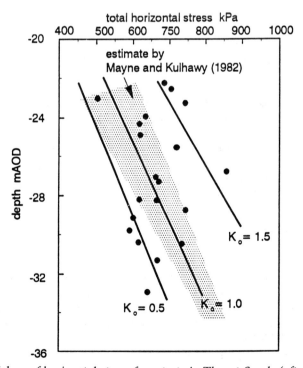

Fig. 3. *Values of horizontal stress from tests in Thanet Sands (after Clarke, 1990, and Newman, 1991)*

enough to reduce the horizontal stress acting on the instrument and, since the rock does not relax onto the instrument as clay does, the horizontal stress may reduce to zero.

It is unlikely that an SBP is ever drilled in without some effect on the soil adjacent to the instrument. Thus, the assessment of horizontal stress will always be subjective. Of all the techniques available the inspection method is probably the best provided the data is evaluated within a framework based on established correlations and theories. This can include an estimation of horizontal stress from the known geological conditions of the site as shown in Fig. 3.

Deformation modulus

In an elastic medium, the radial expansion of a cylindrical cavity is related to the pressure increment, for small strains, by

$$G = \frac{1}{2} \frac{a \Delta p}{a_0 \Delta \varepsilon_c} \quad (2)$$

where G is the shear modulus, p the applied total pressure, a the current cavity radius, a_0 the original cavity radius and ε_c the cavity strain.

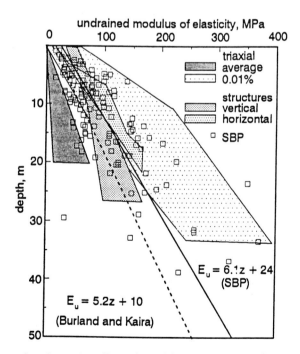

Fig. 4. *Values of undrained stiffness from fifteen sites in London Clay*

The disturbance created during installation implies that the initial slope of a pressuremeter test is a measure of the stiffness of the disturbed annulus of soil. Further, in normally consolidated soils, the initial expansion includes plastic strain. Therefore, the modulus is more often taken from an unload–reload cycle. It is assumed that the slope is independent of the installation technique and therefore must be independent of the instrument type. This only applies for large expansions of the cavity. Equation (2) can be reduced to

$$G = \frac{1}{2} a \frac{\Delta p}{\Delta a} \qquad (3)$$

Therefore, the original cavity dimensions do not affect G. Thus stiffnesses determined from SBP and PBP tests should be similar. Powell (1990) and Clarke et al. (1990) show that results from SBP and PBP tests are similar although no conclusions could be drawn since the strain ranges over which they were measured were not necessarily the same.

Figure 4 shows values of stiffness taken from tests in London Clay compared to those backfigured from observations of structures (St John, 1979). The stiffnesses are the best fit to complete unload–reload cycles where the membrane was unloaded by a pressure approximately equal to the undrained shear strength. The data fall within the bounds set by observations of horizontal and vertical movements of structures.

The modulus from an SBP test is an average modulus which will be referred to as a cavity modulus to distinguish it from the Menard pressuremeter modulus.

The deformation of the ground is dependent on an average modulus. Most design calculations involve changes in vertical loads, therefore a 'vertical' modulus should be used unless the ground is isotropic. Lee and Rowe (1989) show theoretically that ignoring anisotropy has little effect on settlement. Thus, it is possible to use a cavity stiffness, which is measured in the horizontal direction, in displacement calculations based on simple elastic theories. This is supported by the comparisons shown in Fig. 4.

Stiffnesses are dependent on strain range and mean effective stress. Clarke and Wroth (1985) demonstrated that stiffnesses obtained from tests in sands corrected for stress level reduced the scatter in the data. O'Brien and Newman (1989) and Newman et al. (1991) suggested that stiffnesses should be quoted over a constant strain range. Robertson and Hughes (1985) and Newman et al. (1991) showed that stiffnesses should be corrected for stress and strain level. Newman (1991) used the work of Bellotti et al. (1989) to account for the variations in stress and strain level for tests in Thanet Sands. All these authors used a single value of cavity stiffness from the pressuremeter test.

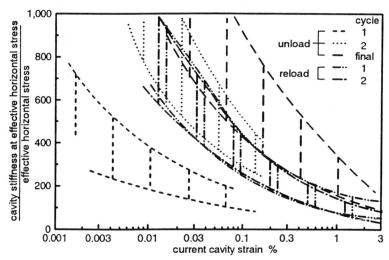

Fig. 5. The variation in stiffness with cavity strain for tests in Thanet Sands

The variation in stiffness with strain can be assessed from the initial loading, the final unloading and both the unloading portion and reloading portion of a unload–reload cycle. The shape of the initial loading curve is sensitive to the installation technique and therefore is unlikely to be of use in the interpretation.

Figure 5 shows the variation in stiffness with strain for twenty tests in Thanet sands between 25 and 40 m below ground level. The strain is expressed in terms of either the radius at the start of loading or the maximum radius on unloading. The cavity stiffnesses, G_{ur}, have been converted to the stiffnesses, G_{ur0}, at the in situ effective horizontal stress by

$$G_{ur0} = G_{ur}(\sigma'_{h0}/\sigma'_{av})^n \quad (4)$$

n has been taken as 0.4 (Bellotti et al., 1989). The average horizontal stress σ'_{av} at the start of unloading for these tests is equal to

$$\sigma'_{av} = \sigma'_{h0} + 0.2(\sigma'_{max} - \sigma'_{h0}) \quad (5)$$

where σ'_{max} is the maximum applied effective stress at the start of unloading, and the constant 0.2 is an average for all the tests. This is the same as observed by Bellotti et al. for their chamber tests and tests in a medium dense sand.

The current cavity strain is restricted to the range 0.001 to 1% which covers medium to large shear strains (Bellotti et al., 1989). The resolution of the displacement measuring systems is equivalent to 0.0005% cavity strain but the resolution of the transducers in the pressuremeter is probably 0.01% because of system compliance. It is noted that for cavity strains less than 0.01% the data are unreliable.

The normalised reloading stiffnesses for the two cycles are similar whereas the unloading stiffnesses are different from each other and the loading stiffnesses. It was noted that the unloading stiffnesses divided by the mean effective stress at the cavity wall assuming critical state conditions are reached were similar.

At the start of unloading the membrane continues to expand due to creep of the sand (Robertson and Hughes, 1986). Thus during unloading, even if the membrane is retracting, there could be a component of creep which combined with the unloading gives a greater value of stiffness than that for pure unloading. This may account for the greater variation in unloading stiffness with depth and, for the second cycle and final unloading cycle, the difference between the unloading and reloading stiffnesses. The first unloading cycle may be dominated by a disturbed zone created during installation.

The ranges of strain for the unload moduli are limited by the extent of unloading. Typically it may take up to five minutes to unload the membrane during a cycle; this is in excess of the time required to ensure that no further creep is taking place following the loading phase. Therefore it may be preferable to determine a stiffness from the reloading curve. This has the additional advantage that the upper limit to the strain range is increased allowing a better definition of the nonlinear stiffness profile. The strains in Fig. 5 can be converted to an average shear strain, γ_{av}, using the relationship

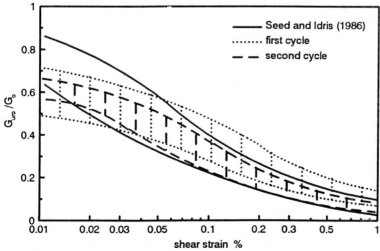

Fig. 6. *Variation in stiffness with shear strain for Thanet Sands*

CLARKE

$$\gamma_{av} = \beta * \Delta\varepsilon_c \qquad (6)$$

Bellotti et al. found that β was equal to 0.5 for medium dense sand but for these tests in a very dense sand β varied from about 0.6 to 0.2 as the applied pressure increased.

The cavity stiffness at the in situ horizontal stress can be related to a material stiffness, G_0, at small strains using a hyperbolic relationship such as

$$\frac{G_{ur0}}{G_0} = \frac{1}{1 + (G_0 \gamma_{av}/(\sigma'_{h0} \sin \phi'))} \qquad (7)$$

The data from the reload cycles in Fig. 5 have been replotted in Fig. 6 to show this relationship; the results agree with the proposal of Seed et al. (1986). The results from the second cycle give a better agreement than those from the first cycle perhaps because the first cycle is dominated by installation effects.

Figure 7 shows data from tests in London Clay. The effective stress is taken as the in situ effective horizontal stress since this represents the stress to which the soil is consolidated prior to undertaking a test. The mean effective stress changes very little once yield has occurred (Wroth, 1982) hence the stiffness will not vary significantly as the membrane expands. This is the case for the reloading stiffnesses. The unloading stiffnesses are different for all the cycles and are different from the reloading stiffnesses for the same cycles. This is the same as the observations for the tests in sand. It may be due to consolidation taking place during unloading.

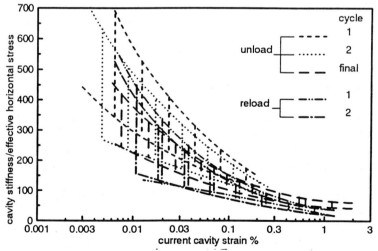

Fig. 7. The variation in stiffness with cavity strain for tests in London Clay

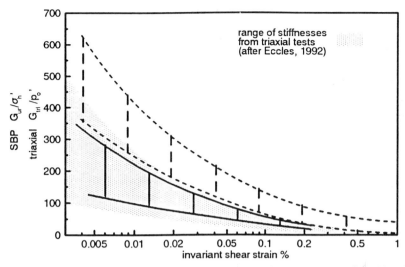

Fig. 8. The variation in stiffness with shear strain for tests in London Clay

Muir Wood (1990) and Jardine (1991) suggest that the cavity stiffness from an unload–reload cycle in clay can be converted to a secant stiffness as measured on an element of soil for example in a triaxial test. Data from tests in London Clay at two sites are plotted in Fig. 8. The cavity strain has been converted to a shear strain using the proposal of Jardine. The range of stiffnesses are similar but the values differ.

Undrained shear strength

Baguelin et al. (1972) and others developed an analysis from which a precise shear stress strain curve could be derived. The shear stress is a function of the slope of the cavity pressure/deformation curve. The change in slope is not uniform because of fluctuations produced by the data recording system. Various curve fitting methods have been developed to smooth the recorded data to simplify the derivation of the shear stress strain curve (for example, Denby et al., 1982; Prevost, 1979). They tend to give different shear stress strain curves and show significant peak values of shear stress. Baguelin et al. (1978) suggest that the peak values are due to disturbance. It is unlikely that a typical test can ever be correctly modelled because of the uncertainty of installation effects.

The undrained shear strength is normally taken from the latter part of the loading curve using the method originally proposed by Gibson and Anderson (1961). It is commonly accepted that a plot of applied pressure against volumetric strain to a logarithmic scale produces a straight line once yield has occurred (ignoring unload–reload cycles). However, in

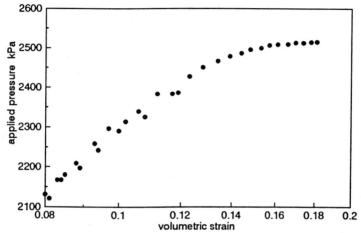

Fig. 9. An example of the change in strength with volumetric strain for a test in London Clay

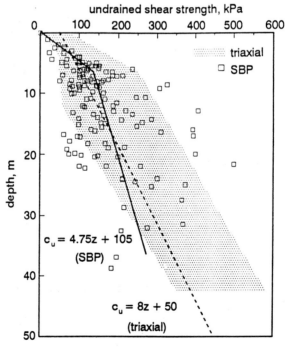

Fig. 10. Variation in undrained shear strength from fifteen sites in London Clay

many tests in stiff clays this is not the case. The data produce two straight lines which normally intersect at between 6 and 7% cavity strain; an example of this is shown in Fig. 9. The slope of the first part of the curve, up to 7%, is greater than the slope of the latter part of the test. The reduction in strength for many tests is, on average, 30%. This may be a function of the assumptions made in the analysis or it may be an effect due to the post-rupture behaviour of the soil. It is necessary, as with the cavity stiffnesses, to quote the strain range over which the strength was measured.

It is not expected that shear strengths derived from pressuremeter tests should be similar to those measured in triaxial tests (Wroth, 1984). Therefore, it is unwise to use pressuremeter strengths in design methods based on triaxial tests. In soft clays the pressuremeter tends to give greater values of strength than the triaxial tests, whereas in stiff clays the results are similar as demonstrated in Fig. 10. These results are taken from the latter part of the test curves and have not been corrected for membrane length using the method proposed by Yeung and Carter (1990).

Angle of friction
The angle of friction can be found for cohesionless soils from a plot of logarithm of effective cavity pressure against logarithm of current cavity strain. This plot is linear after yield and Hughes et al. (1977) have shown that the slope is related to the peak shear stress and angle of dilation.

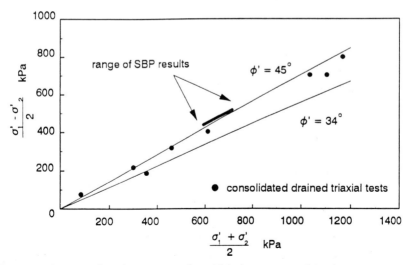

Fig. 11. *A comparison between angles of friction measured in the triaxial and pressuremeter tests (after Newman, 1991)*

Robertson and Hughes (1986) showed that this method produced low values of friction for loose sands because of the limited expansion of the pressuremeter. They produced charts to correct for this problem. Manassero (1989) proposed a numerical solution based on the two methods which incorporates a complete non-linear volume and stress change relationship. Figure 11 shows a comparison between the results of triaxial tests and pressuremeter tests on Thanet Sand.

Conclusions

Pressuremeter results are used either directly in design rules or to produce fundamental parameters. There is considerable experience with the former. This approach only applies to instruments and tests which conform to a standard, and is used to overcome the problems of installation. The SBP has the potential of causing no disturbance to the ground but experience has shown that the installation of the SBP does cause changes in the properties of the soil adjacent to the instrument. Therefore, the interpretation of these tests is in part subjective. There are numerous theories of cavity expansion, some of which attempt to take into account changes that take place during installation. However, in practice the interpretation of a test is based on a few theories using various curve fitting routines to smooth the data and soil mechanics principles to assess the quality of the data. Interpretation of tests which have been carried out under controlled conditions should give reliable, consistent results. This approach, to develop a consistent framework within which test data is analysed, can be used to isolate the fundamental parameters from effects of installation and varying effective stress.

Average cavity stiffnesses from tests in London Clay are similar to those backfigured from observations of structures which suggests that they may be used directly in classical settlement techniques.

The variation of cavity stiffnesses with shear strain for tests in dense sands and stiff clay have been assessed. It was found that reloading cavity stiffnesses gave more consistent results and covered a greater strain range. Hence they give a better definition of the cavity stiffness strain profile. There is good agreement between secant stiffnesses deduced from the reloading cavity stiffnesses and typical values from tests on elements of soil.

The strength deduced from tests in these stiff materials agrees with data from other tests but, as with stiffness, there has to be a defined range over which they are measured since stress–strain curves do not strictly conform to the theories used. This is particularly the case for tests in stiff clays.

References

BAGUELIN, F., JEZEQUEL, J.-F. LEMEE, E. AND LE MEHAUTE, A. (1972). Expansion of cylindrical probes in cohesive soils. Proc. Am. Soc. Civ. Engrs-J. Soil. Mech. Found. Div., Vol. 98, No. SM11, pp. 1129–1142.

BAGUELIN, F., JEZEQUEL, J.F. AND SHIELDS, D.H. (1978). The pressuremeter and foundation engineering. Trans. Tech. Publ.

BELLOTTI, R., GHIONNA, V., JAMIOLKOWSKI, M., ROBERSTON, P.K. AND PETERSON, R. (1989). Interpretation of moduli form self-boring pressuremeter tests in sand. Geotechnique, Vol. 39, No. 2, pp. 269–292.

CLARKE, B.G. (1990). Assessment of in situ properties at Port East using a self-boring pressuremeter. Report for Ove Arup and Partners.

CLARKE, B.G., NEWMAN, R. AND ALLAN, P. (1990). Experience with a new high pressure self-boring pressuremeter in weak rock. Ground Engng., Vol. 22, No. 5, pp. 36–39, No. 6, pp. 45–51.

CLARKE, B.G. AND SMITH, A. (1992). Self-boring pressuremeter tests in weak rocks. Proc. Engngn. Group of Geological Soc., to be published.

CLARKE, B.G. AND WROTH, C.P. (1984). Analysis of Dunton Green retaining wall based on results of pressuremeter tests. Geotechnique, Vol. 34, No. 4, pp. 549–561.

CLARKE, B.G. AND WROTH, C.P. (1985). Discussion on 'Effect of disturbance on parameters derived from self-boring pressuremeter tests in sand'. Geotechnique, Vol. 35, No. 2, pp. 81–97.

DENBY, G.M. AND CLOUGH, G.W. (1980). Self-boring pressuremeter tests in clay. Proc. Am. Soc. Engrs-J. Geotech. Div., Vol. 106, No. GT12, pp. 1369–1387.

ECCLES, C.S. (1992). Some aspects of the effective shear strength and stiffness of the London Clay at depth in the Central London area. Submitted to the British Geotechnical Society for the Cooling Prize.

FAHEY, M. AND JEWELL, R. (1990). Effect of pressuremeter compliance on measurement of shear modulus. Proc. 3rd Int. Symp. on Pressuremeters, pp. 115–124, Thomas Telford, London.

GIBSON, R.E. AND ANDERSON, W.F. (1961). In situ measurements of soil properties with the pressuremeter. Civ. Engng. Public Wks Rev., Vol. 56, No. 658, pp. 615–618.

HUGHES, J.M.O., WROTH, C.P. AND WINDLE, D. (1977). Pressuremeter tests in sands. Geotechnique, Vol. 27, No. 4, pp. 455–477.

JARDINE, R.J. (1991). Discussion on 'Strain dependent moduli and pressuremeter tests'. Geotechnique, Vol. 41, No. 4, pp. 621–626.

JEZEQUEL, J.-F., LEMASSON, H. AND TOUZE, J. (1968). Le pressiometre Louis Menard: Quelques problemes de mise en oevre et leur influence sur les valeurs pressiometriques. Bulletin de LCPC, No. 32, pp. 97–120.

LACASSE, S. AND LUNNE, T. (1982). In situ horizontal stress from pressuremeter tests. Proc. Int. Symp. on the Pressuremeter and its Marine Applications, Paris, pp. 187–208.

WROTH, C.P. AND HUGHES, J.M.O. (1973). An instrument for the in situ measurement of the properties of soft clays, Proc. 8th Int. Conf. on Soil Mechanics and Foundation Engineering, Moscow, Vol. 1.2, pp. 487–494.

YEUNG, S.K. AND CARTER, J.P. (1990). Interpretation of the pressuremeter test in clay allowing for membrane and effects and material non-homogeneity. Proc. 3rd Int. Symp. Pressuremeters, pp. 199–208.

An assessment of seismic tomography for determining ground geometry and stiffness

C.R.I. CLAYTON, M.J. GUNN and V.S. HOPE, Department of Civil Engineering, University of Surrey

Seismic tomography has been used for the past 20 years in the field of oil exploration, for the assessment of sub-surface structure. Its use in civil engineering ground investigation is attractive in that there is the potential not only for the identification of subsurface geometry (including the detection of cavities), but also for the determination of the spatial variability of ground stiffness. But the application of geophysics to civil engineering problems is never as straightforward as might be supposed. This paper briefly describes the tomographic method, before examining in some detail those factors which affect its predictive accuracy. Using examples, the difficulties of detecting voids and recovering the correct values of seismic velocity are demonstrated. Conclusions are drawn regarding both the design of tomographic surveys and the ability of the technique to detect anomalies of different types.

Introduction

The process of routine ground investigation involves vertical profiling, classification, and parameter measurement for the ground beneath a site, in order to determine the geometry of the subsoil, its variability, and its properties. Seismic techniques may be used as part of ground investigation, and can provide valuable information not only on the lateral and vertical variability of the ground, but also its stiffness.

Tomography (from the Greek, *tomos*: a slice or section) is a method whereby an image of the distribution of a physical property within an enclosed region is produced without direct access to the zone. Seismic velocity tomography is a geophysical sectioning technique which determines the spatial seismic velocity distribution within a given area of interest.

Seismic velocity tomography is potentially extremely useful in geotechnical ground investigations, for two reasons. Firstly, tomography can give information on the general variability (i.e. of seismic velocity) beneath a site, and by inference allow a qualitative assessment of the variability of properties such as stiffness and strength. Features such as voids, fractures, rock layers and soft spots are often difficult to detect using more conventional (boring and drilling) techniques, because only a minute proportion of the subsoil is sampled and tested by direct methods of ground investigation. Geophysical techniques may be helpful in detecting such features.

Secondly, the technique can be used to provide values of the 'very small strain' stiffness (G_0, or G_{max}) of the ground, since this is uniquely linked to seismic shear wave velocity (V_s) through the equation

$$G_0 = V_s^2 \cdot \rho \qquad (1)$$

where ρ is the bulk density of the soil. Depending upon the ground conditions and the strain levels imposed by construction, this stiffness may be more or less relevant to the operational stiffness required for the analysis of geotechnical displacement problems. In many ground conditions (for example fractured rock, or granular soils) the alternative methods of estimating stiffness may be sufficiently expensive or inaccurate as to make seismic methods attractive.

In this paper we examine the seismic tomographic technique, as applied to ground investigation, and based upon our experience consider some of the difficulties that may arise.

The tomographic method

The form of tomography that is, perhaps, most familiar is the technique of computer assisted tomography (CAT), as used in diagnostic medicine. The method of seismic tomography has been in use for some years: amongst the first applications of geotomography was a survey by Bois et al. (1972) in an oil field. In seismic velocity tomography, an image of the distribution of the seismic propagation velocity properties within a region of the ground is deduced from measurements of the transit times of artificially induced seismic waves crossing the zone. The process can be divided into a number of activities, each of which requires careful attention if useful results are to be achieved:

(a) data acquisition and reduction
(b) reconstruction of velocities
(c) interpretation.

The field geophysical method involves the input of seismic energy (from mechanical hammers, or sparkers) at discrete points down boreholes, and possibly along the ground surface, and the acquisition of seismic records of the incoming wave energy (via geophones or hydrophones) at as many other discrete points around the area of interest as is feasible. The seismic traces are then inspected to determine the first arrival of the wave type of interest. In many rock investigations

SEISMIC TOMOGRAPHY FOR GROUND INVESTIGATION

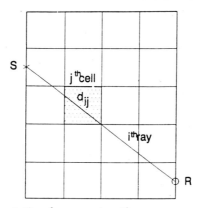

Fig. 1. *Definition of notation for reconstruction*

this may be the onset of primary, compressional (P) waves, but in weaker sediments such as saturated soils, it may be necessary to determine the arrival time of secondary, shear (S) waves. The data required from the field, before reconstruction processing can commence, are the travel times from each source to every receiver, and the coordinates of the source and receiver positions.

Tomographic reconstruction (as applied here) is the mathematical process by which velocities at different points within the ground are calculated. The principles of tomographic reconstruction are well established (Radon, 1917). In general, the region to be imaged is divided into discrete rectangular areas or cells (Fig. 1). Let v_j be the seismic propagation velocity within the jth of n reconstruction cells. This velocity value applies uniformly across the full area of the cell. The travel time, t_i, for the ith ray across the grid of cells is given by:

$$t_i = \sum_{j=1}^{n} \left(\frac{d_{ij}}{v_j} \right) \quad (2)$$

where d_{ij} is the length of the ith ray within the jth cell. In a survey which incorporates m travel times, acquired for rays at various positions and orientations across the region, there will be m such summations. These can be expressed in matrix form as:

$$\mathbf{t} = \mathbf{Dw} \quad (3)$$

where $\mathbf{D}$ is an $m \times n$ array having elements of the form d_{ij}; $\mathbf{w}$ is an n-element column vector containing the current estimate of the reciprocal velocities ('slownesses'); and $\mathbf{t}$ is a column vector of the m travel times calculated across the discretized velocity field.

Let $\mathbf{p}$ be the column vector of the m observed travel times from a field survey. For reasons to be discussed in detail later, it is usually not possible to find a velocity field which is exactly consistent with the measured travel times. Thus the process of reconstructing a velocity field is equivalent to the minimization of a residual vector, **e**, defined by the equation:

$$\mathbf{e} = \mathbf{p} - \mathbf{D}\,\mathbf{w} \qquad (4)$$

Commonly used strategies for computerized geotomographic reconstruction include the damped least squares method (Bois et al., 1971), the Back Projection Technique (BPT – Kuhl and Edwards, 1963; Neumann-Denzau and Behrens, 1984), the Algebraic Reconstruction Technique (ART – Kaczmarz, 1937; Peterson et al., 1985), and the Simultaneous Iterative Reconstruction Technique (SIRT – Dines and Lytle, 1979; Gilbert, 1972), although many others are available (for example, Gordon and Herman, 1974).

The acquisition diagram in Fig. 1 shows straight rays. In practice, rays may deviate from straight paths as waves undergo, for example, refraction in an heterogeneous velocity field. Therefore the path followed by a ray, between a source and receiver, is not known. **D** is a function of **w**: the elements of **w** *and* **D** are unknown. Thus reconstruction involves not only the distribution of errors along ray paths, but also the determination of an appropriate position for that ray path.

The product of reconstruction is a single velocity for each cell. These velocities may be contoured and displayed as colour or grey scales in a *tomogram*. A seismic velocity tomogram is, necessarily, an approximation. It is a two-dimensional, discrete estimate of a continuously varying, three-dimensional function – that is, the distribution of the seismic velocity properties of the ground. The accuracy of a tomographic estimate of the subsurface and, hence, its usefulness as a predictive tool is influenced by many diverse factors. These affect the ability of the technique to determine the size, form and seismic velocity (and hence stiffness) of features in the ground.

Factors affecting the predictive accuracy of geotomography
Idealizations of wave behaviour
Seismic energy travels through the ground as *waves*. There are a number of different physical theories associated with the description of wave behaviour. Each particular theory or idealization may be useful in one context, but it is often necessary to appeal to more than one theory to understand fully how seismic energy can be transmitted through the ground. For example, a source of seismic energy at a point in a homogeneous isotropic medium will produce a set of spherical wavefronts and these can, as a simplification, be represented as a set of rays emanating from the source (Fig. 2(a)). The ray approximation is convenient for use in geotomographic reconstruction because of the line

integral relation between propagation velocity and travel time assumed in eqn. (2). According to Huygens (who first put forward a wave theory for light), each point on a wavefront can be regarded as a possible secondary source. This concept leads to Fermat's principle of stationary travel times, which identifies the ray path as that giving the minimum travel time. A corollary of Fermat's principle is Snell's law of refraction, which governs the deviation of a ray at a velocity interface. Snell's law is restrictive because it only admits of energy propagation normal to a wavefront. Ray paths for reconstruction determined using Fermat's principle will allow for reflection, refraction, diffraction, and also head waves.

Figure 1, with its straight-ray paths, gives an oversimplified picture of energy passing from seismic source to receiver. In situations where there is not much variation in seismic velocity in the region of interest, the straight-ray assumption may be reasonable. ISRM (1988) suggests that this is the case when velocity contrasts are less than 20%.

When there are greater velocity contrasts, straight-ray reconstruction may lead to unacceptably inaccurate tomograms. Wave energy may be refracted (Fig. 2(b)) or diffracted (Fig. 2(c)) around obstacles, and head waves may form along velocity interfaces (Fig. 2(d)). Attempts to reconstruct the distribution of seismic velocities will be doomed to failure if the mathematical model being used assumes that rays can only bend according to Snell's law, but diffraction or head waves dominate.

To illustrate these points consider the two tomographic surveys outlined below. The first was designed to locate a tunnel crossing the tomographic plane. If ray paths are assumed to be those permitted by Snellian refraction then there should be receivers which are unable to detect any energy emitted by the source (Fig. 3(a)). In practice, all receivers detected incoming wave energy, albeit at various reduced amplitudes. The presence of the water table produced a higher velocity layer a short distance below the tunnel and gave rise to head waves which provided a route around the void (Fig. 3(b)). Even without diffraction and head waves, if refraction is invoked then many rays will travel around the anomaly (Fig. 3(c)). As might be expected from the

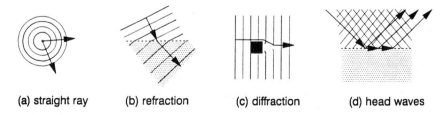

(a) straight ray   (b) refraction   (c) diffraction   (d) head waves

Fig. 2. Wave propagation

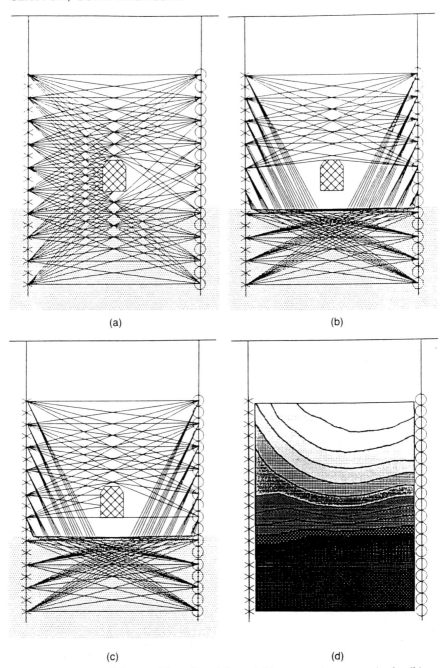

Fig. 3. Unsuccessful tunnel location: (a) straight ray—note non-arrivals, (b) diffracted rays—note lack of ray coverage, (c) refracted rays, and (d) tomogram from field data

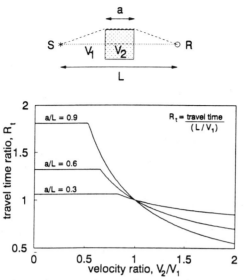

Fig. 4. *Normalized travel time ratio as a function of velocity contrast*

discussion above, a P-wave survey failed to show any signs of the tunnel (Fig. 3(*d*)).

An idealization of tunnel geometry is shown in Fig. 4. Simple mathematics can demonstrate that for any given size ratio (a/L) there is a critical velocity ratio (V_2/V_1) below which an increase in velocity contrast has no effect on travel time. At lower velocity ratios the velocity (and hence stiffness) of the inclusion (V_2) cannot be recovered correctly during reconstruction, because first arrival energy does not travel through the lower velocity material. For the simple geometry shown, it is obvious that the maximum effect that a low-velocity inclusion may have is to double the travel time. When the inclusion occupies as much as 30% of the distance between the source and receiver, the travel time will increase by no more than 6% of the value when no low velocity inclusion is present, regardless of the velocity contrast. The detectability of high-velocity inclusions will not be influenced by this effect.

As a second example of problematic wave phenomena, Fig. 5 shows two adjacent shear wave tomograms that were obtained from the London Clay. The diagonal features which appear in these images result from picking first-arrival shear-wave energy arising from tube waves propagating in the water-filled source boreholes. It is emphasized that these are not a feature of the ground. Virtually identical features can be synthesized by assuming that a wave propagates down the hole in the borehole fluid, generating a shear wave upon its impact with the base.

Finally, there are other aspects of wave behaviour that are important. For example, most tomographic reconstruction is based on the assumption of plane structures in the ground. It should always be remembered that seismic waves are three-dimensional in form and that first arriving seismic events may have followed paths outside the plane of the boreholes. In addition, features in the ground which are smaller in dimension than about one wavelength of the signal will not be capable of being resolved by tomographic reconstruction. In our experience, P-waves typically will be expected to have wavelengths between 1 and 5 m, while for S-waves the wavelength is of the order of 1 to 3 m.

Influence of data errors

Data errors can arise due to inaccurate seismograph triggering, as a result of mis-picking travel times, and also from locating the seismic source and receiver stations incorrectly. The latter class of error will usually be negligible in cases where a borehole deviation survey is available.

Any estimate of travel time between source and receiver will involve some level of error. This error is a function of the accuracy with which the source trigger time is known, and the certainty with which first arrival events can be identified on the seismic records. This latter is largely a function of signal to noise ratio.

The influence of observational errors in the travel time data set, **p**, on the spatial resolution of a survey can be quantified as follows. Let t_{err} be the empirically determined uncertainty on these data. An inclusion would be imperceptible if the increase in travel time due to its presence were less than the travel time error. Thus the following condition must be satisfied:

$$\frac{t_{err}}{t} < R_t \qquad (5)$$

where t is the observed travel time for the wave and R_t is the ratio of travel times for the feature, for example, as defined in Fig. 4. R_t is a function of the geometry of the problem and the velocity ratio; values for a simple square inclusion can be obtained from Fig. 4. If eqn. (5) is not satisfied, data errors will effectively mask the presence of the feature. In practice, estimates of t_{err} can be based on previous field experience, allowing for signal degradation caused by ambient noise levels and attenuation due to increased borehole spacing. For example, in a survey across a 15 m span of London clay, the authors have observed travel time and errors of the order of 10 ± 0.2 ms ($t_{err}/t = 2\%$) and 75 ± 0.8 ms ($t_{err}/t = 1\%$) for P- and S-waves, respectively. These values suggest that, to be detectable, a void must have dimensions greater than about 13% to 18% of the borehole spacing.

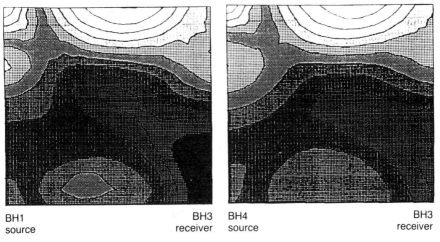

Fig. 5. Artefacts produced by tube-wave propagation

As has been indicated, in a typical geotomographic survey, the matrix **D** does not have full rank. Furthermore, **D** is ill-conditioned: these systems tend to magnify the effects of data errors (Jackson, 1972). This problem can be reduced by identifying and excluding outliers from the data set and also by 'damping out' the influence of the smaller eigenvalues in the tomographic system. Nevertheless, to some degree, observational errors will, unavoidably, affect a tomogram that is derived from field data. An effective empirical method of assessing the influence of travel time errors on a reconstructed image is given by the following procedure. A numerical model of the suspected velocity field is simulated. Using a suitable ray tracing algorithm, a set of travel times for theoretical rays across the field are calculated. A second data set is generated by adding random 'error' values, limited in magnitude by t_{err}, to the travel times in the first set. Tomograms are reconstructed from both data sets. Subtracting cell by cell, the reconstructed velocity values in each image will result in a 'difference' tomogram. This image indicates the effect of data errors in a particular tomographic system, as processed by the chosen reconstruction algorithm. The image can be interpreted thus: If cell velocities in the difference tomogram show a deviation of, say, 5% from the velocities within adjacent cells, then such fluctuations in a field tomogram should, perhaps, be attributed to observational error rather than genuine variations in the properties of the surveyed region.

Mathematical properties of the tomographic equations

In an earlier section it was pointed out that reconstructing velocity

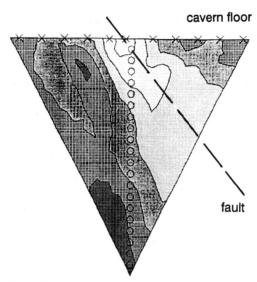

Fig. 6. Successful application of geotomography; detection of a fault in weak rock

values is equivalent to minimizing the residuals **e** defined by:

$$e = p - D w \qquad (4)$$

and that usually one cannot find a solution, **w**, which is both exact (i.e. one corresponding to **e** = 0) and unique. Many of the reasons for this can be shown to be due, when the standard cross-hole tomographic geometry is used, to the mathematical properties of the matrix **D**. For the sake of simplicity, consider first the case where velocity contrasts are relatively small, so that it is reasonable to assume that ray paths are straight.

Recall that the number of rows in **D** is equal to the number of rays (m) and the number of columns is equal to the number of cells (n). In the general case, it is unlikely that m will equal n, but again for simplicity assume that in a particular case there are k sources and k receivers, in two parallel boreholes, and that a k by k grid of cells is to be used for reconstruction. Here, $m = n = k*k$ and the matrix **D** is square. Even in this apparently favourable case there is no straightforward solution to the equations. This is because the matrix **D** does not have full rank. In other words, the number of *independent* rows or columns in **D** is less than the number of rows/columns in the matrix. In effect the matrix equations are simultaneously underdetermined and overdetermined. A consequence of this is that one can change certain combinations of velocities in the grid without altering the travel times, **t**, across the grid.

Mathematically, one says that such changes in velocity lie in the nullspace of **D** and satisfy the equation **Dy** = 0.

A non-trivial nullspace has important consequences. For example, with the standard geometry of tomography between two vertical boreholes, it is possible to add a constant value to the slownesses of all the cells in one column of the grid and to subtract the same value from the slownesses of all the cells in another column of the grid without changing the resulting travel times. A practical consequence is that the *structural resolution* of such a tomogram is poor; one would not expect geotomography to identify, for example, a vertical velocity anomaly which extended the entire height of the tomographic plane.

A second result which follows from the mathematical properties of **D** is that $(m - r)$ independent relations are implied between the travel times where r is the rank of **D**. Due to inconsistencies, for example caused by travel time errors, it is not possible to satisfy the equation **Dw** = **p** exactly. A mathematically equivalent statement is that **p** does not necessarily lie in the column space of **D**. It follows, therefore, that a number of solution strategies may be used, and that each can yield a different result.

The same general conclusions apply when **D** is not square, as is usual in practice. The acquisition of additional, more vertically oriented rays (for example, through the use of sources or receivers on the ground surface) will increase the rank of **D** and assist with the identification of vertical velocity features. To achieve full rank for **D**, however, sources and receivers are required which surround the region to be imaged. This is, of course, not usually feasible for practical reasons.

It is important to emphasize that, for an ill-determined and inconsistent tomographic system (see eqn. (4)), there will be no unique solution vector, **w**. The solution vector that is generated will depend on the reconstruction algorithm that is chosen for the task. Worthington (1984) has presented a review of various reconstruction algorithms that are used in tomography. Each category of solution vector may display certain valuable characteristics. Nevertheless, no one class of solution vectors can be declared to be *correct*. If the original velocity field is not a member of the class of vectors generated by a reconstruction algorithm, then this field cannot be reconstructed accurately.

In practice, one would apply several different algorithms to a field data set. Features that are found to be common in the resulting tomograms can, perhaps, be interpreted with some confidence.

Discussion and conclusions

As with many new and complex techniques, there is a danger that tomography will become discredited as a result of thoughtless mis-use.

As we have demonstrated above, there are potentially many reasons why the technique might be expected not to succeed, given particular site conditions. If tomography is to be successful then a number of basic criteria should be applied during the planning and design of the work.

The preliminary design of a tomographic survey should consider:

(a) The type of wave to be used (i.e. P or S). This decision should be made on the basis of the expected velocity contrasts in each case, and the predicted wavelength in relation to any target.

(b) The expected signal-to-noise ratio, and the repeatability with which the seismograph can be triggered by the source. These will influence the travel time errors.

(c) The size and geometry of the required tomogram, based on the amount of ground to be investigated, and the expected size of the target.

It is essential that shallow tomographic surveys are planned on the basis of a reasonable knowledge of the likely range of ground stiffnesses (and hence seismic velocities), and the information required from the survey. If tomography is intended to detect a 'target' (for example, a fault or a cavity) then the possible orientations and sizes of the target should be estimated. Following this, synthetic travel times should be generated for a number of possible survey geometries (i.e. borehole separation, and the downhole source and receiver spacings) and processed using a range of reconstruction techniques. In this way, the design of the survey may be optimized and the viability of the technique assessed.

On the basis of our experience to date, it may be stated that:

(a) Low velocity anomalies (such as cavities) appear more difficult to detect than high velocity anomalies (e.g. hard inclusions).

(b) The absolute values of seismic velocity recovered from a survey may be regarded as indicative of ground stiffness variations, but should not be used in an absolute way in engineering calculations.

(c) Velocity variations are likely to be most reliably reconstructed when velocity contrasts are low.

(d) Planar, or approximately planar, features (such as faults) can be located relatively successfully, provided that they strike normal to the tomographic plane. An example of the successful reconstruction of real data, over a fault in an oil-storage cavern floor, is shown in Fig. 6.

References

BOIS, P., LAPORTE, M., LAVERGNE, M. AND THOMAS, G. (1971). Essai de determination automatique des vitesses sismiques par mesures entre puits. Geophysical Prospecting, Vol. 19, pp. 42–83.

DINES, K.A. AND LYTLE, R.J. (1979). Computerised geophysical tomography. Proc. I.E.E.E., Vol. 67, No. 7, pp. 1065–1073.

GILBERT, P.F.C. (1972). The reconstruction of a three-dimensional structure from projections and its application to electron microscopy: II Direct methods. Proc. R. Soc. Lond., Vol. B-182, pp. 89–102.

GORDON, R. AND HERMAN, G.T. (1974). Three-dimensional reconstruction from projections: a review of algorithms. Int. Rev. of Cytology, Vol. 5, No. 38, pp. 111–151.

INTERNATIONAL SOCIETY FOR ROCK MECHANICS (1988). Suggested methods for seismic testing within and between boreholes. Commission on Testing Methods. Int. J. Rock Mech. Min. Sci. & Geomech. Abst., Vol. 25, No. 6, pp. 447–472.

JACKSON, D.D. (1972). Interpretation of inaccurate, insufficient and inconsistent data. Geophys. J. R. Astr. Soc., Vol. 28, pp. 97–109.

KACZMARZ, M.S. (1937). Angenaherte Auflosung von Systemen lineareer Gleichungen. Bull. Int. de l'Academie Polonaise des Sciences et des Lettres, Classe des Sciences Mathematiques et Naturelles, Serie A Science Mathematiques, pp. 335–337.

KUHL, D.E. AND EDWARDS, R.Q. (1963). Image separation radioisotope scanning. Radiology, Vol. 8, pp. 653–662.

NEUMANN-DENZAU, G. AND BEHRENS, J. (1984). Inversion of seismic data using tomographical reconstruction techniques for investigations of laterally inhomogenous media. Geophys. J. R. Astr. Soc., Vol. 79, pp. 305–315.

PETERSON, J.E., PAULSSON, B.N.P. AND MCEVILLY, T.V. (1985). Applications of ART to crosshole seismic data. Geophysics, Vol. 50, No. 10, pp. 1566–1580.

The behaviour of granular soils at elevated stresses

M.R. COOP, City University, London, and I.K. LEE, formerly City University, London

The behaviour of three diverse granular soils has been investigated over a wide range of stresses (50 kPa–58 MPa) allowing a general framework to be developed for the behaviour of granular soils. Characteristic features such as a normal compression and critical state lines may be identified and the description of the mechanics of these soils is therefore similar in many respects to the critical state framework developed for clay soils. For granular soils, the principal means of volumetric compression is through particle crushing and this gives rise to some unusual features, notably that for these soils the critical state is not at the apex of the state boundary surface. Particle breakage has been quantified and unique relationships identified between the degree of breakage and the stresses applied, both for isotropic compression and at the critical state.

Introduction

The behaviour of soils and rocks at elevated stresses has generated interest in recent years largely as a result of greatly increased foundation depths, particularly for offshore piling, the construction of deep galleries for the storage of radioactive waste and because of the need of petroleum engineers to understand the mechanics of oil bearing strata. In response to this, over a five year period a high pressure stress path facility has been developed at City University which is comprised of three stress-path triaxial apparatus with full computer control, covering a variety of stress ranges up to a maximum of 70 Mpa. A wide range of soils have been tested during this period, and in particular a number of different granular soils have been investigated in detail. It is the aim of this paper to draw together the general characteristics of the behaviour of granular soils which may be inferred from this work. In each case tests at high pressures have been complemented by others at more usual engineering stresses and the data presented cover a range of stresses from 50 kPa to 58 MPa. The three soils which are discussed here are a carbonate sand (Dogs Bay sand), a decomposed granite and a silica sand (Ham River sand), the gradings of which are shown in Fig. 1.

Dogs Bay sand was one of several carbonate sands tested by Coop (1990) as part of an extensive investigation for British Petroleum of the

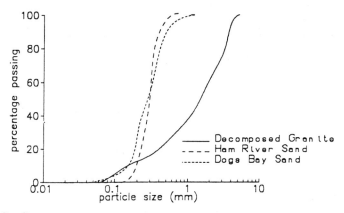

Fig. 1. Gradings

behaviour of piles in these soils, work which was prompted by the problems of low pile capacity encountered at the North Rankin offshore platform (King and Lodge, 1988). The soil is a biogenic carbonate sand, consisting predominantly of mollusc and foraminifera shells, which being relatively unbroken and angular in nature result in a poorly graded soil with high initial void ratios.

Lee (1991) investigated the intrinsic mechanics of a residual soil through the use of reconstituted samples. The soil is a completely weathered granite, well graded and consisting primarily of angular to sub-angular particles of quartz and feldspar with some mica and kaolinite.

The third soil discussed in this paper is Ham River sand, a poorly graded quartz sand the behaviour of which was investigated by Bishop et al. (1965), who tested this soil in a triaxial apparatus at confining pressures up to 6.8 MPa, enabling them to conclude that the principle of effective stress was still valid at these pressures.

In each case the samples were generally prepared in as loose a state as possible. For the poorly graded soils this was achieved by pluviation of the soil directly into a membrane held in place on the triaxial platen. The few denser samples tested were tamped in layers during this process. The particularly high voids ratios for some of the Ham River sand samples were achieved by freezing the samples after pluviation so that they could be set up in the triaxial without further disturbance. Pluviation would result in particle segregation for the well-graded decomposed granite and so these samples were created by lightly compacting the soil unsaturated into a mould, again freezing the loosest samples for installation in the triaxial. In all cases samples were saturated through back-pressure before continuing the test, and further details of the test procedures and apparatus are given in Coop (1990) and Lee (1991).

Isotropic compression

Isotropic compression data for the three soils are given in Fig. 2. In each case both loose and dense soils tend towards a unique normal compression line when plotted in v:ln p' space. Similar normal compression lines at high pressures have been observed by Miura and Yamanouchi (1975) for Toyoura sand and been interpreted by Atkinson and Bransby (1978) from data presented by Vesic and Clough (1968) for

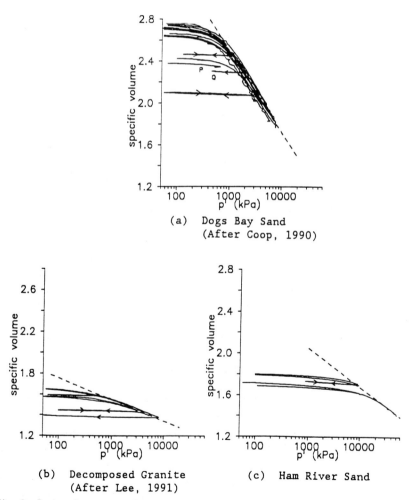

Fig. 2. *Isotropic compression*

Table 1. Summary of critical state parameters

Soil	N	Γ	λ	κ	M	ϕ'_{cs}
Dogs Bay Sand	4.80	4.35	0.34	0.008	1.65	40°
Decomposed Granite	2.17	2.04	0.09	0.005	1.59	39°
Ham River Sand	3.17	2.99	0.16	0.014	1.28	32°

Chatahoochee river sand. The normal compression line may be characterised by:

$$v = N - \lambda \ln p'$$

and the values of these parameters are shown in Table 1. It will be shown later that the normal compression line results from crushing of the soil particles. The stress level at which this breakage starts and the normal compression line is encountered depends not only on the strength of the soil particles but also on the density of the soil. For the Dogs Bay sand, the relatively weak shell fragments combined with the low density of the loose samples and hence high contact stresses gives yield at around 800 kPa. This compares to around 10 000 kPa for samples of Ham River sand prepared in a similar way, resulting from the greater strength of the solid silica particles which comprise this soil, together with its greater density. Dense samples of each soil also showed higher yield stresses than loose samples, again resulting from the greater number of particle contacts in the dense soil and hence lower contact stresses.

In each case this first loading of either dense or loose samples results in a poorly defined yield point which Coop (1990) attributed to the fact that the onset of particle breakage is a gradual process. This behaviour is therefore quite distinct from that in unload–reload for which the soil behaviour is very stiff and elastic resulting from the fact that the particles once crushed cannot be 'uncrushed'. Since the structure of the soil and individual particle contacts are not disturbed by unloading, upon reloading crushing of particles begins again exactly at the previous maximum stress producing a very much better defined yield point than for first loading.

It is interesting to note that the location of the normal compression line for the Ham River sand is very similar to that for Chatahoochee River sand (N = 3.28, λ = 0.175) and Toyoura sand (N = 3.57, λ = 0.198), which probably results from similar crushing characteristics for these quartz sands. In contrast, Coop (1990) showed that for carbonate sands, the diversity of geological origins and consequent variety of particle sizes, shapes and strengths gave a wide range of normal compression line locations.

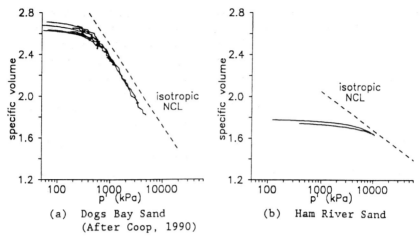

Fig. 3. *One-dimensional compression*

One-dimensional compression

Two of the soils were loaded in one-dimensional compression, the data being shown in Fig. 3. For the Dogs Bay sand it is possible to firmly identify a K_0 normal compression line in $v: \ln p'$ space which lies parallel to the isotropic loading line. The data for the Ham River sand also tend towards a compression line at lower volumes than under isotropic loading. The values of K_0 of 0.51 for Dogs Bay sand and 0.57 for Ham River sand were found not to vary significantly with stress level and so did not change when the soil was taken beyond yield. Both values are significantly greater than those which would be obtained from $K_0 = 1 - \sin \phi'$ as suggested by Jaky (1944) for normally consolidated soils.

Shearing

End of test points for each soil are plotted in $v: \ln p'$ space in Fig. 4. In each case a critical state line may be identified which is approximately parallel to the normal compression line, and which may be defined by

$$v = \Gamma - \lambda \ln p'$$

where values of λ and Γ are given in Table 1. It will be shown later that the critical state line is again governed by the crushing of the soil particles. Been and Jefferies (1985) identified similar 'steady state' lines for Kogyuk sand. As the grading of the sand became less uniform, through the addition of silt sized particles both the gradient and intercept of the critical state line reduced. Coop and Atkinson (1992) found that the addition of fines to the Dogs Bay sand had the effect of reducing N, Γ and λ as the initial density of the samples increased. They also observed that the spacing of the critical state and normal compression lines (i.e. N–Γ) reduced. These features can again be seen from a comparison of the three soils shown in Figs. 2 and 4. The well-graded decomposed granite has much lower values of λ, N, Γ and (N–Γ) than either of the other two soils which are poorly graded.

In q':p' space the critical state lines were found to be straight with a constant gradient, M, at all stress levels, the values of which are given in Table 1 along with corresponding critical state friction angles (ϕ'_{cs}). Where curvature of the failure envelope has been suggested in the past, particularly for the carbonate sands, Coop (1991) suggested that this resulted from incomplete testing at the high pressures where the soils

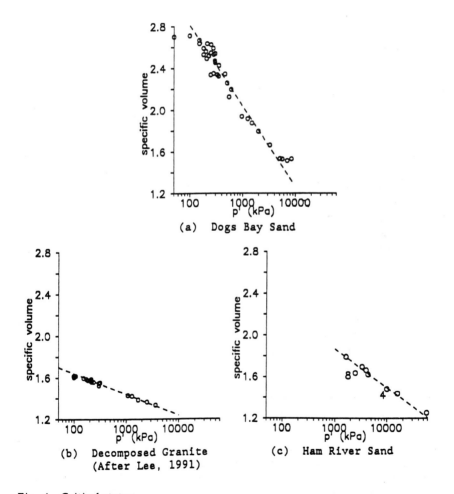

Fig. 4. Critical states

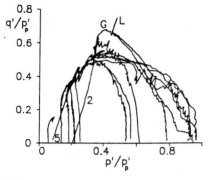

(a) Dogs Bay Sand - Drained Tests (After Coop, 1990)

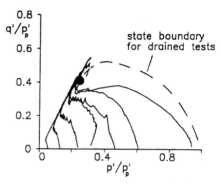

(b) Dogs Bay Sand - Undrained Tests (After Coop, 1990)

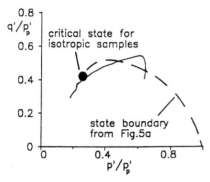

(c) Dogs Bay Sand - Undrained Test on K_o Compressed Sample

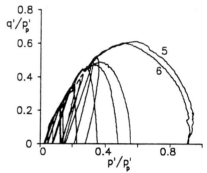

(d) Decomposed Granite - Drained Tests (After Lee, 1991)

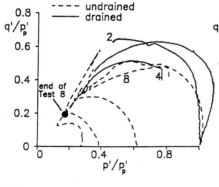

(e) Ham River Sand - Drained and Undrained Tests

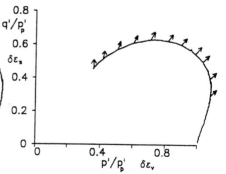

(f) Ham River Sand - Plastic Strain Increment Vectors

Fig. 5. Normalised stress paths

need to be taken to very high strains in order to reach a critical state.

Figure 5 shows the stress paths for the shearing stages normalised with respect to a current preconsolidated pressure (p'_p) defined as:

$$p'_p = \exp[(N - v - \kappa \ln p')/(\lambda - \kappa)]$$

This procedure is similar to using an equivalent pressure on the normal compression line but gives a section of the state boundary surface along an elastic wall rather than a constant volume section. Been and Jefferies (1985) have produced similar normalised plots for Kogyuk sand but as their tests did not reach sufficient pressures to identify a normal compression line they normalised with respect to an equivalent pressure on the critical state line. To avoid the scatter in data which would result from the difficulty in obtaining highly accurate water content data for sands, where the soil has been taken past yield to the normal compression line, the specific volume of the sample after primary compression has been adjusted to fit exactly on that line.

In each case, the drained test data can be used to identify a state boundary surface. On the wet side of critical these have a peak, so that, unusually, the critical state is not at the apex of the state boundary surface. This is, however, a feature which Chandler (1985) predicted for soils with deformable grains, and as a consequence, at the critical state the plastic strain increment vector cannot be perpendicular to the yield curve. This is illustrated for a drained test on Ham River sand in Fig. 5(*f*).

From the critical state data shown in Fig. 4 it can be seen that at the very highest pressures there is some evidence that the critical state line begins to curve and it clearly cannot continue straight in $v:\ln p'$ space indefinitely. Tests G and L for Dogs Bay sand and Test 2 for Ham River sand do not reach the critical state lines identified by the other tests and this is reflected in Fig. 5 by the normalised paths for these tests showing poor agreement with the state boundaries identified by tests at lower stresses. The use of a logarithmic volume scale rather than the conventional linear one as suggested by Butterfield (1979) might give a longer straight critical state line.

For each soil two types of drained stress path were used, some with the usual constant radial stress condition and some holding a constant value of p'. The normalised stress paths of the latter are initially vertical whereas those of the standard drained tests (e.g. Tests 2 and 5 for Dogs Bay sand) are inclined to the vertical. The normalised stress paths of the undrained tests have a characteristic 'S'-shape, again showing the dependence of the normalised stress path on the type of test. For all three soils the normalised stress paths for undrained tests on normally compressed samples curve inside the state boundary surface identified from the drained tests and this is illustrated in Fig. 5 for Ham River sand

and Dogs Bay sand. Similarly for the Decomposed Granite the constant p' test (Test 6) gives a boundary surface apparently inside that of the standard drained test (Test 5). Potts and Gens (1982) have identified other soils, in particular Cowden Till for which Rendulic's principle does not hold and the normalised paths do not give a unique boundary surface. They suggested that the true state boundary surface could only be identified with tests conducted with constant q'/p' ratio.

A further departure from conventional critical state theory can be seen from the data for one-dimensionally normally compressed samples of Dogs Bay sand and Ham River sand, as shown in Fig. 5. While a constant p' drained test on a one-dimensionally normally compressed sample (Test 4 on Fig. 5(e)) gives a stress path which identifies the same critical state as the isotropically loaded samples, the undrained tests (Fig. 5(c) and Test 8 on Fig. 5(e)) give a very brittle response, rapidly unloading to much lower p' values than expected. Although their data were not normalised, the same feature can be seen for the tests of Bishop et al. (1965). It is probable that a particular soil structure is set up during anisotropic loading which gives rise to this feature, while for drained loading the large shear and volumetric strains taken to reach a critical state are sufficient to erase the influence of such a structure.

Compacted and overconsolidated soils

Been and Jefferies (1985) emphasised the importance of the state of a sand relative to the critical state line in $v : \ln p'$ space, showing that as for clay soils this is a principal factor determining the stress–strain behaviour. They defined a state parameter as the difference between the current voids ratio and that on the critical state line at the same pressure. If the normal compression line has been determined, as is the

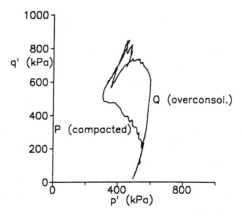

Fig. 6. Dogs Bay Sand – undrained tests on compacted and overconsolidated samples (after Coop (1990))

case here, then there is little need to use the state parameter as an apparent overconsolidated ratio analogous to that for clay soils may be defined:

$$R = p'_p/p'$$

While clay soils reach a given state predominantly through preconsolidation, for sands the principal mechanism in changing its state is compaction, and a comparison must be made between a soil which has been truly overconsolidated to reach a given state and one which has reached the same state through compaction. Such a comparison is made for Dogs Bay sand in Figs. 2(a) and 6. Under undrained loading the overconsolidated sample is initially much stiffer undergoing less particle crushing than the compacted sample since some breakage has already occurred during the preconsolidation.

Particle breakage

Hardin (1985) suggested that the amount of particle breakage could be quantified by the area between the gradings curves before and after a test, considering only that part which was above the 74 μm sieve size since the fines were unlikely to contribute to the overall particle breakage. This he termed total breakage (B_t). Hardin also recognised that the amount of breakage would be influenced by the initial grading of the soil, and defined a breakage potential (B_p) as the area between the original gradings curve of the soil and a vertical at 74 μm. He suggested that comparisons between various soils should be made by normalising B_t with respect to B_p giving relative breakage B_r, and values of B_r for each soil have been plotted in Fig. 7 against the value of p' to which they were subjected. For the sheared samples this is p' at the critical state.

From the critical state data it can be seen that at low pressures B_r tends towards zero and as observed earlier the onset of crushing is a gradual process. At higher stresses the relationships between B_r and ln p' can be approximated as linear. The stress level at which significant breakage starts is clearly related to the particle strength, ranging from around 100 kPa for Dogs Bay sand to around 1500–2000 kPa for Ham River sand. In each case the relationship between B_r and stress is unique, being independent of the stress path taken to reach a critical state. Figure 7(b) also shows that B_r at the critical state is independent of whether the soil was compacted or overconsolidated.

The linearity of these plots suggests a linear relationship between B_r and the specific volume of the soil at the critical state, and a similar relationship might be expected between B_r and the value of p' during isotropic compression. Only in the case of Dogs Bay sand are there sufficient data through which a tentative line has been drawn parallel to

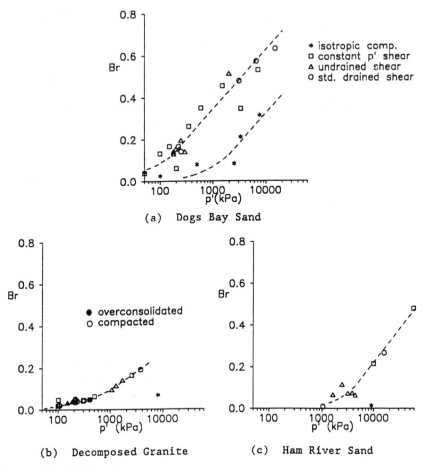

Fig. 7. Particle breakage

that for the critical states. At any given p', the value of B_r for normally compressed states clearly must be less than that after shearing, and the separation of the two lines corresponds to the separation of normal compression and critical state lines in $v: \ln p'$ space.

Conclusions

The behaviour of granular soils at high stresses is governed by the breakage of particles, this being the principal mechanism for plastic volumetric compression. Although this is clearly quite a different process to that for clay soils, the framework of critical state soil mechanics, as developed for the latter, may still be applied as a useful framework within which to examine the soils' behaviour. A number of features appear to result from the breakage of the soil particles, in particular a peak in the state boundary surface on the wet side of critical. Rendulic's principle is also not strictly obeyed and it is therefore not possible to identify uniquely the state boundary surface through the normalisation of shearing data. The amount of particle crushing may be quantified using relative breakage as defined by Hardin (1985), and unique linear relationships may be identified between B_r and $\ln p'$ which correspond to the normal compression and critical state lines.

Acknowledgements
Thanks are due to the following for their help in conducting the experiments presented in this paper: Dr D.M. Dewaikar, Ms M. Ho, Mr D. Nicolau and Ms T. Cuccovillo.

Nomenclature

B_r	relative breakage
B_p	breakage potential
B_t	total breakage
Γ	volume intercept of critical state line at $p' = 1$ kPa
e	voids ratio
κ	gradient of swelling line in $v : \ln p'$ space
K_0	earth pressure coefficient for one-dimensional loading ($= \sigma'_a/\sigma'_r$)
λ	gradient of normal compression and critical state lines
M	gradient of critical state line in $q' : p'$ space
N	volume intercept of isotropic normal compression line at $p' = 1$ kPa
N_0	volume intercept of K_0 compression line at $p' = 1$ kPa
p'_p	equivalent preconsolidation pressure
p'	mean normal effective stress ($= \frac{1}{3}(\sigma'_a + 2\sigma'_r)$)
q'	deviator stress ($= \sigma'_a - \sigma'_r$)
R	apparent overconsolidation ratio
σ'_a	effective axial stress
σ'_r	effective radial stress
ϕ'_{cs}	critical state friction angle [$= \sin^{-1}(3M/(6+M))$]
v	specific volume ($= 1 + e$)

References

ATKINSON, J.H. AND BRANSBY, P.L. (1978). The Mechanics of Soils. London: McGraw-Hill.

BEEN, K. AND JEFFERIES, M.G. (1985). A State Parameter for Sands. Géotechnique, Vol. 35, No. 2, pp. 99–112.

BISHOP, A.W., WEBB, D.L. AND SKINNER, A.E. (1965). Triaxial Tests on

Soil at Elevated Cell Pressures. Proc. 6th Int. Conf. on Soil Mech., pp. 170–174.

BUTTERFIELD, R. (1979). A Natural Compression Law for Soils. Géotechnique, Vol. 29, No. 4, pp. 469–480.

CHANDLER, H.W. (1985). A Plasticity Theory without Drucker's Postulate Suitable for Granular Materials. J. Mech. Phys. of Solids, Vol. 33, pp. 215–226.

COOP, M.R. (1990). The Mechanics of Uncemented Carbonate Sands. Géotechnique, Vol. 40, No. 4, pp. 607–626.

COOP, M.R. AND ATKINSON, J.H. (1992). The Mechanics of Cemented Carbonate Sands. Géotechnique. To be published.

HARDIN, B.O. (1985). Crushing of Soil Particles. Journal of Geotechnical Engineering, ASCE, Vol. 111, No. 10, pp. 1177–1192.

JAKY, J. (1944). The Coefficient of Earth Pressure at Rest. J. Soc. Hungarian Archit. Engnrs, Vol. 22, pp. 355–358.

KING, R. AND LODGE, M. (1988). North West Shelf Development. The Foundation Engineering Challenge. Proc. Int. Conf. on Calcareous Sediments, Perth, Australia, Vol. 2, pp. 333–342.

LEE, I.K. (1991). Mechanical Behaviour of Compacted Decomposed Granite Soil. PhD Thesis, City University.

MIURA, N. AND YAMONOUCHI, T. (1975). Effect of Water on the Behaviour of a Quartz-Rich Sand under High Stresses. Soils & Foundations, Vol. 15, No. 4, pp. 23–34.

POTTS, D. AND GENS, A. (1982). A Theoretical Model for Describing the Behaviour of Soils not Obeying Rendulic's Principle. Proc. Int. Symp. on Numerical Models in Geomechanics, Zurich, pp. 24–32.

VESIC, A.S. AND CLOUGH, E.W. (1968). Behaviour of Granular Materials under High Stresses. Proc. Am. Soc. Civ. Engrs, Vol. 94, No. SM3, pp. 661–688.

Geotechnical characterization of gravelly soils at Messina site: selected topics

R. CROVA, Studio Geotecnico Italiano, Milano,
M. JAMIOLKOWSKI, R. LANCELLOTTA and
D.C.F. LO PRESTI, Technical University of Torino

The paper summarizes the experience gained in the geotechnical characterization of sand and gravel deposits, in connection with the design of the one span suspended bridge over the Messina Strait. The discussion is focused on the development of a dynamic penetration test using a spoon sampler larger than the one employed in Standard Penetration Tests (SPT) with the aim of investigating the possible influence of gravel particles on the measurements with the SPT. The second part of the paper is devoted to a discussion on the influence of the geological age of granular deposits on the penetration resistance and stiffness.

Introduction

Due to the extreme difficulties and high costs connected with undisturbed sampling and with the difficulty of performing many important in situ tests, the characterization of gravelly soils for design purposes still represents an uneasy task of the experimental soil engineering. Therefore, the writers will illustrate some of the problems encountered in the geotechnical characterization of gravelly deposits along the proposed crossing of the Messina Strait with a one span suspended bridge.

The Messina Strait is located in one of the regions with highest seismicity where both atmospheric and marine conditions are extremely hostile. See as an example Table 1 which shows the earthquake and wind conditions that have been considered in designing the bridge.

The subsoil conditions of both sides of the Messina Strait consist of gravelly deposits of holocene and pleistocene age underlain by soft rocks of pliocene and miocene age. Especially on the Sicilian shore, sand and gravel deposits extend to a depth beyond 180 m below the existing GL.

This paper deals with the geotechnical characterization of such deposits on the Sicilian side, see the cross-section in Fig. 1. The geotechnical investigation for the preliminary design consisted in the operations summarized in Table 2 in addition to the evaluation of the

Table 1. Design values for earthquake and wind

	GROUND ACCELERATION (Free field condition)	WIND VELOCITY u(10) (*) (m/s)	RETURN PERIOD (years)
1st LEVEL	0.08g	36	50
2nd LEVEL	0.29g	42	400
3rd LEVEL	0.64g (**)	46	2000

(*) at + 10 m above m.s.l.
$u(z) = u(10) (z/10)^{0.135}$ for $10\ m \leq z < 300\ m$
$u(z) = u(300)$ for $z > 300$
(**) g = gravity acceleration; g = 9.81 m/s²;
z = elevation above m.s.l.

Fig. 1. Geological cross-section on Sicilian shore

Table 2. Soil investigation program

Type of test	CALABRIA		SICILY	
	Foundation	Anchor	Foundation	Anchor
Boring (1)	4	5	2	3
SPT (2)	7	6	7	6
LPT (2)	5	5	5	5
PLT (3)	NONE	3	NONE	3
Pumping test	1	NONE	1	NONE
CPT	1	NONE	3	NONE
CH (4)	2	2	2	1
SASW	1	1	1	1

(1) Depth = 100 m, in addition to a number of geological borings
(2) Depth = 50 m, with rod energy measurements
(3) In 2.5 m diameter cased well, depth = 18 m, in situ density measured
(4) Depth = 100 m

BH	= boring with sampling	CPT	= static cone penetration test
SPT	= standard penetration test	CH	= shear wave velocity measurement by cross-hole methods
LPT	= large penetration test		
PLT	= plate load test	SASW	= spectral analysis of surface waves

index properties and some laboratory tests performed on samples reconstituted with the assumed in situ density.

The discussion that follows will focus on the following:

(a) Development and validation of Large Penetration Test (LPT).
(b) Evaluation of the influence of the geological age of the deposit on the SPT resistance and on the small strain shear modulus (G_0) inferred from shear wave velocity (V_s) measured using cross-hole techniques.

Subsoil conditions

Figure 1 shows a cross-section of the subsoil conditions of the Sicilian shore. As far as the location of the anchor block is concerned, from the ground surface down to the maximum explored depth (100 m), the soil consists of sand and gravel of medium pleistocene age (500 000 to 600 000 years old), which, based on geological borings and geophysical investigations, extend to the depth of at least 180 m below the existing GL. This deposit is locally called Messina Gravel Formation (MGF).

At the location of the bridge tower, the upper part of the subsoil profile consists of sand and gravel of holocene age (6 000 to 15 000 years old), having a thickness ranging between 35 and 67 m. Such deposit is locally called Coastal Plain Deposit (CPD).

Figure 2 shows the range of gradings of the MGF and of the CPD as inferred from soil samples obtained from boreholes (BH), using a sampler having internal diameter ID = 100 mm, from LPTs (ID = 100 mm) and from SPTs (ID = 35 mm).

Table 3. Grain size composition of sands and gravels on Sicilian shore

			Gravel (%)		Sand (%)		Fines (%)		d50 (mm)		d60/d10		d max (mm)		Number of tests
			Mean	St. dev.	Mean	St. dev.	Mean	St. dev.	From	To	From	To	Mean	St. dev.	
Tower foundation	Holocene (CPD)	BH	58	22.1	34.4	18.4	8.6	6.7	0.53	9.60	3.4	223.7	47.0	17.1	46
		LPT	65.5	22.5	31.7	21.6	2.8	2.2	0.60	21.30	2.1	126.9	57.9	24.1	73
		SPT	59.6	17.4	36.4	16.7	4.0	2.1	0.60	15.00	3.1	103.3	30.1	8.1	103
	Pleistocene (MGF)	BH	57.1	16.3	36.3	16.4	6.6	3.3	0.92	44.00	4.6	400.0	56.3	22.2	35
		LPT	85.7	13.2	12.7	12.2	1.7	1.2	7.00	20.00	1.5	31.7	73.3	26.6	6
		SPT	56.3	11.9	39.0	12.6	4.7	1.0	1.00	11.00	11.3	67.6	38.3	7.5	6
Anchor block	Pleistocene (MGF)	BH	56.9	14.9	33.5	11.9	9.6	6.7	0.20	12.00	4.0	335.0	39.9	12.4	74
		LPT	67.0	11.5	30.4	9.6	4.0	2.8	0.54	36.00	3.3	200.0	57.0	22.1	115
		SPT	57.1	9.8	36.5	9.3	6.4	2.1	0.95	7.50	4.4	371.4	31.3	8.2	117

GRAVELLY SOILS AT MESSINA

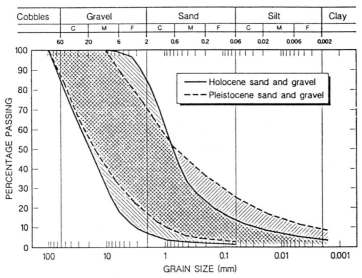

Fig. 2. *Gradation curves of sands and gravels on Sicilian shore*

Larger soil samples obtained during excavation of the 2.5 m diameter and 15 m deep shaft in the MGF and from the 0.8 m diameter and 40 m deep water well installed in the CPD indicated gradings similar to those of LPTs. A more precise perception of the comparison between gradings of CPD and MGF can be inferred from the data reported in Table 3.

To give a more complete picture of the geotechnical features of the gravelly deposits under discussion, the following information is given:

(a) The coefficient of permeability for horizontal flow (k_h) which has been deduced from pumping tests and which represents the overall mass permeability of CPD within the depth of 40 m below the GL, resulted in a value of 5×10^{-3} m/s.

(b) The bulk density measured in situ at the location of the anchor block during excavation of the exploratory shaft for the plate loading tests gave a value of γ ranging from 18.6 to 20 kN/m^3. Such values correspond to the in situ density of the MGF above the ground water level. The range of γ from 18.64 to 19.6 kN/m^3 has been inferred for CPD.

(c) A limited number of electrical static cone penetration tests (CPT) performed both in the CPD and in the MGF reaching a depth ranging between 25 and 30 m have yielded:

(i) values of the cone resistance q_c ranging from 10 to 15 MPa at shallow depth up to 15 to 30 MPa at a depth beyond 10 m

(ii) a mean value of the ratio q_c/N_{SPT} equal to 6.25 and 5.30 in CPD and MGF respectively.

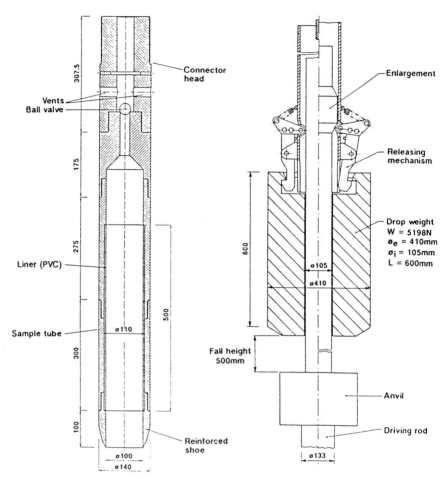

Fig. 3. Details of LPT

Table 4. Dynamic penetration tests

		STANDARD PENETRATION TEST	LARGE PENETRATION TEST (1)	LARGE PENETRATION TEST (2)	BECKER DRILL HAMMER (3)
DRIVE METHOD		FALL WEIGHT	FALL WEIGHT	FALL WEIGHT	DIESEL HAMMER
WEIGHT		623 N	981 N	5592 N	–
DROP HEIGHT		760 mm	1500 mm	500 mm	–
IMPACT ENERGY		473 Nm	1472 Nm	2796 Nm	344 to 6512 Nm
DRIVE LENGTH		450 mm	450 mm	450 mm	300 mm
SAMPLER	OD	51 mm	73 mm	140 mm	165 mm
	ID	35 mm	50 mm	100 mm	CLOSED END

(1) YOSHIDA et al, 1988; (2) USED IN ITALY, 1989; (3) HARDER & SEED, 1986

Large penetration tests

Considering the limitations imposed by the gravelly nature of the soil, the Standard Penetration Test (SPT) has been the only practical test. However, considering the dimensions of the SPT spoon sampler (OD = 51 mm; ID = 35 mm) it was also decided to develop a Large Penetration Test (LPT) in order to investigate a possible influence that the gravel particles might have on the SPT blow/count.

In the present case, the sampler shown in Fig. 3 was adopted as penetration tool. It has ID = 110 mm and OD = 140 mm and contains a plastic liner 5 mm thick. It is driven into the soil using rods having ID = 100 m and OD = 133 mm.

Similarly, as it is done for the SPT, the LPT sampler is driven 450 mm into the soil from the bottom of a borehole and counting the number of blows necessary for penetrating every 150 mm. Table 4 summarizes the characteristics of the above described LPT and, for comparison, those of LPTs used in North America (Harder and Seed, 1986; Harder, 1988) and recently developed in Japan (Yoshida, 1988; Yoshida et al., 1988).

Both SPT and LPT were carried out in cased holes having respectively ID = 75 and 200 mm. The boreholes were always filled with bentonite mud with the aim of preventing hydraulic inflow and consequent loosening of the material at the bottom of the hole.

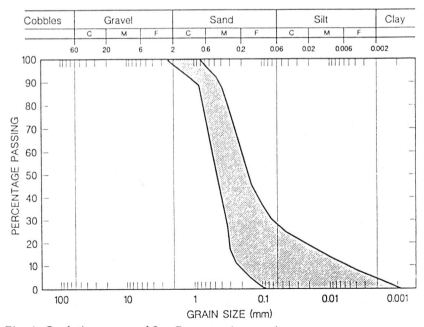

Fig. 4. Gradation curves of San Prospero river sand

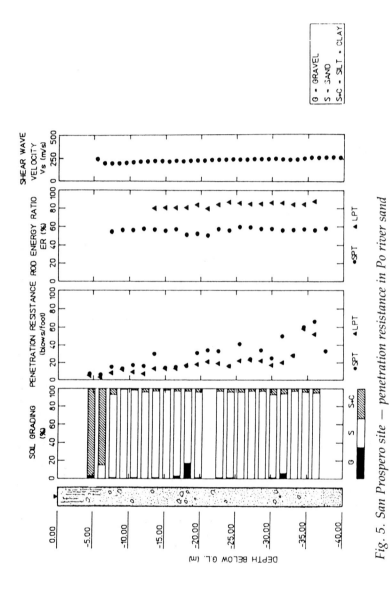

Fig. 5. San Prospero site — penetration resistance in Po river sand

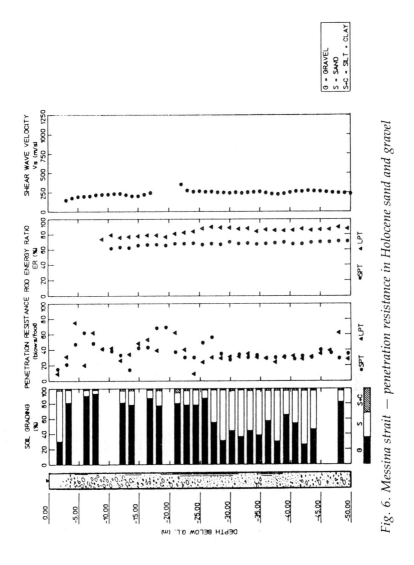

Fig. 6. Messina strait – penetration resistance in Holocene sand and gravel

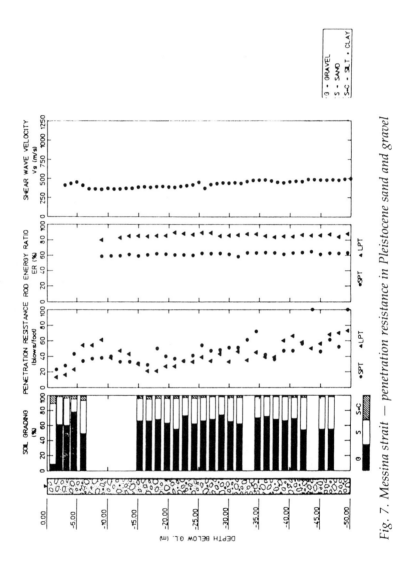

Fig. 7. Messina strait — penetration resistance in Pleistocene sand and gravel

GRAVELLY SOILS AT MESSINA

During the execution of both SPT and LPT the energy delivered to the driving rod was measured using load cells consisting of 1 m-long rod segment instrumented with strain gauges (Schmertmann and Palacios, 1979).

In order to calibrate the newly developed LPT and compare its results with those of the SPT, preliminary tests with both devices were performed at the San Prospero site in the well-investigated Po river sand whose grading can be inferred from Fig. 4.

After the calibration of the LPT in Po river sand, both SPT and LPT were performed with systematic measurement of ER during both tests in sand and gravel deposits at the Messina Strait site.

Examples of N_{SPT}, N_{LPT}, ER and V_s values measured in Po river sand, CPD and MGF are reported in Figs. 5–7.

Table 5 summarizes the obtained values of ER and $N_{1(60)}$, which is the penetration resistance normalized to the ER = 60% and the value of the effective overburden stress $\sigma'_{vo} = 98.1$ kPa, see Skempton (1986).

The statistical distribution of the following data: N_{SPT}, N_{LPT}, $[N_{1(60)}]_{SPT}$ and $[N_{1(60)}]_{LPT}$ have been computed and the results are also summarized in Table 5. It results that the frequency distributions of both measured and normalized penetration resistances are definitely non-symmetrical. Therefore, the mean is higher than the normal as made clear by the data reported in Table 5.

The positive skewness of the aforementioned distributions suggests that the highest values of N_{SPT} and N_{LPT} are due to systematic errors. In these conditions, in the design, the use of the normal rather than the mean of N_{SPT} and N_{LPT}, is advisable as the mean value tends to overestimate the penetration resistance.

Figure 8 shows the ratio between $N_{1(60)}$ from SPT and $N_{1(60)}$ from LPT, as a function of mean grain size (d_{50}).

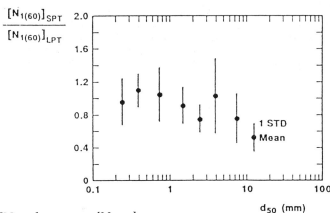

Fig. 8. $[N_{1(60)}]_{SPT}$ versus $[N_{1(60)}]_{LPT}$

Table 5. N_{SPT} versus N_{LPT} in granular deposits

Site	Deposit	[N₁(60)]SPT				[N₁(60)]LPT				Number of tests	$\frac{N_{SPT}}{N_{LPT}}$ ± St. d.	$\frac{[N_{1(60)}]_{SPT}}{[N_{1(60)}]_{LPT}}$ ± St. d.	d_{50} (mm)
		ER ± St. d.	Mean ± St. d.	Norm.	Skew.	ER ± St. d.	Mean ± St. d.	Norm.	Skew.				
SAN PROSPERO	Po River sand	56 ± 3	20 ± 8	15	0.63	84 ± 3	18 ± 7	17	0.21	35	1.41 ± 0.46	1.14 ± 0.40	0.2 to 0.6
MESSINA STRAIT	Holocene sand and gravel (CPD)	66 ± 3	24 ± 10	16	0.85	85 ± 6	30 ± 12	26	0.42	97	1.13 ± 0.52	0.89 ± 0.40	1 to 15
MESSINA STRAIT	Pleistocene sand and gravel (MFG)	61 ± 3	27 ± 7	25	0.29	85 ± 6	29 ± 9	23	0.67	62	1.38 ± 0.45	1.02 ± 0.36	1 to 5

GRAVELLY SOILS AT MESSINA

This ratio (see also Table 5) is, on average, close to 1.0, i.e. in the present case the capability of LPT is close to that of the SPT.

This is true despite the increased driving energy and efficiency of the LPT driving system, because the delivered energy is necessary to overcome the base resistance and the shaft friction (Schmertmann, 1979) when driving a larger sampler.

Influence of age of deposit on test results

The circumstances that both formations CPD and MGF have been deposited in a similar environment and exhibit similar density grading

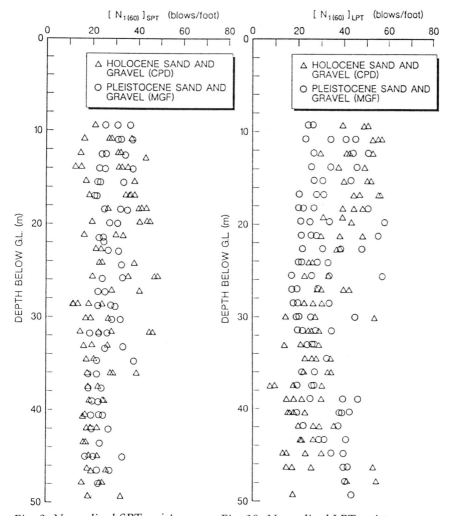

Fig. 9. Normalized SPT resistance   *Fig. 10. Normalized LPT resistance*

and mineralogical composition but substantially different age, offers the possibility of analysing the influence of this latter factor on the results of penetration and seismic tests. As far as penetration test results are concerned, Figs 9 and 10 report comparisons between $[N_{1(60)}]_{SPT}$ and $[N_{1(60)}]_{LPT}$, respectively, as obtained in holocene (CPD) and pleistocene (MGF) deposits.

It appears that on average the penetration resistances resulting from both SPT and LPT in CPD are on average 10 to 20% lower compared to those obtained in much older MGF. To some extent, this finding contradicts those by Skempton (1986) and Mesri et al. (1990) who have found a larger influence of the geological age on N_{SPT} and q_c respectively.

A very different picture emerges when comparing the shear modulus G_0 inferred from the shear wave velocity V_s measured in the two deposits. Figure 11 reports the comparison between the modulus number K_2 of G_0 as defined by Seed et al. (1986) in CPD and MGF respectively. This figure refers to a spot where the contact between the

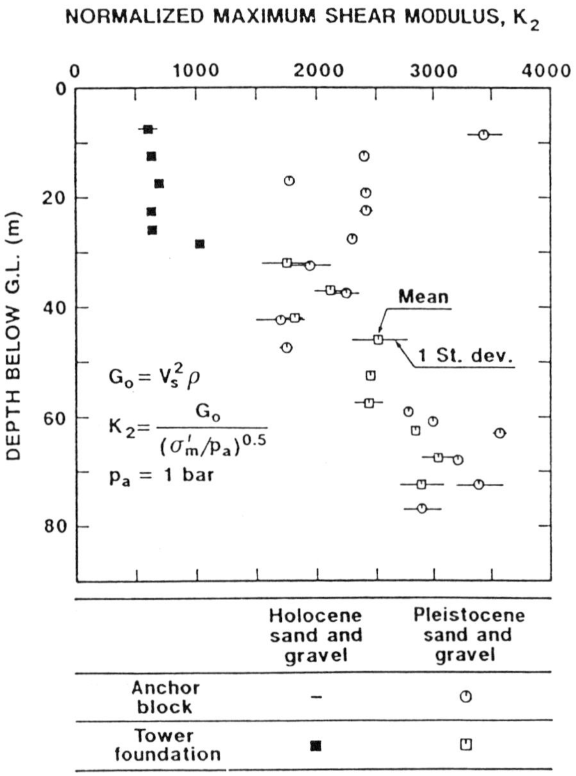

Fig. 11. Normalized shear modulus of sand and gravel from seismic tests

two formations was located by geologists at a depth ranging from 36 to 38 m. The difference of G_0 in holocene and pleistocene sand and gravel appears evident. Based on the already indicated similarities between the two deposits, it is postulated that this difference should be attributed to their different age.

To understand the reasons for the differences in stiffness between CPD and MGF, a number of outcrops of MGF existing in the immediate vicinity of the construction site have been investigated (Bosi, 1990) leading to the conclusion that within the pleistocene sand and gravel formation:

(a) weak to occasionally strong bonding caused by cementing agents like calcium carbonate and iron oxides, together with signs of early diagenesis exist

(b) no sign of liquefaction phenomena which might have occurred in the past were encountered.

In contrast, an absence of cementation phenomena has been ascertained in holocene sand and gravel based on the information collected from shallow excavations in CPD and from the re-examination of the cores of the borings. Within this deposit, there are also historical records (Baratta, 1910) of the occurrence of liquefaction phenomena during the 1908 earthquake.

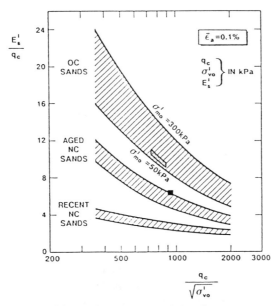

Fig. 12. *Evaluation of drained Young's modulus from CPT in silica sand*

Based on the above, it emerges that penetration tests which induce a large straining in the surrounding soil obliterate the effects of soil structure due to processes like ageing, light cementation and other phenomena of early diagenesis as described by Daramola (1978), Dusseault and Morgenstern (1979), Mitchell (1986), Palmer and Barton (1987), Barton and Palmer (1989), Mesri et al. (1990) and Schmertmann (1991). In contrast, the above-mentioned phenomena strongly influence soil stiffness at small strains.

This poses some doubts about the reliability of correlations between deformation moduli and penetration resistance in soils of the same age when one refers to the results of laboratory tests performed on

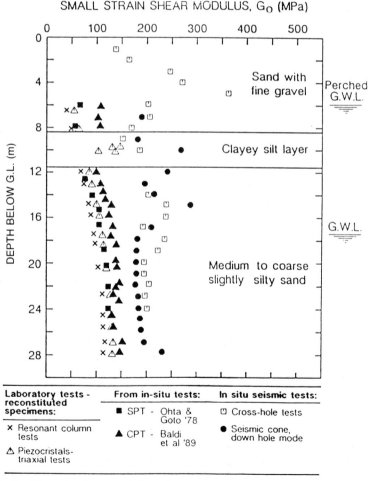

Fig. 13. Small strain shear modulus of Ticino sand at Pavia University site

reconstituted samples. Further confirmation is found in the experience gained by the writers in testing other granular soils.

In this context, some additional experimental evidence is reported here confirming the different response of the natural granular soils if compared, at the same void ratio and effective ambient stress to that of freshly deposited samples, where stiffness parameters are concerned.

Figure 12 shows the correlation between q_c and the secant Young modulus E' at axial strain equal to 0.1%. The ranges given for NC and for mechanically OC sands have been obtained on pluvially deposited specimens from Calibration Chamber (CC) tests (q_c) and drained K_0-consolidated triaxial compression tests E' (Baldi et al., 1989).

The intermediate range for natural sand deposits has been postulated by Baldi et al. (1989) based on a limited number of screw plate loading tests performed in virtually NC Po river holocene sand at the San Prospero site. The same figure reports also the values of E'/q_c ratio obtained by Pasqualini (1989) from the back analysis of field performance of slightly silty sand compacted by vibratory rollers. The above values fall in the range for mechanically OC sand that resulted from CC tests.

Figure 13 shows the variation of G_0 versus depth as obtained by Ghionna (1991) at the Pavia site in Ticino river holocene sand 10 000 to 20 000 years old. Referring to depths below 17 m*, it can be noticed that the values of G_0 inferred from in situ seismic tests are 25 to 40% higher than those obtained from laboratory tests (resonant column and piezo-cristals mounted in triaxial cell) on reconstituted samples and from the correlations between q_c or N_{SPT} and G_0 or V_s valid for freshly deposited sands.

Finally, Fig. 14 from the work by Berardi (1992) compares the E' at $\varepsilon_a = 0.1$ versus $D_R = f\ [N_{1(60)}]_{SPT}$ as obtained from CK_0D triaxial compression tests on reconstituted samples of two silica sands and those inferred (Berardi and Lancellotta, 1991; Berardi et al., 1991) from the field performance of shallow foundations placed in sand and gravel deposits, case records from Burland and Burbidge (1985), properly reanalysed.

The range of E' back-figured from the field performance of predominantly NC deposits falls above the trend resulted for NC freshly deposited Ham river sand.

The three above examples together with that of the Messina Strait suggest that:

(a) The in situ stiffness of natural granular deposits tends to be higher than that of samples reconstituted in laboratory to the same void ratio and subject to the same in-situ effective stresses.

*To the depth below the static GWL to which the following comments apply.

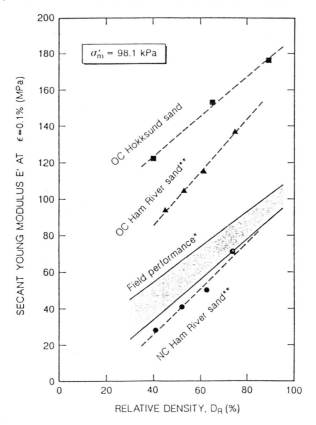

(*) Case records from Burland and Burbidge (1985), see Berardi et al (1991)

(**) Adapted from Daramola (1978)

Fig. 14. *Young's modulus of cohesionless-field performance versus constituted specimens*

(b) It is postulated that this is linked to the complex structure-forming processes developing in natural deposits with time, such as ageing, light cementation, increase of the area of the interparticles contacts, etc., which cannot be replicated in the laboratory.

(c) Penetration resistance versus stiffness correlations based on moduli obtained in laboratory tests on reconstituted samples, tend to underestimate the deformation moduli of natural sand and gravel deposits. This is attributed to the relevant influence that the soil structure developing in geological time has on stiffness, while its impact on the penetration resistance appears much more modest.

References

BALDI, G. ET AL. (1989). Modulus of Sands from CPT's and DMT's. Proc. 12th ICSMFE, Rio de Janeiro.

BARATTA, M. (1910). La Catastrofe Sismica Calabro-Messinese. 25 dicembre 1908. Ristampa Anastatica, Arnaldo Forni Editore.

BARTON, M.E. AND PALMER, S.N. (1989). The Relative Density of Geologically Aged, British Fine and Fine-Medium Sands. Quarterly J. of Engineering Geology, Vol. 22.

BELLOTTI, R., GHIONNA, V.N., JAMIOLKOWSKI, M., ROBERTSON, P.K. AND PETERSON, R.W. (1989). Interpretation of Moduli from Self-Boring Pressuremeter Tests in Sand. Geotechnique, No. 2.

BERARDI, R., JAMIOLKOWSKI, M. AND LANCELLOTTA, R. (1991). Settlement of Shallow Foundations in Sands. Selection of Stiffness on the Basis of Penetration Resistance. Proc. Geotechnical Engineering Congress, Boulder, Col., ASCE Geotechnical Special Publication, No. 27.

BERARDI, R. AND LANCELLOTTA, R. (1991). Stiffness of Granular Soils from Field Performance. Geotechnique, No. 1.

BERARDI, R. (1992). Fondazioni Superficiali su Terreni Sabbiosi — Parametri di Progetto. Ph.D. Thesis, Technical University of Torino.

BOSI, C. (1990). Studio sullo Stato di Aggregazione delle Ghiaie di Messina. Report by SITEC, s.r.l., Roma.

BURLAND, J.B. AND BURBIDGE, M.C. (1985). Settlement of Foundations on Sand and Gravel. Proc. I.C.E. Part 1, 78.

DARAMOLA, O. (1978). The Influence of Stress-History on the Deformation of a Sand. Ph.D. Thesis, Univ. of London.

DUSSEAULT, M.B. AND MORGENSTERN, N.R. (1979). Locked Sands. Quarterly J. of Engineering Geology, Vol. 12.

GHIONNA, V.N. (1991). Personal communication.

HARDER, L.F. AND SEED, H.B. (1986). Determination of Penetration Resistance for Coarse-Grained Soils Using the Becker Hammer Drill. Report no. UCB/EERC-86/06. University of California, Berkeley, Ca.

HARDER, L.F. (1988). Use of Penetration Tests to Determine the Cyclic Loading Resistance of Gravelly Soils During Earthquake Shaking. Ph.D. Thesis, University of California, Berkeley, Ca.

MESRI, G., FENG, T.W. AND BENAK, J.M. (1990). Postdensification Penetration Resistance of Clean Sands. JGE, ASCE, No. 7.

MITCHELL, J.K. (1986). Practical Problems from Surprising Soil Behaviour. JGE, ASCE, No. 3.

OHTA, Y. AND GOTO, N. (1978). Empirical Shear Wave Velocity Equations in Terms of Characteristic Soil Indexes. Earthquake Engineering and Structural Dynamics, Vol. 6.

PALMER, S.N. AND BARTON, M.E. (1987). Porosity Reduction, Microfabric and Resultant Lithification in UK Uncemented Sands. British Geological Society. Special Publication, No. 36.

Pasqualini, E. (1989). Personal communication.

Schmertmann, J.H. (1979). Statics of SPT. JGE, ASCE, No. GT5.

Schmertmann, J.H. and Palacios, A. (1979). Energy Dynamics of SPT. JGE. ASCE, No. GT8.

Schmertmann, J.H. (1991). The Mechanical Ageing of Soils. 25th Terzaghi Lecture. JGE, ASCE, No. 9.

Skempton, A.W. (1986). Standard Penetration Test Procedures and the Effects in Sands of Overburden Pressure, Relative Density, Particle Size, Ageing and Overconsolidation. Geotechnique, No. 3.

Seed, H.B., Wong, R.T., Idriss, I.M. and Tokimatsu, K. (1986). Moduli and Damping Factors for Dynamic Analyses of Cohesionless Soils. JGE, ASCE, GT11.

Yoshida, Y. (1988). A Proposal on Application of Penetration Tests on Gravelly Soils. Abiko Research Laboratory. Rep. No. U 87080 (in Japanese).

Yoshida, Y., Kokusho, T. and Motonori, I. (1988). Empirical Formulas of SPT Blow-Counts for Gravelly Soils. Proc. ISOPT-1, Orlando, Fla.

A critical state constitutive model for anisotropic soil

M.C.R. DAVIES and T.A. NEWSON, School of Engineering, University of Wales, Cardiff

The development of a constitutive model for anisotropically consolidated clay is described. The model, based on critical state soil mechanics concepts, has a rotational hardening non-associated flow rule. To assess the model, data from stress path triaxial tests conducted as part of the research programme and reported in the literature have been predicted. This has resulted in good correlations which significantly improve on predictions using Modified Cam Clay, an isotropic associated model.

Introduction

One of the major assumptions made in the derivation of the earliest elasto–plastic models which describe the constitutive behaviour of soils has been structural isotropy. This assumption was made to avoid excess complication of the models; effects of an anisotropic fabric being considered secondary to a sound fundamental description of the stress–strain behaviour of soil. However, the natural formation of soil, with its accompanying anisotropic stress state, rarely results in a material displaying an isotropic fabric.

Constitutive models which treat soil as an isotropic material can be used to model soils with an anisotropic stress history – provided it is assumed that the stress history does not cause an anisotropic structure to develop. Thus when using, say, one of the family of critical state models to simulate an increase in deviator stress following one-dimensional compression, the resulting strains will be different than those when the soil is subjected to the same change in deviator stress from an isotropic stress state. Comparison of the predicted response of anisotropic soil with experimental measurements indicates the limitation of this approach. Significant variations between measured behaviour of anisotropic soils and that predicted using 'isotropic' models has been observed by a number of workers, e.g. Bondok (1989).

It is desirable that elasto–plastic models should more closely model the stress–strain behaviour of soil by including the effects of anisotropy without resource to greater complication by the inclusion of a large number of extra parameters, e.g. Almeida et al. (1986). With the increased sophistication of numerical modelling in geotechnical engineering, this to some extent is becoming a requirement. The constitutive model constitutes a vital component of a numerical analysis program and if the program is to make accurate predictions then it must use the most appropriate constitutive model for the soil.

Described herein is a programme of work that has been conducted at the School of Engineering at Cardiff with the aim of producing a model based on the critical state concept to accurately describe the stress–strain behaviour of anisotropically consolidated soil. The study has drawn on investigations previously conducted at Cardiff (Stipho, 1978), Cambridge (Parry and Nadarajah, 1973) and elsewhere (e.g. Tavenas and Lerouiel (1977); Graham et al. (1983); Atkinson et al. (1987)). In order to derive models it is clearly necessary to have access to suitable experimental data. To this end the initial phase of this work was to design and commission a triaxial apparatus appropriate for conducting stress probes on anisotropic consolidated specimens. Utilising data from experiments conducted using this apparatus and from other sources in the literature, in the second phase of the study, a non-associated anisotropic constitutive model has, and continues to be, evolved. Assessment of the model has been made by comparison of predictions with the results of laboratory tests.

Experimentation

Apparatus and experimental procedure

The design philosophy of the apparatus was to produce a computer controlled testing facility based on standard triaxial cell and loading frame equipment. This allows both stress and strain controlled stages to be followed in a test. Axial load during stress control is applied to the cell loading ram by a double acting jack connected to the reaction beam on the loading frame. When a strain controlled stage is required the piston in the jack is restrained by a lock nut, thus fixing the location of the loading ram. The base of the cell is then raised or lowered using the loading frame mechanism in the conventional manner. An additional benefit of this system is that since the entire loading system is independent of the triaxial cell, cells of varying size may be used in the apparatus. Feedback is achieved using instrumentation which monitors cell pressure, back pressure, axial force, axial strain, volumetric strain and lateral strain. The last of these is measured using non-contact displacement transducers, and provides feedback allowing strain controlled one-dimensional consolidation loading paths to be followed.

Speswhite kaolin, LL = 59% and PL = 27%, has been the clay used in the testing conducted to date. Samples are prepared by one-dimensionally consolidating a slurry at twice the liquid limit in a specially constructed consolidometer to a vertical effective stress of 130 kPa. Cubical specimens are trimmed from the resulting clay cake. Having been placed in the triaxial apparatus the anisotropic stress history is further imposed by consolidating the specimen either using strain control (i.e. zero lateral strain to achieve K_0 consolidation) or by following a predetermined anisotropic stress path. It is important that the effective stress state is known throughout loading stages. Consequently, the rate of stress change has to be slow enough to prevent significant rises in excess pore water pressure. Close attention has been given to the selection of appropriate rates of testing throughout the testing programme (Davies, 1988; Bondok, 1989).

Experimental results

The primary objective of the experiments was to observe the yielding behaviour of the clay. Following one-dimensional consolidation of the specimens, stress path tests were conducted from either normally consolidated or overconsolidated stress states. The overconsolidated specimens were subjected to stress probes in order to locate yield points and hence find the shape and orientation of the yield locus. Yield points were defined from plots of p' versus volumetric strain, ε_v, q versus shear strain, ε_s, and dissipated strain energy, W, versus length of stress probe, s (Tavenas et al., 1979; Graham et al., 1983). Experimental yield points are shown in Fig. 1 together with the yield locus proposed by Bondok (1989). In common with observations of other workers it may be seen that this yield locus is of the form of a distorted ellipsoid orientated around the K_0 consolidation line.

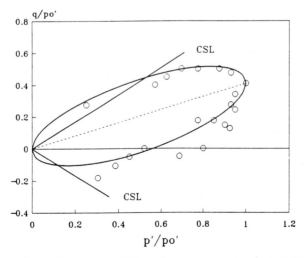

Fig. 1. Comparison of proposed yield surface and experimental yield points

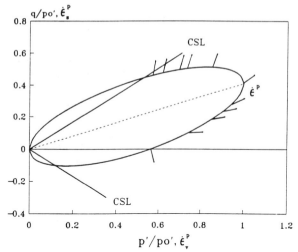

Fig. 2. *Experimental plastic strain increment vector directions*

Closed stress cycles were performed during the stress probes in order to distinguish the plastic strains and hence the directions of the plastic strain increment vectors. These are superimposed on the proposed yield surface in Fig. 2. The results clearly indicate that the plastic strain increment vectors are not normal to the yield surface. The relationship between the yield surface and the plastic potential is therefore non-associated.

Rotational and isotropic hardening of the yield surface were also observed during the stress path tests. The results were used to formulate the rotational hardening rule presented below.

Proposed constitutive model

The proposed model is based on the critical state concept for soils (Roscoe, Schofield and Wroth, 1958). The major features are an empirical yield locus, obtained from the experimental programme, a theoretical plastic potential function originally derived by Dafalias (1987) for an associated anisotropic model and an isotropic and rotational hardening rule. The plastic potential function was also observed to agree with experimental data obtained in this study. Prior to yield the soil is modelled as a linear elastic material.

The yield and plastic potential functions are plotted together in Fig. 3. As has been previously stated the yield function is a rotated distorted ellipse orientated around the K_0 consolidation line. The expression for this yield surface is:

$$(1 - \eta_0^3/M^3) \cdot (q - \eta_0 \cdot p')^2 - (M - \eta_0)^2 (p_0' - p')p' = 0 \qquad (1)$$

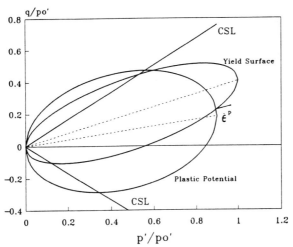

Fig. 3. The CARMEL anisotropic critical state model

where $\eta_0 = q_0/p'_0$, which is the gradient of the K_0 consolidation line.

Similarly the plastic potential is also a rotated distorted ellipse, but in this case it is orientated around a constant stress ratio line ($\eta = q/p'$) which has a gradient, α_0, half that of the K_0 consolidation line. The expression for the plastic potential is:

$$p'^2 - p' \cdot \beta + (q^2 - 2 \cdot \alpha_0 \cdot q \cdot p' - \alpha_0^2 \cdot p' \cdot \beta)/M^2 = 0 \quad (2)$$

where $\alpha_0 = \eta_0/2$ and β is the value of p' at the apex of the plastic potential.

When applying this model the initial anisotropy need not necessarily result from K_0 consolidation; although this would be the most common application in engineering practice. Different forms of anisotropy can be accommodated by changing the value of the parameter η_0 to the appropriate constant stress ratio during consolidation. Indeed, if $\eta_0 = 0$, which corresponds to isotropic consolidation, the model devolves to Modified Cam Clay.

The evolution of the parameter η_0 which defines the rotation of the ellipses, is controlled by the following experimentally derived incremental expression:

$$d\eta_0 = \pm[1 - (\eta_s/M)^2] \cdot \Delta p^* \cdot \exp(\eta_s - \eta_{0s}) \quad (3)$$

where

$\Delta p^* = (p'_s/p'_{0s}) - (p'_f/p'_{0\,iso})$
η_s = stress ratio at start of increment
η_{0s} = rotation of ellipse at start of increment

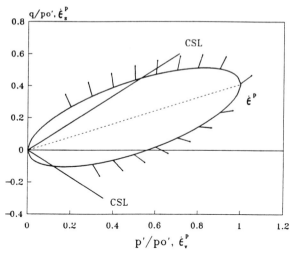

Fig. 4. *Plastic strain increment vector directions predicted by CARMEL*

in which p'_s and p'_f are the values of mean normal effective stress at the start and end of the increment of loading respectively, p'_{0s} is the value at the apex of the ellipse at the start of the increment and $p'_{0\,\mathrm{iso}}$ is the location of the apex of the ellipse at the end of the increment if no rotation occurs (i.e. only isotropic hardening). Equation (3) is positive for increase in q and negative for decreases in q. The model is currently formulated in triaxial space and q is negative when lateral stresses are greater than axial stresses.

During anisotropic normal consolidation (where the stress path is continuously located at the apex of the yield locus) eqn. (3) indicates that an increment of loading will result in the value of Δp^* being zero. It therefore follows that $d\eta_0$ is equal to zero and there is no rotation of the yield locus.

Based on the original concept of simplicity, the model requires only one additional independent parameter to those of the critical state family of models. In common with earlier models the extra parameter η_0 has a physical meaning, i.e. the gradient of the K_0 consolidation line in q:p' space. The rotation of the plastic potential surface α_0 is a function of η_0. Due to the double functioned nature of the model it is known as CARMEL, i.e. the CARdiff Multiple ELlipse model.

Performance of model

The model has been assessed by predicting the results of single element laboratory stress path tests. The results of one such test, using a specimen of anisotropically consolidated kaolin clay, and predictions of

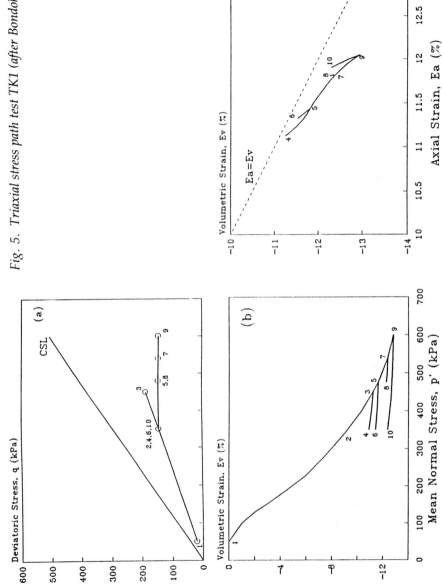

Fig. 5. Triaxial stress path test TK1 (after Bondok (1989))

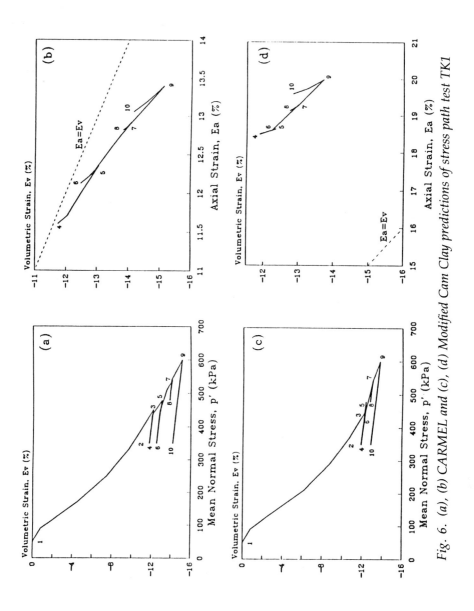

Fig. 6. (a), (b) CARMEL and (c), (d) Modified Cam Clay predictions of stress path test TK1

this test are shown in Figs. 5 and 6, respectively. The predictions of volumetric and axial strain have been made using both the CARMEL and Modified Cam Clay models. In this test (TK1) the specimen, having initially been one-dimensionally normally consolidated to p' = 450 kPa, was allowed to swell to p' = 350 kPa at the same stress ratio, prior to a series of loading and unloading stress probes being applied at constant deviatoric stress, see Fig. 5(a). Comparison of the experimental results and the predictions indicate that throughout the complex stress path test similar volumetric strains were predicted by both models, which correlated reasonably well with experimental observations. Consideration of the values of axial strain reveals that whereas the CARMEL model correlates reasonably well with the experimental values, Modified Cam Clay consistently overpredicts values of axial strain.

Since both models utilise very similar criteria for hardening (i.e. a change in the location of the apex of the yield surface in q:p':e space) comparable volumetric strains would be expected. The difference in axial strain predicted by the two models is due to the differing plastic strain increment vector directions, that result from alternative plastic potential functions. These results also highlight the requirement for non-associated plasticity, since the use of an associated flow rule with the proposed (experimentally observed) yield surface would have resulted in negative axial strains.

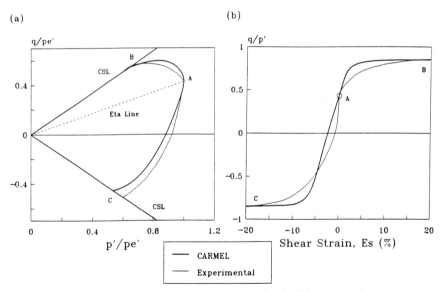

Fig. 7. *CARMEL predictions of K_0 consolidated triaxial compression and extension test (experimental results after Atkinson et al. (1987))*

Assessment of predictions using the proposed model has also been conducted by predicting the results of stress path triaxial tests conducted on anisotropically normally consolidated clay specimens reported by Atkinson et al. (1987). In these experiments following one-dimensional consolidation, reconstituted Speswhite kaolin clay specimens were transferred to a Bishop and Wesley stress path triaxial cell, and radial and axial stresses applied approximately equivalent to those in the oedometer at the end of consolidation. The specimens were sheared in both drained and undrained compression and extension tests and in constant p' tests.

The observed normalised stress paths for these tests are shown in Fig. 7(a), together with the stress paths predicted using the CARMEL model for both undrained compression and extension. Comparison reveals close correlation between experimental and predicted behaviour. It is interesting to note that the CARMEL model predicts the reduction in stress ratio q/p'_e as the stress path approaches critical state which would not be the case for predictions using Modified Cam Clay. Measured and predicted shear strains during undrained shear are compared in Fig. 7(b). Although at small strains the predicted stiffness of the soil is greater than measured for compression tests and less stiff than measured in extension, the experimental trends are adequately modelled.

Conclusions

A programme of research is being conducted to develop a constitutive model for anisotropically consolidated clay to improve methods of predicting engineering processes involving such soils. The achievements of the study to date are:

> (a) The stress path test data necessary for the development of a constitutive model has been obtained using a computer controlled triaxial testing apparatus specifically developed for this application.
>
> (b) A non-associated constitutive model, known as the CARMEL model, has been developed which has an empirically observed yield function and a theoretically derived plastic potential function. It has been demonstrated that predictions using this model correlate reasonably well with the results of laboratory test data and significantly more closely than similar predictions using Modified Cam Clay.

Work on the development of the CARMEL model is currently continuing to more closely model the stress–strain behaviour of anisotropically consolidated clay. In particular, attention is being concentrated on the modelling of large stress reversals. Further, as the next phase of the study, this model is being incorporated into the CRISP finite element package enabling predictions of complex soil mechanics boundary value problems.

Acknowledgements

The authors wish to acknowledge Dr Abdelaziz Bondok for his valuable contribution to this study. The second author is sponsored by the Science and Engineering Research Council.

References

ALMEIDA, M.S.S., BRITTO, A.M. AND PARRY, R.H.G. (1986). Numerical modelling of a centrifuged embankment on soft clay. Canadian Geotechnical Journal, Vol. 23, pp. 103–115.

ATKINSON, J.H., RICHARDSON, D. AND ROBINSON, P.J. (1987). Compression and extension of K_0 normally consolidated clay. ASCE Journal of Geotech. Eng., Vol. 113, No. 12, pp. 1468–1482.

BONDOK, A.R.A. (1989). Constitutive relations for anisotropic soils. Ph.D. Thesis, University of Wales, College of Cardiff.

DAFALIAS, Y.F. (1987). An anisotropic critical state clay plasticity model. Constitutive laws for engineering materials: theory and applications, C.S. Desai et al. (eds.), Vol. 1, pp. 513–521.

DAVIES, M.C.R. (1988). Stress path triaxial testing using a computer controlled apparatus. Proceedings, New Concepts in Laboratory and in-situ testing in Geotechnical Engineering, Rio de Janeiro, Brazil, pp. 71–84.

GRAHAM, J., LEW, K. V. AND NOONAN, M.L. (1983). Yield states and stress–strain relationships in a natural plastic clay. Canadian Geotech. Journal, Vol. 20, No. 3, pp. 502–516.

PARRY, R.H.G. AND NADARAJAH, V. (1973). Observations on laboratory prepared lightly overconsolidated specimens of kaolin. Geotechnique, Vol. 24, pp. 345–358.

ROSCOE, K.H., SCHOFIELD, A.N. AND WROTH, C.P. (1958). On the yielding of soils. Geotechnique, Vol. 8, pp. 22–53.

STIPHO, A.S. (1978). Experimental and theoretical investigation of the behaviour of anisotropically consolidated kaolin. Ph.D. Thesis, University of Wales, College of Cardiff.

TAVENAS, F., DES ROSIERS, J.-P., LEROUIEL, S., LA ROCHELLE, P. AND ROY, M. (1979). The use of strain energy as a yield and creep criterion for lightly overconsolidated clays. Geotechnique, Vol. 29, pp. 285–303.

TAVENAS, F. AND LEROUIEL, S. (1977). The effects of stresses and time on the yielding of clays. Proceedings 9th ISSMFE Int. Conference, Tokyo, Vol. 1, pp. 319–326.

The bearing capacity of conical footings on sand in relation to the behaviour of spudcan footings of jackups

E.T.R. DEAN, Cambridge University Engineering Department,
R.G. JAMES, Andrew N. Schofield & Associates Ltd,
A.N. SCHOFIELD, Cambridge University Engineering Department and Andrew N. Schofield & Associates Ltd,
F.S.C. TAN, formerly Cambridge University Engineering Department, now Shell Eastern Petroleum Ltd, and
Y. TSUKAMOTO, Cambridge University Engineering Department

A simple idealised model of a jackup unit allowing estimations of foundation loads is described. Results of theoretical and experimental investigations of the behaviour of spudcan foundations under combined loads on sand are presented. A simple equation describing the limiting yield surface for such foundations under combined loads is presented, with experimental evidence from centrifuge model tests that show the shape is approximately correct.

Introduction

This paper examines the problem of the interaction of an idealisation of a 3-leg independent leg jackup unit (ILJU) and its spudcan foundations under fully drained conditions on sand. The idealisation considers the foundations to behave as if they had full shear and vertical fixity with an elastic response to moment. Since the behaviour of the foundation is a critical component in relation to the structure, and since most spudcan foundations are conical to some extent, the limiting behaviour of conical footings is examined under conditions of pure vertical loading and of combined loading comprising a vertical component, a moment component, and a lateral shear component. The influence of pore water pressure generation, which may modify the foundation behaviour dramatically, causing reduction of vertical, shear, and moment stiffness, is not considered here.

CONICAL FOOTINGS ON SAND

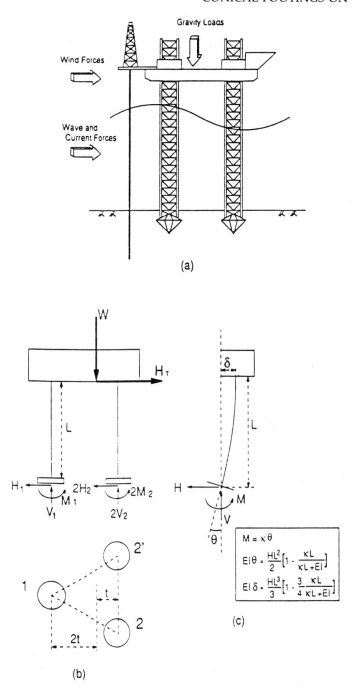

Fig. 1. Schematic of jackup and idealisation.

Structure–foundation interaction

Introduction

Figure 1(a) is a schematic of an independent leg jackup unit (ILJU) subject to wind, wave, and current loading. Whilst the detailed analysis of such a structure is complex, some of the essential features of the interaction of the structure and its foundations may be approximated by the idealisation shown in Fig. 1(b). In the idealisation the three legs of the ILJU are modelled as simple beam columns and the deck as a rigid beam. The three foundations are modelled as rotationally compliant restraints, and vertical settlements and shear displacements of the foundations are considered to be sufficiently small so that complete vertical and shear fixity is assumed. Such an idealisation is considered by Murff et al. (1991).

Considering a single leg as shown in Fig. 1(c) as a simple beam (ignoring P–δ and shear displacement effects), with flexural rigidity EI and length L, subjected to a lateral load of H at the top of the leg, with a foundation fixing moment M, where M is directly related to a secant stiffness k and a rotation θ as follows:

$$M = k \cdot \theta \tag{1}$$

It may be shown that:

$$EI\theta = \frac{H \cdot L^2}{2}\left(1 - \frac{kL}{kL + EI}\right) \tag{2}$$

$$EI\delta = \frac{H \cdot L^3}{3}\left(1 - \frac{3}{4}\frac{kL}{kL + EI}\right) \tag{3}$$

where θ is the rotation of the foundation and δ is the lateral displacement of the top of the beam relative to the foundation. Substituting for θ in eqn. (2) in terms of the fixing moment M and rearranging one may obtain the relationship between M and H for a given secant rotational stiffness K:

$$M = \frac{HL}{2}\left(\frac{kL}{kL + EI}\right). \tag{4}$$

Thus if k, L, and EI are given, the relation between the moment and shear load on the foundation is prescribed.

Considering moment and force equilibrium of the model shown in Fig. 1(b), and taking account of the horizontal deflection δ of the hull relative to the foundations and the eccentricities M_1/V_1 and M_2/V_2 of the vertical reactions at the foundations, the vertical reactions under legs 1, 2, and 2' are obtained as:

$$V_1 = \frac{W \cdot (t - \delta + ((M_2/V_2)) - H_T L}{3t + ((M_2/V_2) - (M_1/V_1))} \tag{5}$$

CONICAL FOOTINGS ON SAND

$$V_2 = \frac{W(t + (\delta/2)) - ((M_1/2V_1)) + (H_T L/2)}{3t + ((M_2/V_2) - (M_1/V_1))} \quad (6)$$

where:
$$W = V_1 + 2V_2 \quad (7)$$
$$H_T = H_1 + 2H_2 \quad (8)$$

and the symbols have meanings as indicated in Fig. 1(b). Thus if secant rotational stiffnesses K_1 and K_2 are prescribed, employing the above equations one may obtain the combined loads acting at each foundation consequent upon the application of a given lateral load H_T.

It is of interest initially to consider two simple situations: (a) the foundations are pinned, and (b) the foundations are encastre or fixed such that $M = HL/2$. For both of these situations the lateral deflection δ will be considered to be small in relation to the footing spacing t ($\delta \ll t$ such that $\delta \approx 0$).

Case A: Pinned
From eqns. (5) and (6) with $M_1 = M_2 = 0$:

$$V_1 = (Wt - H_T L)/3t \quad (9)$$
$$V_2 = (Wt + ((H_T L/2)))/3t. \quad (10)$$

For initial vertical loads of $V_0 = W/3$ one has changes of vertical load $\Delta V = (V - V_0)$ of:

$$\Delta V_1 = -(H_T/3)(L/t) \quad (11)$$

and

$$\Delta V_2 = (H_T/3)(L/2t). \quad (12)$$

For the situation where the secant rotational stiffnesses K_1 and K_2 are equal then:

$$H_1 = H_2 = H_T/3. \quad (13)$$

These equations show that the changes in the vertical reactions are directly related to the lateral load H_T and the parameter (L/t).

Case B: Encastre
For the encastre case, by considering equilibrium of the hull together with the upper half of the legs (noting that the moments in the legs at midheight are zero), and again taking $\delta = 0$, it may be shown that:

$$V_1 = (Wt - (H_T L/2))/3t \quad (14)$$
$$V_2 = (Wt + (H_T L/4))/3t \quad (15)$$

and hence that:
$$\Delta V_1 = -(H_T/3) \cdot (L/2t) \quad (16)$$
$$\Delta V_2 = +(H_T/3) \cdot (L/4t). \quad (17)$$

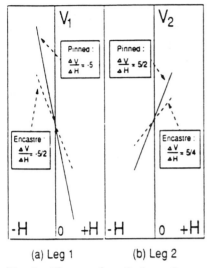

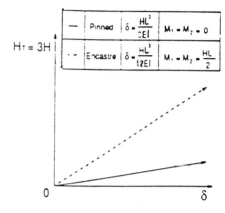

Fig. 2. Change of vertical reactions at legs 1 and 2, for L/t = 5.

Fig. 3. Lateral deflection of hull versus lateral force.

For $K_1 = K_2$ then:

$$H_1 = H_2 = H_T/3 \qquad (18)$$
$$M_1 = M_2 = HL/2. \qquad (19)$$

Equations (16) and (17) show that for the encastre case the changes in the vertical reactions for a given lateral load H_T are one half of those associated with the pinned condition (eqns. (11) and (12) above).

Typical load paths for (L/t) = 5 in V–H space are shown for leg 1 in Fig. 2(a) and leg 2 in Fig. 2(b). The wide variations of vertical load that occur consequent upon the lateral load H for each leg and the contrast between the pinned and encastre conditions are seen from this figure.

Figure 3 shows the relationship between the lateral displacements δ of the rigid deck and the lateral force H_T for the two idealisations from which it may be seen that the lateral displacements for the pinned condition are four times larger than for the encastre condition: from the point of view of the dynamic response for these two idealisations the implication is that the natural period for the pinned condition is twice that for the encastre condition, see Murff et al. (1991).

Loadpaths and limiting loads

It is of interest to plot the above equations as loadpaths and to relate them to yield loci of limiting behaviour of the foundation under combined loads. There is a wide variety of possible equations in use to describe such limiting behaviour. Most are based on Terzaghi's theory of bearing capacity with suitable modifications to account for the influence of lateral and moment loads, footing geometry, and depth of penetration. A limiting yield locus for a flat plate on clay under constant vertical load V and moment M was drawn by Roscoe and Schofield (1956). For the purposes of this paper we will consider an empirical equation of the form:

$$\left[\left(\frac{M}{BV_M}\right)^2 + \beta^2\left(\frac{H}{V_M}\right)^2\right]^{1/2} = \alpha \frac{V}{V_M}\left(1 - \frac{V}{V_M}\right) \qquad (20)$$

where V_M is the current footing vertical bearing capacity (when moment M and lateral load H are both equal to zero), V is the current vertical load, and α and β are factors which are related to footing geometry, roughness, degree of penetration, and the shear strength parameters of a given soil.

The three-dimensional limiting yield locus for V, M, and H on sand as described by eqn. (20) is shown in Fig. 4(a) for $\alpha = 0.35$ and $\beta = 0.625$ which are considered as possible values for a shallow coned semi-rough spudcan on loose sand (for fully drained behaviour).

The loadpaths for leg 1 for the two idealisations are shown in Figs. 4(b) and 4(c) for a preload condition of 2, i.e. $V/V_M = 0.5$ prior to the application of lateral load H. Note that for convenience of plotting the yield locus shown in Fig. 4(b) is transformed to circular sections in plan. Paths OA and OA' in these figures relate to the encastre condition, and are drawn for $L/2B = 5/2$. Paths OB and OB' relate to the pinned condition. It can be seen that the limiting value of lateral capacity H is quite different for the two situations. The lateral capacity for the pinned condition is considerably greater than that of the encastre condition due primarily to the absence of moment (i.e. $M = 0$).

In experiments on jackup rig models it is found that the foundation follows neither of these two idealised loadpaths but may start close to the encastre condition for low values of H and move gradually towards the pinned condition for high values of H, as sketched as paths OCD and OC'D'.

Vertical foundation loads and rotational stiffness
The purpose of the above preamble has been to demonstrate that the manner in which the loads are applied to the foundations of an ILJU is critically dependent upon the rotational stiffness response of the individual foundations which in turn changes with the applied load path. The above simple analysis assumes a fully drained foundation and no vertical or horizontal movements of the foundation after preload. In the case of large prototype spudcans on sand significant excess pore water pressures may be generated and the assumptions with respect to

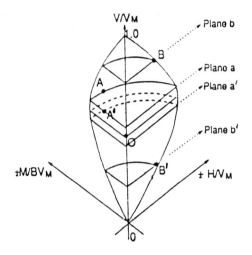

(a) 3-dimensional Yield Locus

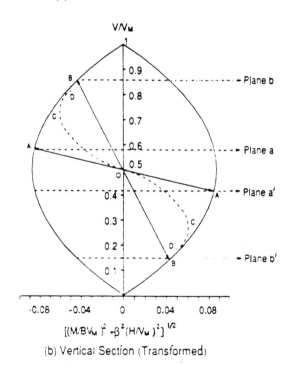

(b) Vertical Section (Transformed)

Fig. 4 (above and facing page). Foundation load paths with respect to an outer limiting yield locus.

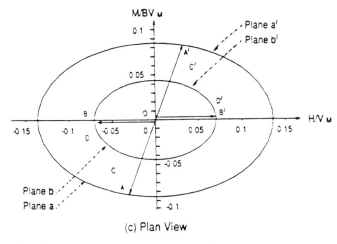

(c) Plan View

Fig. 4—continued

moment, vertical, and shear fixity may require review (see Dean (1991)). Note also that significant movements of the foundations may be expected to occur under cyclic loads that approach the limiting yield surface.

Bearing capacity of conical footings
Introduction
Although as indicated in the previous section the rotational response of spudcan foundations is a major factor it cannot be considered in isolation since all aspects of the foundation behaviour are important. With this in mind, theoretical and experimental studies of conically shaped foundations have been undertaken by Cambridge University Engineering Department (CUED) and by Andrew N. Schofield and Associates Limited (ANS&A) over a period of several years. Work has been reported in Cambridge M.Phil and Ph.D theses, in ANS&A contract research reports, and in various other publications. Some

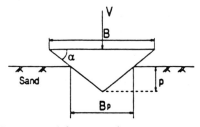

Fig. 5. Conical footing at partial penetration p.

references are provided at the end of this paper. Although some work has been conducted with clay and some work has been done with oil-saturated fine sand, most of the work is on the fully drained behaviour of foundations on sand, generally under conditions of combined loading.

Vertical bearing capacity
Figure 5 shows a conical footing of diameter B partially penetrated in sand to a depth p. The vertical bearing capacity is given by:

$$V = A_p(\tfrac{1}{2}N_\gamma \gamma' B_p) \tag{21}$$

where B_p is the embedded diameter, A_p is the plan contact area, γ' is the soil effective unit weight, and N_γ is an axisymmetric self-weight bearing capacity factor. Thus:

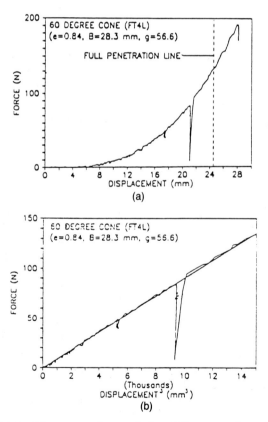

Fig. 6. (a) Initial cubic nature of curve before full penetration, after Tan (1990). (b) Plot of force against the cube of the displacement, after Tan (1990).

$$V = \frac{\pi}{4}(\tfrac{1}{2}N_\gamma \gamma')B_P^3. \tag{22}$$

V varies as the embedded diameter cubed. For cone angle α (not to be confused with the different factor α in eqn. (20)), the depth of penetration and the embedded diameter are related by $B_p = 2p\cot\alpha$. Restating eqn. (22) in terms of penetration p one obtains:

$$V = \pi N_\gamma \gamma' p^3 \cot^3\alpha. \tag{23}$$

Thus for a given cone angle α, V is directly related to the cube p^3 of the penetration, if N_γ is approximately constant (see Fig. 6(a)) so that experimental data should be well fitted by a straight line relationship between V and p^3 the slope of which is related to N_γ. Such a straight line relationship is shown in Fig. 6(b) for a 60° cone (Tan, 1990) on fine Leighton Buzzard sand tested in the drum centrifuge at 56.6 g. A series of tests of cones of angle 0° (flat), 13°, 25°, 45°, and 60° was conducted on sand at 28.3 mm diameter model footings at 56.6 g, representing 1.6 m prototypes, in the drum centrifuge. Typical relationships for semi-rough footings on loose sand are shown in Fig. 7. The results of a set of stress characteristic calculations for the same cone angles on smooth ($\delta = 0$, where δ in this context represents friction angle between the cone surface and the soil), semi-rough ($\delta = 13°$), and rough ($\delta = \phi$) cones for ϕ values of 30°, 35°, and 40° for axisymmetric stress fields are presented in Fig. 8. Typical N_γ values calculated from experimental results using

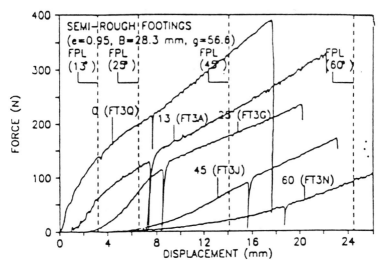

Fig. 7. *Effect of cone angle on force–displacement curves (loose sand), after Tan (1990).*

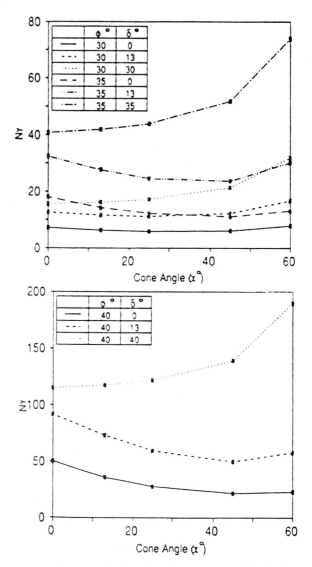

Fig. 8. Theoretical axi-symmetric N values for cones on cohesionless soil, after Tan (1990).

CONICAL FOOTINGS ON SAND

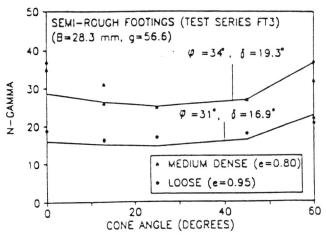

Fig. 9. Effect of cone angle on N_γ of semi-rough footings (theoretical and experimental comparison), after Tan (1990).

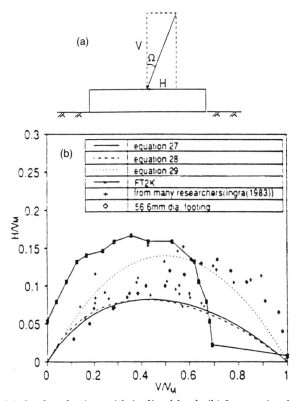

Fig. 10. (a) Surface footing with inclined load. (b) Interaction loci for surface footings with inclined load ($M = 0$).

eqn. (21) are shown in Fig. 9. It appears that the N_γ values vary in a manner similar to that calculated and that they correspond to a ϕ value in the region of 31° and $\delta = 17°$ for loose sand (voids ratio 0.95) and $\phi = 34°$ and $\delta = 19°$ for medium dense sand (voids ratio 0.80).

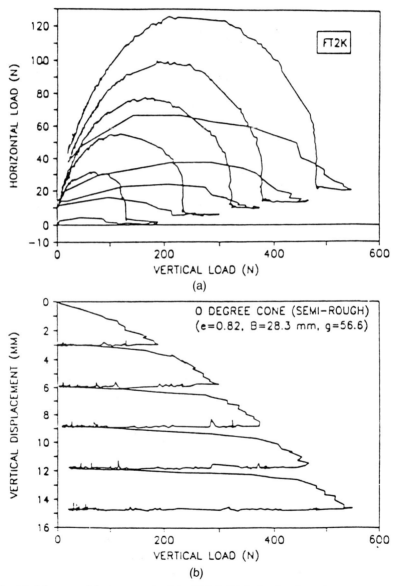

Fig. 11 (above and facing page). (a), (b) Multiple loadpaths, after Tan (1990). (c) Effect of depth of overburden on H_p/V_m, after Tan (1990).

Combined loading for fully penetrated surface footings
Lateral loading

The bearing capacity of a surface strip foundation is normally reduced by a factor i_γ to allow for the effect of load inclined at angle Ω to the vertical as shown in Fig. 10(a), giving:

$$V = A(\tfrac{1}{2}\gamma' BN_\gamma i_\gamma) \tag{24}$$

Meyerhof (1953) suggested that i_γ be taken as:

$$i_\gamma = \left(1 - \frac{\Omega}{\phi}\right)^2 \tag{25}$$

and Hansen (1961) suggested i_γ be taken as:

$$i_\gamma = \left(1 - \frac{H}{V}\right)^4 \tag{26}$$

Hansen (1970) further suggests $i_\gamma = (1 - 0.7(H/V))^5$ and includes a shape factor $s_\gamma = (1 - 0.4 i_\gamma)$ which is discussed later (see also Hambly (1992)). Taking maximum load $V_M = A(1/2)(\gamma' BN_\gamma)$ it is possible to obtain an interaction diagram in terms of (H/V_M) and (V/V_M).

For Meyerhof's i_γ:

$$\frac{H}{V_M} = \frac{V}{V_M} \tan\left[\left(1 - \left(\frac{V}{V_M}\right)^{0.5}\right)\phi\right]. \tag{27}$$

For Hansen's i_γ:

$$\frac{H}{V_M} = \frac{V}{V_M}\left[1 - \left(\frac{V}{V_M}\right)^{0.25}\right]. \tag{28}$$

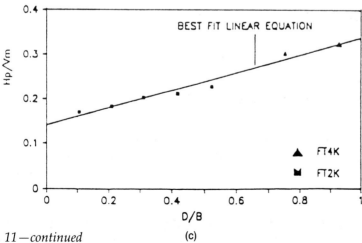

Fig. 11—continued   (c)

These interaction loci are shown in Fig. 10(b) for $\phi = 31.5°$. Also shown is an interaction locus based upon the empirical equation:

$$\frac{H}{V_M} = \left(\frac{\alpha}{\beta}\right)\left(\frac{V}{V_M}\right)\left(1 - \frac{V}{V_M}\right) \qquad (29)$$

where the constant α/β is taken as 0.56. It is well known that the

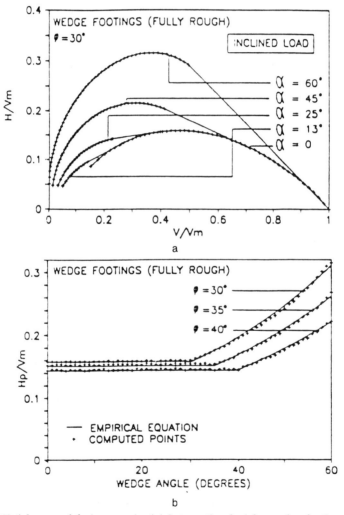

Fig. 12 (above and facing page). (a) Interaction loci for wedge footings, after Tan (1990). (b) Effect of wedge angle on peak of interaction locus. (c) Effect of cone angle on 'corrected' $(H_p/V_m)_0$, after Tan (1990). (d) Effect on depth of overburden on peak of interaction locus, after Tan (1990).

Meyerhof and Hansen loci are in general conservative in relation to the available 1 g data for strip and circular foundations (Ingra and Baecker, 1983). Also plotted in Fig. 10(b) are some of the results of a special drum centrifuge test (known as a 'shear sideswipe') for test FT2K at about 10% beyond full penetration (D/B = 0.1, where D is the depth of overburden above the full diameter of the foundation). Most of the experimental points lie outside of eqns. (27), (28), and (29).

The type of test referred to as a shear sideswipe is performed at constant penetration. That is, having preloaded the footing it is locked vertically such that it can no longer settle vertically. The immediate

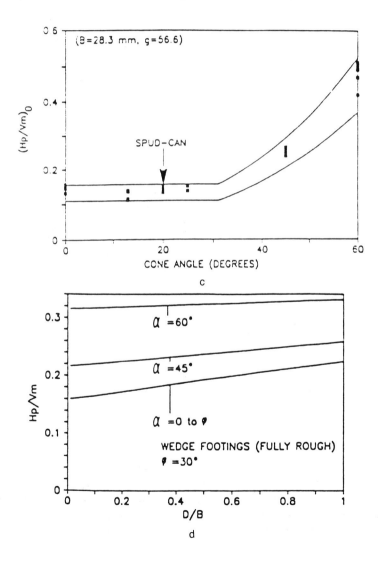

consequence of this is that there is some reduction in vertical load, due to relaxation in the experimental apparatus. A shear displacement is then applied during which the shear load increases until yield is reached; as the soil yields in shear the vertical load continues to decrease. With further shear displacement the shear load increases reaching a maximum at about $V/V_M = 0.5$ and then reduces with further increase in shear displacement.

The results of a series of such tests conducted at 5 different penetrations are shown in Figs. 11(a) and (b). On looking at these sideswipe test data, C.P. Wroth commented that, if the sand foundation were regarded as elastically incompressible under vertical load then, by analogy with the undrained test paths of Granta-Gravel (Schofield and Wroth, 1968), these sideswipe test paths could be regarded as close approximations to yield loci, in which case it can be seen from these curves that the peak lateral resistance H_p increases with penetration. The paper in this conference by G.T. Houlsby and C.M. Martin considers similar loadpaths for foundations on clay. The experimental relationship between H_p/V_M and relative depth D/B is shown in Fig. 11(c).

The above results show the influences of the ratio V/V_M and penetration on the lateral capacity of foundations. In the absence of a satisfactory method of calculation for the axisymmetric case a series of stress field calculations for the lateral loading of plane wedges were carried out. Typical interaction loci are shown in Fig. 12(a). The variation of H_p/V_M with wedge angle is shown in Fig. 12(b) and experimental results corrected for penetration are shown in Fig. 12(c). Figure 12(b) implies that the peak lateral resistance does not vary significantly for wedge angles less than 30°, which appears to be in accord with the experimental observations for conical footings for cone angles up to 30° (Fig. 12(c)). However for cone angles greater than 30° there is a significant increase in lateral capacity. The calculated influence of overburden depth on lateral capacity is shown in Fig. 12(d) for $\phi = 30°$,

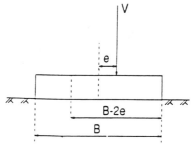

Fig. 13. *Eccentrically loaded footing.*

CONICAL FOOTINGS ON SAND

from which it may be seen that this effect is about one third of that observed experimentally, shown in Fig. 11(c).

Moment loading

The conventional approach is one proposed by Meyerhof (1953) which considers a footing of effective width (B − 2e) where e is the eccentricity of the applied load as shown in Fig. 13. Thus the vertical bearing capacity of a surface strip footing of width B and unit length is:

$$V = \tfrac{1}{2}(B - 2e)^2 \gamma' N_\gamma. \qquad (30)$$

Taking $V_M = (\tfrac{1}{2})\gamma' B^2 N_\gamma$ and the moment $M = Ve$, then:

$$\frac{M}{BV_M} = \frac{1}{2}\frac{V}{V_M}\left[1 - \left(\frac{V}{V_M}\right)^{0.5}\right]. \qquad (31)$$

It is well established that this equation gives a reasonable lower limit to most of the available 1 g experimental data for small model footings on sand, as indicated in Fig. 14 (Ingra and Baecker, 1983). Also shown is the empirical relationship:

$$\frac{M}{BV_M} = \alpha \frac{V}{V_M}\left(1 - \frac{V}{V_M}\right) \qquad (32)$$

for $\alpha = 0.35$, which is less conservative.

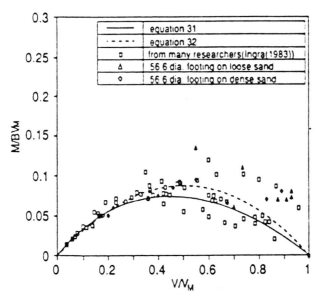

Fig. 14. Interaction loci for surface footings with moment load (H = 0).

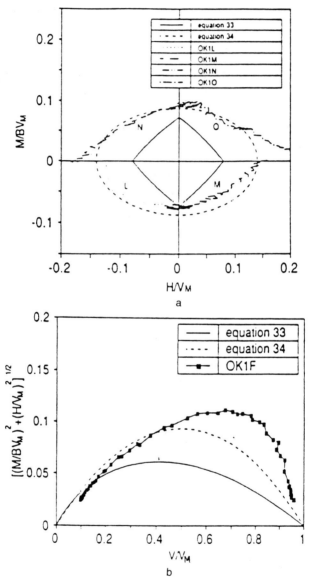

Fig. 15 (above and facing page). (a) Moment–horizontal force interaction loci for surface footings ($V/V_m = 0.5$). (b) Interaction loci for surface footings on a plane $\zeta = -2$, where $M/BV_M = \zeta H/V_M$. (c) Interaction loci for surface footings on a plane $\zeta = 0$, where $M/BV_M = \zeta H/V_M$. (d) Interaction loci for surface footings on a plane $\zeta = 0$, where $M/BV_M = \zeta H/V_M$.

Lateral and moment loading combined

Following Meyerhof's formulation the effect of eccentric and inclined load for a strip surface footing are combined such that:

$$V = \tfrac{1}{2}(B - 2e)^2 N_\gamma \gamma' i_\gamma. \tag{33}$$

For ϕ close to 30°, the Meyerhof and Hansen formulae for i_γ (eqns. (25) and (26) above) give very similar results. Using eqn. (26) (Hansen), one obtains eqn. (33) in a different form:

$$\frac{V}{V_M} = \left(1 - \frac{2M}{BV}\right)^2 \left(1 - \frac{H}{V}\right)^4.$$

Such a locus is plotted for $V/V_M = 0.5$ in Fig. 15(a).

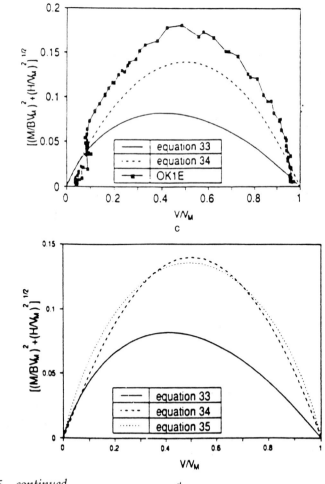

Fig. 15—continued

Recently, over a period of several years, a Joint Industry Study coordinated by Noble Denton Associates (NDA) has been considering many aspects of jackup stability including the questions of soil-structure interaction. A report by NDA included sections originally provided by Houlsby and James relating to the interaction of spudcans respectively with clay and with sand, see Osborne, Trickey, Houlsby, and James (1991). The drum centrifuge data quoted in the present paper form part of an extensive body of model test data, only part of which has as yet been interpreted. A revised version of the original yield locus for sand – which is consistent with test data so far examined, and which can be compared with Meyerhof's and Hansen's curves in Fig. 15 and is plotted for the same value of V/V_M – is the empirical equation:

$$\left[\left(\frac{M}{BV_M}\right)^2 + \beta^2 \left(\frac{H}{V_M}\right)^2\right]^{1/2} = \alpha \frac{V}{V_M}\left(1 - \frac{V}{V_M}\right) \tag{34}$$

where $\alpha = 0.35$, $\beta = 0.625$ (giving $\alpha/\beta = 0.56$).

For purposes of comparison the data from four drum centrifuge tests on a flat spudcan are presented (Tests OK1L, M, N, and O) in which the limiting yield locus for M/BV_M and H/V_M were explored at $V/V_M = 0.5$ on loose (voids ratio 1.0) Leighton Buzzard sand (BS 100/170). These tests were performed using a 57.8 mm dia. flat spudcan at 264 g and are considered representative of a 15.26 m dia. prototype. The penetration of the spudcan corresponded to a D/B of about 10%. The figure indicates that the locus of eqn. (33) is conservative and that the locus of eqn. (34) which lies closer to the observations, is in general slightly unconservative.

Fig. 15(b) shows the loci of eqns. (33) and (34) in V/V_M space on a plane for $\zeta = -2$, i.e. for $M/B = \zeta H$. Also plotted in this figure are the results of a combined moment and shear sideswipe in drum centrifuge test OK1F in which ζ was approximately equal to -2. This test, which was also conducted at 264 g, represented the same prototype as in tests OK1L-O except that the footing penetration corresponded to D/B of about 25%. It is seen in this case that both eqns. (33) and (34) give a conservative result for $V/V_M \geq 0.4$. This would be expected since they apply to a footing level with the sand surface and would therefore be expected to underpredict the combined load capacity at D/B = 25%. However, for $V/V_M \leq 0.4$ eqn. (34) becomes slightly unconservative and for $V/V_M \leq 0.2$ eqn. (33) also becomes slightly unconservative.

Fig. 15(c) shows the loci of eqns. (33) and (34) in V/V_M space for $\zeta = 0$. Also included are the results for a shear sideswipe for $\zeta \approx 0$ from a drum centrifuge test OK1E under similar conditions to OK1F, i.e. 264 g and 25% penetration. In this case eqns. (33) and (34) are again both conservative for $0.1 \leq V/V_M \leq 0.95$.

For completeness Hambly's (1992) interpretation of Hansen (1970) is included as eqn. (35).

$$\frac{V}{V_M} = \left(1 - \frac{2M}{BV}\right)^2 \frac{1}{0.6}\left[\left(1 - 0.7\frac{H}{V}\right)^5 \left(1 - 0.4\left[1 - 0.7\frac{H}{V}\right]^5\right)\right] \quad (35)$$

For purposes of clarity this equation has been omitted from Figs. 15(a), (b), and (c), but is included with eqns. (33) and (34) in Fig. 15(d) for the case of $\zeta = 0$, i.e. $M = 0$. Figure 15(d) shows that the locus represented by eqn. (35) is very similar to the empirical yield locus of eqn. (34).

Conclusions

The foundation loads of an ILJU are a complex function of the interaction of the foundations and the structural characteristics of the ILJU. The bearing capacity of the foundations are also a complex function of the eccentricity and inclination of the loads applied to them by the structure. This paper has outlined a simplified method for estimation of the foundation loads for an idealised structure with complete vertical and shear fixity but variable moment fixity. Limiting interaction yield loci for conical foundations have been examined. By consideration of elastic behaviour within an inner yield locus and work hardening behaviour to an outer limiting yield locus it is possible to develop a calculation procedure for certain idealised situations, viz. complete vertical and shear fixity and fully drained behaviour. However, for prototype ILJUs the assumption of fully drained behaviour may be entirely inappropriate and thus all of the assumptions above may require review as a consequence of the generation of excess pore water pressures.

Acknowledgements

The authors would like to acknowledge the cooperation of T.S. Chilton and the generous support of Esso Exploration and Production UK Limited (EEPUK) and the cooperation of Dr. J.D. Murff of Exxon Production Research (EPR) in the course of an extensive series of drum centrifuge tests within which tests OK1-F, L, M, N, and O quoted in this paper were performed by Prof. O. Kusakabe during a period of leave from Hiroshima University. The Joint Industry Study organised by Noble Denton Associates supported some of the initial pilot 1 g studies and the beam centrifuge studies undertaken by Dr. Q. Shi. While acknowledging the support of these sponsors the authors note that the data obtained are still in the process of interpretation and the work set out in this paper is only an interim interpretation presented for purposes of open discussion: our sponsoring organisations are still in the process of preparing their interpretation of the data. Our thanks are also due to the technicians and colleagues who have worked on the development and operation of the 2 m drum centrifuge in the Geotechnical Centrifuge Centre of Cambridge University Engineering Department.

References

DEAN, E.T.R. (1991). Some Potential Approximate Methods for the Preliminary Estimation of Excess Pore Pressures and Settlement-Time Curves for Submerged Circular Spud Foundations subjected to Time-Dependent Loading. Technical Report CUED/D-Soils/TR240, Cambridge University Engineering Department.

DEAN, E.T.R., JAMES, R.G. AND SCHOFIELD, A.N. (1990, 1991). Contract EP-022R Task Order 1 Phase 1 and Phase 2 Reports. By Andrew N. Schofield & Associates Limited to Esso Exploration and Production UK Limited.

DEAN, E.T.R., JAMES, R.G., SCHOFIELD, A.N. AND TSUKAMOTO, Y. (1991). Contract EP-022R Task Orders 2 and 3, Phase 3, Phase 5 Volume 1, and Phase 5 Volume 2 Reports. By Andrew N. Schofield & Associates Limited to Esso Exploration and Production UK Limited.

DEAN, E.T.R., JAMES, R.G., SCHOFIELD, A.N. AND TSUKAMOTO, Y. (1992). Combined Vertical, Horizontal, and Moment Loading of Circular Spuds on Dense Sand Foundations: Data Report for Drum Centrifuge Model Tests YT1-1L-A thru -G and YT2-1L-G thru V. Technical Report CUED/D-Soils/TR244, Cambridge University Engineering Department.

HAMBLY, E.C. (1992). Jackup Spudcan Sliding/Bearing Resistance on Sand. BOSS 92, Vol. 2, pp. 989-1000.

HANSEN, J.B. (1961). A General Formula for Bearing Capacity. Danish Geotechnical Institute Bulletin, No. 11.

HANSEN, J.B. (1970). A Revised and Extended Formular for Bearing Capacity. Bulletin 28, Danish Geotechnical Institute.

INGRA, T.S. AND BAECKER, G.B. (1983). Uncertainty in Bearing Capacity of Sands. ASCE Vol. 109, No. 7, pp. 899–914.

MEYERHOF, G.G. (1953). The Bearing Capacity of Foundations under Eccentric and Inclined Loads. Proc. 3rd Int. Conf. Soil Mech., Vol. 1.

MURFF, J.D., HAMILTON, J.M., DEAN, E.T.R., JAMES, R.G., KUSAKABE, O. AND SCHOFIELD, A.N. (1991). Centrifuge Testing of Foundation Behaviour using Full Jackup Rig Models. Paper OTC6516, Offshore Technology Conference, Houston, pp. 165–178.

MURFF, J.D., PRINS, M.D., DEAN, E.T.R., JAMES, R.G. AND SCHOFIELD, A.N. (1992). Jackup Rig Foundation Modelling. Paper OTC6807, Offshore Technology Conference, Houston.

OSBORNE, J.J., TRICKEY, J.C., HOULSBY, G.T. AND JAMES, R.G. (1991). Findings from a Joint Industry Study on Foundation Fixity of Jackup Units. Paper OTC6615, Offshore Technology Conference, Houston.

ROSCOE, K.H. AND SCHOFIELD, A.N. (1956). The Stability of Short Pier

Foundations in Sand. British Welding Journal, August, pp. 343–354.

SCHOFIELD, A.N. AND WROTH, C.P. (1968). *Critical State Soil Mechanics*, McGraw-Hill, London.

SHI, Q. (1988). Centrifugal Modelling of Surface Footings subject to Combined Loading. Ph.D. thesis, Cambridge University.

SILVA PEREZ, A.A. (1982). Conical Footings under Combined Loads. M.Phil thesis, Cambridge University.

TAN, F.S.C. (1990). Centrifuge and Theoretical Modelling of Conical Footings on Sand. Ph.D. thesis, Cambridge University.

TANAKA, H. (1984). Bearing Capacity of Footings for Jackup Platforms. M.Phil thesis, Cambridge University.

Co-rotational solution in simple shear test

G. DE JOSSELIN DE JONG, retired from University of Technology, Delft

Simple shear tests by Ladd and Edgers were quoted by Wroth to support the existence of the various failure modes predicted by the double sliding-free rotating (DSFR) model of this author. These tests exhibit a peculiar curvature of the stress path in the prefailure stage. In a 1988 paper it was shown that the stress paths observed after failure correspond to predictions obtained by use of Vermeer's elasto–plastic version of the DSFR model. However, in the elastic stage the computed stress paths are straight lines. By application of corotational stresses in the prefailure stage, the peculiar curvature of the observed stress paths appeared in the solution. Since this correspondence pleased Peter Wroth the details of the computation are shown in this paper.

Introduction

In his article with Randolph (1981), in his Rankine Lecture (1984) and later (1987) Wroth considered the results of simple shear tests on clay as obtained by different investigators. He mentioned observations by Stroud (1971), Borin (1973), Ladd and Edgers (1972) and Airey (1984) showing that the deformation at the onset of failure is apparently not exclusively the one with horizontal failure planes, Fig. 1. In addition to that usually accepted mode another failure mechanism is observed to occur on vertical planes of maximum stress obliquity and in addition to sliding on those vertical planes a rigid body rotation is executed to meet boundary conditions. This second mechanism, shown in Fig. 2, can be visualized as a row of books toppling over sideways, when the left-hand support is removed.

In his papers Wroth referred to this Author's 1972 prediction of the possibility that these two modes are to be expected. This prediction was based on the double sliding, free rotating (DSFR) model for materials with internal friction, proposed in 1959 and specified mathematically in various later papers (de Josselin de Jong, 1971, 1977a, 1977b). Fascinated by the predictive capacity of the model Wroth asked for an explicit description of the basic concepts of the mechanism, which resulted in a paper in *Géotechnique* (de Josselin de Jong, 1988). In that article it is shown how the DSFR model can be deduced from the laws of friction. Relative displacements are considered to take place along planes where shear resistance is exhausted and in a direction such that energy is only dissipated. The resulting constitutive equations are developed, including the case of dilatancy. In addition the stress–strain history of a simple shear test is solved explicitly. A minor improvement of that solution forms the subject of this present paper.

Uniqueness of collapse

In 1972 the possibility of two different failure modes was encountered with disbelief, because in the theory of ideal plasticity there exists the rule that collapse modes are unique. A comprehensive treatment of the uniqueness rules is given by Koiter (1960, Ch.IV, §5), who shows that "the basic property of a limit load system, i.e. the possibility of plastic collapse under constant loads is independent from previous load history of the body ..." and that "the stress distribution at the limit load is essentially unique in those parts of the body, where non-vanishing strain rates occur in the actual collapse mechanism." This uniqueness theorem is based on Melan's (1938) paper, which reveals that he considered uniqueness from the opposite direction. In fact he determined the material properties at yield in such a manner as to ensure uniqueness at collapse. The result of his analysis is that plastic strain rates are then to be associated with the yield surface, creating the concept of ideal plasticity.

This shows that uniqueness is not a consequence of the plastic flow rule, but that the flow rule of ideal plasticity is a consequence of the requirement of uniqueness. In physics there exists no rule that requires uniqueness at collapse, so the associated flow rule is not a physical necessity. The more general case is the non-associated flow rule, which occurs in soil mechanics because of the internal friction. Therefore it is permissible in soil mechanics for several modes of failure to be encountered.

Uniqueness in the simple shear test

Another point, however, is that a particular soil sample will collapse in one particular manner when loaded to the limit. The question of how to determine this choice, which necessarily is unique, could not be answered in 1972 since the DSFR model was incapable of predicting it. This created an uncertainty which was attributed to an incompleteness in the model, thus prohibiting its practical use.

It was recognized by Vermeer (1980, 1981) that the incompleteness of the model is due to its rigid-plastic nature. By adding some elasticity in

Fig. 1. Failure mode with horizontal planes

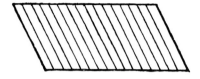

Fig. 2. Failure mode as toppling bookrow

the pre-failure stage he created an elasto-plastic version of the DSFR model which produces unique solutions at failure. Applying this version the sample can be shown to select between the various modes according to its original stress state. In this respect the DSFR model, which is based on internal friction, differs from ideal plasticity, where the unique response at failure is independent of the initial stress state.

Elastic part of the elasto-plastic DSFR model

In the above mentioned 1988 paper it is demonstrated how Vermeer's elasto-plastic version of the DSFR model can be used to develop an explicit solution for the stress history in a simple shear test. This solution demonstrates that a passive original stress state (i.e. horizontal compression larger than vertical) induces the horizontal failure planes of Fig. 1. An active original stress state (i.e. horizontal compression smaller than vertical) leads to the toppling bookrow mechanism, Fig. 2. These findings agreed with test results mentioned by Wroth.

In applying the elasto-plastic version of the DSFR model it is necessary to attribute stress–strain relations to the soil before the limit state is reached. Although perfect elasticity is not a good approximation for soils, it was adopted in the paper because the formulation of elastic stress–strain relations is well known and undisputed.

In the paper these stress–strain relations are expressed in terms of objective co-rotational (Jaumann) stress rates, $\overset{\triangledown}{\sigma}'_{xx} \ldots$ etc. These are related to the time derivatives $\dot{\sigma}'_{xx} \ldots$ etc. by

$$\left.\begin{aligned}\overset{\triangledown}{\sigma}'_{xx} &= \dot{\sigma}'_{xx} + 2\dot{\omega}\sigma_{xy} \\ \overset{\triangledown}{\sigma}'_{xy} &= \overset{\triangledown}{\sigma}'_{yx} = \dot{\sigma}'_{xy} - \dot{\omega}(\sigma'_{xx} - \sigma'_{yy}) = \dot{\sigma}'_{yx} - \dot{\omega}(\sigma'_{xx} - \sigma'_{yy}) \\ \overset{\triangledown}{\sigma}'_{yy} &= \dot{\sigma}'_{yy} - 2\dot{\omega}\sigma_{xy}\end{aligned}\right\} \quad (1)$$

where $2\dot{\omega}$ is the material rotation $(V_{y,x} - V_{x,y})$. In order to be consistent with the 1988 paper, the same notation is used here for all relevant variables. Corresponding equations will be referred to, here, by the same numbers preceded by 88.

In developing the solution for the simple shear test, the difference between the Jaumann stress rates $\overset{\triangledown}{\sigma}'_{xx} \ldots$ etc. and $\dot{\sigma}'_{xx} \ldots$ etc. were

CO-ROTATIONAL SOLUTION IN SIMPLE SHEAR TEST

disregarded for simplicity. It was however mentioned that the more complicated, co-rotational solution existed. Intrigued by this remark Peter Wroth asked the Author, during his sabbatical leave in Australia, to send him this solution in order to study it. Shortly after, he wrote back that he was excited by the result, because "it fits the experimental observations even better". Inspired by his positive reaction the result, which improves the solution slightly, is reproduced in this paper.

Co-rotational solution

In the simple shear test the velocity components are according to 88(28) $V_x = -\dot{\beta}y$; $V_y = 0$ with $\dot{\beta} > 0$ and this gives for the strain rates 88(29): $V_{x,x} = V_{y,x} = V_{y,y} = 0$: $V_{x,y} = -\dot{\beta}$. So in the elastic stage the material rotation is given by (see 88(10))

$$2\dot{\omega} = V_{y,x} - V_{x,y} = \dot{\beta} \tag{2}$$

The elastic constitutive equations express the relation between strain rates and Jaumann stresses. In plane strain they are 88(21)

$$-V_{x,x}^\varepsilon = [(1-\nu)\overset{\triangledown}{\sigma}'_{xx} - \nu\overset{\triangledown}{\sigma}'_{yy}]/2G$$

$$-(V_{x,y} + V_{y,x})^\varepsilon = \overset{\triangledown}{\sigma}'_{xy}/G = \overset{\triangledown}{\sigma}'_{yx}/G$$

$$-V_{y,y}^\varepsilon = [-\nu\overset{\triangledown}{\sigma}'_{xx} + (1-\nu)\overset{\triangledown}{\sigma}'_{yy}]/2G.$$

Using the above mentioned strain rates gives

$$\overset{\triangledown}{\sigma}'_{xx} = \overset{\triangledown}{\sigma}'_{yy} = 0; \qquad \overset{\triangledown}{\sigma}'_{xy} = \overset{\triangledown}{\sigma}'_{yx} = \dot{\beta}G. \tag{3}$$

Introducing (2) and (3) in (1) gives

$$\left.\begin{aligned}\dot{\sigma}'_{xx} &= -\dot{\beta}\sigma_{xy}\\ \dot{\sigma}'_{xy} &= +\dot{\beta}G + \tfrac{1}{2}\dot{\beta}(\sigma'_{xx} - \sigma'_{yy}) = \dot{\sigma}'_{yx}\\ \dot{\sigma}'_{yy} &= +\dot{\beta}\sigma_{xy}.\end{aligned}\right\} \tag{4}$$

This system can be solved by differentiating the second with respect to time, T, giving

$$\ddot{\sigma}_{xy} = \tfrac{1}{2}\dot{\beta}(\dot{\sigma}'_{xx} - \dot{\sigma}'_{yy}) = -\dot{\beta}^2\sigma_{xy} \tag{5}$$

with the solution

$$\sigma_{xy} = C_1 \cos(\dot{\beta}T) + C_2 \sin(\dot{\beta}T) = \sigma_{yx}, \tag{6}$$

C_1, C_2 being integration constants. Using (6) in the first and third of (4) gives after integration

$$\left.\begin{aligned}\sigma'_{xx} &= -C_1 \sin(\dot{\beta}T) + C_2 \cos(\dot{\beta}T) + C_3\\ \sigma'_{yy} &= +C_1 \sin(\dot{\beta}T) - C_2 \cos(\dot{\beta}T) + C_4\end{aligned}\right\} \tag{7}$$

with C_3, C_4 integration constants. Using (7) in the second of (4) and comparing the result with (6), shows that

$$C_3 - C_4 = -2G. \tag{8}$$

The integration constants can be determined from the stress state at the start of the test, when the time T is T_0. At that time the stresses are according to 88(32)

$$\sigma'_{xx}(0) = q_h; \quad \sigma'_{yy}(0) = q_v; \quad \sigma_{xy}(0) = \sigma_{yx}(0) = 0$$

and this applied to (6), (7), (8) gives

$$C_1 = [+\tfrac{1}{2}(q_v - q_h) - G]\sin(\dot{\beta}T_0); \quad C_3 = \tfrac{1}{2}(q_v + q_h) - G;$$
$$C_2 = [-\tfrac{1}{2}(q_v - q_h) + G]\cos(\dot{\beta}T_0); \quad C_4 = \tfrac{1}{2}(q_v + q_h) + G.$$

So the solution for the stresses becomes

$$\left.\begin{array}{l}
\sigma'_{xx} = [-\tfrac{1}{2}(q_v - q_h) + G]\cos(\dot{\beta}(T - T_0)) + [\tfrac{1}{2}(q_v + q_h) - G] \\
\sigma_{xy} = [-\tfrac{1}{2}(q_v - q_h) + G]\sin(\dot{\beta}(T - T_0)) = \sigma_{yx} \\
\sigma'_{yy} = [+\tfrac{1}{2}(q_v - q_h) - G]\cos(\dot{\beta}(T - T_0)) + [\tfrac{1}{2}(q_v + q_h) + G]
\end{array}\right\} \tag{9}$$

This solution is shown in the Mohr diagram of Fig. 3. The stress path for σ'_{yy}, σ_{yx} is the heavy, curved line upwards, and for σ'_{xx}, σ_{xy} the line downwards. The small dotted circle is the Mohr circle for the stress state at the beginning of the test at $T = T_0$, when $\sigma'_{xx} = q_h$ and $\sigma'_{yy} = q_v$. Its centre lies at a distance $\tfrac{1}{2}(q_v + q_h)$ from the origin. The heavy, curved

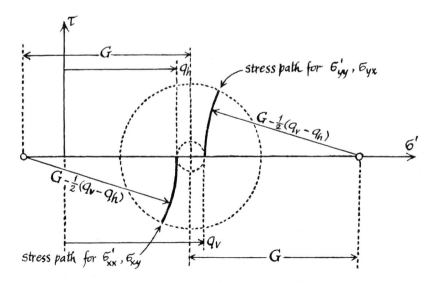

Fig. 3. Stress paths for co-rotational solution in elastic stage

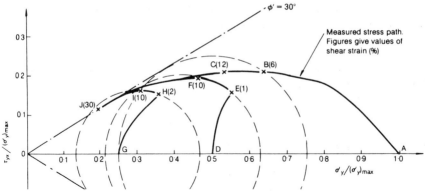

Fig. 4. Stress paths from simple shear tests on Boston blue clay. (Ladd and Edgers, 1972)

lines are arcs of circles with radius $G - \frac{1}{2}(q_v - q_h)$, that are centred around points at a distance G on either side of the centre of the dotted circle.

It may be recalled here, that the use of linear elasticity with a constant shear modulus of magnitude G is not a good approximation for soils and may therefore produce rather unrealistic results. Figure 3 represents a simplified version of reality and may be unappropriate for determining the magnitude of a supposed G.

However, the difference between co-rotational, Jaumann stress rates and the time derivatives of the stresses being ignored in the 1988 paper, the solution for the elastic stage resulted in vertical straight lines for the stress paths (see Figs. 13, 15, 17, 18, 19 in that paper). Taking account of the co-rotational complication leads to a line that is curved in the manner as observed by Ladd and Edgers, see Fig. 4. That agreement pleased Peter and is therefore the reason for this contribution.

References

AIREY, D.W. (1987). Some observations on the interpretation of shearbox results. Report CUED/D-SOILS TR 196. University of Cambridge.

BORIN, D.L. (1973). The behaviour of saturated kaolin in the simple shear apparatus. PhD thesis, University of Cambridge.

DE JOSSELIN DE JONG, G. (1959). Statics and kinematics in the failable zone of a granular material. Dr. Thesis, Univ. of Techn. Delft.

DE JOSSELIN DE JONG, G. (1971). The double sliding, free rotating model for granular assemblies. Géotechnique, Vol. 21, No 2, pp. 155–163.

DE JOSSELIN DE JONG, G. (1972). Discussion. Session II, Roscoe Memorial Symposium. Stress-strain behaviour of soils (ed. Parry, R.M.G.), pp. 258–261. Cambridge: Foulis.

DE JOSSELIN DE JONG, G. (1977a). Mathematical elaboration of the double sliding, free rotating model. Archives of mechanics, Vol. 29, No 4. pp. 561–591. Warszawa.

DE JOSSELIN DE JONG, G. (1977b). Constitutive relations for the flow of a granular material in the limit state of stress. Proc. 9th Int. Conf. Soil Mech. & Found. Eng., Tokyo, pp. 87–95.

DE JOSSELIN DE JONG, G. (1988). Elasto-plastic version of the double sliding model in undrained simple shear tests. Géotechnique, Vol. 38, No. 4, pp. 533–555.

KOITER, W.T. (1960). General theorems for elastic-plastic solids. Progress in Solid Mechanics, Vol. I, pp. 165–221. North-Holland.

LADD, C.C. AND EDGERS, L. (1972). Consolidated undrained direct simple shear tests on saturated clays. Research report T 72-82, Massachusetts, Institute of Technology.

MELAN, E. (1938). Zur Plastizität des räumlichen Kontinuums. Ingenieur Archiv IX, pp. 116–126. Wien (Vienna).

RANDOLPH, M.F. AND WROTH, C.P. (1981). Application of the failure state in undrained simple shear to the shaft capacity of driven piles. Géotechnique, Vol. 31, No 1, pp. 143–157.

STROUD, M.A. (1971). The behaviour of sand at low constant stress levels in the simple shear apparatus. Ph.D. Thesis, University of Cambridge.

VERMEER, P.A. (1980). Double sliding within an elasto-plastic framework. L.G.M. mededelingen, Vol. 21, pp. 323–340.

VERMEER, P.A. (1981). A formulation and analysis of granular flow. Proc. Int. Conf. Mech. Behaviour Struct. Media. B., pp. 325–339.

WROTH, C.P. (1984). The interpretation of in situ soil tests. Rankine Lecture. Géotechnique, Vol. 34, No 4, pp. 449–489.

WROTH, C.P. (1987). The behaviour of normally consolidated clay as observed in undrained direct shear tests. Géotechnique, Vol. 37, No 1, pp. 37–43.

Parameter selection for pile design in calcareous sediments

M. FAHEY, R.J. JEWELL, M.F. RANDOLPH, The University of Western Australia, and M.S. KHORSHID, Woodside Offshore Petroleum Pty Ltd

This paper discusses the site investigation and laboratory testing requirements for the selection of parameters for pile design in calcareous sediments. A series of field tests on drilled and grouted piles was undertaken in South Australia by Woodside Offshore Petroleum and Joint Venture partners in the North-West Shelf project. A detailed site investigation was carried out in order to relate the pile response to soil properties, and hence provide a basis for future pile design in similar soil types. Results of the investigation are summarized and the engineering parameters deduced from the various in situ and laboratory tests are compared with each other and are also linked with the performance of the test piles. Recommendations are then given for in situ and laboratory testing requirements for pile design in calcareous sediments.

Introduction

The calcareous sediments found off the coasts of Australia, Brazil, India and other parts of the world represent an extreme soil type in terms of their compressibility, which results from high in situ void ratios. These soils present a particular challenge for offshore pile design. During shearing, they show a tendency for the skeleton to collapse, leading to reduced radial effective stresses around the pile shaft, and thus the potential for significant degradation of load transfer during large monotonic displacement of the pile, or under the action of cyclic loading (Murff, 1987; Randolph, 1988).

It has long been realized that the shaft friction for piles driven into calcareous soils can be extremely low (5–40 kPa, even at depths of 50–100 m), while that of piles that are grouted in place is often an order of magnitude higher (up to 400 kPa in calcarenite at comparable depths). For both types of pile, there is a wide range in the shaft friction that has been measured in field and laboratory pile load tests, and the design engineer thus faces the problem of identifying which properties of the soil are critical in determining the shaft capacity, and how these properties may be evaluated from the site investigation and laboratory testing.

The difficulties posed by pile design in calcareous sediments have led to the development of new in situ and laboratory tests, modelling the action of a pile, in an effort to reduce uncertainty over key engineering parameters. However, the key features of a full-scale pile are not necessarily reproduced in smaller 'element' tests. A good example of this was the Steel Friction Test, a 2.5 m long by 60 mm diameter probe that was jacked into place and then loaded to measure the average shaft friction during the site investigation for the North Rankin 'A' platform on the North-West Shelf of Australia. This test gave values of peak and residual shaft friction that appear consistently higher than values obtained from tests on conductor pipes (Poulos et al., 1988).

In the planning stage of the Goodwyn platform, which is due to be installed on the North-West Shelf during 1992, a major programme of tests on drilled and grouted pile sections was undertaken at Overland Corner, near the town of Waikerie in South Australia, where there are deposits of weak limestone that have similar geological and engineering characteristics to the offshore calcareous deposits. These tests were accompanied by an extensive site investigation, and thus provide an ideal opportunity to:

(a) compare the results of different forms of in situ and laboratory tests, particularly in respect of engineering parameters deduced from the tests

(b) compare 'Class A' predictions of the pile performance with the actual performance, and review the pile design parameters deduced from the load test with those derived from the site investigation data

(c) assess the apparent merit of different tests in deducing parameters for pile design

(d) assess a methodology for future design of pile foundations in calcareous sediments, without the use of site-specific pile tests at field scale.

Before addressing the above issues, it is helpful to review the key engineering parameters that are required for offshore pile design.

Pile design parameters

For offshore drilled and grouted piles, any support at the pile base is generally ignored, owing to the difficulty of ensuring a clean base to the augered hole, free of debris. The pile shaft performance is generally analysed through simple one-dimensional ('beam-column') models of the pile, with the soil modelled as a set of non-linear springs offering resistance in the axial direction. Each soil spring is specified in terms of a 'load transfer curve' which relates displacement to the average traction acting between the soil and the adjacent pile element. A general form of load transfer curve for axial loading is shown in Fig. 1 (Randolph and

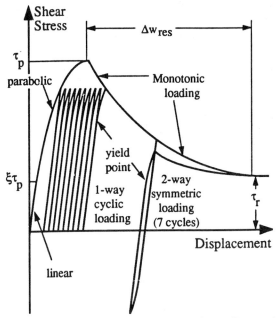

Fig. 1. Load transfer curve in RATZ (Randolph and Jewell, 1989)

Jewell, 1989), and this curve has been incorporated into a load-transfer computer program RATZ (Randolph, 1986). The key load transfer parameters that affect the pile performance are:

(a) the peak and residual values of shaft friction (τ_p and τ_r)
(b) the initial and unload–reload gradients of the load transfer curve
(c) the rate of degradation under monotonic and cyclic loading
(d) the level of cyclic loading threshold below which no degradation will occur.

Evaluation of the key parameters listed above requires a rather different range of laboratory and in situ tests than might be appropriate for other types of foundation. In particular, conventional cone profiles and triaxial test data may prove of limited value except in a qualitative sense, and as 'index' tests to relate soil conditions from different sites. The following sections describe the range of site investigation testing that was undertaken at Overland Corner to support the field pile tests, and discuss the relative consistency and usefulness of the data for pile design.

Site investigation data from Overload Corner

The site investigation and associated laboratory testing for the Overland

Corner site was devised to provide data for the analysis of the pile tests at the site and for the design of the reaction systems for these tests, and to allow the soil properties to be compared with those from the Goodwyn site to facilitate extrapolation of the results to that site. For the site investigation, the philosophy was to try to duplicate the types of test performed offshore, though obviously with the possibility of achieving greater control than is possible with offshore tests.

The pile sections were to be grouted into a formation known as the Morgan Limestone in a zone between 33 and 48 m depth. In this zone, the Morgan Limestone is remarkably similar to the calcarenites found on the North West Shelf. The site investigation included the following:

(a) Drilling and coring in the calcarenite using triple tube coring equipment.
(b) Pull-out tests on 140 mm diameter 3.2 m long grouted prestressing strand anchors; the two of interest were at nominal depths of 35–38 m, and 42–45 m.
(c) Hydraulic fracture tests to determine the maximum allowable grouting pressures.
(d) Electric friction cone (CPT) tests at 4 locations.
(e) Self-boring pressuremeter (SBP) tests at 1 m depth intervals (Fahey, 1988).
(f) Borehole calliper logging to provide accurate values of borehole diameter.

The significant results from the in situ tests in the zone of interest can be summarized as follows:

(a) The CPT q_c values were generally between 10 and 15 MPa, with slightly stronger layers at 31.5 m, 36.0 m and 44.0 m depth.
(b) The τ_p values in the anchor tests were 614 kPa for the test at 35–38 m, and 526 kPa for the test at 42–45 m, with τ_r values of 235 and 137 kPa, respectively.
(c) Hydraulic fracture initiation pressures ranged from 1600 kPa to 2300 kPa.
(d) The G_{ur} values from the SBP tests were generally in the range of 100–120 MPa.†

---

†It was subsequently discovered that because of excessive compliance in the measuring system in the SBP, these values seriously underestimated the true stiffnesses — refer to Fahey and Jewell (1990) for a discussion of the compliance problem. This was not discovered until after the pile testing at Waikerie. When a correction for this compliance was subsequently applied — a process of dubious accuracy — the best estimate of G_{ur} increased to about 350 MPa in zone below 40 m, and to some much greater value above this zone.

(e) When interpreted using the sub-tangent method, the SBP tests gave peak shear strengths of the order of 600–700 kPa, with residual strengths of about 200–300 kPa (see Fahey (1988) for a discussion on the relevance of this interpretation for this material).

(f) Some of the SBP tests incorporated 'pressure holding' tests at different stages. No pore pressures were generated during expansion. However, these tests showed creep-type deformations if the test was conducted at pressures greater than about 1600 kPa (i.e. above the initial linear 'elastic' range) but practically no creep deformations if conducted at pressures below this elastic limit.

Laboratory testing

The laboratory testing programme included fabric studies, oedometer, triaxial and simple shear tests, constant normal stiffness (CNS) direct shear tests, and rod shear tests (RST). Some typical results are presented to illustrate the results obtained.

Consolidation and triaxial tests

The triaxial testing programme included CIU and CID tests, and CK_0U 'active' and 'passive' tests. The mean and axial consolidation stresses in the isotropic and anisotropic tests respectively, were set to the in situ value of σ'_v (300–400 kPa). One dimensional compression tests showed that the apparent preconsolidation pressure p'_c was some 2 to 3 times greater than this range.

Peak shear strengths at very small strains from the CIU and CID tests were about 400–500 kPa, with post-peak strain softening being clearly evident in the undrained tests, though generally with residual strengths still greater than about 60% of peak values. After the initial peak in the CID tests, some slight strain hardening was evident at larger strains. The peak strengths in the 'active' CK_0U tests were slightly lower (300–450 kPa), perhaps reflecting the lower mean consolidation stress. The critical state friction angle was about 45°. Initial Young's Modulus values (E_i) from the isotropically consolidated tests, both drained and undrained, were of the order of 200–280 MPa, with slightly lower values being found for the anisotropic tests. These values were obtained using external rather than internal displacement measurement.

CNS direct shear tests

The constant normal stiffness (CNS) direct shear apparatus is similar to a standard direct shear apparatus except that the normal stress can be varied during shearing in response to the dilation or contraction of the soil according to some predetermined stiffness function (Johnston et al., 1987). It can be used to test the behaviour of the soil (as in this case), or the behaviour on a soil–pile interface. Based on the cavity expansion analogy, the spring stiffness for interface testing is chosen such that the normal stress on the interface σ_n varies according to the relation:

$$\frac{\Delta \sigma_n}{\Delta h} = \frac{2G}{r_{pile}} \qquad (1)$$

where Δh is the outward movement of the top of the sample, and r_{pile} is the radius of the field-scale pile being investigated. The normal stiffness in CNS testing for pile design is therefore chosen to reflect both the soil stiffness and the diameter of the pile of interest.

Johnston et al. (1988) discuss testing of North Rankin calcarenites in the CNS direct shear devices at Sydney and Monash Universities, and present typical results obtained for both monotonic and cyclic loads. The CNS testing for the Overland Corner investigation was conducted at Sydney University (University of Sydney, 1987).

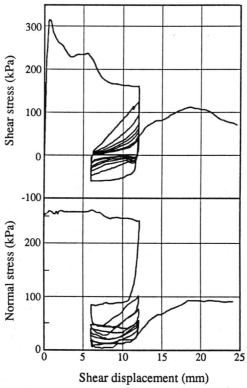

Fig. 2. CNS test on intact core (University of Sydney, 1987)

The results of a typical CNS test, with an initial normal stress of 250 kPa and a normal stiffness of 500 kPa/mm, are shown in Fig. 2. After monotonic shearing to a displacement of 12 mm, a total of 10 cycles of displacement between 12 and 6 mm displacement were applied, before continuing monotonic displacement to 24 mm. The upper part of Fig. 2 shows that a peak shear stress of more than 300 kPa was reached at very small displacement, which reduced to about half that value at the point of first reversal of loading.

For piles of 0.4 m diameter (the most common test piles at Overland Corner), the normal stiffness employed in this test corresponds to an assumed shear modulus G of 100 MPa, according to eqn. (1). The upper part of Fig. 2 shows that the normal stress remained practically constant during the initial monotonic phase, which means that there was no tendency for volume change to occur. Thus, even if an appropriate normal stiffness was used, this would have had no effect on the peak strength or on the post-peak monotonic response. However, these tests were on intact material, rather than on grout–soil interfaces, and the behaviour in the latter type of test might be significantly different, as claimed by Johnston et al. (1988).

At the start of displacement in the reverse sense, a sharp reduction in normal stress occurred, and the maximum shear stress reached in each subsequent cycle continued to decline to very low values. However, in the final monotonic phase, when the displacement was taken outside

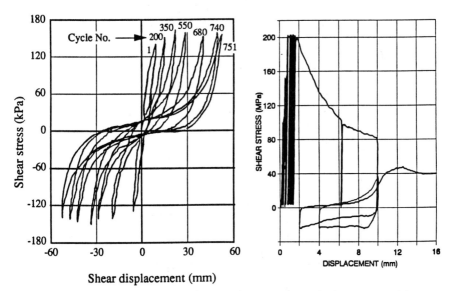

Fig. 3. CNS test on intact core with pre-peak cycling

Fig. 4. Rod shear test with pre-peak and post-peak cycling

the bounds used in the cyclic tests, the shear stress and normal stress gradually recovered to rejoin the projected monotonic response.

Tests with 'pre-peak' cycling were also carried out, including both 'one-way' and 'two-way' tests. The shear stress versus displacement behaviour in a typical two-way test, with cycling between shear stress limits of ±152 kPa, is shown in Fig. 3. As cycling progresses, more and more displacement is required to reach the set shear stress limit, with the normal stress and shear resistance being very low over most of this displacement range.

Rod shear tests

The rod shear test (RST) is a small-scale model pile test carried out in an undisturbed sample of soil, with the pile being grouted into a pre-drilled hole through the sample. The sample is either confined in a triaxial cell, or grouted into a rigid PVC sleeve or jacket, as described by Fahey and Jewell (1988) and Jewell and Randolph (1988). For the Overland Corner tests, both 'triaxial' and 'jacket' types of confining condition were used. Both peak and residual strengths tended to be higher in the 'triaxial' than in the 'jacket' tests, and the post-peak softening was also slightly less severe in the former.

Figure 4 shows the result of one of the tests with the jacket confining condition. In this test, one-way pre-peak cyclic loading was applied, consisting of 'packets' of 50 cycles over shear stress ranges 0–50 kPa, 0–100 kPa, and 0–150 kPa, resulting in a total displacement of 0.5 mm. Failure then occurred after 38 cycles while cycling in the range 0–200 kPa, with a further 1 mm displacement just before failure in the 38th cycle. After a total displacement of 10 mm, the average shear stress had fallen to about 80 kPa. Reversal of the direction of loading at this

Table 1. *Typical values of friction and stiffness parameters*

	CNS	RST	Anchor	SBP	Design	Pile tests
τ_p (kPa)	200–300	300–600† 100–300‡	526, 614	500–800¥	400, 600	400–440
τ_r (kPa)	50–110	100–300† 40–120‡	137, 235	200–300¥	130, 200	#
G (MPa)	¶	¶	¶	100–120§	100, 120	500

Results for † 'triaxial' and ‡ 'jacket' confining conditions.
#Depends on displacement.
¶Not measured.
¥ Shear strength.
§Affected by instrument compliance — see text.

stage resulted in very low shear stress (<20 kPa) being measured in both directions, but some recovery of shear resistance (up to about 50 kPa) was seen when the displacement was taken outside of this cyclic range.

Pile design parameters from test results

Before the pile testing programme was undertaken at the Overland Corner test site, 'Class A' predictions of the performance of the piles were made. The key data for these predictions were chosen on the basis of the evidence from the laboratory and field tests, as summarized in Table 1, and as described in the following sections.

Skin friction

The peak shear stress τ_p was chosen on the basis of CNS, RST and anchor pull-out tests. Broadly, the τ_p ranges were 160–320 kPa for CNS tests, 300–600 kPa for RSTs, and 500–600 kPa for the anchor tests.

In contrast to the laboratory RSTs, in which the insert was grouted in place with only atmospheric pressure on the grout, the field pile sections were to be grouted at a grout pressure of at least 300 kPa above ambient water pressure. This would suggest that the field values of τ_p should be greater than in the RSTs. Thus, two design values of τ_p were chosen: an 'optimistic' value, 600 kPa, and 'pessimistic' value, 400 kPa.

The RST, CNS and anchor tests all indicate significant strain softening of the skin friction under monotonic loading to residual values τ_r which were typically 30 to 40% of τ_p. Thus, τ_r was assumed to be ⅓ of τ_p, or 200 and 130 kPa for τ_p values of 600 and 400 kPa, respectively.

Displacements to mobilize peak and residual friction values

In the monotonic response, the displacements required to mobilize τ_p and to degrade to τ_r are key parameters. The displacement to τ_p depends partly on the shear modulus G of the soil. The SBP tests gave very consistent values of G_{ur} in the range of 100 to 120 MPa (though these were later found to be incorrect, as explained earlier). These values were much higher than deduced from the RST results, which gave values of 10–20 MPa, but were similar to triaxial initial values (E_i of 200 to 280 MPa). Values of 120 and 100 MPa were chosen on the basis of the SBP results, corresponding to the high and low values chosen for τ_p. The displacement to peak was then estimated on the basis of a parabolic load–transfer curve, with an initial slope G/ζ, where ζ depends on the aspect ratio of the section (Randolph, 1986).

It was anticipated that the displacement after peak required to reach τ_r (displacement Δw_{res} in Fig. 1) would be related to diameter, but in the absence of experimental evidence on which to base the relationship, it was taken to be 100 mm, which is of the order of 10 times that observed in the RST and CNS tests.

Effect of cyclic loading

The effects of cyclic loading depend on whether it occurs pre- or post-peak, on whether it is one- or two-way, and on the magnitude of the loading. Randolph and Jewell (1989) discuss the way in which degradation under cyclic loading is handled in the RATZ program. The value assumed for G is critical in this method; as already discussed, two values of G of 100 and 120 MPa were used.

A feature of the RST (and CNS) results is the very low skin friction which is reached when cycling within a displacement range, though the skin friction can increase again when displacements outside this range occur. To allow for this, a cyclic skin friction limit of 50 kPa was assumed for cyclic loading within a previous displacement range.

For the Overland Corner piles, the question of a fatigue limit — the cyclic shear stress level below which no degradation occurs — did not arise, though this is a key issue for offshore piles.

Re-assessment of model parameters after pile testing

The sections tested at Overland Corner consisted of a number of short pile test sections ('elements') and one long pile section. The elements were of diameters 400, 940 and 2000 mm, with lengths nominally 6 times the diameter for the 400 and 950 mm sections, and 3 times the diameter for the 2000 mm section. Four elements of 400 mm diameter were tested, and one of each of the other elements. The long pile section was 400 mm diameter and 15.6 m long.

The testing sequences consisted of either monotonic loading followed by large-displacement post-peak cycling, or pre-peak two-way cyclic loading at increasing load levels until failure was reached. The results of one of the former types of test are shown in Fig. 5. The features of this curve are: very small displacement to a peak load of 1.6 MN (an average shear stress of 440 kPa), followed by post-peak softening; very stiff unload–reload response if the load is not reversed; very low capacity in reverse loading and in cycling within previous displacement limits; and a slow recovery almost to previous levels when loading is taken outside of the previous limits. The overall behaviour is remarkably similar to that observed in the laboratory RSTs in particular (refer to Fig. 4).

Figure 6 shows the behaviour in a test incorporating pre-peak cycling, with 'packets' of about 50 cycles of two-way loading being applied at ±33, ±43, ±52 and ±61% of τ_p, followed by 68 cycles at ±71%, which caused failure of the pile. As failure approaches, the typical low resistance over the central part of each cycle is also evident here.

PILE DESIGN IN CALCAREOUS SEDIMENTS

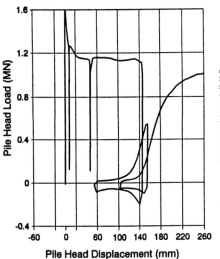

Fig. 5. Load test on pile section (2.4 m long, 400 mm diameter) at Overland Corner

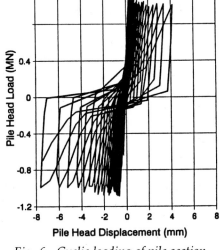

Fig. 6. Cyclic loading of pile section (2.4 m long, 400 mm diameter) at Overland Corner

After the tests, the predicted and measured responses were compared. The RATZ model parameters were then adjusted to achieve the best fit to the results of both monotonic and cyclic tests, and the pre- and post-test parameters were compared.

The first conclusion was that the soil stiffness had been grossly underestimated. A detailed finite element study of the unloading–reloading behaviour in the pile tests was therefore undertaken and a value of G of about 500 MPa was obtained — about five times greater than that measured in the SBP tests and assumed in the predictions. (It was this finding that prompted the re-examination of the compliance of the SBP which showed that the measured values were severely affected by compliance effects.)

It was found that the measured value of τ_p was generally close to the assumed 'pessimistic' value of 400 kPa, but that the exact value depended on section diameter: τ_p values of 300, 400 and 440 kPa were measured for the 2000, 950 and 400 mm sections, and 600 kPa in the 140 mm diameter anchor test at this depth. Part of the dependency is due to the effect of pile diameter on dilation-induced normal stress increase, as expressed in eqn. (1). A full treatment of this aspect is contained in Randolph (1988). In essence, he suggests that in the absence of dilation, the skin friction can be assumed to be:

$$\tau_p = \sigma_g \tan \delta \qquad (2)$$

where σ_g is the grout pressure, and δ is the grout–soil friction angle. The effect of dilation is to increase the normal stress during shearing above this value, but the increase becomes less as the diameter increases. For the grout pressures of 340–400 kPa used on site, and assuming δ of 39°, eqn. (2) would give τ_p of 270 to 335 kPa, which is close to the value observed in the 2000 mm section. For small diameters, the effect of dilation is greater, and τ_p increases accordingly.

In spite of this apparent diameter effect, the laboratory RSTs conducted under triaxial stress conditions appear to give a good prediction of τ_p (and τ_r) values. This may be simply fortuitous, due to the greater relative effect of dilation being cancelled by a cell pressure (250 kPa) and an initial grout pressure ($\sigma_g = 0$) which were both lower than the field grouting pressure (350–400 kPa).

As the pile diameter reduces, the dilation effect cannot cause σ_n to increase indefinitely, because eventually an elastic limit or 'crushing pressure' in the soil will be reached. This would occur at an effective radial stress of about 1100–1300 kPa, according to the behaviour observed in the SBP tests (Fahey, 1988). To give τ_p of just over 600 kPa in one anchor test, σ_n would have to increase to only 740 kPa (assuming δ is 39°), which is well below this elastic limit. Such a limit perhaps could be reached in an RST, especially if grouted under pressure, and this would limit the dilation effect.

If conducted at the appropriate initial normal stress level, CNS tests in theory should be capable of reproducing the correct response. This probably requires that the tests be conducted on interfaces rather than intact core, and that the correct value of G be used in choosing the normal stiffness. In these tests, the initial normal stress used was only 250 kPa; increasing this to 350 kPa could increase the measured τ_p from 200–300 kPa to 280–420 kPa if τ_p is assumed to be proportional to σ_n.

Though τ_r was correctly predicted by the RSTs, the rate at which the friction degraded during cyclic loading was much slower than predicted, and this rate appears to be dependent on section diameter. This necessitated a change in parameters governing the post-peak monotonic behaviour.

The cyclic loading tests confirmed the very low cyclic friction values obtained in the laboratory tests. These values are important in modelling the large displacement cycles shown in Fig. 5, and the behaviour close to failure in pre-peak cycling as in Fig. 6, for the element tests, and in matching the gradual degradation of total capacity in full-scale piles. In this case, even where pre-peak loading is well below the static capacity, the local capacity near the top of the pile can be degraded to these low cyclic values, thereby increasing the cyclic stresses lower in the pile. This can lead to a gradual reduction in total capacity.

The fitting exercise carried out to match the field test behaviour

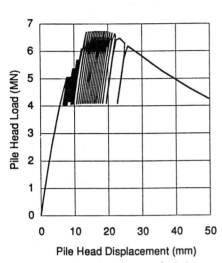
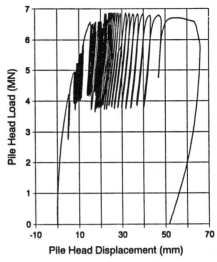

Fig. 7. Predicted response of 15.6 m long, 400 mm diameter section

Fig. 8. Measured response of 15.6 m long, 400 mm diameter section

resulted in some significant changes being made to the parameters in the RATZ model. Nevertheless, the quality of the 'Class A' predictions deserves some comment. The predictions (using the worst-estimate parameters of $\tau_p = 400$ kPa, $G = 100$ MPa etc.) of the load–deformation performance of the long section subjected to cyclic loading are shown in Fig. 7. The applied loading was modelled on a typical storm sequence, and consisted of: 50 cycles from 4 to 5 MN; 5 cycles from 4 to 5.3 MN; 5 cycles from 4 to 5.6 MN; 1 cycle from 4 to 6.0 MN and 1 cycle from 4 to 6.6 MN. This sequence was then followed by 25 cycles at the maximum loading (4 to 6.6 MN) to find out how many cycles at the maximum wave loading would be required to cause failure. In this case, the pile was predicted to survive the storm sequence, but to fail towards the end of the final 25 cycles at maximum load.

The measured response in the field test shown in Fig. 8. The loading sequence applied in the prediction exercise was not followed exactly in the pile test (the final cyclic loading range was 3.8–6.8 MN), but nevertheless it can be seen that the prediction is remarkably accurate. However, though the differences between the predicted and observed behaviour appear subtle, these differences become magnified when applied to the whole loading history of a typical long offshore pile, and are thus of considerable importance.

Application to analysis of offshore piles

Though the pile testing at Overland Corner enabled some of the key features of the RATZ model to be tested, some important areas of uncertainty remained. These involved in particular the question of the effects of large numbers of low-level cycling, and on the numbers of cycles to cause failure at different cyclic stress ratios — in effect, the shape of the whole S–N curve (Airey and Fahey, 1991). Prediction of the effects of such cycling is critically dependent on what if any fatigue threshold exists. It would not be feasible to apply sufficiently large numbers of cycles in a field-scale trial to investigate this problem, so investigation of this aspect must rely on laboratory cyclic tests. For the investigation for the Goodwyn piles, this testing included cyclic triaxial tests (up to 10^6 cycles), resonant column tests (10^8 cycles), and cyclic CNS tests. Though no tests failed below cyclic shear stresses ratios of 37%, which supported the 'most likely' limit of 33% originally assumed, a much lower fatigue limit was adopted to account for present uncertainty. However, this is clearly an issue which requires further work to establish that such a fatigue limit exists, and to determine appropriate values.

Applying the model to offshore piles also involves the question of how to determine the peak, residual and cyclic residual shear stresses. For the design of the Goodwyn piles, the strategy adopted was to express the Overland Corner τ_p results as a ratio τ_p/q_c, and then use this ratio with the q_c data at Goodwyn to determine τ_p. The residual value τ_r was based on a conservative fit to the laboratory RST data, and the displacements to peak and residual were taken directly from the Overland Corner tests.

Discussion

The prediction exercise and subsequent pile testing at Overland Corner show what aspects of behaviour had been correctly predicted from the laboratory and field testing, and what aspects had not been anticipated. This provides some indication of what needs to be done in similar circumstances in future offshore pile design projects which do not have the advantage of a large-scale on-land pile testing programme.

One of the crucial parameters in the analysis was found to be the shear modulus of the soil. It was found that a conservative analysis of the effects of cyclic loading requires the use of the highest likely value of G. The SBP used to obtain G values of Overland Corner proved inadequate to measure the very high stiffnesses involved, though with recent improvements to the instrument, it would now be capable of providing more accurate measurements. Pressuremeter testing is possible offshore using full-displacement or push-in instruments, though the accuracy and resolution of these instruments would have to be very carefully assessed in soils of similar stiffness to the Overland Corner calcarenites. The wire-line SBP developed by the French Petroleum

Institute (Faÿ and Le Tirant, 1990) might be a viable alternative.

This seems to be one class of problem where dynamic methods of shear modulus measurement appear warranted, since the highest value of G possible is required in 'worst case' analyses of pile performance. The seismic cone method, pioneered at the University of British Columbia (Robertson et al., 1985), seems the most appropriate method to use, particularly now that offshore versions have been developed (see for example de Lange et al., 1990). The seismic cone provides a direct measurement of shear wave velocity, from which a dynamic or 'small strain' shear modulus G_0 can be obtained directly.

For the Overland Corner test piles, the peak skin friction τ_p was quite well predicted by the laboratory RSTs, though it is believed that this may be somewhat fortuitous. It would appear that the RST should overpredict τ_p for field-scale piles, because τ_p was found to be dependent on pile diameter, probably because the effect of dilation on σ_n depends on the pile diameter. A logical step would be to carry out RSTs with the grout pressure (and perhaps the cell pressure) chosen to match the intended grout pressures for the field piles, rather than the in situ stress. The parameters obtained would still need to be adjusted for full-scale piles to allow for radius effects, perhaps along the lines outlined by Randolph (1988). This approach appears preferable to relying on what is perhaps a fortuitous agreement between small-scale and full-scale tests, as in this case.

The CNS direct shear test also appears to offer a useful means of measuring peak and residual frictions. However, it appears that CNS testing should be carried out on soil–grout interfaces rather than intact soil samples if the interface behaviour is to be correctly modelled. It is also very important to use the appropriate value of normal stiffness, which depends on both shear modulus and pile diameter.

Remarkably, many of the detailed features of the response of the field-scale piles were reproduced at least qualitatively in the laboratory CNS and RST tests. One of the most important of these was the behaviour of cyclic loading: very low resistance during large displacement post-peak cycling and in pre-peak cycling when close to failure; and recovery of capacity when subsequently loaded monotonically outside the cyclic displacement range. The RATZ analysis of the Overland Corner cyclic loading tests showed that this behaviour must be modelled to reproduce the observed behaviour, and it has also been found to have a significant influence on the predicted behaviour of long offshore piles under storm loading.

In the design of the offshore piles for Goodwyn, a determining factor was the assumption made about the fatigue threshold. It would not be feasible to apply the very large numbers of cycles of load required to investigate this aspect (many millions of cycles) to a field-scale test pile, and therefore laboratory tests are in practice the only source of data. With modern loading machines with hydraulic actuators, it is possible to apply very large numbers of cycles in triaxial or rod shear tests in a relatively short time if very fast loading frequencies are used (a frequency of 10 Hz has been used at UWA, permitting 10^6 cycles to be applied in 28 hours). Providing the question of the effect of such high frequency loading rates can be quantified, this type of testing may offer the best means of resolving this important question of fatigue threshold.

Some of the more standard laboratory and in situ tests — particularly standard triaxial tests and the CPT — may appear to have little relevance to the pile design process described in this paper. However, in practice, the q_c value from the CPT is the most direct measurement to allow scaling of the various skin friction limits (τ_p, τ_r), and for comparing the properties at different sites. The triaxial test also fulfils this function, as it allows the behaviour of the intact material at different sites to be compared.

Conclusion

Predicting the performance of offshore piles in calcareous soils, particularly allowing for the effects of cyclic loading, has been shown to be a process which is very sensitive to the assumptions made about certain parameters. These relate in particular to the soil stiffness, the peak and residual skin friction values, the displacements required to reach these values, and the detailed performance under cyclic loading. The Overland Corner pile testing exercise has shown that reasonable predictions of grouted pile performance in calcareous soil can be made using the results of tests devised specifically to investigate interface behaviour. These tests may need to be developed further to investigate how the size effect can be allowed for in the laboratory tests.

Acknowledgements

The authors wish to thank the Management of Woodside Offshore Petroleum Pty Ltd and the Joint Venture Participants for permission to publish this paper. The Participants in the North West Shelf Joint Venture are: Woodside Petroleum Ltd (through subsidiaries), Shell Development (Australia) Pty Ltd, BHP Petroleum Pty Ltd, BP Developments Australia Ltd, Chevron Asiatic Ltd, and Japan Australia LNG (MIMI) Pty Ltd. The testing described in this paper was carried out for Woodside Offshore Petroleum by a number of organizations. The Geomechanics Group at The University of Western Australia carried out parts of the field and laboratory testing programmes, participated on the Steering Committee and in the running of the pile tests, and carried out review and interpretation of the results.

References

AIREY, D.W. and FAHEY, M (1991). Cyclic response of calcareous soil from the North-West Shelf of Australia. Géotechnique Vol 41, No. 1, pp. 101–121.

DE LANGE, G., RAWLINGS, C.G. and WILLET, N. (1990). Comparison of shear moduli from offshore seismic cone tests and resonant column and piezoceramic bender element laboratory tests. Oceanology International '90, Brighton, 13–20 March.

FAHEY, M. (1988). Self-boring pressuremeter tests in a calcareous soil. Engineering for Calcareous Sediments. R.J. Jewell & D.C. Andrews (eds). Vol. 1, pp. 165–172. Balkema, Rotterdam.

FAHEY, M. and JEWELL, R.J. (1988) Model pile tests in calcarenite. Engineering for Calcareous Sediments. R.J. Jewell & M.S. Khorshid (eds). Vol. 2, pp. 555–564. Balkema, Rotterdam.

FAHEY, M. and JEWELL, R.J. (1990). Effect of pressuremeter compliance on measurement of shear modulus. Pressuremeters: Proceedings of the Third International Symposium on Pressuremeters ISP3, Oxford, pp. 115–124, Thomas Telford Ltd.

FAŸ, J.B. and LE TIRANT, P. (1990). Offshore wireline self-boring pressuremeter. Pressuremeters: Proceedings of the Third International Symposium on Pressuremeters ISP3, Oxford, pp. 55–64, Thomas Telford Ltd.

JEWELL, R.J. and RANDOLPH, M.F. (1988). Cyclic rod shear tests in calcareous sediments. Engineering for Calcareous Sediments. R.J. Jewell & D.C. Andrews (eds), Vol. 1, pp. 215–222.

JOHNSTON, I.W., LAM, T.S.K. and WILLIAMS, A.F. (1987). Constant normal stiffness direct shear testing for socketed pile design in weak rock. Géotechnique, Vol. 37, No. 1, pp. 83–89.

JOHNSTON, I.W., CARTER, J.P., NOVELLO, E.A. and OOI, L.H. (1988). Constant normal stiffness direct shear testing of calcarenite. Engineering for Calcareous Sediments. R.J. Jewell & M.S. Khorshid (eds). Vol. 2, pp. 541–553. Balkema, Rotterdam.

MURFF, J.D. (1987). Pile capacity in calcareous sands: State of the art. Journal Geotechnical Engineering Division, ASCE, Vol. 113 (GT5), pp. 490–507.

POULOS, H.G., RANDOLPH, M.F. and SEMPLE, R.M. (1988). Evaluation of pile friction from conductor tests. Engineering for Calcareous Sediments. R.J. Jewell & M.S. Khorshid (eds). Vol. 2, pp 599–605. Balkerma, Rotterdam.

RANDOLPH, M.F. (1986). RATZ — Load transfer analysis of axially loaded piles. Report No. Geo: 86033, Department of Civil Engineering, The University of Western Australia.

RANDOLPH, M.F. (1988). The axial capacity of deep foundations in calcareous soil. Engineering for Calcareous Sediments, R.J. Jewell and

M.S. Khorshid (eds), Vol. 2, pp. 837–857, Balkema, Rotterdam.

RANDOLPH, M.F. and JEWELL, R.J. (1989). Axial load transfer models for piles in calcareous soil. Proc. 12th Int. Conf. on Soil Mech. and Found. Eng., Rio de Janeiro, Vol. 1, pp. 479–484.

ROBERTSON, P.K., CAMPANELLA, R.G., GILLESPIE, D. AND RICE, A. (1985). Seismic CPT to measure in situ shear wave velocity. Journal of Geotechnical Engineering, ASCE, Vol. 112, No. 8, August, pp. 789–803.

UNIVERSITY OF SYDNEY (1987). Static and cyclic direct shear testing under conditions of constant normal stiffness of core samples from the Waikerie test site, South Australia. University of Sydney, School of Civil and Mining Engineering, Investigation Report No. S612 (in Document No A1820NG002, Woodside Offshore Petroleum Pty Ltd).

Mobilisation of stresses in deep excavations: the use of earth pressure cells at Sheung Wan Crossover

R.A. FRASER, G. Maunsell & Partners

A 30 m deep excavation built as part of the Hong Kong Mass Transit Railway Island Line, utilising diaphragm walls has been monitored using jackout earth pressure cells, inclinometers, piezometers, strut load cells and settlement points. Measurements of horizontal effective stress at the diaphragm wall face indicate that passive pressure in the completely decomposed granite is mobilised more quickly than thought for dense cohesionless material and points to the need for consideration of cohesion or passive stress relief in similar excavations.

Introduction

The initial stress in the ground adjacent to a diaphragm wall after installation will be modified from the in situ stress, and may be either reduced or increased.

The design of deep embedded walls supporting excavations requires knowledge of the in situ horizontal stress acting on the wall after installation and the rate and magnitude of change of horizontal stress as the wall translates either into or out of the excavation. This information can be obtained by installing pressure cells in the face of the walls. Few such walls with embedded pressure cells have been constructed — most

Table 1. Soil parameters used in Sheung Wan Crossover excavation

Soil type	Density (kN/m³)	ϕ (degrees)	K_a	K_p	Cohesion (kPa)
Fill	18.0	35	0.27	5.85	0
Marine deposit	18.0	32	0.31	4.80	5
Alluvium	20.0	36	0.25	6.20	5
CDG (N ≤ 70)	20.0	38	0.24	7.10	5
CDG (N > 70)	21.5	40	0.22	8.10	15

have been investigated by instrumentation installed immediately adjacent to the wall.

The Sheung Wan Crossover was constructed as part of the Island Line development for the Hong Kong Mass Transit Railway in 1982–4 (Fig. 1) and was designed to serve the functions of providing access for tunnelling works, accommodation for plant rooms and to act as partial support for the future development above.

The results of measurement of movement with inclinometers and settlement points and of total stress and porewater pressure with pressure cells installed in diaphragm walls during construction of the 32 m deep excavation at the Crossover, are discussed.

Soil properties

A typical cross-section through the Crossover is given in Fig. 2, incorporating the soil stratification at section AA in Fig. 1.

The parent rock to the site is dominantly slightly decomposed granite (SDG), which occurs at a depth of approximately −40 mPD at the east end, rising to about −21 mPD towards the west. The bedrock is covered by a layer of completely decomposed granite (CDG) consisting of medium dense to very dense silty sand with gravels. The CDG is in turn overlain by alluvium, marine deposits and fill of varying thickness.

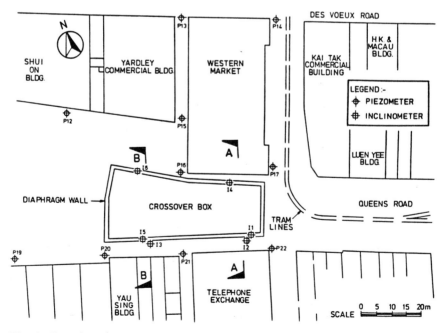

Fig. 1. Location plan

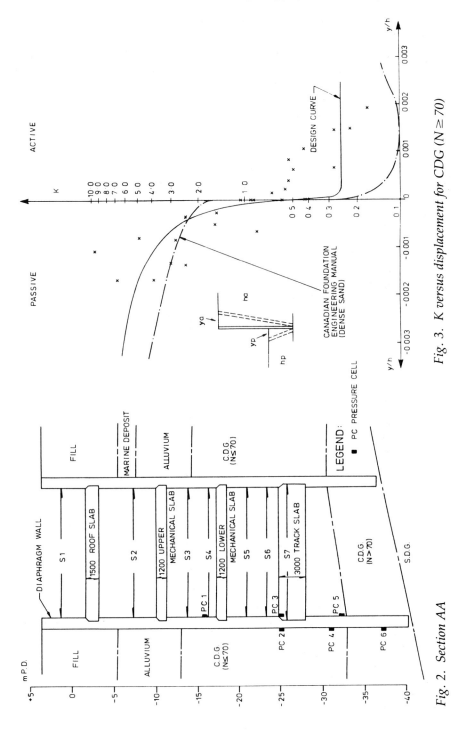

Fig. 3. *K versus displacement for CDG (N ≥ 70)*

Fig. 2. *Section AA*

Groundwater level is affected by tides as the site is close to the sea, and in general fluctuates between +2.0 mPD and +3.0 mPD. A more detailed discussion on the properties of CDG is given in Lumb (1965).

The other materials are generally granular and free draining, with a small component of cohesion. Table 1 presents the soil parameters adopted in the design of the Crossover, which were formulated as a result of laboratory tests on soil samples and previous experience in similar ground conditions.

Analysis methods

The need to consider soil-structure interaction in the design of embedded walls is quite well established as the divergence in bending moments calculated from limit equilibrium methods and that measured for flexible walls has long been known to cause over-design. On the other hand, Potts and Fourie (1985) have shown that limit equilibrium methods are not conservative for stiffer walls wished into place in over-consolidated materials. Soil-structure interaction can be modelled by subgrade reaction models or a full finite element analysis. The simplest model is a subgrade reaction model, and as the process of wall design is essentially a structural optimisation process, the inevitable need for refinements during design development of the structure rules out the routine application of a full finite element analysis of an excavation.

Linear and non-linear beam and spring finite elements are used in a subgrade reaction model, and quick assessments can be made of the effects of changing wall stiffness, wall thickness, temporary strut positions, creep of concrete, yielding supports (often a non-linear response, as at Bell Common), and moment restraint at supports. Each of these aspects can influence the economy and viability of a scheme.

With regard to geotechnical aspects, often sites are not level, and nor are the geotechnical properties uniform. There is a need for analyses to provide the ability to efficiently optimise structural alternatives and construction sequencing for a variety of soil conditions. This again leads to the choice of a subgrade reaction model.

Gudehus (1991) has pointed out that subgrade reaction models require three assumptions:

(a) initial stress state as produced by wall placement
(b) non-linear change of pressure with lateral displacement
(c) limitation by active or passive earth pressure.

The calculations for the Sheung Wan Crossover were undertaken by means of the computer program 'DIANA' (Maunsell Systems Centre Ltd, 1982), which was specifically developed to carry out DIaphragm wall ANAlysis incorporating a non-linear subgrade reaction analytical model. Development was based on back analysis of embedded walls designed and monitored during earlier phases of the Hong Kong Mass Transit (Fraser, 1980).

The prediction of stress in, and deformation of, the diaphragm walls during excavation was carried out by modelling the wall as a series of beam elements rigidly connected at nodes, and the soil materials as a series of non-linear nodal springs. The initial state of the soil after wall placement is discussed in more detail later, but the computer program allows the input of any adopted profile for K.

From the initial value of K, movement of the wall into or away from the soil will mobilise pressure from a non-linear soil spring. The non-linear change of pressure can be related to the lateral wall displacement based on previous studies as in Fig. 3 (Canadian Geotechnical Society, 1985). The soil pressures in the retained soil above the excavation level will drop towards values limited by active pressure coefficients. Similarly, the pressure in front of the wall is represented by a non-linear soil spring which reaches a limit defined by a log spiral passive failure criterion. The program iterates until convergence is reached, following a predefined mobilisation curve such as that given in Fig. 3 for CDG with $N \leq 70$, where the earth pressure coefficient varies as the parameter y/h varies, which is an approximation to half the shear strain for a rigid wall rotating about the toe. This curve was adopted from back analysis of previous excavations in similar soil (compared with data from this study). During analysis h is taken as the dimension from the toe of the wall to the relevant node of the beam element and y is the calculated displacement.

Water pressures due to seepage were assumed for water levels inside and outside the excavation. The sequence of excavation and installation of props was analysed by removal or addition of relevant springs.

Following structural optimisation an assessment of the likely magnitude of other construction effects (e.g. long-term loading, overall deformation of an anchored wall, heave and settlement) may be required. To do this a finite element analysis of a typical section can be carried out. Allowance for groundwater flow and development of long-term conditions can be accommodated in such a model, but because of the assumptions and simplifications the results should be considered to be a qualitative assessment of the ground response.

Despite the different model, the influence of wall placement on in situ stresses is just as important for finite element models as for sub-grade reaction models.

Construction of Sheung Wan station

The Sheung Wan Crossover site is located in an urban area and is in close proximity to a number of buildings (Fig. 1). The majority of buildings adjoining the site were large multi-storey reinforced concrete framed structures founded on either piled foundations or on concrete rafts constructed on grout stabilised soil. However, the Western Market and a few older buildings were 3- to 4-storey structures founded on isolated spread footings.

The envisaged construction sequence of the Crossover required the top down method, incorporating perimeter diaphragm walling followed by the construction of internal permanent concrete slabs and installation of temporary steel struts as excavation progressed. The need for these temporary struts was largely due to the insistence by the Building Authority on the use of cohesionless soil parameters derived by testing

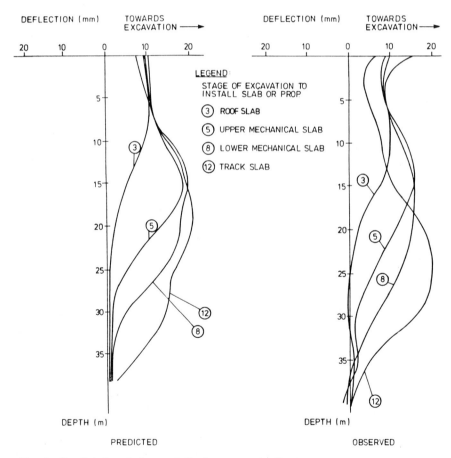

Fig. 4. Predicted and observed displacement; inclinometer 15

back saturated samples of CDG (where the soil structure has been destroyed).

The construction work started in September 1982 with the installation of the perimeter diaphragm wall involving 34 panels. Panel excavation was carried out using grabs, with boulders or other forms of obstruction being overcome by chiselling. However, large initial settlements of the Western Market (approximately 30 mm) occurred as a result of the chiselling work in adjacent panels. The procedure of stitch-drilling the boulders was then adopted, and the corresponding settlements at the Western Market were significantly reduced.

The toe of the diaphragm wall is typically 0.5 m to 8 m above the SDG surface. In order to provide an extension of the hydraulic cut-off between the toe of the diaphragm wall and bedrock material, contact grouting was carried out upon completion of diaphragm wall installation. The grout materials comprised a cement bentonite mix and a low-viscosity silicate mix, and were injected into the soil in two stages beneath the toe of the diaphragm wall using the Tube-a-Manchette method.

The main excavation work for the Crossover commenced at the end of January 1983, and dewatering work for the excavation started later in February 1983 by means of 10 pumping wells located inside the box. At the western section the diaphragm wall terminated above the track slab level due to a high rock surface, and underpinning of the wall was implemented by bottom up construction, with the exposed rock face being supported by temporary strutting. The final excavation level was reached (-28.4 mPD), and the track slab cast in mid-December 1983.

During the later stages of the main excavation, a system of recharge wells for the Western Market and the Telephone Exchange Building came into operation, in order to limit settlements of the buildings concerned by replenishing the ground water.

Details of instrumentation

The locations of piezometers and inclinometers are given in Fig. 1. Numerous settlement markers were installed on adjacent buildings and pavements but for clarity are not detailed in Fig. 1. Strut loads were measured by means of vibrating wire strain gauges which were mounted on selected members.

Six pneumatic pressure cells with piezometers were installed in the south panel of section A-A to measure the total horizontal soil pressure and pore water pressure adjacent to the wall. The positions of the pressure cells are indicated in Fig. 2. Piezometers and settlement markers were generally monitored daily, whereas for inclinometers, strain gauges and pressure cells, readings were normally taken weekly during construction.

Installation of earth pressure cells

Measurements of total stress at the face of diaphragm walls are not common. The use of jack out earth pressure cells to record pressures has been rare compared with other forms of instrumentation, but has been described by Lings et al. (1991), Inoue and Katsura (1977), Di Bagio and Roti (1972), Uff (1967). Jack out pressure cells were installed at various sites on the Hong Kong Mass Transit Modified Initial System and Tsuen Wan extension, and expertise was developed in installation techniques.

The equipment for panels at the Crossover essentially consisted of 6 pneumatic earth pressure cells of 200 mm diameter each containing a pneumatic piezometer with hydraulic jacks with 400 mm extension and with swivel joints fitted to the cells and a reaction plate of 200 mm diameter.

Before the installation all the instruments were tested, and following calibration the earth pressure cells and the piezometers were fixed on the reinforcing cage and sunk into the diaphragm wall trench, and the initial readings of the instruments were taken. Then the earth pressure cells were jacked out by the hydraulic jacks. The pressures registered by the cells and piezometers once they were in position were checked against calculated pressures of the bentonite slurry and were found to be satisfactory, taking into account the variability of the bentonite density. The assemblies were then jacked out such that a nominal increase in all the cells readings was obtained, indicating all the cells were able to butt against the soil face before the rams ran out of their stroke.

Between the time when all the rams had been jacked out and the time concrete had embedded the equipment, the cell and piezometer readings were taken every 10 minutes to make sure the pressure in each cell did not fall. In case there was a pressure drop, the hand pump was operated to top up the nominal pressure in the cell.

Monitored performance

Lateral wall movements

In general, all the inclinometers indicated increasing lateral movements as excavation progressed, the range of maximum deflections recorded being 20 mm to 27 mm for the wall, and 20 mm to 23 mm for the adjacent soil. In addition, the deflection profiles shown by the soil inclinometers I2 and I3 closely paralleled the movements of the adjacent wall inclinometers, I1 and I5 respectively.

The predicted and observed deflection profile at the location of wall inclinometer I5 are shown in Fig. 4. The four stages of excavation indicated in Fig. 4 relate to the stage of excavation to install the four slabs indicated in Fig. 2. Comparison of the observed and predicted profiles reveals that the analytical model used gave reasonable predic-

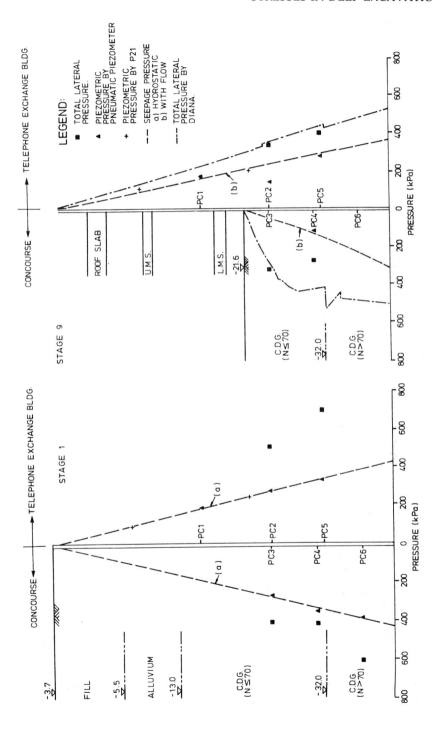

Fig. 5. Predicted and observed piezometric and total pressures — South wall

tions of diaphragm wall movements, particularly the envelopes of maximum deflections.

Earth pressures

The theoretical importance of in situ stress conditions in determining the response of an embedded diaphragm wall wished into place is indicated in Potts and Fourie (1984) and Fourie and Potts (1989), for excavations supported by propped and cantilever walls respectively. Such design methods which assume walls are wished into place show a strong relationship between existing horizontal stress and wall bending moments and displacements. High K_0 values result in very large bending moments and displacements. However, observations of movements during excavation in front of propped walls constructed in over-consolidated materials show a dependency on height, but do not show a distinct relationship with in situ stress. The average displacement observed when excavating in front of diaphragm walls installed in over-consolidated materials is of the order of 0.2% of the retained height (Clough and O'Rourke, 1990). In contrast, the analyses by Potts and Fourie (1984) indicate displacements of about 1% and 0.15% of the retained height for a propped retaining wall in a material with a K_0 of 2.0 and 0.5 respectively.

Powrie (1985) has indicated that due to installation of a diaphragm wall, low in situ values of K_0 (0.5) could be expected to rise to higher values of K within the range of 0.63 to 1.0, and high in situ values of K_0 (2.0) could be expected to fall to lower values of K within the range 1.0 to 1.38.

The observations at Bell Common (Tedd et al., 1984) confirm that K_0 did reduce immediately adjacent to the wall to values of K of the order of 0.6. A back analysis of the Bell Common wall (Higgins et al., 1989) has indicated that if the stresses caused by installation of the secant piles are taken into consideration, reasonable agreement with predicted and observed bending moments and displacements is achieved and that installation of the secant piles could cause a reduction in total horizontal stress. Using the observed piezometric pressures this indicates that K_0 could drop from about 1.5 to a K value of about 0.6. This is the average of an axisymmetric and plane strain analysis of the wall installation. A similar response has been confirmed by the observations of Lings et al. (1991), and Uff (1969) who determined that installation of a diaphragm wall in over-consolidated clay caused a significant reduction in K_0.

From this site, the observed total stresses and pore pressures prior to excavation indicate that a value of K approximately 0.75 has been obtained (Fig. 5), which is above that of about 0.4 assumed for essentially cohesionless materials.

The results of the total horizontal earth pressures measured by the pressure cells prior to excavation and during construction of the track slab are shown in Fig. 5. The pressure cell PC6 was damaged soon after installation and therefore no data are available except for the zero excavation case. The data from pressure cell PC1 are considered erroneous as the total pressure measured by PC1 invariably approximated water pressure. In addition, the predicted variations of total horizontal earth pressures by the diaphragm wall analysis using the DIANA programme are indicated on the plots for comparison. The plots also contain the predicted (from seepage analysis) and observed pore water pressure variations.

A plot of the ratio of horizontal effective pressure mobilised to vertical effective pressure in relation to relative displacement of the diaphragm wall (displacement divided by height of gauge above diaphragm wall toe level) is given in Fig. 3. The design curve has been derived from past back analysis and was used in the diaphragm wall analysis for CDG (N less ≤ 70). Despite the probable errors in the measurements, the assumed design curve compares favourably with the observed data on the passive side but the observed data indicates a stiffer and stronger response. The discrepancy between observations and the design curve on the active side has fewer consequences in design, particularly if water pressure dominates the load on the retained side.

The analysis by Potts and Burland (1983) of the Bell Common tunnel predicted that soil elements on the excavated side of the wall would follow a stress path which caused shear stress increases due to vertical effective stress decrease at approximately constant mean effective stress. This stress path is totally different to that imposed during conventional triaxial testing. Recent specialist testing in London Clay (Fourie and Potts, 1991) modelling such passive stress relief has indicated evidence

Table 2. *Maximum strut loads observed and predicted by computer during excavation*

Strut Number	Maximum Strut Load (tonne)	
	DIANA	Observed
S3 (Section A-A)	226	244
S3 (Section B-B)	230	324
S4 (Section A-A)	212	270
S4 (Section B-B)	240	305
S5 (Section A-A)	460	420
S6 (Section A-A)	400	410
S6 (Section B-B)	250	325
S7 (Section A-A)	390	300

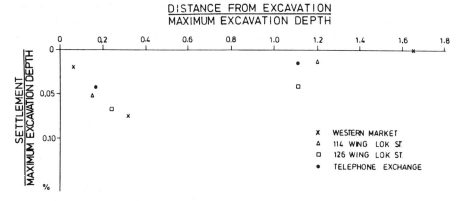

Fig. 6. Nondimensional settlement

of very high mobilised friction angle values (e.g. 35–40° compared with 25° measured in conventional triaxial tests).

The indication of high value of K mobilised on the excavated side of the wall could be related to such effects. Another possible explanation for high K values is that put forward by Howat (1985), who suggested a real cohesion in CDG *'of sufficient magnitude to be of significant use in engineering'*.

Strut loads during excavation

Maximum strut loads predicted by the computer analysis are compared with the maximum observed values in Table 2.

Building settlements due to diaphragm wall installation

The construction of diaphragm wall for Crossover Box was completed in week 52, 1982, when excavation and dewatering activities had not commenced. Therefore, settlements of buildings on shallow footings recorded up to this date are considered to be entirely due to diaphragm wall installation. Settlements of up to 30 mm were recorded, and reduced when stitch drilling was used to overcome boulders rather than relying solely on cheselling.

Building settlements resulting from dewatering and wall deflections

For two components of settlements of buildings occurred simultaneously as excavation progressed. In order to identify the relative magnitude of settlements resulting from the two causes, a theoretical estimation of dewatering settlements was made by using the empirical relationship $1/m_v = 0.8\,N\,\text{MPa}$ between soil compressibility and SPT results (Morton, Leonard & Cater, 1980), and the recorded pore pressure variations.

The settlements due to wall deflections were then computed for these buildings. The data have been plotted in the dimensionless form as shown in Fig. 6. The results indicate an approximate 1 to 1 relationship between maximum building settlement and wall deflection, and settlements of less than 0.1% of the depth of excavation, in line with Clough and O'Rourke (1990).

Conclusions
The results from an instrumented diaphragm wall panel at the Sheung Wan Crossover in Hong Kong indicate that K reached a value of about 0.75 after concreting which is an increase over the theoretical value for cohesionless material. They also indicated a very rapid mobilisation of significant passive resistance with displacement (a high initial stiffness) and point to the possibility of high mobilised values of friction angle in passive stress relief, or a cohesive component of strength for CDG.

References
BRITTO, A.M. AND GUNN, M.J. (1987). Critical State Soil Mechanics via Finite Elements. Ellis Horwood.

CANADIAN GEOTECHNICAL SOCIETY (1985). Canadian Foundation Engineering Manual, Canadian Geotechnical Society, Ottawa.

CLOUGH, G.W. AND O'ROURKE, T.D. (1990). Construction Induced Movements of Insitu Walls. Proc. Conf. Design and Performance of Earth Retaining Structures, Geot. SP25, ASCE, pp. 439–470.

DI BAGIO, E. AND ROTI, J.A. (1972). Earth Pressure Measurements on a Braced Slurry Trench Wall in Soft Clay. Proc. 5th ECSMFE, Madrid, Vol. 1, pp. 473–483.

FOURIE, A.B. AND POTTS, D.M. (1989). Comparison of Finite Element and Limiting Equilibrium Analyses for an Embedded Cantilever Retaining Wall. Geotechnique, Vol. XXXIX, No. 2, pp. 175–188.

FOURIE, A.B. AND POTTS, D.M. (1991). A Numerical and Experimental Study of London Clay Subjected to Passive Stress Relief. Geotechnique, Vol. XLI, No. 1, pp. 1–15.

FRASER, R.A. (1980). Geotechnical Design for Underground Construction. Proc. Conf. Mass Transportation in Asia, Hong Kong, J2/01–J2/13.

GUDEHUS, G. (1991). Displacements and Soil-Structure Interaction of Earth Retaining Structures. Proc. 10th ECSMFE, Florence, Vol. 3, pp. 1065–1074.

HIGGINS, K.G., POTTS, D.M. AND SYMONS, I.F. (1989). Comparison of Predicted and Measured Performance of the Retaining Walls of the Bell Common Tunnel. CR124, TRRL.

HOWAT, M.D. (1985). Completely Weathered Granite – Soil or Rock? Q. J. Eng. Geol., Vol. 18, pp. 199–206.

INOUE, Y. AND KATSURA, Y. (1977). Monitoring of Braced Diaphragm Walling Excavation of Fukoku Life Insurance Head Office in Soft Cohesive Soils, Tokyo. Proc. 9th ICSMFE, Tokyo, Case Studies of Excavation, pp. 303–310.

LINGS, M.L., NASH, D.F.J., NG, C.W.W. AND BOYCE, M.D. (1991). Observed Behaviour of a Deep Excavation in Gault Clay: a Preliminary Appraisal. Proc. 10th ECSMFE, Florence, Vol. 2, pp. 467–471.

LUMB, P. (1965). The Residual Soils of Hong Kong. Geotechnique, Vol. XV, No. 2, pp. 180–194.

MAUNSELL SYSTEMS CENTRE LTD (1982). DIANA — DIaphragm wall ANAlysis Users Manual Version 1.0. Revised 1991 (Version 2.0).

MORTON, K., LEONARD, M.S.M. AND CATER, R.W. (1980). Building Settlements Associated with the Construction of Two Stations of the Modified Initial System of the Mass Transit Railway, Hong Kong. 2nd Conf. on Ground Movements and Structures, Cardiff.

POTTS, D.M. AND BURLAND, J.B. (1983). A Numerical Investigation of the Retaining Walls of the Bell Common Tunnel, TRRL, SR783.

POTTS, D.M. AND FOURIE, A.B. (1984). The Behaviour of a Propped Retaining Wall: Results of a Numerical Experiment. Geotechnique, Vol. XXXIV, No. 3, pp. 383–404.

POTTS, D.M. AND FOURIE, A.B. (1985). The Effect of Wall Stiffness on the Behaviour of a Propped Retaining Wall. Geotechnique, Vol. XXXV, No. 3, pp. 347–352.

POWRIE, W. (1985). Discussion on Performance of Propped and Cantilevered Rigid Walls. Geotechnique, Vol. XXXV, No. 4, pp. 546–548.

TEDD, P., CHARD, B.M., CHARLES, J.A. AND SYMONS, I.F. (1984). Behaviour of a Propped Embedded Retaining Wall in Stiff Clay at Bell Common Tunnel. Geotechnique, Vol. XXXIV, No. 4, pp. 513–532.

UFF, J.F. (1969). In situ Measurements of Earth Pressure for a Quay Wall at Seaforth, Liverpool. Proc. Conf. In situ Investigations in Soils and Rocks, BGS, London, pp. 229–239.

Consolidation of an accreting clay layer: solutions via the wave equation

R.E. GIBSON, Golder Associates, Maidenhead

The dissipation of pore water pressure in a uniform layer of saturated sediment which is accumulating at a constant rate is considered. Problems of this type arise in geotechnical engineering and sedimentology. If the strains are sufficiently small the water pressure will be governed by an equation of heat-conduction type, the discretized form of which can readily be solved numerically. Exact analytical solutions have been obtained to only a few problems and these have required non-standard procedures to be developed, the implementation of which requires considerable ingenuity. A general method of attack from a new direction is developed here which depends on the governing equation being transformed into one of hyperbolic (wave equation) type, the general solution of which is straightforward. Two problems have been chosen to exemplify the method, and it is shown that the success of this depends on the possibility of solving a pair of difference–differential equations.

Introduction

This paper is concerned with the problem of estimating the progress of consolidation in a uniform layer of fully saturated fine-grained soil which is increasing in thickness at a constant rate due to the accumulation of fresh material on its surface. It will be assumed either that the deposition takes place over an area the dimensions of which are large compared with the thickness of the layer, or that the layer is circumscribed by an impervious barrier, so that to a good approximation the flow of pore water and the displacement can be considered to be one-dimensional. Further, if the strains are assumed to be small then the governing equation will be of parabolic type (Terzaghi, 1923).

Some consideration has been given to problems of this type by Terzaghi and Fröhlich (1936) and Olsson (1953) who considered the case of constant rate of deposition and developed some approximate solutions. Some exact solutions (exact, that is, within the framework of the usual assumptions made in Terzaghi's theory) were given by Gibson (1958) who formulated the problem as an integral equation the analytical solution of which turned out, quite fortuitously, to be feasible for the case of a constant rate of deposition. The disadvantage of this approach was its lack of generality. In the absence of a general method of attack (cf. Carslaw and Jaeger, 1959) a solution had to be constructed for each set of boundary conditions by a process of analytical trial and error: for any boundary conditions other than the simplest this could prove difficult (Gibson and Sills, 1990).

A more general approach was devised by Tait (1979) and others, the so-called pseudo-similarity solutions, which made use of, and expressed the solution in terms of, eigen-function expansions (Titchmarsh, 1946). Using this method he derived inter alia the solution to a cosmological problem originally considered by Gibson (1959) and demonstrated the equivalence of the results.

In the present paper the problem with the moving boundary is first transformed into one with fixed boundaries using a mapping employed by Olsson (1953), Gibson (1958) and Tait (1979). It is then shown how the parabolic equation governing consolidation can be transformed into an equation of hyperbolic type,[†] the general solution of which may be written down in terms of two arbitrary functions which must then be found from the boundary conditions. The success of the method depends mainly on the possibility of solving a pair of difference-differential equations for these two functions.

Formulation

We suppose that sedimentation takes place through still water of depth $H(t)$ to a basement stratum, the current thickness of the deposit being $h(t)$. Both $H(t)$ and $h(t)$ are specified functions of the elapsed time t since deposition commenced and the initial thickness $h(0)$ of the layer is taken to be zero. It is worth noting that if $H(t) = h(t)$ this would no longer be a 'sedimentation problem' but a 'layer building problem'.

By considering the net rate of accumulation of pore water in a unit volume of the fully saturated sediment at an elevation x above the base of the layer, the continuity equation is usually derived in the form

$$c_v \frac{\partial^2 u_w}{\partial x^2} + \frac{\partial \sigma'}{\partial t} = 0 \tag{1}$$

where $u_w(x, t)$ is the pore water pressure, $\sigma'(x, t)$ is the vertical effective stress and c_v is the coefficient of consolidation of the sediment. It will be assumed that the permeability and compressibility (and, therefore, c_v) of the sediment are constants and that the pore water and the sediment solids are incompressible: these are the assumptions usually adopted in the small-strain theory of one-dimensional consolidation.

[†]This depends, indirectly, on a connection noted by van der Pol (1950, p. 320).

The total vertical stress

$$\sigma = \sigma' + u_w = \gamma(h-x) + \gamma_w(H-h), \qquad (2)$$

where γ and γ_w are respectively the bulk unit weights of the sediment (assumed to be constant) and water. It follows from (1) and (2) that the pore water pressure in the sediment is governed by the equation

$$c_v \frac{\partial^2 u_w}{\partial x^2} = \frac{\partial u_w}{\partial t} - \gamma_w \frac{dH}{dt} - \gamma' \frac{dh}{dt} \qquad (3)$$

where $\gamma' = \gamma - \gamma_w$. On the moving boundary the water pressure is

$$u_w(h,t) = \gamma_w(H-h). \qquad (4)$$

The excess pore water pressure u will be governed by an equation similar to (3), its form being dependent on exactly how u is defined (Gibson, Schiffman and Whitman, 1989).

Considerable simplification results from working in terms of the vertical effective stress σ' as given in (2). The governing equation (3) then becomes simply

$$c_v \frac{\partial^2 \sigma'}{\partial x^2} = \frac{\partial \sigma'}{\partial t} \qquad (5)$$

while on the moving boundary

$$\sigma'(h,t) = 0. \qquad (6)$$

As mentioned above we take $h(t) = mt$ and restrict discussion to this case. It is also convenient now to introduce non-dimensional quantities and therefore c_v/m will be chosen as the unit of length, c_v/m^2 as the unit of time and $\gamma c_v/m$ as the unit of stress. The existing notation can be retained provided it is understood that non-dimensional quantities are being used. With this understanding (2) to (6) remain unchanged save that c_v must be replaced by unity, γ_w by γ_w/γ and γ' by γ'/γ.

General analysis

A solution to the equation

$$\frac{\partial^2 \sigma'}{\partial x^2} = \frac{\partial \sigma'}{\partial t} \qquad (7)$$

is required over the infinite triangular region $0 \le x \le t \le \infty$ of the (x,t) plane, which meets the boundary conditions on σ' along $x = t$ and $x = 0$.

If the new independent variables

$$\xi = xt^{-1} \quad \text{and} \quad \tau = t \qquad (8)$$

are introduced, (7) becomes

$$\frac{\partial^2 \sigma'}{\partial \xi^2} + \xi\tau \frac{\partial \sigma'}{\partial \xi} = \tau^2 \frac{\partial \sigma'}{\partial \tau} \tag{9}$$

and the domain of solution is now the infinite strip $0 \leq \xi \leq 1$, $\tau \geq 0$ of the (ξ, τ) plane. That is, the moving boundary problem governed by (7) has been replaced by a fixed boundary-value problem governed by the more complicated equation (9).

The dependent variable σ' (ξ, τ) is now replaced by its one-sided Laplace transform

$$\bar{\sigma}'(\xi, p) = \int_0^\infty \sigma'(\xi, \tau) e^{-p\tau} d\tau \tag{10}$$

in terms of which (9) becomes

$$\frac{\partial^2 \bar{\sigma}'}{\partial \xi^2} - \xi \frac{\partial^2 \bar{\sigma}'}{\partial \xi \partial p} = p \frac{\partial^2 \bar{\sigma}'}{\partial p^2} + 2 \frac{\partial \bar{\sigma}'}{\partial p} \tag{11}$$

when account has been taken of the condition

$$\sigma'(\xi, 0) = 0 \quad \text{on} \quad 0 \leq \xi \leq 1. \tag{12}$$

The next step, which is crucial to the method, is to change from the independent variables (ξ, p) to new variables (ζ, s), where

$$s^2 = p + \frac{1}{4}\xi^2 \tag{13a}$$

$$\zeta = \xi \tag{13b}$$

and from the dependent variable $\bar{\sigma}'$ to $\bar{w}$, where

$$\bar{w} = s\bar{\sigma}' \tag{14}$$

It can then be shown, after some algebra, that (11) becomes:

$$4 \frac{\partial^2 \bar{w}}{\partial \zeta^2} = \frac{\partial^2 \bar{w}}{\partial s^2} \tag{15}$$

which is a hyperbolic equation, the so-called wave equation, the general solution of which is (Titchmarsh, 1948):

$$\bar{w} = F_1(2s + \zeta) + F_2(2s - \zeta) \tag{16}$$

and so

$$\bar{\sigma}' = s^{-1} F_1(2s + \zeta) + s^{-1} F_2(2s - \zeta) \tag{17}$$

where F_1 and F_2 are arbitrary functions which must be found, for any particular problem, from the boundary conditions on the base ($\zeta = 0$) and surface ($\zeta = 1$) of the layer. Only linear boundary conditions are considered in the problems which will be examined presently, and then these two functions are determined from a pair of linear equations which may be of differential, difference or difference-differential type (see, for example, Pinney, 1958).

We turn now to the two problems which have been chosen to exemplify the method.

Problem I

This is concerned with consolidation of a layer of fine silty sand accumulating from a sediment settling through still water on to a much more permeable layer from which water is pumped at a constant rate. This problem arose in connection with the construction of off-shore sand islands the stability of which, at any stage, depends on the strength developed in the sediment and this is related in a simple way to the function $\sigma'(x, t)$. An analytical solution was obtained using the integral equation formulation referred to above (Gibson and Sills, 1990). It has been chosen because the solution is known, and also because a

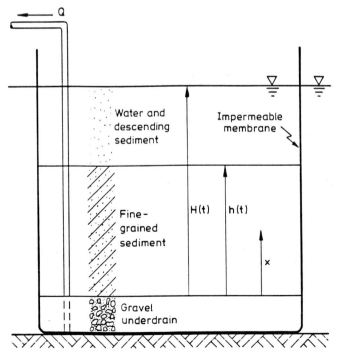

Fig. 1. *Sediment layer on gravel bed.*

comparison could then be made of the relative efficiency of the analyses employed.

Consider a sediment descending at a constant rate through still water of depth H on to the upper surface of a layer (permeability k) of current thickness h = mt, which rests on a gravel bed. The permeability of this gravel bed can be considered to be infinite and its compressibility zero. It is surrounded and underlain by an impervious membrane which rests, in turn, on a firm stratum (Fig. 1). The coordinate x is measured upward from the base of the layer.

In dimensionless variables the vertical effective stress (see eqn. (2)) is given by

$$\sigma' = \frac{\gamma'}{\gamma} t - x + \frac{\gamma_w}{\gamma} H - u_w \qquad (18)$$

and it is governed by eqn. (7). On the surface of the layer

$$\sigma'(t, t) = 0. \qquad (19)$$

Beneath the layer, in the underdrain, water is pumped from cased wells at a constant rate Q. Each well of cross-sectional area A_w serves a base area A_b of the layer. Since the whole system is surrounded by an impermeable membrane it follows that the sum of the rate of pumping Q from each well and the rate of filling of each well (both measured volumetrically) must equal the rate of flow of pore water from the base of the layer to the underdrain. These conditions of continuity lead to the following condition at the base (x = 0) of the layer:

$$\frac{\partial \sigma'}{\partial t} - \lambda \frac{\partial \sigma'}{\partial x} = \lambda \eta \qquad (20)$$

where

$$\eta = \frac{\gamma_w}{\gamma} \frac{Q}{kA_b} + \frac{\gamma'}{\gamma} \left[1 + \frac{mA_w}{kA_b} \right] \qquad (21)$$

$$\lambda = \frac{kA_b}{mA_w}. \qquad (22)$$

When these conditions are expressed in terms of $\bar{\sigma}'(\zeta, s)$ it is found that (19) becomes

$$\bar{\sigma}'(1, s) = 0 \qquad (23)$$

while (20) becomes

$$\lambda \frac{\partial \bar{\sigma}'}{\partial \zeta} + \frac{s}{2} \frac{\partial \bar{\sigma}'}{\partial s} + \bar{\sigma}' = -\lambda \eta s^{-4}. \qquad (24)$$

CONSOLIDATION OF AN ACCRETING CLAY LAYER

From the general solution (17), eqns. (23) and (24) can be expressed in the forms

$$F_1(y+1) + F_2(y-1) = 0, \qquad (25)$$

$$F_1(y) + F_2(y) + (y+2\lambda)F_1'(y) + (y-2\lambda)F_2'(y) = -16\lambda\eta y^{-3}, \qquad (26)$$

where, for convenience, y replaces 2s and the primes denote differentiation with respect to y. To solve these equations the functions

$$\hat{F}_{1,2}(z) = \frac{1}{2\pi i}\int_{Br} e^{yz} F_{1,2}(y)\,dy \qquad (27)$$

are introduced, where Br denotes the Bromwich contour $c - i\infty$ to $c + i\infty$ in the y-plane, in terms of which (25) and (26) become

$$e^{-z}\hat{F}_1 + e^{z}\hat{F}_2 = 0 \qquad (28)$$

$$(\hat{F}_1' + \hat{F}_2') + 2\lambda(\hat{F}_1 - \hat{F}_2) = 8\lambda\eta z \qquad (29)$$

from which it is found, after some algebra, that

$$\hat{F}_1 = 4\lambda\eta e^{z} \operatorname{cosech}^{(1+2\lambda)} z \int_0^z \nu \sinh^{2\lambda}\nu\,d\nu \qquad (30)$$

$$\hat{F}_2 = -4\lambda\eta e^{-z} \operatorname{cosech}^{(1+2\lambda)} z \int_0^z \nu \sinh^{2\lambda}\nu\,d\nu. \qquad (31)$$

The functions $F_{1,2}$ may now be recovered using the inverse of (27), namely

$$F_{1,2}(y) = \int_0^\infty \hat{F}_{1,2}(z) e^{-yz}\,dz, \qquad (32)$$

and reverting to the variables (ξ, p) from (13a, b) and using (17):

$$\bar{\sigma}'(\xi, p) = \lambda\eta \int_0^\infty \frac{\exp[-\beta\sqrt{p+\tfrac{1}{4}\xi^2}]}{\sqrt{p+\tfrac{1}{4}\xi^2}} \frac{\sinh[(1-\xi)\beta/2]}{\sinh(\beta/2)} W(\beta)\,d\beta \qquad (33)$$

where

$$W(\beta) = \operatorname{cosech}^{2\lambda}(\beta/2) \int_0^\beta \mu \sinh^{2\lambda}(\mu/2)\,d\mu. \qquad (34)$$

Finally, the vertical effective stress may be found from (33) using the inversion formula

$$\sigma'(x,t) = \frac{1}{2\pi i} \int_{Br} e^{pt} \bar{\sigma}'(\xi, p) dp$$

and noting that $\xi = x/t$ it is found that in dimensionless variables:

$$\sigma'(x,t) = \frac{\lambda\eta}{\sqrt{\pi t}} \exp(-x^2/4t) \times \int_0^\infty \exp(-\beta^2/4t) \frac{\sinh\left[\left(1-\frac{x}{t}\right)\frac{\beta}{2}\right]}{\sinh(\beta/2)} W(\beta) d\beta \tag{35}$$

and the pore water pressure then follows by substituting (35) in (18).

This recovers the known solution to the problem (Gibson and Sills, 1990, eqn. (36)).

Problem II

In this second problem a clay layer increasing in thickness at a constant rate $\underline{m}$ rests on a rigid semi-permeable base (e.g. concrete) which is underlain by gravel which is free-draining (Fig. 2). Circumstances of this kind might, for example, be encountered in a sludge or tailings containment area, where a forecast of the rate of consolidation might be important.

The case is one in which $h = H$, so that from (18):

$$\sigma' = t - x - u_w \tag{36}$$

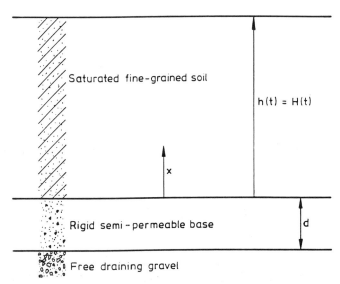

Fig. 2. *Clay layer on rigid semi-permeable base.*

CONSOLIDATION OF AN ACCRETING CLAY LAYER

and σ' is again governed by (7). On the surface of the layer we assume that the effective stress in the placed clay is negligible and that pore water flows out of this surface; but if not, that sufficient rainfall is available to prevent the air–water interface retreating or capillary menisci forming. In these circumstances it follows that (19) is valid.

To find the boundary condition at $x = 0$, the flow of water through the concrete must be considered. The discharge velocity from the clay is, in dimensional variables:

$$-\frac{k_s}{\gamma_w}\left[\frac{\partial u_w}{\partial x} + \gamma_w\right] \quad (37)$$

where k_s is its permeability, while that through the concrete is

$$-\frac{k_c}{\gamma_w}\left[\frac{u_w}{d}(0, t) + \gamma_w\right] \quad (38)$$

where k_c is the permeability of the thickness d of the concrete, the variation of water pressure through the concrete being linear since it is rigid. Continuity of flow at $x = 0$ requires the equality of (37) and (38), from which, using (36), it can be shown that, in dimensionless variables:

$$\frac{\partial \sigma'}{\partial x} - \alpha\sigma' = -\delta - \alpha t \quad (39)$$

$$\alpha = k_c/(k_s d) \quad (40a)$$

and

$$\delta = \left[\frac{\gamma_w k_c}{\gamma k_s} + \frac{\gamma'}{\gamma}\right]. \quad (40b)$$

In this problem, as in the last, the solution (17) requires F_1 and F_2 to be found from the boundary condition equations of which (19) is still valid, while (39) becomes

$$(y + 2\alpha)F_1' - (y - 2\alpha)F_2' - 2\alpha y^{-1}(F_1 + F_2) = -8\delta y^{-2} - 64\alpha y^{-4} \quad (41)$$

The functions $\hat{F}_1$ and $\hat{F}_2$ defined in (27) are given by (28) and from (41) by

$$-(z\hat{F}_1)' + (z\hat{F}_2)' - 2\alpha z(\hat{F}_1 + \hat{F}_2) - 2\alpha \int_0^z (\hat{F}_1 + \hat{F}_2)\,dz = -2\delta z - \frac{8}{3}\alpha z^3 \quad (42)$$

which is a differential equation of the second order. The details of the solution of (28) and (42) will not be given here, but it is found that

$$\hat{F}_1 = 2e^z \operatorname{sech}^{(1+2\alpha)} z \int_0^z (\delta + 2\alpha\nu^2) \cosh^{2\alpha}\nu \, d\nu \quad (43)$$

$$\hat{F}_2 = -2e^{-z} \operatorname{sech}^{(1+2\alpha)} z \int_0^z (\delta + 2\alpha\nu^2) \cosh^{2\alpha}\nu \, d\nu \quad (44)$$

from which F_1 and F_2 can be found using (32), and then from (13a,b), (17) and the inversion integral from (ξ, p) to the (x, t)-plane it is found that

$$\sigma'(x, t) = \frac{1}{\sqrt{\pi t}} \exp(-x^2/4t) \times \int_0^\infty \exp(-\beta^2/4t) \frac{\sinh\left[\left(1 - \frac{x}{t}\right)\frac{\beta}{2}\right]}{\sinh(\beta/2)} V(\beta) \, d\beta$$

(45)

where

$$V(\beta) = \operatorname{sech}^{2\alpha}(\beta/2) \int_0^\beta (\delta + \tfrac{1}{2}\alpha\mu^2) \cosh^{2\alpha}(\mu/2) \, d\mu \quad (46)$$

and the pore water pressure, if this is required, can then be found by setting (45) in (36). Dimensional expressions can be recovered from (36), (45) and (46) by replacing σ' by $m\sigma'/\gamma c_v$, x by mx/c_v, t by m^2t/c_v, and α by $c_v\alpha/m$.

Finally, it is perhaps worth noting some earlier results which may be derived from this example. For instance, in the limit $d = 0$ ($\alpha = \infty$) the case of free drainage from beneath an accreting embankment (Gibson, 1958, p. 178) is recovered. When $k_c = 0$ ($\alpha = 0$) the solution for an impermeable base is found. Similar reductions can be made from Problem I.

Concluding remarks

The purpose of this paper has been to present a general procedure for obtaining closed form solutions to a class of linear problems in consolidation theory which involve a specified moving boundary. Unlike the integral equation approach it can be applied directly and dispenses, in particular, with the need to construct the integral equation in the first place. This analysis would supplement a direct numerical attack on problems of this kind (Abbott, 1967) and the two examples given would provide benchmarks against which numerical solutions could be compared and which can be used directly in other contexts.

References

ABBOTT, M.B. (1967). A note on one-dimensional consolidation of a clay layer that increases in thickness with time. LGM Mededelingen (Delft), XI, p. 4.

CARSLAW, H.S. AND JAEGER, J.C. (1959). Conduction of heat in solids. Oxford University Press. (2nd Ed.)

GIBSON, R.E. (1958). The progress of consolidation in a clay layer increasing in thickness with time. Géotechnique, Vol. VIII, pp. 171–182.

GIBSON, R.E. (1959). A heat conduction problem involving a specified moving boundary. Q. appl. Math., Vol. 16, pp. 426–430.

GIBSON, R.E., SCHIFFMAN, R.L. AND WHITMAN, R.V. (1989). On two definitions of excess pore water pressure. Géotechnique, Vol. XXXIX, pp. 169–171.

GIBSON, R.E. AND SILLS, G.C. (1990). Consolidation due to underpumping an accreting layer of sediment. Q. J. Mech. appl. Math., Vol. 43, No. 3, pp. 335–346.

OLSSON, R.G. (1953). Approximate solution of the progress of consolidation in a sediment. Proc. 3rd Int. Conf. Soil Mech., Vol. 1, pp. 38–42.

PINNEY, E. (1958). Ordinary difference-differential equations. University of California Press.

TAIT, R.J. (1979). Additional pseudo-similarity solutions of the heat equation in the presence of moving boundaries. Q. appl. Math., Vol. 37, pp. 313–324.

TERZAGHI, K. (1923). Die Berechnung der Durchlässigkeitsziffer des Tones aus dem Verlauf der hydrodynamischen Spannungserscheinungen. Stiz. Akad. Wiss. math.-naturw., Abt. IIa, 132 Bd., 3.u.4.H.

TERZAGHI, K. AND FRÖHLICH, O.K. (1936). Theorie der Setzung von Tonschichten. Deuticke, Leipzig.

TITCHMARSH, E.C. (1946). Eigenfunction expansions associated with second-order differential equations. Oxford University Press.

TITCHMARSH, E.C. (1948). Introduction to the theory of Fourier integrals. Oxford University Press (2nd Ed.).

VAN DER POL, B. AND BREMMER, H. (1950). Operational Calculus. University Press, Cambridge.

The prediction of surface settlement profiles due to tunnelling

M.J. GUNN, Department of Civil Engineering, University of Surrey

Recent work has shown the importance of modelling the small strain stiffness of soil when predicting ground movements adjacent to excavations. This paper describes a simple constitutive relation for the undrained behaviour of overconsolidated clays. A power law expression determines the non-linear elastic stiffness of the soil at small strains. When the undrained shear strength is mobilised the soil yields according to the Tresca criterion. This model has been implemented in the geotechnical finite element program CRISP. The paper describes the application of this model to the prediction of ground surface movements caused by tunnelling. Comparisons with analyses which do not model small strain stiffness demonstrate a dramatic improvement in the predictions. Comparisons with empirical relations based on field data indicate that although the 'ground loss' prediction is realistic, the overall settlement profile predicted is still too wide, the maximum settlement being about half that observed in practice.

Introduction

During the period 1975 to 1982 a series of contracts from the Transport and Road Research Laboratory (TRRL) sponsored research at Cambridge into various aspects of tunnelling. Part of my brief (initially under the technical supervision of Peter Wroth) was to develop and improve the finite element software that the group was using at that time. Work by myself and others led to the CRISP program (Britto and Gunn, 1987) which was first officially released into the wider world in 1982. Peter Wroth always had a very clear vision of how numerical techniques could benefit the geotechnical profession, and he gave their development his support. More importantly, however, he insisted that the techniques should be used with intellectual rigour and discrimination.

A major concern in the construction of any tunnel in an urban environment is the ground movement that inevitably takes place. Engineers responsible for the design and construction of tunnels must have some technique for estimating the ground movements so that they can demonstrate that neighbouring buildings will not be subject to excessive differential settlements. The finite element technique seems to be ideally suited to make these predictions, but nearly all calculations carried out using CRISP (and its precursors) on the TRRL research contracts were disappointing. The finite element calculations predicted a distribution of surface settlements that was much shallower and wider than those seen in model tests or real tunnelling. The reason for the poor quality of these predictions was quickly identified as being the elastic part of any of the constitutive models that were used to represent the stress–strain behaviour of the soil.

On the other hand, significant advances were made in devising new stability calculations for tunnels (Davis et al., 1980). It was also shown that although finite elements predicted a poor distribution of settlements, the prediction of 'ground loss' (defined below) was quite good when using the modified Cam-clay soil model. These findings were combined with the empirically observed shape of settlement profiles above tunnels to produce a practical method of making predictions of ground movements (Mair et al., 1981).

Although the method just mentioned is soundly based, it has a major deficiency in that it only produces estimates of settlements for a 'green field' site. In practice we often need to estimate ground movements adjacent to structures on piles or rafts where the influence of the stiffness of piles or raft can be expected to have an important influence. This can only be done through the use of a technique such as finite elements with the use of an adequate constitutive model for the soil.

This paper describes how improved estimates of the ground movements due to tunnelling can be obtained through the use of a constitutive model based on measurement of the small strain stiffness of the soil.

The 'error curve' method for settlements

According to Peck (1969) the distribution of surface settlements due to a single circular tunnel is given by the equation

$$s = s_{max} \exp(-x^2/2i^2) \tag{1}$$

where s is the vertical settlement of a point which is a distance x from the vertical plane containing the tunnel axis (see Fig. 1). s_{max} is the settlement of the point directly above the tunnel and i is a parameter which defines the width of the settlement profile. When $x = i$ the slope of the ground surface is at its maximum and $s = 0.61 s_{max}$ and when $x > 3i$, $s < 0.012 s_{max}$ and the settlement can be disregarded. This is the equation of a normal distribution, and is called the 'error curve' because it corresponds to the expected distribution of measurements of a physical quantity when measurement errors are random. Remarkably (perhaps) Peck found that eqn. (1) fitted all the field records that he was

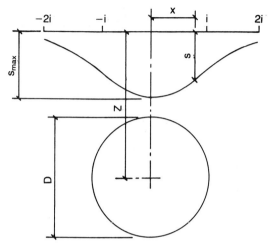

Fig. 1. Surface settlements represented by the error curve (after Peck (1969))

able to collect for tunnels at various depths in both clays and granular deposits.

If the settlements occur with no change in volume of the soil (i.e. the behaviour is undrained) then the volume of soil between the settlement profile and the original ground surface,

$$V_s = \sqrt{(2\pi)} \, i \, s_{max} \qquad (2)$$

represents an extra volume of soil (above the finished volume of the tunnel) that the tunnellers must excavate. This is termed the 'ground loss' and is normally expressed as a percentage of the volume of the tunnel. Peck (1969) also produced a dimensionless plot of the observed widths of settlement profile, compared to the depth of the tunnel axis for different types of soil (Fig. 2). A relationship that gives a reasonable fit to Peck's data for tunnels in clays was given by Schmidt (1969):

$$(2i/D) = (Z/D)^{0.8} \qquad (3)$$

where D is the diameter of the tunnel and Z is the depth of the tunnel axis from the ground surface.

These observations and relationships can be used to estimate settlements for a real tunnel as follows:

(a) the engineer estimates a figure for ground loss on the basis of experience with similar tunnelling techniques in similar soils

(b) a value of i/D is assumed, based on Peck's chart (Fig. 2) or eqn. (3) or a similar relationship

(c) eqn. (2) is now used to find s_{max} and hence eqn. (1) can now be used to predict the surface settlement at any point.

SURFACE SETTLEMENT PROFILES

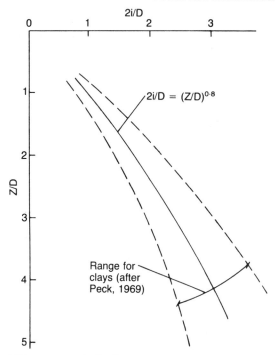

Fig. 2. *Profile width versus axis depth*

Mair et al. (1981) replace step (*a*) by finding the ground loss from a plot of ground loss versus load factor. Load factor is defined as N/N_{TC} where N is the tunnel stability number and N_{TC} corresponds to the stability number for collapse (Davis et al., 1980). When the tunnel is unsupported $N = \gamma Z/c_u$ where γ is the bulk unit weight of the soil and c_u is the undrained shear strength. The plot of ground loss versus load factor could be obtained from a finite element analysis, but in practice would probably be obtained from a published plot such as that in the original paper. Although this approach replaces guessing by a soundly based engineering calculation there are a couple of problems. Firstly, there is not a great deal of experimental data to support any relationship for ground loss versus load factor for soils apart from kaolin. Secondly, the calculation of a load factor is not straightforward as it must be based on an actual distribution of c_u with depth, while the original stability calculations (Davis et al., 1980) were for constant c_u. Unfortunately the ground loss is quite sensitive to the value of load factor, particularly as the load factor approaches 1.

In the author's experience, engineers in the UK make reference to the Mair et al. (1981) paper but in practice use something closer to the original step (*a*) above. When calculating subsurface movements, it is often assumed that the above relationships can be used. The soil above the level where the settlements are calculated is ignored, and movements are calculated on the basis of a reduced depth for the tunnel axis. There is very little published data to support or contradict this procedure.

A small strain model for soil

Simpson et al. (1979) put forward an elasto-plastic stress–strain model for soil which included two regimes of elasticity. Essentially bilinear elasticity was assumed with initial elastic stiffnesses ten times those normally measured in laboratory tests. After a threshold strain (typically 0.02%) the elastic stiffnesses were reduced by a factor of ten and these properties were then operative until plastic yielding occurred. The threshold strain was represented by a kinematic yield surface in strain space and so a change in the direction of loading would lead to a certain amount of straining with the initial higher moduli. The use of this model led to greatly improved predictions of ground movements next to excavations, but (at the time the proposal was made) the model seemed rather complex and many of the assumptions seemed to be rather speculative.

Jardine et al. (1984) published laboratory measurements of soil stiffnesses using local strain measurement devices that could resolve mean axial strains as low as 0.002%. Their data show undrained 'elastic' moduli continually reducing from strains as (typically) low as 0.005%

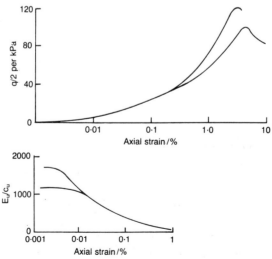

Fig. 3. Jardine et al.'s (1984) data for the low strain stiffness of London clay

SURFACE SETTLEMENT PROFILES

until failure is approached. Figure 3 summarises their data for two intact samples of London Clay taken from depths of 5.3 m (LC1) and 7.5 m (LC2). The measurements of Jardine et al. (1984) generally support the basic approach of Simpson et al. (1979), but give a rather different detailed picture of soil stiffnesses at low strains.

In the work reported here it was decided to use the simplest possible approach to representing soil stiffness and yielding consistent with the data presented by Jardine et al. (1984). The undrained non-linear 'elastic' response of the soil is given by:

$$q = a\varepsilon^n \qquad (4)$$

where q is the deviator stress, ε is the deviator strain and a and n are soil parameters obtained as described below. This power law expression can be manipulated in a very simple way to give expressions for the

$$\sec E_u = a\varepsilon^{n-1} \qquad (5)$$

and the

$$\tan E_u = na\varepsilon^{n-1} \qquad (6)$$

The parameters a and n for the model are recovered from the secant Young's modulus measured at two strain levels in an undrained triaxial test:

$$E_{u1} = a\varepsilon_1^{n-1}$$
$$E_{u2} = a\varepsilon_2^{n-1}$$

(obtained by substituting directly into eqn. (5)). Then:

$$n = 1 + \log(E_{u1}/E_{u2})/\log(\varepsilon_1/\varepsilon_2)$$

and

$$a = E_{u1}\varepsilon_1^{1-n}$$

For example if we adopt values from Fig. 3 of $c_u = 100$ kPa,

$$\sec E_u/c_u = 1000 \text{ for a strain of } 0.01\%$$

and

$$\sec E_u/c_u = 400 \text{ for a strain of } 0.1\%,$$

then the above equations give values of a and n very close to 2500 kPa and 0.6 respectively.

The model adopted here assumes that the soil is elastic, perfectly plastic, yielding according to the Tresca yield condition. The value of undrained shear strength can vary with depth and (consistent with the presentation of data in Jardine et al., 1984) the parameter a varies in proportion to c_u. There is a value of strain (ε_L) below which the stiffness is taken to be constant (and equal to the secant stiffness at ε_L), so the actual stress strain curve and variation of modulus with strain is as

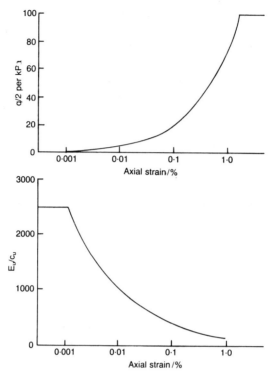

Fig. 4. *Jardine et al.'s (1984) data fitted by* $q = a\varepsilon^n$

shown in Fig. 4 (with a value of $\varepsilon_L = 0.001\%$ assumed). The assumptions made in deriving the model, and the details of its implementation into a version of CRISP mean that there are some important limitations on its use. Firstly, the model has been developed on the basis of data from undrained tests, and so it should just be used to predict undrained deformations. Secondly, the model is designed for modelling situations where the loading is monotonic. Contours of constant elastic shear modulus are circles in the π-plane of principal stress space, centred upon the point corresponding to the stresses at the start of the analysis. In other words, the stiffness in unloading is just the same as the stiffness in loading at the equivalent strain level.

Before eqn. (4) was adopted, previously proposed non-linear elastic relationships for soil due to Naylor (Naylor et al., 1981) and Duncan and Chang (1969) were considered and rejected. Both of these models have two parameters describing a non-linear stress–strain curve and the parameters can be obtained in a similar fashion to the procedure described above, fitting the non-linear curves at low strain values. If this is done, for example with the data quoted above, both of these models predict that the soil fails at a value of q about one third of the value actually seen. In practice one would use these models with the maximum value of q correctly represented, but this will be at the cost of a poor representation of stiffness at some values of low strains. In contrast, eqn. (4) gives a reasonable fit for stiffnesses until plastic yielding starts (at a strain of about 1.5%).

In fact the model described here has some similarity to that described by Jardine et al. (1986). The main difference seems to be that their equation matches the data more precisely at the cost of some extra complexity in the form of the equation and the derivation of material parameters.

Hillier (1989) made use of the soil model reported here to analyse the data of plate loading tests. He discovered from an examination of the load deformation response measured in a plate loading test that one can easily determine a value of n which (when inserted in a finite element computation) will reproduce the test data. Thus in principle one could determine the value of n from an in situ test.

Tunnel analyses

The results of twelve separate analyses, identified by the letters A to L, are reported here. For each analysis the dimensionless ratio Z/D and the properties assumed are given in Table 1. In each case the diameter of the tunnel was taken as 5 m and the mesh extended a distance 3Z to the

Table 1. Summary of analyses performed

Identifying letter	$\dfrac{Z}{D}$	Properties
A	2	Homogeneous elastic E_u = 100 MPa
B	2	Homogeneous EPP, E_u as A, c_u = 100 kPa
C	2	Homogeneous NLEPP, c_u as B, a = 2500 kPa, n = 0.6
D	4	Same properties as A
E	4	Same properties as B
F	4	Same properties as C
G	2	Elastic, E_u = 40 + 10y MPa
H	2	EPP, E_u as G, c_u = 40 + 10y kPa
I	2	NLEPP, c_u as H, a = 2500 kPa (for y = 6), n = 0.6
J	4	Same properties as G
K	4	Same properties as H
L	4	Same properties as I

Note: EPP = elastic, perfectly plastic with Tresca yield criterion; NLEPP = non-linear elastic perfectly plastic; y is equal to the depth below the ground surface; undrained Poisson's ratio = 0.49 for all analyses

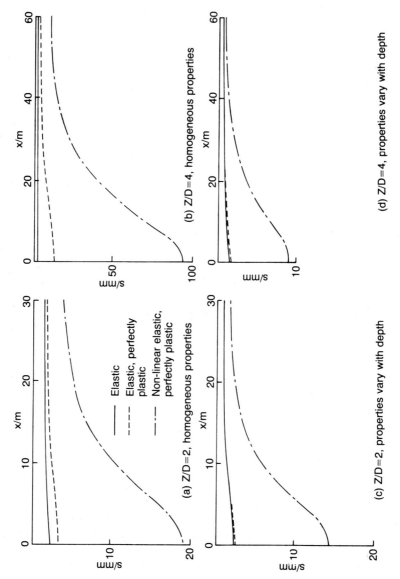

Fig. 5. *Surface settlement profiles*

SURFACE SETTLEMENT PROFILES

right of the tunnel axis and a distance 2Z below it. Symmetry was assumed and the vertical boundaries of the mesh were free of vertical tractions (i.e. on rollers), while the bottom boundary was assumed to be pinned. The in situ total stresses were $\sigma_x = \sigma_y = 20y$ where y is the depth below the ground surface. The excavation of the tunnels is modelled by removing the elements originally occupying the location of the tunnel.

The material properties were chosen to be broadly representative of London clay and at the same time allow a comparison of the various assumptions and their influence on the computed results. Analyses G to L where the properties vary with depth are regarded as more realistic, but comparison with analyses A to F (where homogeneous properties were assumed) allow an assessment of the effect of inhomogeneity. Only one aspect of the computed results is considered here: the shape of the surface settlement profile.

Figures 5(a) and 5(b) compare the surface settlement profiles for the analyses with homogeneous properties where $Z/D = 2$ and $Z/D = 4$ respectively. The first point to note is that the elastic settlement profiles are wide and shallow with the settlements at the edge of the mesh being approximately 70% in each case of the settlement above the tunnel axis. The addition of plastic yielding increases settlements by 50% above the tunnel axis when $Z/D = 2$ and 500% when $Z/D = 4$. In both cases there is a proportionally smaller increase in settlements at the edge of the mesh. The different effect of plasticity on the two depths of tunnel can be explained by noting that the load factor for $Z/D = 2$ is approximately 0.63 while for $Z/D = 4$ it is approximately 0.89 (i.e. the latter tunnel is much closer to failure). Finally we can see that the addition of non-linear elastic properties apparently leads to a significant improvement in the shapes of the settlement profiles in each case.

Figures 5(c) ($Z/D = 2$) and 5(d) ($Z/D = 4$) show the settlement profiles computed in the analyses with properties varying with depth. Broadly, the picture here is similar to the analyses with homogeneous properties except that there is hardly any difference made by the addition of plastic yielding. It is worth noting here that the load factors are approximately 0.69 ($Z/D = 2$) and 0.63 ($Z/D = 4$). The addition of inhomogeneity has led to some improvement in the elastic (and elastic, perfectly plastic analyses) with the settlements at the edge of the mesh now being 40% of the settlement above the tunnel axis. Of course, these results are still unrealistic, the predicted settlement profiles being much too wide.

To compare the four settlement profiles from non-linear elastic, perfectly plastic analyses with Peck's error curve the following procedure is followed. For each analysis the ground loss is calculated by integrating the area above the computed settlement profile. Then eqns. (3) and (2) can be used, as described above, to obtain four values of i and

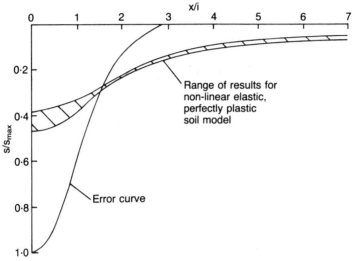

Fig. 6. Equal ground loss settlement profiles normalised by s_{max}

s_{max}. These values are used to normalise the computed settlement profiles which are plotted together with Peck's error curve in Fig. 6. The above procedure has produced settlement troughs with equal volumes and so Fig. 6 allows one to make a direct comparison of the shapes of the settlement profiles. It is clear that the settlement profiles are approximately twice as wide (defining the width by the point at which the settlement is 61% of that above the tunnel axis) as would be expected from the existing empirical approach.

Figure 7 superimposes the data for ground losses and load factors calculated here with the data originally presented by Mair et al. (1981). The new data is fairly consistent with the original data but of course we should bear in mind that the original data refers to tests on model tunnels in kaolin and the material properties used here describe London clay.

Discussion and conclusion

It is interesting to note that the ground loss figures for the analyses with properties varying with depth are quite close to those actually seen for tunnels in London clay (analysis H, Z/D = 2, V_s = 1.6%; analysis L, Z/D = 4, V_s = 2.0%). Although, as demonstrated above, the settlement profiles are too wide, the actual settlements are relatively small as one moves away from the tunnel (say 10% of the maximum expected settlement). The present results suggest that an analysis with the non-linear elastic, perfectly plastic model would lead to a sensible figure for ground loss and maximum settlements that are about 45% of those

SURFACE SETTLEMENT PROFILES

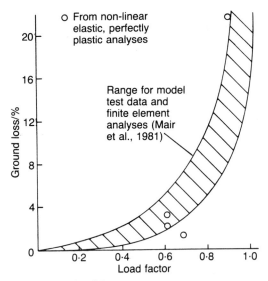

Fig. 7. *Ground loss versus load factor*

that could be expected in practice. This situation is not entirely satisfactory, but represents a definite improvement over what can be achieved using existing models which incorporate elastic moduli that are not strain dependent.

When considering the practical implications of these results, it is important to remember the drastic nature of some of the idealisations that have been made. For example, during excavation most tunnels are supported quite close to the face. The present analyses have assumed that the tunnel is infinitely long and unsupported. Thus real tunnels should have lower load factors than those analysed here and cause smaller ground movements. Three dimensional analyses are required to examine the importance of this point. Secondly, the soil model used is rather simple and perhaps should be modified to allow for plastic hardening/softening, different properties for compression/extension stress paths and anisotropic elastic moduli. It has been recently suggested (Lee and Rowe, 1989) that the use of anisotropic elastic properties should improve the prediction of ground deformations above tunnels. In the author's experience this is not the case when the anisotropic elastic moduli appropriate to London clay are used. The advantage of the model used here is its simplicity and the fact that it is based on directly measured soil properties.

Acknowledgements

The implementation of the soil model described here was originally done in a version of CRISP in 1986 at the request of David Yong of Mott MacDonald. To avoid confusion I should emphasise that it has never been available in any of the standard versions of CRISP issued by Cambridge University. Jason Kenward first applied the model to tunnelling (and demonstrated its potential) in his final year undergraduate project.

References

BRITTO, A.M. AND GUNN, M.J. (1987). Critical state soil mechanics via finite elements, Ellis Horwood, Chichester.

DAVIS, E.H., GUNN, M.J., MAIR, R.J. AND SENIVIRATNE, H.N. (1980). The stability of shallow tunnels and underground openings in cohesive material, Geotechnique, Vol. XXX, No. 4, pp. 397–416.

DUNCAN, J.M. AND CHANG, C.Y. (1969). Non-linear analysis of stress and strain in soils, J. Soil Mech. Found. Div., Proc. ASCE, Vol. 96, pp. 1629–1653.

HILLIER, R.P. (1989). Personal communication.

JARDINE, R.J., POTTS, D.M., FOURIE, A.B. AND BURLAND, J.B. (1986). Studies of the influence of non-linear stress-strain characteristics in soil-structure interaction, Geotechnique, Vol. XXXVI, No. 3, pp. 377–396.

JARDINE, R.J., SYMES, M.J. AND BURLAND, J.B. (1984). The measurement of soil stiffness in the triaxial apparatus, Geotechnique, Vol. XXXIV, No. 3, pp. 323–340.

LEE, K.M. AND ROWE, R.K. (1989). Deformations caused by surface loading and tunnelling: the role of elastic anisotropy, Geotechnique, Vol. XXXIX, No. 1, pp. 125–140.

MAIR, R.J., GUNN, M.J. AND O'REILLY, M.P. (1981). Ground movements around shallow tunnels in soft clay, Proc. 10th Int. Conf. Soil Mech. and Found. Eng., Vol. 2, pp. 323–328.

NAYLOR, D.J., PANDE, G.N., SIMPSON, B. AND TABB, R. (1981). Finite elements in geotechnical engineering, Pineridge Press, Swansea.

PECK, R.B. (1969). Deep excavations and tunnelling in soft ground, Proc. 7th Int. Conf. Soil Mech. and Found. Eng., State of the Art Volume, pp. 226–290.

SCHMIDT, B. (1969). Settlements and ground movements associated with tunnelling in soil, PhD Thesis, University of Illinois.

SIMPSON, B., O'RIORDAN, N.J. AND CROFT, D.D. (1979). A computer model for the analysis of ground movements in London clay, Geotechnique, Vol. XXIX, No. 2, pp. 149–175.

Predicted and measured tunnel distortions associated with construction of Waterloo International Terminal

D.W. HIGHT, K.G. HIGGINS, R.J. JARDINE, D.M. POTTS, Geotechnical Consulting Group, London, and A.R. PICKLES, E.K. DE MOOR, Z.M. NYIRENDA, Sir Alexander Gibb and Partners, Reading

Predictions have been made of the distortions and changes in clearance in the Bakerloo Line tunnels where they pass beneath the excavation for the Waterloo International Terminal. To make the predictions a comprehensive site investigation was carried out at the site and aimed, inter alia, at determining the in situ stress state and non-linear small strain stiffness of the soils. The effect on the in situ stress state of the construction history at Waterloo, including driving of the tunnels, and of groundwater flow towards the tunnels was taken into account in the modelling. The predicted distortions and the measurements made during construction are compared.

Introduction

An International Terminal for Channel Tunnel trains has been constructed at British Rail's Waterloo Station in London. The presence of London Underground Limited's (LUL's) structures beneath the site of the Waterloo International Terminal (WIT) has constrained both the foundation design[†] and construction activities. It has been necessary to ensure that distortion of the underground structures did not exceed prescribed limits, both during construction and in the long term. The limits were set to prevent damage to tunnel linings and to maintain clearances in station and running tunnels.

As essential input to the design process was, therefore, the prediction of ground movements and tunnel distortions caused by demolition of existing structures on the site, excavation to form the basement of WIT

[†]Sir Alexander Gibb and Partners Ltd were responsible for the design of the foundations and basement of WIT. The Geotechnical Consulting Group acted as sub-consultants on the analysis of tunnel movements.

Predictive soil mechanics. Thomas Telford, London, 1993

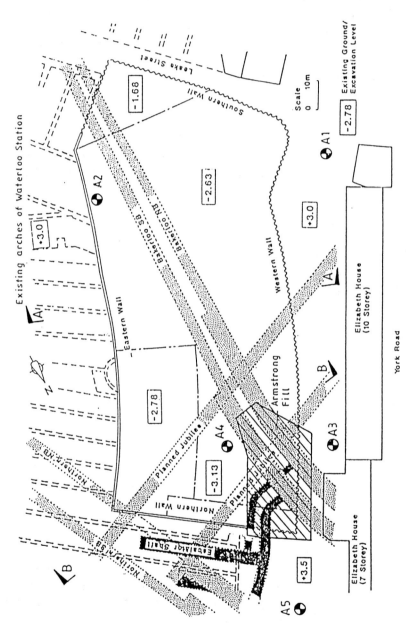

Fig. 1. General layout of Waterloo International Terminal

and loading from the new terminal. The predictions have been made using the finite element program ICFEP and soil models which take into account the non-linear stress–strain characteristics of the Thames Gravel, London Clay and Woolwich and Reading Beds at the site.

This paper describes the investigations made to determine the requisite soil parameters and compares the predictions with field measurements of tunnel movements.

Description of the site

The site for the Waterloo International Terminal is at the western side of the existing Waterloo Station in London (Fig. 1). The site is bounded to the south by Leake Street, to the north by the approach road to Waterloo Station and to the west by York Road. To the east, platforms and track to Waterloo Station will be retained on existing brick arches.

The WIT site was previously occupied by sidings and platforms, mostly carried on brick arches. An area of the sidings next to the Armstrong Lift was carried on 7 m of fill (the Armstrong Fill). Passing beneath the site are station and running tunnels for LUL's Bakerloo and Northern Lines. British Rail's Waterloo and City Line lies immediately to the north east of the site. The alignment of the proposed tunnels for the Jubilee Line extension also passes beneath the site.

Description of WIT

WIT has been constructed as an International Terminal for Channel Tunnel trains for opening in July 1993. It is a reinforced concrete structure comprising a single level of basement, an arrivals floor, a departures floor and a steel and glass roof (Fig. 2). It is founded on a reinforced concrete base slab, approximately 160 m by 60 m. The cross-section through WIT in Fig. 2 shows the Bakerloo Line tunnels approximately 5 m below the underside of the base slab.

Demolition of the arches and excavation for the basement resulted in an average stress reduction of 180 kN/m^2 and a maximum reduction of 260 kN/m^2 below the Armstrong Fill. The final reduction in stress above the tunnels is approximately 100 kN/m^2.

Site investigation

Reliable predictions of ground movement must take account of the dependence of soil stiffness on strain level, effective stress level and direction of shearing. A major effort was made in the site investigation, therefore, to establish both in situ stresses and the non-linear stress–strain properties of the soils.

The investigation was concentrated at five locations, shown as A1 to

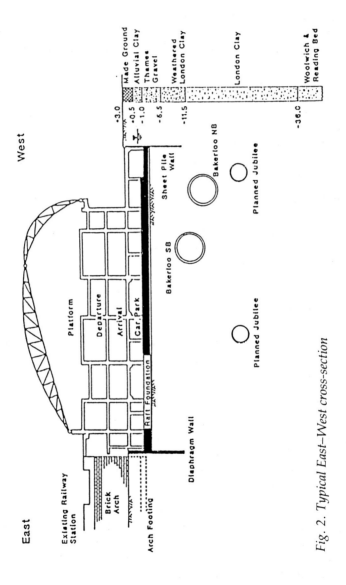

Fig. 2. Typical East–West cross-section

A5 in Fig. 1. At each location a combination of the following activities was carried out:

(a) Pushed thin-wall sampling, to obtain high quality samples of London Clay for measurement of suction and strength.

(b) Rotary coring, to obtain total volume samples for detailed logging of soil fabric and for measurement of non-linear stress–strain properties and of strength.

(c) Self-boring pressuremeter testing, to determine in situ total horizontal stress, σ_{ho}, and in situ shear stiffness characteristics.

(d) Dilatometer testing, to calibrate against the pressuremeter and to increase the quantity of data available on in situ stress and strength.

(e) Piezocone testing, to obtain information on soil profile, in particular the presence of sand seams, and where dissipation tests were carried out, to determine coefficients of consolidation, C_h.

Each of the sampling sites was selected to be at a different location relative to the Bakerloo Line tunnels and the existing brick arches. Location A1 was furthest from the tunnels and arches, and probably least affected by previous construction activities; data from this location was taken as representing initial conditions prior to any construction at the site.

Soil profile and macrofabric

The sequence of soils revealed at each location is shown in Fig. 3. Made ground, approximately 2.5 m thick, overlies 2 m of alluvial clay and approximately 4 m of Thames Gravel. The upper part of the London Clay is weathered throughout the site over a depth of 4.6 m to 7.2 m. The elevation of the base of the London Clay varies between −34.3 m OD and −37.4 m OD.

Particular attention was paid in the site investigation to identifying the presence of sand/silt seams within the London and Woolwich and Reading Bed Clays, noting their location, thickness, spacing and assessing their persistence across the site. These seams were considered to have a potential influence on:

(a) The suction that would be measured in samples which contain the seams; on stress relief the sands/silts desaturate, giving up water to the surrounding clay and lowering the overall suction that is observed; samples containing desaturated sand are difficult to re-saturate in triaxial tests, and, as a consequence, B values are low and estimates of suction are unreliable.

(b) The pattern of groundwater flow towards the tunnels.

(c) The amount of immediate heave accompanying excavation. Where pore pressures in these layers are taken below atmospheric by

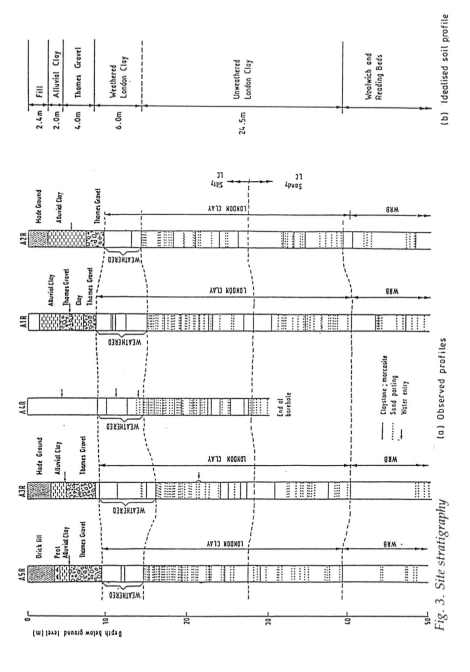

Fig. 3. Site stratigraphy

TUNNEL DISTORTIONS AT WATERLOO

undrained excavation, there is a risk of desaturation and loss of water to the clays, which can, therefore, swell.

A summary of the observations relating to the sand/silt seams is presented in Fig. 3.

Groundwater conditions

Water level in the Thames Gravel was at -0.3 m OD and was not influenced by tidal changes.

The London Clay at the WIT site was known to be underdrained, on the basis of observations reported by Skempton and Henkel (1957) for the adjacent Shell Centre. They reported a water level in the Thanet Sands of -61.5 m OD and a water level in a sand layer at the base of the London Clay (at -38.5 m OD) of -24.6 m OD. The piezometric profiles shown in Fig. 4 were established from vibrating wire piezometers in the design phase. The profile shown in Fig. 4(a) applied remote from the tunnels. Seepage analyses illustrate that the observed piezometric profile at A1 matches that predicted assuming vertical downward seepage to the sands of the Woolwich and Reading Beds, where pore pressures were zero, and assuming a fifteenfold reduction in permeability through the London and Woolwich and Reading Bed Clays.

Following Ward and Pender (1981), it was appreciated that the tunnels would be acting as drains, modifying the local pore pressure regime. Measurements made in the vicinity of the tunnels, of which those shown at A3 in Fig. 4(b) were typical, confirmed such modifications. The observed drawdown towards the tunnels was shown to be reasonable on the basis of seepage analyses which modelled the tunnels as drains, assumed a fifteenfold reduction in vertical permeability, k_v, with depth and a horizontal permeability, k_h, of $2k_v$, and which allowed for enhanced horizontal permeability where silt seams were interpreted as being persistent. The pore pressure distribution predicted from such an analysis is shown in Fig. 4(c) and predicted piezometric profiles adjacent to the tunnel compare favourably with those measured, Fig. 4(b).

In situ stresses

Three approaches were used to estimate the in situ stress state, and its variation with depth and location across the site:

(a) the measurement of suction in specimens cut from thin-wall samples pushed into the London Clay
(b) the direct measurement of σ_{ho}, via lift-off pressures in the self-boring pressuremeter test, and
(c) an indirect assessment of K_0 ($= \sigma'_{ho}/\sigma'_{vo}$) using correlations published for the Marchetti Dilatometer.

The suction existing in a sample of clay after it has been taken from

HIGHT ET AL.

the ground is related to the mean effective stress of the clay in situ, at the depth from which it was taken. The relationship depends on the stress history and type of clay from which the sample is taken. Furthermore, the suction may be modified by the sampling process and by the procedures followed in making the measurement (Hight, 1985). For the thin wall samples of London Clay, it was assumed that the mean effective stress existing in the sample, p'_k, was equal to mean effective

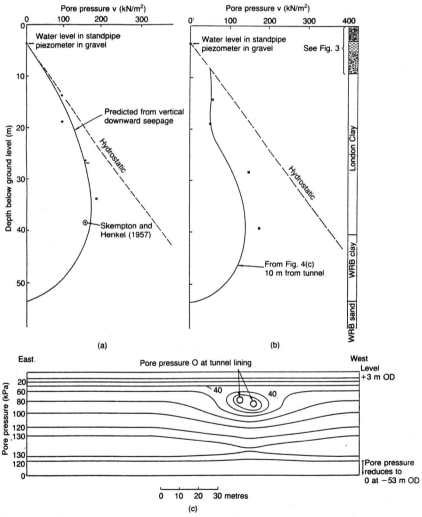

Fig. 4.(a) Observed pore pressures at A1. (b) Observed pore pressures at A3. (c) Predicted pore pressure regime around tunnels

stress in situ, p'_0; this is equivalent to assuming that A_s, the pore pressure parameter for sampling, was equal to one third.

The value of p'_k in the samples was estimated using the filter paper method (Chandler and Gutierrez, 1986), and by installing each 100 mm diameter by 200 mm high triaxial specimen, applying a cell pressure, σ_{ci}, under undrained conditions, and monitoring the pore water pressure until an equilibrium value, u_i, was reached. For an initially saturated specimen and measuring system (B = 1), the mean effective stress measured in the specimen ($p'_i = \sigma_{ci} - u_i$) is equal to p'_k. Due to deviations of B from unity, p'_i is likely to exceed p'_k.

The data for each A location were assembled and a mean line for p'_0 was established. In producing the mean line, measurements on samples thought to contain sand seams and on samples taken from damaged tubes were neglected. Values of horizontal effectives stress were derived from the trend in p'_0 with depth, and the resulting trends in $\sigma'_{h0}/\sigma'_{v0}$ with depth are shown in Fig. 5 for a site remote from the tunnels, A1 (Fig. 5(a)) and a site 8 m from the Northern Line southbound tunnel, A5 (Fig. 5(b)). An effect of the tunnels on $\sigma'_{h0}/\sigma'_{v0}$ can be seen.

Assessment of σ_{h0} from the results of the pressuremeter tests was particularly difficult. Very few of the cavity pressure–strain relationships for individual arms showed the anticipated response of strain increasing from zero at a well-defined pressure. The difference between arms was large and, in two holes the difference was consistent, suggesting that the instrument was poorly adjusted. Estimates of lift-off pressure were made, however, taking a major deviation of the strain from zero (positive or negative) to coincide with σ_{h0}. These estimates of σ_{h0} are compared with the values deduced from suction measurements in Fig. 5.

To assess the reasonableness of the horizontal stresses deduced from suction measurements and pressuremeter tests, comparisons were made with values obtained using a procedure described by Burland, Simpson and St John (1979), in which allowances could be made for vertical loading from the alluvial clay, gravel, fill and existing foundations. The values for $\sigma'_{h0}/\sigma'_{v0}$, from Burland et al.'s approach, have been added to Fig. 5 from which it is apparent that:

(a) At location A1, remote from the tunnels, there is good agreement below 16 m between estimates from suction measurements and Burland et al.'s procedure. (Divergence may be expected at depths in the London Clay where the effects of underdrainage become significant at WIT.)

(b) At location A5, there is a divergence between estimates from the two approaches. Suction measurements lead to higher values of $\sigma'_{h0}/\sigma'_{v0}$ at shallow depths and to lower values deeper in the London

Clay. These divergences may reflect the influence of the tunnels.

(c) Estimates of $\sigma'_{h0}/\sigma'_{v0}$ from the pressuremeter generally lead to higher values at shallow depths.

(d) Estimates based on the dilatometer were found to be in reasonable agreement with those based on suction measurements.

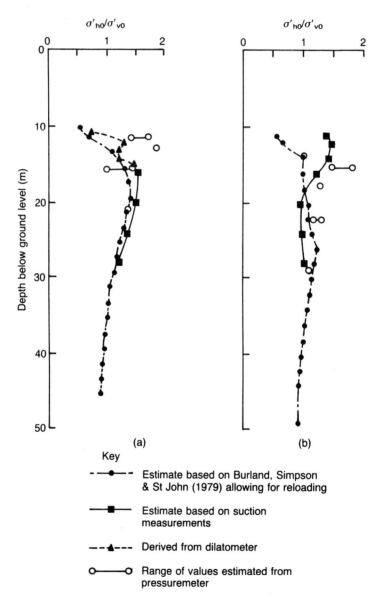

Key

— • — Estimate based on Burland, Simpson & St John (1979) allowing for reloading

— ■ — Estimate based on suction measurements

— ▲ — Derived from dilatometer

○——○ Range of values estimated from pressuremeter

Fig. 5. (a) $\sigma'_{h0}/\sigma'_{v0}$ at location A1. (b) $\sigma'_{h0}/\sigma'_{v0}$ at location A5

TUNNEL DISTORTIONS AT WATERLOO

Small-strain stiffness characteristics

The stress–strain properties of the Thames Gravel were measured by Surrey Geotechnical Consultants in locally instrumented triaxial tests on 100 mm diameter by 200 mm high specimens, prepared by static compaction of material less than 15 mm to relative densities of 0.4, 0.6 and 0.8. On two specimens (1 and 2), drained compression tests were carried out with the total axial stress, σ_a, constant and the total radial stress, σ_r, reduced in steps to simulate the stress path behind a retaining

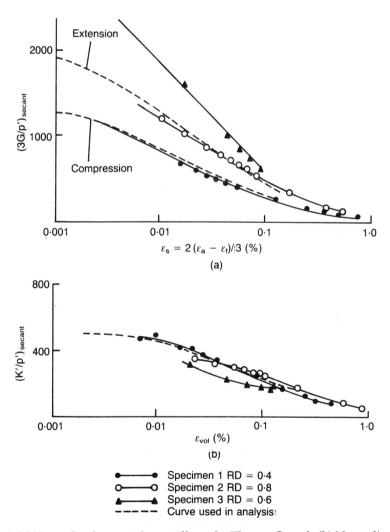

Fig. 6.(a) Normalized secant shear stiffness for Thames Gravel. (b) Normalized secant bulk stiffness for Thames Gravel

327

wall during excavation. Specimen 3 was subject to a conventional drained compression test. The observed variations in secant shear stiffness and secant bulk stiffness with strain are shown in normalized form in Figs. 6(a) and (b). The shear stiffness at small strains can be seen to be sensitive to relative density, while the bulk stiffness is much less so.

The small-strain stiffness characteristics of the London Clay were determined in two series of locally instrumented triaxial tests, one carried out at Imperial College (IC) and one carried out at City University. In the IC tests, samples from relatively shallow depths (11.85 to 15.8 m) were subject to undrained triaxial compression and extension after following reconsolidation paths which traced the soil's recent stress history (i.e. unloading due to erosion and reloading due to deposition and construction) and which brought samples to their estimated in situ effective stress state. The stress–strain data was analysed to provide information on the variation in normalised secant stiffness with strain level (Fig. 7(a)).

Samples from the deeper unweathered London Clay (23.8 to 26.6 m) were tested at City University. Two specimens were subject to un-

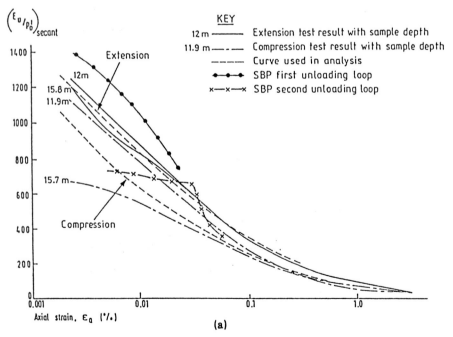

Fig. 7 (facing page and above). (a) Normalised secant shear stiffness for weathered London Clay. (b) Normalised secant shear stiffness for unweathered London Clay. (c) Normalised secant bulk stiffness for unweathered London Clay

TUNNEL DISTORTIONS AT WATERLOO

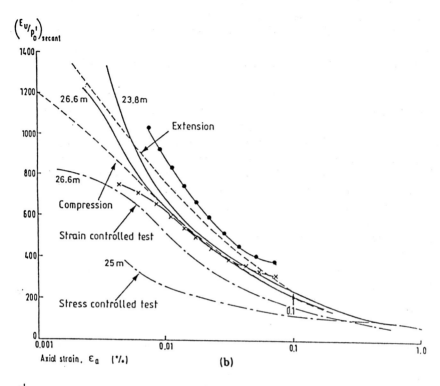

(b)

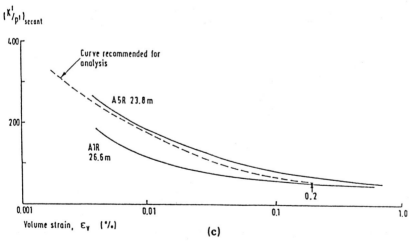

(c)

drained triaxial compression and two to undrained triaxial extension. The secant stiffness data interpreted from these tests is shown in Fig. 7(b). In these tests the reconsolidation paths, which again traced the recent stress history of the soil, were followed sufficiently slowly for pore pressures to equalise throughout the specimen. Data from the reconsolidation paths could be analysed, therefore, to provide information on secant bulk stiffness and this is shown in Fig. 7(c).

Secant shear stiffnesses derived from triaxial tests carried out at City University, particularly from triaxial compression tests, are lower than anticipated. This probably results from the fact that the tests were run initially under stress control, leading to strain rates that were 1/10–1/20 times the rate in the IC tests.

Small-strain stiffnesses have been interpreted from the pressuremeter tests, using a method described by Jardine (1992), and data from the appropriate depths has been added to Figs. 7(a) and (b). These characteristics are generally similar to those measured in the triaxial tests, despite different shearing directions and differences in pre-shear creep rates.

Prediction of distortions

A series of plane strain finite element analyses was carried out using the ICFEP program to make predictions of ground movements and tunnel distortions caused by construction of WIT, modelling the construction sequence described below. The sections which have been analysed are shown in Fig. 1. Section A-A is a typical section through the running tunnels at the site. Section B-B is taken through the station tunnels where they are currently overlain by fill adjacent to the Armstrong Lift.

The Thames Gravel, London Clay and Woolwich and Reading Beds were modelled as non-linear elastic perfectly plastic materials, obeying the Mohr Coulomb failure criterion. A summary of the parameters that

Table 1. *Summary of material properties*

	γ (kN/m^3)	c' (kN/m^2)	ϕ'	ν	K_0
Made Ground/Fill	18	0	30°	0	0.5
Alluvial Clay	17	0	24°	0	0.5
Thames Gravel	20	0	35°	17.5°	0.5
London Clay					
(a) Weathered	20	5	23°	11.5°	Varying
(b) Unweathered	20	15	25°	12.5°	with
Woolwich and Reading Beds	22	200	27°	0	depth

Table 2. Constants for non-linear stiffness expressions

Layer	A'	B'	C' (%)	α'	γ'
A. BULK STIFFNESS					
Thames Gravel – I (Compression)	275	225	2E-3	0.998	1.044
Thames Gravel – I (Extension)	275	225	2E-3	0.998	1.044
Thames Gravel – II	330	270	2E-3	0.998	1.004
All London Clay	245	203	5E-4	1.662	0.555

Layer	A	B	C (%)	α	γ
B. 3G – COMPRESSION					
Thames Gravel – I	622	640	5E-4	0.974	0.94
Thames Gravel – II	1104	998	5E-4	0.974	0.94
Lower London Clay	695	620	5E-4	1.368	0.6964
Upper London Clay	1310	1322	7E-5	1.471	0.4352
C. 3G – EXTENSION					
Thames Gravel – I	933	960	5E-4	0.974	0.94
Thames Gravel – II	as compression	as compression	as compression	as compression	as compression
Lower London Clay	776	753	1E-3	1.571	0.60
Upper London Clay	980	950	5E-5	0.8695	0.8079

I – outside extent of excavation
II – within excavated area
Woolwich and Reading Beds, as lower London Clay, but with A', B', A and B factored by 2.0.

were adopted is given in Table 1. The non-linear elastic properties were described by the following two equations, given by Jardine et al. (1986):

$$3G/p' = A + B \cos(\alpha[\log(\varepsilon_s/C)]^\gamma)$$

$$K/p' = A' + B' \cos(\alpha'[\log(\varepsilon_v/C')]^{\gamma'})$$

Values for the constants A, B, C, α, γ, A', C', α' γ' are presented in Table 2. The variation in normalised stiffness, $3G/p'$ and K/p', which these expressions and constants describe are shown in Figs. 6 and 7 as the 'curves used in analysis'. For the lower London Clay, the characteristics that were adopted were stiffer than those measured in the triaxial tests and closer to those derived from the pressuremeter tests.

For the Woolwich and Reading Beds, the non-linear elastic stiffness was assumed to be twice that of the unweathered London Clay. The fill and alluvial clay were modelled as linear elastic plastic materials obeying the Mohr Coulomb failure criterion and having the properties given in Table 1.

The tunnel lining segments were modelled using beam elements which had an equivalent EI to the cast iron linings, based on Young's modulus for the cast iron of 91×10^6 kN/m^2. Articulation of the linings was not permitted in these particular analyses because the depth of cover provided sufficient hoop stress to maintain thrusts within the width of the flange.

The starting point for the analyses was a 'green-field' site having the in situ stress state shown in Figs. 4(a) and 5(a), which was based on measurements made remote from previous construction events, i.e. at A1. The following essential features of the site's history prior to construction of WIT were then modelled in sequence:

(a) construction of the brick arches carrying Waterloo Station and placement of the Armstrong Fill

(b) construction of the Bakerloo line tunnels, northbound then southbound

(c) dissipation of excess pore pressures generated by tunnel construction to equilibrium values obtained from seepage analyses described above (Fig. 4(c))

(d) excavation for and construction of Elizabeth House, which lies alongside the site.

It was necessary to follow this procedure because of the kinematic nature of soil stiffness, i.e. its dependence on recent stress history, and its dependence on current stress state. The predicted values of mean effective stress, p', in the ground at this stage, i.e. prior to construction of WIT, are shown in Fig. 8 and serve to illustrate the variation in stress state and to emphasise the difficulty in interpreting measurements of p' made at this site.

Construction of WIT was modelled with the following steps:

(a) Installation of the boundary retaining walls.
(b) Demolition of the brick arches and removal of the Armstrong Fill.
(c) Dewatering of the Thames Gravels.
(d) Excavation in layers over the full width to a depth of −2.55 m OD. This followed guidelines for construction sequencing which had been established to minimise tunnel distortions and required excavation to be carried out over a maximum feasible width, removing 2–3 m in each pass. Props to support the retaining walls were installed at +0.5 m OD.
(e) Casting of a 1.6 m thick concrete raft.
(f) Application of the building loads.

During construction of WIT it was assumed that the London Clay would remain undrained; allowances were made subsequently for the potential effects of cavitation in sand layers within the clay during unloading and for some dissipation of negative excess pore pressures during construction.

Final dissipation of negative excess pore pressures was carried out to two different long-term seepage conditions. In both, the tunnels were assumed to continue acting as drains and the boundary retaining walls were assumed to be ineffective cut-offs. In one, continuing underdrainage of the London Clay was assumed; in the other, pore pressures in the London Clay were assumed to return to hydrostatic ('rising groundwater level').

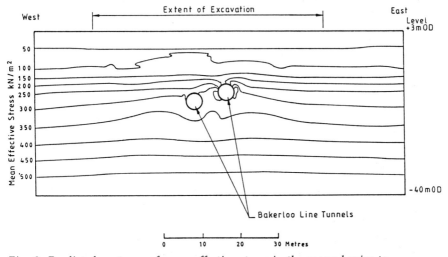

Fig. 8. *Predicted contours of mean effective stress in the ground prior to construction of WIT*

HIGHT ET AL.

The predicted changes in vertical and horizontal tunnel diameters and vertical movements of the tunnel crowns for the case of the Bakerloo Line are related to construction in Fig. 9. Presentation of the predicted thrusts and bending moments in the tunnel lining and the predicted changes in clearance are outside the scope of this paper.

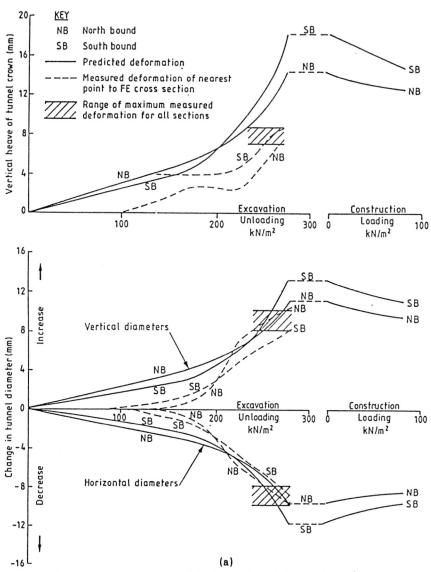

Fig. 9 (above and facing page). (a) Deformation of station tunnels.
(b) Deformation of running tunnels

TUNNEL DISTORTIONS AT WATERLOO

Monitoring of distortions
The monitoring of the LUL underground tunnels has comprised:

(a) precise levelling of points (eyebolts) installed in the crown of the Bakerloo and Northern Line running and station tunnels

(b) measurement of the distances between eyebolts installed in the cast iron lining at selected cross sections; measurements were made

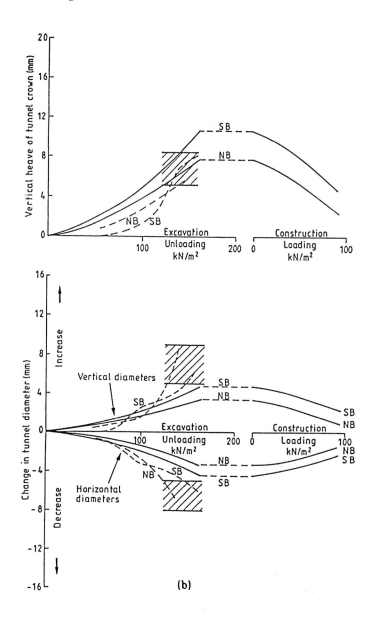

(b)

using a Soil Instruments Mark II tape extensometer; the monitored sections extended 50 m beyond the site boundaries

(c) establishing the coordinates of eyebolts at the crown of the tunnels at 10 m intervals to determine changes in horizontal alignment.

The monitoring showed that the majority of the tunnel movements occurred during the final excavation phase, which was carried out in bays of approximately 30 m by 30 m. This introduced important three dimensional effects to which the movements were sensitive and which made interpretation of the monitoring difficult. Distortions varied according to the location of the monitored section relative to the boundaries of a bay.

Diameter changes and crown heaves which indicate their development with unloading have been added to Fig. 9. The maximum values recorded beneath each bay are also indicated. The following comments may be made:

(a) the measurements confirm the non-linear nature of the build-up in distortion with unloading

(b) the measurements indicate that distortions are less than predicted at low levels of unloading

(c) the maximum predicted distortions are reasonable for the case of the station tunnels

(d) the maximum predicted distortions are lower than those measured in the running tunnels

(e) predictions of crown heave are reasonable for the running tunnels.

Predicted crown heaves for the station tunnels are overestimates. This may be related in part to the location of the station tunnels relative to the boundaries of the site (Fig. 2) which made it impossible to select a reliably representative cross-section that could be analysed assuming plane strain conditions. Predictions made for Section B-B were bound to overestimate displacements, assuming all other aspects were correctly modelled.

Concluding remarks

The original scheme for WIT, involving a two storey basement, had many features in common with the construction of the Shell Centre, on the W. side of York Road (Williams, 1957; Measor and Williams, 1962). The monitoring carried out at the Shell Centre by BRE has been invaluable for checking the predictive methods described in this paper and the monitoring was updated for the WIT project.

State of the art techniques described by Williams (1957) for predicting undrained displacements involved treating the London Clay and Woolwich and Reading Beds as linear elastic materials, ascribing an average elastic modulus determined from back-analysis of three previously monitored excavations in London and measurements made at other locations on the Shell Centre site. Reasonable predictions of surface heaves could be made. However, reliable predictions of displacements at depth and of the interaction between the tunnels and ground were not feasible.

Special measures were taken at the Shell Centre to protect the Bakerloo Line running tunnels and these included:

(*a*) the maintenance during excavation of an earth bund over the tunnels, which was only removed over its full depth is short sections, allowing a capping raft to be constructed over the tunnels, held down on each side by a line of under-reamed hand dug caissons
(*b*) the loosening of tight circle joint bolts inside the boundaries to the excavation.

Our current appreciation of the non-linear nature of soil stiffness indicates that full depth excavation in short sections does not necessarily minimise distortions. Different construction guidelines were set, therefore, for WIT to ensure that excavation for the basement limited distortion of the tunnels. The guidelines were sufficiently broad for the contractor to have flexibility in the phasing of the excavation works. As a result, the construction sequence adopted for the modelling differed in a number of respects from the actual sequence on site.

The construction controls that were imposed at WIT were successful in keeping distortions and changes in clearance within acceptable limits. No damage occurred to the tunnel linings. Three dimensional effects were significant, however, because of the particular construction sequence that was followed. While these effects may help explain why measured values were sometimes smaller than predicted values they emphasise the need for a capability for 3D analysis of non-linear geotechnical problems.

The precedent of the Shell Centre construction and the sophisticated analyses carried out for WIT gave both the designers and London Underground Limited confidence to allow construction of the WIT to proceed without any special precautions being taken within the running and station tunnels, other than regular monitoring.

Acknowledgements
British Rail and Sir Alexander Gibb and Partners have kindly agreed to details of this work being published. Mark Sharrock carried out the seepage analyses described in the paper. Ivan Chudleigh of LUL and John Paterson of Charles Haswell and Partners gave valuable advice.

References

BURLAND, J.B., SIMPSON, B. AND ST. JOHN, H.D. (1979). Movements around excavations in London Clay. Proc. 7th European Conf. on Soil Mech. and Found. Engng, Brighton, Vol. 1, pp. 13–29.

CHANDLER, R.J. AND GUTIERREZ, C.I. (1986). The filter paper method of suction measurement. Géotechnique, Vol. 36, No. 2, pp. 265–268.

HIGHT, D.W. (1985). Laboratory testing: assessing BS 5930. Geological Society. Engineering Geology Special Publication No. 2, pp. 43–52.

JARDINE, R.J. (1992). Non-linear stiffness parameters from undrained pressuremeter tests. Can. Geotech. Jnl, Vol. 29, No. 3, pp. 436–447.

JARDINE, R.J., POTTS, D.M. FOURIE, A.B. AND BURLAND, J.B. (1986). Studies of the influence of non-linear stress-strain characteristics in soil-structure interaction. Géotechnique, Vol. 36, No. 3, pp. 377–396.

MEASOR, E.O. AND WILLIAMS, G.M.J. (1962). Features in the design and construction of the Shell Centre, London. Proc. Instn. Civ. Engrs, Vol. 21, pp. 475–502.

SKEMPTON, A.W. AND HENKEL, D.J. (1957). Tests on London Clay from deep borings at Paddington, Victoria and South Bank. Proc. 4th Int. Conf. Soil Mech. and Found. Engng, London, Vol. 1, pp. 100–106.

WARD, W.H. AND PENDER, M.J. (1981). Tunnelling in soft ground – General Report. Proc. 10th Int. Conf. Soil Mech. and Found. Engng, Stockholm, Vol. 4, pp. 261–276.

WILLIAMS, G.M.J. (1957). Design of the foundations of the Shell Building, London. Proc. 4th Int. Conf. Soil Mech. and Found. Engng, London, Vol. 1, pp. 457–461.

Modelling of the behaviour of foundations of jack-up units on clay

G.T. HOULSBY and C.M. MARTIN, Department of Engineering Science, Oxford University

Theoretical and experimental investigations of the behaviour of spudcan foundations of jack-up units under combined loadings are described. The behaviour of the foundations is described by a work-hardening plasticity theory which bears a close resemblance to the concepts of critical state. Theoretical calculations of the variation of vertical bearing capacity are closely matched by experimental results. A simple form of the yield surface under combined vertical, horizontal and moment loading is suggested. Experiments designed to explore the shape of the surface confirm that the simple shape is approximately correct.

Introduction

In the currently adopted designs of platforms for offshore oil and gas exploration and production, there are three types of foundation which can be treated as approximately circular, shallow foundations. These are gravity bases, the mudmats used for temporary support of piled structures and the spudcan foundations of jack-up units. The design of these structures is complicated by a number of factors.

Firstly and most importantly they are subjected to large horizontal and moment forces, due to environmental loading, as well as vertical loads. The environmental loads are of course applied cyclically to the structure, although this aspect will not be pursued further here. By comparison, the design of shallow foundations onshore is dominated usually by purely static vertical loading.

Secondly, the base of the foundation is often embedded to a significant depth below the ground surface, typically by up to about half a diameter for a gravity base or mudmat, and by up to two diameters for a spudcan. Unless this embedment is taken into account, the design will be excessively conservative, since even a small embedment can significantly increase the foundation capacity. The development of design methods to account for embedment is not, however, straightforward.

Thirdly the quality and level of detail of the site investigation data, especially in the case of a mobile jack-up unit, may be much poorer than for an onshore design.

Although much of the following is also relevant to gravity bases and mudmats, the purpose of this paper is to concentrate on the problem of spudcan foundations on clay. This problem is of particular interest because the mobility of jack-up units means that they are being continually reassessed for different sites. The tendency to use jack-ups in increasingly harsh environments has led to the need for a more detailed understanding of the influence of the foundation on the performance of the structure.

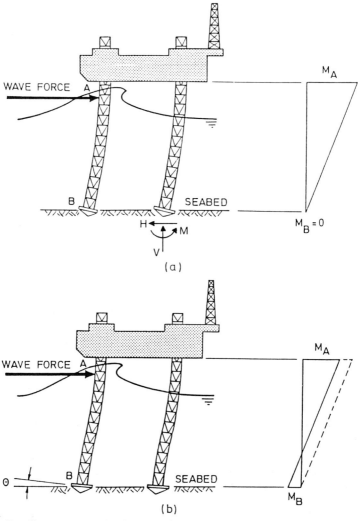

Fig. 1. Bending moments in the legs of jack-up units (a) without moment fixity and (b) with moment fixity (after de Santa Maria, 1988)

FOUNDATIONS OF JACK-UP UNITS

There are two principal concerns in the assessment of a spudcan foundation. The first is the estimation of spudcan penetration under preload conditions. This requires an accurate estimate of the variation of bearing capacity with the penetration depth. This is not usually a problem in coarse materials, where penetrations are typically very small (see for example Dean et al., 1992). In soft clays, where penetrations can be very large (30 m penetration of a 15 m diameter spudcan is not uncommon), the problem is often further complicated by the increase of strength with depth of the clay.

The second concern is the performance of the spudcan under storm loading conditions. It has been conventional to assume, conservatively, that the spudcan offers no moment restraint, in which case the assessment needs to account only for the interaction of horizontal and vertical loading. In extreme conditions, however, the critical feature of the structure is often the bending moment in the leg at the lower leg guide, see Fig. 1. If some account is taken of the moment developed at the spudcan, then the calculated moment at the leg guide is reduced. There is therefore also a need to understand the full interaction between vertical load, horizontal load and moment.

At small deflections the response of a spudcan foundation can be treated as elastic, and the estimation of appropriate elastic stiffness factors has been the subject of a previous paper (Bell, Houlsby and Burd, 1991). The purpose of this paper is to examine the problem of the shape of the yield locus when larger combined loads are applied.

Vertical loading

The calculation of the variation of the bearing capacity with depth is important if a reasonably accurate estimate of spudcan penetration is to be made. This has been approached in two ways: purely theoretical calculations of bearing capacity have been made, and experiments carried out in which small scale model spudcans were pushed into soft clay.

Theoretical modelling

Theoretical calculations of bearing capacity were carried out using the method of characteristics as described by Houlsby (1982), Houlsby and Wroth (1982, 1983). Calculations were made for the bearing capacity of cones of angle 30° to 180° (flat plate) at embedments from zero to 2.5 diameters, Fig. 2. The full range of possible footing roughness values was considered. The soil strength was assumed to increase linearly with depth, as defined by $s_u = s_{u0} + \rho z$. The range of values of $\rho D/s_{u0}$ was from zero to 5.0. Full details of the calculations, together with the tabulated values, are given by Martin (1991).

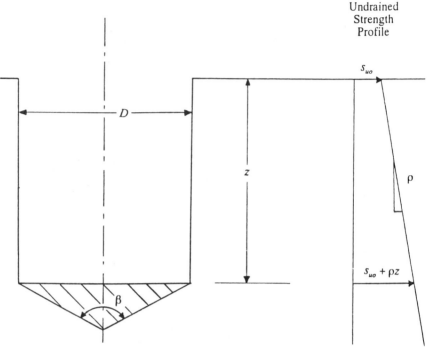

Fig. 2. Definition of cone geometry

The treatment of the effect of embedment can only be approximate when using this method. At the surface the field of stress characteristics (usually interpreted also as the slip planes) is as shown in Fig. 3(a). At small embedments the field shown in Fig. 3(b) is used. This gives a continuously increasing capacity with depth, and at large depths a cut-off value determined by the field shown in Fig. 3(c) is used. The method of integration along stress characteristics gives a lower bound solution to the incremental collapse problem, but it should be noted that the solutions presented here are not strict lower bounds, in that the field of characteristics has not been extended throughout the soil.

The theoretical calculations have been employed in the following way to determine the profile of bearing capacity with depth for a spudcan. A series of depths of penetration is considered, and the undrained strength determined from the design profile at the foundation level, and half a diameter below this level. From these values the local rate of increase of strength with depth is determined. Making use of the foundation geometry (the shape of the spudcan is treated as equivalent to a cone enclosing the same volume) and an assumed footing roughness, the bearing capacity factor is then interpolated from the

FOUNDATIONS OF JACK-UP UNITS

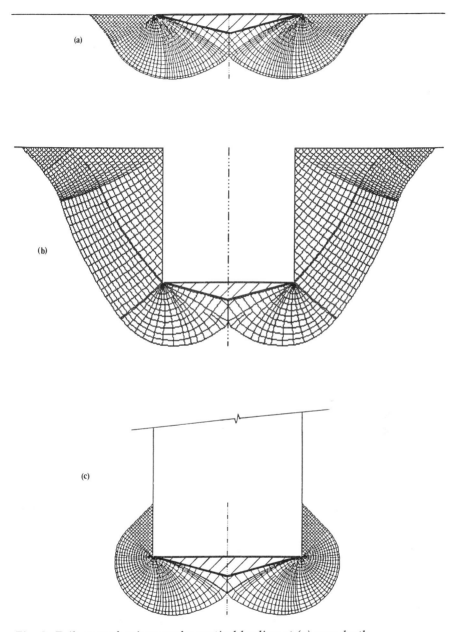

Fig. 3. Failure mechanisms under vertical loading at (a) zero depth, (b) intermediate depths and (c) large depths

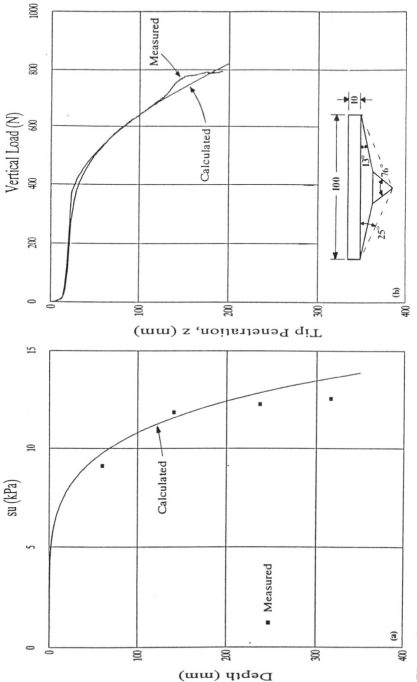

Fig. 4. Comparison of calculated and measured profiles of (a) undrained strength and (b) vertical load on a model spudcan

FOUNDATIONS OF JACK-UP UNITS

calculated values. Proper account is taken of the changing spudcan geometry at partial penetration, and also for shear developed on any vertical upstand at the side of the spudcan.

Experimental modelling

Vertical bearing capacity experiments were conducted using a model spudcan footing and carefully prepared samples of soft clay. The footing was 100 mm in diameter, with a basic cone angle of 154° and a sharper 76° conical tip (see Fig. 4(b)). The clay samples were prepared by one-dimensional consolidation of a speswhite kaolin slurry (PL ≈ 34%, LL ≈ 65%, $\lambda \approx 0.25$, $\kappa \approx 0.04$, $M \approx 0.9$), using tanks of 450 mm inside diameter. Drainage, through porous filter sheets, was allowed at both the top and bottom of the samples. The consolidation pressure was increased incrementally to a maximum stress of 200 kPa, then unloaded in three stages (100, 50 and 0 kPa). The resulting clay samples were about 400 mm deep, with a moisture content of approximately 50%. A simple motorised driving device was used to push the model spudcan footing vertically into each clay sample at a speed of 0.33 mm/s. The vertical load was measured with a 1000 N load cell, and the vertical displacement with a long stroke LVDT. The load and displacement data were automatically logged using a computer equipped with an analogue-to-digital converter card.

Figure 4(a) shows the variation in undrained strength with depth for a typical clay sample. As would be expected for a heavily overconsolidated sample, the strength increase with depth is highly non-linear. Each of the measured points represents an average of four miniature shear vane readings taken at that depth in the sample. The calculated strength profile is derived from the theoretical relation between undrained strength and OCR derived from critical state theory, Wroth and Houlsby (1985). For speswhite kaolin and one-dimensional consolidation conditions, the equation takes the form:

$$\frac{s_u}{\sigma'_v} = 0.17(\text{OCR})^{0.8} \tag{1}$$

This may be rearranged to give:

$$s_u = 0.17(\sigma'_{vmax})^{0.8}(\sigma'_v)^{0.2} \tag{2}$$

where $\sigma'_{vmax} = 200$ kPa and $\sigma'_v = \gamma' z$. As shown in Fig. 4(a), the observed strength values correlate well with the theoretical equation, particularly in the top 250 mm of the sample where footing penetration occurs. The calculated strength profile is most useful for estimating undrained strength in the upper 50 mm of the clay, where it is not possible to make accurate measurements with the shear vane.

Figure 4(b) shows the profile of spudcan bearing capacity with depth, as observed in one of the scale model experiments. Also shown is the theoretical prediction, using the calculated undrained strength profile from Fig. 4(a). The two bearing capacity curves agree very well, and close agreement was also observed in the other experiments. The shape of the plot of bearing capacity with depth may be explained as follows. As the sharp conical tip of the spudcan first penetrates the clay, the bearing capacity remains low because of the small contact area. Once the main conical part of the footing begins to penetrate the clay (at z = 14 mm), the contact area increases rapidly with further penetration, so the bearing capacity rises sharply. After full penetration has been reached (at z = 23 mm) there is no further increase in contact area, but the bearing capacity continues to rise as a result of embedment below the soil surface, and because of increasing clay strength with depth.

Combined loading
Theoretical modelling

As the foundation penetrates the clay under vertical load, the clay is continually yielding. If the vertical load is reduced a much stiffer elastic unloading is observed, and on reloading the response is elastic until maximum past load is reached, when plastic deformation once again occurs. The yield point which is observed is just one point on a yield surface in (V,H,M) space. The situation is analogous to the consolidation of a clay, in which the preconsolidation pressure observed in an isotropic compression test is one point on a yield surface in (p',q) space.

One problem is the determination of the shape of the three-dimensional yield surface. In the next section experiments designed to explore the shape of the surface will be described, but first it is worth considering theoretically the shape which might be expected.

Consider first the problem of combined vertical load and moment. An alternative is to treat this case as an eccentrically applied vertical load, and this has led to the 'effective area' concept. The point of application of the eccentric load is taken as the centroid of the effective area. For a strip footing of width B on a uniform clay this leads to a parabolic shape of yield surface in the (V,M) plane:

$$4\frac{V}{V_0}\left(1 - \frac{V}{V_0}\right) - \frac{M}{M_0} = 0 \qquad (3)$$

where V_0 is the bearing capacity under purely vertical loading and $M_0 = BV_0/8$. Although this theory is not strictly applicable to a circular footing, it will be assumed for simplicity that the section of the yield surface in this plane is a parabola. This means that the maximum moment occurs at exactly half the vertical load. If the average pressure on the effective area is taken as the same as that on the full footing, it

FOUNDATIONS OF JACK-UP UNITS

may be assumed that a semicircular area is in contact with the clay. The centroid of this area is at $4R/3\pi$ from the centre, so that $M_0 = (4R/3\pi)(V_0/2) = 2V_0R/3\pi \approx 0.21V_0R$. By comparison, if the more complex theory suggested by Hansen (1970) were to be used, a peak moment of $0.20V_0R$ would be obtained at a value $V/V_0 = 0.48$.

Considering now only vertical and horizontal load, again an exact theory for strip footings on homogeneous clay exists. The maximum horizontal load is $H_0 = s_u A$, and sliding at this load occurs if $V \leq V_0/2$. For larger vertical loads V and H are related by

$$\frac{V}{V_0} = \frac{1 + \pi - \sin^{-1}(H/H_0) + \sqrt{1 - (H/H_0)^2}}{2 + \pi} \quad (4)$$

which approximates quite closely to a parabola. The large shear load which may be developed at low vertical loads depends on there being no loss of contact, which may occur if any tipping of the foundation takes place. It will therefore be assumed that the yield locus in the (V,H) plane is also a parabola, similar in form to that in the (V, M) plane. The value of H_0 will be taken as $s_u A + 2s_u A_h$, where A_h is the area of the spudcan in elevation. The second term takes some account of the development of passive pressure on the side of the spudcan.

Finally the shape of the yield locus in the (H, M) plane must be considered. For simplicity it will be assumed to be elliptical, although there is evidence (Noble Denton and Associates, 1987) that a more rectangular shape may be more appropriate. Bell (1991) carried out finite element analyses which give further evidence about the shape in this plane. His results suggest that the shape is indeed approximately elliptical, but with the axes of the ellipse rotated with respect to the (H, M) axes. For preliminary purposes, however, the yield surface will be assumed to take the form:

$$16\left(\frac{V}{V_0}\right)^2 \left(1 - \frac{V}{V_0}\right)^2 - \left(\frac{H}{H_0}\right)^2 - \left(\frac{M}{M_0}\right)^2 = 0 \quad (5)$$

It should be mentioned that the type of theoretical model described here is rather similar in structure to that employed by Nova and Montrasio (1991) for modelling the behaviour of strip footings on sand, and also to the methods used by Tan (1990) and Dean et al. (1992) to model spudcan footings on sand.

After the foundation has penetrated to a given depth, a yield surface of a certain size will be set up. If the foundation is then displaced horizontally with no further vertical displacement, then the sum of the subsequent vertical elastic and plastic displacements will be zero. If the vertical elastic stiffness is very high, then the elastic displacements are very small, and so too are the equal and opposite plastic displacements.

The result is that only a very small expansion of the yield locus occurs, and the load path follows a track across the yield surface appropriate to the fixed depth. Tan (1990) carried out such tests on sand. The test is closely analogous to an undrained triaxial test on a normally consolidated clay. The stress path in such a test resembles the yield locus, although the deviation is somewhat larger because the ratio of elastic to plastic stiffness is smaller.

By conducting a test in which vertical penetration is interrupted at various depths for a horizontal displacement to be applied, the evolution of the shape and size of the yield locus can be determined. The interruption must only be for a brief period, so that the assumption of undrained loading remains valid, but excursions of horizontal displacement do not take long. Further information is obtained by replacing the horizontal movement by a pure rotation. The shape of the yield surface at low vertical stresses can be explored by unloading the vertical force before displacing horizontally or rotating.

Experimental modelling

An apparatus has been designed and built at Oxford, specifically for an experimental investigation of model spudcan behaviour under combined loads. A system of three computer-controlled stepper motors allows precise and fully independent control of the vertical displacement, horizontal displacement and rotation of the spudcan. The linear displacements are generated by ballscrew and nut systems, while the rotation is generated by a curved rack and pinion arrangement. Both horizontal movement and footing rotation take place in the same vertical plane, with the centre of rotation fixed at the level of the footing 'shoulder'. The model spudcan has a diameter of 100 mm, and is similar to the one used in the vertical bearing capacity tests described above.

The three displacement components are measured using LVDTs, with the step counts from the motors as an independent check. A specially designed Cambridge type stress transducer (Bransby, 1973), placed between the driving shaft and the model footing, allows simultaneous measurement of all three load components: vertical, horizontal and moment. The measured load and displacement data are logged by an 8 channel analogue-to-digital converter unit with 256 kByte of RAM. The data acquisition is carried out independently of the host computer, although channel readings may be sampled at any time. A compiled BASIC program is used to issue commands to the stepper motor control unit and the data logger via two RS232 interfaces. For the reasons outlined above, the rig has primarily been designed to subject the model footing to pre-determined displacement paths. However, it is envisaged that some relatively simple software development, employing feedback

FOUNDATIONS OF JACK-UP UNITS

control, will allow pre-determined load paths to be followed in future tests, if required.

Typical test procedure

At the time of writing, three experiments have been conducted with the new apparatus. All have been broadly based on the experimental procedure employed by Tan (1990) for investigating the combined load behaviour of model spudcans on sand. Tan's apparatus was not able to impose rotation on the spudcan, but only vertical and horizontal displacements, so his tests were limited to a study of the yield surface in (V,H) space. The new rig for tests on clay also allows pure rotation of the footing, or a simultaneous application of horizontal movement and rotation in any ratio. This means that a complete exploration of the yield surfaces in (V,H,M) space is possible. Of the three experiments carried out to date, two have involved pure horizontal movement of the footing (tests 1 and 3), and one pure rotation (test 2). Test 2 will be used to illustrate a typical experiment using the new apparatus.

It is useful to follow the progress of the test on three separate graphs, see Fig. 5. Figure 5(*a*) shows the profile of vertical load on the spudcan with depth; Fig. 5(*b*) shows the interaction between vertical load and moment; Fig. 5(*c*) is a plot of moment versus footing rotation. The test began at Point A, with all loads zero, and the tip of the model spudcan just above the surface of the clay sample. The spudcan was driven vertically into the clay at 0.33 mm/s, stopping when the shoulder of the footing reached the level of the clay surface (corresponding to a tip penetration of 23 mm). At this stage the vertical load had reached 383 N (Point B), but when penetration was stopped, the load immediately dropped by 11% to 342 N (Point C). This loss of load is thought to be due to the rapid dissipation of transient pore pressures in the soil just beneath the footing, and also due to creep behaviour of the soil skeleton. It was possible to restore the load to its original peak value, and maintain it at this level, by using a simple feedback control loop to generate additional vertical penetration (Path CD). The penetration rate required was initially about 0.05 mm/s, but quickly decayed to less than 0.01 mm/s.

When the vertical penetration rate had settled to a steady value, rotation of the footing was started at a constant rate of 0.17°/s (Path DE). The vertical penetration speed was linearly reduced to zero over the first 2.5 s of rotation, ensuring that the rotation path DE was effectively carried out at a constant vertical embedment (Fig. 5(*a*)). In (M,θ) space (Fig. 5(*c*)) the moment value increased sharply, but subsequently reached an almost constant value. The path DE in (V,M) space (Fig. 5(*b*)) shows that the vertical load initially decreased rapidly, but it too reached an almost constant value at Point E. This point represents a state of

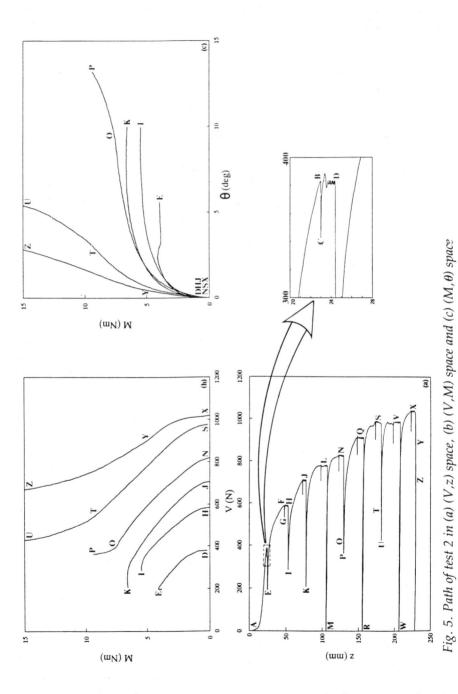

Fig. 5. Path of test 2 in (a) (V,z) space, (b) (V,M) space and (c) (M,θ) space

constant vertical and moment load, which was maintained despite further rotation of the footing. When this constant load state had been reached, the direction of rotation was reversed and the footing was returned to its initial position ($\theta = 0°$). The load and displacement paths during this repositioning have been omitted from Fig. 5 for clarity.

Vertical driving of the footing was then resumed at a rate of 0.33 mm/s (Path EF), stopping when the footing shoulder was 25 mm below the clay surface ($z = 48$ mm). Once again the vertical load immediately dropped by about 10% from its peak value, and the same feedback control loop was employed to restore and maintain constant vertical load (Path GH). Rotation of the footing at constant vertical displacement produced another curved load path in (V,M) space (Path HI), but this time with a higher value of peak moment. The footing was rotated back to its central position and driven another 25 mm into the clay (Path IJ). A third rotation at constant vertical displacement traced out the load path JK in Fig. 5(b). The peak moment value was again greater than those developed by rotation at shallower depths.

The vertical driving of the footing (Path KL) was next interrupted when the shoulder was 75 mm below the clay surface, i.e. when $z = 98$ mm. After the vertical load had been held constant for a few minutes, the footing was moved upwards at 0.17 mm/s until the vertical load reached zero (Path LM). Downward vertical driving of the footing then resumed at 0.33 mm/s (Path MN). Two more unload–reload loops were performed at deeper penetrations (QR and VW in Fig. 5(a)). The unload–reload process allows computation of an approximate elastic shear modulus for the clay. As explained below, a reasonable estimate of shear modulus is essential if the predictions of a theoretical combined load model are to be compared with experimental results. The shear modulus may be calculated from the unload–reload slope in (V,z) space and the following expression for a rigid circular footing on an infinite elastic continuum:

$$\Delta V = K_1 G R \Delta z \qquad (6)$$

The stiffness factor K_1 was obtained from the finite element analysis of Bell (1991), taking into account the angle of the footing and the depth of embedment. A correction was applied to account for the finite size of the testing tank. This method gives a rigidity index $I_r = G/s_u$ of approximately 46 for the clay samples used—a value which is comparable with that obtained from other tests on the same clay.

A further three rotations were performed at footing shoulder depths of 100 mm, 150 mm and 200 mm, represented by the paths NOP, STU and XYZ in Fig. 5. Points O, T and Y mark the onset of a distinct upward curvature in Figs. 5(b) and (c), which is not exhibited for shallower depths. This may be explained as follows. The Cambridge stress transducer is a fragile instrument which must be protected from clay and moisture when the footing has penetrated into the sample. This protection is provided by a watertight metal case which is fixed to the upper side of the spudcan footing, and extends to 250 mm above the footing level. When the footing is rotated at deep footing penetrations, the protective case begins to contact the side of the hole above the footing. Because the case is rigidly connected to the footing, the load transducer not only registers any load on the footing itself, but also any load applied to the protective case. The force exerted on the case as it contacts the clay sidewall thus creates a moment additional to that arising from spudcan–clay interaction. For future experiments a slightly larger footing will be used. This will create a larger hole above the penetrated footing, and therefore allow a larger rotation before the protective case contacts the sidewall. In these circumstances it would be reasonable to expect the load paths in (V,M) space to produce a shape closely resembling the paths DE, HI and JK in Fig. 5(b), where no sidewall contact occurred.

Horizontal loading test results
Test 1 was very similar to test 2 except that pure horizontal displacement, rather than pure rotation, was applied at a variety of vertical penetrations. The load paths traced out by this test in (V,H) space are shown in Fig. 6, and a strong similarity with Fig. 5(b) is apparent. As with rotation at large depths, horizontal movement at large depths also caused the protective casing to dig into the clay sidewall, producing a misleading upward curvature of the load paths in Fig. 6. Note that, in this test, no horizontal displacement was applied at a footing shoulder depth of 25 mm, where an unload–reload loop was performed instead.

Test 3 also involved pure horizontal displacement at a series of vertical penetrations. In this test, however, the vertical load was reduced to a very small value, 10 N, before imposing the horizontal movement. The results of this test are shown in Fig. 6. The horizontal movement caused both an increase in horizontal load and an increase in vertical load, with larger horizontal loads being attained at greater depths (as in test 1). Once again the impact of the protective casing on clay above the footing level has distorted the (V,H) load paths obtained at greater depths.

Significance of test results
In the section on theoretical modelling above, the combined load yield locus of the footing is assumed to expand with further plastic penetration. A consequence of this is that any horizontal displacement or rotation applied while the vertical penetration is held constant will produce a load path which actually tracks along the (V,H,M) yield surface applicable to that particular embedment. Figs. 5(b) and 6 indicate

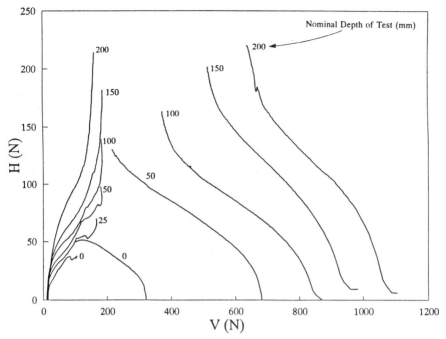

Fig. 6. Yield surface sections from tests 1 and 3

that, as vertical penetration increases, larger horizontal loads and moments are developed on the footing. This confirms that the (V,H,M) yield surface expands with increasing vertical penetration. The figures also suggest that, despite this expansion with depth, the shape of the yield surface remains reasonably constant in both (V,M) and (V,H) space.

It should be pointed out that in test 2 (the pure rotation test), small horizontal loads were developed on the foundation in addition to the moments. Similarly, in tests 1 and 3, small values of moment were recorded, despite the fact that only horizontal displacement was imposed on the foundation with out rotation. This means that Figs. 5(*b*) and 6 represent projections, onto the (V,M) and (V,H) planes, of various tracks across the (V,H,M) yield surface. Future tests involving simultaneous application of both horizontal displacement and rotation, in various ratios, will allow a much more complete definition of the full (V,H,M) yield surface in three dimensions. For the time being, however, it will be assumed that Fig. 5(*b*) defines the yield surface for $H = 0$, and Fig. 6 defines the yield surface for $M = 0$.

Normalisation of results and comparison with theory

In the theoretical model outlined above, the yield locus in (V,H) space at

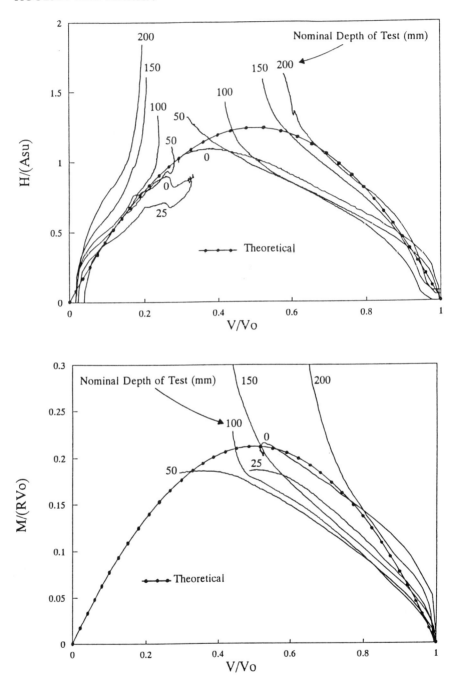

Fig. 7. Normalised yield surfaces

any embedment is assumed to be parabolic, with a maximum value H_0 given by $H_0/s_u A = 1 + 2A_h/A \approx 1.24$ for the model spudcan used. Thus, the experimental results may be normalised with respect to $s_u A$, and compared with the theoretical prediction of the shape and size of the yield locus in normalised (V,H) space. The results of tests 1 and 3 have been normalised in this way in Fig. 7(a), taking s_u to be the undrained strength at the level of the footing shoulder. It is encouraging to note that the experimental results from different embedments all trace out similar paths in normalised load space. The simple parabola adopted in the present theory also gives a reasonable approximation to the experimentally determined normalised yield surface.

The theoretical model suggests that the results of test 2 be plotted as M/RV_0 versus V/V_0. Again a parabolic variation is assumed in the theory, with a maximum moment value of $M_0/RV_0 = 2/3\pi \approx 0.21$.

Figure 7(b) shows both the experimental results and the theoretical prediction in normalised (V,M) space. Once again there is good agreement between the experimental curves, and the theoretical parabolic yield locus provides a fairly good description of the observed load paths.

Implications for design

The type of theoretical model described above bears a striking similarity to the Cam-Clay type of model in critical state soil mechanics. The vertical load is analogous to the mean stress, the horizontal load and moment analogous to the shear stress and the vertical penetration analogous to the specific volume. Such comparisons have been exploited very successfully by Tan (1990) for describing the behaviour of spudcan foundations on sand.

The analogy with critical state soil mechanics can be carried further. Critical state concepts can be used either to construct qualitative conceptual models for the behaviour of clays, or they can be used to construct precise theoretical models for the solution of boundary value problems, based on the theory of work hardening plasticity. The long term objective is to construct such models to describe the behaviour of spudcan foundations under all possible histories of loading. The necessary ingredients for such a model are as follows:

(a) A definition of the elastic response of the spudcan. This is thought to be adequately modelled by the stiffness factors developed by Bell, Houlsby and Burd (1991). The major uncertainty lies in the choice of the appropriate value of the shear modulus. It is well recognised that this is shear strain dependent, and it is difficult to assess the appropriate shear strain to represent the wide variety of stress and strain conditions beneath the footing.

(b) A definition of the load–penetration curve under purely vertical loading (the equivalent of the normal consolidation curve). As demonstrated in the section on vertical loading, a very satisfactory modelling of this aspect has been achieved. Although the shape of the curve is not represented by a mathematical expression (as in the case of a Cam-Clay model), it is defined numerically and can be employed in a numerical computation.

(c) A definition of the shape of the yield surface. As discussed above in the section on combined loading, the yield surface conforms quite closely to the simple mathematical form which has been chosen for preliminary calculations. Further experiments should define the shape of the surface with greater accuracy. A feature of the presently chosen surface which is different from Cam-Clay models is that it changes shape as well as size as penetration proceeds. This is because, although M_0/V_0 is chosen to be a constant, H_0/V_0 is not.

(d) A definition of the flow rule. For preliminary purposes this has been assumed as associated with the yield surface. This is likely to prove a correct assumption on the 'wet' side of critical (high vertical loads) but may well be inadequate on the 'dry' side, where there are features of the behaviour of the foundation which may be expected to lead to non-associated flow.

Caution should be exercised in carrying the analogy with critical state soil mechanics too far. For instance, during horizontal displacement the physical gouging out of a track by the spudcan leads to a strong anisotropy of the yield locus on reversal of the loading. Modelling of such phenomena is beyond the scope of the present work, as is the modelling of cyclic loading response. It should not be expected that a model based solely on the concepts of critical state will be able to capture all the features of the foundation behaviour.

Although the type of model described is a long term research goal, it is unlikely that numerical calculations using a rather complex work-hardening plasticity model are likely to be used in routine assessments of jack-up units in the foreseeable future. It is in the form of simpler qualitative models for the foundation behaviour that the analogy with critical state models is likely to prove valuable in the short term.

Most jack-up units operate with a preload of about twice the operating vertical load, so that the ratio V_0/V is about 2.0 before any lateral load is applied. This quantity is analogous to the overconsolidation ratio. The leeward leg(s) then experience horizontal and moment loads combined with an *increase* in vertical loads. The size of the yield locus can be used to estimate the loads that will cause yield, and since this yield occurs on the 'wet' side of critical it will be accompanied by an increase in vertical penetration. A good estimate of the relationship between the additional penetration and the magnitude of any rotation and horizontal displacement may be possible by assuming an associated flow rule. The acceptability of the unit may depend on whether it can tolerate this additional penetration.

On the other hand the windward leg(s) experience a *reduction* of vertical load as horizontal and moment loads are applied. Again the loads at which yield will occur can be estimated from the yield locus, but this time the yield occurs on the 'dry' side of critical, so additional penetration is not expected. There are good reasons to expect that the flow rule on the dry side of critical may be non-associated (because of problems such as contact breaking), and the detailed movements are hard to predict at present.

Conclusions

Experimental and theoretical studies of the behaviour of spudcan foundations on clay indicates that this can be modelled in terms of work-hardening plasticity theory. The shape of the yield surface in (V,H,M) space, and its evolution with additional penetration of the spudcan, can be described in a form which bears a striking similarity to the structure of critical state theory. Some details of the model are already well established, for instance the overall shape of the yield locus, its increase of size with foundation penetration and the elastic behaviour within the surface. Further work is required to clarify other aspects of the theory such as the details of the flow rule and special features associated with cyclic loading.

References

BELL, R.W. (1991). The Analysis of Offshore Foundations Subjected to Combined Loading. M.Sc. Thesis, Oxford University.

BELL, R.W., HOULSBY, G.T. AND BURD, H.J. (1991). Finite Element Analysis of Axisymmetric Footings Subjected to Combined Loads. Proceedings of the International Conference on Computer Methods and Advances in Geomechanics, Cairns, Australia, May 6–10, Vol. 3, pp. 1765–1770.

BRANSBY, P.L. (1973). Cambridge Contact Stress Transducers. CUED/C-Soils LN2, Cambridge University Engineering Department.

DEAN, E.T.R., JAMES, R.G., SCHOFIELD, A.N., TAN, F.S.C. AND TSUKAMOTO, Y. (1992). The Bearing Capacity of Conical Footings on sand in Relation to the Behaviour of Spudcan Footings of Jackups. Proc. Wroth Memorial Symposium, Oxford.

DE SANTA MARIA, P.E.L. (1988). Behaviour of Footings for Offshore Structures Under Combined Loads. DPhil Thesis, Oxford University.

HANSEN, J. BRINCH (1970). A Revised and Extended Formula for Bearing

Capacity. Bulletin No. 28, Danish Geotechnical Institute, Copenhagen, pp. 38–46.

HOULSBY, G.T. (1982). Theoretical Analysis of the Fall Cone Test. Géotechnique, Vol. 32, No. 2, June, pp. 111–118.

HOULSBY, G.T. AND WROTH, C.P. (1982). Direct Solution of Plasticity Problems in Soils by the Method of Characteristics. Proceedings of the 4th International Conference on Numerical Methods in Geomechanics, Vol. 3, Edmonton, June, pp. 1059–1071.

HOULSBY, G.T. AND WROTH, C.P. (1983). Calculation of Stresses on Shallow Penetrometers and Footings. Proceedings of the International Union of Theoretical and Applied Mechanics (IUTAM)/International Union of Geodesy and Geophysics (IUGG) Symposium on Seabed Mechanics, Newcastle upon Tyne, September, pp. 107–112.

MARTIN, C.M. (1991). The Behaviour of Jack-up Rig Foundations on Clay. First Year Research Report, Department of Engineering Science, Oxford University.

NOBLE DENTON AND ASSOCIATES (1987). Foundation Fixity of Jack-up Units: A Joint Industry Study. Confidential report, Noble Denton and Associates, London.

NOVA, R. AND MONTRASIO, L. (1991). Settlement of Shallow Foundations on Sand. Géotechnique, Vol. 41, No. 2, pp. 243–256.

TAN, F.S.C. (1990). Centrifuge and Theoretical Modelling of Conical Footings on Sand. Ph.D. Thesis, Cambridge University.

WROTH, C.P. AND HOULSBY, G.T. (1985). Soil Mechanics — Property Characterization and Analysis Procedures. Proc. 11th Int. Conf. Soil Mech. Found. Eng., San Francisco, Vol. 1, pp. 1–50.

Development of the cone pressuremeter

G.T. HOULSBY and N.R.F. NUTT, Department of Engineering Science, Oxford University

Methods of interpretation of the cone pressuremeter in clay and in sand are discussed. A method of analysis for the test in clay had previously been compared with results in stiff overconsolidated clay, and is compared here with results in soft clay at Bothkennar. Both shear strength and shear modulus can be measured successfully, and a discussion is presented of comparisons of shear modulus measured by different techniques. In sand interpretation methods are based on correlations determined from calibration chamber tests. The correlations can be used to estimate horizontal stress and relative density, and unload–reload loops can be used to determine stiffness. The important influence of horizontal stress is emphasised, as is the benefit of using relative density rather than voids ratio to quantify sand density. Comparisons are made with the self-boring pressuremeter and with the cone penetrometer, considering both costs and the quality of the information about soil properties which can be achieved with the tests.

Introduction

The development of a new in situ testing device involves the technical development of hardware and testing procedures, and also the establishment of analysis and interpretation procedures. The cone pressuremeter is a device for which the former process was relatively straightforward, since the instrument consists of a standard 15 cm^2 cone, with a pressuremeter mounted behind it. The pressuremeter is very similar to the Cambridge self-boring pressuremeter, differing principally in the much smaller diameter and larger length:diameter ratio (Withers, Schaap, Dalton, 1986). The establishment of analysis procedures has, however, been much less straightforward, since the cone pressuremeter subjects the soil to a more complex stress history than the self-boring pressuremeter. A considerable amount of research has been completed on interpretation methods, but the use of the new device is in its infancy as far as commercial use is concerned.

In situ tests can be used for predictive purposes broadly in two ways. Firstly the design values required can be correlated directly with the results of the test. This method is exemplified by the application of Menard pressuremeter tests in French practice to the design of foundations (Baguelin, Jezequel and Shields, 1978) or by the direct use of cone penetration test (CPT) results for the design of piles (Heijnen, 1974). Such methods are valuable in that they provide simple and often quite reliable solutions to standard problems such as pile capacity design. The main disadvantages are that they rely heavily on totally standardised test procedures, and they do not provide solutions to novel problems. Because they are based on locally derived empirical information they should only be used in new areas with the utmost caution. Establishment of new tests to be interpreted in this way involves a very considerable collection of field data and case histories.

The second approach is to use the in situ test to derive soil parameters such as strength and stiffness, and use these in the design process. This method has the advantage that it can be applied to novel as well as familiar problems, and is more readily adapted to new locations. Again there is a significant reliance on standardised test procedures if parameters are to be repeatable. The major task in establishing a new test is to demonstrate how the measured values correspond to those from existing methods, and demonstrate the confidence which can be attached to the new values. In principle it should be possible to demonstrate this confidence with rather less collection of empirical data than is needed for directly interpreted tests. The indirect approach offers therefore a faster route to acceptance of new methods. The indirect approach has also been the one most commonly adopted in UK geotechnical practice, and is the method used to interpret the cone pressuremeter.

The CPT is an outstanding device for soil profiling. It also provides quite reliable evidence of soil strength, although precise strengths of clay rely on locally determined values of N_{kt}, and the estimation of strength of sands is somewhat approximate. The CPT does not give a reliable estimate of either stiffness or in situ stress. The main rationale behind the development of the cone pressuremeter test (CPMT) was to provide a measure of the soil stiffness. Any additional information it provides may be regarded as a bonus.

The cone pressuremeter in clay

An analysis has been developed for the CPMT in clay, and this has been compared to test results obtained in stiff clay at Madingley (Houlsby and Withers, 1988). In this paper the analysis is also applied to tests in soft clay at Bothkennar.

Analysis
The analysis of the CPMT is based on cylindrical cavity expansion theory. The principal difference from the analysis of the SBPM is that

DEVELOPMENT OF CONE PRESSUREMETER

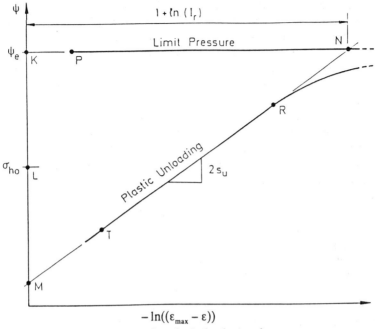

Fig. 1. Graphical representation of analytical solution for cone pressuremeter test in clay

the installation of the pressuremeter is itself regarded as a cavity expansion from zero radius. This of course only approximates the actual installation process. The result is that the entire expansion is predicted as occurring at a limit pressure of:

$$\psi_1 = \sigma_{ho} + s_u(1 + \log_e I_r) \qquad (1)$$

Where $I_r = G/s_u$, the rigidity index. After an initial unloading at a stiffness of $2G$ to a pressure of $\psi_1 - 2s_u$, the unloading curve is then given by:

$$\psi = \psi_1 - 2s_u \left[1 + \log_e \left(\frac{\sinh(\varepsilon_{max} - \varepsilon)}{\sinh(1/I_r)} \right) \right] \qquad (2)$$

This leads to the simple geometric construction in Fig. 1 for the determination of s_u, G, and σ_{ho}. The shear modulus may alternatively be determined from small unload–reload loops, although the values obtained by this method can be obscured by the effects of partial drainage.

Tests in stiff clay
The analysis was originally calibrated against tests in stiff clay at

Madingley (Houlsby and Withers, 1988), demonstrating that the above method gave both strength and stiffness values similar to those from other tests. The horizontal stress values determined using the above procedure were unreasonably high, and a suspicion that this was related to the finite length of the pressuremeter was confirmed by a finite element analysis by Yu (1990).

Tests in soft clay
Twelve tests using the 15 cm^2 cone pressuremeter were recently carried out at the SERC soft clay site at Bothkennar in Scotland. Three soundings were completed 6 m apart on a triangular grid with the maximum depth of penetration ranging from 12 m to 18 m. Before inserting the pressuremeter into the ground it was necessary to push a dummy cone through a firm crust at the top 1.5 m of each probe location.

The cone pressuremeter was pushed into the ground at the standard rate of 2 cm/s, with continuous readings of cone resistance and friction ratio. The centre of the pressuremeter was located one metre behind the cone tip, and so penetration was stopped with the tip that distance deeper than the test location. The time between the end of the cone penetration and the start of the pressuremeter inflation varied between 4 and 7 minutes.

The membrane was inflated at a rate of 2% per minute until one of the strain arms registered a radial expansion of 50%. At three intervals during the test, inflation was stopped and the strain held for one to two minutes, followed by a rapid unload–reload loop. Upon reaching maximum expansion, the pressuremeter was then unloaded at a rate of 2% per minute until the unloading slope became close to horizontal at which point the pressuremeter was vented.

The measured pressure–expansion curves have to be corrected to account for two effects

(*a*) the stiffness of the membrane, which can be obtained by carrying out an expansion in air, and

(*b*) the compliance of the pressuremeter, which can be determined by carrying out a test in which the pressuremeter is enclosed in a steel cylinder.

In soft materials it is the first of these calibrations which is most important, since the stiffness of the instrument is high compared to the stiffness of soft soils. The pressure required to inflate the membrane in air is, however, significant by comparison with the pressures obtained during tests in soft soils. At the shallowest depths the maximum pressures obtained in air are up to 35% of the limit pressures obtained in the soil.

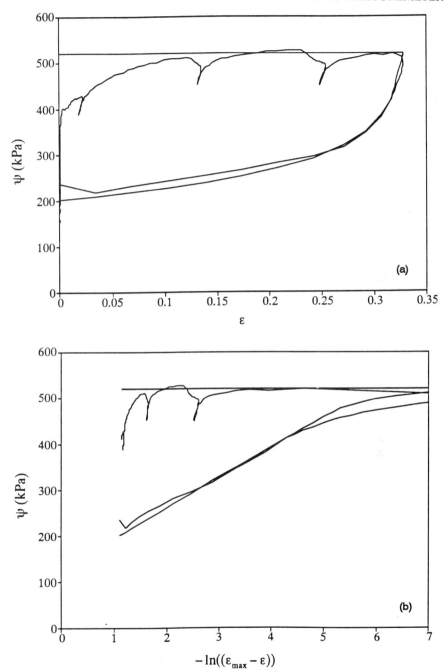

Fig. 2. (a) Measured and fitted pressure-expansion curve for test B1T5 at Bothkennar. (b) Measured and fitted logarithmic plot of test B1T5

Comparisons have been made between the results obtained from the CPMT and the results of the initial investigations at Bothkennar, reported by Nash et al. (1992). A wide range of field and laboratory tests including self-boring pressuremeter tests, dilatometer tests, field vane tests, seismic cone penetration tests and laboratory tests on specimens recovered using a piston sampler and the Laval sampler, have been carried out at Bothkennar, so that the ground conditions there are very well documented.

The results from the cone pressuremeter have been reduced using the above analysis and estimates of the undrained strength, s_u, shear modulus, G_{cc} (from the cavity contraction curve) and lateral horizontal stress, σ_{ho} have been made. The pressuremeter curve from Test B1T5 at a depth of 15 m is shown in Fig. 2(a), with the theoretical curve from the Houlsby and Withers analysis superimposed. A close fitting of the theory to the test is evident, and as shown in Fig. 2(b) it is relatively simple to determine the best straight line fit and limit pressure from the plot of pressure against $-\ln(\varepsilon_{max} - \varepsilon)$. Performing this construction on all of the pressuremeter curves resulted in the soil parameters listed in Table 1. Also given in this table are the values of ψ_1, the limit pressure from the pressuremeter test and q_t, the tip resistance from the cone penetration test.

The undrained shear strength, s_u, has been compared to estimates made from field vane tests, undrained unconsolidated triaxial tests and SBPM tests as reported by Nash et al. (1992), and shown in Fig. 3(a). A consistent trend of increasing s_u with depth is evident, and the CPMT estimates are close to the values from other tests, although slightly lower.

Table 1. Results of cone pressuremeter tests at Bothkennar

Test	Depth (m)	ψ_1 (kPa)	q_t (kPa)	s_u (kPa)	G_{cc} (MPa)	σ_{ho} (kPa)	I_r
B1T1	3	169.6	298	14.5	1.71	85.9	117.8
B2T1	4	174.1	324	14.2	1.38	95.0	97.3
B3T1	5	197.0	364	16.9	1.56	103.4	92.1
B1T2	6	244.3	410	19.3	1.43	141.7	73.9
B2T2	7	257.1	448	17.6	1.92	156.8	109.0
B3T3	8	287.2	484	21.2	1.73	172.5	81.5
B1T3	9	325.8	513	21.5	2.27	204.2	105.7
B2T3	10	347.2	556	23.6	2.64	212.4	112.0
B1T4	12	436.2	630	27.1	4.22	272.4	155.9
B2T4	13	449.0	680	26.3	3.91	291.0	148.4
B1T5	15	518.9	807	32.9	4.46	324.4	135.5
B2T5	16	544.9	800	34.8	6.12	330.2	175.9

DEVELOPMENT OF CONE PRESSUREMETER

Estimates of σ_{ho} made from the CPMT are given in Table 1. The value of K_0 estimated from the lift-off pressure from SBPM tests, and from empirical correlation from DMT tests is typically in the range 0.7 to 1.0 throughout the deposit. By comparison the interpretation of the horizontal stresses from the CPMT given in Table 1 would indicate a value of K_0 of about 1.5, even after accounting for the fact that the correction suggested by Yu (1990) has been used to account for the finite length of the pressuremeter. This reinforces the conclusion made from the results at Madingley, that at present it is not possible to measure the horizontal stress in clay with the cone pressuremeter. The indicated variation of horizontal stress does, however, vary consistently with depth, and in future it may be possible to get some indication of K_0 by empirical correlation.

The estimates of shear modulus from the CPMT are plotted in Fig. 3(b) along with values from the seismic cone test (SCPT) and the SBPM test. The obvious variation of increasing G with depth is clear, and the large difference in the magnitude of G between the in situ devices is due to the fact that the tests measure the stiffness by imposing very different strain amplitudes on the soil. When normalised by the values from the SCPT as shown in Fig. 3(c) it can be seen that both the CPMT and SBPM show a similar magnitude of variation. Small unload–reload loops from the SBPM give a shear modulus about one quarter of the value measured at very small strain by the seismic cone, and the interpretation of the unloading curve of the CPMT gives a modulus about half as small again.

Table 2. Unload–reload moduli at Bothkennar

Test	Depth (m)	Loop 1 G_{ur} (MPa)	Loop 1 $\Delta\varepsilon$	Loop 2 G_{ur} (MPa)	Loop 2 $\Delta\varepsilon$	Loop 3 G_{ur} (MPa)	Loop 3 $\Delta\varepsilon$
B1T1	3	—	—	1.1	0.0102	0.9	0.0046
B2T1	4	1.3	0.0048	1.0	0.0104	0.8	0.0161
B3T1	5	1.7	0.0059	1.2	0.0093	1.2	0.0075
B1T2	6	2.1	0.0037	1.7	0.0090	1.1	0.0145
B2T2	7	2.1	0.0070	1.6	0.0091	1.2	0.0163
B3T3	8	1.6	0.0070	1.9	0.0078	1.5	0.0106
B1T3	9	2.5	0.0054	1.9	0.0061	1.4	0.0075
B2T3	10	3.0	0.0044	2.9	0.0036	2.3	0.0056
B1T4	12	4.1	0.0034	3.0	0.0053	2.3	0.0055
B2T4	13	2.7	0.0113	2.3	0.0188	1.6	0.0236
B1T5	15	4.9	0.0023	4.1	0.0023	3.3	0.0044
B2T5	16	5.3	0.0071	3.7	0.0103	3.3	0.0121

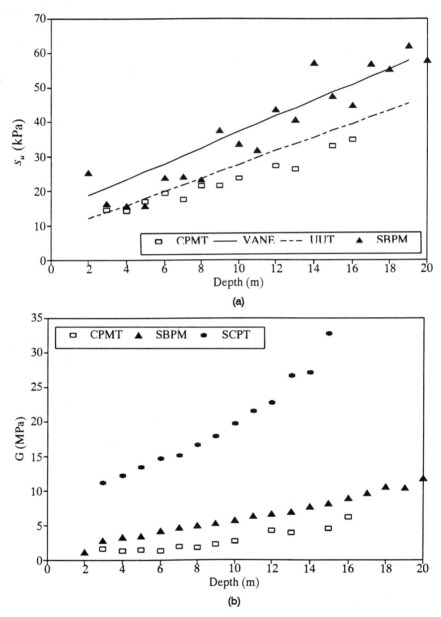

Fig. 3 (above and facing page). (a) CPMT estimates of shear strength compared with vane, triaxial and SBPM data. (b) CPMT estimates of shear modulus compared with SBPM and seismic cone data. (c) CPMT and SBPM shear moduli normalised by SCPT moduli

DEVELOPMENT OF CONE PRESSUREMETER

Shear moduli have also been determined from the unload–reload curves obtained at Bothkennar, and the results are presented in Table 2. The modulus has been determined both by using the slope of the chord through the extreme points of the unload–reload loop (the figures given in Table 2) and also by using a least-squares fit to the whole loop (Schnaid and Houlsby, 1991). The two methods agree closely, with the least-squares method giving consistently lower modulus values by 14%. It is convenient to normalise the shear moduli by the estimated mean normal stress, and the normalised moduli are then expected to be a function of the strain amplitude. Figure 4 confirms that this is the case, with a slight trend observed of lower normalised modulus at higher strain amplitude.

The range of the normalised values from the SBPM tests is also shown at the strain amplitude reported for these tests. The moduli measured by the CPMT seem to be rather lower than those measured by the SBPM at the same strain amplitude, although more detailed information about the SBPM tests is required. The difference may be due to the additional disturbance caused by the CPMT, and this is supported by the fact that the moduli measured by unload–reload loops systematically decrease from loop 1 to loop 3 (see Table 2). When the moduli measured by the SCPT are included as representative of a very small strain amplitude (equivalent to less than 0.000025 cavity strain), the variation of modulus with strain amplitude is broadly similar to that reported by Georgiannou et al. (1991).

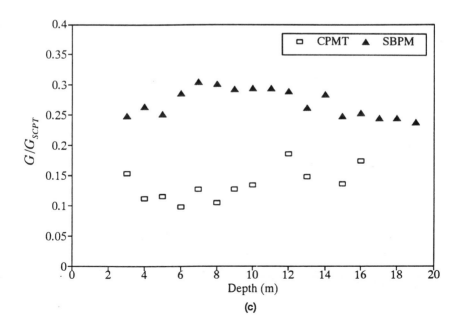

(c)

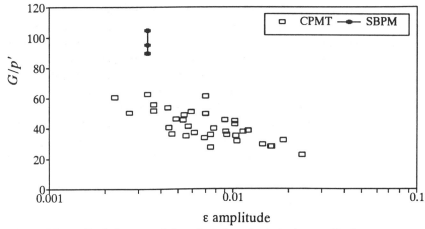

Fig. 4. *Normalised shear modulus plotted against strain amplitude*

The shear moduli G_{cc} measured from the cavity contraction curve correspond broadly to the values obtained from unload–reload loops with a strain amplitude of about 0.3% to 0.4%.

The cone pressuremeter in sand
Analysis
The interpretation of the cone pressuremeter is not as straightforward as the interpretation of the SBPM, particularly in dilatant materials. This is due primarily to two aspects. Firstly, there exists a zone of disturbed material surrounding the pressuremeter due to the penetration of the cone, whereas this zone should not be present around the SBPM, and secondly the cone pressuremeter is capable of larger radial membrane expansions than its self-boring counterpart. The analyses appropriate to the self-boring pressuremeter are not therefore appropriate for the cone pressuremeter. In clay, the analysis by Houlsby and Withers (1988) has modelled the behaviour of the cone pressuremeter expansion and contraction test successfully assuming an incompressible, elastic, perfectly plastic soil. In sand, however, the phenomenon of dilatancy must be taken into account. Hughes et al. (1977) account for this in an analysis of the self-boring pressuremeter expansion test, assuming the soil to be failing in conditions of plane strain at constant stress ratio and constant rate of dilation. This small strain analysis is not applicable to the cone pressuremeter test because of the large shear strains induced in the plastic zone. No closed form solution incorporating large strains and dilatancy exists for pressuremeter expansion or contraction in sand, although recent work by Yu (1990) has provided an analysis expressed in terms of a single power series expansion.

Even more inadequate, however, are the current theories for analysing the cone penetration resistance. Bearing capacity theories are inadequate because they fail to apply correctly the boundary conditions for the test, while spherical cavity expansion underpredicts the observed values of the cone factor. Finite element methods fail to make proper consideration of the steady state nature of deep cone penetration.

Understanding of the behaviour of the cone pressuremeter in sands is lacking not only in a rigorous theoretical analysis, but is also without the support of any large quantity of field data. Emphasis therefore is placed on calibration chamber tests, where control of relative density, stress level and stress ratio can easily be maintained. The calibration chamber at Oxford allows such control and has been used for a large number of tests in two very different types of sand.

Tests in silica sand

Schnaid (1990) carried out a series of 34 tests in the Oxford calibration chamber using 5 cm^2, 10 cm^2 and 15 cm^2 cone pressuremeters in Leighton Buzzard 14/25 sand, a uniformly graded silica sand. By correlating measured values of CPMT limit pressure ψ_l and cone resistance q_t with the relative density, stress level and stress ratio, a set of empirical relationships were determined which could then be used to obtain independent estimates of relative density and horizontal confining stress σ'_{ho}. The same series of tests showed that good estimates of the elastic shear modulus could also be made from the gradient of the pressuremeter unload–reload loops.

Tests in carbonate sand

More recently, a carbonate sand has been used to study the applicability of the findings of Schnaid to different frictional materials. A series of 30 tests was carried out in the Oxford calibration chamber with a 10 cm^2 cone pressuremeter using sand from Dogs Bay on the west coast of Ireland. Details of the test procedure are reported by Nutt and Houlsby (1991).

The test results verify the studies made by Houlsby and Hitchman (1990) and Schnaid (1990) which showed that it is the horizontal confining stress σ'_h and not the vertical confining stress σ'_v which controls the behaviour of both the cone and pressuremeter tests. Figure 5(a) shows the relationship between the limit pressure normalised by the horizontal stress and the relative density. Here the results from both the silica sand and carbonate sand tests have been plotted together and fitted by the relationship:

$$\frac{\psi_l - \sigma_{ho}}{\sigma'_{ho}} = 2.21 + 19.35 R_d \qquad (3)$$

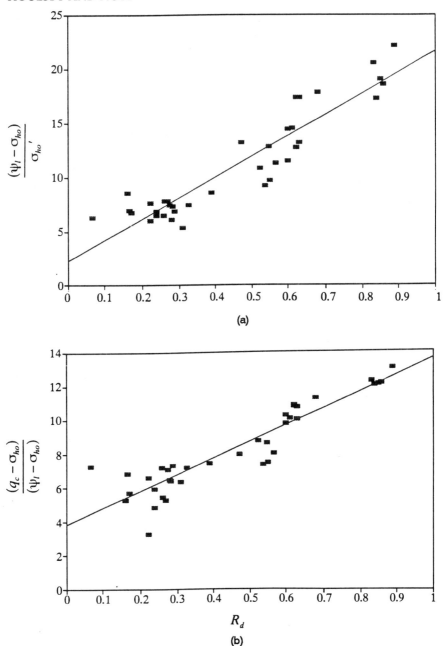

Fig. 5 (above and facing page). (a) Correlation between CPMT limit pressure and relative density. (b) Correlation between cone resistance normalised by limit pressure and relative density. (c) Correlation between cone resistance and relative density

Similarly the ratio of cone resistance to limit pressure is shown in Fig. 5(b), and the relative density estimated by the relationship:

$$\frac{q_t - \sigma_{ho}}{\psi_1 - \sigma_{ho}} = 3.80 + 9.84 R_d \quad (4)$$

where R_d is expressed as a ratio between 0.0 and 1.0.

It seems reasonable therefore to combine these two expressions to obtain an estimate of the cone factor $(q_t - \sigma_{ho})/\sigma'_{ho}$ against relative density. Figure 5(c) shows that the parabolic relationship which results is in fact a better fit to the data than the linear relationship which was assumed by Schnaid (1991). A back-analysis procedure then makes it possible to obtain estimates of R_d and σ'_{ho} which are compared to the actual measured values of R_d and σ'_{ho} measured in each test. In Fig. 6 the measured and estimated values of R_d are compared, where it can be seen that for most of the tests the relative density is predicted to within 10% of the measured value.

The importance of interpreting the cone pressuremeter data in terms of relative density as opposed to voids ratio is illustrated by the plots of Figs. 7(a) and (b). In addition to the cone results from the Dogs Bay and Leighton Buzzard sand (which have been normalised with respect to the mean effective stress), results from Almeida et al. (1991) have also been

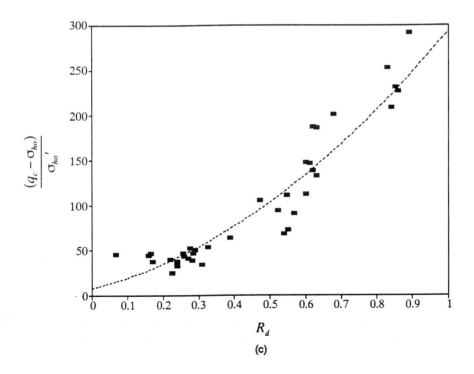

(c)

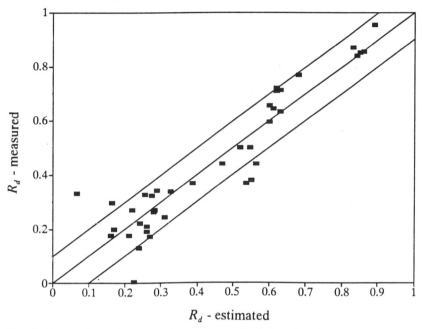

Fig. 6. *Comparison of measured and observed relative density*

included. Figure 7(a) shows that no unique relationship exists between voids ratio, e, and normalised cone resistance, each sand type falling within a different range in the plot. In contrast, Fig. 7(b) shows that when plotted against relative density, a more consistent trend is evident. These results are in contrast to Semple (1988), who suggests that the behaviour of carbonate sand can best be explained by its initial voids ratio. The benefits of using relative density rather than voids are clear, in that the former provides a normalised measure of density which at least in part accounts for the different characteristics of different sands. The disadvantage is that it is less directly defined, depending on standard methods for measuring maximum and minimum voids ratios. The adoption of standard procedures for these density measurements is an important priority.

The use of unload–reload loops in the tests in carbonate sand have also yielded estimates of elastic shear modulus. It has been found, however, that these estimates can be affected by creep strains occurring during the test. It is necessary to carry out pressuremeter tests with a procedure that allows for the creep strains first to decrease to a negligible rate. This is best done by holding the pressure constant prior to unloading, a procedure also reported by Withers et al. (1989).

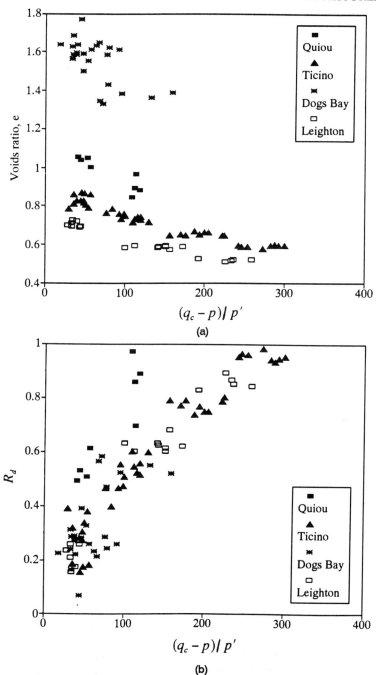

Fig. 7. (a) Correlation of voids ratio with cone resistance. (b) Correlation of relative density with cone resistance

The rôle of the new test

The cone pressuremeter test logically invites comparison with the cone penetration test and the self-boring pressuremeter. What information in addition to the cone test does it provide, and at what extra cost? Can it provide comparable information to the self-boring pressuremeter test, but at reduced cost? Experience is at present limited, but some preliminary conclusions can be drawn which suggest a promising future for the CPMT. Comparisons in terms of equipment needs and operational time are preferred to a direct examination of costs. The following comparisons should, however, be regarded as approximate.

The piezocone will give adequate profiling capability in most materials, and allow identification of broad material types. In clay it should give a reasonable estimate of strength, and (through the pore pressure readings) an indication of OCR. Dissipation tests may provide an estimate of c_h. In sands only a qualitative indication of strength is possible ('dense' or 'loose'), and any estimate of the angle of friction is subject to a wide margin of error. In clays the CPMT could add to this

(a) a second, and more reliable measure of the undrained strength, and

(b) an indication of stiffness, either from the unload–reload loops or from the overall position of the unloading curve.

The interpretation of this latter stiffness is complicated by the difficulty in choosing the appropriate strain magnitude for comparison with other tests. The additional information is based on a theoretical analysis and field tests.

In sand the interpretation of the CPMT is based principally on large calibration tests. The interpretation would yield

(a) an indication of the relative density, providing an angle of friction almost certainly of greater accuracy than that obtained from the CPT,

(b) an indication of horizontal stress, and

(c) measurements of stiffness from unload–reload loops.

The detailed information in sand is therefore much greater from the CPMT. Field experience has also shown the CPMT to be reliable and relatively trouble-free.

Cone testing involves a significant investment in capital equipment to drive the cone and log the data. In addition the cone pressuremeter requires very little. Apart from the device itself and pressure development equipment, only additional datalogging capacity is needed. After initial site mobilisation a CPT to 20 m can be completed in about 1.5 hours, say six tests per working day. Each pressuremeter test takes approximately one hour, so a CPMT sounding with six pressuremeter tests would also take about one day. A programme of 12 CPT tests and one CPMT would therefore take 50% longer in time than the CPT tests alone. In clay the additional confidence in the s_u readings and the estimation of stiffness may not be worth the cost involved, but in sand the additional quantitative information on strength, stiffness and horizontal stress would amply reward the cost in many cases.

The self-boring pressuremeter (SBPM) is used principally where there is already knowledge of the soil profile. It can be used reliably to determine strength, stiffness and horizontal stress in clay, although strength measurements are known often to be high, and the interpretation of horizontal stress requires considerable experience. In sand reliable measurements of stiffness can be made, but both the strength and horizontal stress are very difficult to determine reliably. Problems related to the drilling of the device into sand may be the root cause of the difficulties, and recent advances in drilling procedures and interpretation methods are allowing greater confidence in this area.

In clay the cone pressuremeter could give strength and stiffness measurements as reliably as the SBPM, but would not give horizontal stress. In sand the stiffness measurements would be comparable to those from the SBPM, and the indications of horizontal stress and of strength would in most cases also be as good as could be obtained by the SBPM.

The equipment needed to install the SBPM is simple and relatively inexpensive. The drilling process is, however, relatively slow, so that a 20 m hole with six tests may take two to three days, i.e. at least twice as long as the CPMT. In clays the reliability of determination of horizontal stress may make the additional time worthwhile, but in sands the CPMT must be seen as a strong competitor.

It is important to note that the CPMT would act as a supplement to CPT tests, but as a substitute for SBPM tests. Although possibly justified in some cases, it is likely that in clay the information additional to the CPT is too small to justify the cost, whilst the SBPM would be needed for horizontal stress measurements. In sands the cost additional to the CPT would be well justified, especially as the CPMT can probably provide the necessary information as reliably as the SBPM.

The above conclusions are all based on comparisons for onshore testing. Offshore the situation is quite different, and operational differences mean that the CPMT would be very much easier to use than the SBPM. For both sands and clays the CPMT must therefore have an important rôle to play offshore. One of the principal purposes of current testing is to establish testing procedures onshore before proceeding to offshore use.

Conclusions
Methods of interpretation of the cone pressuremeter in both sand and

clay have been discussed. The quality of the data that can be obtained in clay for undrained strength and shear modulus, and in sand the relative density, shear modulus and horizontal stress, amply justify the development of this new test.

References

ALMEIDA, M.S.S., JAMIOLKOWSKI, M. AND PETERSON, R.W. (1991). Preliminary results of CPT tests in Calcareous Quiou sand. Proceedings of the First International Conference on Calibration Testing, Potsdam, New York, June 28–29, pp. 41–54.

BAGUELIN, F., JEZEQUEL, J.F. AND SHIELDS, D.H. (1978). The Pressuremeter and Foundation Engineering. Trans Tech Publications.

DALTON, J.C.P., SCHAAP, L.H.J. AND WITHERS, N.J. (1986). The Development of a Full Displacement Pressuremeter. Proc. 2nd Int. Symp. on The Pressuremeter and Its Marine Applications, Texas A & M University, pp. 38–56.

GEORGIANNOU, V.N., RAMPELLO, S. AND SILVESTRI, F. (1991). Static and Dynamic Measurements of Undrained Stiffness on Natural Overconsolidated Clays. Proc. X ECSMFE, Vol. 1, pp. 91–96.

HEIJNEN, W.J. (1974). Penetration Testing in the Netherlands. Proc. Eur. Symp. on Penetration Testing, Stockholm, Vol. 1, pp. 79–84.

HOULSBY, G.T. AND HITCHMAN, R. (1988). Calibration Chamber Tests of a Cone Penetrometer in Sand. Géotechnique, Vol. 38, No. 1, March, pp. 39–44. Reply to discussion: Géotechnique, Vol. 39, No. 4, December 1989, pp. 729–731.

HOULSBY, G.T. AND WITHERS, N.J. (1988). Analysis of the Cone Pressuremeter Test in Clay. Géotechnique, Vol. 38, No. 4, December, pp. 575–587.

HOULSBY, G.T. AND YU, H.S. (1990). Finite Element Analysis of the Cone-Pressuremeter Test. Proceedings of the 3rd International Symposium on Pressuremeters, ISP3, Oxford, April 2–6, pp. 221–230.

NASH, D.F.T., POWELL, J.J.M. AND LLOYD, I.M. (1992). Initial Investigations of the Soft Clay Test Bed Site at Bothkennar. Géotechnique, Vol. 42, June, 163–181.

NUTT, N.R.F. AND HOULSBY, G.T. (1991). Calibration Tests on the Cone Pressuremeter in Carbonate Sand. Proceedings of the First International Conference on Calibration Testing, Potsdam, New York, June 28–29, pp. 265–276.

SCHNAID, F. AND HOULSBY, G.T. (1990). Calibration Tests of the Cone-Pressuremeter in Sand. Proceedings of the 3rd International Symposium on Pressuremeters, ISP3, Oxford, April 2–6, pp. 263–272.

SCHNAID, F. AND HOULSBY, G.T. (1992). Measurement of the Properties of Sand by the Cone-Pressuremeter Test. Géotechnique, Vol. 42, No. 4, pp. 587-602.

WITHERS, N.J., HOWIE, J., HUGHES, J.M.O. AND ROBERTSON, P.K. (1989). Performance and Analysis of Cone Pressuremeter Tests in Sands. Géotechnique Vol. 39, No. 3 September, pp. 433–454.

YU, H.S. AND HOULSBY, G.T. (1991). Finite Cavity Expansion in Dilatant Soils: Loading Analysis. Géotechnique, Vol. 41, No. 2, pp. 173–183.

Predicting the effect of boundary forces on the behaviour of reinforced soil walls

R.A. JEWELL, H.J. BURD and G.W.E. MILLIGAN,
Department of Engineering Science, University of Oxford

Practical design methods for reinforced soil retaining walls are often based on limit-equilibrium methods of analysis. It is often found that displacements and reinforcement forces measured after wall construction are substantially lower than those implied in the design analysis. It is suggested that in order to make satisfactory predictions of wall behaviour it is necessary to include the effects of boundary forces which are not incorporated directly in the current design methods. A method of calculating displacements in reinforced soil walls in the absence of boundary forces is described and an analysis is developed in which the effect of boundary forces on the wall displacements is included. This analysis is used to discuss the effects of boundary forces on the measurements made during recent published reinforced soil wall tests.

Introduction

Most current design procedures for reinforced soil walls are based on limit-equilibrium methods of analysis in which an assessment is made of the forces that need to be supplied by the reinforcement to prevent failure. Limit-equilibrium design methods involve the study of wall failure and the choice of reinforcement to ensure that the possibility of failure is sufficiently small. It is often found after construction, however, that the displacements occurring in reinforced soil walls are surprisingly small. This has raised the question as to whether current design methods are over-conservative.

The factors which contribute to this apparent conservatism in the design of reinforced soil walls are the subject of this paper, and touch on the three main themes of the Wroth Memorial Symposium, namely soil properties and their measurement, prediction and performance, and design. A significant factor is the high peak angle of friction in compact granular soil at low effective stress, a case which would apply in many reinforced soil walls. The appropriate value of peak friction angle for a particular backfill soil is often not identified by standard test methods and interpretation procedures. A further important factor is the presence of external boundary forces mobilised at the base of the wall face and, for the case of a test wall, frictional forces mobilised on the sidewalls of the test chamber. These boundary forces may reduce the reinforcement forces by a significant amount and this has a corresponding effect on the magnitude of front wall displacement. This paper is concerned primarily with the analysis of the effect of these boundary forces on wall displacements and reinforcement forces.

In order to make realistic *predictions* of deformations and forces in reinforced soil walls it is important to include all of the relevant details, described above, which may influence wall behaviour. Some of these factors, however, would normally not be considered reliable when calculating the *design* strength of the wall. Inadequate compaction, for example, can reduce markedly the peak angle of friction in granular soil, and a shallow trench dug at the toe of a wall (to install or repair a utility for example) would usually suffice to eliminate the forces acting at the base of the wall face. In order to *predict* the behaviour of a reinforced soil wall it is necessary to consider the expected conditions that govern wall performance; in a *design* calculation it is necessary to consider the less likely but possible conditions which might lead to collapse.

The above points are discussed in detail with reference to the recently published results of two sets of test walls. The first set of test walls were part of a prediction symposium and are described by Bathurst et al. (1988). Two walls from this test series are described in this paper; an unreinforced wall and a wall with four layers of relatively stiff geogrid reinforcement. The results of the unreinforced test are used to study the effects of sidewall friction on the active forces developed in a conventional retaining wall; the second test is used as the basis of a discussion of the prediction of reinforced soil wall displacements under working conditions. The second set of tests, carried out using the same sand backfill in the same test facility but with a relatively flexible reinforcement is described by Bathurst and Benjamin (1990) and Bathurst et al. (1992). In these tests a surcharge was applied to the top of the wall to induce failure.

The predictions of reinforced soil wall performance described in this paper are based on the limit-equilibrium method described by Jewell and Milligan (1989). This method, which does not include the effect of boundary forces, may be used to derive a series of design charts for wall displacements which are given by Jewell and Milligan (1989). This approach is extended here to include the effects of boundary forces in the analysis. Further studies in which this extended analysis is shown to compare satisfactorily with the results of finite element studies and the measured performance of full scale walls are given in Jewell and Burd (1993) and Jewell and Milligan (1993).

This paper is concerned primarily with the prediction of front wall

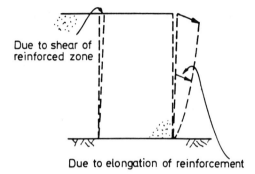

Fig. 1. *Wall displacements*

EQUILIBRIUM IN REINFORCED SOIL WALLS
Equilibrium without boundary forces

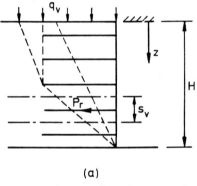

Fig. 2. *Distribution of reinforcement force*

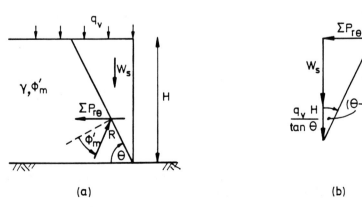

Fig. 3. *Force equilibrium*

displacements in reinforced soil walls built on rigid foundations. These displacements, in this case, consist of two components as shown in Fig. 1. Forces generated in the reinforcement cause elongations which contribute to front wall movement. It is also possible that the lateral stresses in the unreinforced backfill may cause additional shear displacements in the reinforced zone. This effect is normally only significant when a short reinforcement length is used and is not considered in this paper.

Equilibrium in reinforced soil walls
Equilibrium without boundary forces

The study of equilibrium in reinforced soil walls is based on many of the features of classical active earth pressure theory. Instead of providing the necessary stabilising forces externally, however, the required forces for equilibrium are transmitted back into the fill by the reinforcement in order to maintain equilibrium on all potential failure mechanisms. The maximum reinforcement forces required for equilibrium may be estimated using a stress analysis in which the soil is assumed to be in a state of active failure as described below with reference to the wall shown in Fig. 2(a). The self-weight of the backfill soil is γ and the surcharge applied to the top of the wall is q_v. The horizontal stresses required for equilibrium, σ_r, for the case of a smooth front wall are assumed to be $K_a(q_v + \gamma z)$ as shown in Fig. 2(b) where K_a is the coefficient of active earth pressure. The required force in a particular layer of reinforcement may therefore be assumed to be $\sigma_r s_v$ where s_v is the local reinforcement spacing as shown on Fig. 2(a). For the wall to be stable, sufficient reinforcement must be provided to sustain this force and the reinforcement must extend sufficiently far into the backfill soil to provide equilibrium on all potential failure mechanisms as shown in Fig. 2(a). Further details of this active stress analysis are given by Jewell (1990). An equivalent calculation of the reinforcement force required for stability may be performed using a limit-equilibrium analysis of potential wedge failure mechanisms as described below. Consider the equilibrium in a wall subject to uniform vertical surcharge as shown in Fig. 3(a). An equation for the total required force for equilibrium, $\Sigma P_{r\theta}$, may be found from a conventional active wedge analysis as illustrated in Fig. 3(b) to give

$$\Sigma P_{r\theta} = \left(W_s + q_v \frac{H}{\tan \theta} \right) \tan(\theta - \phi'_m) \qquad (1)$$

where θ is the angle of inclination of the assumed failure plane, W_s is the weight of the failing soil block, and ϕ'_m is the mobilised angle of friction of the soil. The inclination of the assumed failure plane, $\theta_{\text{crit.}}$, for which the required reinforcement force is maximised, may be found analytically and is given by $(45° + \phi'_m/2)$. In this case the maximum total reinforcement force, ΣP_r, may be shown using eqn. (1) to be

$$\Sigma P_r = K_a \frac{\gamma H^2}{2}\left(1 + \frac{2q_v}{\gamma H}\right) \qquad (2)$$

The active earth pressure coefficient, K_a, is given by:

$$K_a = \tan^2\left(45° - \frac{\phi'_m}{2}\right) \qquad (3)$$

In this case the active stress and limit-equilibrium analyses described above give the same result for the total maximum required reinforcement force.

Equilibrium allowing for external boundary forces
The limit-equilibrium analysis described above may be extended to the case where boundary forces are included as discussed below. Consider a wall with a continuous facing of weight W_f which is subjected to vertical and horizontal external forces at the base of the facing panel, P_v and P_h, as shown in Fig. 4(a). In the case of experimental walls, an additional external force, F, associated with sidewall friction, may act on the sides of the failing soil to resist the displacement in the soil. The sidewall force is shown acting at an angle of θ_f with respect to the assumed failure plane in Fig. 4(a). An appropriate choice for this angle would be $\theta_f = \psi$, where ψ is the angle of soil dilation.

The force polygon for a trial wedge, allowing for all the boundary forces, is shown in Fig. 4(b). The total required reinforcement force in this case is found to be

$$\Sigma P_{r\theta} = (W_s + W_f + q_v H \cot\theta - P_v)\tan(\theta - \phi'_m) - P_h - F\frac{\cos(\theta_f - \phi'_m)}{\cos(\theta - \phi'_m)} \qquad (4)$$

A numerical search is necessary in order to find the angle, θ, of the critical failure plane which requires the maximum reinforcement force for equilibrium.

Analysis of sidewall friction
Sidewall friction can influence the measured performance of large-scale experimental walls and is a source of uncertainty for the back-analysis of large scale tests, such as those built at the Royal Military College (RMC), Canada (Bathurst et al. (1988) and Bathurst and Benjamin (1990)).

Until recently, the main study of the effect of sidewall friction on the results of retaining wall tests was that of Bransby and Smith (1975). In this study a series of active and passive unreinforced retaining walls was

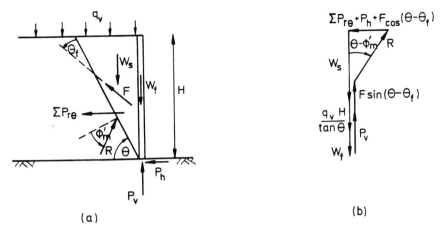

Fig. 4. *Force equilibrium including boundary forces*

examined using a two-dimensional method of characteristics analysis to obtain the stresses in the zone of plasticity behind the wall. The effect of the sidewalls was included by the application of body forces equivalent to the assumed shear stresses mobilised between the deforming soil and the sidewalls. Unfortunately, only a few combinations of wall height to width ratio, angle of internal friction of the soil and sidewall roughness were investigated.

To assess the influence of sidewall friction on the RMC test walls that are discussed in detail later in this paper, Bathurst and Benjamin (1988) and Jewell (1988) developed limit-equilibrium methods which provided a satisfactory match with the results given by Bransby and Smith (1975). A detailed analysis of sidewall friction, which is a development of these preliminary studies, is given below.

Calculation of sidewall friction force

The total frictional force generated by the sidewalls, F, is assumed to oppose the movement of the backfill with respect to the sidewalls as shown in Fig. 4(a). The magnitude of the force depends on the normal stress on the sidewalls, their roughness and the area of contact with the failing soil. The effect of this sidewall force on the total required reinforcement force may be deduced from eqn. (4).

The state of stress in the failing wedge of soil is not precisely known because of the influence of external boundary forces, P_v and P_h, and because of sidewall friction itself. Burd (1993) suggests that it is sufficient to estimate the magnitude of the sidewall force from an idealised stress equilibrium in the failing wedge in which the soil stresses are calculated neglecting the effect of boundary forces. This approach is further supported by the back-analysis described in this paper.

The magnitude of the sidewall force per unit width of the wall, F, is evaluated from the integral

$$F = \frac{2}{w} \int \sigma_2 \tan \phi'_{sw} \, dA \quad (5)$$

where σ_2 is the direct stress acting normal to the sidewall, ϕ'_{sw} is the angle of sidewall friction and the factor 2/w allows for two sides of a reinforced soil wall of width w. The integral is taken over the triangular area of the failure wedge.

The direct stress acting normal to the sidewall is assumed to be given by the following function of the vertical stress, σ_v, and the horizontal stress, σ_h,

$$\sigma_2 = K_2 \frac{(\sigma_v + \sigma_h)}{2} \quad (6)$$

where the coefficient, K_2, is a constant (Stroud, 1971; Symes, 1983). The vertical stress is assumed to be given by

$$\sigma_v = q_v + \gamma z \quad (7)$$

Note that in this expression the small reduction in vertical stress due to sidewall friction is neglected. The horizontal stress is found by assuming active conditions in the failing wedge

$$\sigma_h = K_a \sigma_v \quad (8)$$

It is sufficient in this case to use the value of active earth pressure coefficient, K_a, appropriate for a smooth front wall. The above equations lead to an expression for the total force due to sidewall friction per unit width of the wall

$$F = \frac{K_2 H}{2w} \left(q_v + \frac{\gamma H}{3} \right) (1 + K_a) H \cot \theta \tan \phi'_{sw} \quad (9)$$

Equations (4) and (9) together define the total required reinforcement force for equilibrium. The critical failure plane inclination must be obtained numerically.

The measurements of wall displacement and reinforcement force in the reinforced RMC tests were all carried out using a central panel of width 1 m which was independent of two side panels of width 0.7 m in order to reduce the effects of sidewall friction on the measured results. It is likely that this procedure would indeed reduce the influence of sidewall friction although there is considerable uncertainty about the precise mechanisms involved. A detailed analysis of this effect is difficult (although it is attempted by Burd (1993)) and is not included here.

Back-analysis of forces at failure in RMC test walls

The analysis described above is used to examine the forces at wall failure measured in an unreinforced test (Bathurst et al., 1988) and a reinforced test (Bathurst and Benjamin, 1990) at RMC.

The granular fill used in the RMC tests was an angular sand containing feldspar. The peak angle of friction for the sand, at the stress level and soil density relevant to conditions in the walls at failure, was measured to be in the range $53° \geq \phi'_p \geq 51°$, and the critical state angle of friction was found to be of the order of $\phi'_{cv} = 38°$ (Bathurst and Benjamin, 1988; Jewell, 1988). The unit weight of the compacted sand was $\gamma = 17.6 \text{ kN/m}^3$. Direct shear tests were carried out to measure the likely angle of sidewall friction in the RMC test facility; this was found to be $17.5° \geq \phi'_{sw} \geq 15°$ (Bathurst and Benjamin, 1988).

The relative displacement across the wedge shear surface is assumed to occur at the angle of dilation in the soil, i.e. $\theta_f = \psi$ as shown in Fig. 4(a). The angle of dilation, ψ, depends on the mobilised angle of friction, ϕ'_m, and in this analysis the empirical relationship relation suggested by Bolton (1986) is adopted

$$\psi = \frac{\phi'_m - \phi'_{cv}}{0.8} \tag{10}$$

Data for simple shear tests carried out on Leighton Buzzard sand and reported by Stroud (1971) suggest that the coefficient K_2 in eqn. (6) is approximately 0.74 and a similar value can be deduced from Symes (1983). In fact, it is likely that the value of K_2 depends on both the dilational and frictional characteristics of the sand. A brief study has been carried out assuming that the sand may be represented by a simple elasto-plastic model (Burd et al., 1989). This study suggests that for sand with a higher critical state angle of friction such as the backfill used in the RMC tests, the value of K_2 would be expected to be of the order 0.8 to 0.9. In view of this a value of 0.8 is adopted for K_2 in the analyses presented in this paper.

The RMC tests were all conducted in a test chamber for which the wall height was 3 m and the wall width 2.4 m. Although the fill was confined between two lubricated sidewalls the measured angle of sidewall friction was found to be significant, as described above. The retaining walls were constructed with props applied to the full-height facing panel; after construction the props were removed and the wall loaded by the application of a uniform vertical surface. The weight of the front panel, W_f, was 2 kN/m.

Back-analysis at failure of an unreinforced test wall
An unreinforced test wall was built at RMC using a propped, full-height

facing with the aim of assessing the sand properties and the influence of sidewall friction on the active forces developed in the test (Bathurst and Benjamin, 1988). A vertical surcharge of 22 kN/m² was applied to the top of the wall and then the wall prop was released to allow the face to rotate outward about the toe until failure was judged to have occurred (i.e. a minimum force was achieved). The prop force and the vertical and horizontal forces at the base of the facing panel were measured. At failure the total horizontal and vertical forces applied to the wall were found to be 10.4 kN/m and 6.7 kN/m respectively. Assuming a value of W_f of 2 kN/m, this corresponds to an angle of front wall friction, δ, of 24°.

It is possible to adapt eqn. (4) to deal with the case of an unreinforced test by interpreting the total reinforcement force, $\Sigma P_{r\theta}$, as the force supplied externally by the front wall prop. In this case it is convenient to define a new parameter, the total horizontal force acting on the soil, P_H, where $P_H = \Sigma P_r + P_h$.

The horizontal and vertical forces at failure calculated using eqns. (4) and (9) for the expected range of soil and sidewall friction properties are

Table 1. Gross calculated forces at failure in the unreinforced RMC test

	$\phi'_{sw} = 15°$		$\phi'_{sw} = 17.5°$	
	P_H kN/m	P_v kN/m	P_H kN/m	P_v kN/m
$\phi'_m = 51°$	11.1	6.94	10.4	6.61
$\phi'_m = 53°$	9.82	6.37	9.13	6.07

Table 2. Critical wedge angle and percentage reduction in the gross horizontal required force for equilibrium due to the external boundary forces (see Table 1)

	$\phi'_{sw} = 15°$		$\phi'_{sw} = 17.5°$	
	θ_{crit} (°)	Reduction (%)[1]	θ_{crit} (°)	Reduction (%)[1]
$\phi'_m = 51°$	72.3	39%	72.9	43%
$\phi'_m = 53°$	73.4	40%	74.0	44%

[1] Percentage reduction is given by $(P_H^* - P_H) \times 100/P_H^*$ where P_H^* is the active force for the case where F and P_v are both zero

listed in Table 1. The range in Table 1 corresponds well with the forces that were measured during the test.

The corresponding critical wedge angles are given in Table 2 and do not differ by more than a few degrees from the result for no external forces, $\theta_{crit.} = 45° + \phi'_m/2$.

An important feature of these data is that the effect of boundary forces is shown to reduce the expected value of horizontal force required for equilibrium by about 40% as shown in Table 2. A net reduction of about 30% in the required horizontal force is caused by sidewall friction, with the remainder due to forces applied at the base of the facing panel.

Back-analysis at failure of a reinforced test wall

The RMC test described by Bathurst and Benjamin (1990) and Bathurst et al. (1992) was constructed using a relatively weak reinforcement which allowed the wall to be taken to failure during the test. The external forces at the base of the facing were measured, as were the strains along the four reinforcement layers.

The 100 hour isochronous load–displacement curve for the grid reinforcement at 20°C is thought to be representative of the reinforcement properties in the test and this may be approximated by the isochronous stiffness, J, of 60 kN/m over the range of strains of interest (Bathurst et al., 1992). The strain measured at the end of the test in the four reinforcement layers at positions corresponding to an assumed failure plane were, from the top down, $\varepsilon_r = 4\%, 5\%, 5\%$ and 0.5% respectively. This corresponds to a total reinforcement force at failure, ΣP_r, of 8.7 kN/m.

The vertical surcharge load at failure was 80 kN/m²; at this stage in the test the external forces at the toe of the wall face were measured to be $P_v = 27$ kN/m and $P_h = 6$ kN/m. The other parameters are as described previously and analysis using eqns. (4) and (9) gives the predictions summarised in Table 3. The measured value of ΣP_r falls within the range of the predictions.

Table 3. *Gross calculated reinforcement force at failure for the reinforced test wall at RMC (Bathurst and Benjamin, 1990)*

	$\phi'_{sw} = 15°$		$\phi'_{sw} = 17.5°$	
	ΣP_r kN/m	θ_{crit} (°)	ΣP_r kN/m	θ_{crit} (°)
$\phi'_m = 51°$	11.6	71.3	9.4	71.9
$\phi'_m = 53°$	8.7	72.3	6.6	72.9

Table 4. *Influence of boundary forces on the gross required reinforcement force*

Assumptions: $\phi'_m = 52°$, $\phi'_{sw} = 16.25°$	ΣP_r kN/m	Percentage reduction[1]
No external boundary forces	38.5	0%
Including the vertical toe force, P_v	29.8	23%
Including the horizontal toe force, P_h	23.8	38%
Including sidewall friction, F	9.1	76%

[1]Percentage reduction given by $(38.5 - \Sigma P_r)/38.5$

Although the test wall represents an extreme case, with a narrow width and a highly frictional fill, it is nevertheless useful to examine the effect of the various boundary forces on the equilibrium. The results in Table 4 give the total required horizontal force for equilibrium for the cases

(*a*) with no boundary forces,
(*b*) including a vertical force at the toe of the wall, P_v, of 27 kN/m
(*c*) including a horizontal force, P_h, of 6.0 kN/m, and
(*d*) including sidewall friction.

In all of these cases the value of W_f was taken to be 2 kN/m.

Prediction of wall displacements

A detailed description of an analysis method, based on a limit-equilibrium approach, to predict the deformations in a reinforced soil wall in the absence of boundary forces is given by Jewell (1988). This analysis is based on idealised solutions, termed the *ideal length* and the *truncated length* cases, which are intended to provide bounds to the expected wall displacements. The method is further developed by Jewell and Milligan (1989), to include the effects of soil dilation and to allow for the calculation of vertical settlement. This latter analysis results in a series of design charts, given by Jewell and Milligan (1989) from which wall deformations may be estimated. This method is developed further here in order to include the effects of boundary forces.

A detailed analysis of the displacements in two of the RMC test walls is given below. The first of these was the wall taken to failure and analysed above. The second wall, described by Bathurst et al. (1988), was constructed using a relatively strong reinforcement and the wall was not loaded to failure.

Application of the displacement charts (Jewell and Milligan (1989))
The displacement charts apply to reinforced soil walls with any

reinforcement spacing arrangement. The horizontal displacement, δ_h, at the level of each reinforcement layer is expressed in terms of the wall height, δ_h/H, the maximum force mobilised in a layer of reinforcement, P_r, and the reinforcement secant stiffness, J. The reinforcement length is assumed to satisfy the stability requirements for the wall, and does not explicitly enter the analysis.

The non-dimensional parameter $\delta_h J/HP_r$ is evaluated from the charts at each reinforcement level. The magnitude of this parameter depends on the mobilised angle of friction in the soil, and the associated angle of dilation. These parameters are implicitly linked using eqn. (10). The maximum force in the reinforcement at each level is calculated using the idealised stress equilibrium outlined in Fig. 2(b). The reinforcement is assumed only to support the required stresses in the soil for half the spacing to the next reinforcement layer above and below. In order to modify the method to include the effects of boundary forces an analysis is carried out using eqns. (4) and (9) to find the total required reinforcement force for equilibrium both *with* and *without* any anticipated external boundary forces. The maximum reinforcement force in each layer found using this idealised stress equilibrium is then reduced in proportion to the ratio of the calculated required forces. These procedures are illustrated below.

Back-analysis of wall displacements at failure

In this section an analysis of the RMC wall described by Bathurst and Benjamin (1990) and discussed previously is presented. The number, depth and spacing of the four reinforcement layers are given in Table 5. The force in each reinforcement layer ignoring any external boundary forces is found from the reinforcement spacing and the distribution of maximum required stress as illustrated in Fig. 2.

The displacement charts given in Jewell and Milligan (1989) apply for

Table 5. *Displacements for a geogrid reinforced soil wall at failure. Analysis for $\phi'_m = 53°$ and $\phi'_{sw} = 17.5$. Truncated length solution (Fig. 5(b))*

Layer	z (m)	S_v (m)	P_r† (kN/m)	P_r‡ (kN/m)	z/H	$\dfrac{\delta_h J}{HP_r}$	δ_h (mm)
4	0.50	0.875	8.6	1.6	0.17	0.60	48
3	1.25	0.750	8.6	1.6	0.42	0.45	36
2	2.00	0.750	9.7	1.8	0.67	0.31	28
1	2.75	0.625	8.9	1.6	0.92	0.14	12

†Ignoring the boundary forces
‡Allowing for boundary forces

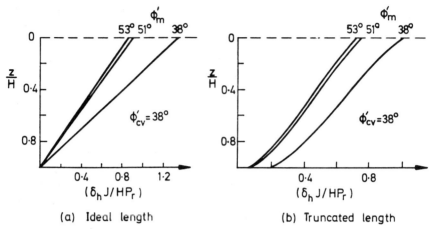

Fig. 5. Design charts. (a) Ideal length. (b) Truncated length

three angles of dilation, $\psi = 0°$, $10°$ and $20°$. In order to use this method for the back-analysis described here a set of charts have been derived (Fig. 5) for the friction angles of interest in the RMC walls using dilation angles calculated from eqn. (10). The analysis for the case $\phi'_m = 53°$ and $\phi'_{sw} = 17.5°$ is described in detail below using the truncated length solutions given in Fig. 5(b). The relevant value for $\delta_h J/HP_r$ at each depth

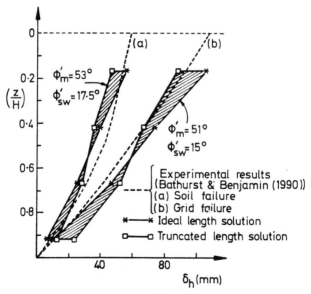

Fig. 6. Comparison between experimental results and limit-equilibrium analysis

found from Fig. 5(b) is listed in Table 5. The corresponding displacement (for the case where boundary forces are absent) may be evaluated using the appropriate value of required reinforcement force P_r.

An assessment of the effect of boundary forces on the total required horizontal force may be carried out using the limit-equilibrium analysis described previously. In this case the total required reinforcement force in the absence of boundary forces is found from eqn. (2) to be 35.7 kN/m. When the boundary forces are included, the total required reinforcement force reduces to 6.6 kN/m as reported in Table 3; a reduction of 81.5%. In order to include the effect of boundary forces on the expected wall displacements, the assumption is made that the force in each individual reinforcement, and hence the corresponding displacement, is reduced by 81.5%. Details of the calculated reinforcement forces and corresponding displacements are given in Table 5.

The wall face displacements calculated from the truncated length solution given in Table 5 are plotted in Fig. 6. The calculated displacements for the ideal length solution (Fig. 5(a)) are plotted in Fig. 6 for comparison. The predictions for the combinations $\phi'_m = 51°$, $\phi'_{sw} = 15°$ and $\phi'_m = 53°$, $\phi'_{sw} = 17.5°$ (chosen to bound the likely combinations of parameters) are also shown together with the measured displacements reported at soil failure and grid rupture in the test (Bathurst and Benjamin, 1990). Given the uncertainties in evaluating the parameters for the analysis, the results obtained using this simple calculation procedure agree well with the observed behaviour.

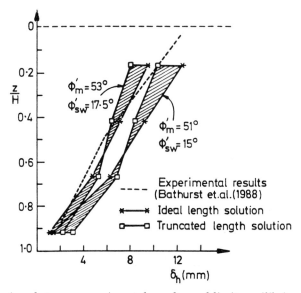

Fig. 7. Comparison between experimental results and limit-equilibrium analysis

Back-analysis of displacements under working conditions
Data from a geogrid reinforced soil wall at working equilibrium are available from a test at RMC using a much stronger and stiffer reinforcement material and described by Bathurst et al. (1988). In this case the surcharge was applied to the structure for a sufficiently long period for the 1000 hour isochronous stiffness of the geogrid (550 kN/m) to be selected for the back-analysis. In this test no measurements were made of the boundary forces acting at the base of the facing panel. It is therefore necessary to estimate values of these forces in order to carry out a back-analysis. Since the facing wall was allowed to slide outwards at the base a value of P_h of zero is adopted. In order to estimate P_v a value of δ of 50° is assumed. This value is justified on the basis that a similar value of δ was measured in the test carried out to failure described in the previous section.

The maximum vertical surcharge used in the test was $50 \, kN/m^2$. The calculation of wall displacement using the displacement charts in Fig. 5 follows the same procedure as described above. The calculated displacements are compared with those measured in the test in Fig. 7.

Conclusions

In order to make satisfactory predictions of behaviour of reinforced soil walls it is necessary to include the effects of boundary forces in the analysis. The effects of forces at the base of the facing panel and, for the case of a test wall in a narrow chamber, the friction acting on the sidewalls are shown to be significant. A limit-equilibrium method in which boundary forces are included is shown to provide wall displacements that are in broad agreement with values measured in tests carried out with two different reinforcement stiffnesses.

The forces acting at the base of the facing panel depend on the stiffness of the facing and cannot be deduced from limit-equilibrium methods alone. It is possible, however, to use limit-equilibrium methods as described in this paper to estimate the effect of known values of force applied to the base of the facing on the wall performance. It is also possible to make realistic predictions of the frictional force generated on the sidewalls of a test chamber and an analysis is presented in this paper. It is suggested that for design purposes the influence of boundary forces should be neglected since changes in construction procedures and design details could have a marked effect on their values. For performance prediction, however, these effects should be included in order to obtain realistic results.

References

BATHURST, R.J. AND BENJAMIN, D.J. (1988). Preliminary assessment of sidewall friction on large scale wall models in the RMC test facility. Proc. NATO Advanced Research Workshop on Application of Polymeric Reinforcement in Soil Retaining Structures, Kingston, Ontario, Canada, pp. 181–192.

BATHURST, R.J. AND BENJAMIN, D.J. (1990). Failure of a geogrid reinforced soil wall. Transportation Research Record 1288, Washington D.C., pp. 109–116.

BATHURST, R.J., KARPURAPU, R. AND JARRETT, P.M. (1992). Finite element analysis of a geogrid reinforced soil wall. Proc. ASCE Symp. on Grouting, Soil Improvement and Geosynthetics, New Orleans, USA, pp. 1213–1224.

BATHURST, R.J., WAWRYCHUK, W.F. AND JARRETT, P.M. (1988). Laboratory investigation of two large-scale geogrid reinforced soil walls. Proc. NATO Advanced Research Workshop on Application of Polymeric Reinforcement in Soil Retaining Structures, Kingston, Ontario, Canada, pp. 71–126.

BOLTON, M.D. (1986). The strength and dilatancy of sands. Geotechnique, Vol. 36, No. 1, pp. 65–78.

BRANSBY, P.L. AND SMITH, I.A.A. (1975). Side friction in model retaining-wall experiments. J. of Geo. Eng. Div., ASCE GT7, July, pp. 615–632.

BURD, H.J. (1993). Finite element analysis of reinforced soil walls. Delft University of Technology, Dept of Civil Engineering report no. 332.

BURD, H.J., YU, H.-S. AND HOULSBY, G.T. (1989). Finite element implementation of frictional plasticity models with dilation. Proc. Int. Conf. on Constitutive Laws for Engineering Materials, China, pp. 783–787.

JEWELL, R.A. (1988). Analysis and predicted behaviour for the Royal Military College trial wall. Proc. NATO Advanced Research Workshop on Application of Polymeric Reinforcement in Soil Retaining Structures, Kingston, Ontario, Canada, pp. 193–238.

JEWELL, R.A. (1990). Strength and deformation in reinforced soil design. Proc. 4th Int. Conf. on Geotextiles, Geomembranes and Related Products, The Netherlands, Vol. 3, pp. 913–946.

JEWELL, R.A. AND BURD, H.J. (1993). Predicting the behaviour of soil walls reinforced by geotextiles part 1; theoretical basis. Oxford University Internal Report No. 1929/93.

JEWELL, R.A. AND MILLIGAN, G.W.E. (1989). Deformation calculations for reinforced soil walls. Proc. 12th Int. Conf. on Soil Mech. and Fdn. Eng., Rio de Janeiro, pp. 1257–1262.

JEWELL, R. A. AND MILLIGAN, G.W.E. (1993). Predicting the behaviour of soil walls reinforced by geotextiles part 2; practical application. Oxford University Internal Report No. 1930/93.

STROUD, M.A. (1971). The behaviour of sand at low stress level in the simple shear apparatus. PhD Thesis, University of Cambridge.

SYMES, M.J.P.R. (1983). Rotation of principal stresses in sand. PhD thesis, Imperial College, London.

Some thoughts on the evaluation of undrained shear strength for design

F.H. KULHAWY, Professor of Civil/Geotechnical Engineering, Cornell University

Initial observations are made on the rationale for total stress analyses and the use of the undrained shear strength, s_u. It is shown that different s_u values are appropriate for different field loading conditions and that there are many uncertainties in s_u as a material property. Another call is made to adopt the CIUC test as the standard test of reference for evaluating s_u. Results of comprehensive studies are presented that show the relative comparisons among the CIUC and other major test types. Finally, illustrative comparisons are presented to show the relative undrained strength ratios for some different field loading conditions.

Introduction

Evaluation of the undrained shear strength of fine-grained soils is no trivial task. This fact has long been known within the research community that focuses on soil properties, but it is only slowly being realized in many other sectors of practice. During the past fifteen years, a number of major overview papers have been written that have traced the steady accumulation of knowledge on soil properties (e.g. Ladd et al., 1977; Wroth, 1984; Wroth and Houlsby, 1985; Jamiolkowski et al., 1985; Jamiolkowski et al., 1991), and a major design manual on estimating soil properties also has been prepared (Kulhawy and Mayne, 1990). These documents have shown an increasing sophistication in all types of property evaluation, specifically including a careful matching of test and prototype variables and a direct awareness of variability (or uncertainty) in the property evaluation process.

Peter Wroth was a major player in the development of this knowledge, and he was a strong proponent of modelling test and prototype conditions as realistically as possible and of developing minimum standards of reference in testing. In the many discussions (or mini-debates?) that we had on these subjects, Peter always impressed upon me the need to focus on these basic issues. Continuing in the spirit of these discussions, this paper focuses on some key issues in evaluating the undrained shear strength of fine-grained soils for design purposes. For simplicity, the soil is assumed to be saturated and relatively

UNDRAINED SHEAR STRENGTH FOR DESIGN

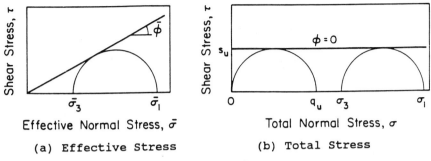

Fig. 1. Idealized Coulomb–Mohr failure envelopes

unstructured. Special behavioural issues dealing with sensitive, cemented, and other structured soils are beyond the scope of this paper.

Basic characterization

In geotechnical engineering analyses involving fine-grained soils, either effective stress or total stress methods can be used. Total stress methods normally are adopted because of (implied) simplicity. However, the failure of all soils actually occurs on the effective stress failure envelope shown in Fig. 1(a). Loading generates excess pore water stresses (Δu) that change the original effective stresses and, in turn, influence the stress state relative to the envelope defined by the effective stress friction angle ($\bar{\phi}$). Since the total stress loading path and the developed excess pore water stresses (Δu) may not be known with confidence, a total stress analysis with $\phi = 0$ and s_u = undrained shear strength, as shown in Fig. 1(b), provides a simple and idealized analysis alternative. However, it must be remembered that s_u incorporates both $\bar{\phi}$ and Δu, and it varies with the initial or in-situ effective stress level.

The undrained shear strength may very well be the most widely used parameter for characterizing fine-grained soils. In some circles, it is even portrayed as a fundamental material property, which it isn't. Instead, it is a measured soil response during undrained loading that assumes zero volume change. As such, s_u is affected by the mode of testing, boundary

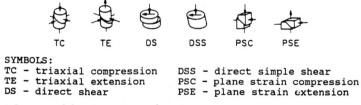

SYMBOLS:
TC - triaxial compression
TE - triaxial extension
DS - direct shear
DSS - direct simple shear
PSC - plane strain compression
PSE - plane strain extension

Fig. 2. Common laboratory strength tests

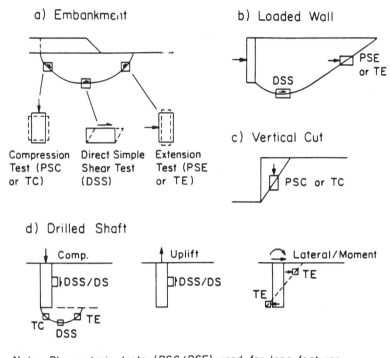

Fig. 3. Relevance of laboratory strength tests to field conditions

conditions, rate of loading, initial stress level, and other variables. Consequently, s_u is and should be different for different test types. Consider the test representations depicted in Fig. 2. All could be used to estimate a soil strength, but the results should be different. The overview references cited previously cover these issues very well.

These same points carry over into the field, where different boundary conditions, stress paths, etc. also will apply. Figure 3 illustrates a few common cases related to embankments, walls, slopes, and drilled shaft foundations. It is clear from this figure that no one type of test usually addresses the actual field conditions. Instead, a combination commonly is warranted.

The issues raised above are not new, and they do not represent an academic exercise. They are real, and they have important implications in practice because, if the 'wrong' test is used to characterize a particular field situation, there could be significant implications on the actual factor of safety in contrast with the perceived value.

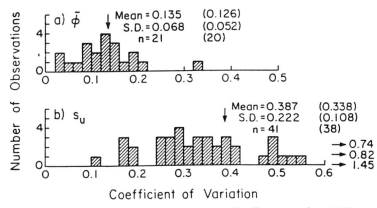

Fig. 4. Histograms of soil strength parameters (Kulhawy et al., 1991)

Uncertainties in strength parameters

Uncertainties are always introduced in any evaluation process. Some represent the inherent material or property variability, some represent measurement errors, and some represent modelling inadequacies or inaccuracies. A complete evaluation of these issues is well beyond the scope of this modest paper and our general state of knowledge at this time. However, a general framework for assessing these uncertainties is given by Kulhawy (1992).

As a first-order assessment of uncertainty, one can evaluate the coefficient of variation, COV (standard deviation/mean), of soil strength properties. Figure 4 summarizes data for both $\bar{\phi}$ and s_u, as reported in the literature. The databases for these studies have been highly variable, ranging from as few as 5 samples in some cases to as many as 295 in others. For $\bar{\phi}$, the range was 5 to 81; for s_u, the range was 10 to 295. A further complicating factor is the lack of control of data in any literature survey. Undoubtedly, there is mixing of test types and testing procedures in these data, so the summary in Fig. 4 is likely to be a bit on the high side in addressing the variability. For each parameter, the mean, standard deviation (S.D.), and number of samples (n) are given. The parenthesized values represent the mean, S.D., and n without the several high values that appear to be outside of the main populations. For comparison, the COV for the compressive strength of concrete and the tensile strength of steel is about 6% (Harr, 1977).

As can be seen, the COV for $\bar{\phi}$ is relatively low and is about double that for concrete or steel. However, the COV for s_u is quite large, necessarily indicating more uncertainty in the property. This significant difference does not necessarily imply that there is more uncertainty in total stress analyses. Consider, for instance, the geotechnical prediction model shown in Fig. 5. To make a prediction from a given load, the

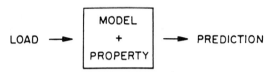

Fig. 5. *Components of geotechnical prediction*

model and property must be considered or calibrated together. For total stress analyses, the models may be simple, with little uncertainty, but the properties have significant uncertainty. The reverse tends to be true with undrained effective stress analyses. Therefore, with total stress analyses, it is much more important that the properties replicate the prototype conditions.

Importance of a standard 'test of reference'

As can be seen in Fig. 3, quite a number of different types of tests and equipment might be needed for a particular design condition. However, this level of testing is likely to be excessive in common and routine design cases. Therefore, it is both appropriate and convenient to establish a standard 'test of reference' that would be applicable in some design cases and would be simple and expedient from a commercial testing standpoint. The test that was recommended by Wroth (1984) and others is the isotropically consolidated, triaxial compression test for undrained loading (CIUC). This test is logical for high-quality field samples because it satisfies the above criteria, re-establishes a state of stress in the soil that is approximately consistent with the overburden stress, minimizes the sampling disturbance effects, and includes a reconsolidation phase.

It should be noted that most soils in situ actually will be consolidated anisotropically. This difference in consolidation stresses has no appreciable influence on $\bar{\phi}_{tc}$, the effective stress friction angle in triaxial compression (e.g. Kulhawy and Mayne, 1990), but it does influence s_u, as will be shown shortly.

There also are simpler forms of triaxial test that are available, such as the unconsolidated, undrained (UU) triaxial and unconfined compression (UC) tests. However, many detailed studies (e.g. Ladd et al., 1977; Tavenas and Leroueil, 1987) have shown that the UU and UC tests often are in gross error because of sampling disturbance effects, incorrect initial shear stress level, and omission of a reconsolidation phase. Based on studies such as these, the CIUC test also should be considered to be the minimum quality laboratory test for evaluating s_u.

With the CIUC test as the standard reference, the results of all other tests can be compared simply and conveniently. Since s_u is stress-dependent, its value typically is normalized by the vertical effective overburden stress ($\bar{\sigma}_{vo}$) at the depth where s_u is evaluated. The result is the undrained strength ratio $(s_u/\bar{\sigma}_{vo})_{CIUC}$.

Based on an evaluation of analytical expressions and a detailed comparison of available undrained strength data for the major test types, Kulhawy and Mayne (1990) developed the mean normalized undrained strength ratios shown in Fig. 6. For this figure, the reference strength ratio is given, to sufficient accuracy (S.D. ≈ 0.05), by the modified Cam clay model as follows (e.g. Wroth and Houlsby, 1985):

$$(s_u/\bar{\sigma}_{vo})_{CIUC} = 0.5\,M\,(0.5)^\Lambda \qquad (1)$$

in which $M = 6 \sin \bar{\phi}_{tc}/(3 - \sin \bar{\phi}_{tc})$ and Λ = critical state parameter. This relationship works rather well for relatively unstructured soils. For sensitive, cemented, and other structured fine-grained soils, eqn. (1) tends to be a lower bound. Figure 6 represents an illustrative comparison of a very extensive set of databases, presented here for the specific

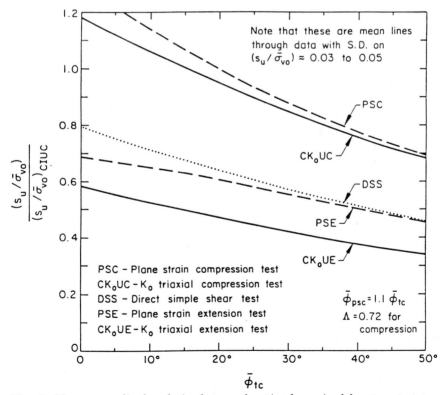

Fig. 6. *Mean normalized undrained strength ratios for major laboratory tests (Kulhawy and Mayne, 1990)*

cases of $\bar{\phi}_{psc}$ and Λ shown. The detailed comparisons are given by Kulhawy and Mayne (1990), along with appropriate modifications for testing rate, overconsolidation, and other test specifics.

As can be seen in Fig. 6 on a relative basis, considerable variation occurs among the test types. For the typical range of $\bar{\phi}_{tc}$ from 20° to 40°, the CIUC value will always be greater than all of the others, and therefore it may be unconservative to use directly. In addition, the other test values in compression are roughly double those in extension.

Although not recommended for any future work because of the problems cited previously, the UU test is the only strength documentation for many sites evaluated in the past. For this reason, it is of interest to examine the interrelationships between the UU and CIUC tests, as shown in Fig. 7. In this figure, the data were grouped by overconsolidation ratio, OCR, as follows: normally consolidated, NC (1.0 < OCR < 1.3); lightly overconsolidated, LOC (1.3 < OCR < 3.0); moderately overconsolidated, MOC (3 < OCR < 10); and heavily overconsolidated, HOC (OCR > 10).

Figure 7 represents reasonably homogeneous soil deposits and 'well-conditioned' data. As such, it may be interpreted as near the upper bound in quality (i.e. minimum S.D.) in the interrelationships. Following this preamble, it is clear that there is a well-defined relationship between the UU and CIUC s_u values. In the NC range, UU values may only be ½ the CIUC values. However, for the HOC range, UU values can exceed the CIUC values. This general behaviour can be predicted by

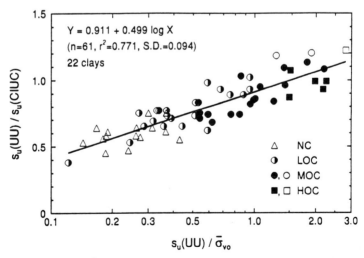

Fig. 7. Comparison of undrained strengths from UU and CIUC Tests (Chen and Kulhawy, 1993)

Comparison of s_u for different field loading conditions

Figure 3 portrayed the applicable undrained strength tests for a variety of field loading conditions, and Fig. 6 demonstrated the relationships between specific laboratory strength tests and the CIUC reference test. By integrating these two data sets, a comparison of the field loading conditions can be made with the CIUC test by field case type, as given in Table 1. This table shows that the normalized undrained strength ratio for design is always less than one. Therefore, if the CIUC results are used directly in design, then the results will be unconservative because the actual operative strength is less than the CIUC value.

However, if the analyses are based on UU test results instead of CIUC test results, then the situation changes. For NC soils, $s_u(UU)/s_u(CIUC) \approx 0.6$. Dividing the results in Table 1 by 0.6 would give a value of $(s_u/\bar{\sigma}_{vo})/(s_u/\bar{\sigma}_{vo})_{UU}$ closer to 1.0 for 5 of the 6 loading conditions portrayed, indicating compensating errors leading to an apparently acceptable result, as long as the UU data are representative and reliable.

For rational design, compensating errors of this type should not be relied upon, and lower grade tests simply are not acceptable for modern (informed) practice. Direct recognition of the boundary conditions and their relationship to the CIUC test results should be introduced into the design explicitly.

Summary

Evaluation of the undrained shear strength (s_u) of fine-grained soils should be based on sound geotechnical principles. Ample evidence

Table 1. *Illustrative undrained strength ratio comparisons computed for different field loading conditions*

Field Loading Condition	$(s_u/\bar{\sigma}_{vo})/(s_u/\bar{\sigma}_{vo})_{CIUC}$		
	$\bar{\phi}_{tc} = 20°$	30°	40°
Long Embankment (PSC + DSS + PSE)	0.75	0.67	0.60
Long Wall (DSS + PSE)	0.62	0.57	0.51
Short Vertical Cut (CK$_o$UC)	0.94	0.85	0.75
Shaft Bearing Capacity (CK$_o$UC + DSS + CK$_o$UE)	0.68	0.62	0.55
Shaft Side Resistance (DSS)	0.64	0.58	0.51
Shaft Lateral Load (CK$_o$UE)	0.47	0.42	0.38

exists that s_u varies greatly with the test conditions and other factors. A standard 'test of reference' should be adopted for all future work, and this test should be the CIUC triaxial test. Correlations to other site, geometry, or load-specific conditions can be made through simple correlations developed from extensive research studies. Prior usage of the UU test gave compensating errors that nearly 'corrected' the strength results adequately. However, the UU results are subject to many vagaries, and they cannot be depended on. The CIUC should be the minimum quality of test.

Acknowledgements

The concepts expressed were developed largely during geotechnical studies for the Electric Power Research Institute, Palo Alto, California. V.J. Longo was the EPRI Project Manager. Many discussions with H.E. Stewart and P.W. Mayne crystallized the concepts, and H.E. Stewart provided many useful review comments on this paper. L. Mayes prepared the text, and A. Avcisoy prepared the figures.

References

CHEN, Y.-J. and KULHAWY, F.H. (1993). Undrained shear strength interrelationships among CIUC, UU, and UC tests. J. Geotech. Engng ASCE. In press.

HARR, M.E. (1977). Mechanics of particulate media. McGraw-Hill, New York.

JAMIOLKOWSKI, M., LADD, C.C., GERMAINE, J.J. and LANCELOTTA, R. (1985). New developments in field and laboratory testing of soils. Proc. 11th Intl. Conf. SM & FE (1), San Francisco, pp. 57–153.

JAMIOLKOWSKI, M., LEROUEIL, S. and LOPRESTI, D.C.F. (1991). Design parameters from theory to practice. Proc. Geo-Coast '91, Yokohama.

KULHAWY, F.H. (1992). On the evaluation of static soil properties. Stability and Performance of Slopes and Embankments – II (GSP 31), ASCE, New York, pp. 92–115.

KULHAWY, F.H. and MAYNE, P.W. (1990). Manual on estimating soil properties. Report EL-6800, Electric Power Res. Inst., Palo Alto.

KULHAWY, F.H., ROTH, M.J.S. and GRIGORIU, M.D. (1991). Some statistical evaluations of geotechnical properties. Proc. 6th Intl. Conf. Applic. Stat. & Prob. in Civil Eng., Mexico City, pp. 705–712.

LADD, C.C., FOOTT, R., ISHIHARA, K., SCHLOSSER, F. and POULOS, H.G. (1977). Stress-deformation and strength characteristics. Proc. 9th Intl. Conf. SM & FE (2), Tokyo, pp. 421–494.

LAMBE, T.W. and WHITMAN, R.V. (1969). Soil Mechanics. Wiley, New York.

TAVENAS, F. and LEROUEIL, S. (1987). State-of-the-art on laboratory and in-situ stress-strain-time behavior of soft clays. Proc. Intl. Symp. Geotech. Eng. Soft Soils, Mexico City, pp. 1–46.

WROTH, C.P. (1984). Interpretation of in-situ soil tests. Geotechnique, Vol. 34, No. 4, pp. 449–489.

WROTH, C.P. and HOULSBY, G.T. (1985). Soil mechanics — property characterization and analysis procedures. Proc. 11th Intl. Conf. SM & FE (1), San Francisco, pp. 1–55.

Attempts at centrifugal and numerical simulations of a large-scale in situ loading test on a granular material

O. KUSAKABE, Hiroshima University, Y. MAEDA, Japan Highway Public Cooperation, M. OHUCHI, Shiraishi, Co. Ltd, and T. HAGIWARA, Gunma University, Japan

This paper describes a large-scale in situ loading test on a scoria, a granular material produced by volcanic eruptions, conducted in a pneumatic caisson. Presented also are test results on undisturbed scoria samples procured from the loading test site. For the loading test, two types of simulation were attempted: centrifugal simulation using undisturbed samples and FEM simulation of the CRISP package using the element test data. It was found that centrifugal simulations using undisturbed samples predicted the yield load, and elastic deformations close to the prototypes, but clear slip lines were not developed and settlements at yield load were much larger than those in prototype. It was inferred that the observations may be partly due to particle size effect. Difficulty in the selection of soil parameters in the FEM analysis was encountered and uncertainty in the determination of K_0 values for very high overconsolidation ratios was recognized.

Introduction

Two types of modelling are available for geotechnical problems: physical modelling and numerical modelling. The validation of these modellings requires well documented prototype behaviour, with which the effectiveness of the modellings can be directly compared. This study focuses on two particular modelling techniques: the centrifuge test and CRISP FEM programme, both of which have been developed at Cambridge University.

Direct comparisons between centrifuge modelling and prototype behaviour have been relatively rare in number, compared with FEM versus prototype comparisons. For clays, Lyndon and Schofield (1970, 1978) reported successful centrifuge simulations of failures on natural slopes of London clay and Lodalen landslide using undisturbed block samples. Basset and Horner (1979) also attempted such direct comparisons with prototypes of clay foundations.

Well documented field data are scarce. Detailed information on soil element behaviour and boundary conditions are needed for proper comparison with prediction. Since these details can be determined for well controlled centrifuge tests, this had led to such tests being regarded as prototypes. Many centrifuge modellers have attempted direct comparisons between centrifuge tests of this type and FEM predictions using the CRISP package (Britto and Gunn, 1987). Works by Basset et al. (1981), Kimura et al. (1984), Almeida et al. (1984), Davies and Parry (1985) are good examples. All these works have dealt with the problems of clay foundations and they seem to share the view that FEM analysis using the CRISP package can predict centrifuge results with acceptable accuracy. Quite recently Indraratna et al. (1992) utilized the CRISP package to analyse a field test embankment behaviour on a clay foundation.

In contrast, for granular materials there still exists scepticism about the accuracy of predictions by centrifuge tests and about the CRISP analysis. Only limited information is available for us to reach any conclusion about the validity of centrifuge modelling and FEM predictions compared with prototype behaviour. Fujii et al. (1988) compared large-scale in situ field loading tests of shallow foundations on a pumice flow deposit called Shirasu with centrifuge tests using undisturbed samples. They concluded that the centrifuge tests could predict ultimate bearing capacities fairly well, but overestimated settlement behaviour by a factor of 2 to 3. They inferred that the difference in settlement behaviour might have stemmed from sampling disturbance.

Recently Tatsuoka et al. (1991) conducted a direct comparison of loading tests of shallow footings between a 1G model and corresponding centrifuge tests using dry Toyoura sand. They claimed that the particle size effect cannot be ignored for centrifuge simulations for granular materials. Bolton and Lau (1988) performed centrifuge tests on particle size modelling, from which they stated that the particle size modelling may cause another complication due to the fact that smaller particles have higher crushing pressures.

This paper describes a large-scale in situ loading test on a scoria, a granular material produced by volcanic eruptions, conducted in a pneumatic caisson. Presented also are test results on undisturbed scoria samples procured from the loading test site. For this particular loading test, two types of simulation were attempted: centrifugal simulation using undisturbed samples and FEM simulation of the CRISP package using the element test data. Direct comparisons with the prototype are presented in terms of load–settlement curves, failure mechanism and characteristics of particle crushing.

It should be mentioned here that the scoria is very different from the Kaolin and the standard sand generally used in simulations. While the

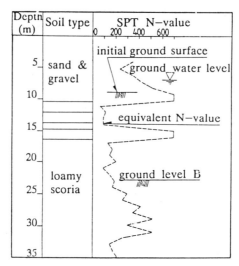

Fig. 1. Soil profile and elevation of loading test

result of this study is that the simulations are of 'predictive' value, nevertheless care should always be exercised in extending modelling techniques to new materials.

In situ loading tests
Test site and soil properties
A series of loading tests was conducted in the base chamber of a caisson under construction, of dimensions 16 m in diameter, and 23 m in height when completed. Figure 1 presents the soil profile of the test site with the level of ground water, consisting of a thick layer of dark brown scoria mixed with loam from 16.8 m below the ground surface and overlying layers of gravel and sandy deposits. The loading test selected for the direct comparison in this study was the one conducted at 14 m below the ground level, called B ground hereafter. The SPT-N values of the scoria layer are over 70 at the B ground.

The scoria belongs to the eruptions of Old Hakone Volcano deposited in the Quaternary period and has high angularity and numerous small voids. Prior to the loading test, undisturbed samples were manually cut from the test ground level in the form of blocks of 0.4 m length, 0.3 m width and 0.3 m height. These block samples were used for both element and centrifuge tests.

The soil test programme included physical tests, isotropic compression tests and triaxial compression tests under drained conditions (CID test). The grain size distribution curve of the B ground level showed that the scoria is a granular material with coefficient of uniformity U_c of 4.69, gravel content of 40.0%, sand content of 57.7% and fines content of 2.3%. Other physical properties are as follows: the specific gravity $G_s = 2.85$, maximum and minimum void ratios $e_{max} = 1.74$, $e_{min} = 1.17$, in situ void ratio $e_0 = 1.05$, saturated unit weight $\gamma_{sat} = 18.6 \text{ kN/m}^3$, and natural water content $W_n = 37.0\%$, average grain size $D_{50} = 1.57$ mm, maximum grain size $D_{max} = 19.0$ mm. It should be emphasized that the relative density in situ is as large as 121%.

The blocks of the scoria were frozen in unsaturated conditions and then cut out by a diamond cutter from different angles (δ) relative to the horizontal plane to obtain the samples with $\delta = 0, 30, 60, 90, -60$ and -30 degrees for the mechanical tests, which enabled us to examine the degree of anisotropy of the soil. It should be added here that the effect of freezing and thawing was found to be negligible in the stress–strain curves (Kusakabe et al., 1991).

Stress–strain curves of the CID triaxial tests for various confined pressures are shown in Fig. 2 for the cases of $\delta = 90°$. In common with cemented sands, the soil behaviour changes from that of a dilatant brittle material to that of a plastic material showing volume reduction during shearing as the confining pressure increases; as if the scoria behaves like a loose sand or a normally consolidated clay under high pressures. By defining the failure state at the axial strain of 15%, when the peak value was not observed, the failure envelopes of Mohr's Circles at failure as well as at residual states can be shown in Fig. 3, from which $c_d = 191$ kPa, $\phi_d = 33.3°$, and $c_r = 76$ kPa, $\phi_r = 33.9°$ were obtained. It is clear that the scoria is cemented and structured.

The data of $(\sigma'_1 - \sigma'_3)_f/\sigma'_3$ at failure for the samples with different angles

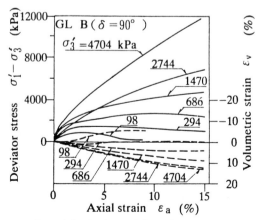

Fig. 2. Deviator stress, volumetric strain–axial strain curves

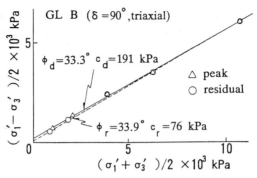

Fig. 3. Failure envelope of Mohr's Circles at failure and residual state

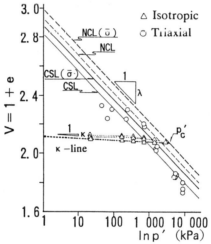

Fig. 4. Specific volumes at residual states versus mean effective stress

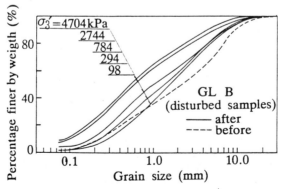

Fig. 5. Change in grain-size distribution curves before and after triaxial tests

showed that for this particular scoria, the assumption of isotropy may be adopted for the analysis (Kusakabe et al., 1992). In Fig. 4 specific volumes at residual states in CID tests are plotted against mean effective stresses at failure. The data seem to lie in a straight line, giving V = 2.826–0.112 ln p'. The data from the isotropic compression tests are also plotted in Fig. 4, from which $\kappa = 0.00573$ is obtained.

Another feature of the scoria is its high crushability. Figure 5 shows grain-size distributions before and after the triaxial compression tests. It clearly shows that the fines content after the test increases with increasing confining pressure.

Loading test programme and test procedures

The in situ loading test programme included eight tests with different dimensions of footings. For this particular study, a 400 mm square footing (B = 400 mm) was selected for comparison. A model footing was composed of rigid steel frames and was placed onto the excavated ground surface after having placed a thin layer of cement paste on the surface ground in order to have a rough condition at the footing base. Four hydraulic jacks were used to apply the pressure onto the footing, and a pin joint was implemented between the footing and the jacks. The measurements of the footing displacement were made at the four corners of the footing attached to reference beams.

In order to carry out the loading test on a saturated ground, the ground water level was maintained at the horizontal surface of the base of the excavated ground level by adjusting the air pressure fed to the base chamber of the caisson. The change in the ground water level was

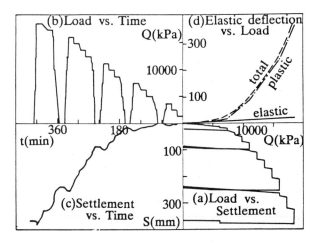

Fig. 6. Loading test results

monitored during the loading test by a float in a small pit dug out in the ground.

The loading was of the load control type, applying the loads in a step loading; at the end of each loading step, the applied force was held constant for 15 min. Prior to the loading test, an elastic consolidation analysis using the CRISP package had been performed with the value of the coefficient of permeability ($k = 3.2 \times 10^{-3}$ cm/s) obtained from in situ permeability tests to ensure that 15 min was sufficient for excess pore water pressure to dissipate and achieve the drained condition. More detailed descriptions and the results of the whole loading tests are reported elsewhere (Kusakabe et al., 1992).

Loading test results

The test results are plotted in Fig. 6, in terms of

(a) load (Q) against settlement (S)   (Q–S curve)
(b) load against time (t)   (Q–t curve)
(c) settlement against time   (S–t curve)
(d) load against elastic deflection (recoverable deflection upon unloading) (Se)   (Q–Se).

By plotting Q and S in double log-scale the yield load was determined as a deflection point. The ultimate bearing capacity was defined as a point where a linear portion appeared in the Q–S curve plotted in normal scale.

The soils were excavated after the test to investigate failure surfaces. Figure 7 shows a view of the zone beneath the footing after the footing

Fig. 7. Visible slip lines developed beneath footing

was penetrated to an S/B of as large as 0.9, in which the zone beneath the footing is heavily compressed and a series of slip lines developing from both edges of the footing is clearly shown by the change in tone on the excavated surface. Apart from the radial slip lines, no indication of clear shear bands was observed in other zones.

Centrifugal simulation

Centrifuge and test system
Two centrifuge tests with a reduced model (denoted US-2 and US-4) were performed using undisturbed block samples. The centrifuge used was the machine at Utsunomiya University with effective radius 1.18 m. A rectangular strong steel box (inner dimensions 262 mm in width, 299 mm in depth, and 498 mm in length) housed the model ground. The loading system, comprising an AC motor of 0.1 kW, a load cell and a dial gauge, was mounted directly on the strong box. The footing had a concave loading point to simulate the pin condition.

Model preparation and test procedures
The size of model footing was determined to be 1/13.3 of the prototype (30 mm square) by considering that the strong box boundaries do not interfere with possible failure zones. Consequently the centrifugal acceleration was 13.3, giving a prototype footing width (B) of 400 mm. The line of effective radius is located at 1B depth below the ground surface.

Two boxes of the undisturbed samples procured from the B ground

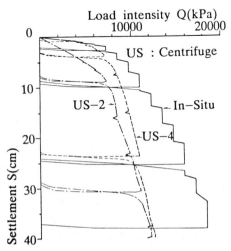

Fig. 8. Comparison of centrifugal simulations with in situ loading test

were used for the centrifuge tests. Model preparation consisted of four stages: water feeding, freezing, trimming and defrosting. Having sprayed water over the undisturbed sample, the sample was kept in a refrigerator at a temperature of −20°C for two days. The block sample was then trimmed down to a size of 180 mm in thickness, 200 mm in width, and 400 mm in length by using the diamond cutter fitted into a wooden liner for handling purposes. The trimmed sample was again placed in the refrigerator for another two days. One day before the loading test, the sample was weighed and its dimensions were measured. The model ground in the liner was lowered carefully into the strong box. A gap between the container and the sample was filled with wet Toyoura sand, which was then compacted to achieve the relative density of over 90%.

After placing cement paste on the surface of the model ground, the model footing made of steel was set on that location to simulate prototype conditions as precisely as possible. Water was then gradually fed to the sample from the bottom of the container for two hours and the container was hung from the centrifuge arm. Water was further added from the top surface of the sample. The centrifuge was accelerated to 13.3 G in 4 min and maintained for 15 min before the loading sequence was started.

Because of the system available at that time, the loading was of the settlement control type and was therefore different from the prototype condition. According to the $1/n^2$ similitude for time, an equivalent rate of loading and unloading of 4 mm/min was adopted and the load holding time was 5 s, corresponding to 15 min in prototype.

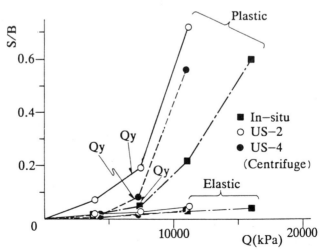

Fig. 9. Elastic and plastic deflections plotted against load intensity

Results of centrifugal simulations and comparisons

The load–intensity settlement curves obtained in the centrifuge tests are superimposed on Fig. 6(a), as is presented in Fig. 8. Curves of the two centrifuge tests are somewhat different in the early stage and become very close to each other in the later stage. What is noticeable in direct comparison with the prototype is that the centrifugal simulations underestimate both the coefficient of initial subgrade reaction K_{Vi}, and the yield load intensity Q_y and ultimate bearing capacity Q_u. The ratios of centrifuge result to prototype are 0.98 (US-4), 0.23 (US-2) for K_{Vi}, 0.93 (US-4), 0.89 (US-2) for Q_y, and 0.76 (US-4), 0.68 (US-2) for Q_u, respectively. Here, the determination of the yield load and ultimate load were the same as those adopted in the field test. Similarly, the ratio of centrifuge to prototype about the settlement/footing width ratios (for yield S_y/B; for ultimate S_u/B) are 1.05, 0.80 for the case of US-4.

Figure 9 shows the relationship between the elastic deflection and the load intensity for the case of US-4. It is clear that the centrifugal simulation predicts fairly well the value of elastic deflection even up to beyond the yield loads of the tests, suggesting that the centrifuge tests results in larger plastic deflections.

Settlement controlled loading was used in the centrifuge, whereas the prototype loading tests were of load control type. The match between the centrifuge simulation and prototype in terms of the settlement–time relationship is compared in Fig. 10. It is seen that up to 1.15 times the prototype yield load both are quite close to each other. Large discrepancy after yield would not affect other comparisons, provided that both prototype and centrifuge tests had been conducted under drained conditions.

Figure 11 compares the failure mechanism. Centrifuge models did not exhibit clear visible slip lines, although general patterns of the failure mechanism are in conformity with the prototype observations.

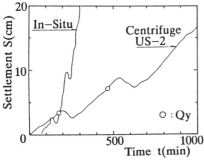

Fig. 10. Settlement–time relationship

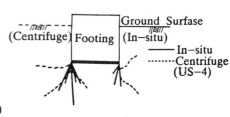

Fig. 11. Comparison of failure mechanism

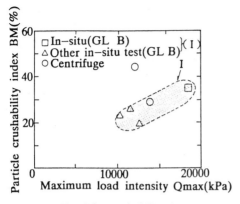
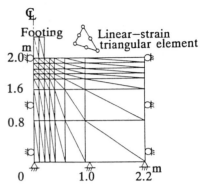

Fig. 12. Particle crushability index versus maximum load intensity

Fig. 13. FEM discretization of loading test

It was pointed out earlier that the scoria has a tendency for particle crushing. Both in the prototype and centrifuge tests, the two grain size distributions were measured after the tests; one from directly beneath the footing and the other from a point away from the loading point. Figure 12 summarizes the results of particle crushing in terms of crushability ratio defined by Marsal (1965), who plotted the crushability ratio against load intensity including the data of three other loading tests conducted on the B ground. There seems a trend that the crushability ratio increases with increasing load intensity. The degree of particle crushing is strongly related to the pressure level that the ground will experience, as is seen in Fig. 12. The centrifuge test does simulate the phenomenon of particle crushing, but the degree of particle crushing is underestimated because of the underestimation of the ultimate bearing capacity.

Numerical simulation
CRISP programme and analysis conditions
The CRISP programme is readily available in Japan with a Japanese user's manual (Akaishi et al., 1989). One of the attractive features of the CRISP programme is that it can be used on a personal computer. The personal computer used for this analysis was a 32-bit NEC PC-H98 S-model with a 100 Mb hard-disk memory.

The original Cam-clay model (Schofield and Wroth, 1964) was purposely selected as a constitutive model. For granular materials, alternative soil models are of course available and analyses with sophisticated constitutive soil models could have been done, but requires a mainframe computer, which was not our intention for this particular study.

The 400 mm square footing was modelled by a circular footing having the same cross-sectional area as the square footing and a drained axisymmetric analysis was carried out. FEM discretization of the loading test is shown in Fig. 13. It has 102 cubic strain triangular elements and 67 nodes. A depth of 5 B and a width of 5.5 B were considered adequate for the purpose of analysis. The boundary conditions of displacement imposed in the analysis are also shown in Fig. 13. It was assumed that the footing is effectively rigid and its base is perfectly rough. The number of loading steps was 1402 with a load intensity of 7415 kPa, which was restricted by the capacity of the personal computer used. The execution time was typically about 18 h.

The necessary input critical state parameters (κ, λ, M, Γ, p'_c) were selected mainly from the data of the triaxial CID and the isotropic compression tests. Namely, κ was determined from the rebound portion of the isotropic compression tests. λ was taken from Fig. 4. The value of M was estimated from the residual friction angle of 33.9° from Fig. 3. In order to satisfy the relationship of the Cam clay between the above parameters.

$$N = \Gamma + \lambda - \kappa,$$

the critical state line was shifted to the right by one standard deviation ($\bar{\sigma}$) and then N and Γ values were determined as 2.99, 2.89, respectively. Consequently the p'_c value became 3840 kPa.

Poisson's ratio (ν) was evaluated to be 0.3 from two elastic soil parameters K and G obtained from the initial portions of volumetric strain–deviator stress curve, and deviator strain–stress curves in the CID tests, respectively. The values of relevant soil parameters are summarized in Table 1. The ground water level was set at the ground surface. Initial vertical effective stresses were assumed to be of a trapezoidal shape due to selfweight with a small surcharge of 1 kPa.

Since K_0 values were not measured in situ, three different methods were used to estimate the K_0 values. The process of one-dimensional loading to p'_c value of 3840 kPa and unloading to the in situ effective vertical stress was simulated by a CRISP calculation using the soil parameters listed in Table 1, which gave $K_{nc} = 1.0$ and $K_0 = 17$. Wroth (1975) and Parry (1982) proposed the followed equations for K_0: $K_0 = OCR \times K_{nc} - \{\nu(OCR - 1)/(1 - \nu)\}$,

$$K_0 = K_{nc} \times (OCR)^{\phi'}.$$

Table 1. Soil parameters used in CRISP

κ	λ	Γ	M	ν	γ (kN/m³)
0.00573	0.112	2.89	1.37	0.3	18.6

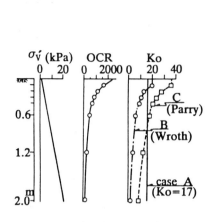

Fig. 14. In situ stress distribution in analysis

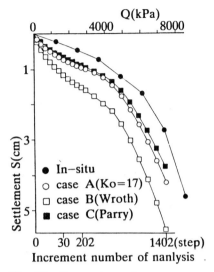

Fig. 15. Comparison of pressure curves for different in situ stress conditions

Using Jaky's equation $K_{nc} = 1 - \sin \phi'$ for OCR = 1, the vertical preconsolidation stress was calculated to be 3190 kPa. In situ stress distributions assumed in the analysis are shown in Fig. 14. Note that OCRs are extremely high in the zone shallower than 1B depth; consequently extremely high values of K_0 are obtained. Although these are outside the range of values considered reasonable, it should be mentioned here that none of the effective horizontal stresses calculated by these three methods exceeds the Rankine passive earth pressure using the residual strength parameters.

Results of numerical simulations and comparisons
Figure 15 is a comparison of the load-intensity–settlement curves against the prototype for various K_0 values, which clearly indicates that the shape of the initial portion of the curves is strongly influenced by the selection of K_0 values. The best fit curves up to the yield load among the three are the ones with the K_0 distributions predicted by the CRISP and suggested by Parry, although they overestimate the settlements by a factor of about 1.4 at a prototype load intensity of 7415 kPa.

Development of failure was examined by plotting the three zones (hardening, critical state and softening) at two different loading stages, as shown in Fig. 16(a). The slip lines observed in the prototype are also drawn. Figure 15(b) presents the state paths of two particular elements, one beneath the footing, the other in the passive zone. It is noted that

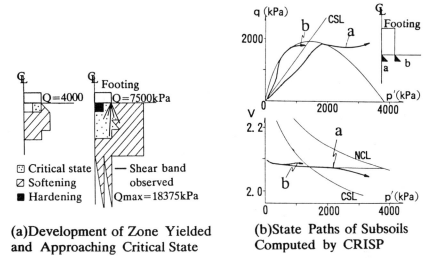

Fig. 16. Plastic zone and state paths

the zone of critical state is well matched with the area where the slip lines were observed, whereas the area beneath the footing is compressed and continuously hardening as is seen in the behaviour of the element (a), suggesting that particle crushing may be probable. This result is in conformity with the observation in the field.

Discussion

The centrifugal simulation carried out in this study was a relatively straightforward exercise, but assumed that the freezing and thawing process provided undisturbed samples of high quality. Although the yield load was quite accurately predicted, the difference between centrifuge and prototype did exist in two aspects:

(1) the underestimation of the coefficient of initial subgrade reaction K_{Vi}, and the ultimate load Q_u, and
(2) the overestimation of plastic settlement.

Both of these may be on-the-safe-side predictions from a practical point of view. The underestimation of K_{Vi}, Q_u may have arisen from the difference in loading system, the disturbance of samples and, in particular, the deterioration of the cohesion term of the cemented scoria. The overestimation of plastic deformation may have resulted from the fact that the ratio of particle size to footing width (B/D_{50}) was as small as 8.82–13.6. This may be compared with the case of the loading tests on Shirasu, of which the B/D_{50} value was about 6 times larger than the above. It is obvious that further research is required into whether centrifuge tests for natural granular deposits give predictions on the safe side or unsafe side, and into how seriously particle size affects the test results of granular materials. But the authors' experience in this attempt was that centrifuge modelling did work well in estimating the bearing capacity and in producing crushing phenomenon.

In FEM simulations some difficulties were experienced, mostly the selection of soil parameters and K_0 value. Some freedom was used to determine values of N and Γ consistent with the Cam clay model, which may imply that the Cam clay may not be the best choice for modelling the scoria dealt with in this study. It was demonstrated that the initial portion of the load–settlement curve was very sensitive to the selection of the K_0 value, and the use of a very high K_0 was needed to get reasonable agreement up to the yield load. This poses two questions:

(a) how to select appropriate elastic parameters for modelling the scoria, constant shear modulus G, or K varying with p', as is the Cam clay model

(b) how to estimate K_0 values for extremely high over-consolidation ratios, as is the case of this study.

These are the problems associated with soil model rather than those of numerical scheme.

As to the comparison of the two modellings it must be emphasized that the centrifuge tests readily provided the results without any information about soil element behaviour (best soil model and proper values for parameters) and in situ stress conditions.

Concluding remarks

A series of large-scale loading tests was successfully carried out in a pneumatic caisson which provided a body of information about the prototype behaviour of shallow footing on cemented and structured granular materials susceptible to particle crushing. It was found that centrifugal simulations using undisturbed samples predicted the yield load to be 89–93% of the corresponding prototype, and elastic deformations close to the prototypes. In the models, although the phenomenon of the particle crushing was observed, clear slip lines were not developed and settlements at yield load were larger than those in prototype. It was inferred that the two observations may be partly due to the particle size effect, which requires further research.

Attempts at numerical simulation for a scoria, naturally deposited, cemented granular material were made by the CRISP programme with the Cam clay model. Difficulty in the selection of soil parameters was encountered and the uncertainty in the determination of K_0 values for very high overconsolidation ratios was recognized.

References

AKAISHI, M., KUSAKABE, O., KOGO, T. AND TAKAGI, N. (1989). CRISP manual, Tokai University Press (in Japanese).

ALMEIDA, M.S.S., BRITTO, M. AND PARRY, R.H.G. (1986). Numerial modeling of centrifuged embankment on soft clay. Canadian Geotechnical Journal, Vol. 23, pp. 103–114.

BASSET, R.H. AND HORNER, J. (1979). Prototype deformations from centrifuge model tests. Proc. of 7th European Regional Conf. on SMFE, Vol. 1, pp. 1–9.

BASSET, R.H., DAVIES, M.C.R., GUNN, M.J. AND PARRY, R.H.G. (1981). Centrifuge models to evaluate numerical methods. Proc. 10th ICSMFE, Vol. 1, pp. 557–562.

BOLTON, M.D. AND LAU, C.K. (1988). Scale effects arising from particle size, Centrifuge '88, edited by Corte, Balkema, pp. 127–131.

BRITTO, A.M. AND GUNN, M.J. (1987). Critical state soil mechanics via finite elements, Ellis Horwood.

DAVIES, M.C.R. AND PARRY, R.H.G. (1985). Centrifuge modelling of embankments on clay foundations, Soils and Foundations, Vol. 25, No. 4, pp. 19–36.

FUJII, N., KUSAKABE, O., KETO, H. AND MAEDA, Y. (1988). Bearing capacity of a footing with an uneven base on slope: Direct comparison of prototype and centrifuge model behaviour, Centrifuge '88, edited by Corte, Balkema, pp. 301–306.

INDRARATNA, B., BALASUBRAMANIAM, A.S. AND BALACHANDRAN, S. (1992). Performance of test embankment constructed to failure on soft marine clay. ASCE Journal of Geotechnical Engineering, Vol. 118, No. 1, pp. 12–33.

KIMURA, T., KUSAKABE, O. AND SAITOH, K. (1984). Undrained deformation of clay of which strength increases linearly with depth. Proc. of Symp. Application of centrifuge modelling to geotechnical design, pp. 315–335.

KUSAKABE, O., MAEDA, T., OHUCHI, M. AND HAGIWARA, T. (1991). Strength-deformation characteristics of an undisturbed scoria and effects of sampling disturbance (in Japanese), Proc. of Japan Society of Civil Engineers. No. 430/III -15, pp. 97–106.

KUSAKABE, O., MAEDA, Y. AND OHUCHI, M. (1992). Large-scale loading tests of shallow footings in a pneumatic caisson. ASCE, Journal of Geotechnical Engineering, Vol. 118, No. 11, pp. 1681–1695.

LYNDON, A AND SCHOFIELD, A.N. (1970). Centrifugal model test of a short-term failure in London clay, Geotechnique, Vol. 20, 440–442.

LYNDON, A. AND SCHOFIELD, A.N. (1978). Centrifugal model tests of the Lodalen landslide. Canadian Geotechnical Journal, Vol. 5, pp. 1–13.

MARSAL, R.J. (1965). Soil properties–shear strength and consolidation, Proc. 6th Int. Conf. SMFE, Vol. 3, pp. 310–316.

PARRY, R.H.G. (1982). Quoted by Britto and Gunn (1987), p. 184.

SCHOFIELD, A.N. AND WROTH, C.P. (1964). Critical State Soil Mechanics, McGraw-Hill.

TATSUOKA, F., OKAHARA, M., TANAKA, T., TANI, K., MORIMOTO, T. AND SIDDIQUEE, M.S.A. (1991). Progressive failure and particle size effect in bearing capacity of a footing on sand, Geotechnical Engineering Congress, ASCE, Vol. II, pp. 788–802.

WROTH, C.P. (1975). In-situ measurements of initial stresses and deformation characteristics, Proc. Spec. Conf. on In-situ measurement of soil properties, ASCE, pp. 181–230.

The behaviour of a displacement pile in Bothkennar clay

B. LEHANE and R.J. JARDINE, Imperial College of Science & Technology

The paper describes a series of six experiments conducted with extensively instrumented displacement piles at the SERC soft clay test bed site at Bothkennar, Scotland. Reliable measurements of pore-water pressures, radial total stresses, local shear stresses, temperature and pile axial force were made at various levels on the shaft of a steel, closed-ended, jacked pile during installation, equalisation and load testing to failure. The research sheds light on the fundamental processes that govern displacement pile behaviour in clays; the test series also evaluates the influence on capacity of parameters such as pile slenderness ratio, displacement rate during installation, set-up time, drainage conditions during loading and shaft capacity in tension versus compression.

Introduction
The Bothkennar instrumented pile tests form part of a series of investigations into the behaviour of displacement piles in sands and clays; see Bond and Jardine (1990), Lehane and Jardine (1992a), Lehane (1992). Measurements are presented in this paper of the radial effective stresses and shear stresses developed in a soft sensitive clay during pile installation, equalisation and load testing. These data provide fresh insights into the processes that control displacement pile capacity in soft clays. The Bothkennar piling research has also included strain path analyses and other theoretical studies (Whittle, 1991; Lehane, 1992).

Site description
The Science and Engineering Research Council (SERC) has set up a national soft clay test bed at Bothkennar, Scotland, on the southern bank of the Forth estuary. The ground conditions have been investigated extensively using state-of-the-art sampling, laboratory testing and in situ test techniques. Details of these investigations are reported by Hawkins et al. (1989), Hight et al. (1992), Smith et al. (1992) and others.

Soil profile
The Bothkennar clay was deposited in a stable shallow marine environment and, within the depths penetrated by the piles (1–6 m), comprises

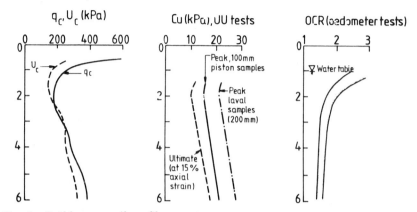

Fig. 1. Bothkennar soil profile

soft black silty mottled clay with some local silt laminae; a 1 m thick weathered firm crust has formed above the soft clay. Plasticity indices (PI) increase from ≈25% at 1.5 m to ≈50% between 4 and 6 m. Paul et al. (1991) have shown that these indices are enhanced by the moderate proportion (≈3%) of organic residues bound to the clay minerals and that the PI reduces to between 20 and 25% when this is removed. The clay fraction increases from ≈15% at 2 m to ≈40% at 6 m and consists of kaolinites, illites and rock flour. Quartz is the principal mineral in the silt fraction (50–75%). Post-depositional chemical alterations have led to some local *bonding* between silt particles (Paul et al., 1991). The soil is moderately sensitive and has liquidity indices between 0.5 and 0.9.

The profiles obtained with piezocone (CPTU) tests and quick undrained (UU) triaxial compression tests are shown in Fig. 1. Below 2 m, the CPTU end resistance (q_c) and pore pressure (u_c) both increase slowly with depth in a similar way to the undrained shear strength (c_u) profiles. The peak values of c_u vary with the sample type; 200 mm diameter Laval samples (La Rochelle et al., 1981) gave strengths typically 40% greater than those of 100 mm diameter piston samples. This dependence on sample size and quality is typical of sensitive clays (e.g. Smith, 1992). The reference undrained strength (c_{u0}) used as a normalising parameter to calculate adhesion factors (α) in the pile tests is the peak strength of the more conventional 100 mm diameter samples.

The apparent overconsolidation ratios (OCRs) at Bothkennar (Fig. 1) were estimated from oedometer tests which followed the standard incremental load procedure and showed that the OCR reduces from ≈1.9 at 2 m to ≈1.5 at 6 m. In situ horizontal stresses were measured in self-boring pressuremeter tests. These measurements and standard relationships between OCR and K_0 suggested that K_0 reduced from ≈0.65 at 2 m to ≈0.52 at 6 m.

Behaviour in undrained shear

Undrained triaxial tests on intact samples of Bothkennar clay, which were re-consolidated to in situ (K_0) stress conditions, showed the following characteristics (see Hight et al. (1992)):

(a) Significant excess pore pressures were not developed until a large strain yield surface was reached. Samples tested in compression exhibited pronounced brittleness and reducing mean effective stress when strained beyond this point. The degree of brittleness was more significant in tests on 'undisturbed' block samples than in tests on piston samples. Brittleness was absent in extension tests.

(b) The mean large strain (critical state) friction angle (ϕ'_{cv}) was 35.5° for piston samples and ≈38° for block samples; the latter may have retained a degree of bonding even after developing large shear strains. Smith (1992) suggested that the clay's high ϕ'_{cv} values are associated with the significant proportion of rock flour and the colloidal organic content.

Behaviour in interface shear

Soil on soil and soil–steel interface ring shear tests were performed on Bothkennar clay, as described in Lehane and Jardine (1992b). The experiments used interfaces of the same material and roughness as those of the piles and modelled the displacement history of soil elements adjacent to a pile during installation.

Samples were subjected to an initial relative displacement of 1.2 m in 8 fast shearing pulses (at 500 mm/min) and subsequently tested in slow (drained) shear. The peak drained resistances corresponded to friction angles (= $\tan^{-1}(\tau/\sigma'_n)$) of 33° ± 1° in soil on soil shear and 31° ± 2° in soil–steel shear. Post-peak reductions in τ/σ'_n were observed in about one third of all tests, whilst little or no loss of strength was recorded in the remainder. This unstable response led to ultimate angles which ranged from 25 to 32°.

Table 1. Pile and load tests details at Bothkennar

Pile	L(m)	t_{eq} (days)	Load test	Peak τ_{av} (kPa)	α	Disp. at peak (mm)
BK1T	6.00	0.83	Tens.	15.3	0.90	12.0
BK2C	6.00	4.20	Comp.	16.7	0.98	3.75
BK3C/1	3.15	0.08	Comp.	9.2	0.61	2.05
BK3C/2	5.90	3.90	Comp.	17.4	1.02	4.90
BK4C	3.15	4.20	Comp.	17.8	1.19	3.70

Pile experiments
Testing programme
The pile tests performed at Bothkennar are summarised in Table 1. Four instrumented 102 mm diameter cone-ended steel, tubular piles were jacked into the ground in a series of 220 mm stroke pushes at a rate of 500 mm/min. These pushes were separated by pause periods of typically 5 min duration. Piles were installed to a depth of either ≈ 3.2 m or ≈ 6 m and were load tested after an equalisation period (t_{eq}) of ≈ 4 days, with two exceptions:

(1) Pile BK3 (which was installed first to 3.15 m) was load tested after just 2 h (BK3C/1). This pile was then jacked on to a final penetration of 5.9 m and re-tested after the standard period of 4 days (BK3C/2).

(2) Pile BK1T experienced a minor leak and was load tested after a shortened equalisation of 20 h so as to limit instrument damage.

All piles, except BK3C/2, were load tested to failure by applying 12 load increments over a period of ≈ 1 h. The excess pore pressures generated by loading did not decrease appreciably between load increments and it suggested that this testing procedure gave essentially undrained conditions at the pile shaft. At failure, attempts were made to maintain the maximum load by pumping the loading ram continuously. For the special case of pile BC3C/2, loading was slowed down to ensure that all excess pore pressures dissipated between load increments; this drained test took more than 24 h to complete.

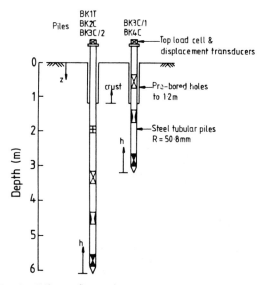

Fig. 2. Pile configuration

Instrumentation

The piles were equipped with three main instrument clusters, as shown in Fig. 2. Each cluster comprised a load cell to measure axial load (P), two independent probes to measure pore pressures (u) and a surface stress transducer (SST) which recorded radial stress (σ_r), shear stress (τ_{rz}) and temperature. Measurements made by each device are referred to by their distance from the pile tip (h) normalised by the pile radius (R) (i.e. h/R). Additional axial load cells were used at the pile head and at h/R = 77 on piles installed to ≈6 m. A full description of all instruments is given by Bond and Jardine (1990).

Excepting difficulties associated with the leakage encountered in pile BK1T, all instruments remained stable, showing virtually no zero shift or change in sensitivity during the programme.

Pile installation
Shaft resistance

The average shaft shear stresses (τ_{av}) measured during a typical pile installation at Bothkennar are shown in Fig. 3. The resistance is seen to reduce steadily from a maximum at the start of a pile push (τ_{avp}) to a minimum at the end of the push. A similar trend was observed during fast shearing pulses in ring shear tests on Bothkennar clay (Lehane and Jardine, 1992b). Considerably higher values of τ_{avp} are mobilised in jacking stages which were preceded by extended pause periods (of greater than ≈7 min). However the minimum resistances measured by all piles tended towards a consistent lower bound value of ≈5 kPa. It

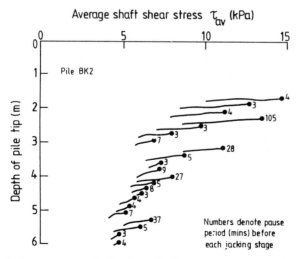

Fig. 3. *Shaft shear stresses during installation*

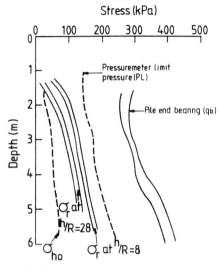

Fig. 4. *Installation radial total stresses*

appears that τ_{av} would remain at 5 kPa, which is equivalent to an α value of ≈ 0.3, if the piles were pushed continuously (with no pause periods) to their final penetration depth. Such a condition of 'steady' penetration is assumed by the strain path method when modelling the installation process (Baligh, 1985).

Radial total stress
Envelopes to the radial total stresses (σ_r) recorded during all pile installations are shown in Fig. 4. These lie between the initial undisturbed horizontal stress (σ_{h0}) and the limit pressure measured in self-boring pressuremeter tests (p_{lim}), demonstrating that cylindrical cavity expansion calculations overpredict the stresses developed against displacement piles. The radial stresses recorded at h/R = 8 are typically $\approx 0.8 p_{lim}$ but reduce (within a given soil horizon) to $\approx 0.6 p_{lim}$ when the pile tip penetrates further to h/R = 28. The tendency for σ_r (as measured at fixed depths) to reduce with h/R appears to stabilise when h/R > 28. Similar tends were shown by the local shear stress measurements (τ_{rz}).

Pore pressure
The installation pore pressures are summarised in Fig. 5. It is notable that the pressures recorded while the piles were moving (u_m) are less than the stationary values (u_s) measured during the pause periods in between jacking stages. This phenomenon has been observed in other instrumented pile tests in a wide range of clays (e.g. Coop and Wroth

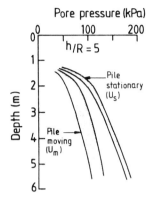
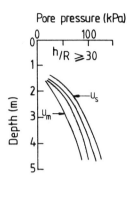

Fig. 5. *Installation pore pressures*

(1989), Lehane and Jardine (1992a)) but appears to occur only when the pore pressure sensors are located within material that has been pre-sheared in a previous jacking stage (see Lehane (1992)).

As with the σ_r measurements, pore pressures (u_m and u_s) increase with depth and reduce with h/R. Pore pressures recorded with a piezocone (for which h/R ≈ 1) are about twice the values of u_s recorded at h/R = 5.

Radial effective stress

Although similar radial total stresses are measured in the moving and stationary conditions, the discrepancies between u_m and u_s lead to large differences between the effective stresses, σ'_r, computed during jacking and in pause periods. σ'_r appears to fall rapidly during the first minute after jacking (as pore pressures rise). The strong vertical hydraulic gradients between the probes at h/R = 5 and h/R = 30 must be accounted for when calculating σ'_r at the leading instrument position; if not, apparently negative values are obtained.

Interface friction angles

Interface friction angles may be calculated from the effective stress measurements made during jacking as $\delta = \tan^{-1}[\tau_{rz}/(\sigma_r - u_m)]$. Accepting the u_m values at face value leads to δ values between 12° and 18°. This range falls well below the ultimate residual angle (δ_{ult}) measured in ring shear tests. As will be shown, the subsequent load tests also proved far higher and more credible δ values. From this and other evidence, it appears that pore pressure measurements at the pile shaft are anomalous when displacement rates exceed ≈1 mm/min and do not represent those acting on the principal displacement shear, which probably existed a short distance from the shaft in the Bothkennar tests. More plausible values of pore pressure acting in the failure zone may be estimated from the installation shear and radial total stress records by assuming that the interface friction angles proven (at very slow rates of displacement) during load tests also apply during installation. The pore pressures deduced in this way lie *between* the recorded values of u_m and u_s.

Equalisation
Radial total stress (σ_r)

The changes in radial total stress (σ_r) which took place during equalisation are summarised in Fig. 6 for all instrument positions, using the non-dimensional ratio $H/H_i = (\sigma_r - u_0)/(\sigma_{ri} - u_0)$, where σ_{ri} is the measurement made at the end of installation and u_0 is the hydrostatic water pressure. A consistent pattern is seen; H/H_i remains constant for the first few minutes, but then reduces steadily, reaching an ultimate value of less than 0.5. If the material adjacent to the pile was elastic (and isotropic) throughout the equalisation process, no such reduction would

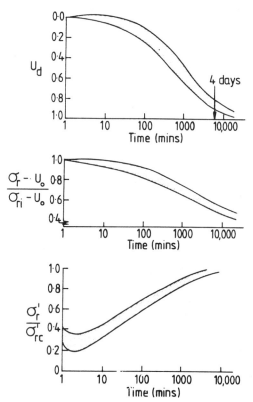

Fig. 6. Equalisation data

Pore pressure

As witnessed during installation pause periods, pore pressures rose rapidly to reach maxima (u_{max}) about 1 min after the end of installation. The subsequent variation of the pore pressure dissipation factor (U_d) defined as $(u_{max} - u)/(u_{max} - u_0)$, during equalisation is shown in Fig. 6 which combines data measured by all instrument clusters. It is evident that the average time taken for 50% excess pore pressure dissipation is ≈750 min. After 4 days, U_d values are in excess of 0.8. The spread in these data reflects the more three dimensional drainage conditions near to the tip. Dissipation at $h/R = 53$ was consistently almost twice as slow as that at $h/R = 5$ and about seven times slower than at $h/R = 1$ (as inferred from piezocone dissipation tests).

However, dissipation curves for the same h/R – but different soil depths – were closely comparable, indicating that local variations in soil permeability and compressibility had no significant effect. Two dimensional uncoupled consolidation analyses indicated radial coefficients of consolidation (c_h) of ≈30 m²/year, which falls in the range measured in laboratory tests for small-strain swelling or re-compression (Smith, 1992).

Radial effective stress (σ_r')

The variations of σ_r' during equalisation are summarised in Fig. 6 by normalising σ_r' values recorded at each instrument position by their

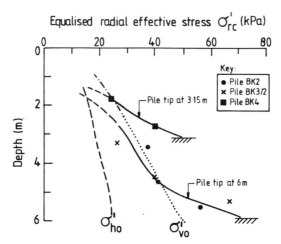

Fig. 7. Equalised radial effective stress profiles

respective final equilibrium value (σ'_{rc}). It is seen that after about 2 min, σ'_r values increase rapidly with time, reaching 85% of fully equalised values after 1000 min and full equilibrium after 4 days. This rate of equalisation exceeds that for pore pressure decay because pore pressure and radial stress changes are practically equal during the latter stages of dissipation. σ'_{rc} values are between three and five times the minima measured shortly after installation.

The profiles of σ'_{rc} for the two pile lengths investigated are shown in Fig. 7. σ'_{rc} increases with depth in both cases and lies within the range $[0.8-1.4]\sigma'_{v0}$. However, the magnitudes of stresses acting on the shorter pile appear to be about 50% greater than those developed at the same depths by the longer piles, suggesting that σ'_{rc} also varies with the relative depth of the pile tip (h/R). This dependence is comparable to that shown by the installation radial stresses (σ_{ri}) as values of $[\sigma'_{rc}/(\sigma_{ri} - u_0)]$ measured at *all* instrument positions were relatively constant at 0.44 ± 0.04. Comparisons with published solutions for other clays shows that, except near the pile tip, the σ'_{rc} values fall well below the predictions of cylindrical cavity expansion theory.

Load testing

The compression load tests performed at t_{eq} = 4 days all gave similar initial load displacement curves. Peak shaft capacities were mobilised; the mean typically, after displacements of $\approx$4 mm (see Table 1). However the tension test (BK1T) (and also tension retests, if performed after compression tests) exhibited a significantly softer response, suggesting that the stiffness is related to the principal stress rotations which the soil adjacent to the pile experiences during loading.

Excluding the short term test (BK3C/1), the peak average shaft resistances (τ_{avp}) are within 10% of the mean value of 16.6 kPa and corresponds to an α value of $\approx$1. This narrow range suggests that the unit capacities, mobilised more than $\approx$1 day after installation, are relatively insensitive to the pile depth (from 3 to 6 m), the direction of loading and the pile loading rate (at <1 mm/min). Although the length of the equalisation period has little influence on capacity after one day has elapsed, it has a critical influence over the first few hours: capacities after full equalisation are typically three times those measured during installation (see Fig. 3).

The variations of local stresses during the loads test are described most concisely by plotting τ_{rz} against σ'_r at each instrument position. These 'stress paths' are shown in Fig. 8 and reveal two consistent features:

(a) Radial effective stresses reduce during pile loading and at peak shear stress are typically $\approx$0.85 of the pre-loading stress.

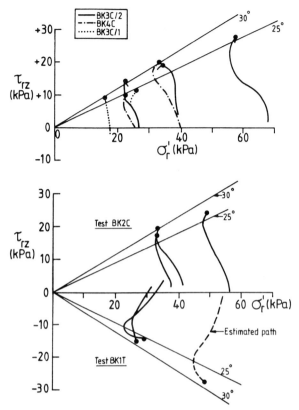

Fig. 8. τ_{rz} versus σ'_r, variations during load tests

(b) The peak obliquities (δ_p) lie within a narrow range of 25° to 30°, with a mean value of 28.5°.

It appears that the values of δ_p and the relative reductions in σ'_r are insensitive to the direction of loading, the rate of loading and the length of the equalisation period. In Bothkennar clay, differences in capacity are associated (primarily) with variations in the pre-loading radial effective stresses.

Only the data recorded up to, and including, the peak shear stress are shown for the standard 'undrained' pile tests in Fig. 8. In these tests, the piles accelerated from ≈0.2 mm/min to ≈5 mm/min post-peak as attempts were made to maintain the maximum applied load. Anomalously low post-peak pore pressures were recorded (as during installation) which led to apparent ultimate obliquities of only ≈15°. However, these low angles were only measured at high velocities and disappeared when the rate was dropped to below ≈1 mm/min. Furthermore, reload tests gave comparable δ_p values to values measured in first-time tests, proving that the apparent large post-peak reductions in δ were not associated with the development of a low strength soil fabric adjacent to the pile.

The complete (pre- and post-peak) stress paths of the drained test (BK3C/2) are shown on Fig. 8. The pre-failure paths of this test are comparable to those of undrained tests, suggesting that the soil adjacent to this pile was sheared at nearly constant volume conditions with reductions in σ_r replacing the pore pressure increases seen in the undrained tests. Post-peak, the test shows only a slight reduction in δ and little variation in σ'_r.

Comparison of load test data with laboratory tests

The kinematic boundary constraints close to a pile during undrained loading impose a constant volume simple shear mode of deformation on the adjacent soil elements. With this in mind, Randolph and Wroth (1981) proposed that data from constant volume direct simple shear (DSS) tests on (K_0) normally consolidated clay might be used to predict the shaft failure criteria. It was also suggested that failure might start on horizontal surfaces. The obliquity at peak local shear stress in this instance is given by $(\tau_f/\sigma'_{rf}) = [\sin \phi' \cos \phi']/[1 + \sin^2 \phi']$, which for Bothkennar clay with $\phi'_{cv} \approx 35°$, is equal to 0.32. Azzouz et al. (1990) noted that discrepancies may exist between pile loading stress paths and those measured in DSS tests (at OCR = 1) because of the difference between the consolidation stress conditions operating in the DSS apparatus and those adjacent to the pile. However, they also suggested that DSS tests could provide a useful tool for predicting the effective stress changes and δ angles applying to undrained pile loading. In their analyses, DSS tests performed at an OCR = 1.2 provided the best model for expected behaviour.

Figure 9 shows the typical normalised DSS stress paths for Bothkennar clay at OCR = 1 and 1.2. These paths have been extrapolated from measurements made in DSS tests on intact Bothkennar clay at in situ stress conditions (Lehane, 1992). The mean trend stress path established for the drained and undrained pile tests is plotted on the same diagram. We note:

(a) The pile tests show smaller reductions in σ'_r up to peak than would DSS tests on soil at OCR $\leq$ 1.2.

(b) The peak obliquity coincided with the peak shear stress in the pile tests; this is not true in DSS tests.

(c) The dramatic post-peak reduction in shear and normal effective stress seen in DSS tests was not evident in the pile tests.

(d) The measured peak obliquity is not compatible with the hypothesis that failure starts on a horizontal surface.

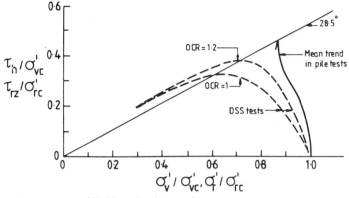

Fig. 9. Comparison of DSS and pile test data

Although pile loading involves a simple shear mode of deformation up to peak shear stress, DSS tests on K_0 consolidated samples (at OCR ≤ 1.2) overestimate the changes in σ'_r and underestimate δ. There appears to be a departure from continuum behaviour when peak local shear stresses are mobilised. Under these conditions, ring shear interface tests provide the best means of determining peak and ultimate obliquities (δ_p and δ_{ult}); these tests gave δ values that agree well with the mean of the peak angles recorded in all pile tests (28.5°) and the mean ultimate angles measured under drained loading conditions.

Conclusions

The most conclusive result to emerge from the instrumented pile tests at Bothkennar is that the peak local shear stresses are controlled by the criterion $\tau_f = \sigma'_{rf} \tan \delta_f$, where δ_f is the peak residual angle measured in soil–interface shear. The values of σ'_{rf} depend on the soil consistency, the degree of equalisation and the relative depth of the pile tip (h/R) and are $\approx 15\%$ lower than the equalised radial effective stresses (σ'_{rc}). The measurements do not follow the patterns predicted by any existing simple theory.

Acknowledgements

The Bothkennar pile research was funded by: The Science and Engineering Research Council (through MTD Ltd), AMOCO (UK) Exploration Co., Building Research Establishment, CONOCO (UK) Ltd., Exxon Production Research Co., Health & Safety Executive, Mobil Research and Development Corp., Shell (UK) Ltd. The support of the sponsors and the efforts of all the staff at Imperial College are gratefully acknowledged.

References

Azzouz, A.S., Baligh, M.M. and Whittle, A.J. (1990). Shaft resistance of piles in clay. J. Geotech. Eng. Div., ASCE, Vol. 116, No. 2, pp. 205–221.

Baligh, M.M. (1985). Strain path method. J. Geotech. Eng. Div. ASCE, Vol. 111, No. 9, pp. 1108–1136.

Bond, A.J. and Jardine, R.J. (1990). Research on the behaviour of displacement piles in an overconsolidated clay, UK Dept. of Energy OTH Report, OTH89 296.

Coop, M.R. and Wroth, C.P. (1989). Field studies of an instrumented model pile in clay. Geotechnique, Vol. 39, No. 4, pp. 679–696.

Hawkins, A.B., Larnach, W.J., Lloyd, I.M. and Nash, D.F.T. (1989). Selecting the location, and initial investigation of the SERC soft clay test bed site. Quart. J. Eng. Geology, Vol. 22, pp. 281–316.

Hight, D.W., Bond, A.J. and Legge, J.D. (1992). Characterisation of Bothkennar clay: an overview. Geotechnique, Vol. 42, No. 2, pp. 303–348.

La Rochelle, P., Sarrailh, J., Tavenas, F., Roy, M. and Leroueil, S. (1981). Causes of sampling disturbance and design of a new soil sampler for sensitive soils. Can. Geotech. J. Vol. 18, No. 1, pp. 85–107.

Lehane, B.M. (1992). Experimental investigations of pile behaviour using instrumented field piles. PhD thesis, Univ. of London (Imperial College).

Lehane, B.M. and Jardine, R.J. (1992a). The behaviour of displacement piles in glacial till. Proc. Conf. on the Behaviour of Offshore Structures, BOSS 1992, Vol. 1, pp. 555–568. BPP Technical Services, London.

Lehane, B.M. and Jardine, R.J. (1992b). The residual strength characteristics of Bothkennar clay. Geotechnique, Vol. 42, No. 2, pp. 363–368.

Paul, M.A., Peacock, J.D. and Wood, B.F. (1991). The engineering geology of the soft clay at the national soft clay research site Bothkennar. Dept. of Civ. Eng., Heriot-Watt University, Edinburgh.

Randolph, M.F. and Wroth, C.P. (1981). Application of the failure state in undrained simple shear to the shaft capacity of driven piles. Geotechnique, Vol. 31, No. 1, pp. 143–157.

Sills, G.C. (1975). Some conditions under which Biot's equations of consolidation reduce to Terzaghi's equation. Geotechnique, Vol. 25, No. 1, pp. 129–132.

Smith, P.R. (1992). Properties of high compressibility clays with reference to construction on soft ground. PhD Thesis, University of London (Imperial College).

Smith, P.R., Jardine, R.J. and Hight, D.W. (1992). On the yielding of Bothkennar clay. Geotechnique, Vol. 42, No. 2.

WHITTLE, A.J. (1991). Predictions of instrumented pile behaviour at the Bothkennar site. Private report to Dept. of Civ. Engineering, Imperial College.

Three-dimensional tests on reconstituted Bothkennar soil

P.I. LEWIN and M.A. ALLMAN, City University

Tests on reconstituted Bothkennar soil were carried out in a conventional triaxial apparatus and in a true triaxial cube apparatus. All samples were initially subjected to one-dimensional normal consolidation to the same stress point. They were then tested under different three-dimensional conditions of drained or undrained loading. They are grouped in five sets of tests – axisymmetric, constant mean effective stress, plane lateral strain, plane vertical strain and undrained – and it is shown how these sets interrelate.

Introduction

A conventional triaxial apparatus, testing a cylindrical sample enclosed laterally by a membrane, is limited in that the two lateral principal stresses are necessarily equal. Field situations, however, are rarely so simple. In an analysis of a soil mechanics problem, if it is intended to use parameters derived from a triaxial test then it is still necessary to have some appreciation of the influence of an intermediate principal stress lying somewhere between the major and minor values. Certainly an important requirement for any successful constitutive modelling process is that it includes the ability to predict for plane strain or plane stress conditions – conditions which were of particular interest to Wroth (1984) in his Rankine Lecture.

As a help towards the understanding of three-dimensional problems, this paper looks at a normally consolidated reconstituted soil and, by using a true triaxial apparatus, investigates five particular surfaces in stress space to see how the stress–strain behaviours on these surfaces can be interrelated. Each surface is generated by a family of stress paths where, for each family, a single condition is allowed to vary over a wide spectrum. Particular attention is paid to the conditions that give plane strain.

Some of the experiments described here were carried out in the course of an investigation, sponsored by the SERC, into the mechanical behaviour of Bothkennar soil (Allman and Atkinson, 1992).

Apparatus

The triaxial apparatus was an hydraulic stress path cell with computer control and logging facilities (Atkinson, Evans and Scott, 1985) and was fitted with standard pore pressure and cell pressure transducers, internal load cell, IC type volume gauge and an external displacement transducer to measure axial strains.

The cube apparatus has been briefly described by Abbiss and Lewin (1990) and was basically of the type developed by Ko and Scott (1967) with pressurized flexible membranes on each face of a 60 mm cubic sample. Drainage was allowed by filter strips in contact with the sample which conveyed the water through channels in the cell frame either to a pore pressure transducer or to a volume measuring device and back-pressure. Deformations were measured at the centre of each face by means of a thin steel pin, mounted on the membrane, which could slide through an O-ring seal and so emerge from the cell where it registered against an external LVDT. The computer control and logging facilities were similar to those used for the triaxial apparatus, although necessarily more complex. Strain control was achieved by the computer comparing the measured strain with the required strain and then instructing the appropriate pressure to adjust accordingly; this method generally worked well but occasionally produced hunting.

A problem with membranes is that they allow the face of the sample to bow instead of remaining flat or straight, so that strain is not strictly uniform. In a triaxial apparatus (which is a mixed membrane and platen device) the sample tends to barrel in compression and the true lateral strain is then greater than that calculated from the assumption of a right cylinder. In the cube apparatus, the measured drainage in the volume transducer was generally 20% less than the sum of the strains measured by the LVDTs during consolidation – but this characteristic could be reduced or reversed in direction during swelling, according to circumstances. The basis for calculations made here is

(a) that the triaxial sample deforms as a right cylinder
(b) the strains in the cube are based on the deformations measured at the face centres by the LVDTs even if in an undrained test, when the drainage is shut off, these strains ($\varepsilon_x + \varepsilon_y + \varepsilon_z$) do not add up to zero.

Sample preparation

All samples were prepared from slurry mixed with saline water to a water content of around 90% ($w_L = 67$, $w_P = 27$). The slurry was placed in 'floating-ring' formers (38 mm or 100 mm dia. for the triaxial and 60 mm square for the cube). The vertical stress was taken to 50 kPa for the triaxial and 180 kPa for the cube. (A higher value in the cube was required since further available strain in the apparatus was limited.) The samples were extruded and placed in the cells. Drainage was allowed against a back-pressure of 200 kPa. The effective stresses were raised at a ratio of lateral/vertical effective stress of 0.5 (to give approximately K_0 conditions) until they reached $\sigma'_z = 300$ and $\sigma'_x = \sigma'_y = 150$ kPa (marked A in the figures) where the vertical direction is designated z. Before testing, the stresses were left at A until the vertical deformation rate had dropped to 0.01% per hour (about 24 hours).

Results

Because of the initial axisymmetric preparation of the samples, each cube test can be considered to have a mirrored version in which the x and y directions are transposed; such putative tests will be labelled with an # as, for example, 1212#. Tests performed in the triaxial apparatus are prefaced with a B as, for example, B33.

A summary description of all the tests is given in Table 1 which shows to which set or sets each test belongs and what the effective stress loading rates were when stress paths were specified. The same test may appear in more than one set.

Table 1. Sets and loading rates

Test No.	Set 1 $\sigma_x = \sigma_y$	Set 2 const p'	Set 3 ε_x or ε_y = 0	Set 4 $\varepsilon_z = 0$	Set 5 drain shut	Rates – kPa/h σ'_x	σ'_y	σ'_z
Cube								
611			y			−2	0.75§	2
619		√§	y			2	0§	−2
628	√	√				1	1	−2
722		√				1	−1	0
919	√		xy§			−1.28	−1.28	−4
1007	√					−1.25	1.25	−5
1014		√		√§		4	−2	−2
1104			x			1.5§	−2	4
1111		√		√§		4	−1.5	−2.5
1212		√	y			−2.3§	0.3§	2
1310			x	√		1§	4	−1.4§
1320	√	√				−2	−2	4
1328	√					3	3	−2
1406			y		√	0.7§	−0.9§	−3.5§
1417			y		√	−1.9§	−0.4§	0§
Triaxial								
B33	√	√				−2	−2	4
B58	√	√				2	2	−4
B75	√				√	−2§	−4§	0§
B90	√				√	0§	0§	−4§

§ denotes a condition that was not directly stipulated

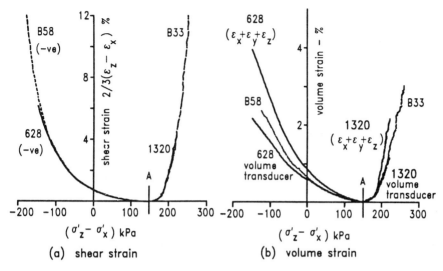

Fig. 1. *Constant p' tests comparing triaxial and cube. (a) Shear strain, (b) volume strain.*

Set 1. Triaxial plane, $\sigma'_x = \sigma'_y$

Since these tests were all confined to the axisymmetric triaxial plane, where $\sigma'_x = \sigma'_y$, they could be performed in either the triaxial apparatus or the cube apparatus. The opportunity was therefore taken to duplicate the constant p' tests in both kinds of apparatus. This was done for both compression (B33 and 1320) and extension (B58 and 628). The shear strains (Fig. 1(a)) compared very well. The volume strains (Fig. 1(b)) compared quite well when looking at the actual drainage measured by the volume transducer but were not so good when looking at the sum of the strains ($\varepsilon_x + \varepsilon_y + \varepsilon_z$); this is another manifestation of the membrane-bowing phenomenon already mentioned.

The stress paths for all tests in this set are shown in Fig. 2(a). The undrained compression and extension tests – B75 and B90 – were performed in the triaxial apparatus. These two tests, alone of this group, have stress paths which are curved since they are strain-ratio controlled tests. However the initial portions of their stress paths are straight and the results from these straight portions may therefore be considered to belong to the same family as the rest of the tests in this group which were carried out at controlled stress ratios and therefore necessarily have straight stress paths.

The path marked K_0 was not actually performed but is an extrapolation of the consolidation stage up to state A.

Tests 919 and 1007 were carried out with stress ratios known to

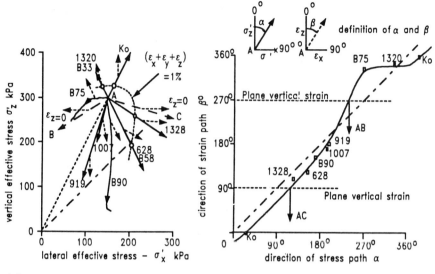

Fig. 2. Axisymmetric triaxial plane where $\sigma'_x = \sigma'_y$. (a) Stress paths and initial strain directions, and (b) initial stress and strain angles.

produce near K_0 swelling. Test 1328 was along a consolidation path with a reducing vertical stress.

Some strain paths were reasonably straight while others had a very slight curve. The direction of the first part of each strain path (between the values 0.01% and 0.05%) has been superimposed a short arbitrary distance along the appropriate stress path in Fig. 2(a). This clearly illustrates the difference in the direction of the strain vector at the start of loading in relation to the associated stress vector and how this difference changes as the stress vector rotates through 360°. This information is presented in an alternative form in Fig. 2(b) which shows the angle β of the strain vector plotted against the angle α of the stress vector (defined in Fig. 2(b), inset). From Fig. 2(b) it is possible to pick off the conditions for plane vertical strain (where $\beta = 90$ and 270°). This gives respective values for α of 115 and 240°, shown AC and AB in Fig. 2(a).

In Fig. 2(a) the position on each stress path is marked at the point where the volume strain reached 1%. These points have been joined up to form a contour and it will be noted that this contour reflects the shape of the undrained test paths. Further contours can be constructed for higher values of volumetric strain and this knowledge, together with the direction of the strain paths given in Fig. 2(b), could be used to predict (following principles enunciated by Rendulic and later amplified by Henkel (1959)) the strains for any direction of stress path radiating outwards from A and being 'outside' the undrained test paths. (The method of constructing strain contours for families of similarly related tests was used by Wroth and Loudon (Wroth, 1984) when examining the effects of different degrees of overconsolidation.)

Set 2. Constant mean effective stress, $p' = const$

This family consists of tests for which it was arranged that the mean effective stress p' should remain constant. The stress paths are shown in Fig. 3(a) as the octahedral view in stress space looking in a direction normal to that plane. The vertical axis is an axis of symmetry so mirror images of the stress paths, marked with #, are shown on the opposite side.

An equilateral outer triangle shows where the constant p' octahedral plane intersects the three planes for zero σ'_x, σ'_y and σ'_z. Stress states outside this triangle would involve negative effective stresses. The inner 'rounded triangle' represents the Lade-Duncan (1975) failure criterion where the product of the normalised principal effective stresses reaches a critical value $(\sigma'_x \cdot \sigma'_y \cdot \sigma'_z)/(p')^3 = C$. Here the value of C is taken as 0.6 which is appropriate to the failure condition for the triaxial compression test B33 where, at 20% shear strain, $\sigma'_x = \sigma'_y = 114$ kPa and $\sigma'_z = 372$ kPa. (This suggests that at failure $(\sigma_z - \sigma_x)/p' = 1.29$, although a more representative figure taken from other tests was 1.4.)

The axisymmetric cube tests 628 and 1320 and their triaxial counterparts B58 and B33 (represented only by their end points) have already been reported in Set 1. Test 722 was carried out at constant vertical stress and the stress path is therefore horizontal. This is a test where, despite the fact that the vertical stress was not being increased, the vertical direction still produced a large positive strain because it was still the major stress.

Test 1014 was carried out by increasing one of the lateral stresses at a rate of 4 kPa/h while reducing both the vertical stress and the other lateral stress at 2 kPa/h so that $\sigma'_x : \sigma'_y : \sigma'_z = 4:-2:-2$. Test 1111 was a very similar test except with $\sigma'_x : \sigma'_y : \sigma'_z = 4:-1.5:-2.5$.

In test 1212 (lateral plane strain compression)

(a) the vertical stress σ'_z was raised at a set rate (2 kPa/h)
(b) one lateral stress σ'_y was instructed to maintain zero strain $\varepsilon_y = 0$
(c) the first two principal stresses were read and the third stress σ'_x then calculated so as to maintain a constant mean effective stress p'.

For test 619 (lateral plane strain extension) a slightly different regime was used in which the vertical stress σ'_z was reduced by 2 kPa/h, one lateral stress σ'_x was increased by 2 kPa/h and the other lateral stress σ'_y

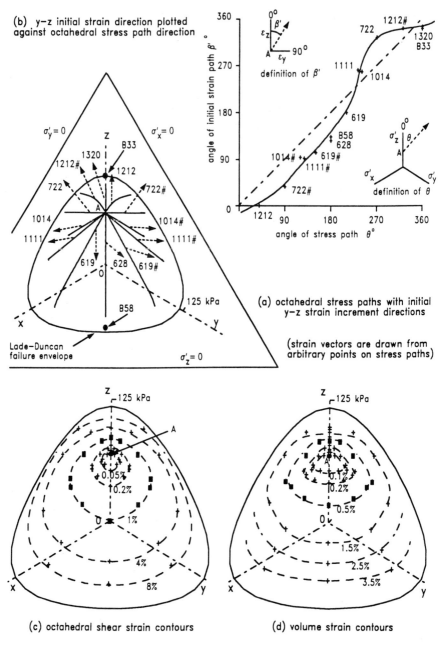

Fig. 3. Constant p' tests. (a) Octahedral stress paths with initial y–z strain increment directions (strain vectors are drawn from arbitrary points on stress paths). (b) y–z initial strain direction plotted against octahedral stress path direction. (c) Octahedral shear strain contours. (d) Volume strain contours.

was instructed to maintain zero strain $\varepsilon_y = 0$. Although p' was not controlled to remain constant, it did in fact do so within 1 kPa until nearly the end of the test. The curves in Fig. 3(a) for 619 (left-hand-side) and 1212 (right-hand-side) are thus the two curves for lateral plane strain $\varepsilon_x = 0$. However, although there appears to be a smooth transition at A this does not follow necessarily. For these two tests 1212 and 619 both stress path and strain path were curved but were reasonably straight for early stages of the test.

The octahedral strain increment directions have not been superimposed onto the stress paths in Fig. 3(a) because, unlike the stress paths which are fully contained within their octahedral plane, these strains contain a component of volume change which means that the vectors cannot be contained within a single octahedral plane. The initial strain directions are therefore presented instead in Fig. 3(a) as vectors showing just the y and z components ε_y and ε_z. These vectors are seen to rotate with the rotation of the stress path but they should not be expected to have any symmetry with respect to the z-axis. This point is emphasised by Fig. 3(b) which shows the direction of the x–z strain vector compared with the direction of the octahedral stress path vector. (It will be noted that each test provides two points on this plot since the mirrored test gives an extra point for the alternative lateral direction.) The significance of Fig. 3(b) lies principally in the confirmation it gives to the stress directions which produce zero vertical or lateral strain ($\beta' = 0$ or 180°, and $\beta = 90°$ or 270° respectively); even if the direct plane strain tests had not been made, the required stress path directions could still have been deduced by interpolation.

Shown in Fig. 3(c) are the points at which the octahedral shear strain $\gamma_{OCT} = 2/3[(\varepsilon_x - \varepsilon_y)^2 + (\varepsilon_y - \varepsilon_z)^2 + (\varepsilon_z - \varepsilon_x)^2]^{0.5}$ reached values of 0.05, 0.2, 1, 4 and 8%. This has enabled the relevant contours to be drawn, the relative spacing of which gives an immediate indication of the change of stiffness. The wider apart the contours are, the stiffer the behaviour in terms of the length of the octahedral stress path. As the stress states approach failure so the shape of the contours gets closer to the shape of the limiting envelope. Unfortunately, with the cube apparatus there is a limit to the strain (around 10%) to which the samples, or rather the membranes, can be taken. This particularly applies to the vertical compression tests where K_0 consolidation has already taken up much of the permissible vertical strain. It may be noted here that the pattern in Fig. 3(c) is very similar to the closely related pattern of stress curves derived from strain ratios as discussed by Lewin (1972).

A similar presentation is used in Fig. 3(d) to show contours of volume strain ($\varepsilon_x + \varepsilon_y + \varepsilon_z$) for the values 0.1, 0.2, 0.5, 1.5, 2.5 and 3.5%. Here again the contours approach the shape of the limiting envelope when the stress states get close to failure. However, this is a characteristic that is peculiar to normally consolidated samples. With overconsolidated samples, the contours might start reducing in value, or even to become negative if the samples were tending to dilate as failure is approached. This sort of characteristic was seen with results from cube tests on Tokyo sand (Lewin, Yamada and Ishihara, 1982).

Although there is an apparent similarity between the contour patterns for shear strain and volume strain (Figs. 3(c) and 3(d)), when considered three-dimensionally they are very different in character. Take the case of an infinite number of straight stress paths radiating in all three-dimensional directions from A. Suppose points were to be marked on these paths corresponding to particular levels of shear strain. One would then observe a set, or nest, of funnel-like surfaces which would narrow down towards the origin but open out for the higher stresses. However, if the points were marked for levels of volume strain then one would observe a completely different set of curved surfaces, geometrically similar in shape, which were transverse to the isotropic stress space axis. Their section would be that of the 1% volume strain contour sketched in the triaxial plane in Fig. 2(a). (If the material were isotropic and linear elastic then the shear strain surfaces would be concentric circular tubes lying parallel to the stress space diagonal while the volume strain surfaces would be flat planes, equally spaced apart for regular strain increments, which would be normal to the diagonal.)

Set 3. Plane lateral strain $\varepsilon_y = 0$

The stress paths for this family are shown in Fig. 4(a) as plots of σ'_z against σ'_x. The paths marked K_0 and 919 (K_0 swelling) have already been presented in Set 1, while the paths for 619 and 1212 were from tests in Set 2 where the stress paths were confined to the constant p' plane.

There are five new tests here. In test 611, lateral plane strain $\varepsilon_y = 0$ was stipulated while raising σ'_z and lowering σ'_x at equal rates. There were two undrained lateral plane strain tests 1417 (compression) and 1406 (extension). Test 1104 is shown as 1104# to indicate that the x and y directions have been interchanged. This was a vertical compression test similar to the axisymmetric constant p' test 1320 using the same effective stress loading for the minor and major principal stresses but substituting plane strain for the intermediate direction. Finally, there is test 1310 which was a test in which the vertical strain ε_z and one of the lateral strains ε_y were both held at zero while the stress in the third direction σ'_x was slowly increased. It may be seen from Fig. 4(a) that it would be a long time before the stress path for 1310# would reach the 'lateral K_0' condition at which $\sigma'_z/\sigma'_x = 0.5$. (When aiming to produce an isotropically consolidated characteristic from an anisotropically consoli-

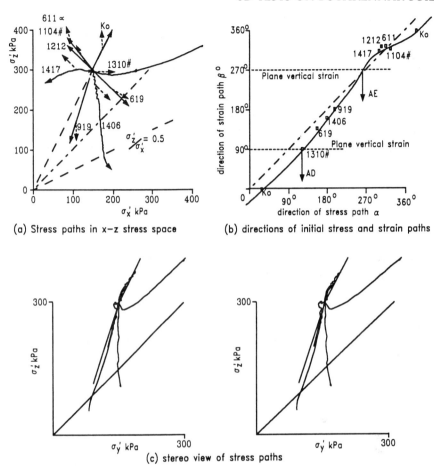

Fig. 4. Tests at plane lateral strain $\varepsilon_y = 0$. (a) Stress paths in x–z stress space. (b) Directions of initial stress and strain paths. (c) Stereo view of stress paths.

dated sample it is usually regarded as necessary at least to double the mean effective stress in order to overcome the effects of the previous stress history; but here it is obviously going to take a lot longer to switch from one anisotropic K_0 system to the other.)

Here again the initial directions of the strain vectors (which do not have a y component) have been indicated on the x–z stress paths shown in Fig. 4(a). Figure 4(b) shows the angle of the initial strain path plotted against the angle of the stress path, where α and β are still the same as defined in Fig. 2(b) (inset). The arrows marked AD and AE, drawn from where the plot cuts the values $\beta = 90°$ and $270°$ indicate the pair of initial stress path directions, $\alpha = 125°$ and $260°$ respectively, which could

445

be expected to give not only plane strain in one lateral direction but at the same time plane strain in the vertical directions.

The stress paths for all the tests in this set are re-presented in Fig. 4(c), this time plotted on a y:z projection. The left-hand plot is a true projection while the right-hand plot has been rotated 3° about the z axis. This provides a stereo view of the paths in three-dimensional stress space. This view emphasizes the point that the stress paths for lateral plane strain lie on a surface which is not flat and is certainly not confined to the vertical plane for plane stress σ'_y = constant.

Set 4. Vertical plane strain $\varepsilon_z = 0$

This set covers all test conditions which would produce zero vertical strain. Although not strictly plane strain, two of the constant p' tests (1014 and 1111) may be considered as belonging to this set. These were tested with slightly different stress ratios $\delta\sigma'_x : \delta\sigma'_y : \delta\sigma'_z$ of $4:-2:-2$ and $4:-1.5:-2.5$ respectively. Both these tests gave initially low vertical strain rates as shown in Fig. 3(b). It will be noted that in each case the effective vertical stress change was negative and this is demonstrated by the downward slope of their stress paths in the octahedral view of the constant p' plane shown in Fig. 3(a). This illustrates quite clearly that, for the initial part of the loading path for anisotropically consolidated soils, plane vertical strain should not be associated with plane vertical stress.

A particularly interesting pair of tests in this set would be those for the axisymmetric condition where the two lateral stresses are increased or decreased equally and the vertical stress adjusted to maintain zero vertical strain. These two tests have not been performed but their stress

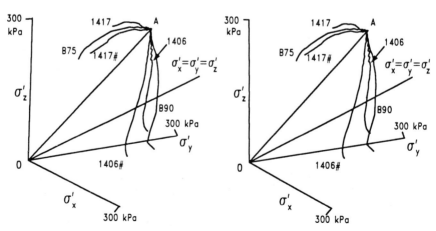

Fig. 5. Undrained tests – stereo view of stress paths.

paths can be deduced by interpolation in Fig. 2(b) where the stress path angles can be read off from the graph against the strain path angles $\beta = 90°$ and 270°. The respective stress path angles are seen to be $\alpha = 115°$ and 240° and stress paths corresponding to these values have been marked in Fig. 2(a) as dotted lines AC and AB. These stress paths may be considered to form a back-bone for a flattish conical surface in stress space for the family of all paths emanating from A which produce zero vertical strain.

Set 5. Undrained tests
The final set to be considered is that of the four undrained tests B75, B90, 1406 and 1417 together with the two mirrored tests 1406# and 1407#. The stress paths (already given in Fig. 2(a)) have been used to construct a stereo presentation of the undrained surface in three-dimensional stress space (Fig. 5). This surface has similarities with that shown in Lewin, Yamada and Ishihara (1982) which was generated from undrained cube tests on Tokyo sand where all the tests started from an hydrostatic stress state. The anisotropy evident in that surface derived from the original deposition of the sand in the apparatus.

Conclusions

The paper shows how it is possible to portray the three-dimensional character of stress–strain behaviour by establishing patterns of behaviour for five different surfaces in stress space. By conducting tests across a spectrum of conditions it is shown how one can deduce results for particular conditions, such as plane strain and plane stress, and that it is erroneous to assume that plane strain and plane stress are necessarily mutually associated. The difference in the directions of the stress increment vector and the initial strain increment vector has been demonstrated for various conditions. Contours have been drawn for shear strain and volume strain for drained tests conducted at constant mean effective stress and the three-dimensional implications have been discussed drawing attention to the difference in character between these two sets of contours. The contours showed a high degree of anisotropy as is to be expected from soil samples that had been initially anisotropically consolidated.

References

ABBISS, C.P. AND LEWIN, P.I. (1990). A test of the generalized theory of visco-elasticity in London Clay. Geotechnique, Vol. XL, No. 4, pp. 641–646.

ALLMAN, M.A. AND ATKINSON, J.H. (1992). Mechanical properties of reconstituted Bothkennar soil. Geotechnique, Vol. XLII, No. 2, pp. 289–301.

ATKINSON, J.H., EVANS, J.S. AND SCOTT, C.R. (1985). Developments in stress path testing equipment. Ground Engineering, Vol. XVIII, No. 1, pp. 15–22.

HENKEL, D.J. (1959). Characteristics of saturated clays. Geotechnique, Vol. IX, No. 3, pp. 119–135.

KO, H.-Y. AND SCOTT, R.F. (1967). A new soil testing apparatus. Geotechnique, Vol. XVII, pp. 40–57.

LEWIN, P.I. (1972). 3D anisotropic consolidation of clay and the relationship between plane strain and triaxial test data. RILEM Symposium 'Deformation of solids subjected to multiaxial stresses'. Cannes.

LEWIN, P.I., YAMADA, Y. AND ISHIHARA, K. (1982). Correlating drained and undrained 3D tests on loose sand. IUTAM Symp. 'Deformation and failure of granular materials'. Delft.

WROTH, C.P. (1984). The interpretation of in-situ soil tests. Rankine Lecture. Géotechnique, Vol. XXXIV, No. 4, pp. 447-489.

Prediction of clay behaviour around tunnels using plasticity solutions

R.J. MAIR, Geotechnical Consulting Group, and
R.N. TAYLOR, City University

Simple plasticity solutions can be used to predict ground deformations and pore pressure changes as a tunnel is constructed in clay. Field measurements of sub-surface vertical and horizontal ground deformations in the plane perpendicular to the tunnel axis are shown to be reasonably consistent with predictions of a cylindrical contracting cavity in an idealised linear elastic–perfectly plastic soil. Movements ahead of an advancing tunnel heading are better predicted assuming spherically symmetric conditions. Measured pore pressure changes within the plastic zone around tunnels can be reasonably predicted by the simple linear elastic–perfectly plastic model. Poor predictions are obtained of pore pressures in the surrounding elastic zone: improved predictions are obtained assuming non-linear elastic soil behaviour in this zone.

Introduction

Tunnel construction in clay is usually sufficiently rapid that the clay behaviour around the heading is undrained. By making a number of simplifying assumptions, plasticity solutions can be used to model the general behaviour of the clay around a tunnel heading. Predictions can then be made of sub-surface ground deformations and pore-pressure changes as the tunnel is constructed. The following assumptions are made:

(a) The tunnel heading is idealized as in Fig. 1(a). The tunnel is circular, of diameter D, and the lining is installed at a distance P behind the face.
(b) Axisymmetric conditions are assumed; the initial stress state is isotropic.
(c) The clay is assumed to behave as a linear elastic-perfectly plastic soil.

The behaviour of the clay around the advancing tunnel heading will be idealized either in terms of the unloading of a spherical cavity, as shown in Fig. 1(b), or in terms of the unloading of a plane strain cylindrical cavity, as shown in Fig. 1(c). Neither idealization is strictly correct; for some aspects of tunnel behaviour, one idealization is more

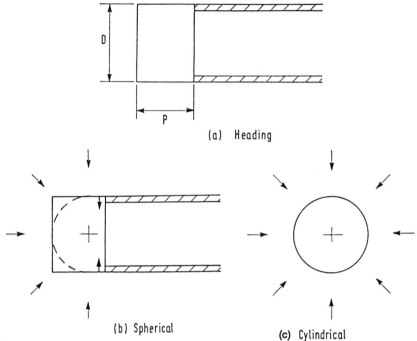

Fig. 1. Idealisation of a tunnel heading

appropriate than the other, and in some circumstances a combination of the two idealizations can be used. The purpose of this Paper is to compare predictions from the idealizations with field measurements, and to highlight how the simple concept of the unloading of a spherical or cylindrical cavity can be a useful predictive tool.

Deformations
Plasticity solutions
Closed form solutions exist for the response of a linear elastic–perfectly plastic continuum around an expanding cavity (Hill, 1950; Gibson and Anderson, 1961). Similar solutions can be derived for a contracting cavity and, for a fully unloaded spherical cavity, it can be shown that

$$\frac{\delta}{a} = \frac{S_u}{3G}\left(\frac{a}{r}\right)^2 \exp(0.75N^* - 1) \qquad (1)$$

where

δ = radial movement at radius r

a = inner radius of the cavity (tunnel)
N* = stability ratio = σ_0/s_u
σ_0 = initial total stress at cavity boundary
s_u = undrained shear strength
G = elastic shear modulus (for isotropic conditions, the undrained Young's modulus $E_u = 3G$)

For tunnels in clay without internal temporary support, the stability ratio is usually defined as σ_v/s_u, where σ_v is the total vertical stress at the level of the tunnel axis prior to tunnelling. Eqn. (1) is derived from the assumption that the continuum extends to a radius significantly larger than that of the plastically deforming zone. For a deep tunnel it is reasonable to assume that $\sigma_v \approx \sigma_0$ and therefore that N* in eqn. (1) corresponds to the usual definition of stability ratio in tunnelling.

Tunnel excavation in clay can also be considered in terms of a fully unloaded cylindrical cavity (see, for example, Schmidt, 1969, and Lo et al., 1984). Using the same notation as defined for eqn. (1), it can be shown that

$$\frac{\delta}{a} = \frac{s_u}{2G}\left(\frac{a}{r}\right)\exp(N^* - 1) \qquad (2)$$

From eqn. (1) it can be seen that for unloading of a spherical cavity δ/a is proportional to $(a/r)^2$, as depicted in Fig. 2(a). Correspondingly, for unloading of a cylindrical cavity eqn. (2) shows that δ/a is proportional to a/r, as depicted in Fig. 2(b). It should be noted that the linear relationships between δ/a and $(a/r)^2$ or (a/r) follow directly from the

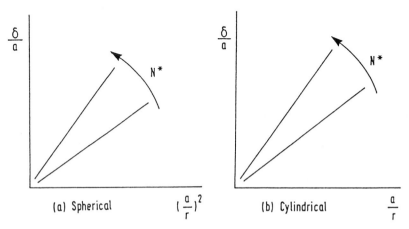

Fig. 2. Radial deformation associated with unloading a cavity in an elastic–perfectly plastic continuum. (a) Spherical, (b) Cylindrical

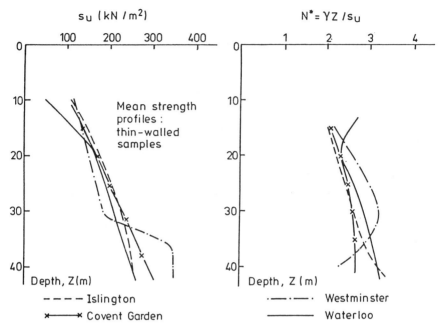

Fig. 3. *Shear strength and stability ratio profiles for London Clay*

constant volume condition associated with undrained behaviour. These apply irrespective of the soil model.

For the linear–elastic, perfectly plastic soil model assumed, eqns. (1) and (2) show that the slopes of the lines in Fig. 2 increase exponentially with increasing stability ratio N^*. If N^* is less than 4/3, in the case of the spherical cavity, or 1 in the case of the cylindrical cavity, the clay behaves entirely elastically. For these cases, eqns. (1) and (2) can still be used by substituting $N^* = 4/3$ and 1 respectively and replacing s_u with σ_0. For larger N^* values, a plastic zone develops around the tunnel, and this strongly influences the magnitude of the deformations.

Sub-surface deformation measurements around tunnels

For most tunnels in London Clay at depths between about 20 m and 40 m, the stability ratio typically varies between about 2.5 and 3. This is illustrated in Fig. 3, which shows profiles of undrained strength obtained from four London Clay sites. The corresponding stability ratio profiles are also shown in Fig. 3. The strength profiles are mean lines of results from unconsolidated undrained triaxial compression tests on 100 mm diameter specimens obtained from high quality thin-walled samples. The stability ratio profiles were derived assuming a bulk unit weight for the clay of 20 kN/m^3.

Fig. 4. *Deformations in front of an advancing tunnel heading in London Clay (Fig. 4(a) after Ward, 1969)*

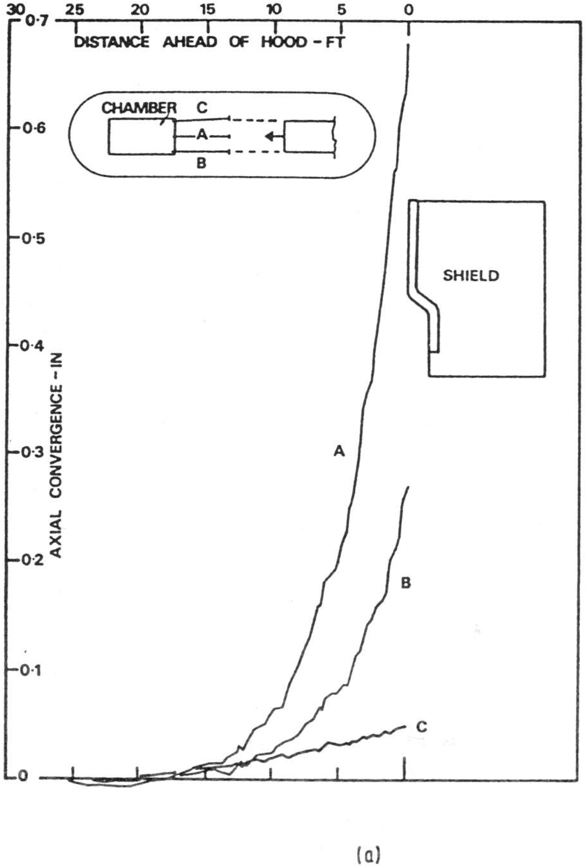

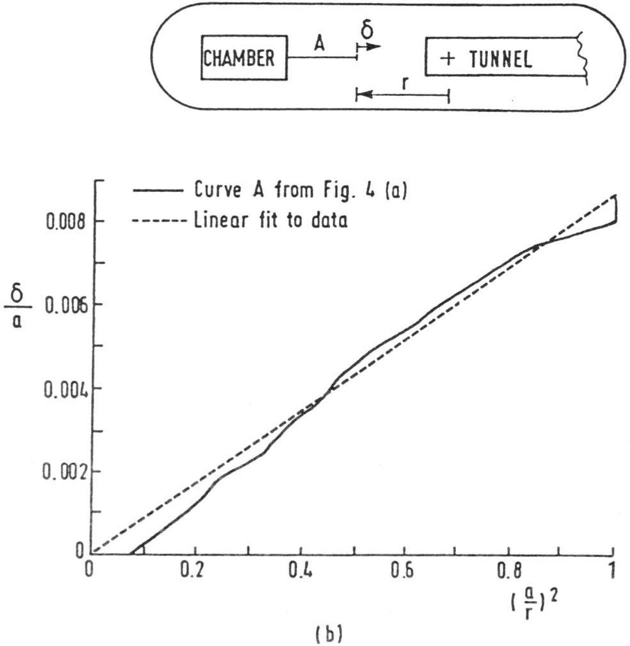

Figure 4(a) shows the results of detailed measurements of ground movements as a tunnel heading at a depth of 24 m in London Clay at Brixton was advanced towards an underground chamber (Ward, 1969). Of particular interest are the measurements of the ground movements along the tunnel axis parallel to the direction of tunnelling (i.e. curve A in Fig. 4(a)). These data are replotted in Fig. 4(b) in the form shown in Fig. 2(a). A reasonably linear plot is obtained, indicating that the unloading of the spherical cavity is a reasonable approximation for predicting the behaviour of the clay around the tunnel heading. Detailed strength measurements are not available for the London Clay at the Brixton site, but it is likely that at a depth of 24 m N^* is about 2.5 (see Fig. 3). Using the slope of the straight line fit to the data in Fig. 4(b) in conjunction with eqn. (1), a G/s_u ratio of 96 is obtained for $N^* = 2.5$.

Figure 5 shows field measurements of sub-surface vertical ground movements (δ_v) above the centre-line of tunnels constructed in London Clay for London Underground Ltd. The measurements were made on tunnels of 4.1 m excavated diameter beneath Green Park at a depth of 29 m (Attewell and Farmer, 1974) and Regent's Park at depths of 20 m

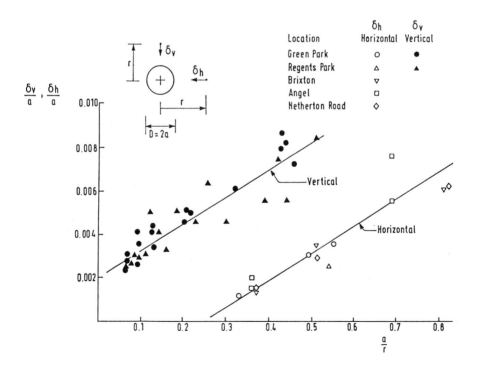

Fig. 5. Vertical and horizontal sub-surface movements in the vicinity of tunnels in London Clay

and 24 m (Barratt and Tyler, 1976). Also shown in Fig. 5 are measurements of horizontal ground movements (δ_h) at the level of the tunnel axis in a direction perpendicular to the direction of the tunnel. Measured horizontal ground movements from Green Park and Regent's Park are shown, together with data from Brixton (Ward, 1971) and Netherton Road (Smyth-Osbourne, 1971, and Ward, 1971) and from the recent Angel Station Reconstruction project. All the measurements are plotted in the normalized form shown in Fig. 2(b), corresponding to the idealization of an unloaded cylindrical cavity.

The vertical and horizontal ground movement measurements from the different London Clay sites in Fig. 5 are reasonably consistent and are in general agreement with linear plots. This suggests that the ground movements above and to the side of the tunnel can be reasonably predicted on the basis of an unloaded cylindrical cavity. The two plots are almost parallel, but they do not pass through the origin. This reflects the non-axisymmetric conditions observed in reality: at any radius r, the vertical movements above the tunnel (settlements) are greater than the horizontal movements at the same radius to the side of the tunnel at axis level. Even at large distances above the tunnel (small values of a/r) measurable settlements are observed, whereas no measurable horizontal movements are observed to the side of the tunnel beyond a radius of 4a (i.e. for a/r values of less than 0.25). Nevertheless the data in Fig. 5 indicate that the idealization of axisymmetric conditions around a cylindrical cavity provides a valuable framework for prediction of ground movements around tunnels; such a framework has the merit of simplicity.

Under normal tunnelling conditions a lining is installed at a distance P behind the face, as shown in Fig. 1(a); usually the ratio P/D is less than 1 and often close to zero. As the tunnel face moves away from the installed lining, pressure from the ground rapidly builds up on the lining under undrained conditions. (Subsequently, the lining pressure increases as pore pressures dissipate.) For tunnels in London Clay, for example, the average total stress imposed by the ground on the lining as soon as the face has advanced about two to three tunnel diameters is generally about 30% of the total overburden stress (Barratt and Tyler, 1976; Ward and Thomas, 1965; Thomas, 1976). When considering the plane strain cylindrical cavity model, illustrated in Fig. 1(c), it is therefore more appropriate to view it in terms of partial unloading to the stress provided by the lining support, σ_{li}, rather than a complete unloading to zero internal support. Equation (2) can then be rewritten as:

$$\frac{\delta}{a} = \frac{s_u}{2G}\left(\frac{a}{r}\right)\exp[N^*(1-n)-1] \qquad (3)$$

The same notation is used as defined for eqn. (1) and the equivalent stability ratio N is defined as:

$$N = \frac{\sigma_0 - \sigma_{li}}{s_u} \quad (4)$$

i.e.

$$N = N^* - \frac{\sigma_{li}}{s_u} \quad (5)$$

and $\quad N = (1 - n)N^* \quad (6)$

where $n = \sigma_{li}/\sigma_0$ and $N^* = \sigma_0/s_u$. The parameter n represents the proportion of the total overburden pressure ($\sigma_v \approx \sigma_0$) imposed by the ground on the lining. It is reasonable to assume $N^* \approx 2.5$ and $n = 0.3$ for tunnels in London Clay, as discussed earlier. Using these values, and taking the average gradient of the lines through the data in Fig. 5, eqn. (3) gives $G/s_u = 87$. This is consistent with the value of G/s_u of 96 obtained earlier from the back-analysis of the data in Fig. 4 using the analogy of unloading a sphere and eqn. (1). For tunnels in London Clay,

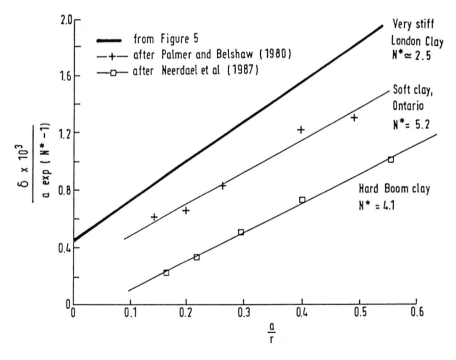

Fig. 6. Normalized vertical movements for tunnels in clay

therefore, it appears that a linear elastic stiffness given by the ratio $G/s_u = 100$ is reasonably representative for the purposes of predicting ground movements in terms of a linear elastic–perfectly plastic soil model.

Sub-surface settlement measurements from a number of tunnels in clays other than London Clay are shown plotted in Fig. 6 in terms of the simple cylindrical unloading analogy. Measurements are shown for a tunnel in a soft clay in Ontario (Palmer and Belshaw, 1980) and a hard clay (Boom clay) in Belgium (Neerdael et al., 1987). The line through the London Clay settlement data from Fig. 5 is also shown for comparison. To account for the large differences in stability ratio, the data in Fig. 6 have been normalized by $\exp(N^* - 1)$, as suggested by eqn. (2). The reasonable linearity of the plots indicates that the concept of a partially unloaded cylindrical cavity under axisymmetric conditions has general application to prediction of sub-surface settlements above tunnels in clays.

Surface settlements can be predicted by assuming a settlement trough in the shape of an error function (Peck, 1969); this assumption has been shown to be reasonable by comparison with field measurements (e.g. O'Reilly and New, 1982). It is often assumed that similar shaped error function settlement troughs develop between the ground surface and the tunnel. However, very few sub-surface field measurements are presently available and, consequently, less is known about the shape and magnitude of settlement troughs between the ground surface and the tunnel, particularly at relatively small distances above the tunnel. The simple concepts described in this paper provide an alternative means of predicting sub-surface ground movements.

Pore pressure changes

Assumptions of the linear elastic–perfectly plastic soil model gives the following expression for the pore pressure change at a radius r within the plastic zone around an unloaded cylindrical cavity under axisymmetric conditions:

$$\frac{\Delta u}{s_u} = 1 - N + 2\log_e\left(\frac{r}{a}\right) \quad (7)$$

The plastic zone extends to a radius $r = R_p$, given by

$$\frac{R_p}{a} = \exp\left(\frac{N-1}{2}\right) \quad (8)$$

The analysis shows that in the elastic zone, beyond the plastic zone, zero pore pressure change is predicted by the linear elastic–perfectly plastic model.

Figure 7 shows measurements of pore pressure changes in a centrifuge test by Mair (1979) on a model tunnel in soft clay. The cover-to-diameter ratio was 3.1, and the internal support pressure was progressively reduced until collapse of the tunnel occurred. Measurements are

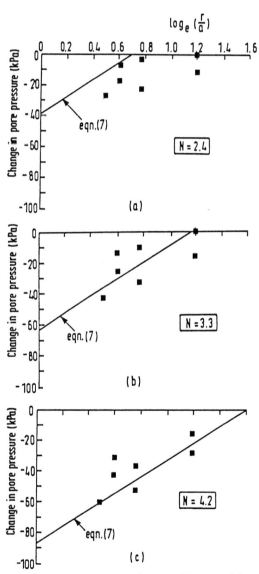

Fig. 7. Changes in pore pressure around a centrifuge model tunnel in clay ($s_u = 27$ kPa)

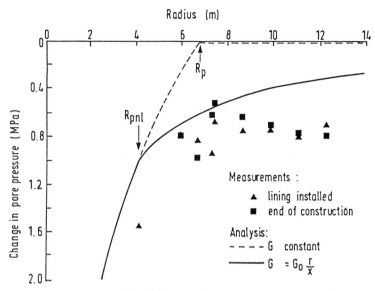

Fig. 8. *Measured and predicted changes in pore pressure near a deep tunnel in Boom clay*

shown for three different stages of the test, corresponding to stability ratios (N) of 2.4, 3.3 and 4.2 (Figs. 7(*a*), (*b*) and (*c*) respectively). Good agreement is obtained with eqn. (7) for the measurements within the plastic zone. Beyond the plastic zone, however, significant pore pressure changes were observed, as shown in Fig. 7(*a*).

Measurements of pore pressure changes around a deep tunnel in Boom clay are shown in Fig. 8. The tunnel is of 4.7 m excavated diameter and is at a depth of 223 m (Neerdael and De Bruyn, 1989). At this depth it is probable that reasonably axisymmetric conditions were operating. Also shown plotted in Fig. 8 is the predicted distribution of pore pressure changes, derived from the solution for an unloaded cylinder given in eqn. (7) (Mair et al., 1992). The parameters assumed for the prediction were $N^* = 4.46$, $s_u = 1$ MPa and $n = 0.3$; the values for s_u and n were based respectively on laboratory triaxial tests and in situ measurements of lining loads. It is evident that substantial pore-pressure changes were observed in the elastic zone beyond the plastic zone (the latter being predicted as having a radius of 6.8 m). This is in conflict with the prediction of zero pore pressure change in the elastic zone.

One possible reason for the difference between predictions and observations of pore pressure changes in the elastic zone both for the centrifuge tests (Fig. 7) and for the tunnel in Boom clay (Fig. 8) lies in the assumption of isotropic soil behaviour. A second possible reason is the assumption of axisymmetric conditions, although this is felt to be reasonable for the deep tunnel in Boom clay. The third, and probably most significant, reason lies in the assumption of linear elastic soil behaviour.

The influence of non-linear elastic behaviour (Jardine et al., 1985) on pore-pressure changes can be qualitatively illustrated by making the simple assumption that the secant stiffness G increases linearly with radius:

$$G = G_0 \left(\frac{r}{x} \right) \qquad (9)$$

where G_0 is the stiffness at $r = x$. This is equivalent to a non-linear elastic soil model in which the stiffness increases with decreasing strain level, since the strain level around a tunnel decreases with increasing radius. For a cylindrical cavity, strain is proportional to $1/r^2$ and therefore eqn. (9) implies that G is proportional to $(strain)^{-1/2}$, giving a parabolic stress–strain curve. With this assumed function for stiffness variation, the change in pore pressure within the plastic zone is unaltered from that determined using a linear elastic–perfectly plastic model for clay behaviour. In the elastic zone, the change in pore pressure is given by:

$$\frac{\Delta u}{s_u} = - \frac{R_{pnl}}{r} \qquad (10)$$

R_{pnl} is the radius of the plastic zone assuming non-linear elasticity, as defined in eqn. (9), and differs from R_p given by eqn. (8) assuming linear elasticity. R_{pnl} is given by:

$$\frac{R_{pnl}}{a} = \exp\left(\frac{N}{2} - 1 \right) \qquad (11)$$

The predicted distribution of pore pressure changes for the tunnel in Boom clay is shown in Fig. 8; the same values for N^*, s_u and n were used as for the previous prediction with the assumed linear elastic–perfectly plastic soil model. Significant changes in pore pressure are predicted in the elastic zone.

The influence of non-linear elasticity on pore pressure changes is illustrated by Fig. 9. This shows stress paths for an element of clay around a circular tunnel being unloaded under axisymmetric conditions. The effective stress path A'B'C' is vertical, i.e. the mean normal effective stress, s', is constant during undrained loading and when the response is elastic (for an isotropic soil). If the soil behaviour is *linear* elastic and isotropic, the total mean normal stress remains constant and conse-

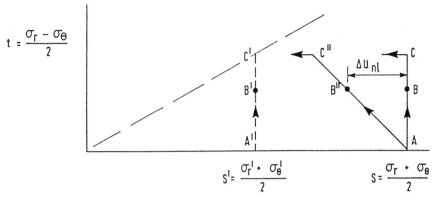

Fig. 9. *Predicted stress paths using simple linear and non-linear elastic–perfectly plastic models*

quently the total stress path ABC is vertical. When the effective stress path reaches C', yield occurs and the total stress path remains at constant shear stress. If the soil behaviour is *non-linear* elastic and isotropic, the total stress path is no longer vertical, but follows the path AB"C". At the point B", the reduction in pore pressure is then Δu_{nl}, rather than zero for the linear elastic case at point B.

The simple analysis using eqn. (9) does not accurately predict the observations, but it illustrates how non-linear elastic soil behaviour leads to significant pore pressure reduction in the elastic zone surrounding tunnels, even under conditions of axial symmetry. Similar behaviour is predicted by finite element analysis incorporating a more realistic non-linear tangent stiffness soil model (Mair et al., 1992). This is important, because long-term behaviour of clay around tunnels is strongly influenced by the dissipation of excess pore pressures generated during tunnel construction.

Conclusions

1. Comparison with field data indicates that the plasticity solution for unloading of a spherical cavity is a reasonable approximation for prediction of the ground movements in front of an advancing tunnel heading. Field data of sub-surface vertical and horizontal movements above and to the side of tunnels indicate that the unloading of a cylindrical cavity provides a useful framework for prediction.

2. Back-analysis of field measurements around 4 m diameter tunnels in London Clay indicates that an elastic stiffness given by the ratio $G/s_u = 100$ is reasonable for prediction of ground movements in terms of a linear elastic–perfectly plastic soil model. Plasticity has an important

influence on ground movement prediction, and this can be taken into account by use of the solutions discussed in this paper. For most tunnels in London Clay at depths between about 20 m and 40 m the stability ratio is typically between about 2.5 and 3.

3. Predicted pore pressure changes within plastic zones around tunnels during their construction in clays are in reasonable agreement with centrifuge test data and with field measurements. Discrepancies are evident in the elastic zone, and these can be explained in terms of non-linear elastic soil behaviour.

4. By making a number of simplifying assumptions, the plasticity solutions discussed in this paper are useful for predicting sub-surface deformations and pore-pressure changes in clay around tunnels, particularly in regions close to the tunnels.

Acknowledgements

The authors are grateful to Dr W.H. Ward for many stimulating discussions concerning deformations around tunnels. They also acknowledge London Underground Ltd for their permission to present material relating to the Angel Station Reconstruction project.

References

ATTEWELL, P.B. AND FARMER, I.W. (1974). Ground deformations resulting from shield tunnelling in London Clay. Canadian Geotechnical Journal, Vol. 11, pp. 380–395.

BARRATT, D.A. AND TYLER, R.G. (1976). Measurements of ground movement and lining behaviour on the London Underground at Regent's Park. Transport and Road Research Laboratory Report LR 684.

GIBSON, R.E. AND ANDERSON, W.F. (1961). In-situ measurement of soil properties with the pressuremeter. Civ. Engng Pub. Wks Rev. 56, pp. 615–618.

HILL, R. (1950). The mathematical theory of plasticity. Oxford University Press.

JARDINE, R.J., FOURIE, A., MASWOSWE, J. AND BURLAND, J.B. (1985). Field and laboratory measurements on soil stiffness. Proc. 11th Int. Conf. on Soil Mechanics and Foundation Engineering, San Francisco, Vol. 2, pp. 511–514.

LO, K.Y., NG, M.C. AND ROWE, R.K. (1984). Predicting settlement due to tunnelling in clays. Proc. Geotech. '84. Tunnelling in Soil and Rock, Atlanta, Georgia. Ed. Lo, K.Y. ASCE, pp. 46–76.

MAIR, R.J. (1979). Centrifugal modelling of tunnel construction in soft clay. Ph.D. Thesis, Cambridge University.

MAIR, R.J., TAYLOR, R.N. AND CLARKE, B.G. (1992). Repository tunnel construction in deep clay formations. Commission of the European Communities Report EUR 13964 EN.

NEERDAEL, B., DE BRUYN, D., VOET, M., ANDRE-JEHAN, R., BOUILLEAU, M., OUVRY, J.F. AND ROUSSET, G. (1987). In-situ testing programme related to the mechanical behaviour of clay at depth. Conference on Numerical Methods in Geomechanics, Kobe, Japan.

NEERDAEL, B. AND DE BRUYN, D. (1989). Geotechnical research in the test drift of the HADES underground research facility at Mol. Proceedings of CEC Technical Session 'Geomechanics of clays for radioactive waste disposal', Commission of the European Communities Report EUR 12027 EN/FR, pp. 83–94.

O'REILLY, M.P. AND NEW, B.M. (1982). Settlements above tunnels in the United Kingdom – their magnitude and prediction. Tunnelling '82, IMM, London, pp. 173–181.

PALMER, J.H.L. AND BELSHAW, D.J. (1980). Deformations and pore pressures in the vicinity of a precast, segmented, concrete-lined tunnel in clay. Canadian Geotechnical Journal, Vol. 17, pp. 174–184.

PECK, R.B. (1969). Deep excavations and tunnelling in soft ground. Proc. 7th Int. Conf. on Soil Mechanics and Foundation Engineering, Mexico, State of the Art Report, Vol. 3, pp. 225–290.

SCHMIDT, B. (1969). Settlements and ground movements associated with tunnelling in soil. Ph.D. thesis, University of Illinois.

SMYTH-OSBOURNE, K.R. (1971). Discussion on Muir Wood and Gibb. Proceedings of the ICE, Vol. 50, pp. 190–196.

THOMAS, H.S.H. (1976). Structural performance of a temporary tunnel lined with spheroidal graphite cast iron. Proceedings of the ICE, Part 2, Vol. 61, pp. 89–108.

WARD, W.H. (1969). Yielding of ground and the structural behaviour of linings of different flexibility in a tunnel in London clay. Proc. 7th Int. Conf. on Soil Mechanics and Foundation Engineering, Mexico, Vol. 3, pp. 320–325.

WARD, W.H. (1971). Some field techniques for improving site investigation and engineering design. Proc. Roscoe Memorial Symposium, Cambridge, pp. 676–682.

WARD, W.H. AND THOMAS, H.S.H. (1965). The development of earth loading and deformation in tunnel linings in London clay. Proc. 6th Int. Conf. on Soil Mechanics and Foundation Engineering, Toronto. Vol. 2, pp. 432–436.

Settlement predictions for piled foundations from loading tests on single piles

A. MANDOLINI and C. VIGGIANI, Università di Napoli Federico II

The paper reports a case history of settlement of a piled foundation. The observed settlement is compared with that predicted by means of the superposition method developed by Poulos, based on the axial compliance of the single pile obtained by pile loading tests. Further evidence on the same topic, collected for a number of other structures, is presented. It is concluded that a simple linear superposition technique is surprisingly effective in predicting the amount and distribution of settlement.

Introduction

The settlement of piled foundations is generally rather small and is not considered a major problem in foundation engineering. Settlement observations and analyses, however, may contribute to the development and validation of satisfactory models of the soil-structure interaction, in view of a more rational and economical design of piled foundations (Burland et al., 1977; Cooke, 1986; Bilotta et al., 1991; Caputo et al., 1991).

In the opinion of the authors, simple linear elastic models still deserve some attention in engineering applications, provided they are properly used (Poulos, 1981; Viggiani, 1981). In this paper it is shown how linear elastic analyses may be employed to predict the settlement of a piled foundation, starting from the observed settlement of individual test piles.

The case history of two tall buildings founded on pyroclastic soils by means of auger piles is presented in some detail; furthermore a number of data relating to other case histories are reported.

Method of analysis

Poulos and Davis (1980) suggest that an evaluation of the settlement of a free standing pile group can be obtained by assuming linear elastic behaviour for both the soil and the piles, and either full flexibility or infinite stiffness of the slab connecting the pile heads; the behaviour of

SETTLEMENT PREDICTION FOR PILED FOUNDATIONS

the real foundation being actually intermediate between these two limit situations.

As is well known, the settlement w_s of a single elastic pile of length L and diameter d in an elastic soil, under the action of a vertical load Q, may be expressed as:

$$w_s = \frac{Q}{E_s d} I_w = w_1 Q \tag{1}$$

where E_s is the Young's modulus of the soil, I_w an influence factor depending on the slenderness L/d of the pile, on its stiffness relative to the soil E_p/E_s (E_p = Young's modulus of the pile), on the Poisson's ratio ν_s of the soil and on the subsoil model, and w_1 is the axial compliance of the pile.

The settlement w_p of a pile belonging to a pair of identical piles, with a spacing i from axis to axis, both acted upon by a load Q, may be expressed as:

$$w_p = w_1(1+\alpha)Q \tag{2}$$

where α is a superposition factor depending on the same parameters influencing I_w plus the ratio i/d.

Values of I_w and α for a number of simple subsoil models (homogeneous half space, Gibson soil, finite layer, etc.) have been published by Banerjee (1978), Poulos and Davis (1980) and others. When the subsoil conditions prevailing at a site are such that none of these simple schemes applies, it is often reasonable to model the subsoil as consisting of several horizontal layers with different values of Young's modulus and Poisson's ratio. Simple computer programs are widely available for the determination of I_w and α values in these cases (see, for instance, Caputo and Viggiani (1985)).

Poulos (1968) suggested evaluating the settlement w_i of the ith pile in a group of n piles by superimposing the coefficients α

$$w_i = \sum_{j=1}^{n} w_{1i} \alpha_{ij} Q_j \tag{3}$$

where α_{ij} is the superposition factor between the piles i and j. If the piles i and j have different diameter and/or length, w_1 and α are evaluated as if both piles had the diameter and length of the loaded pile j (Poulos and Hewitt, 1985).

In the case of a fully flexible raft, the load Q_j acting upon each pile is known, and eqn. (3) may be used directly to evaluate the settlement of each pile. In the case of an infinitely stiff raft, the settlement consists of a rigid motion that is fully defined by the vertical displacement w_0 of the centre and by two rotations β_x and β_y.

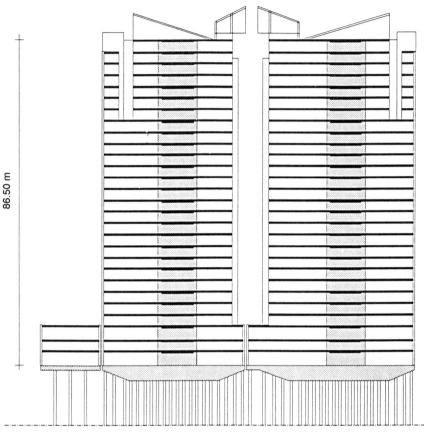
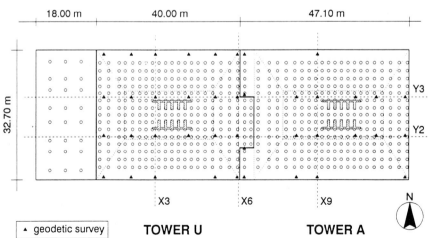

Fig. 1. Plan of the foundations and cross-section of the towers

The settlement of the pile i (x_i, y_i) must be compatible with such a displacement; hence:

$$\sum_{i=1}^{n} w_{1j}\alpha_{ij}Q_j = w_0 + \beta_x y_i + \beta_y x_i \quad (4)$$

Furthermore, if Q is the total load acting upon the raft with eccentricities e_x, e_y, the equilibrium requires that:

$$\sum_{i=1}^{n} Q_i = Q; \quad \sum_{i=1}^{n} Q_i x_i = Qe_x; \quad \sum_{i=1}^{n} Q_i y_i = Qe_y \quad (5)$$

The system of n compatibility conditions (eqn. (4)) plus three equilibrium conditions (eqn. (5)) gives the n unknown loads Q_i plus the values of w_0, β_x and β_y.

The buildings and the subsoil

The two tall buildings whose settlement is reported are at present under construction in Napoli; they will be referred to as Tower A and Tower U. A plan and elevation of the buildings are reported in Fig. 1.

The towers are provided with reinforced concrete slabs with thicknesses varying between 1.2 and 3.5 m and resting respectively on 323 and 314 auger piles of the 'PressoDrill' type (Mascardi, 1985; Viggiani, 1989a) with diameter 0.6 m and length 20 m. The two plates are in contact but separated by a joint. The structures consist of steel frames with r.c. stiffening cores. The total weight of each tower is around 300 MN, including the weight of the raft and accounting for the buoyancy; the average permanent load per pile is nearly 1 MN; the maximum design load under wind or seismic action is 1.2 MN on the most heavily loaded pile.

As of February 1992, Tower A had reached the 17th floor and Tower U the 11th floor, in a total of 28 floors. The construction of Tower A started 4 months before that of Tower U; the low-rise building adjacent to Tower U will be constructed after completion of the two towers.

The building site belongs to the new Directional Centre of Napoli, located in the eastern area of the town. The subsoil of the whole area has been thoroughly investigated by a number of authors (Croce and Pellegrino, 1967; Rippa and Vinale, 1982; Vinale, 1988; Viggiani, 1989b) and its general features are well known.

According to Caputo et al. (1991), starting from the ground surface and moving downwards the following soils are typically found:

(a) made ground
(b) volcanic ashes and organic soils

(c) stratified sands
(d) pozzolana, cohesionless or slightly indurated
(e) volcanic tuff.

The made ground is 3–4 m thick, and consists essentially of rubbles and tuff fragments, embedded in a brown silty matrix of pyroclastic nature.

Ashes and organic soils represent the bottom of the marsh that occupied this zone until less than a century ago. Ashes have a silty texture, and exhibit a slight plasticity; organic soils are generally pyroclastic sediments with diffused organic matter, but sometimes real peat layers are found.

Stratified sands are alluvial soils of volcanic origin (sand, pumices, lapilli, ashes), washed out from the surrounding hills and deposited

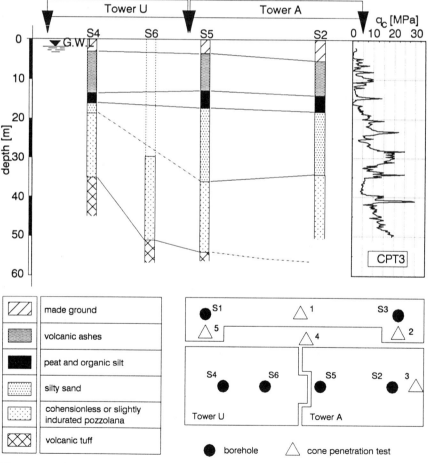

Fig. 2. *Constitution of the subsoil*

SETTLEMENT PREDICTION FOR PILED FOUNDATIONS

under lagoonal or marsh environment. The thickness of each layer ranges between some tens of centimetres and a few metres.

In some instances a coarse sand fills erosion gullies in the underlying pyroclastic formation. The latter consists of pozzolana (cohesionless or slightly indurated) and tuff; they belong to the same geological formation, the only difference being the degree of diagenesis. The upper part of the formation, starting from a depth of 25–30 m, generally consists of cohesionless pozzolana; indurated pozzolana and/or tuff are subsequently found.

The tuff is a soft rock with a compressive strength ranging between 2 and 10 MPa; the thickness of the layer is usually of some tens of metres. Depending on the conditions prevailing at the time of the deposition, at some locations the tuff is missing and the pyroclastic formation consists entirely of cohesionless and slightly indurated pozzolana.

The groundwater table is situated at a shallow depth below the ground surface. At the site a number of boreholes, SPT and CPT, have been carried out; the subsoil may be described as shown in Fig. 2. The thickness of the upper soils (made ground, volcanic ashes and organic soils) is about 16 m; at this depth, a peat layer has been found, where the q_c values are very small.

The stratified sands have a thickness increasing from 5 to 20 m moving from the west side (Tower U) towards the east side (Tower A). Such an increase follows the deepening of the top of the pyroclastic formation from 22 to 37 m below the ground surface. Within the pyroclastic formation, the thickness of the cohesionless and indurated pozzolana also increases eastwards; accordingly the top of the tuff layer is found at a depth of 35 m at the west side, goes down to 54 m at the centre of the area and is even deeper under Tower A.

The separation among the various soils has been based on a comparative examination of the cores from the boreholes and of q_c profiles. Some typical CPT results are reported in Fig. 3, with an indication of the related soils. The upper soils exhibit relatively low q_c values, with some peaks due to the stony fragments; a similar trend is found in the ashes, but with slightly higher values. The peat layer is characterized by a low and fairly constant cone resistance; the stratified sands by the typical very variable trend. The cohesionless pozzolana has a fairly uniform cone resistance, while the indurated pozzolana shows q_c values increasing with depth.

In Fig. 4 a plot of the SPT blowcount N against depth is reported. The overall trend is that of an increase with depth; no significant differences among the different soils are noticeable, due also to the scatter of the data.

A statistical analysis of all the available CPT results shows that, within each layer, the cone resistance may be assumed as constant and equal to

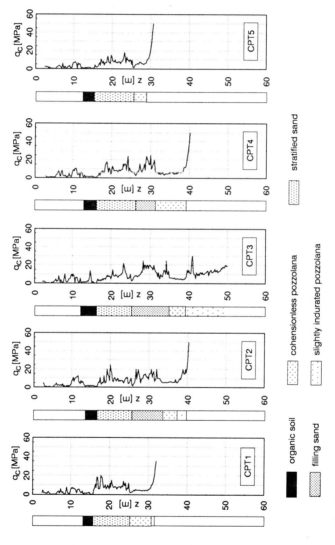

Fig. 3. Results of the CPT

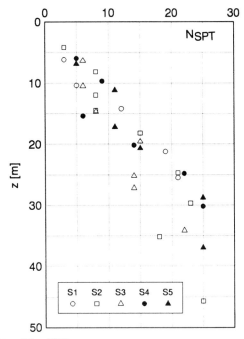

Fig. 4. Results of the SPT

the mean value of the same layer; no statistically significant differences have been found among different locations within the area, apart from the different thickness of some layers.

Observed and predicted settlement

The settlement of the two buildings has been observed at a number of points on the foundation slabs, whose location in plan is reported in Fig. 1. The survey has been performed by a high precision optical instrument, referring to a local datum fixed to a nearby building, built some years before and founded on piles end bearing on the tuff. Since the concreting of the slab has been carried out in two lifts, the levelling points have been first installed at the surface of the lower lift and then moved to the upper one, in order to measure the settlement due to the weight of the second lift of concrete.

The total load acting on the foundations is reported, as a function of the time, in Fig. 5; as may be seen, the slab of Tower A has been concreted 4 months before that of Tower U. As a consequence, the settlement induced by the weight of the former in the area of Tower U has not been measured; this fact has been accounted for in the comparison between predicted and observed settlement.

As of February 1992, the total load acting on the foundation equals 294 MN; the self weight of the two foundation slabs accounts for almost 60% of this load.

In the lower part of Fig. 5 the observed average settlement is reported. At present the average settlement amounts to 7 mm for Tower U and 11 mm for Tower A, with a maximum of 17 mm and a minimum of 3 mm.

It may be observed that the concreting of the second lift of the foundation slab of Tower A has produced an average settlement of 2.6 mm, while in Tower U no significant settlement has been observed under the corresponding load.

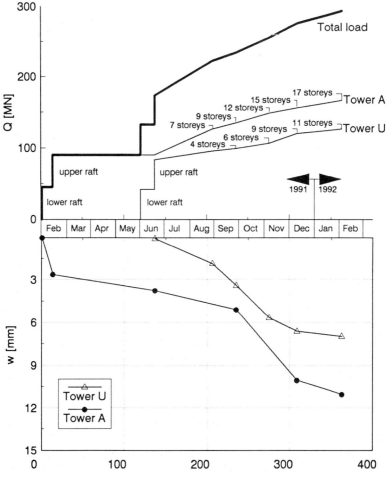

Fig. 5. *Loading and settlement history*

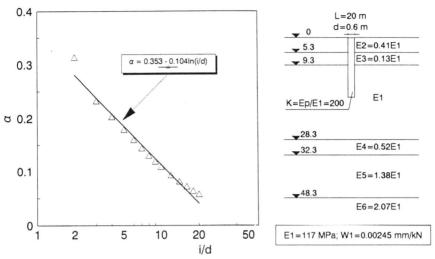

Fig. 6. *Computed values of the superposition factors and subsoil model*

For the purpose of settlement prediction, the subsoil has been modelled as being composed of a succession of horizontal elastic layers of constant thickness; the depth of excavation (6.7 m) has been subtracted from the thickness of the upper layer. The ratio between the moduli of any two layers has been assumed equal to the ratio between the corresponding average q_c values (De Beer and Martens, 1957; Schmertmann, 1970; Bellotti et al., 1986).

The scheme reported in Fig. 6 has been considered sufficiently representative of the average situation of the two towers. The value of the Young's modulus E1 (and, as a consequence, having fixed the ratios Ei/E1, of all other moduli Ei) has been obtained by imposing that the calculated value of the settlement of a single pile $w_s = w_1 Q$ equals the mean value of the settlement measured in the pile loading tests at a load level corresponding to the average load on the foundation piles in the analysed situation.

In Fig. 7 the load–settlement curves for 10 proof loading tests and those for the two instrumented test piles are reported. The value of $w_1 = 0.00245$ mm/kN has been taken on the mean curve at a load of 0.45 MN, which corresponds to the average load per pile in the examined situations.

Having developed a subsoil model that fits at best the available data (CPT, pile loading tests), this has been used to compute the values of the superposition coefficients, α, as a function of the spacing i, by means of the simplified procedure developed by Caputo and Viggiani (1985). The values obtained are plotted against i/d in Fig. 6; for the sake of

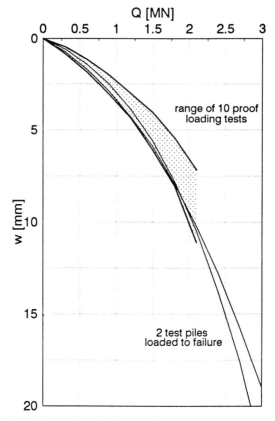

Fig. 7. Results of the loading tests on piles

computational ease the relationship between α and i/d has been interpolated by the logarithmic expression $\alpha = M + N \ln (i/d)$.

On the basis of previous experience (Caputo and Viggiani, 1984), it has been assumed that no interaction occurs ($\alpha = 0$) for i/d $\geq$ 20.

In Fig. 8 the distribution of the axial load along the shaft of the test piles at the different load levels is reported, as determined by means of strain gauge extensometers fixed to the reinforcement bars. An attempt to fit the observed pattern of the axial load by means of the above elastic model has been only partially successful.

A comparison between the observed and predicted settlement at two selected dates is reported in Figs. 9 and 10, where the measured settlement is represented by the dots and the computed one by the solid line. The comparison is limited to a few alignments for reasons of space, but the remaining data follow a very similar trend.

The prediction that best fits the observations is that based on the

SETTLEMENT PREDICTION FOR PILED FOUNDATIONS

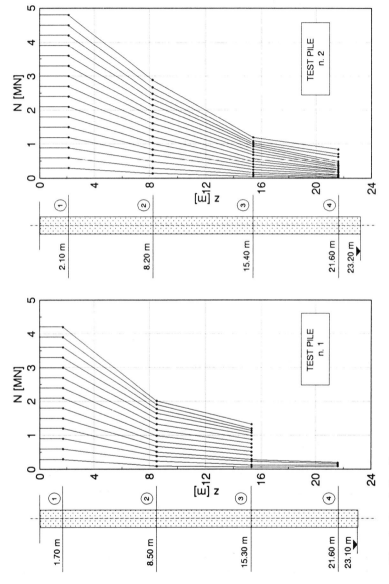

Fig. 8. Load distribution along the shaft of the instrumented piles during the load tests to failure

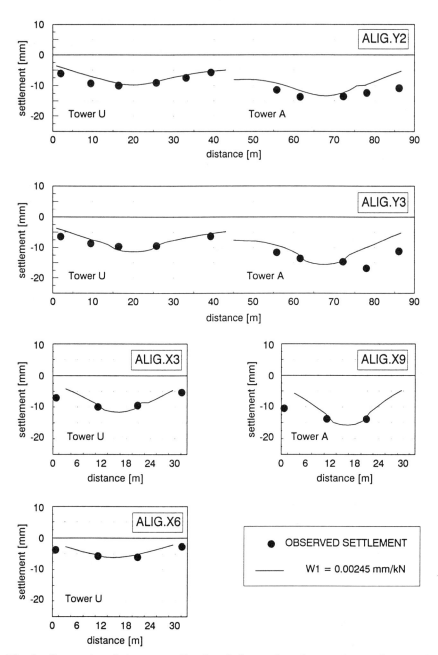

Fig. 9. Comparison between predicted and observed settlement (December 1991)

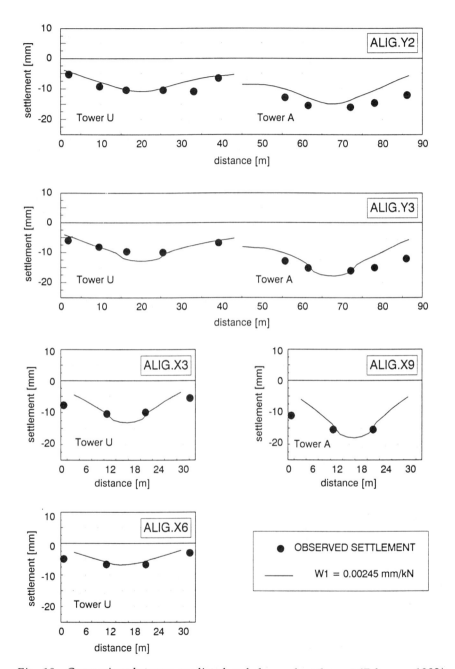

Fig. 10. Comparison between predicted and observed settlement (February 1992)

hypothesis of flexible foundations, with the loads on piles evaluated by their influence area; for the sake of clarity, the prediction with the rigid slab is not reported in the figures.

A first comment is that the subsoil model adopted (Fig. 6) averages the thickness of the different layers, due to the necessity of adopting horizontal constant thickness layers that is intrinsic to the model. For this reason, the model should underpredict the settlement under Tower A, where the thickness of the soil layers overlying the tuff is larger than average, and overpredict it under Tower U. This is indeed the case, especially for Tower A.

Both the magnitude and distribution of the settlement are predicted satisfactorily by the simple model adopted. The largest differences occur at the perimeter of the buildings, due to the fact that the load acting in that zone is practically nil and the settlement computed under the assumption of a fully flexible foundation slab is correspondingly very small.

The discontinuity between the settlement predicted for the two towers is due to the correction made to account for the differences in the period of construction of the two slabs; in other words, the discontinuity equals the calculated settlement produced at the border between the two towers by the weight of the slab of Tower A.

Of course, the settlement observations are continuing and more significant conclusions will be drawn from the data under full loads.

Table 1.

CASE	Ref.	SOIL	FOUNDATION		PILES			
			L (m)	B (m)	type	number	length (m)	dia. (m)
New Law Court	Caputo et al. (1991)	pyroclastic	198.5	52	bored	241	42	1.5-2.2
Mail Bldg. Tower A	Caputo (1991)	pyroclastic	56	38	bored	182	43	1.5
Mail Bldg. Tower C		pyroclastic	64	25.2	bored end-bearing on tuff	136	35	1.5
Re.Di.T Tower U	present paper	pyroclastic	43.6	32.7	auger	314	20	0.6
Re.Di.T Tower A		pyroclastic	47.1	32.7	auger	323	20	0.6
Garigliano viaduct	Mandolini, Viggiani (1992)	soft clay	8.0	8.0	driven through soft clays to gravelly sand	16	44-48	0.36/0.41
			9.7	8.0		18	44-48	0.36/0.41
			8.0	8.0		20	44-48	0.36/0.41
			10.4	8.0		24	44-48	0.36/0.41

Concluding remarks

Terzaghi and Peck (1948) discuss in detail the use of the plate load tests as a means to predict the settlement of a large foundation; they point out that this procedure has a number of shortcomings, mainly connected with the heterogeneities of the subsoil. Above all, the plate load test as a means to predict the settlement of a large foundation is totally unreliable when the properties of the soil vary significantly with depth.

Similar arguments are presented in connection with pile foundations, discussing the reasons why the settlement of a large piled foundation could differ significantly from that inferred from the settlement of a single pile, and presenting examples to substantiate this statement.

The evidence presented in this paper, on the contrary, seems to indicate that a reasonable evaluation of the magnitude and distribution of the settlement may be obtained by a careful extrapolation of the settlement of a single test pile. More results of the same type, collected by the writers, may be quoted. In Table 1 the main features of the available case histories, including the one presented in this paper, are reported. As may be seen, a considerable variety of soil type and of number, size and type of pile have been considered; the table includes foundations ranging from large piled rafts to small pile groups.

Table 2.

CASE		$W1 \times 10^4$ (mm/kN)	DATE	Q_{av} (kN)	calculated settlement (mm)				observed settlement (mm)	
					RIGID		FLEXIBLE			
					average	max diff.	average	max diff.	average	max diff.
New Law Court		4.9-10.1	2/87	5700	19.5-25.7	10.3-13.6	21.6-28.4	25.7-33.8	28.7	21.3
Mail Bldg. Tower A		2.7-4.4	4/90	1450	3.1-6.3	1.0-2.1	3.5-7.5	2.9-6.9	2.95	7.0
Mail Bldg. Tower C		3.1-5.7	6/89	770	1.5-3.6	0.0	1.8-4.3	2.0-4.5	1.5	1.0
			3/90	1770	3.5-9.0	0.0	4.1-10.6	4.7-11.0	5.4	2.4
			5/90	2050	4.1-9.7	0.0	4.7-11.4	5.4-11.9	8.1	10.9
Re.Di.T Tower U		24.5	12/91	250	***	***	6.3	10.3	6.6	7.3
			2/92	270	***	***	6.9	11.8	7.0	8.9
Re.Di.T Tower A		24.5	12/91	340	***	***	8.6	13.0	10.0	14.91
			2/92	380	***	***	9.5	15.1	11.0	13.47
Garigliano viaduct	a	38-58	1990	170 450	1.8-9.5	0.0	***	***	1.6-7.7	0.0
	b	58	1990	70 210	4.9-7.7	0.0	***	***	1.4-4.0	0.0
	c	55	1990	90	0.4-0.9	0.0	***	***	1.4	0.0
	d	58	1990	140 310	5.2-8.5	0.0	***	***	3.5-7.7	0.0

The data refers to: a) 16 piers; b) 2 piers; c) 2 piers; d) 4 piers

In Table 2 the elements for a comparison between predicted and observed settlement are reported. For some cases different dates are considered; in these cases, including that of the present paper, the comparison has been carried out for different loading conditions during the construction.

Only the average settlement and the maximum differential settlement are compared; in some instances, the computed values refer to a range of single pile compliances w_1.

On the whole, the data shown are believed to substantiate the proposed procedure. Similar evidence may be found in the literature (Koizumi and Ito, 1967; Mattes and Poulos, 1971).

The success of the proposed procedure is probably due to a combination of factors. Firstly, all the cases reported refer to subsoil conditions where the variations of the soil properties with depth are small, and in any case the overall soil stiffness increases with depth. In such situations, the load tests on single piles are more significant.

Furthermore, it is becoming more and more evident that the deformations of soils are more concentrated than those predicted by linear elastic theory, due to a number of factors, among which a primary role is played by non-linearity (Jardine et al., 1986). A more concentrated deformation corresponds to an increased significance of the single pile settlement.

Finally, it should be stressed that the proposed procedure is limited to settlement prediction; the settlement results from an integral, and is not very sensitive to the local values of parameters, depending essentially on average values. On the contrary a prediction of the load acting on the piles under a rigid foundation, for instance, would be grossly in error if based on the same linearly elastic model (Caputo and Viggiani, 1984).

As usual in foundation engineering, the simplest models have to be used in conjunction with a maximum of engineering judgement.

References

BANERJEE, P.K. (1978). Analysis of axially and laterally loaded pile groups. In: Developments in Soil Mechanics, C.R. Scott, Editor, Applied Science Publishers, London, pp. 317–346.

BELLOTTI, R., GHIONNA, V.N., HOLTZ, R.D., JAMIOLKOWSKI, M., LANCELLOTTA, R. AND MANFREDINI, G. (1986). Deformation characteristics of cohesionless soils from in situ tests. Proc. Symposium In-Situ 86, ASCE, Blacksburg.

BILOTTA, E., CAPUTO, V. AND VIGGIANI, C. (1991). Analysis of soil-structure interaction for piled rafts. Proc. X. Europ. Conf. Soil Mech. Found. Eng., Firenze, Vol. 1, pp. 315–318.

BURLAND, J.B., BROMS, B.B. AND DE MELLO, V.F.B. (1977). Behaviour of

Foundations and Structures. Proc. IX Int. Conf. Soil Mech. Found. Eng., Tokyo, Vol. 2, pp. 495–546.

CAPUTO, V. (1991). Contribution to the discussion, Session 3b. Proc. X Europ. Conf. Soil Mech. Found. Eng., Firenze. In press.

CAPUTO, V., MANDOLINI, A. AND VIGGIANI, C. (1991). Settlement of a piled foundation in pyroclastic soils. Proc. X Europ. Conf. Soil Mech. Found. Eng., Firenze, Vol. 1, pp. 353–358.

CAPUTO, V. AND VIGGIANI, C. (1984). Pile foundation analysis: a simple approach to nonlinearity effects. Riv. It. di Geotecnica, Vol. 18, No. 1, pp. 32–51.

CAPUTO, V. AND VIGGIANI, C. (1985). Analisi dell' interazione terreno-struttura per fondazioni su pali. XII Conf. di Geotecnica, Torino.

COOKE, R.W. (1986). Piled raft foundations on stiff clays; a contribution to design philosophy. Géotechnique, Vol. 36, No. 2, pp. 169–203.

CROCE, A. AND PELLEGRINO, A. (1967). Il sottosuolo della città di Napoli. Caratterizzazione geotecnica del territorio urbano. Atti VIII Conv. Naz. di Geotecnica, Cagliari, Vol. 3, pp. 233–270.

DE BEER, E.E. AND MARTENS, A. (1957). Method of computation of upper limit for the influence of heterogeneity of sand layers on the settlement of bridges. Proc. IV Int. Conf. Soil Mech. Found. Eng., London, Vol. 1, pp. 275–282.

JARDINE, R.J., POTTS, D.M., FOURIE, A.B. AND BURLAND, J.B. (1986). Studies of the influence of non-linear stress-strain characteristics in soil-structure interaction. Géotechnique, Vol. 36, No. 3, pp. 377–396.

KOIZUMI, Y. AND ITO, K. (1967). Field tests with regard to pile driving and bearing capacity of piled foundations. Soils and Foundations, No. 3, pp. 30–53.

MANDOLINI, A. AND VIGGIANI, C. (1992). Terreni ed opere di fondazione di un viadotto sul fiume Garigliano. Riv. It. di Geotecnica, Vol. 26, No. 2.

MASCARDI, C. (1985). Esecuzione e cenni sul dimensionamento dei pali trivellati con elica continua. XII Conf. di Geotecnica, Torino.

MATTES, N.S. AND POULOS, H.G. (1971). Model tests on piles in clay. Proc. 1st Austr. New Zeal. Conf. on Geomechanics, Melbourne, pp. 254–259.

POULOS, H.G. (1968). Analysis of the settlement of pile groups, Géotechnique, Vol. 18, pp. 449–471.

POULOS, H.G. (1981). Soil–Structure Interaction. General Report, X Int. Conf. Soil Mech. Found. Eng., Stockholm, Vol. 4, pp. 307–334.

POULOS, H.G. AND DAVIS, E.H. (1980). Pile foundations analysis and design. John Wiley & Sons, New York.

POULOS, H.G. AND HEWITT, C.M. (1985). Axial interaction between dissimilar piles in a group, Univ. of Sydney, School of Civil and Mining Eng., Res. Rep. R512.

RIPPA, F. AND VINALE, F. (1982). Experiences with CPT in eastern Naples area. Proc. II ESOPT, Amsterdam, pp. 797–804.

SCHMERTMANN, J.H. (1970). Static cone to compute static settlement over sand. Journ. Soil Mech. Found. Div., Proc. ASCE, Vol. 96, SM3, pp. 1011–1043.

TERZAGHI, K. AND PECK, R.B. (1948). Soil Mechanics in Engineering Practice. John Wiley & Sons, New York.

VIGGIANI, C. (1981). Simple methods for soil-foundation-structure interaction analysis. Proc. X Int. Conf. Soil Mech. Found. Eng., Stockholm, Vol. 4, pp. 704–710.

VIGGIANI, C. (1989a). Influenza dei fattori tecnologici sul progetto dei pali di fondazione. Atti XVII Conv. Naz. di Geotecnica, Taormina, Vol. 2, pp. 83–91.

VIGGIANI, C. (1989b). Terreni ed opere di fondazione della Cittadella Postale nel nuovo Centro Direzionale di Napoli. RIG, XXIII No. 3, pp. 121–146.

VINALE, F. (1988). Caratterizzazione del sottosuolo di un' area campione di Napoli ai fini di una microzonazione sismica. RIG, XXII No. 2, pp. 77–100.

In situ determination of clay stress history by piezocone model

P. W. MAYNE, Associate Professor, Georgia Institute of Technology, Atlanta

A simple piezocone model combines spherical cavity expansion theory and Modified Cam Clay concepts to represent both the cone tip resistance (q_T) and penetration pore water pressure (u_{bt}) measured behind the tip. In closed form, the OCR is shown to be a function of the effective friction angle (ϕ') of the clay and the normalized piezocone parameter, $(q_T - u_{bt})/\sigma'_{v0}$. The approach is supported by laboratory experiments using a series of miniature electric cone and piezoprobe tests on kaolinitic clay that was prestressed in a large fixed-wall chamber. The method is further applied to a variety of soft to stiff to hard, sensitive and insensitive, intact and fissured clays which have been field tested by piezocones. These deposits have known profiles of σ'_p evaluated from laboratory oedometer tests and values of ϕ' determined from triaxial tests.

Introduction

In the last decade, much interest has centered around the use of in situ tests for determining soil properties. In particular, the ability to profile the preconsolidation yield stress (σ'_p) is attractive since it separates elastic from plastic behaviour. Moreover, to benefit from the scaling laws of continuum mechanics, Wroth (1988) suggested the use of dimensionless variables and, in this regard, the use of overconsolidation ratio (OCR = σ'_p/σ'_{v0}) is a preferable means of expressing the variation of stress history at a given site.

Several researchers have pursued the use of piezocone penetration tests (CPTU) for determining the OCR of clays since the cone geometry and steady penetration rate favour a theoretical assessment. Whilst classical interpretations of measured cone tip resistance (q_c) have focused on a total stress analysis, the advent of the piezocone now permits an effective stress analysis, as well as the correction of tip resistance (q_T) for pore water pressures acting on unequal areas of the cone tip (Campanella and Robertson, 1988). This correction is only possible for piezocones where penetration pore water pressures are measured behind the tip (u_{bt}). Moreover, Wroth (1984) explained that u_{bt} measurements were better suited for profiling stress history than for pore water pressures taken at the cone tip (u_t). Consequently, only piezocones with q_T and u_{bt} data will be considered herein.

Model development

The cone tip resistance (q_T) in clay is often expressed in terms of the undrained shear strength (s_u):

$$q_T = N_{kT} s_u + p_0 \qquad (1)$$

where p_0 = total overburden stress and N_{kT} = cone bearing capacity factor. The bearing factor N_{kT} depends upon the specific theory employed. If the spherical cavity expansion theory of Vesić (1977) is invoked, N_{kT} is simply:

$$N_{kT} = (4/3)(\ln I_r + 1) + \pi/2 + 1 \qquad (2)$$

where $I_r = G/s_u$ = rigidity index and G = shear modulus of the soil. Values of N_{kT} from eqn. (2) are comparable to those from more sophisticated strain path analyses reported by Houlsby and Wroth (1989).

Modified Cam Clay can be used to describe s_u, corresponding to triaxial compression loading, in terms of effective stress state and stress history effects (Wroth and Houlsby, 1985):

$$s_u = (M/2)(OCR/2)^\Lambda p_0' \qquad (3)$$

where $M = 6 \sin\phi'/(3 - \sin\phi')$, ϕ' = effective stress friction angle, $\Lambda = 1 - \kappa/\lambda$ = plastic volumetric strain ratio, p_0' = initial effective overburden stress, and κ and λ are the respective swelling and compression indices. Available laboratory strength data indicates that, for natural clays, the parameter $\Lambda \approx 0.75$, 0.80, and 0.85 for compression, simple shear, and extension modes, respectively (Kulhawy and Mayne, 1990). A value $\Lambda = 0.75$ has been adopted herein, corresponding to triaxial compression. As suggested by Wroth (1988), the OCR may be related directly to the normalized cone tip resistance, $(q_T - p_0)/p_0'$. This is accomplished by combining (1), (2), and (3) to give:

$$OCR = 2\left[\frac{(2/M)(q_T - p_0)p_0'}{(4/3)(1 + \ln I_r) + \pi/2 + 1} \right]^{1/\Lambda} \qquad (4)$$

Available data from 83 different piezocone sites (Mayne, et al., 1990) indicate a strong relationship between OCR and normalized cone tip resistance, as shown by Fig. 1. Since the value of p_0' is not actually known, σ_{v0}' has been used for the overburden stress. Therefore, the practical version of the normalized piezocone parameter is $(q_T - \sigma_{v0})/\sigma_{v0}'$, as recommended by Wroth (1988). From Fig. 1, it may be seen that eqn. (4) bounds the piezocone data for typical ranges of $20° < \phi' < 40°$ and $100 < I_r < 500$.

The excess pore water pressures ($\Delta u = u_{bt} - u_0$) generated during piezocone penetration may also be expressed in terms of cavity

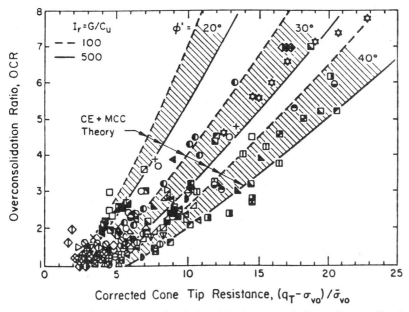

Fig. 1. *Theoretical and measured relationships between OCR and normalized cone resistance (after Mayne (1991))*

expansion and critical-state concepts (Mayne and Bachus, 1988). These excess pore water pressures are due to a combination of changes in octahedral and shear stresses:

$$\Delta u = \Delta u_{oct} + \Delta u_{shear} \qquad (5)$$

Using spherical cavity expansion to describe the octahedral component and Modified Cam Clay for the shear-induced pore water pressures, Δu can be approximated by:

$$\Delta u = (4/3)s_u \ln I_r + p'_0[1 - (OCR/2)^\Lambda] \qquad (6)$$

A similar approach was used by Randolph, Carter, and Wroth (1979) to describe pile installation effects in clay. By introducing (3) for representing strength and stress history, eqn. (6) becomes:

$$OCR = 2\left[\frac{(\Delta u/p'_0 - 1)}{(2M/3)\ln I_r - 1}\right]^{1/\Lambda} \qquad (7)$$

The frictional parameter M (or alternative parameter, ϕ') is a fundamental parameter, easily measured by triaxial shear tests. The practical difficulty with using either (4) or (7), however, is that the proper selection or measurement of the relevant I_r is not straightfor-

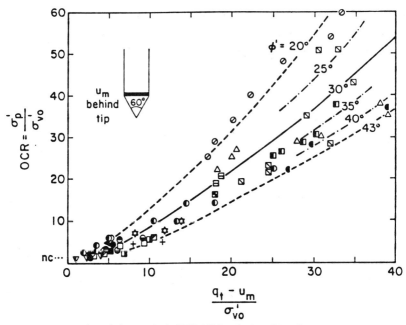

Fig. 2. Measured and theoretical CE/MCC relationships between OCR and normalized piezocone parameter. Individual site symbols given in Mayne et al. (1990)

ward, as noted by Wroth and Houlsby (1985). For piezocones, the problem can be resolved by utilizing both q_T and u_{bt} data in the formulation. Combining (1), (2), (3), and (6), the net cone resistance is:

$$q_T - p_0 = \Delta u + p'_0[(1.95M + 1)(OCR/2)^\Lambda - 1] \tag{8}$$

Introducing a value of $\Lambda = 0.75$ and the substitution $\sigma'_{v0} = p'_0$, the expression for OCR becomes (Mayne, 1991):

$$OCR = 2\left[\frac{1}{1.95M + 1}\left(\frac{q_T - u_{bt}}{\sigma'_{v0}}\right)\right]^{1/33} \tag{9}$$

Although this simple approach does not include the influence of strength anisotropy, initial stress state, stress rotation, strain rate, sensitivity, soil fabric, and other important considerations, Fig. 2 shows that (9) appears to give practical and reasonable first-order predictions of OCR for a variety of clay deposits. The individual symbols refer to the clays cited in the database given by Mayne et al. (1990).

The aforementioned derivation associates OCR with the normalized piezocone parameter, $(q_T - u_{bt})/\sigma'_{v0}$. It is interesting to note that Houlsby (1988) also obtained this parameter using dimensional analysis. Furthermore, this parameter bears an uncanny resemblance to those independently developed by both Konrad and Law (1987) in an effective stress interpretation of piezocone penetration (Robertson et al., 1988) and Senneset et al. (1989) using a plane strain plasticity bearing capacity formulation.

Laboratory study

During an experimental test program involving model foundation testing in a large fixed-wall calibration chamber, an opportunity arose to evaluate (9) in predicting OCR profiles via a series of miniature in situ tests. In this study, a lean kaolinitic slurry (LL = 33, PI = 11, CF = 33%, G_s = 2.65, D_{50} = 0.006 mm, e_0 = 0.91) was prepared at an initial water content w_n = 66% (Kulhawy, 1991). The slurry was pumped into a large cylindrical steel chamber having an inside diameter of 1.37 m and height of 2.13 m. Pneumatic pressure was applied to the top of a rigid piston and the specimen was consolidated one-dimensionally to $\Delta\sigma'_v$ = 48 kN/m² with double drainage. After completion of primary consolidation, the specimen was rebounded to atmospheric conditions. A water reservoir maintained the 'groundwater' level contiguous with the surface of the clay.

A complementary suite of laboratory testing included triaxial, direct simple shear, oedometer, creep, isotropic consolidation, K_0 tests, and fall cone tests on the material. Some test results are reported in

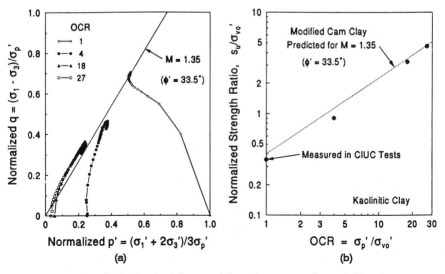

Fig. 3. Results of CIUC triaxial tests: (a) q–p' stress paths and (b) s_u/σ'_{vo} versus induced OCRs

McManus and Kulhawy (1991). Briefly, the critical state parameters are: $\lambda = 0.076$, $\kappa = 0.011$, and $M = 1.35$ (or $\phi' = 33.5°$). Figure 3(a) shows the effective stress paths for four CIUC triaxial tests on the material. The consolidation parameters give a theoretical value $\Lambda = 1 - \kappa/\lambda = 0.85$, while backanalysis of the triaxial series ($1 \leq OCR \leq 27$) and direct simple shear tests ($1 \leq OCR \leq 53$) both indicate $\Lambda = 0.78$, which is consistent with the adopted value $\Lambda = 0.75$. Using this value and $M = 1.35$, Fig. 3(b) indicates that MCC is reasonable in representing undrained strength ratios over a wide range in stress history.

Miniature in situ tests were conducted to evaluate the uniformity and consistency of the prestressed clay deposit. These included vane, cone, two types of piezoprobe, as well as water content determinations. A motorized Wykeham–Farrance vane apparatus was used to perform the vane shear tests with a rectangular blade (12.7 mm diameter by 25.4 mm height). Undrained strengths measured by the vane were essentially constant with depth at $s_{uv} = 8.51 \pm 0.73$ kPa. Consolidated water contents decreased from about 36% at the top to 34% at the bottom of the deposit.

Electric cone penetration tests were performed using a 23.3 mm diameter miniature penetrometer (Fugro-type geometry) with 60° apex to provide measurements of q_c. The cone has a net area ratio $a = 0.88$. Piezoprobe soundings were conducted using brass cones (also with 60° tips) that were fitted with Druck pore water stress transducers and sintered brass porous elements. Two types of piezoprobes were built so

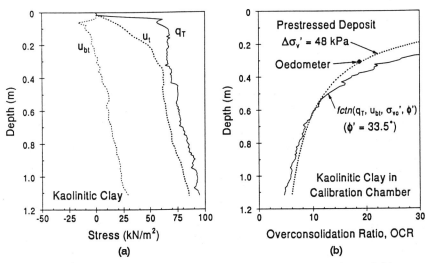

Fig. 4. Results of composite piezocone test: (a) sounding records, and (b) predicted OCR profiles

that penetration pore water pressures could be measured at the tip (u_t) and behind the tip (u_{bt}). Figure 4(a) shows the records of the penetration tests in the prestressed clay deposit. The combined data from the cone and piezoprobes result in the equivalence of a piezocone sounding.

A comparison of the actual induced OCR and predicted OCR profile in the overconsolidated clay is shown in Fig. 4(b). Results from a confirmatory oedometer test on a sampled specimen with an interpreted $\sigma'_p = 50$ kN/m² are also indicated. Considering the simplistic nature of the approach, the overall agreement is quite good.

Applications to field sites

The cavity expansion/critical state approach has also been applied to a number of field cases where piezocones have been advanced into clay deposits and the soundings have been reported in the open literature. At each of these sites, the profile of σ'_p had been determined from companion series of oedometer tests and triaxial compression tests used to determine ϕ'. Measured values of ϕ' span from 20° to 40°.

Average values of index properties of the clay sites are summarized in Table 1 and include: natural water content (w_n), liquid limit (w_L), plasticity index (PI), clay fraction (CF), and sensitivity (S_t). For the twelve sites reviewed, a wide variation in properties is observed: $4 \leq PI \leq 50$; $2 \leq S_t \leq 500$; $23\% \leq w_n \leq 92\%$; $7 \leq CF \leq 61$. The in situ overconsolidation ratios range from 1 to 100+.

These sites have been grouped into three categories for presentation: soft clays ($1 \leq OCR \leq 3$), stiff clays ($3 \leq OCR \leq 15$), and hard clays ($15 < OCR$). At each of these sites, the groundwater table was generally within ±1 m of the ground surface, except for two deposits:

Table 1. Piezocone sites and index properties of clays

Clay Site	w_n	w_L	PI	CF	S_t	OCR
Brent Cross	28	78	50	57	—	20–60
Glava	30	34	12	41	5–8	2–7
Bothkennar	65	73	41	27	4–6	1–3
Haga	35	41	22	45	4–7	2–15
Hamilton AFB	92	88	48	46	4–8	1–3
Madingley	31	73	44	61	—	18–100+
Onsøy	63	65	37	60	6–9	1–4
St. Hilaire	75	55	33	—	14±	1–3
S.-J. Vianney	45	33	10	50	500+	25–50
Taranto	23	60	27	—	—	20–40
Troll East	55	65	42	35	2–4	1–2
Yorktown	31	31	4	7	4–8	4–12

(1) Haga, Norway which is adjacent to but above the Glomma River
(2) Troll East, an offshore deposit located in 300 meters of open water in the North Sea.

Soft clays

The stress history profiles from oedometer tests on four normally to lightly overconsolidated natural clays are shown in Fig. 5. The sites include: an excavation site in Saint Hilaire, Quebec; the Troll East

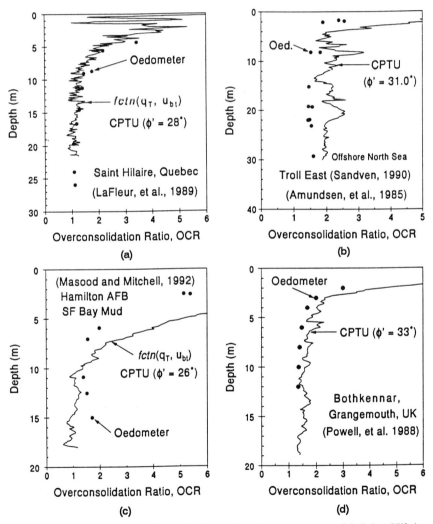

Fig. 5. Measured and predicted OCR profiles for soft clays at (a) Saint Hilaire, (b) Troll East, (c) Hamilton AFB, and (d) Bothkennar

deposit in the Norwegian Trench; San Francisco Bay Mud at Hamilton AFB, California; and the SERC national test site at Bothkennar, Scotland. The reference sources for each site are also noted on the figure. Using the reported value of ϕ' for each site, eqn. (9) was utilized to generate predicted OCR profiles from piezocone data that were re-digitized from sounding records. Saint Hilaire is a somewhat sensitive Champlain Sea clay, while the remaining three clays exhibit low to medium sensitivities, as measured by field vane tests. In each case, the

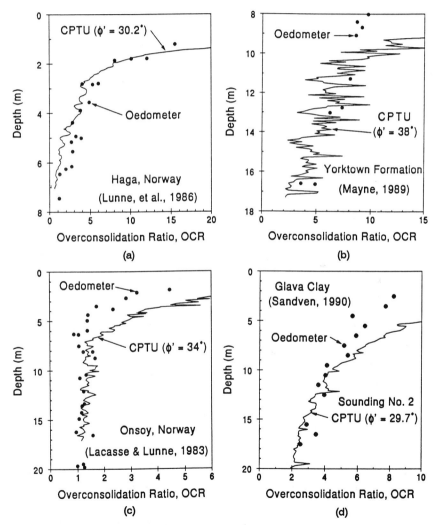

Fig. 6. Measured and predicted OCRs in stiff clays at (a) Haga, (b) Yorktown, (c) Onsøy, and (d) Glava

agreement between measured and estimated OCR is reasonable, although some overprediction is noted in the upper portions of the Hamilton site and lower portion of the Troll site.

Stiff clays

A similar comparison for low to moderately overconsolidated clays is shown in Fig. 6. The clay sites include Onsøy clay, near Fredrikstad, Norway; Haga clay, northeast of Oslo, Norway; sandy clay of the

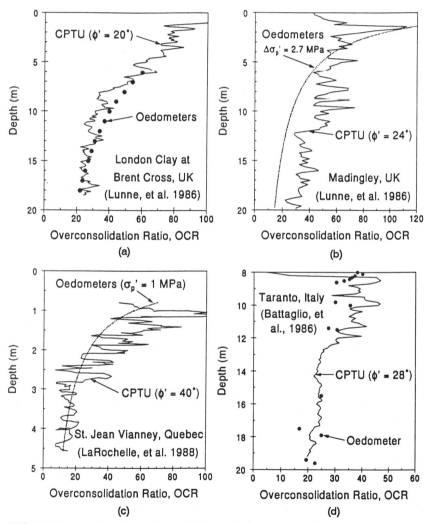

Fig. 7. Measured and predicted OCRs in hard clays at (a) Brent Cross, (b) Madingley, (c) St.-Jean Vianney, and (d) Taranto

Yorktown Formation in Newport News, Virginia; and Glava, near Stjørdal, Norway. Reasonable profiles of OCR are predicted over most of the sounding depths, despite the fact that possible variations in ϕ' may occur vertically. Possibly, the derived methodology would be improved if either an effective cohesion intercept (c') or attraction (a) term were considered (Senneset et al., 1989).

Hard clays
The application of the CE/MCC approach to four heavily overconsolidated clays is shown in Fig. 7. These include two glaciated and highly fissured clays (London clay at Brent Cross and Gault clay at Madingley), a highly sensitive clay associated with landslide instability (St. Jean Vianney), and a hard, cemented and microfissured clay located in southern Italy (Taranto). Overall, the first-order estimates of OCR for these clays is rather good.

Conclusions

The development of a simple hybrid theory based on spherical cavity expansion and Modified Cam Clay has been shown to approximately relate the in situ OCR to the normalized piezocone parameter, $(q_T - u_{bt})/\sigma'_{v0}$, and effective stress friction angle (ϕ') of clays. Preliminary calibration of the model has shown promise in evaluating the stress history of soft to stiff to hard clays ($1 \leq OCR \leq 100+$) over a wide range in frictional characteristics ($20° \leq \phi' \leq 40°$). Ongoing research is underway to evaluate dual and triple element piezocones which provide supplementary pore water pressure data at different positions along the cone penetrometer.

Acknowledgements

The author is currently funded for piezocone research under NSF Grant No. MSS-9108234 for which Dr. Mehmet T. Tumay is the programme director. Appreciation is extended to Fred H. Kulhawy of Cornell University, for providing the opportunity for conducting miniature in situ tests, and to Vito Longo of EPRI, for financial support during this period.

References

AMUNDSEN, T., LUNNE, T. AND CHRISTOPHERSEN, H.P. (1985). Advanced deep-water soil investigation at the Troll East field. Advances in Underwater Technology, 3, Offshore Site Investigation. Graham & Trotman, London, pp. 165–186.

BATTAGLIO, M., BRUZZI, D. AND JAMIOLKOWSKI, M. (1986). Interpretation of CPT and CPTUs. Proceedings, 4th International Geotechnical Seminar, Nanyang Institute, Singapore, pp. 129–143.

CAMPANELLA, R.G. AND ROBERTSON, P.K. (1988). Current status of the piezocone test. Penetration Testing 1988, 1, Balkema, Rotterdam, pp. 93–116.

HOULSBY, G.T. (1988). Piezocone penetration test. Penetration Testing in the U.K. Thomas Telford, London, pp. 141–146.

HOULSBY, G.T. AND WROTH, C.P. (1989). The influence of soil stiffness and lateral stress on the results of in-situ soil tests. Proceedings, 12th Intl. Conference on Soil Mechanics and Foundation Engrg. (1), Rio de Janeiro, Brazil, pp. 227–232.

KONRAD, J.M. AND LAW, K. (1987). Preconsolidation pressure from piezocone tests in marine clays. Géotechnique, Vol. 37, No. 2, pp. 177–190.

KULHAWY, F.H. (1991). Fifteen+ years of model foundation testing in large chambers. Calibration Chamber Testing. A.B. Huang (ed.). Elsevier, New York, pp. 185–196.

KULHAWY, F.H. AND MAYNE, P.W. (1990). Manual on estimating soil properties for foundation design. Rept. EL-6800. Electric Power Research Inst., Palo Alto.

LACASSE, S. AND LUNNE, T. (1982). Penetration tests in two Norwegian clays. Penetration Testing, 2, A. Verruijt (ed.), Balkema, Rotterdam, pp. 607–613.

LAFLEUR, J., SILVESTRI, V., ASSELIN, R. AND SOULIÉ, M. (1988). Behavior of a test excavation in soft Champlain Sea clay. Canadian Geotechnical Journal, Vol. 25, No. 4, pp. 705–715.

LAROCHELLE, P., ZEBDI, M., LEROUEIL, S. AND TAVENAS, F. (1988). Piezocone tests in sensitive clays of eastern Canada. Penetration Testing 1988, 2, Balkema, Rotterdam, p. 831.

LUNNE, T., EIDSMOEN, T., GILLESPIE, D. AND HOWLAND, J. (1986a). Laboratory and field evaluation of cone penetrometers. Use of In Situ Tests in Geot. Engrg. ASCE, New York, pp. 714–729.

LUNNE, T., EIDSMOEN, T., POWELL, J. AND QUARTERMAN, R. (1986b). Proceedings, 39th Canadian Geot. Conf. Ottawa, pp. 209–218.

LUNNE, T., POWELL, J., HAUGE, E. et al. (1991). Correlation of dilatometer readings with lateral stress in clays. Transportation Research Record, 1278, pp. 183–193.

MASOOD, T. AND MITCHELL, J.K. (1993). Estimation of in situ lateral stresses in soils by cone penetration. Journal of Geotechnical Engineering, in press.

MAYNE, P.W. (1989). Site characterization of Yorktown formation for new accelerator. Foundation Engineering: Current Principles and Practices, 1, ASCE, New York, pp. 1–15.

MAYNE, P.W. (1991) Determination of OCR in clays by piezocone tests using CE and CSSM concepts. Soils and Foundations. Vol. 31, No. 2, pp. 65–76.

MAYNE, P.W. AND BACHUS, R.C. (1988). Profiling OCR in clays by piezocone. Penetration Testing 1988, 2, J. DeRuiter (ed.), Balkema, Rotterdam, pp. 857–864.

MAYNE, P.W., KULHAWY, F.H., AND KAY, J.N. (1990). Observations on the development of pore water stresses during piezocone penetration in clays. Canadian Geotechnical Journal. Vol. 27, No. 4, pp. 418–428.

McMANUS, K.J. AND KULHAWY, F.H. (1990). A cohesive soil for large-size laboratory deposits. ASTM Geotechnical Testing Journal. Vol. 14, No. 1, pp. 26–36.

POWELL, J.J.M., QUARTERMAN, R.S.T. AND LUNNE, T. (1988). Interpretation and use of the piezocone test in UK clays. Penetration Testing in the UK. Thomas Telford, London, pp. 151–156.

RANDOLPH, M.F., CARTER, J.P. AND WROTH, C.P. (1979). Driven piles in clay – the effects of installation and subsequent consolidation. Géotechnique. Vol. 29, No. 4, pp. 361–393.

ROBERTSON, P.K., HOWIE, J.A. AND SULLY, J.P. (1988). Discussion. Géotechnique. Vol. 38, No. 3, pp. 455–465.

SANDVEN, R. (1990). Strength and deformation properties of fine grained soils obtained from piezocone tests. PhD Thesis. Norwegian Institute of Technology, Trondheim.

SENNESET, K., SANDVEN, R. AND JANBU, N. (1989). Evaluation of soil parameters from piezocone tests. Transportation Research Record 1235, pp. 24–37.

VESIC, A.S. (1977). Design of Pile Foundations. Synthesis of Highway Practice 42. Transportation Research Board, Washington, D.C.

WROTH, C.P. (1984). The interpretation of in situ soil tests. 24th Rankine Lecture. Géotechnique. Vol. 34, No. 4, pp. 449–489.

WROTH, C.P. (1988). Penetration testing – a more rigorous approach to interpretation. Penetration Testing 1988, Vol. 1. (Proc. ISOPT-1). J. De Ruiter (ed.), Balkema, Rotterdam, pp. 303–311.

WROTH, C.P. AND HOULSBY, G.T. (1985). Soil mechanics – property characterization and analysis procedures. Proceedings 11th Intl. Conf. on Soil Mechanics and Foundation Engineering, 1, San Francisco, pp. 1–55.

Selection of parameters for numerical predictions

D. MUIR WOOD, N.L. MACKENZIE and A.H.C. CHAN,
Department of Civil Engineering, Glasgow University

Data from a test on reconstituted kaolin performed under axially symmetric stress conditions in a true triaxial apparatus are used to generate sets of values of soil parameters for use with the (modified) Cam clay model. First, parameters are chosen by the traditional route, one at a time: slope of normal compression line and slope of unloading line in the compression plane, critical state stress ratio, and elastic property. This fails to take any direct account of the shear strains that occur and yet it is in order to predict the response of a soil to shearing that a model such as Cam clay is normally applied. An alternative procedure adopts an optimisation strategy to produce a simultaneous best fit set of all parameters in order to match any section or sections of the test that are reckoned to be of importance. The values of the parameters thus deduced are rather different, but the model reproduces the soil behaviour more accurately.

Introduction

The Cam clay models have become firmly established in the language of soil mechanics since their first introduction some thirty years ago (Roscoe and Schofield, 1963; Roscoe and Burland, 1968). Over the past two decades, in particular, they have been widely used in numerical analysis of geotechnical structures, especially those involving the loading of soft normally consolidated or lightly overconsolidated clays (Wroth, 1977). The Cam clay models have an important pedagogic role to play in illustrating the way in which rather simple but complete models of soil behaviour can be developed by a logical extension from consideration of ideas of yielding and plastic hardening of ductile metals (Schofield and Wroth, 1968; Muir Wood, 1990). The appeal of the Cam clay models lies in their compactness, in the very small number of soil parameters — five, plus permeability — that are necessary for a complete definition of the models, and in the physical basis of all these parameters. The Cam clay models formed a central element of a number of courses on Critical State Soil Mechanics that were presented in Britain and Europe between 1975 and 1985 (Wroth et al., 1975, 1979, 1981, 1982, 1985) and much was always made of the fact that the five parameters were not really new parameters, but familiar quantities seen in a new (truer) light. Thus the slope M of the critical state line in the $p':q$ effective stress plane is linked with the angle of shearing resistance ϕ'; the slopes λ, κ of normal compression and unload–reload lines in the $v:\ln p'$ compression plane are linked with compression and swelling indices C'_c, C'_s; the location of the critical state line in the compression plane is defined by a reference specific volume Γ which can be linked with liquid limit w_L; and some second elastic property is required such as shear modulus G or Poisson's ratio ν. The model takes care of the rest.

The continuation of this sales tactic is therefore that no special tests are required to determine the values of the soil parameters: testing can continue as before and the five parameters can be picked off one by one.

The fundamental feature of these soil models — and the vital message of critical state soil mechanics — is the importance of volumetric strains and the parallel significance of change in volume and change in effective stress. These models belong to a more general class of volumetric hardening models. The models are driven by the volume changes occurring during normal compression; shear strains are deduced indirectly by introducing a family of plastic potentials (which in the Cam clay models happen to be identical to the yield loci, but which in other volumetric hardening models are not necessarily so (Mouratidis and Magnan, 1983)).

If soil parameters are being chosen in order that the model can be made to give a good general fit to a complete range of laboratory test data — particularly if these data are obtained from tests on reconstituted clays which undergo large volume changes as they are consolidated from slurry — then this volumetric basis for the parameter selection has a certain logic. If, however, the model is to be used for prediction of field response of natural soils then the volumetric response may be much less important than the distortional response, which is hardly considered during the process of parameter selection. Undrained deformation is a purely distortional process — neatly described in volumetric hardening models as the result of balancing equal and opposite recoverable and irrecoverable volumetric changes — and a strong emphasis on volumetric response in selection of parameters may in fact be particularly unhelpful in modelling undrained behaviour.

Equally, the models are strongly governed by the choice of critical state stress ratio M which describes an ultimate condition of infinite distortion. In practice, numerical predictions are required of deformations of geotechnical structures at working loads far removed from collapse conditions.

Numerical modelling is always an extrapolation from the known region of experimental data towards the unknown region of field response. This paper explores the heretical idea that the reputation of the Cam clay models could be improved still further if parameter selection were made a more interactive process, with the selector more consciously choosing experimental data from laboratory (or field) tests with stress levels, stress states and stress paths close to those for which numerical predictions were subsequently required.

Cam clay

For the purposes of this paper, the name Cam clay will be assumed to refer to the modified Cam clay model of Roscoe and Burland (1988) rather than the original Cam clay model of Roscoe and Schofield (1963). Whatever the historical origins of the two models it has been found easier to explain them through an assumed shape of yield locus and coincident plastic potential, rather than through an assumed plastic energy dissipation function and the assumption of normality. It is then natural to start with the (modified) Cam clay ellipse rather than the original Cam clay bullet.

The models are well known and do not require detailed description. Loading and unloading at constant stress ratio $\eta = q/p'$ are associated with linear response in the semi-logarithmic compression plane v:ln p', thus introducing parameters λ and κ (Fig. 1(b)). One-dimensional normal compression in an oedometer is a constant stress ratio loading process so that the validity of the assumption of linearity from which λ emerges can be directly assessed in routine testing. One-dimensional unloading is not a constant stress ratio unloading process so that the selection of κ and its link with swelling index C'_s are less soundly based.

Yield loci in the p':q effective stress plane are assumed to be elliptical, passing through the origin, centred on the p' axis, with the slope to the top of the ellipse given by M (Fig. 1(a)). The assumption of coincident yield loci and plastic potentials, together with the assumption that the soil is volumetric hardening — so that change in size of the yield loci implies irrecoverable plastic volume change — in turn implies that the soil ends with critical states at the stress ratio $\eta = M$. Evidently the existence of critical states can be assessed if laboratory tests are taken to sufficiently large distortions. Certainly failure stress ratios can be determined. Neither the shape of the yield loci, which has an essential but hidden role in all model predictions, nor the coincidence of plastic potentials and yield loci is ever actually investigated in routine testing.

The several assumptions lead to the plastic compliance relationships:

$$\delta\varepsilon_p^p = (\lambda - \kappa)[(M^2 - \eta^2)\delta p' + 2\eta\delta q]/[vp'(M^2 + \eta^2)] \qquad (1)$$

$$\delta\varepsilon_q^p = (\lambda - \kappa)[2\eta\delta p' + \{4\eta/(M^2 - \eta^2)\}\delta q]/[vp'(M^2 + \eta^2)] \qquad (2)$$

which apply whenever the changes in effective stress imply a change in size of the yield locus. The elastic compliance relationships:

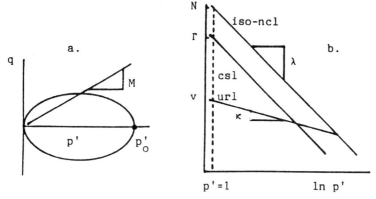

Fig. 1.(a) Elliptical Cam clay yield locus in $p':q$ effective stress plane; (b) isotropic compression line (iso–ncl), critical state line (csl), and unloading–reloading line (url) in $v:\ln p'$ compression plane

$$\delta\varepsilon_p^e = [\kappa/(vp')]\delta p' \qquad (3)$$

$$\delta\varepsilon_q^e = [1/(3G)]\delta q \qquad (4a)$$

or

$$\delta\varepsilon_q^e = [2(1+\nu)\kappa]\delta q/[9(1-2\nu)vp'] \qquad (4b)$$

apply for all changes in effective stress. The symbols for volumetric and distortional increments, $\delta\varepsilon_p$, $\delta\varepsilon_q$, are chosen following Calladine (1963) to indicate work conjugacy with the volumetric and distortional effective stresses p', q.

Shear modulus G, or Poisson's ratio ν, enters as a second elastic parameter, to complete the description of the isotropic elastic properties of the soil. It is recognised that with bulk modulus $K = vp'/\kappa$ proportional to mean effective stress p' it is not thermodynamically acceptable to have shear modulus also proportional to p', as is implied through the selection of a constant value of Poisson's ratio ν (Zytynski et al., 1978). This will in practice cause problems only when predictions are required of response of soils to cycles of loading and unloading.

The fifth soil parameter is required in order to be able to calculate the current specific volume v from the known stress history of the soil. If the size of the current yield locus is given by p'_0 (Fig. 1(a)) then:

$$v = \Gamma - \lambda \ln(p'_0/2p'_\Gamma) + \kappa \ln(p'_0/2p') \qquad (5)$$

where Γ and p_Γ' define a reference point on the critical state line in the compression plane. Conventionally $p_\Gamma' = 1$ measured in whatever units of stress are being used. A stress of 1 kPa is extremely low for most engineering purposes. If $p_I' \simeq 4$ kPa then $\Gamma \simeq 1 + G_s w_L$, indicating the link between this reference volume Γ and liquid limit (Muir Wood, 1990). The choice of this reference stress would not be important if the v:ln p' relationships were indeed linear over the stress range from p_I' to the stresses of engineering interest. Experimental evidence does not always support this (Butterfield, 1979) and since the range of mean effective stress experienced in typical geotechnical structures is not great it might be more rational to choose the value of p_I' to match the ambient mean effective stress and then choose Γ to fit the in situ specific volume.

Experimental data

Data to be fitted with the Cam clay model are obtained from the series of experiments performed in the Cambridge True Triaxial Apparatus (Wood and Wroth, 1972; Airey and Wood, 1988) and described by Wood (1974). These tests were performed on samples of spestone kaolin ($w_P = 0.40$, $w_L = 0.72$, 73% finer than 2 μm) consolidated in the True Triaxial Apparatus from a slurry prepared at a water content equal to twice the liquid limit. The True Triaxial Apparatus permits independent control of three principal stresses (or principal strains) without allowing any rotation of principal axes. Stress paths can be continued without interruption from consolidation to shearing (within the deformation capability of the apparatus).

The (effective) stress path of test L1 is shown in Fig. 2. It consists of anisotropic compression ($\eta = 0.3$) to p' = 150 kPa and unloading (OAB), followed by isotropic reloading (CDF) with two cycles of constant p' loading and unloading (DED) with p' = 100 kPa; FGF with p' = 150 kPa. The whole test was performed with two stresses equal to each other ($\sigma_2 = \sigma_3$) and could in principle therefore have been performed in a conventional triaxial apparatus.

Proceeding incrementally, the value of λ can be deduced from the

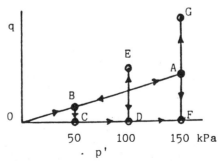

Fig. 2. Stress path of test L1: OABCDEDFGF

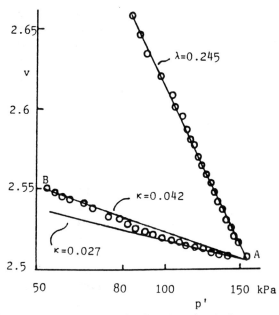

Fig. 3. Anisotropic compression and unloading in $v:\ln p'$ plane

initial anisotropic compression OA (Fig. 3) giving $\lambda = 0.245$. The value of κ can be deduced from the anisotropic unloading AB or the isotropic reloading (CD, DF). Volumetric unload–reload cycles are not the ideally elastic processes that they are assumed to be in Cam clay, and there is room for interpretation in selecting a value of κ. (Kinematic hardening models such as the 'bubble' extension of Cam clay described by Al-Tabbaa and Wood (1989) are introduced precisely to improve the match with the experimentally observed unload–reload hysteresis.) From the data shown in Fig. 3, a value of $\kappa = 0.027$ could be deduced from the initial slope of the $v:\ln p'$ relationship, immediately after the change in loading direction. Alternatively, a value $\kappa = 0.042$ could be chosen as an average slope of the complete unloading or reloading process.

Study of the shape of the deviatoric stress:strain relationship (Fig. 4(a)) suggests that the ultimate stress ratio which would have been reached on section FG if shearing had been continued further would have been about 0.7–0.75. The implied value of M is consistent with values observed in other true triaxial tests reported by Wood (1974).

The stress:strain relationship for the final loading stage FG is shown in Fig. 4(a). The initial section should, according to the Cam clay model, be purely elastic, because of the size of the yield locus which was

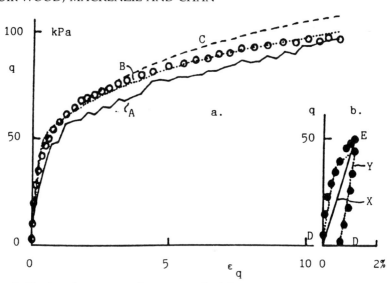

Fig. 4. *Deviatoric stress:strain response for (a) stage FG (○: experiment; A: prediction based on visual selection of parameters; B: optimised fit; C: optimised fit to stage FG for $0 < \eta < 0.5$); and (b) stage DED (●: experiment; X: elastic prediction; Y: optimum fit allowing premature yield)*

established during the initial anisotropic consolidation, and can be used to estimate a value of tangent shear modulus $G_t = 7.5$ MPa. Alternatively a secant shear modulus $G_s = 3.16$ MPa could be calculated for the increase of stress ratio from zero to 0.3, the entire elastic region according to the Cam clay description of this test. It is well known that for most soils the strain range over which the response is truly elastic is extremely small. However, in any situation where the plastic response of the soil is expected to become dominant, as for the soft clay being considered here, the details of the pseudo-elastic response are perhaps less important.

These values of shear modulus can be converted to equivalent values of Poisson's ratio. The measured specific volume at the start of the shearing stage FG was $v = 2.479$, the mean stress $p' = 150$ kPa. With $\kappa = 0.027$ this implies a value of bulk modulus $K = vp'/\kappa = 13.8$ MPa. Poisson's ratio can be calculated from the relationship

$$\nu = (3K - 2G)/(6K + 2G) \tag{6}$$

assuming that the clay is behaving isotropically. The values of Poisson's ratio are then $\nu_t = 0.270$ using the tangent shear modulus G_t, or $\nu_s = 0.394$ using the secant shear modulus G_s. The values of Poisson's

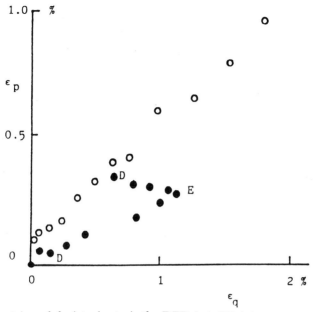

Fig. 5. Volumetric and deviatoric strain for DED (●), FG (○)

ratio would be reduced if higher values of κ were used to calculate correspondingly lower values of bulk modulus. Rather similar values of Poisson's ratio can be calculated from the intermediate loading cycle DED (Fig. 4(b)): the initial tangent stiffness gives $G_t = 5.0$ MPa and $\nu_t = 0.274$, the overall secant stiffness gives $G_s = 1.49$ MPa and $\nu_s = 0.424$.

Examination of the experimental data shows that the supposedly elastic cycle DED and the initial section of the loading FG are accompanied by some volumetric straining, even though the stress changes are entirely distortional ($\delta p' = 0$) ($\delta\varepsilon_p/\delta\varepsilon_q = 0.239$ for DED, $\delta\varepsilon_p/\delta\varepsilon_q = 0.505$ for FG) (Fig. 5). Such response could be described by an anisotropic elastic model such as that proposed by Graham and Houlsby (1983), but that extra refinement has not been considered here, even though the effect is clearly not insignificant.

If it is assumed that the critical state had been reached at point G, the maximum deviator stress applied during the final shearing, then the value of Γ could be calculated from the corresponding specific volume $v = 2.388$ and mean effective stress $p' = 150$ kPa. With a value of $\lambda = 0.245$, this implies $\Gamma = 3.616$. Continuing the argument of the previous section, however, it may be more useful to be able to set the value of the specific volume at the start rather than at the end of shearing. A reference specific volume can be obtained by fixing the

503

location of the isotropic normal compression line in the compression plane:

$$v = N - \lambda \ln p' \qquad (7)$$

According to the Cam clay model:

$$N = \Gamma + (\lambda - \kappa)\ln 2 \qquad (8)$$

Combination of the specific volume and mean effective stress at F with the known stress history, through the Cam clay model with $\lambda = 0.245$, $\kappa = 0.027$, $M = 0.75$, leads to a value of $N = 3.739$ (which in turn implies, from (8), $\Gamma = 3.588$).

Optimisation procedure

As an alternative to direct individual estimation of values of parameters for the Cam clay model the possibility of using an automatic optimisation procedure to produce a simultaneous best fit set of parameters has been explored. Such a procedure can be adapted to ensure that the fit is obtained over the range or ranges of stress change that are expected to be relevant in a particular application — with the emphasis on the ability to match and predict response under working loads which may not approach failure.

The stress:strain response that emerges from a constitutive model is an extremely non-linear function of several model parameters. In only a very few cases will it be possible to obtain an analytical solution to the search for the optimum set of parameters, and a numerical procedure is to be preferred. The procedure adopted here is that proposed by Rosenbrock (1960), and the program used for the optimisation process has been adapted from a program written by Klisinski (1987).

The program searches for the set of n parameters that produces the minimum value of an objective function F which is a measure of the overall difference between experimentally observed and numerically predicted responses. With a given starting set of parameters, the program varies each parameter in turn in order to discover which direction in n-dimensional parameter space leads to the greatest improvement in the value of F. A new set of parameters related to the first by the direction of maximum improvement is then chosen and the procedure is repeated. The process is adaptive in that the direction of maximum improvement will in general involve variation of more than one of the n parameters: a set of n mutually orthogonal directions of progressively decreasing improvement is computed and the search for further improvement makes use of this previously determined set of directions.

The procedure moves through n-dimensional parameter space in the direction of greatest change of the function F, but since it retains

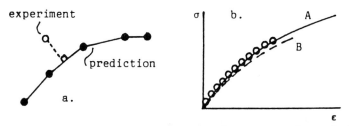

Fig. 6. (a) Objective function: shortest distance between experimental point and predicted curve; (b) alternative fits to experimental data

information from each previous step concerning the relative advantage of moving in each of the n orthogonal directions, it is more robust and more rapidly convergent than the simpler method of steepest descent. It is relatively simple and places no constraint on the nature of the objective function F. It relies on the continuity and smoothness of F but with a very irregular function it may be sensitive to starting point, and may converge on a local minimum rather than the global minimum of the function.

The choice of objective function F which is to be minimised is essentially arbitrary. Some measure of least squares fit is an obvious candidate. Klisinski (1987) uses the square root of the sum of the squares of the shortest distances from each experimental point to the piecewise linear path joining the theoretical prediction points (with appropriate scaling values to allow for the different dimensions of stress and strain) (Fig. 6(a)). This has a potential limitation since the closest experimental and calculated points may correspond to rather different points on the path. For example a curve (A in Fig. 6(b)) which stays close to the shape of the data but which is much 'longer' or 'shorter' than the experimental curve may be a 'better' fit than a curve of the correct form but slightly displaced (B in Fig. 6(b)), even though curve B reproduces the nature of the experimental curve rather better. To try to overcome this difficulty Klisinki adds a term to the objective function equal to the difference between the end points of the experimental and calculated paths. However, this term becomes less important as the number of data points increases, and could perhaps better be made proportional to the number of data points. This definition of objective function may also have difficulties with cyclic paths, where it is not always straightforward to identify, numerically, the relevant closest segment. Such an objective function is, however, useful when the control of the test to be predicted involves a mixture of stress and strain constraints, such as strain control of a specimen tested under conditions of plane stress, or stress control of a specimen tested under conditions of plane strain.

It has been preferred here to define the objective function directly in terms of the differences between corresponding points on the calculated and experimental paths, using experimental values of one set of quantities to control the prediction. For example, in stress driven paths the calculated path is forced to pass through all the stress points, and the objective function is simply the sum of the squares of the strain differences.

Both these objective functions have the disadvantage that sections of the path with widely spaced experimental points will receive less weighting than sections with many points. This could be overcome by weighting each increment of the objective function by some measure of the distance between adjacent points on the experimental path.

The program requires a file of the experimental data points to be fitted; a file containing the control path which provides input for the prediction; and a file which specifies the lower and upper bounds to the n parameters, initial values of these parameters, and an indication of whether each parameter is allowed to be varied as part of the optimisation procedure.

The Cam clay model is most conveniently described in terms of the strain response to changes in effective stress. The model is complete in the sense that it is able to make predictions of response in all regions of strain space — including independent variation of all three principal strains, rotation of all three principal axes. However, the structure of the model implies that not all changes in *stress* are permissible. Any attempt to cause plastic deformations with stress ratio $\eta > M$ leads to collapse of the yield surface: the soil cannot support outward stress increments, and a section of stress space (which depends on the current size of the yield surface) is thus inaccessible. It is therefore preferable to use as the control path the observed strain path even where, as for the true triaxial tests used here, the test has been conceived as a stress-controlled test, because while every strain increment implies a corresponding stress increment the converse is not always true.

Results

Although the primary objective is to improve the prediction of the model during the shear stage FG, it is of interest to observe how the optimisation procedure attempts to cope with other stages of the test. The volumetric data (specific volume and mean effective stress) from the initial anisotropic consolidation OA have been used to obtain a value for $\lambda = 0.245$. The optimisation program prefers a slightly higher value $\lambda = 0.272$, partly because the definition of objective function F implicitly gives greater weight to the data at higher stresses. The optimisation procedure for this stage can also present an opinion on the values of the other parameters because these control the link between stress ratio η

and ratio of distortional to volumetric strain $\delta\varepsilon_q/\delta\varepsilon_p$. Cam clay is not very good at getting this link correct: Muir Wood (1990) notes that Cam clay tends to predict values of earth pressure coefficient at rest K_0 which lie above Jaky's (1948) empirical expression

$$K_0 = 1 - \sin \phi' \tag{9}$$

unless simultaneous low values of both ν and $\Lambda = (\lambda - \kappa)/\lambda$ are assumed, implying dominance of the deformation by low Poisson's ratio elastic response. The optimisation program suggests $\nu = 0.37$ but $\kappa = 0.26$, implying $\Lambda = 0.04$, and $M = 0.86$.

The procedure can also be applied to the anisotropic unloading stage. The average value of $\kappa = 0.04$ for this stage is confirmed, but there is a problem with the search for the optimum value of ν (the only other parameter which has any effect during this elastic unloading). A very small positive shear strain was observed during unloading, while the deviator stress q was reducing. This pattern of response cannot be predicted with an isotropic elastic model. The best the program can do is to set $\nu \simeq 0.5$, making the shear modulus as low as possible, so that the predicted stress path shows no change in q.

The cycle of loading and unloading DED, with $p' = 100$ kPa, is expected to be purely elastic according to Cam clay, with the known stress history. The observed, typically hysteretic, shape of the stress - strain response on this cycle (Fig. 4(b)) clearly conflicts with this expectation, and the program, not surprisingly, is not particularly happy in trying to fit the data varying only G, or κ and ν. (Although this is a purely distortional stress path, both κ and ν are required in order to compute the value of the shear modulus.) The objective function F in this case seems to be rather flat and undulating (a Cambridge-like landscape) with a number of false minima: convergences are obtained with $\nu = 0.12$, $\kappa = 0.15$ but also with $\nu = 0$, $\kappa = 0.34$ (Fig. 4(b): line X).

With $M = 0.75$, the size of the yield locus created by the original consolidation OA is $p'_0 = 174$ kPa. If this known history is ignored then the observed behaviour on cycle DED can be better matched with an elastic–plastic Cam clay prediction with $p'_0 = 130$ kPa, and with $G = 3.25$ MPa, $\kappa = 0.20$, $\lambda = 0.26$, $M = 0.68$ (Fig. 4(b): curve Y). This is a more robust minimum to which the optimisation process is able to converge from several different starting points. Whether such a distortion of the actual history would be acceptable from an engineering point of view is a separate issue. Besides, the value of shear modulus that has been selected implies a negative value of Poisson's ratio.

The final shearing FG is best fitted with the set of parameters $G = 4.09$ MPa, $\kappa = 0.35$, $\lambda = 0.62$, $M = 0.82$ (Fig. 4(a): curve B). The optimisation procedure is happy to choose the size of the yield locus at the start of shearing to be $p'_0 = 178$ kPa, which is surprisingly but gratifyingly close to the value $p'_0 = 170$ kPa calculated from the known history with M = 0.82. This is again a rather robustly convergent set of parameters. It might be suggested that the value M = 0.82 gives a truer estimate of the stress ratio towards which the stress:strain response is actually heading. Again the chosen combination of shear modulus and κ implies a negative value of Poisson's ratio. If it is required to restrict the search to $\nu > 0$ then the optimum fit is obtained with $\nu \approx 0.0$, $\kappa = 0.18$, $\lambda = 0.45$, M = 0.82. It is significant that the value of $(\lambda - \kappa)$ has remained almost the same, while the individual values have changed: it is $(\lambda - \kappa)$ that primarily controls the magnitude of the plastic distortional strain increments $\delta\varepsilon_q^P$ (eqn. (2)). It is particularly the value of $\Lambda = 1 - \kappa/\lambda$ that is being pulled down, indicating that improved fitting is obtained by increasing the contribution of the recoverable component of volumetric deformation. A zero or negative value of Poisson's ratio, implying a low ratio K/G

$$K/G = (2/3)(1 + \nu)/(1 - 2\nu) \tag{10}$$

is also apparently beneficial, but the present Cam clay algorithm does not permit negative Poisson's ratio to be specified.

However, if the optimisation procedure is applied only to the initial part of the shearing FG, up to stress ratio $\eta = 0.5$, then the optimum fit

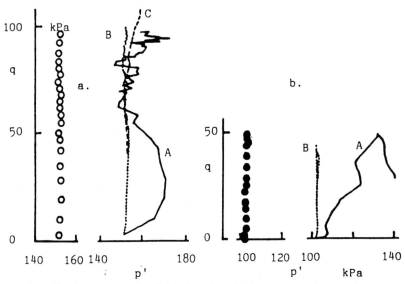

Fig. 7. $p':q$ effective stress paths (a) FG; (b) DED (○, ●: experiment; A: prediction based on visual selection of parameters; B: optimised fits; C: optimised fit to stage FG for $0 < \eta < 0.5$)

is obtained with $G = 5.0$ MPa, $\kappa = 0.43$, $\lambda = 0.58$, $M = 0.78$ and $p'_0 = 172$ kPa (Fig. 4(a): curve C) (or $\nu \simeq 0.0$, $\kappa = 0.13$, $\lambda = 0.28$, $M = 0.80$, again with the same difference $(\lambda - \kappa)$ being chosen) — but these sets of parameters give a worse overall fit to the data of the whole shearing stage. Clearly the choice of parameters can be tuned to match the data over a selected range of interest. Both curves B and C in Fig. 4(a) provide a major improvement over the prediction based on the visual, stepwise selection of parameters (curve A).

The volumetric response has not been mentioned so far. The strain path is used as input to control the prediction; the success of the volumetric response can be judged by comparing the predicted stress paths with the (applied) constant mean stress paths (Fig. 7(a), (b)). In detail these paths are of course very sensitive to erratic changes in direction of the (experimentally derived) strain paths, particularly with the visual stepwise selected parameters (curves A). The optimisation procedure is very successful in matching the actual stress changes (curves B).

Discussion and conclusions

The stress path method (Lambe, 1967) seeks to encourage engineers to match laboratory and field stress paths in order to be able to estimate field deformations in a more rational way. Estimation of field deformations is more readily achieved by numerical analysis than by hand calculation, and consideration of stress paths encourages engineers to be aware of the nature of the extrapolation that is implied in the numerical predictions (Wood, 1984). It is a logical extension then to encourage engineers to make their numerical models match the experimental data over the ranges of stress or strain changes that are actually expected to be important. Clearly this will often be an iterative process, with stress paths that emerge from numerical analyses performed for working loads being used to define the range of stress in laboratory tests over which the optimum set of soil parameters should be assessed.

The possibilities of optimisation in parameter selection have been presented here for just one test, to illustrate some of the problems that may emerge. It would normally be preferable to combine data from several tests, either repeating the response on a single path, to provide some information about reliability of experimental data, or in order to increase the volume of relevant stress hyperspace over which data have been gathered. Different tests can be assigned different weights in the optimisation procedure in order to reflect assessments of test quality or relevance to the particular prototype problem.

It will be noted that no suggested optimum values of Γ or N have been quoted. The Cam clay algorithm used here is typical of those used with finite element programs (see, for example, Britto and Gunn (1987)) in that it expects the initial size p'_{0i} of the yield surface to be specified at the same time as the initial effective stresses. The initial specific volume v_i is required in order that strain increments may be calculated from (1)–(4) but the link between v_i, p'_{0i} and initial mean effective stress p'_i through N, λ and κ is not forced:

$$v_i = N - \lambda \ln p'_{0i} + \kappa \ln (p'_{0i}/p') \qquad (11)$$

the value of p'_{0i} becomes an optimisation variable, whereas comparison and combination of tests with different consolidation histories requires that N or Γ be used instead. This merely requires a minor program modification.

It would of course be quite unwise to use such an optimisation procedure without interaction with an informed user. The process cannot be allowed to become a 'black box'. The user needs to ensure that the parameters that are chosen are indeed reasonable, and needs to be intelligent in choosing data which cover the appropriate stress level and stress and strain ranges. The objective function for a model like Cam clay has many minima, and it is clearly sensible to seed the optimisation process with initial values which have been deduced from visual interpretation of the experimental observations in the traditional manner.

Nevertheless, releasing the Cam clay parameters from their physical origins, and concentrating the prediction on stress changes of prototype interest, improves the performance of the model. It remains a simple model, requiring a small number of soil parameters, and it may be preferable to tune it to give a good local prediction of response, rather than to tune it to the global response of the soil and still to expect it to perform well locally.

Acknowledgements

Some of the work described here was performed by the second author with support from the Science and Engineering Research Council under grant GR/E84167.

References

AIREY, D.W. AND WOOD, D.M. (1988). The Cambridge true triaxial apparatus. In Advanced triaxial testing of soil and rock (eds. R.T. Donaghe, R.C. Chaney and M.L. Silver) ASTM, STP977, pp. 796–805.

AL-TABBAA, A. AND WOOD, D.M. (1989). An experimentally based 'bubble' model for clay. Numerical models in geomechanics NUMOG III (eds. S. Pietruszczak and G.N. Pande) Elsevier Applied Science, pp. 91–99.

BRITTO, A.M. AND GUNN, M.J. (1987). Critical state soil mechanics via finite elements. Ellis Horwood.
BUTTERFIELD, R. (1979). A natural compression law for soils (an advance on e-log p'). Géotechnique, Vol. 29, No. 4, pp. 469–480.
CALLADINE, C.R. (1963). The yielding of clay. Géotechnique, Vol. 13, No. 3, pp. 250–255.
GRAHAM, J. AND HOULSBY, G.T. (1983). Elastic anisotropy of a natural clay. Géotechnique, Vol. 33, No. 2, pp. 165–180.
JAKY, J. (1948). Pressure in silos. Proc. 2nd ICSMFE, Rotterdam Vol. 1, pp. 103–107.
KLISINSKI, M. (1987). Optimisation program for identification of constitutive parameters. Structural Research Series No. 8707, Department of Civil, Environmental, and Architectural Engineering, University of Colorado, Boulder.
LAMBE, T.W. (1967). Stress path method. Proc. ASCE, Vol. 93, SM6, pp. 309–331.
MOURATIDIS, A. AND MAGNAN, J.-P. (1983). Modèle élastoplastique anisotrope avec écrouissage pour le calcul des ouvrages sur sols compressibles. Rapport de recherche, Laboratoires des Ponts et Chaussées No. 121.
MUIR WOOD, D. (1990). Soil behaviour and critical state soil mechanics. Cambridge University Press.
ROSCOE, K.H. AND BURLAND, J.B. (1968). On the generalised stress-strain behaviour of 'wet' clay, in Engineering plasticity (eds. J. Heyman and F.A. Leckie) Cambridge University Press, pp. 535–609.
ROSCOE, K.H. AND SCHOFIELD, A.N. (1963). Mechanical behaviour of an idealised 'wet' clay. Proc. 2nd Eur. Conf. SMFE, Wiesbaden 1, pp. 47–54.
ROSENBROCK, H.H. (1960). An automatic method for finding the greatest or least value of a function. The Computer Journal, Vol. 3, pp. 175–184.
SCHOFIELD, A.N. AND WROTH, C.P. (1968). Critical state soil mechanics. McGraw-Hill.
WOOD, D.M. (1974). Some aspects of the mechanical behaviour of kaolin under truly triaxial conditions of stress and strain. PhD thesis, Cambridge University.
WOOD, D.M. (1984). Choice of models for geotechnical predictions. In Mechanics of engineering materials (eds. C.S. Desai and R.H. Gallagher) John Wiley, pp. 633–654.
WOOD, D.M. AND WROTH, C.P. (1972). Truly triaxial shear testing of soils at Cambridge. Proc. Int. Symp., The deformation and the rupture of solids subjected to multiaxial stresses, Cannes, RILEM Vol. 2, pp. 191–205.
WROTH, C.P. (1977). The predicted performance of a soft clay under a

trial embankment loading based on the Cam clay model. Finite elements in geomechanics (ed. G. Gudehus) J. Wiley, pp. 191–208.

WROTH, C.P. AND WOOD, D.M. (1975). Critical state soil mechanics. Lecture notes for short course given at Chalmers Tekniska Högskola, Göteborg.

WROTH, C.P., WOOD, D.M. AND HOULSBY, G.T. (1981). Critical state soil mechanics. Lecture notes for short course given at Oxford University.

WROTH, C.P., WOOD, D.M. AND HOULSBY, G.T. (1985). Critical state soil mechanics. Lecture notes for short course given at Cambridge University.

WROTH, C.P., WOOD, D.M., HOULSBY, G.T. AND BROWN, S.F. (1982). Critical state soil mechanics. Lecture notes for short course given at Nottingham University.

WROTH, C.P., WOOD, D.M. AND STEENFELT, J. (1979). Critical state soil mechanics. Lecture notes for short course given to Dansk Geoteknisk Forening, Lyngby.

ZYTYNSKI, M., RANDOLPH, M.F., NOVA, R. AND WROTH, C.P. (1978). On modelling the unloading–reloading behaviour of soils. Int. J. for Numerical and Analytical Methods in Geomechanics, Vol. 2, pp. 87–94.

Use of field vane test data in analysis of soft clay foundations

H. OHTA, Kanazawa University, A. NISHIHARA, Fukuyama University, A. IIZUKA, Kanazawa University and Y. MORITA, Kiso-Jiban Consultants, Japan

The use of field vane test data is introduced in specifying the material parameters needed in the Cam clay type of constitutive model. Data from unconfined compression tests are also utilized. Some theoretical reasoning derived from constitutive models is used in the interpretation of the test data. Unconfined compression strength of clay is converted into the undrained compression strength which should be obtained from K_0 consolidated or isotropically consolidated undrained triaxial compression tests. Conversion is made by applying four kinds of correction factor to s_u values. Data from four series of field vane tests give the undrained vane shear strength mobilized on the vertical and horizontal planes as well as the ratio of vertical strength to horizontal strength. From these values, material parameters K_0, M and C_s/C_c are estimated. Thus specified parameters are found to be in good agreement with the failure of a slope and with the settlement of an embankment. This leads to the conclusion that the use of field vane test data in specifying the material parameters is promising from the viewpoint of engineering practice.

Introduction

In the interpretation of any of the laboratory tests for undrained strength, the following 5 effects should be carefully considered:

(*a*) stress release (effective stress change caused by taking the sample from the ground to the surface)
(*b*) sample disturbance during handling and trimming procedures
(*c*) shearing (or loading) rate relative to the shearing rate observed in the actual failure of the soft foundations
(*d*) incompletely undrained conditions (partial drainage or suction)
(*e*) anisotropy of undrained strength (the undrained strength of anisotropically consolidated clay is a function of the principal stress direction and therefore of the inclination of the slip surface; this requires a correction factor to convert the strength obtained from the lab test into the average strength (design strength) which is expected to be mobilized along the whole portion of the slip surface in the field).

Table 1. Undrained strength for various testing methods (after Ohta, Nishihara and Morita (1985), test data reported by Ladd (1973))

$$\frac{S_u}{\sigma'_{vi}} = \frac{OCR^\Lambda(1+2K_0) M \exp(-\Lambda)}{3\sqrt{3}(\cosh\beta - \sinh\beta\cos2\theta)} \quad (1)$$

$$\frac{S}{\sigma'_{vi}} = \frac{OCR^\Lambda(1+2K_0) M \exp(-\Lambda)}{3\sqrt{3}(\sqrt{\cosh^2\beta - \sinh^2\beta\cos^22\omega} - \sinh\beta\sin2\omega)} \quad (2)$$

$$M = \frac{6\sin\phi'}{3-\sin\phi'} \qquad \Lambda = 1 - C_s/C_c \qquad n_0 = \frac{3(1-K_0)}{1+2K_0} \qquad \beta = \frac{\sqrt{3}n_0\Lambda}{2M}$$

Type of test	Reduced equation for specified test on normally consolidated clay	Blue marine clay PI=20 φ'=33° Ko=0.5	
		measured	predicted
(A)	Ko-consolidated Plane strain Comp. KoPUC $\frac{S_u}{\sigma'_{vo}} = \frac{(1+2K_0) M \exp(-\Lambda)}{3\sqrt{3}(\cosh\beta - \sinh\beta)}$	0.34	0.347
	Ko-consolidated Triaxial Comp. KoUC $\frac{S_u}{\sigma'_{vo}} = \frac{1+2K_0}{6} M \exp(\frac{\Lambda n_0}{M} - \Lambda)$	0.33	0.318
	Shear Box Test SBT $\frac{S}{\sigma'_{vo}} = \frac{(1+2K_0) M \exp(-\Lambda)}{3\sqrt{3}}$	—	0.239
(D)	Direct Simple Shear DSS $\frac{S}{\sigma'_{vo}} = \frac{(1+2K_0) M \exp(-\Lambda)}{3\sqrt{3}\cosh\beta}$	0.20	0.224
	Ko-consolidated Plane strain Ext. KoPUE $\frac{S_u}{\sigma'_{vo}} = \frac{(1+2K_0) M \exp(-\Lambda)}{3\sqrt{3}(\cosh\beta + \sinh\beta)}$	0.19	0.165
(P)	Ko-consolidated Triaxial Ext. KoUE $\frac{S_u}{\sigma'_{vo}} = \frac{1+2K_0}{6} M \exp(-\frac{\Lambda n_0}{M} - \Lambda)$	0.155	0.135
	Field Vane FV $\frac{S_h}{\sigma'_{vo}} = \frac{(1+2K_0) M \exp(-\Lambda)}{3\sqrt{3}}$ $\frac{S_v}{\sigma'_{vo}} = \frac{1+2K_0}{3\sqrt{3}} \sqrt{(\frac{MP}{\Lambda P_0} \ln\frac{P}{P_0})^2 - (1-\frac{P}{P_0})^2 n_0^2}$	0.19	0.182

ANALYSIS OF SOFT CLAY FOUNDATIONS

When in situ tests for undrained strength are to be interpreted, care should be taken in correcting the effects of

(a) disturbance caused by the placement of the measuring device in the clay deposit
(b) shearing rate effect
(c) partial drainage
(d) failure mode/anisotropy of undrained strength (in situ tests are boundary value problems, therefore the value of test data obtained should properly be converted into the 'design strength' taking into account both the failure mode seen in an in situ test and the geometry of the slip surfaces expected to exist in the failure of foundations).

Some work has been done on the pressuremeter test, for instance, by Fukagawa et al. (1990) and Ohta et al. (1991). They concluded that the in situ pressuremeter test should be carried out much faster than it is currently performed.

This paper presents a method for correcting the field vane strengths to convert them into theoretically interpretable values from which soil parameters needed in the constitutive model can be derived. The deformation and stability of soft clay foundations which were observed in the field will then be successfully analysed by employing the obtained parameters.

Unconfined compression strength

In order to introduce the concept of how to correct the experimentally obtained strength in the filed, let us briefly review the correction procedure for unconfined compression strength in laboratory tests. Eqns. (1) and (2) in Table 1 give the undrained strength of anisotropically consolidated clay in laboratory tests as a function either of the principal stress direction or of the inclination of the slip surface; see Ohta, Nishihara and Morita (1985). These equations are reduced to the equations shown in the lower rows in Table 1, when the stress and geometrical boundaries which are applied represent each of the listed testing methods. It is assumed, for simplicity, that the overconsolidation ratio is unity. In the right hand columns, the experimentally obtained values of undrained strength of Boston Blue Clay reported by Ladd (1973) are compared with values calculated by using the equations listed in the central column. Theoretical values are in good agreement with experimental values; it was not possible intentionally to get close agreement between the experimental and theoretical values, since all the parameters used in the calculation were specified based on the parameters reported by Ladd.

The constitutive model used in deriving the theoretical equations listed in Table 1 is the one proposed by Sekiguchi and Ohta (1977). The original form of this constitutive model was developed by Shibata (1963). Shibata's model was then expanded and improved by Hata and Ohta (1969), Ohta (1971) and Sekiguchi and Ohta (1977). They took into consideration some of the important findings obtained by Shibata and Karube (1965) and Karube and Kurihara (1966) through a series of very careful experiments on the dilatancy of clays. The development of the constitutive model introduced above went forward along the lines of a study in which the major object was the dilatancy characteristics of clays.

It should be mentioned that this model happened to be essentially identical in its mathematical form with the Cam Clay model proposed by Roscoe, Schofield and Thurairajah (1963) although the basic assumptions employed in developing the model were totally different from the assumptions used in the latter model. The Cam Clay model, as is widely known, is based on an assumption about the energy dissipated during shear, while the model employed here is based on an empirical equation of dilatancy originally found by Shibata through a series of experiments in which clay specimens were sheared keeping the effective mean principal stress constant during shear. The model proposed by Sekiguchi and Ohta (1977) may be distinguished from the Original Cam Clay model when it is applied to K_0-consolidated clays, see Ohta and Sekiguchi (1979).

As seen in Table 1, the undrained strength of K_0-consolidated clay is a function of the principal stress direction θ and hence of the inclination ω of the slip surface. This means that the undrained strength mobilized along the slip surface depends on the inclination of the slip surface. The undrained strength mobilized on a fraction of the slip surface changes its value from fraction to fraction depending on the inclination of the fraction. If the geometry of the slip surface is specified, it is possible to derive the theoretical value of average strength (design strength) mobilized along the slip surface. In the case of a circular slip surface which has right–left symmetry about a vertical line (such as a circular slip surface caused by a load applied on a part of the horizontal ground surface), the ratio of undrained design strength to pre-consolidation pressure s_u/σ'_{v0} ranges in a very narrow band between 0.24 and 0.26 depending on the value of plasticity index, see Ohta, Nishihara and Morita (1985). In calculating the value of the theoretically expected 'design strength', the material parameters M, Λ and K_0, are estimated from the plasticity index by using widely accepted empirical equations which were summarized by Iizuka and Ohta (1987).

Suppose we have the data of undrained strength of undisturbed clay specimens measured in K_0-consolidated undrained triaxial compression tests. Suppose the loading rate during the tests was so slow that there is no need to apply the correction factor μ_R for rate effect. If we know the

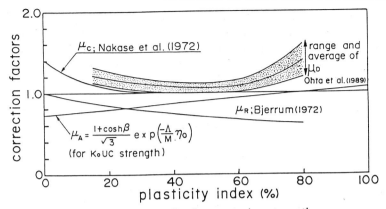

Fig. 1. *Correction factors for unconfined compression strength*

material parameters M, Λ and K_0, it is possible to calculate the ratio of 'design strength' to K_0UC strength. This ratio μ_A should be applied to the K_0UC strength when we need to convert K_0UC strength to the 'design strength'. The correction factor μ_A for strength anisotropy thus obtained is plotted against the plasticity index in Fig. 1, see Ohta et al. (1989). The correction factor for stress release (μ_S) was theoretically found to be unity ($\mu_s = 1$). The correction factors for rate effect (μ_R) and for confining pressure (for incomplete drainage, for partial suction) (μ_C) are also plotted in Fig. 1. These two correction factors are proposed by Bjerrum (1972) and by Nakase, Katsuno and Kobayashi (1972). The use of Bjerrum's correction factor μ_R in correcting the unconfined compression strength may be justified since the time required to achieve the failure state is more or less the same in the field vane tests as in the unconfined compression tests. A band in Fig. 1 indicates the probable range of correction factor for sample disturbance (μ_D). It was estimated by Ohta et al. (1989) based on the experimentally obtained negative pore water pressure remaining in the undisturbed samples collected from 22 sites distributed all over Japan.

Suppose we have data of the unconfined compression test on undisturbed soft clay samples. By multiplying $\mu_S \times \mu_R \times \mu_C \times \mu_D$ by half of the unconfined compression strength $q_u/2$, we get the value of K_0UC strength converted from $q_u/2$. Since we have the theoretical equation for K_0UC strength as seen in Table 1, the converted value of K_0UC strength will provide the information about the material parameters. This implies that q_u values can be a source of information useful in estimating the material parameters needed in the constitutive model of the Cam Clay type. If we need the 'design strength' for symmetric circular slip surfaces, the correction factor for strength anisotropy μ_A should further be multiplied to the K_0UC strength converted from $q_u/2$.

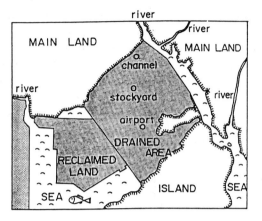

Fig. 2. Sites investigated

The aim of this paper is to show that a principle similar to that introduced above can be applied to the in situ vane test data.

Sites investigated and soil properties

An area of the shallow inland sea located in the western part (called Kasaoka) of Japan was drained during a period from March 1976 to August 1977. The drained area was 861 ha; the sea bed was covered by soft marine clay. From the geotechnical viewpoint, there were three

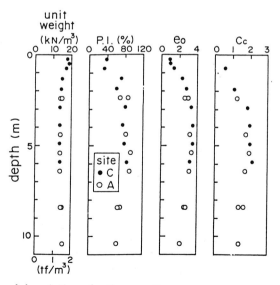

Fig. 3. General description of soil properties

interesting sites in the drained area. The slope of a drainage channel failed during the excavation work in the summer of 1989, site C. A trial embankment was placed as a pre-load for the stockyard facility in 1991, site S. A small airport with an 800 m runway was completed in October 1991, site A. The relative location of the sites is shown in Fig. 2. Since the soft clay was uniform, the subsoil conditions at these three sites were more or less the same. Figure 3 shows the vertical profile of the soft clay layer.

A series of very careful field vane tests were carried out at site C. Dilatometer tests and pressuremeter tests of pre-boring type were also performed at site C. Undisturbed samples were taken from sites A and C, and subjected to unconfined compression tests. In the far left of Fig. 4, the undrained strength obtained from the standard vane (vane height/vane width = H/B = 2) tests is plotted. Since vane torque T is interpreted by using the equation

$$\frac{2T}{\pi B^3} = \frac{H}{B} s_v + \frac{1}{3} s_h,$$

the undrained strength s_v on a vertical–cylindrical side plane can be separated from the undrained strength s_h on horizontal top and bottom planes provided a series of vane test data with different H/B are available (Cading and Odenstad, 1950, see Fig. 7). Values of s_v and s_h are plotted in the middle-left of Fig. 4.

Half of the unconfined compression strength $q_u/2$ obtained at sites A and C is shown in the middle-right of Fig. 4. From $q_u/2$, we get $s_u(K_0UC)$

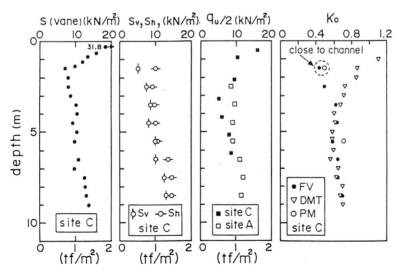

Fig. 4. Vane strength, unconfined strength and K_0-value

through multiplying by correction factors. The ratio $s_u(K_0UC)/s_u(IUC)$ is calculated using the theoretical equation for $s_u(K_0UC)$ in Table 1 as a function of K_0 ($K_0 = 1$ for IUC). Karube's empirical relation $M = 1.75 \Lambda$ (Karube, 1975), is used in the calculation. It is interesting to see that this calculated ratio $s_u(K_0UC)/s_u(IUC)$ is very close to unity (from 0.989 to 1.023 for K_0 values ranging between 0.5 and 1.0). This is probably the reason why the isotropically consolidated undrained compression test could be substituted for the K_0-consolidated undrained compression test in estimating the undrained strength of K_0-consolidated clays.

It now becomes clear that $s_u(IUC)$ is also estimated from $q_u/2$. Since $(s_h/\sigma'_{v0})/(s_u(IUC)/\sigma'_{v0}) = 2(1+2K_0)/3\sqrt{3}$ (see Table 1, and recall that $K_0 = 1$ for IUC) and also since we now have the values both of s_h and $s_u(IUC)$, we can calculate the value of K_0 as shown in the far-right of Fig. 4. As seen in the figure, these estimated K_0 values are in surprisingly good agreement with the K_0 values obtained from dilatometer tests and pressuremeter tests, with the exception of two data points circled by dotted curve. These are the data measured 15 months after failure at a location very close (2 m) to the failure surface in the field, and hence are very likely to be affected by the lateral stress release due to the preceding failure of the slope. It should be noted here that the correction factor for the rate effect is cancelled when we calculate the ratio $s_h/s_u(IUC)$, since both the vane shear test and the unconfined

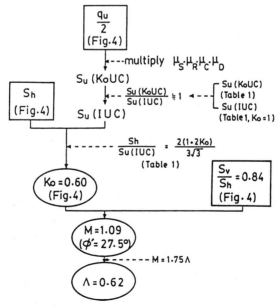

Fig. 5. Use of field vane and unconfined compression strength

compression test are performed within the same order of time period and hence the correction factor for the rate effect may be the same.

It is believed that the deepest part of the slip surface of the channel slope failure reaches a depth of about 5 m. The average of the K_0 values (shown by solid circles in Fig. 4) obtained by the use of field vane tests in the middle of the soft clay layer is 0.60 ($K_0 = 0.60$). It is also possible to calculate the value of M from the ratio s_v/s_h together with $K_0 = 0.60$ and Karube's relation $M = 1.75 \Lambda$, although the procedure is not simple since the value of the effective mean principal stress p at failure in the vane test has to be calculated, see Table 1. The value of M obtained is 1.09 ($M = 1.09$, $\phi' = 27.5°$). The procedure mentioned above is summarized in Fig. 5.

Initial stress state and rate effect

In analysing the behaviour of soft clay foundations, it is essential to estimate the in situ effective stress state prior to the construction works. Figure 6 shows the effective overburden pressure before and long after the draining of the sea bed. The experimentally obtained effective vertical stresses prior to the placement of embankment and excavation of the channel are plotted. The effective stress state indicates that the soft clay ground is still consolidating. The dotted curve represents the effective vertical stress distribution computed by one dimensional FEM using the soil parameters eventually specified through the process

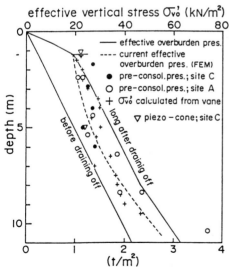

Fig. 6. Effective vertical stress prior to embankment and/or excavation

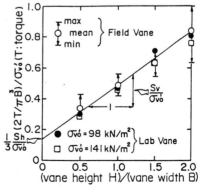

Fig. 7. Separation of strength on vertical and horizontal planes

introduced in this paper. Since all the data points are located around the dotted curve, the material parameters used in the analysis seem to be reasonable. In order to confirm the appropriateness of the material parameter specification, the authors tried to compare the computed rate of settlement of the sea bed after the draining with the monitored settlement. However, unfortunately, there was no settlement record taken after the completion of draining.

Figure 7 shows the results both of lab and field vane tests. The data points represented by solid circles and open squares are the results of the lab vane tests on one-dimensionally reconsolidated clay specimens. The arrows indicating the range of values with the mean value represented by open circles are the data obtained from the field vane. Since the ground is still under consolidation, the effective vertical stress σ'_{v0} is not clearly known. The vane torque T on the vertical axis of Fig. 7 is divided by the σ'_{v0} obtained by 1-D FEM (dotted curve in Fig. 6). Together with the lab vane data, the open circles more or less overlap the lab vane data and form a straight line, as they should, implying that the effective vertical stress σ'_{v0} estimated by FEM was close to the actual value.

Figure 8 shows how the undrained strength experimentally obtained by vane and share box tests is reduced as the rate of shear diminishes (time to failure increases). The data points shown by the open circles and open squares represent the strength of remoulded-reconsolidated samples. As pointed out by Torstensson (1977), the time to failure in the field is generally of the order of 1000 times longer than the time to failure either in vane tests or in shear box tests. The experimentally obtained value of s(vane)/σ'_{v0} (field vane) at time to failure of 10 min is 0.36, while the value of s(vane)/σ'_{v0} at time to failure of 10 000 min is estimated by extrapolation of the experimental data as 0.22. This makes

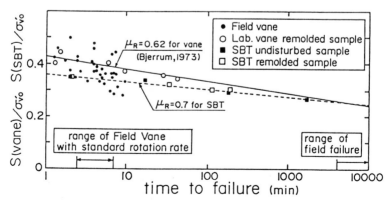

Fig. 8. Rate effects in vane and shear box tests

the correction factor for rate effect to be 0.61. Since Bjerrum's correction factor is 0.62 (for clays of PI = 80%) and is in agreement with the experimental results as seen in Fig. 8, let us use Bjerrum's factor in this investigation. In Fig. 8, the rate effect in the shear box test is also shown in comparison with Bjerrum's factor.

Channel slope slide

At site C, a slide took place during the excavation work of a channel as shown in Fig. 9. It is reported that the slope failed when the excavation depth reached about 2 m, about a week after they started to excavate.

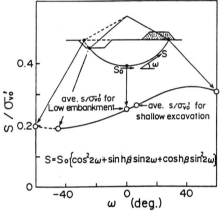

Fig. 9. Average strength for the channel slope slide

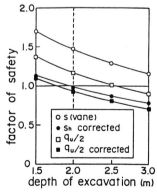

Fig. 10. Factor of safety for the channel slope failure

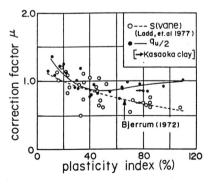

Fig. 11. Correction factor back-calculated from the actual failure records

SETTLEMENT OF PRELOAD

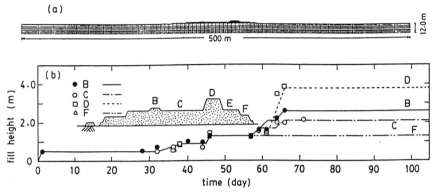

Fig. 12. Finite element model and loading rate

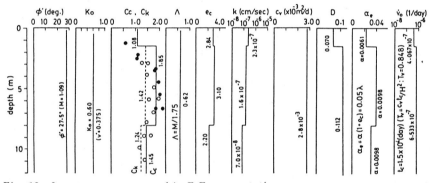

Fig. 13. Input parameters used in F.E. computation

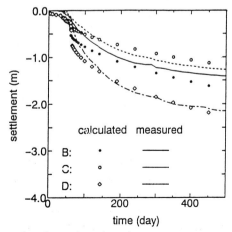

Fig. 14. Computed and monitored settlement

ANALYSIS OF SOFT CLAY FOUNDATIONS

Eqn. (2) in Table 1 is rewritten in a form shown in Fig. 9. The shear strength s_0 along the horizontal plane ($\omega = 0$) is theoretically derived as the average strength for a circular slip surface generated by placing a low embankment, see Ohta, Nishihara and Morita (1985). However, in the case shown in Fig. 9, the average strength mobilized along the slip surface is slightly higher than s_0 because of the lack of slip surface portion shown by the dotted curve in Fig. 9.

The factor of safety calculated from the strength based either on the field vane tests or on the unconfined compression tests is shown in Fig. 10. Apparently the use of raw data of field vane and unconfined compression strength (open circles and open squares) leads to a factor of safety higher than 1.0 at the excavation depth of 2 m indicating that some correction has to be applied to the raw data of strength. The use of strength corrected in a way explained in the previous sections (solid circles and solid squares) seems to give reasonable factors of safety. It should be noted here that the s_h from field vane tests is equal to the strength obtained by shear box tests (see Table 1) and again is equal to s_0 in Fig. 9. Since the correction is consistent with the constitutive model used and the parameters obtained, it may be concluded that the whole story introduced in this paper is in accordance with the observed failure. The critical slip surfaces obtained by the stability analysis are found to be very close to the circle shown in Fig. 9. This is also in accordance with the location of the observed slip surface.

Figure 11 shows the correction factors back-calculated from the observed failure records. Open circles (Ladd et al., 1977) are case records of the use of vane tests, while solid circles are case records of the use of unconfined compression strength collected by the authors. The dotted curve is the curve proposed by Bjerrum (1972) and the solid curve is drawn by the authors in such a way that it passes through the middle of the data points shown by solid circles. Therefore, this curve is only an empirical curve and is not the curve theoretically derived. Ohta et al. (1989) obtained the theoretical curve which locates in the position lower than the solid curve shown in Fig. 11 by an amount 0.1–0.2 of the correction factor. The failure at site C in Kasaoka is represented in Fig. 11 by the data points specified by an arrow indicating that the failure at site C was just one of the common failures.

Settlement of preload

At site S, an embankment was placed as a preload to strengthen the soft clay for the stockyard facilities. The height of the fill was different from place to place as shown in Fig. 12(*a*). Since the loading rate was also complicated as seen in Fig. 12(*b*), the deformation of the soft clay foundation was not uniform. This two dimensional consolidation process is analysed by employing a soil–water coupling finite element programme called DACSAR (Deformation Analysis Considering Stress Anisotropy and Reorientation). The programme was developed by Iizuka and Ohta (1987) using an elasto-viscoplastic constitutive model proposed by Sekiguchi and Ohta (1977). The material parameters are those obtained in the previous sections from field vane tests and unconfined compression tests. The results of oedometer tests are also used. In specifying all the material parameters needed in the analysis, some empirical procedures recommended by Iizuka and Ohta (1987) are also used. The vertical distributions of material parameters are summarized in Fig. 13. The permeability of the soft clay immediately under the embankment was improved by the placing of sand drains, the effect of which is evaluated in the computation by assuming a permeability 20 times higher than the adjacent natural clay.

The settlement records of the different part of the embankment are compared in Fig. 14 with those computed by the DACSAR programme. The computed results seem to be in accordance with the settlements monitored at the different part of the fill. This implies that the two dimensional simulation of consolidation is successfully performed by using the estimated material parameters, mostly specified based on the field vane and unconfined compression strength. In judging whether the computer simulation was successful or not, we would wish to see the lateral movement of the soil as well as the change in pore water pressure. However, regrettably, those data were not available to the authors and could not be compared with the computed results. Since the authors did not make any efforts to modify the material parameters (except the overall permeability of the sand drained area), the agreement seen in Fig. 14 may be seen as a positive support for the use of field vane in estimating the material parameters needed in the constitutive model.

Conclusions

The use of field vane test data in specifying the material parameters needed in the Cam Clay type of constitutive model was introduced in this paper. In the interpretation of the test data, some theoretical reasoning derived from the constitutive model was used. Material parameters thus specified were found to be in good agreement with the failure of a slope and with the settlement of an embankment. This leads to the conclusion that the use of field vane test data in specifying the material parameters is promising from the viewpoint of engineering practice. It may also be concluded that the use of a constitutive model has reached the level of practical applications. In this paper, the effect of partial drainage during the field vane tests was not considered since the tests are usually carried out very quickly. The effect of insertion of the vane blade into the soft clay ground was also ignored in this paper. Whether this is appropriate or not is still open to question.

References

BJERRUM, L. (1972). Embankments on the soft ground, Performance of Earth and Earth-Supported Structures, ASCE Specialty Conference, Vol. 2, pp. 1–54.

FUKAGAWA, R., FAHEY, M. AND OHTA, H. (1990). Effect of partial drainage on pressuremeter test in clay, Soils and Foundations, Vol. 30, No. 4, pp. 134–146.

HATA, S. AND OHTA, H. (1969). On the effective stress paths of normally consolidated clays under undrained shear, Proc. JSCE, No. 162, pp. 21–29 (in Japanese).

IIZUKA, A. AND OHTA, H. (1987). A determination procedure of input parameters in elasto-viscoplastic finite element analysis, Soils and Foundations, Vol. 27, No. 3, pp. 71–87.

KARUBE, D. (1975). Non-standardized triaxial testing methods and their problems, Proc. 20th Symposium on Soil Engineering, JSSMFE, pp. 45–60 (in Japanese).

KARUBE, D. AND KURIHARA, N. (1966). Dilatancy and shear strength of saturated remoulded clay, Trans. JSCE, No. 135.

LADD, C.C. (1973). Discussion, Main Session 4, Proc. 8th ICSMFE, Moscow, Vol. 4.2, pp. 108–115.

LADD, C.C., FOOTT, R., ISHIHARA, K., SCHLOSSER, F. AND POULOS, H.G. (1977). Stress-deformation and strength characteristics, SOA Report, Proc. 9th ICSMFE, Tokyo, Vol. 2, pp. 421–494.

NAKASE, A., KATSUNO, M. AND KOBAYASHI, M. (1971). Unconfined compression strength of sandy clay, Report of Port and Harbour Research Institute, Vol. 11, No. 4, pp. 83–102 (in Japanese).

OHTA, H. (1971). Analysis of deformation of soils based on the theory of plasticity and its application to settlement of embankments, D. Eng. Thesis, Kyoto University.

OHTA, H., IIZUKA, A., NISHIHARA, A., FUKAGAWA, R. AND MORITA, Y. (1991). Design strength Su derived from pressuremeter tests, Proc. 7th Int. Conf. Computer Methods and Advances in Geomechanics, Cairns, Vol. 1, pp. 273–278.

OHTA, H., NISHIHARA, A. AND MORITA, Y. (1985). Undrained stability of K_0-consolidated clays, Proc. 11th ICSMFE, San Francisco, Vol. 2, pp. 613–616.

OHTA, H., NISHIHARA, A., IIZUKA, A., MORITA, Y., FUKAGAWA, R. AND ARAI, K. (1989). Unconfined compression strength of soft aged clays, Proc. 12th ICSMFE, Rio de Janeiro, Vol. 1, pp. 71–74.

OHTA, H. AND SEKIGUCHI, H. (1979). Constitutive equations considering anisotropy and stress orientation in clay, Proc. 3rd Int. Conf. Numerical Methods in Geomechanics, Aachen, pp. 475–484.

SEKIGUCHI, H. AND OHTA, H. (1977). Induced anisotropy and time dependency in clays, Proc. 9th ICSMFE, Specialty Session 9, Tokyo, pp. 229–238.

SHIBATA, T. (1963). On the volume change of normally consolidated clays, Disaster Prevention Research Institute Annuals, Kyoto Univ., No. 6, pp. 128–134 (in Japanese).

SHIBATA, T. AND KARUBE, D. (1965). Influence of the variation of the intermediate principal stress on the mechanical properties of normally consolidated clays, Proc. 6th ICSMFE, Montreal, Vol. 1, pp. 359–363.

TORSTENSSON, B.A. (1977). Time-dependent effects in the field vane test, Proc. Int. Symp. on Soft Clay, Bangkok, pp. 387–397.

Linear and nonlinear earthquake site response

M.J. PENDER, University of Auckland

This paper considers linear and nonlinear aspects of the way in which a soil profile modifies, during transmission, an earthquake motion. A review of published site response data reveals that, at the majority of instrumented sites, the soil profile responds in an approximately linear manner. As many instrumented sites are very stiff it is not surprising that they behave linearly. Nevertheless some important indications of nonlinear behaviour have been obtained recently. The paper concludes with a brief discussion of the process of predicting site response.

Introduction

This paper attempts to clarify aspects of the manner in which the soil profile at a given site modifies the incoming rock motion. It goes on to discuss briefly the process of predicting the response of a given site.

Many recent earthquakes, either by direct measurement (Loma Prieta (1989) and Mexico City (1985)) or by inference (Philippines (1990), Newcastle (1989), Armenia (1988)), have confirmed earlier evidence that soil sites will behave differently from rock sites. For small to modest peak accelerations in the incoming rock motion, it is very clear that the soil layers amplify the response. At larger accelerations there is the possibility that the soil will deform in a nonlinear manner. If this happens weak motion data cannot be extrapolated to give strong motion predictions. The question of nonlinear soil behaviour has been controversial. A substantial part of the paper reviews instrumental data on the response of soil sites. There have been some significant developments in recent years.

Site effects and methods of description

Firstly we need a definition of the term, site effect. This imagines that we have adjacent sites one with rock at the surface and the other with a soil profile overlying the rock. If the sites are adjacent we can assume that the incoming rock motion will be the same for both cases. Then any difference between the motions recorded at the two sites can be attributed to the effect of the soil profile at one of the sites. Such differences are referred to as a site effect. The velocities and accelerations at the surface of the soil may be greater or less than those at the surface of the rock; either case is referred to as a site effect. Site effects might also be indicated by differences in frequency content of spectra. Notice that usually we compare motions at the free surface, rather than the rock motion at the base of the soil column with the motion at the surface.

There are various ways in which this difference might be quantified. The simplest is to take the ratio of the peak ground acceleration (PGA) at the surface of the soil profile to that at the surface of the rock site. The Loma Prieta records in the Oakland area provide an example of this type of effect. The PGA at rock sites was of the order of 0.08 g and those at the surface of soil profiles were up to three times greater. In some cases an adjacent outcrop rock motion is not available but instruments are positioned at various depths in the soil profile and possibly in the underlying rock. The ratio of the PGA at the ground surface to that in the underlying rock can be used as a measure of the site effect, but it is not the same as the ratio of the soil and rock site peak ground surface accelerations.

Both of these are single figure indications of site effects. Measures over the range of frequencies can be obtained by taking ratios of response spectra, that is at each period the ratio of the spectral acceleration for the soil surface spectrum is divided by the corresponding value for the response spectrum of the recorded motion at the rock site. A related method is to take Fourier spectral ratios, either velocities or accelerations, rather than ratios of response spectra. Because of the 'spiky' nature of many Fourier spectra smoothing is usually necessary. These spectral ratio comparisons tend to emphasise the features of the soil profile and smooth out the particulars of the incoming rock motion.

Decreasing spectral ratios with increasing input excitation are indicative of nonlinear soil behaviour.

Evidence for site effects in soil profiles

In this section we will review published information about site effects concentrating on instrumental evidence rather than inference from damage patterns.

Soil profiles which exhibit amplification of rock motion
Chusal Valley, Tadzhikistan. Tucker and King (1984) present data gained from an array of instruments in the Chusal Valley. The sediments in the valley are derived from the surrounding mountainous topography and are very stiff with shear wave velocities ranging from 300 m/s at the ground surface, to 500 m/s at 10 m depth and 1000 m/s at 60 m depth. At the location of the array of instruments the valley is about 370 m in width and the maximum depth of the sediments is approximately 60 m.

An array of four weak motion instruments was located across the valley, the first instrument being located on rock at the edge and the fourth at the centre. Subsequently an array of strong motion instruments was positioned at the weak motion sites. In this way records with peak horizontal ground accelerations ranging from 10^{-5} to 0.2 g were recorded. As the spectral ratios for all events were very similar the authors concluded that strong motion behaviour can be inferred from weak motion data.

Coalinga, California. Strong motion data was recorded at two sites near Coalinga during the 1983 earthquake and aftershocks. The peak horizontal ground accelerations recorded ranged between 0.03 g and 0.72 g. In addition 23 weak motion events were recorded in April and May 1985 with two digital seismometers co-located with the strong motion instruments. The site consists of up to 150 m of alluvium underlain by Cenozoic and Cretaceous sedimentary rocks. The alluvium is said to be unsaturated but no shear wave velocity data are given. Figure 1 has spectral ratios, obtained from smoothed Fourier spectral amplitudes, calculated by Jarpe et al. (1988) for the strong and weak motions. For the 1 to 10 Hz frequency band the average strong motion spectral ratios are similar to the average weak motion ratios, although Jarpe et al. noted that scatter for the strong motions is considerably greater than for the weak motions. Consequently the Coalinga sites seem, on average, to have behaved linearly.

McGee Creek, California. At this site in the Mammoth Lakes region

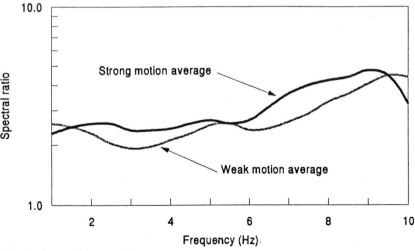

Fig. 1. *Spectral ratios for the Coalinga strong and weak ground motions (after Jarpe et al. (1988))*

strong motion three component accelerometers were located at the surface and at depths of 35 m and 166 m. The site consists of 30 m of glacial till overlying hornfels. The shear wave velocity of the rock is 2800 m/s. The till was represented as two sublayers, the upper 14 m having a velocity of 290 m/s, from 14 to 30 m the velocity is 620 m/s. Seale and Archuleta (1988) report on the response of the instruments to the Chalfont Valley earthquake of 1986 and the Round Valley event of 1984. The PGA recorded at the ground surface in the Chalfont Valley earthquake was 0.10 g and 0.12 g for the Round Valley event. In each case there was an amplification in the range 4 to 5 between the peak acceleration at depth and the ground surface. The authors found that for the Round Valley records the surface response calculated by assuming elastic one-dimensional vertical propagation of the earthquake motion was a good approximation to the recorded surface motion.

Gilroy, California. An array of six recording instruments is maintained across the Santa Clara valley near Gilroy as part of the California Strong Motion Instrumentation Programme (CSMIP). The Gilroy No. 1 instrument is sited on rock and the No. 2 instrument is on stiff alluvium 2 km distant from the No. 1 site. Three strong motion events have been recorded: Coyote Lake in 1979, Morgan Hill in 1984 and Loma Prieta in 1989. The PGAs at the rock site for these events are: 0.11, 0.10 and 0.44 g. The PGAs for the same three events at the alluvium site are: 0.25, 0.21 and 0.35 g. Joyner et al. (1981) have analysed the response to the

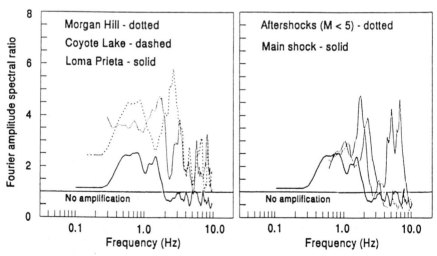

Fig. 2. Spectra for various events recorded at the Gilroy array; (a) mainshock records, (b) Loma Prieta mainshock and aftershock records (after Darragh and Shakal (1991))

1979 Coyote Lake earthquake. By computing the one-dimensional response of the soil site with the rock motion as input they conclude that there is no evidence of nonlinear soil behaviour. Darragh and Shakal (1991) have analysed the data captured during the Coyote Lake, Morgan Hill and Loma Prieta aftershock events. Spectral ratios for these, as well as the mainshocks, are reproduced in Figs. 2(a) and (b). Although there is scatter amongst these ratios it is apparent that they fall in about the same position in the spectral ratio plot so the authors conclude that the behaviour of the alluvium layer at Gilroy No. 2 is linear for incoming rock motions with a PGA up to 0.11 g. (The nonlinear behaviour for the Loma Prieta mainshock with a recorded PGA of 0.44 g is discussed further below.)

Nevada Test site. Hays et al. (1979) discuss recordings of ground motions generated by nuclear explosions at two pairs of sites in Nevada. At the first two sites, separated by 800 m, one being on welded tuff and the other on 100 m of desert alluvium over tuff, two events were recorded with PGAs between 0.03 and 0.10 g. At the second pair of sites 20 m apart, one being on rock and the other having 10 m of alluvium over rock, three events were recorded with PGAs between 0.05 g to 0.54 g. Ratios of the velocity spectra between the site pairs were found to be independent of the PGA in all cases.

Mexico City. The Mexico City response to the 1985 earthquake has been thoroughly investigated. As explained by Whitman (1986) there are several features of the site conditions in Mexico City which are unusual and even unique. One of these is the large strain range over which the volcanic clay behaves in an elastic manner; consequently during the 1985 and other earthquakes the local soils amplified the incoming rock motion elastically (Romo et al., 1989).

Union Bay, Seattle. The Union Bay soil profile, described by Seed and Idriss (1970), is underlain by glacial till at a depth of 33 m. The upper 18 m of the site consists of peat having an undrained shear strength less than 10 kPa. The remainder of the soil profile is clay with strengths rising to about 40 kPa at the top of the till. Recording instruments were located at depths of 3 m, 20 m and 35 m. Accelerations were recorded during a mild earthquake in 1967 and from nuclear blasts in Nevada from 1966 to 1968. The spectral shapes of the recorded rock motions are different as the earthquake was a nearby event and the nuclear explosions distant. For the mild earthquake, which generated a peak horizontal ground acceleration in the till of about 0.002 g, the peak acceleration in the clay was about four times greater, whilst the peak acceleration in the peat was about one quarter of that in the till. On the other hand the nuclear blasts give a typical PGA at the top of the till of about 2×10^{-5} g and this is magnified by a factor of 1.5 to the top of the clay and by a factor of 5.8 to the top of the peat. The large differences between the response of the soil to a nearby earthquake and the distant blast motions are unlikely to be a consequence of nonlinear deformation because the earthquake PGA is only 0.002 g. A more likely explanation is the different frequency characteristics of the two incoming motions as revealed by the different spectral shapes. The earthquake energy was concentrated in the short period range, at periods much less than that of the site. The nuclear blast motions contained longer period energy and had rather flat spectra; such a motion was more likely to excite the natural period of the site and give the large amplification observed.

Oakland and San Francisco. Comparisons between strong motion and nuclear blast data have been made for the San Francisco region. Borcherdt (1991) reports that there is no difference of statistical significance between the average spectral amplification for strong and weak motion data for a range of sites in the San Francisco Bay area. Furthermore the spectral ratios for motions recorded from explosions have many similarities to those from small earthquakes and were found to give a better indication of response than recordings of seismic noise.

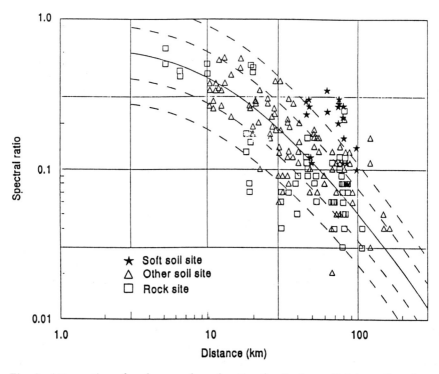

Fig. 3. *Attenuation of peak ground acceleration for the Loma Prieta earthquake plotted with the Seed and Idriss attenuation relationship (after Idriss (1991))*

The greatest influence on the spectral amplification is the site geology, see Borcherdt (1970). An example of this occurs in Fig. 3 which shows how the soft soil sites in Oakland responded differently from other materials during the Loma Prieta event.

Soil profiles which appear not to amplify rock motion
Los Angeles Basin. The San Fernando earthquake that occurred in 1971 resulted in many recorded ground motions. The locations of these instruments covered a range of site conditions. Berrill (1977) analysed data from 71 strong motion accelerograms and concluded that there were no systematic effects of the various site conditions on the recorded motions and in particular that softer sedimentary sites in the Los Angeles region did not on average record significantly stronger or weaker accelerations than the stiffer sedimentary rock sites. It is of interest that Rogers et al. (1984) have extended this work to compare earthquake data with that recorded in the Los Angeles region from nuclear blasts in Nevada. Their conclusion is generally in line with that of Berrill.

Loma Prieta near to source sites. Figure 3 plots the attenuation of PGA with distance from the source of the earthquake. It indicates that the mean and scatter of the PGAs recorded at the soil sites are about the same as those at rock sites at the same distance. This is similar to the conclusion reached by Berrill (1977) for the Los Angeles sites after the San Fernando earthquake, although Berrill looked at complete spectra whereas Fig. 3 is concerned with only one point on the spectrum. The interpretation of Chin and Aki (1991) that this Loma Prieta data can be explained in terms of nonlinear soil behaviour is discussed below.

Sites which behave in a nonlinear manner
In this section sites of liquefaction are not considered as there is no doubt that this involves nonlinear deformation.

The Lotung experimental site, Taiwan. At this site two downhole arrays have been installed at various depths in a soil profile (Chang et al., 1989, 1990). Instruments were installed at the surface and at depths of 6, 11, 17 and 47 m. At the test site the soil profile consists of about 50 m of recent alluvium underlain by 350 m of pleistocene formation. A shear wave velocity profile is available for the upper 50 m or so of the profile. The upper 30 m of the profile consists of silty sand and sandy silt with some gravel. Beneath this the soil is predominantly clayey silt and silty clay. The water table is within 0.5 m of the ground surface. A number of moderate to strong earthquakes have been recorded at the Lotung site with PGAs in the range 0.03 to 0.26 g. As no nearby rock outcrop record is available Chang et al. consider a number of one-dimensional analyses to arrive at changes from the in situ shear wave velocity required to produce the Fourier spectrum of the recorded motion. They compared the measured small strain shear wave velocities for the soil profile with the computed apparent shear wave velocities in the soil profile during the passage of the earthquake. Evidence for nonlinear behaviour is provided by this reduction of shear wave velocity.

Treasure Island, San Francisco Bay. The records for the Treasure Island Yerba Buena Island instrument pair, which are separated by only a few kilometres, have received intense investigation since the Loma Prieta earthquake in 1989 (Jarpe et al. 1989; Idriss, 1990; Hryciw et al., 1991; Dickensen et al., 1991). The soil profile at Treasure Island consists of 13 m of hydraulically placed sand, overlying 18 m of Young Bay Mud, followed by 10 m of dense sand and then 45 m of Old Bay Sediments to bedrock. A shear wave velocity profile was obtained using the Seismic Cone Penetration Test.

The PGA recorded on rock at Yerba Buena Island was 0.06 g and 0.16 g at Treasure Island. The ratio between these two is of the same order as ratios in nearby Oakland but in the case of Treasure Island there is good evidence of nonlinear soil behaviour. Jarpe et al. (1989) and Darragh and Shakal (1991) computed the spectral ratios between the two sites for the main shock and also for aftershock records captured at the same locations. The results are shown in Fig. 4 and provide clear evidence of nonlinear behaviour at the site. This is of interest because the PGA recorded at Treasure Island was nearly three times that at Yerba Buena Island, which suggests that amplification between a soil site and an adjacent rock site is not necessarily evidence for linearity. Liquefaction was observed to occur at Treasure Island after 15 s. The

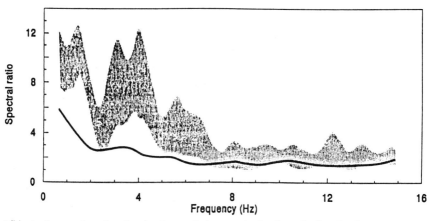

Fig. 4. Spectral ratios for the Loma Prieta mainshock and aftershocks at Treasure Island; solid line: main shock, shaded band: 95% confidence interval on aftershock data (after Jarpe et al. (1989))

strong motion spectral ratios in Fig. 4 were calculated from the first 5 seconds of the S-wave main shock so liquefaction will not have affected the ratios.

Gilroy, California. Returning now to the Gilroy array, Fig. 2 also includes the spectral ratios for the Loma Prieta main shock. These are different from those for the main shocks of the Coyote Lake and Morgan Hill earthquakes and also for the Loma Prieta aftershocks. Darragh and Shakal (1991) accept that the Treasure Island data of the same type is evidence for nonlinear behaviour but in the Gilroy case they are unsure because the spectral ratios for the Loma Prieta main shock differ from those of the other events by a factor of only about 2. They are of the opinion that such variations can be explained by differences in the incoming rock motion or azimuth effects rather than nonlinearity. Nevertheless the spectral ratios do plot as expected for nonlinear behaviour; in addition Chin and Aki (1991) conclude that the site behaved nonlinearly during the Loma Prieta event.

The contribution of Aki and his co-workers. Several recent papers on site effects have come from Aki (Aki, 1988; Aki and Irikura, 1991; Chin, Aki and Martin, 1991; Chin and Aki, 1991). These papers have moved from a position that seismological evidence reveals a good correlation between

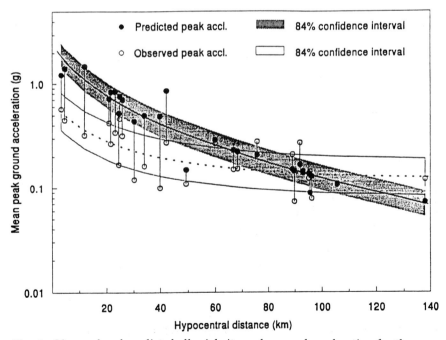

Fig. 5. Observed and predicted alluvial site peak ground acceleration for the Loma Prieta earthquake (after Chin and Aki (1991))

weak and strong ground motion, to a statement that nonlinear soil behaviour may have been more significant than most seismologists have thought, to the presentation of evidence for nonlinear behaviour. Chin and Aki (1991) report on the use of weak motion site amplification factors dervived from coda waves of the Loma Prieta strong motion records. They then used these to predict the PGA for the strong motion event; Fig. 5 shows how this technique overestimates the PGA for distances less than 80 km from the source. Chin and Aki conclude that there exists a pervasive nonlinear site effect in the epicentral region of the Loma Prieta earthquake.

Discussion
The above cases reveal that, at the majority of instrumented sites, the soil profile responds in a linear manner. Many sites, such as Chusal, Coalinga and McGee creek, are very stiff and so it is not surprising that they behave linearly. Nevertheless some important evidence of nonlinear behaviour has been obtained recently.

Parameters characterising site response
In this section we consider the main factors that affect the response of a soil profile in order to gain some insight about factors other than nonlinear soil behaviour.

Soil type and model
Since the Mexico City earthquake of 1985 there has been an awareness that the details of the relation between the apparent shear modulus and shear strain amplitude are influenced by the plasticity index of the soil (Vucetic, 1987; Sun et al., 1988). As the plasticity index increases the clay behaves in an apparently elastic manner over a larger range of shear strain amplitudes. This has a profound effect on the response of a soil profile, and is thought to be the main explanation for the damaging effect of the Mexico City earthquake and also for the response of soft sites in Oakland shown in Fig. 3. Clays and silts of low plasticity and saturated sands, liquefaction aside, are the most likely soils to exhibit nonlinear behaviour. Since the upper 13 m of the Treasure Island profile are sand this may be the reason for the pre-liquefaction evidence of nonlinear response there. It might also be significant that the Lotung profile is in low plasticity soils.

Site period and the characteristic period of the earthquake
The small strain period of the site relative to the period(s) in the incoming earthquake containing significant amounts of energy is particularly significant when the site is behaving in a linear manner. Not only does correspondence of the two periods affect the peak spectral

acceleration markedly but also the PGA at the ground surface. The Union Bay record (Seed and Idriss, 1970), cited earlier, is an example of this effect.

Characteristic acceleration for the profile

If we compare a deep stiff site and a shallow soft site they can be expected to respond to a given earthquake in a different manner even if they have the same small strain period. We need an additional parameter to clarify this difference. For a clay layer of depth D the horizontal acceleration needed to generate a shear stress at the base equal to the undrained shear strength is $s_u/\gamma D$ if the clay is considered rigid. A more sophisticated definition of this acceleration would consider the maximum shear strain in the soil profile; such an approach would also encompass sand profiles. This characteristic acceleration will be small for a soft shallow site and larger for a deep stiff site. Sites will behave linearly for PGAs which are small in relation to the characteristic acceleration. For a set of sites with the same small strain period, the one with the smallest characteristic acceleration will be most susceptible to nonlinear behaviour.

Impedance contrast at the base

The impedance contrast ($\approx$ to the ratio of the shear wave velocities) between the bottom of the soil layer and the underlying rock is very important when the site behaves in a linear manner. As the ratio increases the amplification caused by the soil profile increases.

Characteristics of the incoming earthquake motion

If the response of a site is calculated to a range of earthquakes and the spectral ratios plotted, certain common features emerge but there is scatter. Thus spectral ratios emphasise the contribution of the site characteristics whilst diminishing, but not eliminating altogether, the particulars of individual earthquakes. This is part of the reason Darragh and Shakal were conservative in their assessment of the Loma Prieta and other records captured at the Gilroy array.

Two- and three-dimensional effects

At sites where the lateral extent of the soil profile is considerable in relation to the depth of the soil layer the response can be realised as one dimensional. When this is not the case two- and three-dimensional response may be important. Elastic solutions of some typical cases have been published which show local resonances at and near the edges of a valley. The Chusal valley data of Tucker and King (1984) provide some evidence for edge effects. Surface topography will also affect site response.

Predicting site response

Space precludes a detailed treatment of this topic but general comments on procedures that could be followed might be useful. There are two approaches. One can use empirical attenuation relations such as Katayama (1982), Idriss (1985) or Joyner and Boore (1988). These handle both rock and soil sites by proposing average attenuation relations. In view of the variable nature of the response of the soil sites discussed earlier it may be more useful to follow the suggestion of Idriss (1990) and use an attenuation relation for rock motion and then incorporate specific adjustment for soil effects at the site under consideration. It requires a rock motion attenuation relation suitable for the regional geology and earthquake source mechanisms relevant to the site. The three attenuation relations mentioned above have been derived for interplate earthquakes; they are not appropriate for intraplate earthquakes. If this approach is adopted then the following steps are suggested to estimate the response of a given site:

(a) Estimate the nature of the incoming rock motion, including the PGA and the shape of the response spectrum.

(b) Estimate the small strain period of the site. Possibilities for this include in situ measurement of shear wave velocities, seismic reflection surveys of the site, and use of correlations between penetration resistance and shear wave velocity. The deployment of a sensitive instrument to record small earthquakes might also give useful information.

(c) Estimate the characteristic horizontal acceleration for the site. If the PGA of the incoming rock motion is small in relation to the characteristic acceleration the site will behave in a linear manner, in which case on-site recordings of small earthquakes will give useful information on the site response.

(d) Consider how the apparent shear modulus and equivalent viscous damping ratio will degrade with shear strain amplitude. If clays are present this will require, as a minimum, knowledge of the plasticity index.

(e) Estimate the impedance contrast at the base of the soil layer. This is particularly important if the site is expected to respond in a linear manner.

(f) Consider if the site is likely to be subject to significant surface wave, directional and topographic effects. Then decide if it is appropriate to model the response of the site by simple one-dimensional methods or if two- (or even three-) dimensional response estimates will be needed.

(g) Estimate using an appropriate numerical method the response of the site to the expected incoming rock motion.

Conclusions

This paper has reviewed the evidence for nonlinear site response during earthquake excitation of a soil deposit. Several of the sites discussed are stiff and would not be expected to respond in any other than a linear manner even at quite large values of the PGA. On the other hand there are some cases, most recently from the Loma Prieta event, which provide good evidence of nonlinear behaviour.

A number of factors, in addition to the stress strain properties of the soil, contribute to the response of a site. These interact in various ways, some enhancing site response and others diminishing it. In view of the complex interactions between the various factors it is hardly surprising that nonlinear effects have been difficult to isolate from earthquake motions recorded at various sites. Of the parameters considered it appears that the characteristic horizontal acceleration for the profile acts in a manner independent of the small strain site period and is a guide to whether the site will behave in a linear or nonlinear manner.

Acknowledgements

The helpful suggestions of Dr. T.J. Larkin, of the Civil Engineering Department of the University of Auckland, and Dr. T. Matuschka, of Engineering Geology Ltd. of Auckland, are gratefully acknowledged.

References

AKI, K. (1988). Local site effects on strong ground motion. Proc. ASCE Specialty Conf. Earthquake Engineering and Soil Dynamics II, pp. 103–155.

AKI, K. AND IRIKURA, K. (1991). Characterisation and mapping of earthquake shaking for seismic zonation. Proc. 4th International Conference on Seismic Zonation, Stanford. Vol. 1, pp. 61–110.

BERRILL, J.B. (1977). Site effects during the San Fernando, California, earthquake. Proc. 6th World Conf. on EQ Engineering, Vol. I, pp. 432–438.

BORCHERDT, R. (1970). Effects of local geology on ground motion near San Francisco Bay, Bull. Seis. Soc. of America, Vol. 60, No. 1, pp. 29–61.

BORCHERDT, R.D. (1991). On the observation, characterisation, and predictive GIS mapping of strong ground shaking for seismic zonation. Pacific Conference on Earthquake Engineering, Auckland, Vol. 1, pp. 1–24.

CHANG, C.Y., MOK, C.M., POWER, M.S., TANG, Y.K., TANG, H.T. AND STEPP, J.C. (1990). Equivalent linear versus nonlinear ground response analyses at the Lotung seismic experiment site. Proc. 4th U.S. National Conf. on EQ Engineering, Palm Springs, Vol. 1, pp. 327–336.

CHANG, C.Y., POWER, M.S., TANG, Y.K. AND MOK, C.M. (1989). Evidence of nonlinear soil response during a moderate earthquake. Proc. 12th International Conference on Soil Mechanics and Foundation Engineering, Rio de Janeiro, Vol. I, pp. 1927–1930.

CHIN, B.-Y. AND AKI, K. (1991). Simultaneous study of the source, path and site effects on strong ground motion during the 1989 Loma Prieta earthquake: a preliminary result on pervasive nonlinear site effects, Bull. Seis. Soc. of America, Vol. 81, No. 5, pp. 1859–1884.

CHIN, B.-Y., AKI, K. AND MARTIN, G.R. (1991). The nonlinear response of soil sites during the 1989 Loma Prieta Earthquake, EOS, Vol. 72, p. 339.

DARRAGH, R.B. AND SHAKAL, A.F. (1991). The site response of two rock and soil station pairs to strong and weak ground motion. Bull. Seis. Soc. of America, Vol. 81, No. 5, pp. 1885–1899.

DICKENSON, S.E., SEED, R.B., LYSMER, J.L. AND MOK, C.M. (1991). Response of soft soils during the 1989 Loma Prieta earthquake and implications for seismic design criteria. Proc. Pacific Conf. on EQ Engineering, Auckland, Vol. 3, pp. 191–203.

DOBRY, R. AND VUCETIC, M. (1987). Dynamic properties and response of soft clay deposits. Proc. International Symposium in Geotechnical Engineering of Soft Soil, Mexico City, Vol. 2, pp. 51–87.

HAYS, W.W., ROGERS, A.M. AND KING, K.W. (1979). Empirical data about local ground response. Proc. 2nd U.S. National Conf. on EQ Engineering, Stanford, pp. 223–232.

HRYCIW, R.D., ROLLINS, K.M., HOMOLKA, M., SHEWBRIDGE, S.E. AND MCHOOD, M. (1991). Soil amplification at Treasure Island during the Loma Prieta Earthquake, Proc. Second International Conference on Recent Advances in Geotechnical Earthquake Engineering and Soil Dynamics, St. Louis, pp. 1679–1685.

IDRISS, I.M. (1985). Evaluating seismic risk in engineering practice. Proc. 11th International Conference on Soil Mechanics and Foundation Engineering, San Francisco, Vol. 1, pp. 255–320.

IDRISS, I.M. (1990). Response of soft soil sites during earthquakes, Proc. Seed Memorial Symposium, pp. 273–289.

JARPE, S.P., CRAMER, C.H., TUCKER, B.E. AND SHAKAL, A.F. (1988). A comparison of observations of ground response to weak and strong ground motion at Coalinga, California. Bulletin Seismological Society of America, Vol. 78 No. 2, pp. 421–435.

JARPE, S.P., HUTCHINGS, L.J., HAUK, T.F. AND SHAKAL, A.F. (1989). Selected strong- and weak-motion data from the Loma Prieta earthquake sequence. Seismological Research Letters, Vol. 60 No. 4, pp. 167–176.

JOYNER, W.B. AND BOORE, D.M. (1988). Measurement, characteristics and prediction of strong ground motion. Earthquake Engineering and

Structural Dynamics II — recent advances in ground motion evaluation. ASCE Geotechnical Special Publication No. 20, pp. 43–102.

JOYNER, W.B., WARRICK, R.E. AND FUMAL, T.E. (1981). The effect of quaternary alluvium on strong ground motion in the Coyote Lake, California, earthquake of 1979. Bull. Seis. Soc. of America, Vol. 71, No. 4, pp. 1333–1349.

KATAYAMA, T. (1982). An engineering prediction model of acceleration response spectra and its implication to seismic hazard mapping. Earthquake Engineering and Structural Dynamics, Vol. 10, pp. 149–163.

RODGERS, A.M., BORCHERDT, R.D., COVINGTON, P.A. AND PERKINS, D.M. (1984). A comparative ground response study near Los Angeles using recordings of Nevada nuclear tests and the 1971 San Fernando earthquake. Bulletin Seismological Society of America, Vol. 74, No. 5, pp. 1925–1949.

ROMO, M.P., OVANDO-SHELLEY, E., JAIME, A. AND HERNANDEZ, G. (1989). Local site effects on Mexico City ground motions. Proc. 12th International Conference on Soil Mechanics and Foundation Engineering, Rio de Janeiro, Vol. 3, pp. 2001–2008.

SEALE, S.H. AND ARCHULETA, R.J. (1988). Site effects at McGee Creek, California. Earthquake Engineering and Structural Dynamics II – recent advances in ground motion evaluation. ASCE Geotechnical Special Publication No. 20, pp. 173–187.

SEED, H.B. AND IDRISS, I.M. (1970). Analysis of ground motions at Union Bay, Seattle during earthquakes and distant nuclear blasts. Bull. Seis. Soc. of America, Vol. 60, No. 1, pp. 125–136.

SUN, J.I., GOLESORKHI, R. AND SEED, H.B. (1988). Dynamic moduli and damping factors for cohesive soils. Report No. UCB/EERC-88/15, Civil Engineering Department, University of California, Berkeley.

TUCKER, B.E. AND KING, J.L. (1984). Dependence of sediment-filled valley response on input amplitude and valley properties. Bull. Seis. Soc. of America, Vol. 74, No. 1, pp. 153–165.

WHITMAN, R.V. (1986). Are the soil conditions in Mexico City unique? Proc. International Conference on the Mexico City Earthquakes — 1985 Factors involved and lessons learnt. ASCE, pp. 163–177.

Observed and predicted response of a braced excavation in soft to medium clay

S. RAMPELLO, C. TAMAGNINI and G. CALABRESI, University of Rome 'La Sapienza'

A test section was established next to a 5.5 m braced excavation in made ground consisting mainly of partially weathered clay elements. A sewer trunk line had to be installed at Pietrafitta (Central Italy), where the Italian Electricity Board (ENEL) is building a new thermo-electric plant. Surface ground movements, strut loads, sheet-pile deflections and strains were measured. The relationships between the in situ observations and the construction sequences showed the strong influence of the construction activities on the sheet-pile performance. The measured ground and bracing system responses are compared with results from standard design procedures and finite element analysis. The empirical relationships by Peck (1969) and the method by Mana and Clough (1981) were used to predict the ground response, while the free earth support method and the Peck (1969) earth pressure envelopes were used to evaluate the loads in the bracing system. A plane strain, total stress finite element analysis was carried out by modelling the excavation geometry, the construction sequences and the soil-sheet pile interface. The FEM capability of accounting for the construction procedures provides a much closer evaluation of the deflection profile, if compared with the predictions obtained by the classical design methods. The ground-surface settlements are matched better by the empirical relationships than by the finite element method, probably because of soil non-linearity or strain localisation phenomena not accounted for in the analysis. The standard design analyses appear to underestimate the loads measured in the bracing system, while a better prediction is obtained through finite elements, in spite of the simple soil model selected.

Introduction

The design of a braced excavation in soft clay requires the evaluation of loads in the bracing system and the prediction of the ground movements associated with the excavation process. The former, in order to satisfy the required safety factors against structural failure, the latter to evaluate the excavation effects on nearby structures or existing underground services.

Different methods of analysis are available, standard design procedures and finite element analyses being the most widely used. The standard procedures consider a condition of limit equilibrium for an idealised mechanism of soil–structure interaction, or make use of empirical relationships based on the back-analyses of a large number of documented case histories; in this case, soil movements or earth pressures are related to the main variables of the problem such as excavation geometry, soil characteristics, stiffness of the bracing system, etc. Apart from the classical works by Peck (1969) and D'Appollonia (1971), Mana and Clough (1981) recently proposed a simple method for predicting the ground movements associated with braced excavations in soft to medium clays, by comparing field data and FE parametric studies. They provide non-dimensional charts for estimating magnitudes and distributions of wall movements and surface settlements as a function of the factor of safety against basal heave (as defined by Terzaghi (1943)), soil and bracing system stiffness, struts preload and excavation geometry.

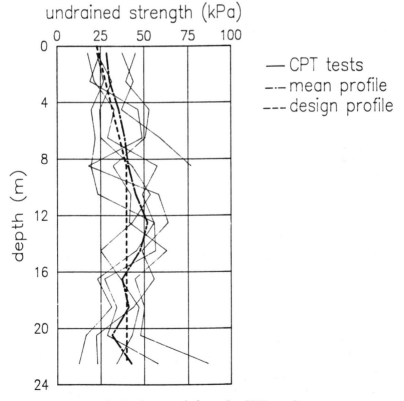

Fig. 1. *Profile of the undrained strength from the CPT results*

The finite element method has been used extensively since the early 'seventies to study the behaviour of braced excavations. Due to its ability to model the construction procedures, the FE analysis has been successfully used to evaluate their influence on the response of the ground and the bracing system. Several comparisons of observed performances and finite element analyses have been published (e.g. Palmer and Kenney, 1971; Clough and Mana, 1976; Mana, 1978; Borja, 1990; Finno and Harahap, 1991; Finno et al., 1991). In the last decade finite element analyses have become increasingly widespread in the design practice, due to the rapid increase in computer power and the consequent diffusion of efficient and versatile FE packages on desktop computers.

The analysis of the braced excavation presented herein was performed using both standard procedures and the finite element method. Among the classical methods, the empirical earth pressure diagrams proposed by Peck (1969) and the free earth support method (FESM) have been selected to evaluate the strut loads and the sheet-pile bending moments. The ground-surface settlements and the sheet-pile deflections were estimated using the method by Mana and Clough (1981) and the normalised surface settlement profiles given by Peck (1969). The CRISP package developed at Cambridge University (Britto and Gunn, 1987) was used for the finite element analysis. The comparison of the predicted and observed response of the case-history to hand allowed for an assessment of the predictive capability, the advantages and the shortcomings of the above-mentioned methods of analysis.

Site description

The excavation was performed in made ground 20 to 25 m thick, mainly consisting of soft to medium clay. The natural clay, originally overlying a lignite layer, was excavated in the sixties for the exploitation of a coal-mine. Once the exploitation activities were concluded, about 15 m of the excavated clay were re-placed in situ in 1964–65, while the remainder was re-placed 10 years later. Although the silty–clayey fraction of the made ground is predominant, it contains inclusions of silty-sand and lignite. The soil heterogeneity is caused by the irregular mixing of the excavated soil, and, according to the activities of the mine exploitation, by the different influences of swelling and weathering processes on the excavated clayey elements which are a few decimetres of average size.

The effects of the above-mentioned mine activities are evident in the soil properties. The unit weight is 16.8 to 20.9 kN/m^3, the mean value being 18.4 kN/m^3; an average water content of 38.5% was measured, the coefficient of variation being as high as 36%. A high scattering characterised the measurements of the undrained strength c_u, the coefficient of variation being 42 to 44% for both laboratory and in situ tests. The undrained strength measured on 54 TX-UU tests was within the range of 10 to 110 kPa, the mode being 30–35 kPa. The high scatter and the lowest c_u values may be attributed to occurred contacts, at the sample scale, between the excavated clay elements. Due to the partial swelling and weathering of the element surfaces, a sort of welding occurs between the clay elements, still undisturbed in the inner part. The made ground thus appears as a continuum deposit, in which however the contacts represent weakness lanes forming an irregular net. From the whole set of available experimental data the made ground may be considered as a heterogeneous, lightly overconsolidated, silty–clayey deposit.

The results of undrained strength measurements from six CPT tests performed close to the excavation are shown in Fig. 1. The undrained strength was evaluated through the cone resistance q_c by the relationship:

$$c_u = \frac{q_c - \sigma_v}{n}$$

with n = 15. The c_u profiles are shown in Fig. 1; the mean and the design profiles are also reported in the figure. The decrease in the undrained strength observed below 12 m of depth can probably be attributed to a consolidation process still in progress in the lower part of the clay layer caused by the two-staged land re-filling operations.

Instrumentation and excavation stages

A plan view and a section of the trench and the installed instrumentation is shown in Fig. 2. The ground instrumentation consisted of two inclinometer casings, a topographic benchmark net and two displacement transducers. The inclinometer TIV1, 12 m deep, was grouted into rings welded to the sheeting, while the inclinometer TIV2, 9 m deep, was installed at 3.5 m south of the southern sheet-pile. Pins A to F gave horizontal and vertical displacements of the ground surface, while pins H, K and Z provided vertical displacements only. The displacement transducers were connected to the top of the south sheet-pile.

The response of the bracing system was monitored by measuring the sheet-pile deflections by means of the inclinometer TIV1, and the strain distribution obtained from strain gauges glued to the sheet-pile, opposite to the excavation side, at regular intervals of 1 m. The strut loads were measured on two struts of the first level by means of strain gauges.

The strain gauge readings were automatically monitored every 60 min while, due to the excavation activities, only three readings were carried out at the inclinometers and the survey net.

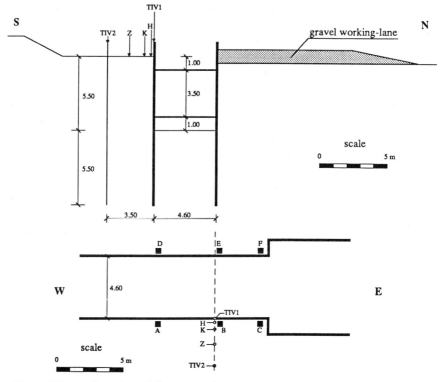

Fig. 2. Plan and section of the excavation

Table 1. Excavation sequences

Date	Stage	Event
23/1/91	A	Excavation face 15 m east of the test section. Drive sheeting in south and north sides. Install the instrumentation.
24/1/91	B	Excavation face 5 m east of the test section, readings of the installed instrumentation.
25/1/91	C	Excavate to a depth of 1 m at the test section.
28/1/91	D	Install the first strut level at a depth of 1 m. Excavate to a depth of 2 m at the test section. Perform inclinometer and survey net readings
29/1/91	E	Excavate to a depth of 5.5 m at the test section. Perform the survey net readings. Install the second strut level at a depth of 4.5 m. Perform the inclinometer readings.
30/1/91	F	Excavation face 5 m west of the test section. Pour the mud slab at the test section. Perform inclinometer and survey net readings.

The excavation sequences are summarised in Table 1. The BU-20 sheet-piles, 12 m long, were driven and the instrumentation installed at the test section on 23 January 1991, the excavation face being about 15 m far east. The test section, 4.5 m wide, was located about 5 m west of a square shaft $7 \times 7\,m^2$. Before driving the sheet-piles, an extended excavation was performed for a depth of 1.5 m, from 5 m south of the southern sheet-pile. A gravel layer about 0.5 m thick was then placed north of the northern sheet-pile as a working-lane basement for the heavy construction equipment, a surcharge being thus applied to the surface. No equipment operated on the south side of the excavation.

Excavation began to the east of the instrumented section and proceeded west as detailed in Table 1. When the excavation reached a depth of 1 m, the first level of wales and struts were installed to brace the sheet-piles. The struts, spaced at an average of 2.7 m, consisted of 220 mm diam × 10 mm pipes, and abutted on HEB-280 wales. The second strut level was installed at a depth of 4.5 m after the completion of the excavation at the full depth of 5.5 m.

Observed behaviour

The overall ground response is described herein by examining the sheet-pile and the ground surface displacements. A significant horizontal displacement toward the area where the excavation had first been made (east) was observed at pins A to F, located on top of the

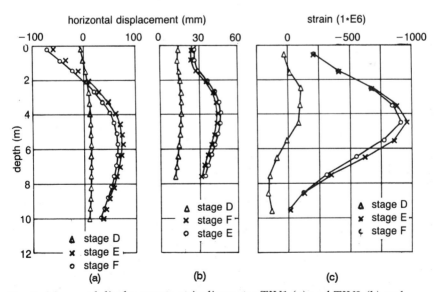

Fig. 3. *Measured displacements at inclinometer TIV1 (a) and TIV2 (b) and strain distribution in the instrumented sheet-pile (c)*

sheet-piles 1 m above the ground surface; it was 8–13 mm on the south side, and 10–14 mm on the north side. The displacement parallel to the excavation axis mainly occurred at stage D, at an excavation depth of 2 m; minor changes were measured thereafter. A different behaviour was observed between the north and the south sheet-piles when considering the horizontal displacement perpendicular to the excavation axis. The north sheet-pile, where the construction equipment was operating, showed maximum inward surface movements of 140 to 330 mm, while the south sheet-pile moved opposite to the excavation at 30 to 150 mm.

The above observations are consistent with the presence of the construction equipment on the north side of the excavation, and the wider shaft 5 m east of the test section. The working-lane surcharge and its higher stiffness were likely to produce the inward movement of the north sheet-pile, even after the strut installation. The shaft was probably the cause of the major displacements measured at pins C and F, and of the significant displacements parallel to the excavation axis.

Ground movements measured by the inclinometer TIV2, 3.5 m south of the southern sheet-pile, showed a maximum displacement of about 40 mm toward the excavation (Fig. 3(b)). The displacements are virtually constant with depth at stage D, while at stages E and F they gently increase from 28.6 mm at a depth of 2 m to about 40 mm at a depth of 3 m, then showing small gradients with depth.

Figure 3(a–c) shows the deflections and the strain distribution of the south sheet-pile. The point of zero displacement, 1 m deep at stage D, moved to a depth of 2 m at stage F. The movements were oriented toward the excavation below it, and opposite to the excavation in the upper few metres. About 0.5 m below the excavation base, a maximum inward movement of $\simeq$71 mm occurred at stage E, before the mud slab was poured. The strain distribution did not show any curvature change when the first strut level was installed.

The strut loads may be conveniently related to the sheet-pile deflec-

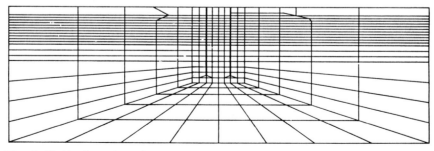

Fig. 4. Finite element mesh

tions, as they increase with them. The east strut was forced when installed at stage D due to the significant inward movements of the sheet-piles next to the wider shaft; a load of about 500 kN was then measured on it, while on the west strut a load of 70 kN was obtained. The strut loads steeply increased at stage E, when the full depth of the excavation was attained, and remained constant at 770 and 670 kN thereafter.

Finite element analysis

The FE mesh adopted in the analysis is shown in Fig. 5; it consisted of 472 8-noded Linear Strain Quadrilateral (LSQ) elements and 42 3-noded beam elements which model the sheet-piles and the struts. Due to the lack of axial symmetry, the mesh was extended to the whole domain on both the excavation sides, to a horizontal distance of ≃8.5 times the excavation width. The bottom boundary of the FE mesh was placed at the top of the lignite layer, assuming this as an infinitely rigid boundary. Both horizontal and vertical displacements were restrained at the lower boundary, while horizontal displacements alone were restrained at the sidewalls. The soil-sheet pile interface was modelled through 8-noded LSQ thin layer elements, with an aspect ratio of 10 to 30 (Pande and Sharma, 1979).

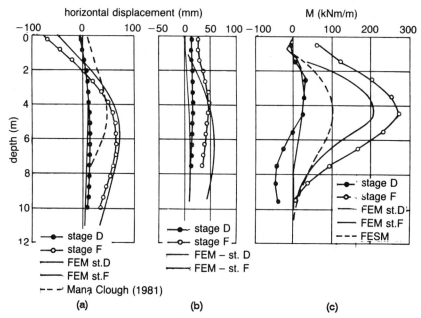

Fig. 5. Measured and predicted displacements: (a) inclinometer TIV1; (b) inclinometer TIV2; (c) measured and predicted bending moments

Due to the low soil permeability, and to the short duration of the excavation process (7 days), undrained behaviour was assumed in the clay layer throughout the analysis, as confirmed by the work of Osaimi and Clough (1979). Among the existing techniques developed to handle problems involving media subjected to incompressibility constraints (e.g., see Borja (1992)), the simple total stress approach was used for a variety of reasons: the difficulty of obtaining a reliable evaluation of the effective stress–strain behaviour, due to the soil heterogeneity; the uncertainties associated with the pore pressure field in the made ground, due to a consolidation process still in progress; the aim of using the FE method as a design tool for a problem the provisional character of which would hardly justify a more refined numerical analysis in terms of effective stresses.

The soil and the soil-sheet pile interfaces were modelled as an elastic perfectly plastic material obeying the Von Mises yield criterion. The undrained strength profile of Fig. 1 was adopted in the analysis; a reduced c_u profile, of a factor 4, was attributed to the interface elements. A modulus multiplier M ($E_u = M \cdot c_u$) of 100 was selected for the undrained stiffness of the soil and the interfaces. Such a value might appear low if compared with those typical of natural soft to medium clay deposits (e.g. see Duncan and Buchignani (1976)). However it reflects the full-scale behaviour of the made ground, which is strongly influenced by the previously described replacement procedure. The gravel working-lane was modelled by an elastic material with a Young modulus of 1×10^5 kPa and a Poisson ratio of 0.25. The initial state of stress in the soil was evaluated by assuming a horizontal to vertical total stress ratio K = 1.

The geometrical and mechanical properties considered in the analysis for the sheet-piles and the struts are summarized in Table 2. Since each sheet-pile element could bend independently from the adjacent ones, due to the poor workmanship in the driving operations, the sheet-pile moment of inertia J was evaluated by assuming a zero efficiency of the joints.

The FE analysis was performed in two separate sections: the removal

Table 2. Properties of the structural elements

Element	$A^{(1)}$ (m²/m)	$J^{(1)}$ (m⁴/m)	E (kPa)
BU-20 sheet-pile	1.86E-2	1.001E-4	2.0E8
Pipe struts	2.19E-3	6.450E-6	2.0E8

(1) per unit length of excavation

Table 3. *Maximum horizontal displacements of the south sheet-pile*

	Stage D		Stage F	
	u_{max} (mm)	error (%)	u_{max} (mm)	error (%)
Measured	14.0	–	65.0	–
Mana and Clough (1981)	8.7	−37.9	49.2	−24.3
FE analysis	10.7	−23.6	74.4	14.5

of a layer 1.5 m thick and the construction of the working-lane were simulated first under undrained conditions; the trench excavation was then modelled under plane strain conditions following the construction sequence shown in Table 1. The excavation was completed in 7 stages by removing subsequent rows of elements 1 m thick, and by applying surface tractions at the excavation boundaries; the tractions are equivalent to the pre-existing state of stress of the removed elements (Mana, 1978).

Comparison of observed and predicted responses
Sheet-pile and ground displacements

Table 3 summarises the maximum horizontal displacements of the south sheet-pile. Comparison between measured and evaluated deflections are considered at stages D and F (2b and 7 in the FE analysis). In applying the method by Mana and Clough (1981) the factor of safety against basal heave was computed for an average undrained strength of 30 kPa. The percentage differences between the computed and the measured displacements are also reported in the table. A fair agreement is obtained by both the methods, the finite element analysis being slightly closer to the measurements.

For the same stages as those above, the deflection profiles as measured at inclinometers TIV1 and TIV2 are compared in Fig. 5(a–b) with the predicted displacements. The deflection profile by the method of Mana and Clough (1981) was obtained by the maximum horizontal displacement and one of the normalised profiles as suggested by Mana (1978). A close agreement is observed for the FE results, the analysis being capable of correctly locating the point of maximum inward movement and the occurrence of large displacements opposite to the excavation side, in the first 2 m of the sheet-pile. The method by Mana and Clough (1981) implies significant errors at the sheet-pile edges and predicts a quite different deflection profile. The difference can possibly be attributed to the presence of the gravel working-lane and the construction equipment acting on the north side of the excavation; the consequent surcharge applied to the surface yields a non-symmetric displacement field oriented towards the south. The central portion of the final deformed mesh next to the excavation zone clearly supports this interpretation (Fig. 6).

The ground-surface settlements, normalised by the excavation height, are compared in Fig. 7 at stage F. The curves proposed by Peck (1969) and the surface settlement envelope suggested by Mana and Clough

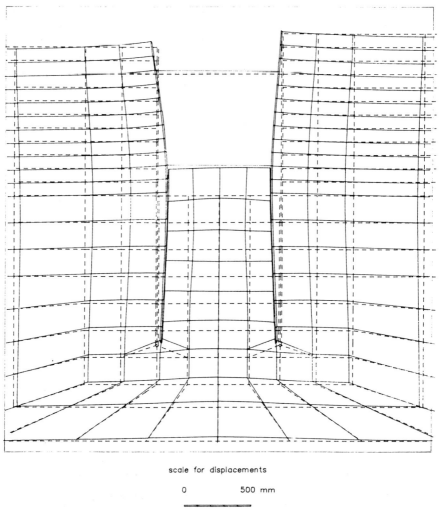

scale for displacements

0   500 mm

Fig. 6. Detail of the deformed mesh; displacements are magnified by a factor of 5

(1981), both on an empirical basis, are also shown in the figure. Except for measurements from pins B and H, next to the sheet-pile, the data points plot close to the boundary between Peck's zones I and II, consistently with the low stability number $N = \gamma H/c_u$ equal to 3.4 (by assuming a constant c_u value of 30 kPa). The normalised curve by Mana and Clough (1981) is also in fair agreement with the measurements, the maximum normalised settlement being about 0.67% against a measured one of 0.45% to 0.89%.

A worst agreement is observed with the profile of the surface settlements obtained by the finite element analysis. Next to the top of the excavation the ground-surface heaves at about 0.75% of the excavation height and the settlement profile extends to a much greater distance than that indicated by the empirical curves. Similar results were obtained by a linear elastic FE analysis of the deep excavation for the New Palace Yard Car Park in London (Ward and Burland, 1973). Simpson et al. (1979) and Jardine et al. (1986) attributed this limitation to the assumed linearity of the soil behaviour in the elastic range, and suggested that the account of soil non-linearity within the yield surface would improve the finite element predictions. Similar conclusions can probably be applied to the problem at hand, since the soil behind the excavation remains in the elastic range throughout the analysis as shown in Fig. 8; the figure shows the extension of the plastic zone at stage F, when the excavation reaches the full depth. The presence of plastic zones in between the sheet-piles may be attributed to the assumed initial stress state and stress history, the stress state in the soil being closer to a condition of passive failure than to an active one.

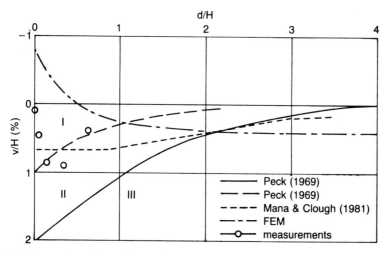

Fig. 7. Ground surface settlements

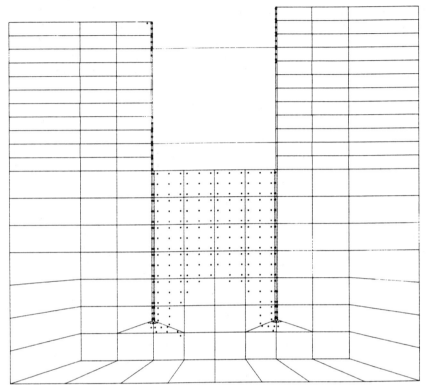

Fig. 8. Development of plastic zones at stage F

However, apart from the limits of the soil model, it is worth noting that the inability of a conventional displacement finite element procedure in modelling strain localisation phenomena could prevent a reliable evaluation of the ground movements after a certain level of strain is reached in the soil mass (Finno and Harahap, 1991).

Loads on sheet-piles and struts

The bending moment distributions computed at stage F by the finite element analysis and the free earth support method are compared with the measurements on the instrumented sheet-pile in Fig. 5(c). The latter were derived by the strain gauges measurements, taking account of the sheet-pile bending stiffness, under the hypothesis of pure bending. The moment of inertia of a single sheet-pile was considered in the evaluation of the bending stiffness as already assumed in the finite element analysis. The analyses appear to underestimate the maximum bending moment on the sheet-pile; values of 104 and 205 kN/m are obtained by the FESM and FEM computations against a measurement of 274 kN/m, at a depth of $\simeq 6$ m, with a percentage error of 62% and 25% respectively. However it is worth stressing that some doubts on the reliability of the strain measurement arise, at least in the upper portion of the sheet-pile, from the observation that the strain measurements were of the same sign even in the zone where an inversion of the bending moment was to be expected.

Table 4 summarises computed and measured loads on the upper strut level. Apart from the evaluations by the free earth support method and the finite element analysis, strut loads were also calculated by the earth pressure diagrams proposed by Peck (1969), in the hypotheses of one or two strut levels. Again, the conventional design methods underestimate the strut loads, the Peck's method giving an error of 37% while the FESM of 75%; a slightly better prediction is obtained by the FE analysis, with an underestimation of about 22%.

The above-mentioned observations are consistent with the results reported by Potts and Fourie (1984) on the behaviour of propped retaining walls. They found that for excavated walls in high K_0 conditions, FEM evaluations of prop forces and bending moments can greatly exceed those computed by the limit equilibrium approach. The discrepancy is to be found in the different earth pressure distributions of the two analyses (Fig. 9). In the free earth support method the earth pressures are determined for a condition of limit equilibrium on both the wall sides, the passive earth pressure being scaled back by a safety factor according to different approaches. In the present analysis, the undrained shear strength was reduced on the passive side by the actual factor of safety. On the contrary, in the finite element analysis the earth pressures are the result of the soil-structure interaction; the progressive deflections of the sheet-pile towards the excavation promote the redistribution of earth pressures in the embedded part of the sheet-pile and the restrained zones, following the well-known mechanism of arching (Bjerrum et al., 1972). In the presence of a large degree of

Table 4. Loads on the first strut level

	strut load[1] (kN/m)	error (%)
Measured	265.1	–
Peck (1969)[2]	83.57 (167.0)	68.5 (37.0)
FESM	66.4	75.0
FEM	206.4	22.1

(1) per unit length of excavation
(2) bracketed figures refer to the case with a single strut level

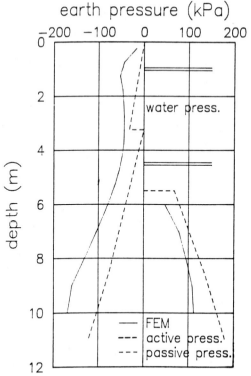

Fig. 9. Horizontal stress distribution

horizontal restraint (as in multi-propped or strutted excavations) or in high K_0 conditions, not even large horizontal movements of the wall are sufficient to bring the earth pressures to the classic active distribution, while passive conditions may be partially or completely reached in front of the sheet-pile wall.

Conclusions

In the present work, the behaviour of a 5.5 m deep braced excavation in soft to medium clay was analysed by means of both standard design procedures and a simple elastic–perfectly plastic, total stress finite element analysis.

From the comparisons of the analysis results with the observed response of an instrumented test section the following conclusions can be drawn:

(a) The maximum lateral movement of the sheet-pile is successfully predicted by both the method by Mana and Clough (1981) and the finite element analysis, the latter matching slightly better with the measurements; however, as far as the deflection profile is concerned, the FE analysis shows its higher predictive capability, the influence on the structure behaviour of factors such as non-symmetric loads and construction sequences being accounted for in the analysis.

(b) A different trend is observed if the comparison of predicted and measured ground-surface settlements are considered. In this case a less satisfactory agreement with the measurements is obtained by the finite element analysis for the maximum displacement and the shape of the settlement trough. This finding is likely to be the result of the initial stress state and the simple soil model adopted in the analysis; the former leads to the elastic behaviour of the soil behind the retaining structure, while the latter does not account for soil non-linearity within the elastic range. A reasonable assessment of the ground movements has been obtained by the normalised surface settlement profiles (Peck, 1969) and by the method proposed by Mana and Clough (1981). This is likely because of the empirical basis of these relationships which include the effects of factors such as soil non-linearity, strain localisation, consolidation, etc., not considered in a simple elastic–perfectly plastic, total stress analysis.

(c) The conventional design methods largely underestimate the loads on the support system for the examined case-history. Although qualitatively in agreement with the measurements, limit equilibrium analysis yields considerably lower bending moment distributions on the sheet-pile. A better prediction is given by the finite element analysis, with an error of about 25% on the maximum bending moment. Similar results are found for the strut loads, for which much closer evaluation is obtained by the finite element computations than by any of the considered standard design procedures. The latter finding may probably be attributed to the high K value selected for evaluating the initial stress state; in similar conditions, Potts and Fourie (1984) found that conventional limit equilibrium approach can severely underestimate both maximum bending moment and strut load.

(d) In agreement with the recent findings of Ho and Smith (1991), the use of the finite elements in the analysis of braced excavations, can capture most of the physical features of the soil-structure interaction process, even performing simple elastic–perfectly plastic total stress analyses by using standard computer packages on desktop computers.

Acknowledgements

The authors gratefully acknowledge the support of the Italian Electricity Board (ENEL) in carrying out the field tests. They are especially indebted to Mr. A. Neri for his invaluable contribution in devising and performing the whole set of measurements on the braced structure. The help of Mr. M. Serini in collecting field data is also gratefully acknowledged.

References

BJERRUM, L., FRIMANN CLAUSEN, C.J. AND DUNCAN, J.M. (1972). Earth pressures on flexible structures – A state-of-the-art report. Gen. Rep. 5th ECSMFE, Madrid, Vol. 2, pp. 169–196.

BORJA, R.I. (1990). Analysis of incremental excavation based on critical state theory. J. Geotech. Engrg. Div., ASCE Vol. 116, No. 6, pp. 964–985.

BORJA, R.I. (1992). Free boundary, fluid flow and seepage forces in excavations. J. Geotech. Engrg. Div., ASCE Vol. 118, No. 1, pp. 125–146.

BRITTO, A.M. AND GUNN, M.J. (1987). Critical state soil mechanics via finite elements. John Wiley and Sons, Inc., New York, N.Y.

CLOUGH, G.W. AND MANA, A.I. (1976). Lessons learned in finite element analyses of temporary excavation. Proc. 2nd Int. Conf. on Num. Meth. in Geomech., Blacksburg, Virginia, Vol. 1, pp. 243–265.

D'APOLLONIA, D.J. (1971). Effects of foundation construction on nearby structures. Proc. 4th Pan American Conf. on Soil Mech. and Found. Engrg, Vol. 1.

DUNCAN, J.M. AND BUCHIGNANI, A.L. (1976). An engineering manual for settlement studies, Geotech. Engrg. Rep., Dept. of Civ. Engrg., Univ. of California, Berkeley.

FINNO, R.J. AND HARAHAP, I.S. (1991). Finite element analyses of HDR-4 excavation. J. Geotechnical Engrg. Div., ASCE, Vol. 117, No. 10, pp. 1590–1609.

FINNO, R.J., HARAHAP, I.S. AND SABATINI, P.J. (1991). Analysis of braced excavations with coupled finite element formulations. Computer and Geotechnics, Vol. 12, pp. 91–14.

HO, D.K.H. AND SMITH, I.M. (1991). Analysis of construction processes in braced excavations. 10th ECSMFE, Florence, Vol. 1, pp. 213–217.

JARDINE, R.J., POTTS, D.M., FOURIE, A.B. AND BURLAND, J.B. (1986). Studies on the influence of non-linear stress–strain characteristics in soil structure interaction. Géotechnique, Vol. 36, No. 3, pp. 377–396.

MANA, A.I. (1978). Finite element analyses of deep excavation behaviour in soft clay. PhD Thesis, Stanford Univ., Stanford, Calif.

MANA, A.I. AND CLOUGH, G.H. (1981). Prediction of movements for braced cuts in clay. J. Geotech. Engrg. Div., ASCE, Vol. 107, No. 6, pp. 759–778.

OSAIMI, A.E. AND CLOUGH, G.W. (1979). Pore pressure dissipation during excavation. J. Geotech. Engrg. Div., ASCE, Vol. 105, No. 4, pp. 481–498.

PALMER, J.H. AND KENNEY, T.C. (1972). Analytical study of a braced excavation in weak clay. Canadian Geotechnical J., Vol. 9, No. 2, pp. 145–164.

PANDE, G.N. AND SHARMA, K.G. (1979). On joint-interface elements and associated problems of numerical hill-conditioning. Int. J. for Num. and Anal Meth. in Geomech., Vol. 3, pp. 293–300.

PECK, R.B. (1969). Deep excavations and tunnelling in soft ground. State-of-the-art report, 7th ICSMFE, Mexico City, pp. 225–281.

POTTS, D.M. AND FOURIE, A.B. (1984). The behaviour of a propped retaining wall: results of a numerical experiment. Géotechnique, Vol. 34, No. 3, pp. 383–404.

SIMPSON, B., O'RIORDAN, N.J. AND CROFT, D.D. (1979). A computer model for the analysis of ground movements in London Clay. Géotechnique, Vol. 29, No. 2, pp. 149–175.

TERZAGHI, K. (1943). Theoretical Soil Mechanics. John Wiley and Sons, Inc., New York, N.Y.

WARD, W.H. AND BURLAND, J.B. (1973). The use of ground strain measurement in civil engineering. Phil. Trans. R. Soc., Vol. A274, pp. 421–428.

Seismic and pressuremeter testing to determine soil modulus

P.K. ROBERTSON and R.S. FERREIRA, University of Alberta, Canada

A method to interpret undrained, self-boring pressuremeter test results in clay and a technique to combine the results with the small strain modulus (G_0) determined from shear wave velocity measurements has been presented. The pressuremeter interpretation incorporates the unloading portion of the pressuremeter test to derive the initial shear modulus and undrained shear strength. The soil response is represented by a hyperbolic relationship between the shear stress and circumferential strain. The method accepts that some level of disturbance may exist for self-boring pressuremeter test results. Pre-bored and full-displacement undrained pressuremeter test results can also be analysed using the proposed method. The proposed interpretation method involves comparison of the measured loading and unloading pressuremeter curves with analytically derived curves. This comparison can be achieved using commercially available microcomputer application software. For this study the software was Kaleidagraph^{TM} (version 2.1) developed and operated for the Macintosh^{TM} microcomputer. Hence, it should be possible for practising engineers to apply this proposed method to SBPT results without the need of special customized software. The proposed interpretation method has been evaluated using high quality self-boring pressuremeter results performed in Fucino clay in Italy. The interpreted soil parameters had reasonable values when compared to other in situ and laboratory test results (Ferreira and Robertson, 1992). The proposed combination of seismic small strain modulus with pressuremeter modulus represents a simple and easy method to evaluate the variation of shear modulus with shear strain and provides a consistent framework to derive other soil parameters from the pressuremeter test. The methodology can be applied to pressuremeter tests which have some disturbance and does not require high resolution in the pressuremeter measurement system. The resulting variation of shear modulus with shear strain appears to be representative when compared with the generalized curves suggested by Vucetic and Dobry (1991).

Introduction

The assessment of soil deformation moduli for application in deformation analyses represents one of the most important but also one of the most difficult problems in geotechnical engineering. Soil response is

TESTING TO DETERMINE SOIL MODULUS

highly non-linear and is influenced by many variables, such as, mineralogy, fabric, structure, stress state, loading conditions and drainage conditions. Sampling techniques have improved considerably in recent years. However, even with the best sampling techniques some disturbance is inevitable due to either mechanical disturbance or disturbance caused by stress changes. Laboratory techniques have improved with the development of local strain measuring methods (Jardine et al., 1984; Atkinson and Evans, 1985). These developments have demonstrated the highly non-linear response of soils and have illustrated the problems associated with sample disturbance. There are many soils that are extremely difficult to sample using conventional techniques. In these soils in situ techniques are required to evaluate soil deformation parameters.

It has been common practice to evaluate soil moduli in the form of an equivalent linear elastic response over a wide range of strains. Research has shown that soil can only be considered essentially linear elastic at very small strains (Hardin, 1978; Jardine et al., 1984). Figure 1 presents a qualitative model of soil behaviour adapted from the work of Jardine et al. (1991), that illustrates the different responses at different strain levels. At strains less than about 10^{-3}% the soil response is essentially linear elastic. From about 10^{-3}% to 10^{-2}% the response can be approximated to non-linear elastic. The stress–strain response within this region remains stiff but highly non-linear and stress path dependent. Creep and rate phenomena generally play a minor role in this

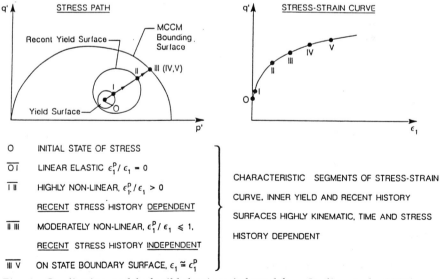

Fig. 1. *Qualitative model of soil behaviour (adapted from Jardine et al., 1991)*

563

region and unload–reload cycles exhibit some hysteresis but plastic strains are generally small. Beyond strains of about $10^{-2}\%$ the plastic strains increase as the stress path approaches the state boundary surface. As soon as the stress path reaches the state boundary surface the soil experiences large plastic strains. The two regions inside the state boundary surface are highly kinematic, i.e. the regions move with the current stress state (Jardine et al., 1991). The shear strain levels that define the boundaries between regions are uncertain and are probably influenced by the soil plasticity (Vucetic and Dobry, 1991).

The selection of relevant soil deformation parameters is also a function of the design application and the project requirements. It is not always feasible to apply complex elasto–plastic soil models with features such as kinematic hardening to small geotechnical engineering problems where budgets are often limited. Hence, there is a need to develop a rational procedure for the selection of appropriate soil deformation moduli and this procedure must be consistent with the analysis technique to be applied. Often engineers prefer to apply simple linear or non-linear elastic solutions to a design problem provided they can evaluate a reasonable equivalent elastic modulus applicable to the induced strain level of the problem. Figure 2 illustrates the possible soil models associated with their applicable strain range.

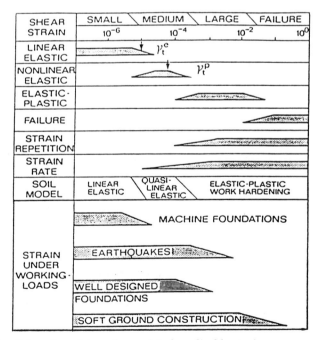

Fig. 2. *Possible soil models and associated applicable strain ranges*

TESTING TO DETERMINE SOIL MODULUS

Ideally, the determination of stiffness models for analysis should involve the interpretation and synthesis of all possible data sources (e.g. lab, in situ testing and previous field experience) within a consistent framework. An overall assessment should then be made to take account of the strengths and weakness of each data source.

The objective of this paper is to describe how the variation in modulus with strain can be evaluated using in situ tests. The paper will concentrate on the use of seismic techniques combined with pressuremeter test results to evaluate the variation of the shear modulus with shear strain for clay soils. However, the basic philosophy described should allow the technique to be applied also to sandy soils.

Small strains

Seismic techniques can be used to determine the small strain (engineering strain, $\gamma < 10^{-3}\%$) stiffness of soil. Based on elastic body wave theory the following moduli can be determined:

Shear modulus $\qquad G_0 = \rho V_s^2 \qquad (1)$

Constrained modulus $\qquad M_0 = \rho V_p^2 \qquad (2)$

where:
G_0 = small strain shear modulus
M_0 = small strain constrained modulus
ρ = mass density of soil
V_s = shear wave velocity
V_p = constrained compression wave velocity

In loose or soft saturated soil the compression wave velocity is dominated by the stiffness of the water and V_p is often close to 1500 m/s. For a small drop in the degree of saturation V_p can decrease significantly and hence, the measurement of V_p can be a highly effective means to evaluate soil saturation.

For a perfectly elastic isotropic soil, the value of the shear modulus is unaffected by the drainage conditions. Hence, the two pairs of elastic constraints are related as follows (Wroth and Houlsby, 1984);

$$\frac{E_u}{2(1+\nu_u)} = G = \frac{E'}{2(1+\nu')} \qquad (3)$$

where: E_u, E' = undrained and drained Young's moduli, respectively.
ν_u, ν' = undrained and drained Poisson's ratio, respectively.

Hence, the small strain Young's moduli (E_{u0}, E'_0) can be determined from seismic technique via G_0 if ν is estimated. Fortunately, the relationship between E_0 and G_0 is not sensitive to the estimated value of ν.

In situ seismic techniques have many advantages for the evaluation of soil moduli. The methods are based on sound, well established elastic theory, the modulus is determined at a known (although small) strain level (usually $\gamma < 10^{-3}\%$), stress levels are essentially unchanged, moduli are measured in situ, the soil is undisturbed and moduli are determined for a large volume of soil. Traditional in situ seismic techniques include surface refraction and reflection, cross-hole, down-hole, and up-hole methods. Recently, the cone penetration test (CPT) has been modified to include seismic measurements (Robertson et al., 1986; Baldi et al., 1988). The seismic cone penetration test (SCPT) represents an ideal tool for soft soils since it combines the logging capabilities of the CPT to record a continuous profile of soil type and shear strength with the downhole seismic capability to determine G_0. A recent development that shows considerable promise is the application of surface waves to determine the shear wave velocity profile and hence, the G_0 profile (Nazarian and Stokoe, 1984; Addo and Robertson, 1992). This technique is based on spectral-analyses-of-surface-waves (SASW) and is non-destructive, requires no boreholes or penetration devices, is fast and therefore economic to perform. The SASW technique can be performed on ground varying from very soft mud through to rock. Seismic techniques, such as the SASW, are ideal for heterogeneous deposits such as fill, colluvium, gravels and waste.

Traditionally, seismic techniques have been used to evaluate G_0 for

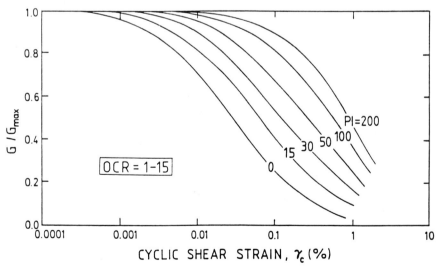

Fig. 3. Shear modulus curves for saturated soil proposed by Vucetic and Dobry (1991)

application to small strain problems, such as vibrating machine foundations and small to moderate earthquake loading. Hardin and Drnevich (1972a, 1972b) proposed that G_0 could be extended to larger strains by assuming a hyperbolic stress–strain relationship with G_0 as the initial tangent modulus. This approach is simple and attractive for engineering applications. However, recent research (Tatsuoka and Shibuya, 1991) have shown that the hyperbolic function is unable to match simultaneously the soil response at both low and high strain levels. The hyperbolic function is effective however, for matching soil response over either the low strain or the high strain range. Vucetic and Dobry (1991) report information available in the geotechnical literature to show how the shear modulus normalized with G_0 varies with shear strain for saturated soils of different plasticity, as shown on Fig. 3. Hence, with a knowledge of G_0 from in situ seismic testing and the soil's plasticity index (PI) it is possible to estimate the secant shear moduli at different strain levels. However, as illlustrated by the generalized soil model in Fig. 1, the soil stiffness at larger strains is more influenced by factors such as stress path, strain rate effects and the distance from the state boundary surface. Therefore, there is a need to determine directly the variation of moduli with strain over the intermediate to large strain region ($\gamma > 10^{-2}\%$).

Intermediate to large strains

The pressuremeter probably represents the best in situ test to evaluate soil stiffness over the intermediate to large strain range. The pressuremeter test has well defined boundary conditions with well established solutions to the boundary value problem and the test can be performed in most soils. The test has developed considerably since its first introduction by Menard in 1956. A number of publications have chronicled this development (Gambin, 1990; Baguelin et al., 1978; Wroth, 1984; Mair and Wood, 1987; Briaud and Cosentino, 1990; Clough et al., 1990). Existing pressuremeter testing can be broadly divided into two main groups: pre-bored and self-bored. The pre-bored pressuremeter test (PMT) is performed in a pre-drilled hole, whereas the self-bored pressuremeter test (SBPT) is self-bored into the soil in an effort to minimize soil disturbance. Recently, full-displacement pressuremeter tests (FDPT) have been developed (Hughes and Robertson, 1985; Withers et al., 1986; Campanella et al., 1990) where the probe is pushed into the ground in a full-displacement manner. These different pressuremeter tests (PMT, SBPT and FDPT) are thought of as distinct and separate in situ testing techniques, with different interpretation methods. The PMT is usually analysed using empirical correlations related to specific design rules. The SBPT is generally performed in relatively soft soils and the results are analysed using theoretical relationships to derive basic soil parameters. The FDPT is relatively new and interpretation techniques are still evolving (Houlsby and Withers, 1988; Withers et al., 1989).

Efforts to minimize soil disturbance led to the development of self-boring pressuremeters in the early 1970s. This resulted in a series of developments related to the theoretical interpretation of the SBPT. However, the process of installation of current SBPs is not always efficient (Clough et al., 1990) and is often subject to problems, resulting in some disturbance especially in very stiff soils. Hence, most SBPT results are subject to some degree of disturbance, the level of disturbance tending to increase with increasing soil stiffness. Soil disturbance during installation of the SBP has the greatest effect on the shape of the initial loading portion of the pressuremeter curve. It is therefore, common practice to put less reliance on the initial loading portion of SBPT results.

The introduction of the SBPT has resulted in substantial research related to various theoretical interpretation techniques of the SBPT results to give basic soil parameters. Almost all of these developments have resulted in interpretation techniques that utilize only the loading portion of the SBPT. Recently, Jefferies (1988) proposed that the interpretation of undrained SBPT results in clay could be extended to include the unloading part of the test. Houlsby and Withers (1988) also incorporated the unloading portion for interpretation of FDPT results in clay. The methods by Jefferies (1988) and Houlsby and Withers (1988) utilize simple elastic–perfectly plastic soil models. Several studies (Hughes and Robertson, 1985; Robertson, 1982; Robertson and Hughes, 1986; Bellotti et al., 1986; Schnaid and Houlsby, 1990) have shown that the unloading portion of pressuremeter tests appear to be insensitive to disturbance caused due to installation. It would, therefore, appear useful and logical to incorporate the unloading portion of pressuremeter tests into any interpretation techniques, especially those related to soil modulus.

Ferreira and Robertson (1992) recently proposed a method for the interpretation of self-boring pressuremeter tests in clay that is an extension of the method developed by Jefferies (1988). However, soil nonlinearity is incorporated assuming the soil stress–strain response can be represented by a hyperbolic function. The method has the following assumptions:

(*a*) The pressuremeter test is performed undrained from the start of expansion to complete contraction.

(*b*) The test is treated as an expansion of an infinity long cylindrical cavity (i.e. radially symmetric plane strain).

(*c*) The vertical stress remains the intermediate principal stress.

(d) The soil stress–strain behaviour can be represented by a hyperbolic function in both loading and unloading.
(e) The ratio of the unloading strength of the clay to the loading strength is known.
(f) Strains are considered to be small.

These assumptions are essentially the same as those made by Jefferies (1988) and Gibson and Anderson (1961), except for the hyperbolic representation of the stress–strain behaviour. The selection of the hyperbolic representation of soil behaviour was made for the following reasons:

(a) The hyperbolic stress–strain model (Kondner, 1963) have proven effective in describing soil behaviour under a variety of monotonic loading conditions (Duncan and Chang, 1970) over a range of strain levels.
(b) The need to keep the soil model simple, as suggested by Wroth (1984), and to avoid generating a method that requires a solution for many unknown parameters.
(c) The parameters that define the soil model have some engineering significance, so that when the interpretation process is completed the parameters derived can be understood and applied in design.

The hyperbolic model for loading is defined in terms of the shear stress (τ) and the cavity strain (ε) as follows:

$$\tau = \frac{\varepsilon}{1/2G_i + \varepsilon/\tau_{ult}} \tag{4}$$

where: τ — mobilized shear stress
G_i — initial shear modulus
τ_{ult} — ultimate undrained shear strength during loading
$\varepsilon = -\varepsilon_\theta = \Delta R/R_0$ — cavity strain at the pressuremeter wall (loading)
ε_θ — hoop strain
$\Delta R = R - R_0$ — change in cavity radius
R — current pressuremeter radius
R_0 — initial pressuremeter radius

The hyperbolic model is usually applied in terms of shear stress (τ) and shear strain (γ). However, for undrained cylindrical cavity expansion the following relationship holds:

$$\gamma = 2\varepsilon \tag{5}$$

Hence, the term $2G_i$ in eqn. (4) stems from the use of cavity strain (ε) instead of the more conventional engineering shear strain (γ).

The complete stress–strain curve of the soil can then be defined using eqn. (4) and the two parameters G_i and τ_{ult}. The parameter G_i represents the initial tangent shear modulus.

The hyperbolic model for unloading is defined in terms of the change of stress and strain from the loaded state, as follows:

$$\tau^* = \frac{\varepsilon^*}{1/2G_i - \varepsilon^*/\tau^*_{ult}} \tag{6}$$

where: τ^* mobilized shear stress
τ^*_{ult} ultimate undrained shear strength during unloading
$\varepsilon^* = \Delta R/R_{max}$ cavity strain at the pressuremeter wall (unloading)
R_{max} maximum pressuremeter radius at the end of the loading stage

The derived relation between ε and ε^* is:

$$\varepsilon^* = \frac{\varepsilon - \varepsilon_{max}}{1 + \varepsilon_{max}} \tag{7}$$

where ε_{max} is the maximum cavity strain at the start of the unloading.

The hyperbolic soil model when combined with the governing equations of equilibrium of stresses, compatibility of strains and the boundary conditions for a pressuremeter produces the following closed form solutions:

(a) loading

$$p = \sigma_{h0} + \tau_{ult} \ln(1 + (2G_i\varepsilon/\tau_{ult})) \tag{8}$$

(b) unloading

$$p = p_{max} + \tau^*_{ult} \ln\left(1 \Big/ \left(1 - \frac{2G_i(\varepsilon - \varepsilon_{max})}{(1 + \varepsilon_{max})\tau^*_{ult}}\right)\right) \tag{9}$$

where: p pressuremeter expansion or contraction pressure exerted on soil after correction for membrane stiffness
σ_{h0} initial horizontal in situ stress
ε_{max} cavity strain at the cavity (pressuremeter) wall at beginning of unloading
p_{max} pressure at the cavity wall at beginning of unloading.

The loading and unloading parts of the pressuremeter test can be fitted using eqns. (8) and (9), respectively, to derive the parameters G_i, τ_{ult} and σ_{h0}. Full details of the derivations are given by Ferreira and Robertson (1992).

TESTING TO DETERMINE SOIL MODULUS

As discussed earlier, the unloading part of the pressuremeter curve is less influenced by disturbance during installation. Hence, it is logical to first analyse the unloading part of the test to derive τ^*_{ult} and G_i. If a value for the ratio $R_\tau (= \tau^*_{ult}/\tau_{ult})$ is assumed, the undrained shear strength in loading (τ_{ult}) is therefore determined. Using these values of τ^*_{ult}, G_i and R_τ the loading part of the test can then be analysed to derive σ_{h0}.

If the loading part of the test is influenced by disturbance, the early part of the loading portion will not agree with the derived curve. Hence, the loading part of the test should only be analysed over the last part of the curve, which will, in general be less influenced by disturbance. Based on this logic, the following steps are prescribed for the proposed interpretation methodology:

Step (1) Use the unloading analytical eqn. (9) to fit the unloading portion of the pressuremeter test. Two parameters are derived from the best fit: τ^*_{ult} and G_i.

Step (2) Assume a value for R_τ and apply the derived values of τ^*_{ult} and G_i to fit the last part of the loading portion of the pressuremeter test to determine σ_{h0}.

This interpretation methodology is shown schematically in Fig. 4. This process accepts that the initial loading portion of the pressuremeter test is influenced by some amount of disturbance. If the self-boring installation process has resulted in very little disturbance the analytical curve

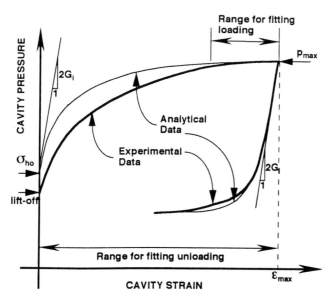

Fig. 4. *Schematic representation of interpretation methodology for self-boring pressuremeters (after Ferreira and Robertson (1992))*

should match closely the entire measured pressuremeter curve, even though the curve fitting is performed using the last portion of the loading curve.

The proposed method (eqns. (8) and (9)) is based on small strains and would not be applicable to large strain pressuremeter expansion. A large strain solution has been developed but for the purpose of this paper the simpler small strain solution is adequate (Ferreira, 1992).

The parameter R_τ is required to obtain the value of the undrained shear strength τ_{ult} and the horizontal in situ initial stress σ_{h0}. Ideally, the parameter R_τ should be obtained from laboratory testing on high quality undisturbed samples, tested under stress paths similar to those experienced during undrained pressuremeter expansion and contraction, as suggested by Jefferies (1988). However, for the initial interpretation of most pressuremeter tests this is not possible, and an estimate of R_τ should be made. Undrained pressuremeter expansion and contraction can be considered to be a plane strain problem where the stress path in unloading is the reverse of loading. Hence, since the plane strain strength envelope is the same in loading and unloading, it is reasonable to assume that R_τ is equal to 2.0. Both Jefferies (1988) and Houlsby and Withers (1988) also assumed that the strength in loading equals the strength in unloading. Therefore, for the soil model used in this methodology the value of the ratio R_τ is assumed to be 2.0.

Application example

The methodology proposed by Ferreira and Robertson (1992) has been applied to high quality SBPT results performed in a uniform clay deposit, reported by Fioravante (1988) and A.G.I. (1991).

Several approaches can be used to fit an equation to experimental data. Rather than develop specific software to perform the curve fitting process using an optimization routine, it is preferable to use available application software developed for personal computers. The software selected for this study was Kaleidagraph™ (version 2.1) developed for a Macintosh™ microcomputer. This application is a powerful tool to perform calculations, graphs and curve fitting. The experimental data, from the field, after being corrected for membrane stiffness, are copied to a Kaleidagraph™ worksheet, so the data become available for analyses. The only manipulation needed is to separate the loading and unloading portions of the test, eliminate any unload–reload data, and arrange both sets of remaining data in ascending order. The data are then ready to be analysed. The least square error curve fitting method was used to fit the general function to a set of experimental data. This is a very simple and well understood method and can be used readily when the data do not present a large scatter and have a defined trend. This is the case of the data from pressuremeter tests. Loading and

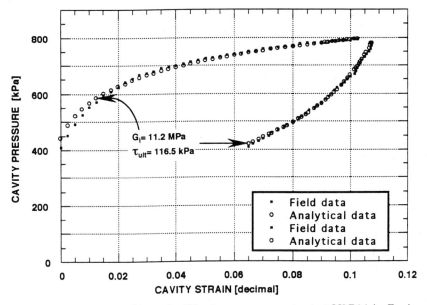

Fig. 5. Computer matching of self-boring pressuremeter test V2P14 in Fucino clay (after Ferreira and Robertson (1992))

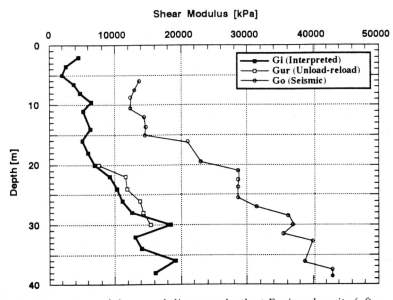

Fig. 6. Comparison of shear moduli versus depth at Fucino clay site (after Ferreira and Robertson (1992))

unloading analytical equations are entered and the curve fitting is performed. As a result, a graph is automatically displayed on the monitor screen, and a visual check of the match between the experimental and analytical curves can be made.

Fioravante (1988) presented 36 self-boring pressuremeter test results performed in Fucino clay in two boreholes, at the same site. A complete geotechnical characterization of the Fucino clay is presented by A.G.I. (1991). The clay deposit, located within the central Apennines, is described as a soft, homogeneous, highly structured $CaCO_3$ cemented, lacustrine clay. The cementation with calcium carbonate plays an important role in the mechanical behaviour of the clay, being responsible for some discrepancies shown by different tests, mainly when disturbance is present.

To illustrate the technique, the interpretation of one high quality self-boring pressuremeter test (V2P14, depth 26 m) will be presented. Figure 5 shows the computer matching of the analytical solution with the measured pressuremeter response and shows that the analytical solution provides an excellent fit to both the measured loading and the measured unloading curves. Figure 6 compares the interpreted initial shear modulus from the proposed interpretation (G_i) with the unload–reload modulus (G_{ur}) and with the small strain shear modulus derived from in situ shear wave velocity measurements (G_0) for all 20 SBPTs in

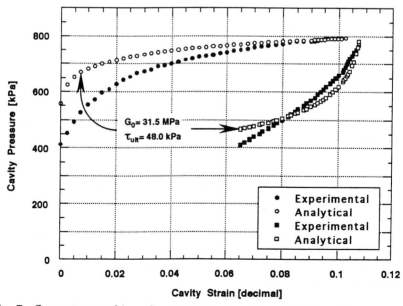

Fig. 7. *Computer matching of test V2P14 using G_0 as initial modulus*

borehole V2 in the Fucino clay. The interpreted G_i values are very close to the unload–reload moduli G_{ur} and are approximately one third the values of G_0.

The shear modulus (G_i) derived from the interpretation is based on an assumed hyperbolic stress–strain relationship. Experience has shown that the hyperbolic expression is a reasonable representation of the stress–strain response of many soils over a variety of strain ranges (Vucetic and Dobry, 1991). Duncan and Chang (1970) suggested that the hyperbolic expression was reasonable to describe the stress–strain response from an initial shear strain of around $10^{-1}\%$ to failure ($>1\%$). The results presented in Fig. 6 support this suggestion.

The interpretation procedure proposed by Ferreira and Robertson (1992) is dominated by the intermediate to large strain response of the pressuremeter. It should be possible to apply the same methodology to the smaller strain response of the pressuremeter by curve fitting the analytical solution to the unload–reload loops or the initial portion of the unloading curve. However, to perform this successfully requires exceptionally high quality pressuremeter data with a large amount of very accurate strain measurements over a small range of cavity strain. These measurements are complicated by the interaction between creep and consolidation which frequently produces continued cavity expansion at constant pressure or during initial unloading. The pressuremeter strain data are therefore unreliable at small strains, even if the pressuremeter instrumentation offers high resolution and good stability. An alternative approach is to combine the proposed pressuremeter interpretation of modulus with the small strain shear modulus derived from in situ shear wave velocity measurements. Figure 7 illustrates the computer matching of test V2P14 using G_0 as the initial modulus. This computer curve matching is clearly unacceptable over the complete strain range, illustrating the limitations of the simple hyperbolic relationship to match the complete soil response. This observation is in agreement with recent laboratory measurements (Tatsuoka and Shibuya, 1991; Jardine et al., 1991). Note also, that the computer matching using G_0 produces a much smaller value for τ_{ult}.

Figure 8 presents the results of the pressuremeter interpretations in the form of normalized shear modulus (G_s/G_0) versus log shear strain, (where G_s is the secant shear modulus at any strain level). Figure 8 compares the variation of G_s/G_0 with shear strain from the pressuremeter interpretation using both the pressuremeter derived modulus G_i and the seismic modulus G_0 as the initial slope of the hyperbolic model. The correct variation of modulus is probably some combination of these two simple relations. The suggested variation of the shear modulus over the entire range of the shear strain is obtained by combining the two plots shown in Fig. 8. The resulting modulus curve is shown in Fig. 9

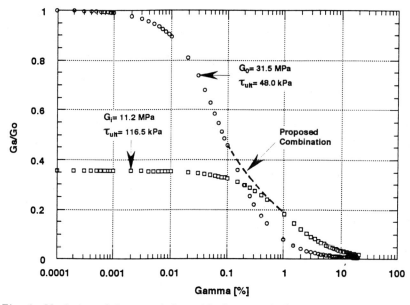

Fig. 8. Variation of shear modulus with shear strain for test V2P14

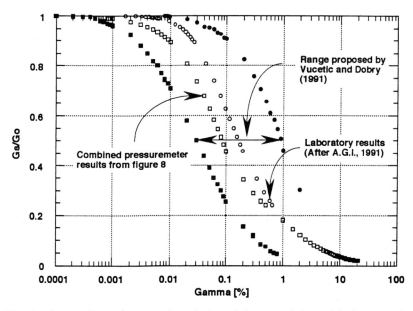

Fig. 9. Comparison of proposed variation of shear modulus with shear strain with curves suggested by Vucetic and Dobry (1991) and results of laboratory tests A.G.I. (1991)

and is compared with results of laboratory tests on undisturbed samples of Fucino clay (A.G.I., 1991). Also shown on Fig. 9 are the curves proposed by Vucetic and Dobry (1991) for PI of 0% and 200%. The PI for the Fucino clay is approximately 70% (A.G.I., 1991). The suggested pressuremeter variation of shear modulus with shear strain is very close to the laboratory results and similar to that suggested by Vucetic and Dobry (1991) for the same PI soil.

Wood (1990) proposed a technique for computer matching of reloading curves from pressuremeter data using the more realistic relationship to model the variation of modulus with strain proposed by Jardine et al. (1986). However, to be able to model the variation of modulus with strain in a more realistic way requires many variables, often each variable has little or no physical meaning. The computer matching is then more difficult and may produce non-unique solutions when many variables are involved. The proposed combination of seismic small strain modulus with pressuremeter modulus represents a simple and easy method to evaluate the variation of shear modulus with shear strain and provides a reasonable and consistent framework to derive other soil parameters from the pressuremeter test. The methodology can be applied to pressuremeter tests which have some disturbance and does not require high resolution in the pressuremeter measurement system.

Acknowledgements

The authors would like to acknowledge the financial support from the Federal University of Santa Catarina, Brazil, and the Brazilian Government (CAPES-MEC). The authors would also like to acknowledge Prof. M. Jamiolkowski for providing the pressuremeter data for Fucino clay.

References

ADDO, K. AND ROBERTSON, P.K. (1992). Shear-wave velocity measurements of soils using Rayleigh waves, Canadian Geotechnical Journal, June.

A.G.I.-ASSOCIAZIONE GEOTECNICA ITALIANA. (1991). Geotechnical characterization of Fucino clay, X ECSMFE, Vol. 1, Florence, Italy.

ATKINSON, J.H. AND EVANS, J. S. (1985). Discussion on: The measurement of soil stiffness in the triaxial apparatus, by Jardine, R. J. Symes, M. J. and Burland, J. B., Geotechnique, No. 3.

BAGUELIN, F., JEZEQUEL, J.F. AND SHIELDS, D.H. (1978). The Pressuremeter and Foundation Engineering. Trans. Tech. Publications.

BELLOTTI, R., GHIONNA, V., JAMIOLKOWSKI, M., LANCELLOTTA, R. AND MANFREDINI, G. (1986). Deformation characteristics of cohesionless soils from in situ tests. In Situ '86 – Use of In Situ Tests in Geotechnical Engineering, ASCE, Blacksburg, pp. 47–73.

BRIAUD, J.L. AND COSENTINO, P.J. (1990). Pavement design with the pavement pressuremeter. Proceedings of the Third International Symposium on Pressuremeters, Oxford, pp. 401–413.

CAMPANELLA, R.G., HOWIE, J.A., SULLY, J.P. AND ROBERTSON, P.K. (1990). Evaluation of cone pressuremeter tests in soft cohesive soils. Proceedings of the Third International Symposium on Pressuremeters, Oxford, pp. 125–135.

CLOUGH, G.W., BRIAUD, J.L. AND HUGHES, J.M.O. (1990). The development of pressuremeter testing. Proceedings of the Third International Symposium on Pressuremeters, Oxford, pp. 25–45.

DUNCAN, J.M. AND CHANG, C.Y. (1970). Nonlinear analysis of stress and strain in soils. Journal of the Soil Mechanics and Foundation Division, ASCE, 96: SM5, pp. 1629–1653.

FERREIRA, R.S. (1992). Soil parameters derived from pressuremeter tests using curve fitting techniques, Ph.D. Thesis, Civil Engineering Department, University of Alberta, Canada.

FERREIRA, R.S. AND ROBERTSON, P.K. (1992). Interpretation of undrained self-boring pressuremeter test results incorporating unloading, Paper under review, Canadian Geotechnical Journal.

FIORAVANTE, V. (1988). Interpretazione delle prove pressiometriche in argille con particolare riferimento alla fase di holding. Dottorato di Ricerca in Ingegneria Geotecnica. Politecnico di Torino, Italy.

GAMBIN, M.P. (1990). The history of pressuremeter practice in France. Proceedings of the Third International Symposium on Pressuremeters, Oxford, pp. 5–24.

GIBSON, R.E. AND ANDERSON, W.F. (1961). In situ measurement of soil properties with the pressuremeter. Civil Engineering and Public Works Review, Vol. 56, pp. 615–618.

HARDIN, B.O. (1978). The nature of stress–strain behaviour for soils. Proc. ASCE Conf. on Earthquake Eng. and Soil Dynamics.

HARDIN, B.O. AND DRNEVICH, V.P. (1972a). Shear modulus and damping in soils: Measurements and parameters effects. Journal of the Soil Mechanics and Foundation Division, ASCE, Vol. 98, pp. 603–624.

HARDIN, B.O. AND DRNEVICH, V.P. (1972b). Shear modulus and damping in soils: Design equations and curves. Journal of the Soil Mechanics and Foundation Division, ASCE, Vol. 98, pp. 667–692.

HOULSBY, G.T., CLARKE, B.G. AND WROTH, C.P. (1986). Analysis of the unloading of a pressuremeter in sand. The Pressuremeter and Its Marine Applications: Second International Symposium, pp. 245–262.

HOULSBY, G.T. AND WITHERS, N.J. (1988). Analysis of the cone-pressuremeter test in clay. Geotechnique, Vol. 38, No. 4, pp. 575–587.

HUGHES, J.M.O. AND ROBERTSON, P.K. (1985). Full displacement pressuremeter testing in sands. Canadian Geotechnical Journal, Vol. 22, No. 3.

JARDINE, R.J., SYMES, M.J. AND BURLAND, J.B. (1984). The measurement of soil stiffness in the triaxial apparatus, Geotechnique, No. 3.

JARDINE, R.J., POTTS, D.M., FOURIE, A.B. AND BURLAND, J.B. (1986). Studies of the influence of non-linear stress–strain characteristics in soil structure interaction, Geotechnique, No. 3.

JARDINE, R.J., ST. JOHN, H.D., HIGHT, D.W. AND POTTS, D.M. (1991). Some practical applications of a non-linear ground model, Proc. X ECSMFE, Florence, Italy.

JEFFERIES, M.G. (1988). Determination of horizontal geostatic stress in clay with self-bored pressuremeter. Canadian Geotechnical Journal, Vol. 25, pp. 559–573.

KONDNER, R.L. (1963). Hyperbolic stress–strain response: cohesive soil. Journal of the Soil Mechanics and Foundation Division, ASCE, Vol. 89, pp. 115–143.

MAIR, R.J. AND WOOD, D.M. (1987). Pressuremeter testing – Methods and interpretation. CIRIA Ground Engineering Report In situ testing. Butterworths, pp. 160.

NAZARIAN, S. AND STOKOE, K.H. (1984). In-situ shear wave velocities from Spectral Analyses of Surface Waves, 8th World Conf. on Earthquake Eng., Vol. III, pp. 31–38.

ROBERTSON, P.K. (1982). In situ testing of soil with emphasis on its application to liquefaction assessment. Ph.D. Thesis, University of British Columbia, Canada.

ROBERTSON, P.K., CAMPANELLA, R.G., GILLESPIE, D. AND RICE, A. (1986). Seismic CPT to measure in-situ shear wave velocity, Journal of Geot. Eng., ASCE, Vol. 112, No. 8, pp. 791–803.

ROBERTSON, P.K. AND HUGHES, J.M.O. (1986). Determination of properties of sand from self-boring pressuremeter tests. The Pressuremeter and Its Marine Applications: 2nd Inter. Sym., pp. 283–302.

SCHNAID, F. AND HOULSBY, G.T. (1990). Calibration chamber test on the cone pressuremeter in sand. Proceedings of the Third International Symposium on Pressuremeters, Oxford, pp. 263–272.

TATSUOKA, F. AND SHIBUYA, S. (1991). Modelling of non-linear stress–strain relations of soils and rocks – Parts 1 and 2, Seisan-Kenkyu, Journal IIS. Univ. of Tokyo. Vol. 44, Nos. 9 and 10.

VUCETIC, M. AND DOBRY, R. (1991). Effect of soil plasticity on cyclic response. Journal of the Geotechnical Engineering Division, ASCE, Vol. 117, pp. 89–107.

WITHERS, N.J., SCHAAP, L.H.J. AND DALTON, C.P. (1986). The development of a full displacement pressuremeter. The Pressuremeter and Its Marine Applications: Second International Symposium, pp. 38–56.

WITHERS, N.J., HOWIE, J., HUGHES, J.M.O. AND ROBERTSON, P.K. (1989). Performance and analysis of cone pressuremeter tests in sands. Geotechnique, Vol. 39, No. 3, pp. 433–454.

WOOD, D.M. (1990). Strain-dependent moduli and pressuremeter tests, Geotechnique, Vol. 40, No. 3, pp. 509–512.

WROTH, C.P. (1984). The interpretation of in situ soil tests. Geotechnique, Vol. 34, pp. 449–489.

WROTH, C.P. AND HOULSBY, G.T. (1985). Soil Mechanics property characterization and analysis procedures, Proc. XI ICSMFE, San Francisco, USA, Vol. 1, pp. 1-55.

Prediction and performance of ground response due to construction of a deep basement at 60 Victoria Embankment

H.D. ST JOHN, Imperial College of Science, Technology and Medicine, D.M. POTTS and R.J. JARDINE, Geotechnical Consulting Group, and K.G. HIGGINS, Imperial College of Science, Technology and Medicine

A 19 m deep excavation for a complex L-shaped basement was carried out using top–down construction techniques at a city site in London. Ground conditions were variable due to the riverside location and detailed investigation was necessary. This included in situ measurements. Constraints to construction are discussed and the final sequence of events is described. Ground movements during construction were predicted based on analyses using a small strain stiffness model for the soils in a finite element analysis and experience elsewhere in London. The predictions are compared with measurements made during construction and conclusions drawn concerning the accuracy of modelling techniques, and the benefits of monitoring both to the project and in terms of data collection of empirical prediction of ground movements.

Introduction

The site at No. 60 Victoria Enbankment has been redeveloped over the period 1987 to 1991. The Building Design Partnership were appointed as Architects and Structural Engineers, while the Geotechnical Consulting Group were appointed by BDP at the start of the project to advise the team on all geotechnical issues.

The development required the construction of a 19 m deep basement on a complex river side site whilst retaining an existing listed building and façades, with the basement passing under one of the latter.

Preliminary studies revealed very variable ground conditions. The site included an area reclaimed from the mouth of a tributary of the Thames, fell within the vicinity of a known deep scour hollow in the London clay and had previously been the site of a gasworks. Because of these geological complexities and the proximity of existing buildings it was considered essential to undertake particularly detailed investigations.

The overall aim was to obtain sufficient information for accurate predictions to be made of the ground movements associated with basement construction, and so enable appropriate construction sequences and methods to be chosen.

Predictions of ground movement around the site were made using the

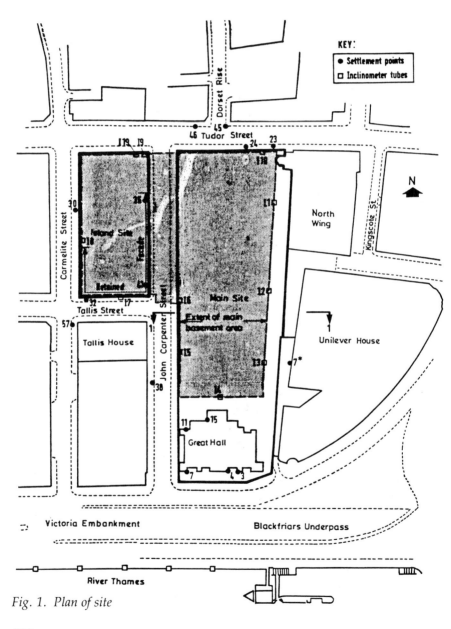

Fig. 1. Plan of site

GROUND RESPONSE AT A DEEP BASEMENT

Imperial College Finite Element Program (ICFEP) with the soil represented by a non-linear elastic model combined with Mohr–Coulomb yield and flow criteria. The parameters for the analysis were obtained directly from the results of laboratory tests undertaken at Imperial College. The results of a number of analyses representing different sections through the site were used together with the results of observations on other similar sites to determine the likely movement of the ground around the basement area.

Detailed measurements were taken of ground surface settlements and lateral movements of the retaining walls during construction. This

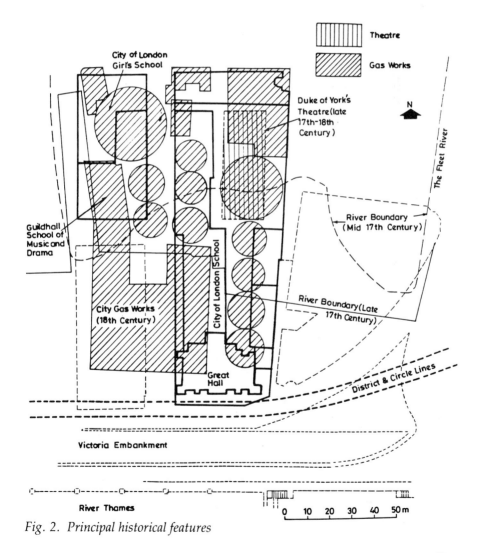

Fig. 2. Principal historical features

enabled potential problems to be recognised at an early stage and was essential under such variable ground conditions. It also enabled the owners of neighbouring properties to be assured that their buildings were not being adversely affected.

This paper describes the main geotechnical issues of the project and the analyses and predictions made of ground behaviour. The observations made during construction are summarised and comparisons are made between these and the original predictions.

Description of the site

Figure 1 shows a plan of the site with the principal features marked. The principal historical features are shown in Fig. 2. The topography varies little across the site with the average ground level being approximately +6 m OD. Immediately to the north the ground rises towards Fleet Street. The shallow District and Circle Underground Line runs just south of the site, with the front steps of the Great Hall resting over it.

The Great Hall of the City of London Boys School at the Southern End of the Main Site was built in the 1880s. It comprises an external masonry structure with a massive timber roof which covers the entire plan area. There are two internal floors beneath the main hall area which are supported on cast iron columns beneath brickwork walls.

The structure was founded on mass concrete footings beneath the main columns and buttresses, which extended to around 2 m below the ground surface. Contemporary records suggested that fill material beneath these footings had been excavated and replaced before the foundations were constructed.

The area to the north west of the development contained the old Guildhall School of Music, three sides of which comprised listed four storey masonry façades. Substantial basements had been constructed in the northern section of this as part of a printing works. This site was named the Island Site. The remaining buildings on the Main Site were to be demolished.

The buildings outside the site also received considerable attention. The most important, Unilever House, lies immediately to the east and was constructed in the 1920s. It has eight storeys, an elaborate curved façade on the corner of the Victoria Embankment and Blackfriars Road and is founded on driven cast in place piles beneath a two level basement. The rear of the building where one wall forms the site boundary, was in a visibly poor condition. Professor C.P. Wroth was appointed as advisor to the party wall surveyor by the owners.

The other surrounding buildings varied from small houses with shallow footings to recently constructed office developments founded on large diameter under-reamed bored piles.

GROUND RESPONSE AT A DEEP BASEMENT

A further matter of concern was the large gas main which ran close to the site boundary along John Carpenter Street and along the northern site boundary.

Description of the development

The main superstructure of the proposed building which was to be a steel frame with concrete floors was designed to maximise the space available on the two sites. The building height was constrained by the St Paul's grid, and consequently deep basements were proposed which involved a general excavation to 19 m below street level. Rather than develop two separate sites, an early decision was taken to excavate the entire site from the rear of the Great Hall, temporarily closing and removing the northern half of John Carpenter Street and supporting the portion of the retained façade over the basement on the main foundation piles.

The outline of the main retaining walls is shown in Fig. 1. The superstructure was to be supported on the walls and columns spaced approximately in a 7 m grid.

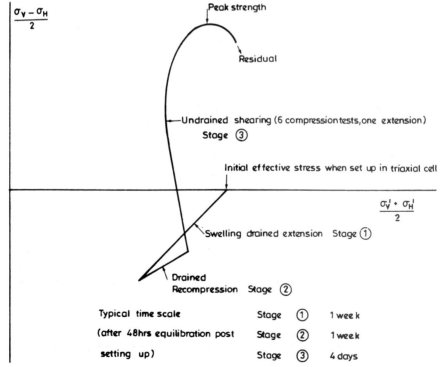

Fig. 3. Typical stress path

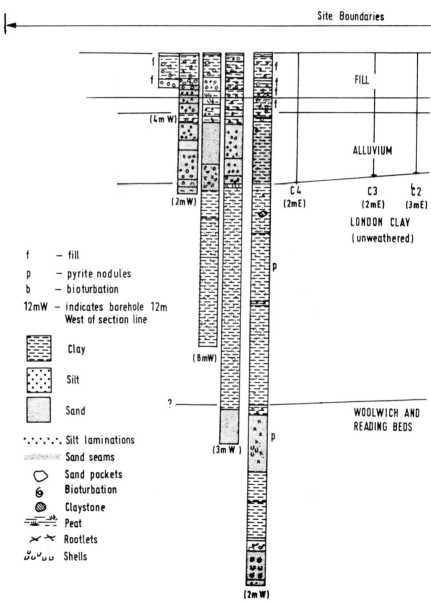

Fig. 4 (above and facing page). Typical section through boreholes North to South

GROUND RESPONSE AT A DEEP BASEMENT

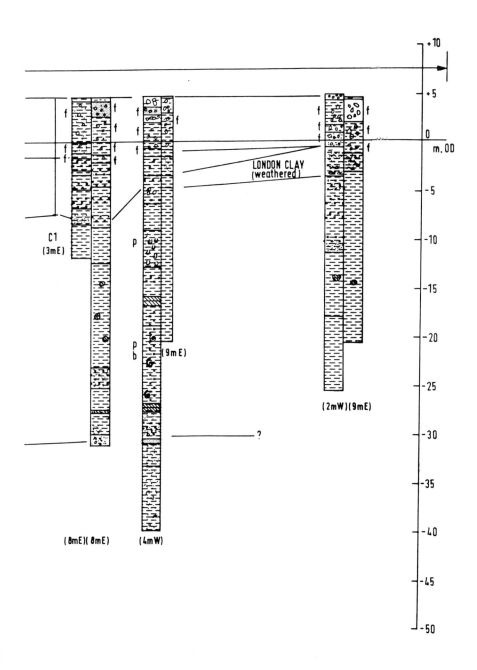

Site investigation strategy

The site investigation was designed to first establish the general ground conditions and secondly provide the detailed information on soil properties required for the proposed finite element analyses. The field work was performed in phases which were supervised directly by the design team, so allowing modifications when appropriate. The principal components were:

(a) Shallow borings to define extent of fill and granular deposits.

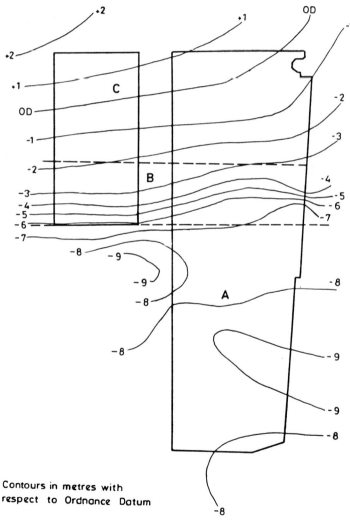

Fig. 5. Contours of London clay surface

(b) Cone penetration tests to define the extent and properties of near surface alluvial materials.
(c) Self-boring pressuremeter tests to determine initial stress conditions.
(d) Deep borings to examine variability of soils at and below founding depths of piles.
(e) Careful sampling for laboratory testing and fabric studies.

Standpipes and piezometers were installed in a number of boreholes;

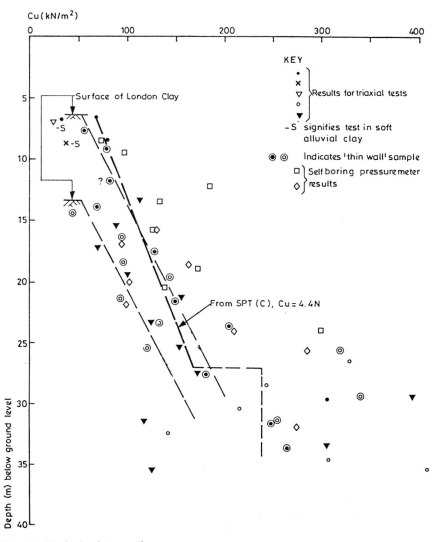

Fig. 6. Undrained strengths

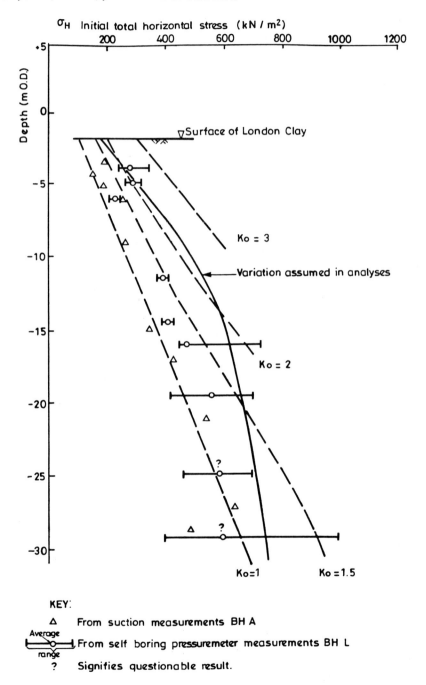

Fig. 7 (above and facing page). Initial total horizontal stresses

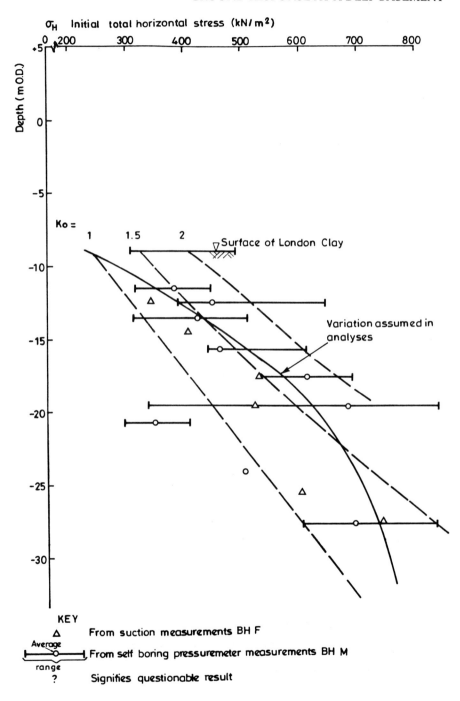

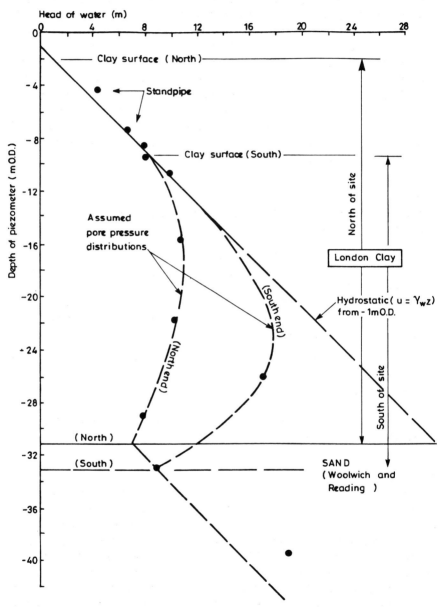

Fig. 8. Pore pressures

constant head permeability tests were carried out in some of these and a pumping test was carried out in the soils just below the London clay.

The sampling operations included driven U102s, pushed in thin walled tubes and rotary coring. The latter was attempted, with limited success, to examine detailed soil fabric. Hight et al. (1992) report how improved techniques gave rather better results in the subsequent Waterloo International Terminal Project. The thin walled tube samples were intended for laboratory triaxial tests and suction measurements.

In addition to the standard laboratory tests, stress path experiments were undertaken at Imperial College with local strain measurements. A typical stress path is shown in Fig. 3. These tests were used to define the non-linear ground model.

Stratigraphy

A typical section through the boreholes is given in Fig. 4.
The following can be seen:

(a) fill is variable and generally 5 to 7 m thick
(b) the thickness and nature of the alluvial deposits varies significantly across the site
(c) the base of the London clay generally occurred at −30 m OD
(d) a water bearing dense fine silty sand was found immediately beneath the London clay

Contours of the top of the London clay are shown in Fig. 5; the overlying alluvium comprised principally a soft silty calcareous clay towards the north east, becoming progressively more granular to the south and west. The silty sand beneath the London clay was not found in the boreholes on the Island site, and was nowhere greater than 7 m thick. Pumping tests revealed that it acted as a confined aquifer.

Soil properties

Testing concentrated on the behaviour of the London clay which was both the main bearing stratum and the main stratum influencing global ground movements. The laboratory tests showed that the London clay could be roughly divided into zones of variable plasticity, an upper zone of higher plasticity (PI = 41–54%) being located above −19 m OD. The results of undrained strength tests on the London clay are summarised in Fig. 6. Two sets of data are presented, one associated with the areas where the clay surface was higher (0 to −2 m OD) and the other where the surface was lower (−6 to −9 m OD). It is clear that strength was related more to depth below the clay surface than total depth.

Figure 7 presents the results of measurements of initial total horizontal stress from boreholes at the two ends of the site. These are compared to the values for assumed K_0 (σ'_h/σ'_v) taking the measured pore pressures given in Fig. 8.

The results of the special triaxial tests are summarised in Fig. 9 in terms of normalised stiffness and bulk modulus against axial strain on a logarithmic scale. The significance of the form of presentation is described by Jardine et al. (1991). The results are comparable with results from tests on London clay at other sites, taking into account the effect of different stress histories. It is clear that over the small to

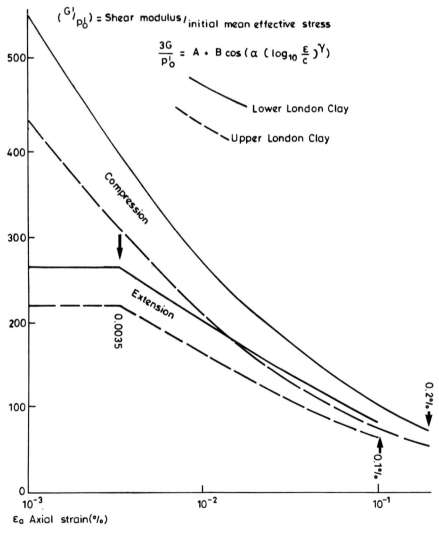

Fig. 9 (above and facing page). Non-linear deformation parameters for London clay

moderate strain range operating during the excavation process, the relationship is steeply non-linear. It can also be seen that the upper London clay appeared to be slightly less stiff than the remainder, and that the clay was less stiff in triaxial extension than compression.

Geotechnical design and sequence of construction

The essence of the substructure design was to limit ground movements whilst minimising the obstructions made to economical construction. A further significant feature was the concern over the possible effects of rising groundwater in the chalk aquifer beneath the site exacerbated by the presence of high permeability layers beneath the London clay. It was anticipated that in this area of London the natural water table might rise to 5 m OD (CIRIA, 1989).

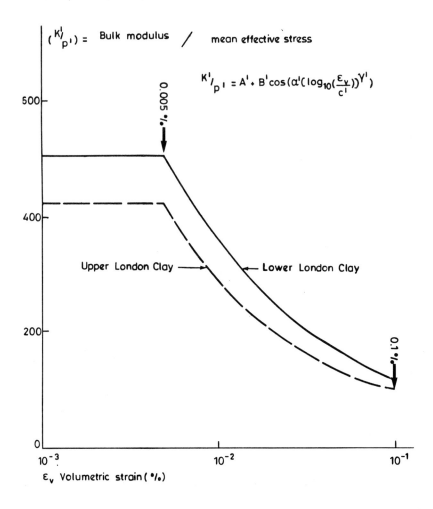

The final choice of method and sequence was guided by the results of the finite element analyses described earlier. The analyses were also used to provide information for the design of the retaining system and piles.

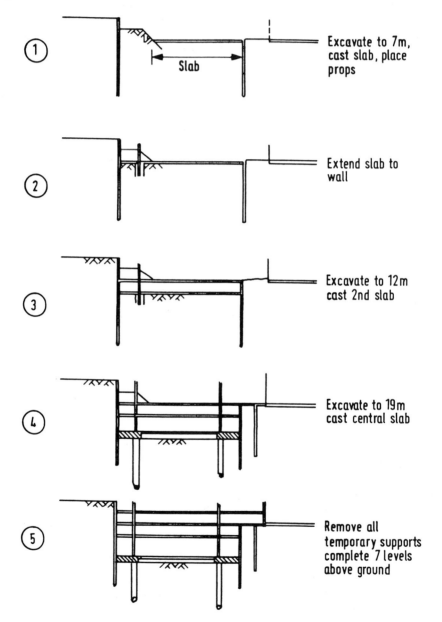

Fig. 10. Basement construction sequence

As a result of these constraints, a top-down construction sequence was adopted with the building being founded on piles placed at an early stage of construction. The piles were cut off at the bottom of the basement and steel stanchions were placed and fixed inside casings, which were subsequently backfilled. The retained façade could therefore be supported on the stanchions prior to the start of excavation. However, the first operation was to install a cast in situ concrete retaining wall which would provide both permanent support to the sides of the excavation and act as a cut-off.

Because of the relatively high water table, the high permeability of the (relatively thick) alluvium and fill and the proximity of the River, exclusion of water was a major consideration. The presence of significant deep obstructions, and the need to provide considerable bending resistance, led to the choice of a secant pile wall comprising 1180 mm diameter piles at 1050 mm centres with steel joists in alternate piles. These piles extended a minimum of 5 m below the excavation and were required to take the loads from the external walls of the structure.

In order to avoid constructing piles that penetrated beyond the London clay, single under-reamed bored piles were used at each column location (up to 1.8 m shaft diamter with up to 4.5 m diameter under-reams). These were terminated at a maximum depth of 3 m above the sandy stratum.

The long-term behaviour of the substructure was checked for the case where the water pressure rose to the anticipated +5 m OD in the sandy stratum. It was found that the whole volume of clay beneath the basement could be taken to a condition of zero vertical effective stress. In order to prevent this unsatisfactory condition a series of open wells were installed into the sandy layer which vented to a void beneath the basement slab. This void also acted as a heave space to prevent pressure developing on the underside of the slab. A further ground bearing slab was placed around the periphery of the basement in order to reduce the risk of inward rotation of the retaining walls beneath the basement due to long-term softening of the clay in front of the walls.

A typical sequence of construction of the basement is shown in Fig. 10. This section is taken across the southern end of the site through the basement of Unilever House. Piles were placed after excavation to a general site level around 4 m below street level on the main site. The general sequence of construction was complex because of the shape of the site and the existence of soft zones along the eastern boundary. This gave rise to out of balance forces across the basement which necessitated the completion of effective floor plates across zones of the basement which could act as waling beams, before excavation started beneath them. The final dig left a 7 m clear span between the lowest slab and the excavated surface. Because of the depth of alluvium at the southern end of the site, excavation adjacent to the retaining walls was carefully controlled and the peripheral slab completely in short sections whilst carefully monitoring ground movement.

Prediction of ground movements

Predictions of ground movement were made using plane strain finite element models of three sections through the site representing typical ground and loading conditions. The most complex of these was that shown in Fig. 10. ICFEP was used to model the construction procedure. A brief description of the program may be found in Potts and Day (1990).

Tables 1 and 2 summarise the principal parameters used in the Finite Element model. The non-linear elastic stiffness for the fill, alluvium and the lower sandy layers were based on laboratory test data reported by Daramola (1978) for soils of the appropriate relative density. The clays of the Woolwich and Reading Beds were assumed to have properties similar to the lower London clay. The London clay inside the excavation below final basement level was given the lower stiffness because of the difference observed between extension and compression response.

The ICFEP calculations were carried out using an effective stress soil model. The volumetric strains in clays during the short-term (undrained) phases are controlled by assigning a high bulk modulus to the pore fluid with the pore pressure changes resulting from construction also being calculated. Excess pore pressure can then be dissipated incrementally to

Table 1. Summary of c', ϕ', ν values for different soil types

Material No.	Material description	c'	ϕ'	ν
1	Granular fill in extension	0	30	0
2	Alluvial clay in extension	0	22	0
3	Sand in extension	0	22	15
4	London clay (Zone II) in extension	5	22	0
5	Granular fill in compression	0	30	0
6	Alluvial clay in compression	0	22	0
7	Sand in compression	0	30	15
8	London clay (Zone II) in compression	5	22	0
9	Woolwich and Reading beds in compression (sand)	0	30	15
10	Woolwich and Reading beds in compression (clay)	10	25	10
11	Woolwich and Reading beds in extension (sand)	0	30	0
12	Woolwich and Reading beds in extension (clay)	10	25	0
13	London clay in extension (Zone I)	5	22	0
14	London clay in compression (Zone II)	5	22	0

Table 2. Summary of constants for stiffness model

Material No.	A	B	C	α	γ	E_{MIN}	E_{MAX}	A'	B'	C' %	α	γ'	ε_{MIN}	ε_{VMAX}	G	K_{MIN}
1	600	450	0.002	0.873	1.107	0.0069	0.69	750	650	0.002	0.759	1.364	0.003	0.25	1000	2000
2	341	319	0.003	1.571	0.608	0.006062	0.1732	520	450	0.001	2.069	0.420	0.005	0.1	1000	2000
3	As for material no. 1															
4	410	383	0.003	1.571	0.608	0.006062	0.1732	624	540	0.001	2.069	0.420	0.005	0.1	1200	2400
5	As for material no. 1															
6	1050	950	0.0001	1.335	0.617	0.001732	0.3464	52	450	0.001	2.069	0.420	0.005	0.1	1000	2000
7	As for material no. 1															
8	1365	1235	0.0001	1.335	0.617	0.001732	0.3464	624	540	0.001	2.069	0.420	0.005	0.1	1000	2000
9	4950	4950	0.0005	1.348	0.58	0.0121	0.866	2150	1800	0.0008	1.462	0.77	0.004	0.2	1000	2000
10	As for material no. 8															
11	As for material no. 9															
12	As for material no. 8															
13	As for material no. 2															
14	As for material no. 6															

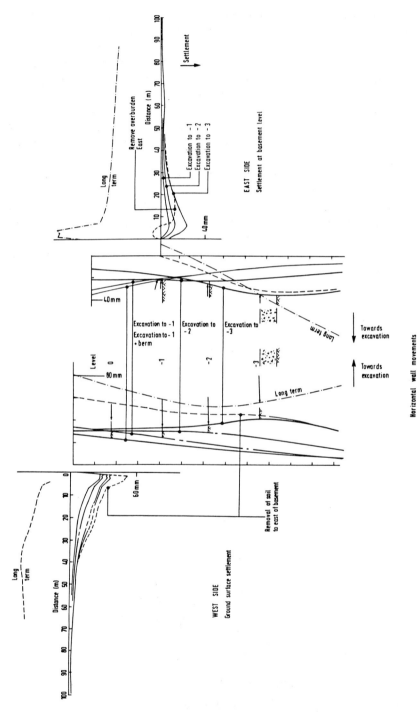

Fig. 11. Summary of results of finite element analysis

a pressure distribution representing a steady state of flow so simulating the effects of long-term post-construction drainage.

The walls were modelled as concrete elements with an elastic Young's modulus of 28 GN/m² in the short term and 14 GN/m² in the long term. Wall installation was modelled by changing the element properties without changing local stress conditions.

Slabs were given stiffness of about 20% of their anticipated full structural stiffness, in order to allow for shrinkage and construction joints based on the results of back analyses of other excavations (St John, 1975). The stiffnesses were increased in the upper floors after the completion of construction in recognition that shrinkage and joint closure would be complete.

The analyses did not model the effect of the piles and applied building loads in restraining ground movement, although this could have been done in an approximate fashion by increasing the unit weight and stiffness of the ground inside the basement. The restraint offered by the peripheral ground bearing slab, and the row of piles nearest to the wall was modelled in an iterative fashion. A vertical spring restraint was applied to the slab to model the pile, the spring's stiffness was

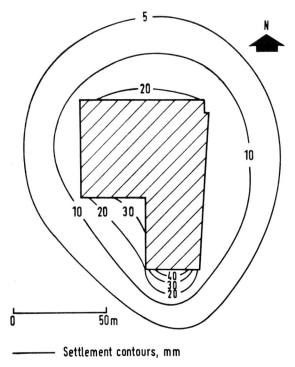

Fig. 12. *Contours of predicted surface settlements*

compatible with the anticipated ground load–displacement relationships of the piles in the field.

Unilever House was represented (in the section through the south portion of the site) by a suitably surcharged 1 m thick concrete slab founded at basement level. This modelled the restraint which the building offered to the ground.

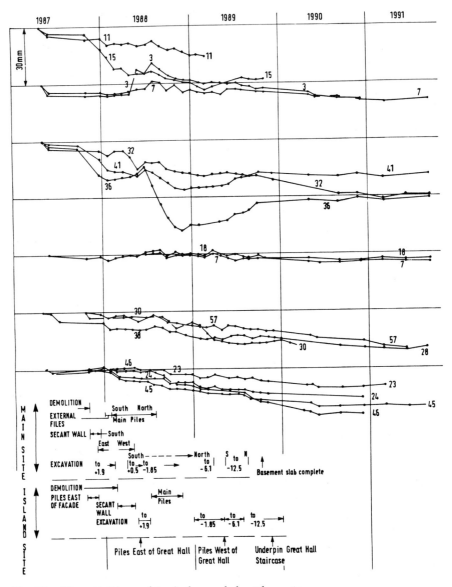

Fig. 13. *Site activities and typical recorded settlements*

The results of the analyses are illustrated in Fig. 11 which summarises the predicted ground movements during and after construction at the critical Unilever House cross-section. It was assumed in this analysis that the ground level to the east of the basement was reduced after completion of the basement. In reality it was removed during the early stages and long-term movements were restrained by the floor slabs acting as walings spanning north to south.

The results of three such analyses were used in combination with observations such as those reported by Burland and Hancock (1977), theoretical analyses described by Burland et al. (1979) and approximate superposition calculations to predict the three-dimensional distributions of movement around the excavation. The results of this exercise for surface settlements are shown in Fig. 12.

Observations during construction

A comprehensive site monitoring system was set up which included 11 wall inclinometers and over 100 settlement points, on and inside buildings, and on the pavements. Distance measurements were set up along an East–West line from Blackfriars Road to across the basement to 100 m beyond Carmellite Street.

Observations were also made of the tilt of the retained façade during critical phases. These measurements were started at the beginning of the site works and were maintained by the Contractor throughout.

Figure 13 shows examples of measured settlements at some typical locations. The relevant site activities are also indicated. A careful examination of this and other data shows that the most significant ground movements occurred in those nearby structures that were unpiled and were associated with local disturbances caused by:

(a) the removal of obstructions to allow the main piling
(b) the installation of the secant pile wall
(c) the installation of grouted mini piles between the Great Hall and Unilever House

Initially, larger than anticipated settlements were noted during the installation of the secant wall. The presence of loose deposits and high ground water table were believed to be the cause of the problem. In order to limit settlement the Contractor was made to control water levels in the casing and control the rate of advance of the auger and casing so that ground was not lost.

Movement adjacent to the Great Hall was apparent when the main grid piles were installed just inside the basement area. Their installation not only resulted in the removal of a substantial volume of soil but also ground loss (in front of the casings) as they were installed through the

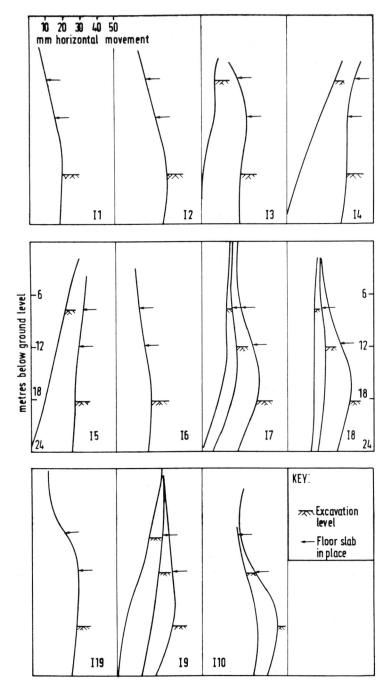

Fig. 14. Summary of inclinometer results

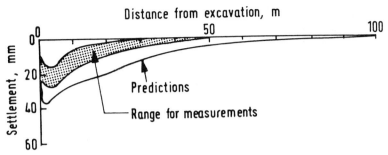

Fig. 15. *Settlements versus distance from excavation*

water bearing overburden. Movement also occurred during the installation of piles around the retained façade. The piles to the east of the façade were installed prior to the demolition of the internal structure and were used to provide a base for an external support to stabilise the façade in the temporary condition. Consequently, when demolition was complete and the internal piles were installed, causing settlement of the façade foundations, the façade tilted in towards the Island site. This tilt was carefully monitored until the façade could be supported by the piles on both sides. A vertical saw cut was made down the corner of the façade in order to reduce the risk of damage due to differential settlement and torsion.

Figure 14 shows some typical results from the inclinometers located around the basement. These seemed to indicate that the most significant movements occurred during the early stages of excavation, whilst the walls were less substantially propped. Large displacements did not develop during the final excavation stages.

In Fig. 15 the range of measured settlements, excluding measurements on piled foundations, are compared to the range of envelope predicted from the analyses.

Discussion

(1) This case study demonstrates that it is possible to predict with reasonable accuracy the ground movements developed around a deep excavation at a complex site.

(2) To do this, careful site investigations and detailed studies of the soil properties are required. A sophisticated finite element code is essential.

(3) Even when these tools are available judgement and experience are necessary to provide a realistic prediction of the ground movements to be expected during construction.

(4) The finite element analyses enabled a range of construction

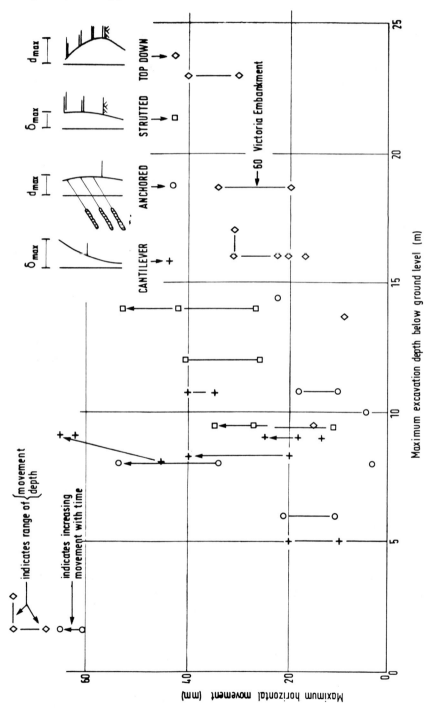

Fig. 16. Observed maximum horizontal wall movement for different depths and types of excavation in London clay

options to be examined and provided detailed information for design of walls and support systems, so proving to be a powerful and versatile design tool.

(5) The value of careful field monitoring was also clearly demonstrated. It enabled initial predictions to be confirmed, problems to be highlighted and their effects minimised, and provided a basis on which to agree the causes and effects of building damage with adjacent owners. The most significant movements occurred due to what could be regarded as peripheral activities adjacent to existing structures e.g. pile installation. In the case of Unilever House a limiting acceptable movement of the secant wall was agreed between the consultants; this limit was not reached.

(6) Collating field data from case histories provides an invaluable way of checking the results of theoretical analyses.

The value of collating the results is indisputable; each new case history adds to the data bank which allows an invaluable way of checking simple predictions for the performance of future projects. Figure 16 is an example of how such collated information can be used to predict horizontal movements of retaining walls embedded substantially in stiff over consolidated (medium to high plasticity) clays. Maximum measured horizontal movement on any part of a wall is plotted against maximum excavation depth based on reported case histories. The type of wall support is indicated for each case.

The range of wall movements measured at 60 Victoria Embankment is indicated on the figure showing that such plots can reasonably estimate the overall ground movement associated with any excavation.

References

BURLAND, J.B. AND HANCOCK, R.J.R. (1977). Underground Car Park at the House of Commons: Geotechnical aspects. Struct. Engrs. Vol. 55, pp. 87–100.

BURLAND, J.B., SIMPSON, B. AND ST JOHN, H.D. (1979). Movements around Excavations in London Clay. ECSMFE Brighton, Vol. 1, pp. 13–29.

CIRIA (1989). The engineering implications of using ground water levels in the deep aquifer beneath London. Special Publication 69.

DARAMOLA (1978). The influence of stress history on the deformation of sand. Ph.D. thesis, University of London.

HIGHT, D.W., PICKLES, A.R., DE MOOR, E.K., HIGGINS, K.G., JARDINE, R.J., POTTS, D.M. AND NYIRENDA, Z.M. (1992). Predicted and measured tunnel distortions associated with Construction of Waterloo International Terminal.

JARDINE, R.J., POTTS, D.M., ST JOHN, H.D. AND HIGHT, D.W. (1991). Some

Practical Applications of a non-linear ground model. Deformation of Soils and Displacement of Structures. XECSMFE, Florence 1991, Vol. 1, pp. 223–228.

POTTS, D.M. AND DAY, R.A. (1990). Use of sheet pile retaining walls for deep excavations in stiff clay. Proc. Instn. Civ. Eng. Part 1 1990, Vol. 88, December, pp. 899–927.

ST JOHN, H.D. (1975). Field and theoretical studies of behaviour of ground around deep excavations in London Clay. Ph.D. thesis, University of Cambridge.

An investigation of bearing capacity and settlements of soft clay deposits at Shellhaven

F. SCHNAID, formerly of Fugro–McClelland Limited,
W.R. WOOD, A.K.C. SMITH, Fugro–McClelland Limited,
and P. JUBB, Shell UK Oil

The paper describes the foundation design for haul routes at the Shellhaven Refinery. Large modules, weighing up to 770 tonnes with their transporters, were to be transported approximately 2 km over soft normally consolidated clay. Preliminary assessment indicated that the problem was one of plane strain bearing capacity of a stiffer crust overlying normally consolidated soil, and that careful design would be necessary to ensure stability of the road at an economic cost. A full scale load test was carried out on a section of road, and finite element analyses carried out to match predicted and observed response of the trial. From these, design recommendations were made for the whole length of the road. Finally, the response of the road was monitored as a module was transported along it, and the measured results compared with the predictions.

Introduction

Shell UK Oil are extending the facilities at their Shellhaven Refinery on the North bank of the Thames in Essex (Fig. 1) by the construction of a Naptha Minus Plant. Some of the major components of the new plant have been fabricated offsite as preassembled units (PAUs). The units, weighing up to 580 tonnes, were transported to the refinery by barge for off-loading at a jetty at the western end of the site. Multi-wheeled transporters (Fig. 2) were used to carry the units across the site to their final location; a distance of about 2 km. The weight of the heaviest PAU and associated transporter was 770 tonnes. Because of the close spacing of the wheels, the transporter can be assumed to transmit a uniform pressure to the ground over its entire base area of approximately 6 m wide by 20 to 30 m long.

Soil conditions at the Shellhaven site consist of soft estuarine clays. A comprehensive site investigation programme, comprising boreholes, sampling, static cone penetration tests and vane tests, was carried out and soil properties were determined both from the in situ tests and from laboratory tests carried out on undisturbed samples. The results indi-

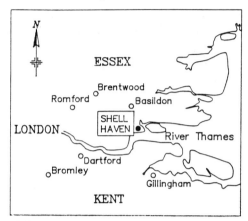

Fig. 1. Location of site

cated that the soil conditions could be modelled as a two layer system, with a relatively stiff crust and softer material below.

Predictions of the stability and settlements of the PAUs during transportation were carried out using simple stability analysis and also two and three dimensional finite element analyses. The predicted ultimate bearing capacity was of the same order as the anticipated applied bearing pressures on the road surface. An instrumented test

Fig. 2. Transporter and PAU

loading was therefore carried out, in which a section of road was loaded to failure, and the displacement and pore pressure response of the soil to the loading was measured.

Further finite element runs were then carried out, examining the stiffness of both the crust and the soft clay beneath, in order to improve the match with the observed load test data. From these runs a maximum allowable bearing pressure was assessed for the PAUs during transportation, and recommendations were made for design of the whole length of the haul route, taking into account variations in soil conditions along the route.

Finally, during construction of the Naptha Minus Plant, the haul road response was monitored as a PAU was transported along it. Survey markers were installed on and beside the road, and were surveyed before and after transportation. Some of the markers on the road were levelled as the PAU passed by them, and the measured transient settlements compared with values calculated from the finite element analyses.

Soil conditions

The soil conditions at the Shellhaven Site were determined from the results of a geotechnical site investigation comprising both 'in situ' and laboratory tests. In situ testing comprised static cone penetration tests,

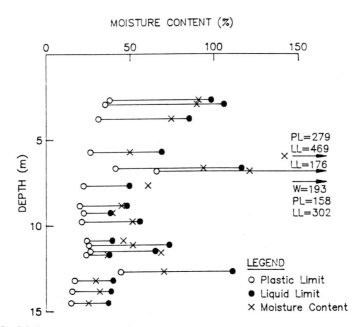

Fig. 3. Moisture content and plasticity profile

piezocone tests and vane shear strength tests. Undisturbed 'GMF' push in samples were taken, and unconsolidated undrained (UU) and isotropically consolidated undrained (CIU) tests carried out. This investigation indicated that the site consists of normally consolidated clays to a depth of 10 to 15 m, generally overlain by a crust up to approximately 1.5 m thick. Dense gravel lies below the clay.

The crust and clay strata govern the design of the haul route. The crustal layer provides valuable load spreading to the underlying clay. Unfortunately, the necessity to excavate inspection pits to search for services limited the amount of cone testing and undisturbed sampling that could be carried out in this material. Data on its strength was

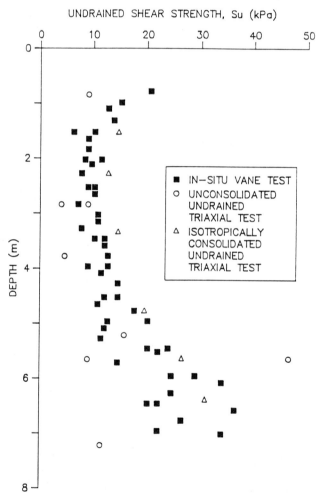

Fig. 4. *Undrained shear strength profile*

obtained from selected CPTs, vane profiles and hand tests carried out in some of the inspection pits. Implied shear strengths in the range 20 to 40 kPa were found, the results from the in situ vane tests generally being lower than from other tests. A profile of the moisture content and plasticity is presented in Fig. 3.

The second significant soil stratum is a layer of very soft to soft silty clay below the firm crust. The undrained shear strength profile is shown in Fig. 4, in which the undrained shear strength s_u is plotted against depth. Results from the in situ vane tests and from unconsolidated-undrained (UU) and isotropically consolidated undrained (CIU) triaxial tests are presented. Results from UU triaxial tests were generally lower than those from vane tests, but results from CIU triaxial tests (consolidated to estimated mean in situ effective stress) were generally higher. In view of the sensitivity of these clays it is considered that UU results may under estimate the in situ strength of the clay. In general, the ratio of shear strength to effective overburden pressure is about 0.20 to 0.30, which gives good agreement with previous studies carried out at the Shellhaven site (Skempton, 1952).

Further information on the shear strength of the soft clay was obtained from static cone penetration tests and piezocone tests. A typical example of a piezocone profile is shown in Fig. 5. This clearly

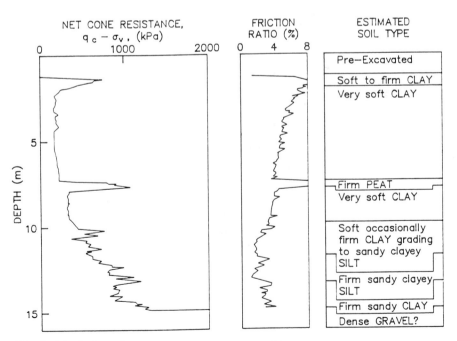

Fig. 5. Typical static cone penetrometer test profile

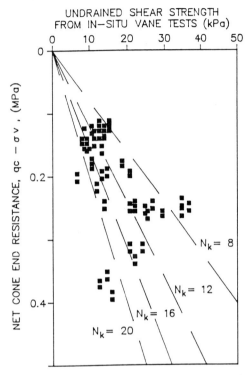

Fig. 6. *Determination of cone factor N_k from in situ vane data*

indicates the inspection pit, the transition from the crust to the soft clay beneath, the gradual increase in strength of the clay (with a band of peat at about 7.5 to 8 m depth) and the termination of the test on the dense gravel at about 15 m depth. The friction ratio generally decreases with depth, indicating an increasing silt content, and corresponding with the general decrease in plasticity with depth (Fig. 3).

The shear strength profile was investigated in detail by means of vane and triaxial tests at only a few locations. The soil strength variation along the land route was investigated by means of cone tests, because of the considerably greater speed and economy of investigation. It was therefore necessary to relate the measured cone resistance, q_c, to the shear strength s_u. This was done by means of the cone factors N_k and N_{kt} (Robertson and Campanella, 1983).

$$N_k = \frac{q_c - \sigma_v}{s_u}$$

$$N_{kt} = \frac{q_t - \sigma_v}{s_u}$$

SOFT CLAY DEPOSITS AT SHELLHAVEN

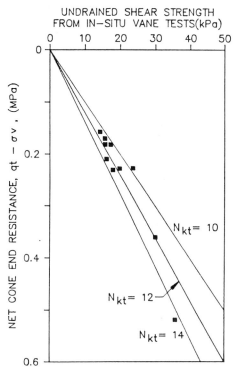

Fig. 7. *Determination of cone factor N_{kt} from in situ vane data*

The value of q_t is computed for piezocone tests, where a correction is made to account for the water pressure that acts on the shoulder of the cone tip at its joint with the sleeve.

The values of N_k and N_{kt} were computed from data presented in Figs. 6 and 7 respectively, in which the net cone resistances, $q_c - \sigma_v$ and $q_t - \sigma_v$, are plotted against in situ vane undrained shear strength. The wide scatter of N_k values is attributed to the variability of the soil, which was indicated by the variation of plasticity index measured across the site.

Bearing capacity assessment

Simple bearing capacity analysis indicated that the ultimate bearing capacity of the soil would be of the order of 60 kPa, compared to an anticipated applied bearing pressure of the order of 65 kPa. It was therefore decided to carry out a trial loading to failure, in which the load–deflection behaviour and the pore pressure response of the soil would also be measured.

It was also decided to carry out a finite element analysis of the trial loading before it was carried out, based upon the site investigation data available, in order to obtain a 'Type A' prediction (Lambe, 1973). Further site investigation was carried out at the trial location so that the analysis could be refined after the trial to obtain a 'Type C1' prediction, which could then be used to predict behaviour of the actual loading over the whole length of the proposed haul route.

Both two-dimensional and three-dimensional finite element analyses were carried out before the trial. The two-dimensional plane strain analysis was performed using a linear elastic–perfectly plastic soil model where the inception of plasticity is defined by the Tresca failure criterion. Eight noded isoparametric plane strain elements were used for the studies reported here. The mesh and the boundary conditions are presented in Fig. 8. Analyses were performed assuming both perfectly flexible and perfectly rigid behaviour of the loaded slab.

A value of 20 kPa was used for the undrained shear strength of the crust, and a value of 20 used for the cone factor N_k for the soft clay. Correlation between the deformation modulus and the undrained shear

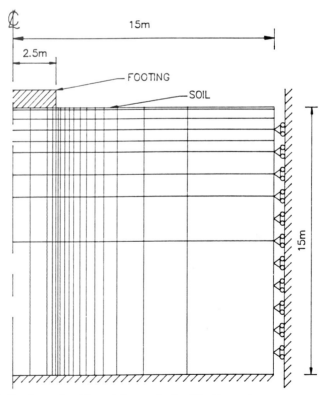

Fig. 8. Finite element grid used for analysis of footing

SOFT CLAY DEPOSITS AT SHELLHAVEN

strength provided the basis for a settlement calculation. Various researchers have back-analysed loading tests on normally consolidated clay sites to establish a value for the ratio E_u/s_u. Values for this ratio in the range 40 to 3000 were tabulated by Simons (1975) including one value of 220 reported by Bjerrum (1964) for a tank test at Shellhaven. The work of Duncan et al. (1976) as quoted in the CIRIA report (Meigh, 1987) gives values from 300 to 600 for the ratio for normally consolidated clays of medium plasticity.

For the two-dimensional analysis a ratio of E_u/S_u of 200 was adopted. Ultimate bearing pressures of 61 kPa and 67 kPa for flexible and rigid footings respectively were obtained.

The three-dimensional analyses were carried out on a finite element mesh 18.5 m by 30.0 m by 14.0 m outside dimensions. This volume was filled with 125 20-noded isoparametric brick elements. The mesh sides were on vertical rollers (no lateral displacements) and the base was fully fixed (no lateral or vertical movements). A linear elastic–perfectly plastic soil model with a Tresca yield criterion has also been used. Soil displacements were derived for these analyses using $E_u = 100 S_u$. Predictions for ultimate bearing pressure for flexible and rigid slabs of 81 kPa and 106 kPa were obtained.

Load test

The load test was carried out on an instrumented selection of an existing road. The road pavement, which consisted of a 175 mm thick by 5 m

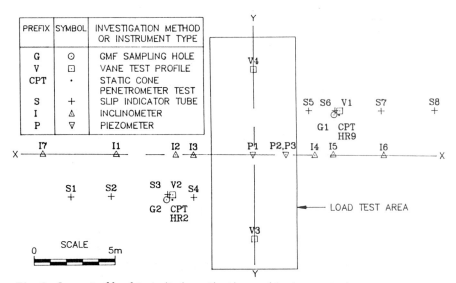

Fig. 9. *Layout of load test site investigation and instrumentation*

wide reinforced concrete slab placed on 225 mm of hardcore, was cut to form an isolated length of 14 m. A trench was excavated all round this slab to the base of the hard core and backfilled with clean sand to ensure no load spread to the adjacent hardcore.

The locations of a range of monitoring points and instruments in relation to the test slab are shown in Fig. 9. A total of 45 survey targets were established to be monitored by two geodi-meter Electronic Distance Measurement Instruments. Each target's movements in three orthogonal directions were measured relative to remote datum targets mounted on disused foundation units.

Seven inclinometer ducts were installed to depths of 8 m below ground. Duct keyway orientation was accurately recorded for most ducts after the test. Profiles of the duct were recorded using a 0.5 m gauge length Slope Indicator inclinometer.

Three-push in electronic piezometers were installed below the slab through holes cut in it. Finally, a total of eight slip indicators were installed. These consisted of flexible plastic tubes set in sand within borings. Rigid 0.5 m long rods placed inside the tubes were connected by cord to ground surface. Any failure plane intersecting and displacing the tube would render the rods impossible to withdraw and the depth of such a slip plane could then be recorded.

All sub-soil instrumentation was pushed into the ground by the hydraulic equipment of a cone truck. This procedure was much faster and probably caused much less soil disturbance than the conventional method of installing instruments in a shell and auger borehole.

Loading of the test bed was carried out by the placement of cubical concrete blocks, each weighing 20 kN. Blocks were placed in a predefined order starting in the centre and moving to the edges of the loading area. A safe method of placement was adopted whereby mobile platforms were used to supervise location and enable sling removal for each block. All plant and block storage was outside the anticipated zone of influence of the test area, such that no surcharge of the failure zone took place.

Monitoring of the movement of the principal targets (the 16 targets around the periphery of the test slab) was made after placements of every 5 blocks (100 kN) and measurements of all 45 targets were taken at 300 kN intervals.

Piezometer readings were taken periodically and full inclinometer records were made on completion of the placement of layers 2 and 3 during a load holding period of 1.5 to 2 h. Those still intact after failure were also recorded.

Vertical deformations of targets immediately adjacent to the centre points of the longitudinal edges are presented as functions of mean bearing pressure in Fig. 10. Significant vertical deflections of about 50 to

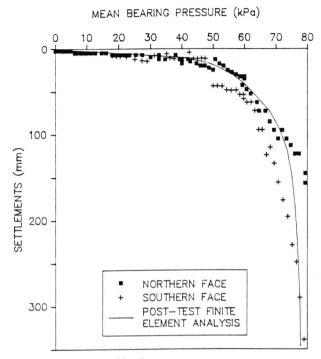

Fig. 10. *Settlement response of load test*

70 mm, indicating some plastic soil behaviour, had occurred by the completion of the third layer (corresponding to a bearing pressure of about 64 kPa). During the latter stages of the placement of the fourth layer movements were occurring without increasing load suggesting a failure condition under sustained load. In view of the nature of the design loading for the road the fourth layer was completed giving a total load of 6000 kN and a mean bearing pressure of 84 kPa. Overall failure occurred 70 minutes later as a rotation of the slab causing toppling of many blocks with consequent damage to some targets and inclinometer tubes (Fig. 11).

Typical results recorded from an inclinometer are presented in Fig. 12 for various stages of loading. Displacements were measured in two diametrically opposite orientations referred to as XX (perpendicular to failure plane) and YY (parallel to failure), as indicated on Fig. 9. Cumulative deflections in the direction YY were generally very small (less than 5 mm), suggesting that failure took place under plane strain conditions. Piezometer readings indicated no discernible dissipation of pore pressure during the test, failure taking place under essentially undrained conditions.

Fig. 11. Load test after failure

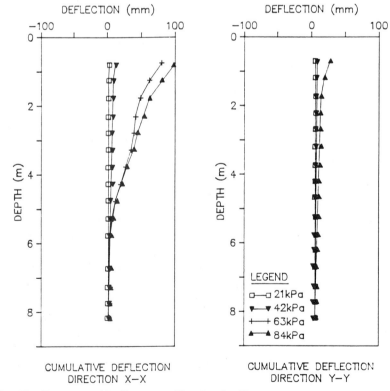

Fig. 12. Response of inclinometer No. 6 to loading

Analysis after trial

After the trial, the results were compared with the finite element predictions. Although it is possible that failure would have occurred at a slightly lower load had it been maintained for longer, the design failure load was taken to be 84 kPa, since the PAU transporter would apply only a transient load at any location. The failure loads predicted by the two-dimensional analysis were 25% and 15% lower than the actual load for flexible and rigid bases respectively, whereas the three-dimensional method gave predictions that were 4% low and 13% high. However, both methods predicted too soft a settlement response. In addition, examination of the surface contours after failure and of the symmetry of the inclinometer data indicated that the failure was essentially two dimensional. Settlement contours of the loaded area after failure were very evenly spaced, implying rigid body rotation of the slab. Consequently, it was concluded that the two-dimensional (plane strain) analysis with a rigid base was most appropriate, but that the soil strength and stiffness parameters needed reassessment to give a better fit to the trial results.

A re-comparison of the cone resistances and the vane shear strengths, and in particular a reassessment of some cone tests that gave particularly low end resistances, indicated that the value of the cone factor, N_k, used for the pre-trial analysis had been too high. The value was revised from 20 to 16, with a proportional increase in the shear strength.

A parameter study using the finite element analysis revealed a considerable influence of the strength of the crust on the soil bearing capacity. The ultimate bearing pressure increased by about 30% for an increase of the crust shear strength from 20 kPa to 35 kPa. A very close matching of behaviour of the load test at ultimate conditions was obtained by adopting a shear strength of 30 kPa for the stiff crust.

In the pre-test analysis the modulus adopted for the linear elastic stage of soil behaviour was derived from published relationships between modulus E_u and undrained shear strength s_u. Predicted settlements were much greater than observed field measurements.

The success of the analysis in predicting undrained settlements depended crucially on the use of appropriate values of undrained modulus E_u. However, E_u is known to be considerably affected by stress level, shear strain amplitude, stress history and soil disturbance (Ladd, 1964; Marsland, 1971; Atkinson, 1973). Since deflections under serviceability conditions were critical in the haul road, a more realistic modulus was obtained from a laboratory triaxial CIU test, carried out on one of the push-in samples obtained at the trial location. Sample strains were measured internally by a local Hall sensor measuring device (Clayton and Khatrush, 1986). Results from both internal and external strain measurements are shown in Fig. 13. It can be seen that the

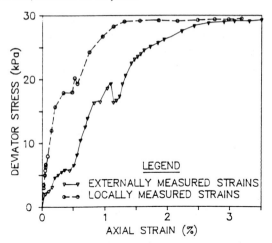

Fig. 13. *Measured strains in consolidated undrained triaxial test*

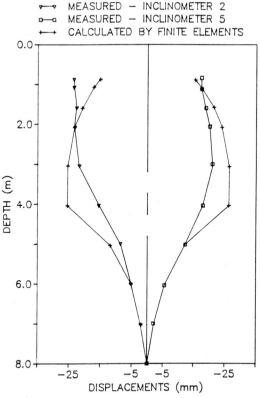

Fig. 14. *Comparison of measured and calculated horizontal displacements immediately before failure*

internal strain measurement response is much stiffer than that measured externally, as errors arising from equipment compliance and sample end effects are removed. The modulus value derived from local measurements yields a value for the ratio E_u/s_u of 400 to 500. This value is within the range of published data and justifies the use of 400 for the test back-analysis.

A comparison between measured and predicted values is presented in Fig. 10, in which the mean vertical displacement is plotted against bearing pressure. An excellent agreement is observed for initial elastic deformations (up to about 20 mm) and good agreement for settlements up to about 100 mm. Such a settlement would be classified as a serviceability failure. Beyond this, the test behaviour became markedly asymmetric as the final gross failure occurred. The finite element model, being symmetric, could not show this; nonetheless, it did indicate rapidly increasing settlement at a load very close to the final failure load.

Predicted and measured lateral displacement are presented in Fig. 14. In order to ensure a valid comparison, the predicted lateral deflections have been reduced by the predicted lateral movements at the bottom of the inclinometer ducts. A good agreement is generally observed, but comparisons are influenced by the asymmetry of the deformations beyond the commencement of yield.

Design of haul route

As design criteria, it was decided that all points on the selected route would have to:

(a) Meet a serviceability criterion of essentially elastic behaviour with settlements less than 25 mm under working bearing pressure.

(b) Provide an adequate factor of safety against bearing failure at working bearing pressure. A factor of safety in excess of 1.5 was believed essential but a value of 2.0 was desirable. The acceptable value was intended to reflect the lateral variability of soil conditions and the available safeguards against the load stopping, for instance because of a breakdown.

Whilst the ultimate pressure applied in the test was 84 kPa, a value of 80 kPa was adopted for design to allow for Hansen's correction (Hansen, 1970) for foundation shape — a simplified method of compensating for end effects in the two-dimensional analysis. A review of test data and corresponding back-analysis indicated markedly linear behaviour up to a settlement of 20 mm, which corresponded to bearing pressures of up to 45 kPa. Therefore, a design criterion was adopted to provide a factor of safety of 1.8 against failure at working bearing pressures up to 45 kPa, at which elastic settlements of less than 20 mm were predicted to occur.

The configurations of the PAUs and transporters were changed in order to reduce the applied load.

Finite element analyses were carried out based on the cone resistance profiles recorded along the route, mainly to assess bearing capacity in areas where weak profiles were identified. In these areas the lower bound line of all cone profiles was analysed. A lower bound is probably marginally weaker than any individual profile and constitutes a minimum design basis. It yielded an ultimate bearing pressure of 60 kPa. The proposed bearing pressure of 45 kPa would not be acceptable for this profile, providing a factor of safety of only 1.3. The prediction again suggests that linear behaviour only takes place up to 20 mm settlement and this would occur at a mean bearing pressure of 35 kPa. A design based on the use of soil reinforcement or concrete slabs to accomplish load spreading was considered appropriate to road lengths in which these weak profiles were identified.

An adoption of a piled option was recommended for localised areas of even weaker soil profiles, which are attributed to local drainage features existing prior to the construction of the refinery. Piles provide holding areas which can support PAU loads for longer periods, in the event of a refinery emergency. Such a piled holding area would not be affected by long-term consolidation settlements.

Monitoring of actual load

As a final test of the analysis, a monitoring exercise was carried out during the early stages of transporting PAUs across the site. Survey markers were installed on and beside the haul route. The first PAUs to be brought to site were comparatively light, and generally applied pressures to the ground much less than the design value of 45 kPa. However, one of the PAUs applied a mean pressure of 37 kPa, and the markers were surveyed before and after it was brought in, to measure any permanent movement. In addition, a selection of the markers were levelled as the PAU passed them, to measure transient settlements. In addition, two pneumatic piezometers were installed below the road surface, and monitored as the PAU passed over them.

Figure 15 shows a typical settlement record of a marker as the PAU went past, together with a schematic profile of the PAU and transporter. The maximum settlement (in this case 6 mm) usually occurred as the rear pair of supports were close to the marker. (It was not possible to take levels as they actually passed because they protruded from the side of the transporter.) At about half the markers, a smaller peak was measured as the front support went past, and a small recovery, of 1 to 3 mm, was measured between the front support and the rear pair. In a few cases a heave of 1 mm was measured before the PAU reached the marker.

SOFT CLAY DEPOSITS AT SHELLHAVEN

The average peak settlement measured was 4.5 mm, and the highest value was 7 mm. No significant difference was observed between areas where the haul road had been reconstructed because of the presence of weaker soils, and areas where the original road remained. A marker on a piled area of the haul route settled a maximum of 2 mm. About two thirds of the markers recovered fully and returned to their original position after the PAU had passed. On the others, 1 mm of residual settlement was measured.

The surveys carried out before and after transportation of the PAU indicated a permanent settlement of 6 mm on one marker, but no other settlements were greater than 3 mm and the average was 1 mm.

As stated above, it was not possible to take levels as the rear supports passed the survey markers. Examination of the settlement curves indicated that the maximum settlement of markers was of the order of 8 to 10 mm. In addition, the readings were taken about 0.5 m from the edge of the PAU. It is likely that settlements beneath it were slightly greater. The steepest observed settlement-distance gradient was about 1 mm per metre. The maximum settlement beneath the middle of the PAU, 3 m away, may therefore have been of the order of 12 mm.

The average surface loading applied by the PAU was approximately

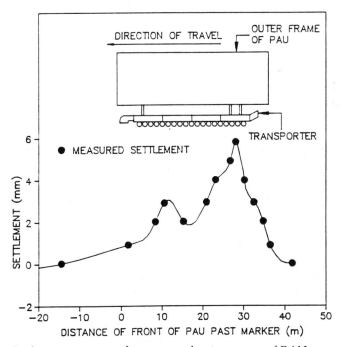

Fig. 15. *Settlement response of survey marker to passage of PAU*

37 kPa. The maximum design loading was 45 kPa. It is considered that this is sufficiently close to 37 kPa, and the observed behaviour was sufficiently close to being elastic, for the maximum settlement to be assumed proportional to the load. On this basis a maximum settlement of the order of 15 mm may be estimated, which compares with a value of 20 mm estimated from the finite element analysis. Measured settlements slightly less than those predicted probably result from the transient nature of the loading and the load spreading provided by the road surface.

This assumption of essentially elastic behaviour up to an applied loading of 45 kPa is also supported by the measurement of permanent settlements and the piezometer response. Readings on both piezometers indicated only small pore pressure responses of 1 kPa as the PAU passed over them. On the basis of pore pressures measured in the loading trial responses of the order of 10 kPa would have been expected. It may be that the short period of about 2 min during which the PAU passed over the piezometers was not long enough for either to respond fully to pore pressure changes in the soil. However, it is considered that the results indicate that pore pressures were not high enough for the soil to be close to a state of failure.

Conclusions

Simple bearing capacity calculations indicated that proposed ground loadings during transportation of PAUs would be close to the ultimate bearing capacity of the soil. This was confirmed by finite element analyses and by a full scale instrumented trial. Finite element analyses carried out before the trial predicted failure pressures which agreed reasonably with that observed, but predicted settlements which were much greater than those measured.

Analyses carried out after the trial were supported by a more detailed geotechnical investigation and, therefore, by a more soundly based choice of parameters to assess soil strength and compressibility. Calculated and measured immediate settlements were in fairly good agreement. The accuracy of post event prediction of ultimate pressure was improved by changing the average strength of the surface crust, within the range of values obtained from vane tests.

Once the model was adjusted to local conditions by the pressure–displacement response of the load test, predictions were made across the site on the basis of lateral soil variability observed on CPT profiles. This method proved to be useful for the road construction design, in which three cases were engineered — normal road construction, soil reinforcement or placement of concrete slabs and piling, according to predicted settlement and stability analyses. Finally, monitoring of an actual load gave measured settlements which agreed reasonably with those predicted, and confirmed that an adequate margin of safety was being achieved.

Acknowledgements

The analyses presented here are based on work contracted by Davy McKee Ltd., acting on behalf of Shell UK Oil. The authors gratefully acknowledge permission to use the technical information obtained in the geotechnical investigation. They would also like to express their appreciation for the technical advice provided by Mr. C. Buijs of Shell International Petroleum, Maatschappij, the Hague, during the project.

References

ATKINSON, J.H. (1973). The deformation of undisturbed London Clay. Ph.D. Thesis. University of London.

BISHOP, A.W. AND HENKEL, D.J. (1962). The measurement of soil properties in the triaxial test. Edward Arnold, London, 2nd Edn.

CLAYTON, C.R.I. AND KHATRUSH, S.A. (1986). A new device for measuring local axial strains on triaxial specimens. Technical note, Geotechnique, Vol. 36, No. 4, pp. 593–598.

DUNCAN, M.J. AND BUCHIGNANI, A.L. (1976). An engineering manual for settlement studies. University of California, Berkeley.

HANSEN, B.J. (1970). A revised and extended formula for bearing capacity. Danish Geotechnical Institute, Bulletin No. 28.

LADD, C.C. (1964). Stress–strain modulus of clay in undrained shear. ASCE Proceedings 90, Sm5, pp. 103–132.

LAMBE, T.W. (1973). Predictions in Soil Engineering. Geotechnique, Vol. 23, No. 2, pp. 149–202.

LEROUEIL, S., MAGHAN, J.P. AND TAVENAS, F. (1990). Embankments on soft clays. Ellis Horwood.

MEIGH, A.C. (1987). Cone penetration testing — methods and interpretation. CIRIA Ground Engineering Report: In situ testing.

MARSLAND, A. (1971). Laboratory and in situ measurements of the deformation moduli of London clay. Proc. Symp. Interaction of Structure and Found., Midland Soil Mech. and Found. Eng. Society. Dept. Civil Eng., University of Birmingham, pp. 7–17.

SIMONS, N.E. (1974). Normally consolidated and lightly over-consolidated cohesive materials. General report, Settlement structures, COSOS, British Geotechnical Society, pp. 500–530.

SKEMPTON, A.W. (1952). The sensitivity of clays. Geotechnique Vol. 3, No. 1, pp. 30–53.

ROBERTSON, P.K. AND CAMPANELLA, R.G. (1983). Interpretation of Cone Penetration Tests, Part 2: Clay. Canadian Geotechnical Journal, Vol. 20, pp. 732–745.

Development and application of a new soil model for prediction of ground movements

B. SIMPSON, Arup Geotechnics

Engineers studying the behaviour of soils in dynamic situations have for many years understood that stiffness reduces in a predictable way as strains become larger. This aspect of behaviour was a main theme of the Florence conference, applied to static as well as dynamic situations. In order to model this behaviour, a simple but novel concept has been proposed in which an analogue in strain space replaces the traditional use of yield surfaces in stress space. The original intention was to model stiff clays at small strains, but the concept has proved surprisingly applicable to other soils, including both sands and normally consolidated clays. The paper describes the details of the model and outlines predictions obtained for pressuremeter tests. Results for three deep excavations will be published separately.

Notation

e = voids ratio
$e_\lambda = e + \lambda \ln(s)$
p = mean normal stress = $(\sigma_1 + \sigma_2 + \sigma_3)/3$
s = mean normal stress in plane strain = $(\sigma_1 + \sigma_2)/2$
t = shear stress = $(\sigma_1 - \sigma_2)/2$
$\varepsilon_1, \varepsilon_2$ = principal strains
ϕ = angle of shearing resistance
σ_1, σ_2 = principal stresses
c_u = undrained strength
E_u = secant Young's modulus
G_t = tangent shear modulus

Other symbols are defined in the text as they occur. Except for E_u and c_u, all parameters refer to effective stresses.

Introduction

Several major cities, including London, are founded on deep deposits of stiff clays which offer opportunities for the construction of raft foundations, deep basements, road cuttings and tunnels. Large structures involving these features may cause ground movements beneath and around them which are sufficient to damage the new or neighbouring structures and services. The prediction of these ground movements is a challenge which is attracting the attention of many engineers and which has demanded improved understanding of the behaviour of the clays.

Stiffness properties can be measured using samples from boreholes in conventional laboratory tests, but if the results are applied in calculations it is clear that displacements are significantly over-predicted; that is, the test results imply stiffnesses which are much lower than those exhibited by the clay in the field situation. Recent research has done much to clarify the very stiff behaviour of clays and other soils at small strains and emphasis has been laid on the importance of the recent stress or strain history of each element of soil. Several methods of representing this behaviour in finite element programs are in use or being developed (Jardine et al., 1985; Al Tabbaa, 1987; Stallebrass, 1990). The model described in this paper was conceived originally as a simple way of representing the behaviour of stiff clays for a wide range of stress paths and which would be easy to implement in finite elements. It soon became apparent that the basic concept might be applicable to soft clays and possibly a much wider range of soils.

Peter Wroth often referred to the parameter e_λ or its more recent equivalents, using it as a parameter to relate stiffness to overconsolidation (Wroth, 1971). Stroud (1971) showed that the same parameter controls the peak angle of shearing resistance available in a sand. An equivalent of this parameter is used in the model to determine stiffness and incidentally increases the angle of shearing resistance above the critical state value for overconsolidated materials.

Basic concept

Engineers working on dynamic behaviour in soils have for many years used the 'S-shaped curve' shown in Fig. 1 to show the way in which stiffness of most soils decreases with increasing shear strain (Seed and Idriss (1970), for example). The same concept has more recently been

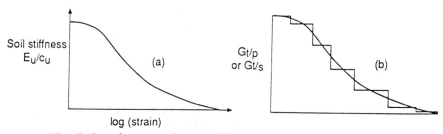

Fig. 1. The 'S-shaped curve' relating stiffness to magnitude of strain

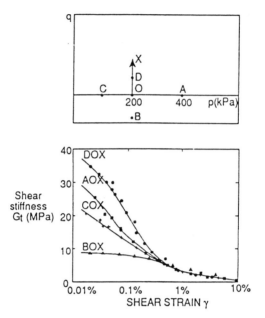

Fig. 2. Stress path tests by Richardson (1988)

adopted for prediction of static behaviour and it featured strongly in the 12th European Conference of the ISSMFE held in Florence (Atkinson and Sallfors, 1991). It is often expressed as E_u/c_u or E_u/p versus $\log_{10}(\gamma)$, using a secant Young's modulus, E_u, as shown in Fig. 1(a). Although this is useful for some forms of computation, a more fundamental form of the same information is obtained by plotting G_t/p or G_t/s versus $\log_{10}(\gamma)$, as shown in Fig. 1(b). The present model was developed for plane strain and uses the curve as G_t/s, assuming that it can be specified adequately by a number of points. If required, an equation such as that proposed by Jardine et al. (1985) could be adopted.

Richardson (1988) studied the behaviour of London clay, measuring small strain effects at abrupt changes in direction of the stress paths. Reconstituted specimens were consolidated isotropically in a triaxial cell to 400 kPa before being returned to points A to D in Fig. 2(a), via point O. From there they were taken in stress space to point O (p = 200 kPa, q = 0) and G_t was then recorded for stress path OX. Figure 2(b) shows how the measured stiffness on path OX varied with the angle of approach of the stress path to point O.

Tests of this type have shown that soils generally offer less resistance to continuation of straining in the direction they were previously following than they do to reversal of direction. Figure 3 shows results from similar tests by Stallebrass (1990) on kaolin. If the stress path turns

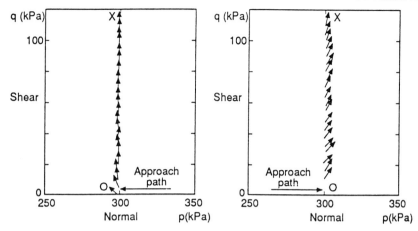

Fig. 3. Tests by Stallebrass (1990) on kaolin

through 90° the strains initially tend to continue in their previous direction before swinging round to follow the stress path.

A physical analogue bears much similarity to this type of behaviour. Imagine a man walking around a room and pulling behind him a series of bricks, each on a separate string. Some possible paths for the man and the strings are shown in Fig. 4. If he walks continuously in one direction the bricks line up behind him and follow him (Fig. 4(a)). If he turns back (Fig. 4(b)) the bricks initially do not move; then the ones on shorter strings start to move, gradually followed by the longer strings (Fig. 4(c)). If he turns through 90°, the bricks initially keep moving in their previous direction but gradually swing round behind him (Fig. 4(d)). The similarity to soil behaviour is obvious, but it is less clear how the analogue could be applied. What do the bricks and strings represent and what are the axes of the 'room'?

Engineers with a background in plasticity might see in this analogue plastic yield surfaces with axes of stress space. However, the equivalence which has been found most useful is to regard the 'room' as *strain* space. The 'man' represents the point in strain space of a soil element and each brick represents a proportion of the element. Movement of a brick represents *plastic* strain, and elastic strain is given by the difference between the movement of the 'man' and the sum of the movements of the bricks, each weighted by the proportion of the soil it represents. In this view, pure elastic behaviour only occurs on the rare occasions when no bricks are moving, i.e. immediately after a reversal of the strain path. It is assumed that only elastic strains cause changes of stress; these will be denoted with subscript e.

The initial aim was to develop a model for plane strain behaviour,

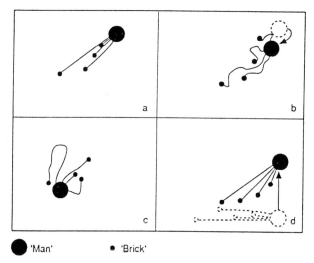

Fig. 4. *The analogue — a man pulling bricks around a room*

though in principle the analogue could be applied in more dimensions. Considering first principal strains (ε_1, ε_2) in fixed directions, a plane strain model for soil which is structurally isotropic should treat each direction equally and the axis system could therefore be (ε_1, ε_2). It is more convenient to use (v, γ), where v is volumetric strain and γ is the diameter of the Mohr's circle of strain. In coordinate axes (x, y), γ is equal to $((\varepsilon_x - \varepsilon_y)^2 + \gamma_{xy}^2)^{1/2}$. Since this has the appearance of the vector sum of two orthogonal components, it is reasonable to adopt (v, $\varepsilon_x - \varepsilon_y$, γ_{xy}) as axes. The validity of this choice will be reviewed later.

Additional assumptions

The concept outlined above could be used in conjunction with a variety of additional equations describing other features of the soil behaviour. A series of assumptions which are well known in soil mechanics will be used here to complete the model.

Volumetric stiffness is assumed proportion to current mean normal stress, so:

$$\delta s \propto s \delta v_e \qquad (1)$$

The constant of proportionality is similar to λ and κ as used in the Cam clay models. However, it relates only to very small (ideally elastic) strains and will be denoted by the previous letter in the Greek alphabet, ι, iota. (The parameters λ, κ and ι used in this paper are equivalent to λ^* and κ^* as discussed by Houlsby and Wroth (1991), since they are defined in terms of volumetric strain v rather than voids ratio.) Thus eqn. (1) becomes

$$\delta s = s \delta v_e / \iota \quad (2)$$

Elastic shear stiffness is derived from elastic volumetric stiffness assuming a constant Poisson's ratio, ν. Thus, for plane strain,

$$\delta t = s \delta \gamma_e (1 - 2\nu) / \iota$$

It is possible that Poisson's ratio should be varied as a function of stress state, but this has not been investigated.

Consolidation and swelling. When the normal stress on a frictional body is increased its ability to strain elastically without slip is also increased. Hence it would be expected that as mean normal stress is increased the soil's capacity to strain elastically and so to accept further normal and shear stresses will be extended. This gives rise to the log–linear compression and swelling lines which are often clearly displayed by soils and which are represented in Cam clay models by the parameters λ and κ. In the present model it is assumed that during normal compression increase of mean normal stress makes available an additional capacity for elastic strain, v_c, equal to $(\iota/\lambda)\delta v$. During unloading and reloading this is modified to $v_c = (\iota/\kappa)\delta v$, making the model display the normal λ, κ behaviour typical of Cam clay, but modified by the Brick effect when the stress/strain path is not straight. Additional capacity for elastic shear is also assumed, given by the formula $(t/s)v_c/(1 - 2\nu)$, which has the effect of ensuring radial straight lines in (t,s) stress space during isotropic normal consolidation.

Effect of overconsolidation. Houlsby and Wroth (1991), Viggiani (1992) and others have suggested that stiffness varies linearly with logarithm of overconsolidation ratio. Houlsby and Wroth pointed out that an equivalent effect could be achieved by assuming stiffness proportional to $p^{n-\nu^1} p_c^n$, where p_c is the preconsolidation pressure. This alternative may be preferable and could readily be incorporated into the Brick concept. The formulation adopted assumes that stiffness is proportional to $(v - v_0 - \lambda \ln (s/s_0))$, which is similar in concept to the e_λ used by Wroth (1971); stiffness is multiplied by $(1 + \beta(v - v_0 - \lambda \ln (s/s_0)))$, where β is a material constant and v_0 and s_0 indicate the initial state relative to which consolidation is to be measured. This should generally be a state of isotropic normal consolidation; in most applications to date it has been taken to represent a slurry.

If there were no 'Brick' effect, the term $(v - v_0 - \lambda \ln (s/s_0))$ would reduce to $(\lambda - \kappa)\ln (\text{OCR})$. One potential advantage of the formulation adopted, however, is that it may be appropriate to coarser granular soils in which density (represented by $v - v_0$) is determined by effects other than precompression.

Derived properties

Conventional parameters such as ϕ and K_0 are not used as data for the model, but follow automatically from the concept and assumptions described above. Suppose the S-shaped curve is plotted as G_t/s versus shear strain γ (not log (shear strain)) and the area under the curve is denoted by A. With the assumptions of eqns. (2) and (3) and $\beta = 0$, it is found computationally that the soil always fails in shear at a constant angle of friction given by $\phi = \sin^{-1}$ (A). This is easily demonstrated for the case s = constant:

$$A = \int (G_t/s)\,d\gamma = (1/s) \int (dt/d\gamma)\,d\gamma = t/s$$

Thus, when γ is large, the area A tends to the limiting value of t/s, i.e. $\sin \phi$.

Thus specification of a complete S-shaped curve, extending to failure, automatically sets the value of ϕ. Alternatively, since the details of the curve may be uncertain at large strain, a known value of ϕ, together with a required strain to failure can be used to ensure that the high strain portion of the curve is sensible. Furthermore, a parameter such as β which changes G_t/s will automatically change ϕ. Thus the multiplier related to overconsolidation $(1 + \beta(v - v_0 - \lambda \ln (s/s_0)))$ will also lead to increased values of $\sin \phi$. This is consistent with Stroud's observation for sands that $\sin \phi$ varies with e_λ. The area under the basic S-shaped curve, for which $(v - v_0 - \lambda \ln (s/s_0)) = 0$, is the critical state angle of shearing resistance, ϕ_{crit}.

The model does not have K_0 as a parameter, but during 1-dimensional normal consolidation the positions of the strain point and bricks are as shown in Fig. 4(a). This implies that the elastic shear strain which will

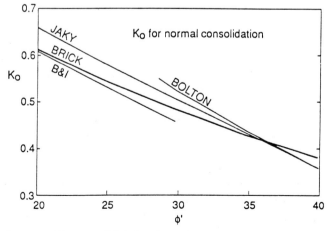

Fig. 5. K_0 for normally consolidated soils

govern shear stress will be $1/\sqrt{2}$ times that developed during failure in pure shear. Hence the angle of friction mobilised in K_0 consolidation will be given by $\sin\phi_{mob} = \sin\phi/\sqrt{2}$. This leads to the following equation for 1D normal consolidation:

$$K_0 = (1 - \sin\phi/\sqrt{2})/(1 + \sin\phi/\sqrt{2})$$
$$= (\sqrt{2} - \sin\phi)/(\sqrt{2} + \sin\phi) \qquad (4)$$

In Fig. 5 this equation is compared with the familiar Jaky's equation

$$K_0 = 1 - \sin\phi$$

Brooker and Ireland (1965) proposed a modified form of Jaky's equation for clays:

$$K_0 = 0.95 - \sin\phi$$

Recently Bolton has proposed another equation based on the study of sands:

$$K_0 = (1 - \sin\phi_{mob})/(1 + \sin\phi_{mob})$$

in which $\phi_{mob} = \phi - 11.5°$.

Both these modified equations are shown in Fig. 5. Equation (4) agrees very well with Brooker and Ireland's equation for the range of ϕ appropriate to clays (20° to 30°) and with Bolton's equation for that appropriate to sands (30° to 40°). It also lies fairly close to Jaky's equation throughout. This close agreement between the prediction of the Brick model and measured values for K_0 gives strong support to the choice of axes in which to use the model which was discussed earlier.

Parameter values

A constant value of Poisson's ratio of 0.2 has been assumed. This is needed in order to derive ι from measured values of shear modulus. If a different value were adopted for ν, ι could be changed to keep the shear modulus correct.

Derivation of parameter values for the model requires results from high quality small strain testing which is now becoming more common. Values relevant to London Clay have been derived from textbooks together with the work of Richardson (1988) and Viggiani (1992). The proportions of the S-shaped curve were derived mainly from the path DOX in Fig. 2, assuming that a complete reversal of the stress/strain path would allow elastic strains equal to double the string lengths. The longer strings were chosen to give an angle of shearing resistance ϕ_{crit} of 20°, which is perhaps a little too low.

From the maximum shear stiffnesses reported by Richardson, the value of ι was determined as 0.0032 and this was confirmed from the

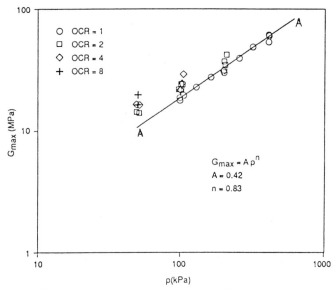

Fig. 6. *Relationship between G_{max}, p and overconsolidation ratio reported by Viggiani (1992)*

work of Viggiani. Figure 6 shows maximum shear stiffnesses, G_{max}, measured in a triaxial apparatus for a range of mean normal stress p and overconsolidation ratio. Viggiani proposes that G_{max} is proportional to $p^{0.83}$, not $p^{1.00}$ as presently assumed in the model. However, the line AA on Fig. 6 lies close to the data for normally consolidated clay and gives $G_{max}/p = 185$. Assuming that this is equivalent to $G_{max}/s = 185$, it follows that $(1 - 2\nu)/\iota = 185$, from which $\iota = 0.0032$ for $\nu = 0.2$.

Viggiani also proposes that stiffness is proportional to $OCR^{0.25}$, and a value of $\beta = 4$ was found to correspond quite closely to this.

Results from the model

All the results presented in this paper have been derived using the parameter values reported above. Although the model was conceived as a representation of stiff clays, it soon became clear that it had wider application. Qualitatively, at least, it reproduces many of the phenomena of real soils but in practical use it would be sensible to fit parameter values to test results which are directly relevant to the situation to be analysed. Nevertheless, it is considered that a model which remains reasonably accurate over a wide range of behaviour is very desirable.

Figure 7 shows the stress paths computed for undrained tests on normally consolidated specimens with isotropic (plane strain) and

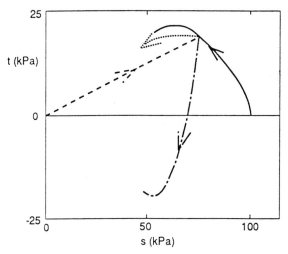

Fig. 7. Predictions for undrained tests on normally consolidated specimens

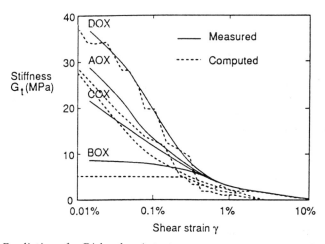

Fig. 8. Predictions for Richardson's tests

anisotropic consolidation. The paths are immediately recognizable as typical behaviour and the specimens fail consistently at $\phi = 20°$. Figure 8 shows computed results for Richardson's tests (Fig. 2), replacing the triaxial mean normal effective stress p by the plane strain value s. The model reproduces the effect of the direction of approach of the stress path at the start of the constant s test and displays appropriate stiffnesses throughout the tests.

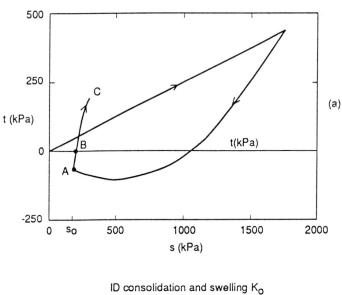

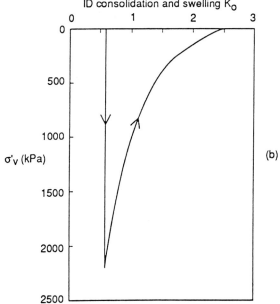

Fig. 9. *One-dimensional loading and unloading followed by undrained shear*

SOIL MODEL FOR PREDICTION OF GROUND MOVEMENTS

Figure 9(a) shows the stress path predicted for 1-dimensional consolidation and unloading followed by undrained shearing. The preconsolidation pressure has been chosen to be equivalent to 200 m of eroded overburden, as proposed for London Clay by Skempton and Henkel (1957). In this process, the model predicts the value for K_0 as a function of overconsolidation ratio. For low OCR, the values derived are mainly dependent on the specified values of Poisson's ratio (0.2) and κ, and are significantly higher than values given by formulae such as $K_0 = (1 - \sin \phi)\, OCR^{0.5}$; higher Poisson's ratio or lower κ would give better agreement with this. Broms (1971) suggested that heavily overconsolidated soils would have $K_0 = 1/K_0(nc)$ and this occurs in the model, so K_0 tends to $(\sqrt{2} + \sin \phi)/(\sqrt{2} - \sin \phi)$. However, the effect of overconsolidation, governed by β, means that ϕ is now larger than during normal consolidation and the values of K_0 derived in this way are therefore relatively large. Figure 9(b) shows the relationship between K_0 and final vertical stress; the results are considered typical of field measurements in London Clay.

In the overconsolidated state, the model in its present formulation displays more strength in the compressive (passive) state than in extension. As a result, the undrained test in Fig. 9(a) reaches failure at $s = 297$ kPa, $t = 193$ kPa, which for $\phi' = 26°$ requires $c' = 70$ kPa, a rather high value for c'.

Figure 10 shows the relationship between the secant undrained Young's modulus, secant E_u, and strain for the undrained test ABC in Fig. 9(a). E_u has been normalized by dividing by s_A, the value of s at

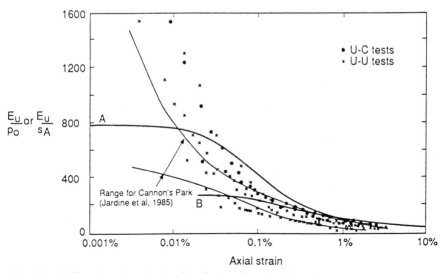

Fig. 10. *Stiffness/strain relationship for 'undisturbed' London clay*

point A. This is compared with results from triaxial tests reported by Jardine et al. (1985) and others carried out on recent projects for Arup Geotechnics. For the triaxial tests, the origin of strain is generally taken at zero shear stress but the details of the recent stress/strain history are unclear. The results from the Brick model can be plotted in different forms, according to the choice of origin of 'axial strain', taken to be $\frac{2}{3}\gamma$. If the origin is taken as point A on Fig. 9(a), the Brick model gives a slightly higher stiffness than measured for the practical range of strains; however, if it is taken at zero shear stress (point Q), the Brick model gives lower stiffness. In general, therefore, the model brackets the observed results in a reasonable manner though it shows lower stiffness at shear strains less than 0.01%. The parameters of the model were derived from tests on reconstituted samples and it may be that this is not appropriate.

Application in a finite element program

The Brick model is ideal for use in a finite element program because, unlike most plasticity models, it derives stress increments from strain increments. The model itself requires about 80 lines of Fortran and additional program is needed to set up initial states of stress and strain, including the initial positions of the 'bricks'. The model has been found to be very easy to use, invariably giving good convergence and smoothly distributed stress and strain fields. It has been implemented for use on a PC in the Arup Geotechnics program SAFE, using 8-node plane strain elements with 4 integration points. Undrained behaviour is modelled as described by Simpson et al. (1979). At the start of each computation, the geological history of the soil is reproduced, considering consolidation from a slurry to maximum overburden and subsequent unloading. This computation takes about 30 s and is carried out only once for each stratum; the results are stored as a function of OCR for use throughout the stratum.

Remembering Peter Wroth's great interest in the pressuremeter, an attempt has been made to reproduce the results of self-boring pressuremeter tests carried out at the British Library site (Simpson et al., 1981). The plane strain model is not ideally suited to this, so the analysis used a horizontal plane strain slice. This was preconsolidated isotropically (in horizontal plane strain) to the horizontal effective stress calculated for 200 m of overburden in the normally consolidated state, and then allowed to swell back to a horizontal stress of 200 kPa. The results are compared with field measurements in Fig. 11, in which measured pressures, ψ, are plotted against radial strain ε_r. The pressures have been normalized by dividing by the undrained shear strength, c_u, appropriate to their depth z(m), assuming $c_u = 17.5z$ for the field data,

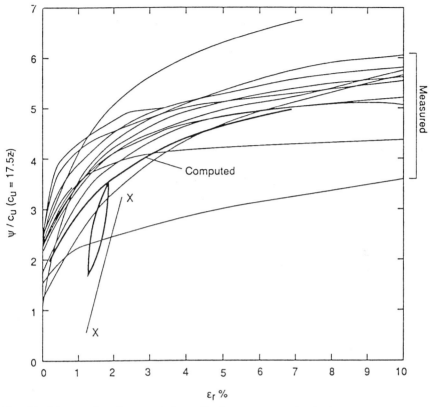

Fig. 11. *Measured and computed results for pressuremeter tests*

which was a best fit through the pressuremeter results; the model result had an undrained strength of 166 kPa. For first loading, the Brick model predicts a slightly lower stiffness than recorded in the tests, possibly reflecting the rather different history and initial state of field and computation. However, line XX in Fig. 11 represents the average stiffness of unload/reload loops in the field tests and this is very similar to the stiffness predicted by the model.

The model has been tested by comparing computed results with field measurements at three excavations, constructed respectively in London clay, Gault clay and Singapore soft marine clay. The detailed results of this work will be the subject of separate publications including Simpson (1992). Unfortunately, site specific, high quality small strain test data were available for none of these sites, but the results indicate that, using parameter values derived from laboratory tests and text books, the model gives acceptable predictions of ground movements.

Concluding remarks

(1) The Brick model is centred around a very simple concept which is capable of many extensions and variants. The main constituent of the theory cannot be represented analytically but requires simple repetitive calculations and so is ideally suited to computing, especially in a finite element program. The 'S-shaped curve' of stiffness versus strain and the importance of recent stress/strain history are key elements.

(2) With the addition of a few equations commonly adopted in critical state soil mechanics the model is able to predict the behaviour of clays in a wide range of states and stress paths, together with credible values for K_0 in both normally consolidated and overconsolidated states.

(3) The requirement to predict increase of stiffness with overconsolidation ratio leads automatically to increased available ϕ or, alternatively, to a c' component.

(4) The model is ideally suited for use in a finite element program and has been implemented in the Arup Geotechnics program SAFE. It is found to be easy to use, in particular showing the smooth convergence which might be expected of a realistic model.

(5) The model has been used to compute the results of pressuremeter tests in London clay and wall movements of deep basement excavations in London clay and Singapore soft clay. In all cases the parameter values were derived from the results of laboratory tests on reconstituted specimens. The results suggest that the model has considerable potential for predicting the field behaviour of clay deposits.

Acknowledgements

The author gratefully acknowledges the assistance of Prof. J.H. Atkinson, Dr S.E. Stallebrass and Miss G. Viggiani of City University and Mr C. Ng of Bristol University. Colleagues at Arup Geotechnics are thanked for their input of time, interest and computer facilities.

References

AL TABBAA, A. (1987). Permeability and stress strain response of speswhite kaolin. PhD thesis, University of Cambridge.

ATKINSON, J.H. AND SALLFORS, G. (1991). Experimental determination of soil properties (stress-strain-time). Proc. 10th Euro. Conf. SMFE, Vol. 3, pp. 915–958.

BROMS, B. (1971). Lateral earth pressures due to compaction in cohesionless soils. Proc. 4th Conf. on Soil Mechanics, Budapest, pp. 373–384.

BROOKER, E.W. AND IRELAND, O.H. (1965). Earth pressures at rest related to stress history. Canadian Geotechnical Journal, Vol. 2, No. 1, pp. 1–15.

HOULSBY, G.T. AND WROTH, C.P. (1991). The variation of shear modulus

of a clay with pressure and overconsolidation ratio. Soils and Foundations, Vol. 31, No. 3, pp. 138–143.
JARDINE, R.J., FOURIE, A., MASWOSWE, J. AND BURLAND, J.B. (1985). Field and laboratory measurements of soil stiffness. Proc. 11th Int. Conf. San Francisco, SMFE, Vol. 2, pp. 511–514.
NEWMAN, R.L., CHAPMAN, T.J.P. AND SIMPSON, B. (1991). Evaluation of pile behaviour from pressuremeter tests. Proc. 10th Euro. Conf. Florence, SMFE, Vol. 2, pp. 501–504.
RICHARDSON, D. (1988). Investigations of threshold effects in soil deformations. PhD thesis, City University.
SEED, H.B. AND IDRISS, I.M. (1970). Soil moduli and damping factors for dynamic response analysis. EERC Report No. 70–10, Berkeley, California.
SIMPSON, B., O'RIORDAN, N.J. AND CROFT, D.D. (1979). A computer model for the analysis of ground movements in London Clay. Geotechnique, Vol. 29, No. 2, pp. 149–175.
SIMPSON, B., CALBRESI, G., SOMMER, H. AND WALLAYS, M. (1991). Design parameters for stiff clays. Proc. 7th Euro. Conf. SMFE, Brighton, 1979, Vol. 5, pp. 91–125.
SIMPSON, B. (1992). Retaining structures — displacement and design. 32nd Rankine Lecture, Geotechnique.
STALLEBRASS, S.A. (1990). Modelling the effect of recent stress history on the deformation of overconsolidated soils. PhD thesis, City University, London.
SKEMPTON, A.W. AND HENKEL, D.J. (1957). Tests on London Clay from deep borings at Paddington, Victoria and the South Bank. Proc. 4th Int. Conf. SMFE, London, Vol. 1, pp. 100–106.
STROUD, M.A. (1971). The behaviour of sand at low stress levels in the simple shear apparatus. PhD thesis, University of Cambridge.
VIGGIANI, G. (1992). Dynamic measurement of small strain stiffness of fine grained soils in the triaxial apparatus. Proc. Workshop on experimental characterization and modelling of soils and soft rocks, Napoli, pp. 75–97.
WROTH, C.P. (1971). Some aspects of the elastic behaviour of overconsolidated clay. Proc. Roscoe Mem. Symp., Foulis, pp. 347–361.

Stability of shallow tunnels in soft ground

S.W. SLOAN and A. ASSADI, University of Newcastle, Australia

This paper examines the undrained stability of a shallow circular tunnel in soft ground using theoretical and experimental methods. Rigorous bounds on the loads required to resist collapse are derived using two numerical methods which are based on finite element formulations of the classical limit theorems. Both of these methods can account for the variation of shear strength with depth and lead to large linear programming problems. To improve the quality of the numerical solutions, the upper bounds from the finite element procedure are compared with those obtained from a number of rigid block mechanisms. The theoretical predictions, which are presented in the form of dimensionless stability charts, are compared against the results from controlled centrifuge experiments and their limitations are discussed.

Introduction

This paper considers the stability of a shallow circular tunnel in a soil whose shear strength increases linearly with depth. The tunnel is of diameter D, has a depth of cover C, and is assumed to be loaded under undrained conditions in a state of plane strain. The idealized problem, shown in Fig. 1, models the construction of a bored tunnel in soft ground where a rigid lining is placed in position as the excavation

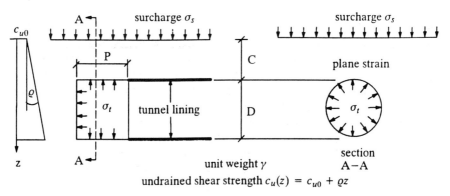

Fig. 1. *Shallow circular tunnel in soil whose shear strength increases with depth*

proceeds and the unlined heading, of length P, is supported by an internal pressure σ_t. Collapse of the heading is triggered by the action of gravity and the surcharge σ_s.

For the geometry shown in Fig. 1, the assumption of plane strain deformation is valid only if the length of the unlined heading P is significantly greater than its diameter D. It should be noted that the stability of a plane strain tunnel is more critical than that of a tunnel heading, and this model thus provides a conservative estimate on the loads that are likely to trigger the true mode of collapse. When considering the stability of deeper tunnels, the possibility of local collapse must also be taken into account. In this paper, the soil is assumed to be normally consolidated or overconsolidated and its undrained shear strength is permitted to vary linearly with depth according to

$$c_u(z) = c_{u0} + \rho z \tag{1}$$

where c_{u0} is the undrained shear strength at the ground surface and $\rho = dc_u/dz$ is the rate of change of shear strength with depth. The overall stability of the tunnel is described conveniently by two load parameters $(\sigma_s - \sigma_t)/c_{u0}$ and $\gamma D/c_{u0}$, which are functions of $\rho D/c_{u0}$ and C/D. As the tunnel of Fig. 1 is excavated, either by a tunnelling machine or (increasingly rarely) by hand, the heading is typically supported by compressed air or a clay slurry. This means that in most design situations the quantities $\gamma D/c_{u0}$, $\rho D/c_{u0}$, and C/D are known and it is necessary to be able to predict the value of $(\sigma_s - \sigma_t)/c_{u0}$ at incipient collapse.

A comprehensive study of the behaviour of tunnels and headings in clays has been conducted at Cambridge over the last two decades. In the thesis of Cairncross (1973), the stability of unlined circular tunnels in overconsolidated kaolin was investigated by testing a number of planar models in a rectangular box. Each model was loaded by an internal tunnel pressure and a surface pressure and the surrounding clay was impregnated with lead shot. Both of the loadings were initially pre-scribed to be equal and applied by lubricated rubber bags in an effort to sustain a uniform pressure. A state of drained collapse was induced by holding the surface pressure constant and reducing the internal tunnel pressure very slowly so that no excess pore pressures were generated. By using X-rays to locate the position of the lead shot, Cairncross (1973) was able to determine the velocity and strain fields at various stages during each test. This type of laboratory experiment has proved invaluable for suggesting likely modes of collapse in theoretical studies. Indeed, as a result of these tests, Atkinson and Cairncross (1973) proposed a limit equilibrium mechanism which appears to give good predictions of the drained collapse load, at least for small values of C/D.

Limit equilibrium solutions suffer from the disadvantage that they are neither rigorous upper bounds nor rigorous lower bounds. Although this limitation makes it difficult to isolate the modelling error from other sources of error, it is rather academic if the theoretical predictions match the experimental data closely.

Following on from the initial work of Cairncross, further experimental investigations of circular tunnel stability in clays were carried out at Cambridge by Orr (1976) and Casarin (1977). The latter study focused on the behaviour of a cylindrical heading, as shown in Fig. 1, and confirmed the intuitively obvious result that the heading is least stable when the ratio P/D is large.

Using essentially the same experimental apparatus as Cairncross (1973), Seneviratne (1979) also examined the deformation and stability of plane strain tunnels in clay. He conducted drained and undrained collapse tests as well as consolidation tests where the internal tunnel pressure was reduced quickly to a stable value and then held constant whilst the excess pore-pressures dissipated. In contrast to the stiff overconsolidated kaolin that was used in the work of Cairncross (1973), Seneviratne (1979) concentrated on the behaviour of soft normally-consolidated clay. He employed classical plasticity theory to derive a number of rigorous upper and lower bound stability solutions for undrained loading and also used modified Cam clay in a displacement finite element analysis to try and predict the load–deformation behaviour of the various tunnel models. Generally speaking, the bound solutions gave good estimates of the collapse load and supported the use of classical limit theory for undrained conditions. The finite element solutions, on the other hand, modelled the observed behaviour less closely for this case and were most accurate when they were used to predict the results of the drained tests and the consolidation tests.

The various studies of tunnel and heading stability conducted at Cambridge during the 1970s culminated in the exhaustive experimental and theoretical investigation of Mair (1979). In contrast to previous work, Mair (1979) used the centrifuge to observe the undrained collapse of two-dimensional circular tunnel sections and three-dimensional cylindrical headings which were constructed in kaolin. A major advantage of centrifuge testing is that it enables geotechnical models to be loaded to failure under different gravity regimes, thus permitting parametric studies to be conducted. It also allows the stress history of the soil to be controlled in a relatively precise manner. During each of Mair's centrifuge experiments, pore pressures and displacements were measured continuously at key locations in the model using transducers. The strain and deformation fields at various stages of loading were also determined by taking in-flight photographs of a grid of silvered perspex balls which were placed in each model prior to testing. All of these tests were performed on models with a uniform undrained strength profile and considerable care was taken to establish accurate estimates of the soil parameters. In an effort to ascertain the effect of the heading geometry, Mair (1979) constructed a number of models to investigate the influence of the ratio P/D on the stability of an unsupported three-dimensional cylindrical heading. His experiments showed that a significant drop in stability occurred as P/D increased from 0 to 1, with the collapse mechanism becoming essentially two-dimensional once this ratio exceeded 3. In practice, the value of P/D is typically between 0 and 1, with the larger values occurring for small diameter tunnels, and thus the approximation of plane strain collapse will always be conservative. To verify the assumptions that are implicit in centrifuge testing, Mair (1979) performed a number of plane strain tunnel tests at two different modelling scales using models which were geometrically similar but of differing sizes. His experimental results not only confirmed the correctness of the scaling procedure but also provided an invaluable source of high-quality data for comparison with theoretical solutions. Following on from Seneviratne (1979), Mair (1979) computed rigorous upper bounds on the undrained collapse load for a plane strain circular tunnel by using a variety of rigid block mechanisms. He also proposed a three-dimensional mechanism for finding a rigorous upper bound for the more complex case of a cylindrical heading with $P/D = 0$. In an effort to try and predict pore pressures and displacements for a number of the plane strain tunnel tests, Mair (1979) used the modified Cam clay soil model and displacement finite element analysis with fully-coupled consolidation. Although these analyses are quite sophisticated and demand considerable care to obtain sensible results, the finite element solutions gave good qualitative estimates of the overall behaviour. Quantitatively, however, the computed pore pressures and displacements tended to predict incipient collapse at tunnel support pressures which were lower than those observed. In addition, the extent of the surface settlement profiles was typically too large. One possible reason for the latter discrepancy is that the nonlinear elastic behaviour of the soil is not well characterized by the modified Cam clay model.

As a result of the continuing research programme at Cambridge, a comprehensive analysis of the stability of various types of underground openings has been given by Davis et al. (1980). This study used a number of the theoretical results accumulated in the above-mentioned theses, together with the classical limit theorems, to present rigorous bounds on the collapse loads for circular plane strain tunnels, plane strain tunnel headings, and three-dimensional cylindrical headings. Other literature on the undrained stability of circular tunnels and headings is surprisingly scarce. Muhlhaus (1985) has given analytic lower bound solutions for both drained and undrained conditions, but the latter are substantially poorer than the numerical lower bounds presented by Seneviratne (1976).

Using a completely different approach, which involves a finite element formulation of the limit theorems, Assadi and Sloan (1991) have obtained tight bounds on the collapse load for a shallow square tunnel in a soil with a uniform undrained shear strength. This technique has proved to be a most successful tool for stability analysis and will be used in this paper to analyse the behaviour of the plane strain tunnel of Fig. 1.

Finite element formulation of the lower bound theorem

This section describes a finite element formulation of the lower bound theorem. The key idea behind the technique is to use a linear variation for the stress field, in conjunction with an internal linear approximation of the Tresca yield criterion, to express the unknown collapse load as a solution to a large linear programming problem. Since the history and detail of the method has been described elsewhere in Sloan (1988a) and Assadi and Sloan (1991), only a very brief outline of its essential features will be given here.

The formulation uses the three types of elements shown in Fig. 2. The stress field for each of these is assumed to vary linearly. Unlike the more familiar types of elements used in the displacement finite element method, each node is unique to a single element and several nodes may share the same coordinates.

To broaden the range of stress fields that are available to a particular mesh, statistically admissible stress discontinuities are permitted at all edges that are shared by adjacent elements, including those edges that are shared by adjacent extension elements. A rigorous lower bound on the exact collapse load is ensured by insisting that the stresses obey equilibrium and satisfy both the stress boundary conditions and the

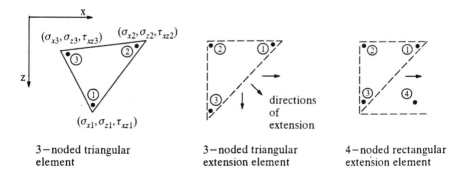

Fig. 2. Elements for lower bound analysis

SHALLOW TUNNELS IN SOFT GROUND

yield criterion. Each of these requirements imposes a separate set of constraints on the nodal stresses. To incorporate the yield condition, the Tresca yield surface, which gives rise to a nonlinear inequality in the nodal Cartesian stresses, is replaced by a series of linear inequalities. This ensures that the yield condition is satisfied rigorously and corresponds to replacing the Tresca surface, which plots as a circular cylinder in stress space, by an n-sided internal prism.

In a typical lower bound calculation, a statically admissible stress field is sought which maximizes an integral of the normal stresses over some part of the boundary. This integral corresponds to the collapse load and can be expressed in terms of the unknown Cartesian stresses to define an objective function.

Once the various constraints and objective function terms have been assembled, the problem of finding a statically admissible stress field which maximizes the collapse load $\mathbf{C}^T\mathbf{X}$ may be written as

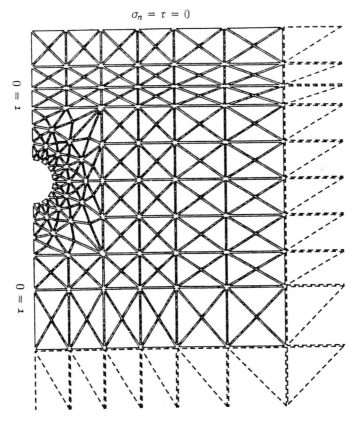

Fig. 3. *Lower bound mesh for circular tunnel with C/D = 3*

$$\begin{aligned}\text{Minimise} \quad & -\mathbf{C}^T\mathbf{X} \\ \text{Subject to} \quad & \mathbf{A}_1\mathbf{X} = \mathbf{B}_1 \\ & \mathbf{A}_2\mathbf{X} \leq \mathbf{B}_2\end{aligned} \qquad (2)$$

where **X** denotes the global vector of unknown nodal stresses. This type of optimzation problem is a classical linear programming problem but is unusual in that all of the variables are unbounded and the overall constraint matrix has many more rows than columns. A detailed discussion of the various alternatives for solving eqn. (2) may be found in Sloan (1988a, 1988b) and will not be repeated here.

A typical lower bound mesh for a plane strain circular tunnel is shown in Fig. 3. This mesh is typical of those that are used to obtain rigorous lower bounds for various values of C/D.

Finite element formulation of the upper bound theorem

An elegant finite element formulation of the upper bound theorem can be derived by adopting a similar approach to that used for the lower bound theorem. By assuming a linear variation for the velocity field, in conjunction with an external linear approximation of the Tresca yield criterion, the rate of internal energy dissipation may be expressed as a solution to a large linear programming problem. A comprehensive history of the method, together with explicit coding details, may be found in Sloan (1989).

The three-noded triangle used in the upper bound formulation is shown in Fig. 4. Each node has two velocity components and each element has n plastic multiplier rates (where n is the number of sides in the linearized yield criterion). The velocities are assumed to vary linearly across each element.

The upper bound grid is similar to one that would be used in a conventional finite element analysis, with most nodes being shared by

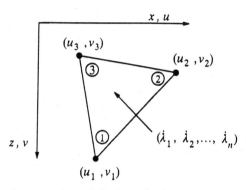

Fig. 4. Element for upper bound limit analysis

more than one element, except that it also incorporates a number of velocity discontinuities. The position and sign of shearing for each discontinuity must be specified *a priori*. Unlike an upper bound calculation which is based on a rigid block mechanism, plastic deformation is permitted to occur throughout the soil mass. To obtain a rigorous upper bound on the collapse load, the rate of work done by the external loads is equated to the rate of internal energy dissipation for a kinematically admissible velocity field. For a given mesh, the best upper bound is found by choosing a mode of deformation that minimizes the power dissipation (or some related load parameter which is the quantity of interest). To ensure that the velocity field is kinematically admissible, it must satisfy both the plastic flow rule and the velocity boundary conditions. Each of these conditions gives rise to separate sets of constraints on the nodal velocities and plastic multiplier rates.

To remove the stress terms from the flow rule equations, and thus provide a linear relationship between the unknown velocities and plastic

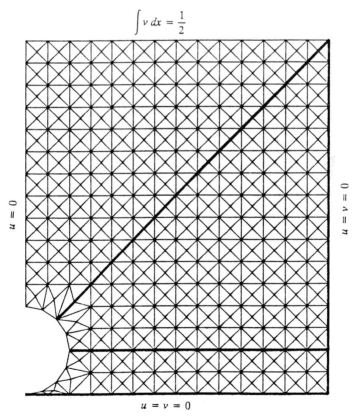

Fig. 5. Upper bound mesh for circular tunnel with $C/D = 3$

multiplier rates, it is again necessary to linearize the Tresca yield surface. To preserve the bounding property of the solution, the upper bound scheme uses an external n-sided prism, to approximate the Tresca circular cylinder.

To define the objective function, the dissipated power (or some related load parameter) is expressed in terms of the unknown velocities and plastic multiplier rates. As the soil deforms, power dissipation may occur in the velocity discontinuities as well as in the triangles.

Once the constraints and the objective function coefficients are assembled, the task of finding a kinematically admissible velocity field, which minimizes the internal power dissipation for a specified set of boundary conditions, may be written as

$$\begin{aligned}
\text{Minimise} \quad & \mathbf{C}_1^T \mathbf{X}_1 + \mathbf{C}_2^T \mathbf{X}_2 \\
\text{Subject to} \quad & \mathbf{A}_{11} \mathbf{X}_1 + \mathbf{A}_{12} \mathbf{X}_2 = \mathbf{B}_1 \\
& \mathbf{A}_2 \mathbf{X}_1 \leq \mathbf{B}_2 \\
& \mathbf{A}_3 \mathbf{X}_1 = \mathbf{B}_3 \\
& \mathbf{X}_2 \geq 0
\end{aligned} \quad (3)$$

where $\mathbf{X}_1$ and $\mathbf{X}_2$ are global vectors of nodal velocities and plastic multiplier rates respectively. A detailed analysis of various strategies for solving this type of linear programming problem may be found in Sloan (1988b, 1989) and will not be covered here.

A typical upper bound mesh for analysing the stability of a plane strain circular tunnel is shown in Fig. 5. Similar grids are used to compute rigorous upper bounds for values of C/D other than 3.

Results and discussion

For the plane strain circular tunnel shown in Fig. 1, the internal pressure σ_t provides resistance against collapse which is driven by the action of gravity and the surcharge σ_s. Under this type of loading, the power dissipated by the external loads is

$$P_{ext} = (\sigma_s - \sigma_t)\left(\int v\, dx\right)_{z=0} + \gamma \iint_A v\, dxdz \quad (4)$$

where A is the area of the soil mass which deforms plastically at constant volume. Equation (4) may also be written in terms of the dimensionless parameters $(\sigma_s - \sigma_t)/c_{u0}$ and $\gamma D/c_{u0}$ according to

$$P_{ext} = \left(\frac{\sigma_s - \sigma_t}{c_{u0}}\right)\left(c_{u0} \int v\, dx\right)_{z=0} + \left(\frac{\gamma D}{c_{u0}}\right)\left(\frac{c_{u0}}{D}\iint_A v\, dxdz\right) \quad (5)$$

These two quantities, which are both functions of C/D and $\rho D/c_{u0}$, provide a concise means for summarising the stability of a plane strain circular tunnel. In practice, the parameters $\rho D/c_{u0}$ and $\gamma D/c_{u0}$ are usually known for a given site and it is necessary to be able to determine the maximum value of $(\sigma_s - \sigma_t)/c_{u0}$ for a prescribed C/D. Once the value of $(\sigma_s - \sigma_t)/c_{u0}$ at the point of collapse is found, an appropriate factor of safety may be applied to it to deduce a safe working load range for tunnelling operations.

Since the aim of this study is to investigate the potentially unstable behaviour of shallow tunnels in soft ground, solutions are presented for C/D ratios in the range of 1 to 5. The methods of stability analysis that have been described could, of course, be applied to deeper tunnels in stiffer materials without difficulty. The experimental evidence reported in Mair (1979), however, suggests that the mechanism of failure predicted by classical plasticity theory is inappropriate once C/D is greater than about 3. In particular, the collapse mechanisms observed in the centrifuge did not propagate through to the ground surface, with failure causing large inward movements of the tunnel walls but only small surface settlements. This suggests that, in practice, the collapse of deeper tunnels is probably caused by a complicated local mechanism involving both elastic and plastic deformation, at least for cases where the undrained shear strength profile is uniform.

For a wide variety of clays, the undrained shear strength is observed to increase linearly with depth. This phenomenon has been described empirically by Skempton (1957), who analysed a range of field data for normally consolidated clays and suggested an expression of the form

$$\frac{c_u}{\sigma'_{v0}} = 0.11 + 0.0037 I_p$$

where σ'_{v0} denotes the initial effective overburden stress and I_p is the plasticity index. Substituting representative values for the plasticity index and submerged unit weight of clay soils, Skempton's equation suggests that ρ may vary anywhere between 0.5 kN/m^3 and 5 kN/m^3. These predictions are in broad agreement with the field strength profiles published in Lambe and Whitman (1979), which give typical values for normally consolidated soft clay deposits of $1–2 \text{ kN/m}^3$. Bearing these values in mind, the dimensionless quantity $\rho D/c_{u0}$ is varied from 0 to 1 in increments of 0.25.

To complete the specification of the parametric study, $\gamma D/c_{u0}$ is selected to have the values 0, 1, 2, and 3. The first case corresponds to a weightless soil and is included to highlight the effects of gravity, whilst the other cases are typical for tunnels in soft ground.

Note that the arrangement of the extension elements in the lower bound mesh of Fig. 3 permits the stress field to be extended throughout the semi-infinite domain of the problem without violating equilibrium, the stress boundary conditions, or the yield criterion. This property ensures that the stress field is complete and guarantees that the corresponding solution for the collapse load is a rigorous lower bound on the exact collapse load. Numerical experiments suggest that at least twelve sides are needed to avoid excessive approximation error in the linearized yield surface and this value is used in all of the bound calculations presented in this paper.

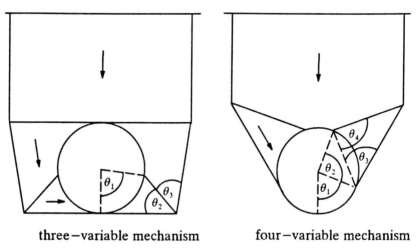

three−variable mechanism   four−variable mechanism

Fig. 6. *Three-variable and four-variable mechanisms for tunnel collapse*

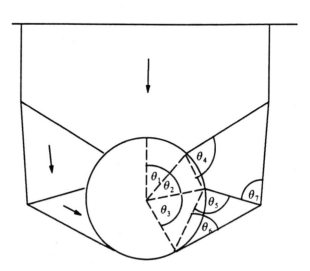

Fig. 7. *Seven-variable mechanism for tunnel collapse*

In the upper bound mesh of Fig. 5, the loading is applied by prescribing the integral of the downward velocity at the ground surface to be equal to unity. Because the plastic deformations associated with the Tresca yield criterion are incompressible, this condition is equivalent to imposing the constraint that the integral of the outward normal velocities over the tunnel face is equal to unity.

To provide additional upper bound solutions, a number of rigid block mechanisms were also considered. The three-variable and four-variable mechanisms, shown in Fig. 6, are taken from Davis et al. (1980) and furnish good upper bounds for shallow tunnels in clay with a uniform undrained shear strength. The seven-variable mechanism, shown in Fig. 7, is considerably more complex to implement than the simpler models but gives substantially better solutions for cases where the strength increases linearly with depth.

For deeper tunnels, where the failure mechanism is unlikely to propagate all the way to the ground surface, it is necessary to check for local collapse. One possible local collapse mechanism, which is defined by two variables and involves a rigid body rotation along a circular failure surface, is shown in Fig. 8. When the undrained shear strength is uniform, the solution from this model is independent of C/D and gives an upper bound of $\gamma D/c_u = 11.72$. Davis et al. (1980) have reported a solution of $\gamma D/c_u = 8.71$ using a similar mechanism which involves only one variable, but this would appear to be incorrect. The relatively high value of $\gamma D/c_u$ required to cause local collapse for the uniform strength case suggests that the mode of failure of Fig. 8 is unlikely to occur for shallow tunnels where the shear strength increases with depth.

Table 1 gives a complete summary of the stability bounds on $(\sigma_s - \sigma_t)/c_{u0}$ for the various values of C/D, pD/c_{u0}, and $\gamma D/c_{u0}$. Because the statically admissible stress fields are determined for a soil model with a convex yield surface and an associated flow rule, the safe load region defined by the dimensionless load parameters $(\sigma_s - \sigma_t)/c_{u0}$ and $\gamma D/c_{u0}$ must be convex. This feature permits a rigorous parametric study to be completed without determining the lower bound solutions for every combination of pD/c_{u0} and $\gamma D/c_{u0}$. Instead, for each value of

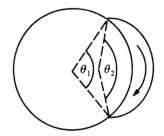

Fig. 8. Two-variable mechanism for local tunnel collapse

Table 1. Stability bounds on $(\sigma_s - \sigma_t)/c_{u0}$ for a shallow circular tunnel

$\dfrac{C}{D}$	$\dfrac{\rho D}{c_{u0}}$	\multicolumn{8}{c}{$\dfrac{(\sigma_s - \sigma_t)}{c_{u0}}$}
		\multicolumn{2}{c}{$\dfrac{\gamma D}{c_{u0}} = 0$}	\multicolumn{2}{c}{$\dfrac{\gamma D}{c_{u0}} = 1$}	\multicolumn{2}{c}{$\dfrac{\gamma D}{c_{u0}} = 2$}	\multicolumn{2}{c}{$\dfrac{\gamma D}{c_{u0}} = 3$}
1	0	2.27	**2.55**	1.08	**1.37**	−0.16	**0.15**	−1.60	**−1.12**
	0.25	2.73	**3.05**	1.60	**1.91**	0.40	**0.73**	−0.98	**−0.47**
	0.5	3.18	**3.53**	2.06	**2.41**	0.90	**1.26**	−0.38	**0.10**
	0.75	3.61	**4.01**	2.50	**2.89**	1.36	**1.76**	0.18	**0.62**
	1	4.04	**4.46**	2.93	**3.35**	1.80	**2.24**	0.70	**1.11**
2	0	3.25	**3.68**	0.97	**1.41**	−1.40	**−0.91**	−3.87	**−3.28**
	0.25	4.54	**5.10**		**2.89**		**0.65**	−2.20	**−1.61**
	0.5	5.79	**6.48**		**4.29**		**2.08**	−0.83	**−0.14**
	0.75	7.02	**7.85**		**5.66**		**3.46**	0.43	**1.26**
	1	8.24	**9.20**		**7.02**		**4.83**	1.68	**2.64**
3	0	3.78	**4.51**	0.47	**1.18**	−2.95	**−2.20**	−6.49	**−5.63**
	0.25	6.05	**7.11**		**3.85**		**0.56**	−3.76	**−2.72**
	0.5	8.22	**9.62**		**6.39**		**3.15**	−1.47	**−0.11**
	0.75	10.38	**12.10**		**8.88**		**5.66**	0.74	**2.42**
	1	12.52	**14.57**		**11.36**		**8.14**	2.92	**4.92**
4	0	4.30	**5.17**	−0.08	**0.80**	−4.57	**−3.61**	−9.11	**−8.08**
	0.25	7.67	**9.15**		**4.86**		**0.56**	−5.20	**−3.77**
	0.5	10.97	**12.99**		**8.73**		**4.46**	−1.78	**0.18**
	0.75	14.24	**16.80**		**12.56**		**8.30**	1.56	**4.03**
	1	17.50	**20.60**		**16.36**		**12.11**	4.84	**7.85**
5	0	4.65	**5.67**	−0.74	**0.30**	−6.20	**−5.10**	−11.80	**−10.60**
	0.25	9.30	**11.25**		**5.94**		**0.61**	−6.61	**−4.74**
	0.5	13.83	**16.59**		**11.30**		**6.01**	−1.99	**0.70**
	0.75	18.32	**21.89**		**16.62**		**11.33**	2.57	**6.04**
	1	22.80	**27.17**		**21.91**		**16.63**	7.10	**11.35**

Notes: 1. **Bold** entries indicate upper bounds from 7-variable mechanism
Other upper bounds from finite element formulation.
2. Blank entries for lower bounds can be found by linear interpolation along rows using values at $\gamma D/c_{u0} = 0$ and $\gamma D/c_{u0} = 3$.

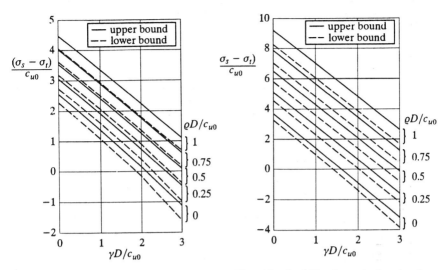

Fig. 9. Stability bounds for circular tunnel with C/D = 1

Fig. 10. Stability bounds for circular tunnel with C/D = 2

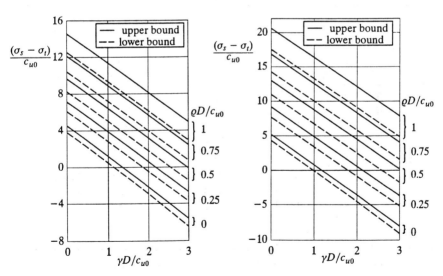

Fig. 11. Stability bounds for circular tunnel with C/D = 3

Fig. 12. Stability bounds for circular tunnel with C/D = 4

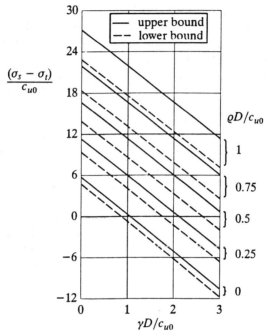

Fig. 13. Stability bounds for circular tunnel with $C/D = 5$

pD/c_{u0}, it is only necessary to compute values of $(\sigma_s - \sigma_t)/c_{u0}$ for $\gamma D/c_{u0} = 0$ and $\gamma D/c_{u0} = 3$, as any intermediate values may be found by linear interpolation. To illustrate this point, a complete set of lower bounds is presented for the case of $C/D = 1$ and for all cases where the undrained shear strength is uniform. These solutions are almost identical to those that are obtained using linear interpolation. The results shown in Table 1 are presented as dimensionless stability charts in Figs. 9 to 13.

As indicated in the notes to Table 1, the seven-variable mechanism of Fig. 7 provides the best upper bounds for all cases where $C/D < 5$. In most instances, the finite element solutions are only a few percent above the upper bounds for this mechanism, with the largest discrepancies occurring for very shallow tunnels with $C/D < 2$. It is not uncommon for rigid block models to give accurate estimates of the undrained collapse load. Indeed, provided an appropriate mechanism is selected, it is usually possible to get to within about ten percent of the true solution for some geometries.

For the special case where the undrained shear strength c_u is uniform, another set of lower bounds has been derived by Seneviratne (1979). He used the method of characteristics and was able to obtain rigorous solutions for a number of the tunnel geometries discussed here. For cases where $C/D \leq 1.5$ and $\gamma D/c_u \geq 3$, however, Seneviratne (1979) could not extend the stress field without violating yield and the solutions are not complete. The lower bounds obtained from the method of characteristics are, for all practical purposes, identical to the lower bounds obtained by the finite element technique. Because it is based on a finite element discretization, the latter method has no difficulty in extending the stress field and can model complicated geometries with ease.

For the case where the undrained shear strength is uniform, the seven-variable mechanism is the best rigid block model for all values of C/D. The solutions from this mechanism are typically two to three percent lower than those from the three- and four-variable mechanisms of Davis et al. (1980).

The undrained stability of a plane strain circular tunnel in a soil with a uniform strength profile has also been investigated by Sloan (1981). He used the displacement finite element method and employed the fifteen-noded triangular element which is known to give accurate estimates of collapse. Special care was taken with this analysis in an effort to minimize the effects of the mesh refinement and load increment size.

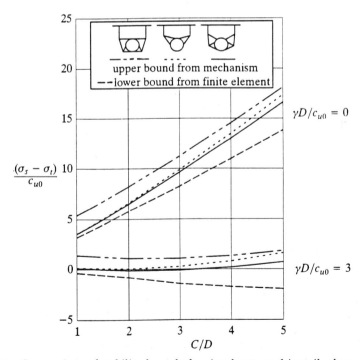

Fig. 14. *Comparison of stability bounds for circular tunnel in soil whose undrained shear strength increases linearly with depth ($\rho D/c_{u0} = 0.5$)*

For the case of a tunnel in a weightless soil with $C/D = 3$, Sloan (1981) estimated that failure occurs when $(\sigma_s - \sigma_t)/c_u = 4.00$. This value is between the lower and upper bounds of 3.78 and 4.51 which are shown in Table 1. Bearing in mind that the displacement finite element method invariably over-estimates the true collapse load, this result suggests that the lower bound solution is probably more accurate than the upper bound solution. In another displacement finite element analysis for the same value of C/D but with $\gamma D/c_u = 3$, the load parameter $(\sigma_s - \sigma_t)/c_u$ at incipient collapse was found to be -6.10. This result again falls between the rigorous bounds of -6.49 and -5.63 that are shown in Table 1 and confirms the likely accuracy of the lower bound solution.

The relative performance of the various upper bound mechanisms for the case where the undrained shear strength increases linearly with depth is illustrated in Fig. 14. Results are shown for two extreme values of the load parameter $\gamma D/c_{u0}$ with a strength profile of $\rho D/c_{u0} = 0.5$. This plot indicates that the four-variable and seven-variable mechanisms give similar results for shallow tunnels, but the latter is clearly superior once $C/D > 3$. The upper bounds from the finite element technique are very close to those of the three-variable mechanism and are not shown for the sake of clarity. Physically, the increasing shear strength is likely to cause the failure mechanism to be relatively narrow, with little movement of the lower walls of the tunnel, so that shearing of the stronger material is avoided. This explains the poor performance of the three-variable mechanism, which gives excellent results for the case where the undrained shear strength profile is uniform, as well as the good performance of the four-variable mechanism for shallow tunnels.

The dimensionless width of the failure mechanism at the ground surface, as predicted by the seven-variable model for various values of $\rho D/c_{u0}$ and C/D, is shown in Fig. 15. This ratio, denoted by W/D, is relatively independent of $\gamma D/c_{u0}$ for all cases where $\rho D/c_{u0} > 0$, and thus results are presented only for a typical value of $\gamma D/c_{u0} = 2$. As expected, W/D decreases with increasing $\rho D/c_{u0}$. Indeed, the value of W/D for cases where $\rho D/c_{u0} = 1$ is typically half that for cases where the undrained shear strength is uniform. The marked dependence of W/D on $\rho D/c_{u0}$ is one possible explanation as to why uniform strength plasticity solutions tend to overestimate the width of the collapse mechanism at the ground surface.

To assess the ability of plasticity theory to predict the stability of tunnels in practice, it is necessary to compare the theoretical predictions with observed behaviour wherever possible. The centrifuge experiments conducted by Mair (1979) are a valuable source of data for this purpose, particularly as they provide some guidance on the limitations of the overall approach. Mair's centrifuge results, which are for cases where the surcharge at the ground surface is zero and the undrained shear

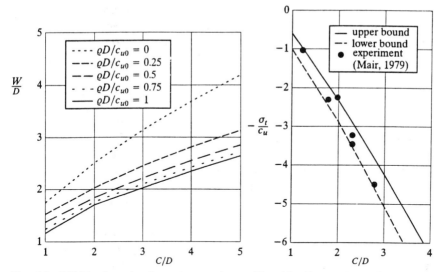

Fig. 15. Width of mechanism at ground surface for seven-variable mechanism ($\gamma D/c_{u0} = 2$)

Fig. 16. Comparison of theory and experiment for tunnel in soil with uniform undrained shear strength ($\gamma D/c_u = 2.6$)

strength is relatively uniform, are shown in Fig. 16. The stability bounds predicted by plasticity theory are in good agreement with the experimental observations which show surprisingly little scatter. It should be stressed, however, that Mair's results also suggest that the mode of collapse for tunnels with $C/D > 3$ is not modelled accurately by a failure mechanism which propagates all the way through to the ground surface. For these cases, collapse is accompanied by large radial deformations of the tunnel walls, with only small settlements taking place at the ground surface. The rigorous upper and lower bounds given in Davis et al. (1980) for the special case of a uniform undrained shear strength, are within a few percent of the limit solutions shown in Fig. 16 and cannot be distinguished on the scale of this plot.

As mentioned in the introduction, the stability solutions presented here must be used with care for tunnels in heavily overconsolidated clays. In these cases, the reduction in ground stress at the tunnel face will lead to the generation of negative excess pore pressures which in turn will result in a loss of strength with time. As the pore pressures dissipate and the clay softens, it is possible that a configuration which is initially stable may eventually become unstable. This phenomenon is known as the 'standup' problem and is not addressed here.

Conclusions

The stability of a plane strain tunnel in a soil whose undrained shear strength varies linearly with depth has been investigated using numerical techniques based on the theory of plasticity. A parametric study has been conducted for a typical range of soil properties and the results have been presented in the form of dimensionless stability charts. The solutions are appropriate for assessing the behaviour of tunnels in soft normally consolidated or overconsolidated deposits. In the latter case, the solutions are only relevant for immediate stability, and care must be exercised in allowing for a possible decrease in strength with time.

For all the cases considered, the numerical solutions bound the true collapse load most closely when the undrained shear strength increases slowly with depth so that $\rho D/c_{u0}$ is small. The evidence available suggests that the lower bound solutions are likely to be more accurate than the upper bound solutions. This is possibly because the lower bound finite element formulation permits large numbers of discontinuities in the stress field and thus admits a very broad range of solutions.

For the special case where the undrained shear strength is uniform, the new limit solutions are only a slight improvement on the bounds presented by Mair (1979) and Davis et al. (1980). Mair's experimental results suggest that the stability charts are appropriate for shallow tunnels where $C/D \leq 3$, but should be used with caution for deeper cases.

References

Assadi, A. and Sloan, S.W. (1991). Undrained stability of a shallow square tunnel. J. Geotech. Div., ASCE, Vol. 117, pp. 1152–1173.

Atkinson, J.M. and Cairncross, A.M. (1973). Collapse of a shallow tunnel in a Mohr–Coulomb material. Proc. Symp. Role of Plasticity in Soil Mechanics, Cambridge, pp. 202–206.

Cairncross, A.M. (1973). Deformation Around Model Tunnels in Stiff Clay. PhD Thesis, University of Cambridge.

Casarin, C. (1977). Soil Deformations Around Tunnel Headings in Clay. MSc Thesis, University of Cambridge.

Davis, E.H., Gunn, M.J., Mair, R.J. and Seneviratne, H.N. (1980). The stability of shallow tunnels and underground openings in cohesive material. Geotechnique, Vol. 30, pp. 397–416.

Lambe, T.W. and Whitman, R.V. (1979). Soil Mechanics – SI Version, Wiley, New York.

Mair, R.J. (1979). Centrifugal Modelling of Tunnel Construction in Soft Clay. PhD Thesis, University of Cambridge.

Muhlhaus, H.B. (1985). Lower bound solutions for circular tunnels in

two and three dimensions. Rock Mech. Rock Engrg., Vol. 18, pp. 37–52.
ORR, T.L.L. (1976). The Behaviour of Lined and Unlined Tunnels in Stiff Clay. PhD Thesis, University of Cambridge.
SENEVIRATNE, H.N. (1979). Deformations and Pore Pressures Around Model Tunnels in Soft Clay. PhD Thesis, University of Cambridge.
SKEMPTON, A.W. (1957). The planning and design of the new Hong Kong airport. Discussion, Proc. Inst. Civil Engineers, London, Vol. 7, pp. 305–307.
SLOAN, S.W. (1981). Numerical Analysis of Incompressible and Plastic Solids Using Finite Elements. PhD Thesis, University of Cambridge.
SLOAN, S.W. (1988a). Lower bound limit analysis using finite elements and linear programming. Int. J. Numer. Anal. Methods Geomech., Vol. 12, pp. 61–77.
SLOAN, S.W. (1988b). A steepest edge active set algorithm for solving sparse linear programming problems. Int. J. Numer. Methods Engrg., Vol. 26, pp 2671–2685.
SLOAN, S.W. (1989). Upper bound limit analysis using finite elements and linear programming. Int. J. Numer. Anal. Methods Geomech., Vol. 13, pp. 61–77.

Sliding resistance for foundations on clay till

J.S. STEENFELT, Danish Geotechnical Institute, Lyngby

In connection with the establishment of the Fixed Link across the Great Belt, Denmark, sliding resistance of clay till turned out to be important for the design of the foundations for bridge piers and anchor blocks. The sliding resistance was investigated experimentally by large scale field tests and more extensively by tests in a purpose-built Large Sliding Box in the laboratory. In a multi-factor test programme the effects from ageing, overconsolidation, pre-shearing, displacement rate and loading rate were investigated for (primarily) remoulded Storebælt clay till. The paper presents the test programme, the basic results and tentative correlations for the undrained sliding resistance.

Introduction

The Fixed Link across the Great Belt, Denmark, is designed, constructed and operated by A/S Storebæltsforbindelsen (Great Belt A.S.). The Fixed Link will connect two of the Danish main isles as shown in Fig. 1. The small island Sprogø, in the middle of the Great Belt, will be the meeting point for the (low level) West Bridge, a 6.6 km twin girder concrete bridge for combined traffic across the Western Channel, two 8 km long bored railway tunnels and the 6.8 km long East Bridge for the motorway across the Eastern Channel. The East Bridge is a suspension bridge with

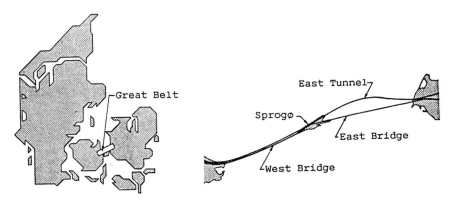

Fig. 1. Location of the Fixed Link across the Great Belt, Denmark

SLIDING RESISTANCE ON CLAY TILL

a main span of 1624 m and an elevation of 65 m above sea level, where the forces from the cables are taken up by two concrete anchor blocks. The foundations for the bridge piers, the pylons and the anchor blocks are raft type concrete caissons, where the majority rests in glacial tills.

To provide soil design parameters for a safe and economical design of the foundations, a number of numerical and experimental studies were carried out. A sliding type failure was found to be important for the anchor block foundations subjected to the large anchor forces and for the West Bridge piers subjected to impact forces from ice or ship collisions. As a consequence, experimental efforts were concentrated on establishing soil parameters for sliding failure, in particular for clay till and for the anchor blocks.

The joint venture CBR, COWIconsult, B. Højlund Rasmussen and Rambøll & Hannemann, has been in charge of the design of the East Bridge and has supervised the experimental programme. The Danish Geotechnical Institute was geotechnical consultant to Great Belt A.S. and in charge of the planning, execution and evaluation of the experimental programme. The results are reported to the Client in a number of project reports and form the basis for the selection of the clay till design parameters.

However, the experimental results also provide a more general insight into the shear strength behaviour of clay till for a sliding type failure. The present paper focuses on this general aspect for remoulded clay till.

Sliding resistance problem

The assessment of the reliability of a gravity foundation on clay till is usually based on the undrained shear strength, τ_u, in the disturbed

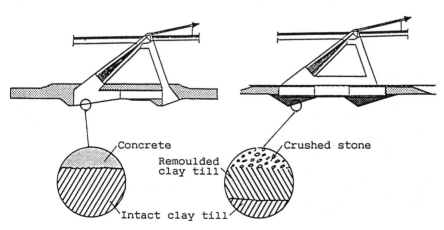

Fig. 2. Interface problem studied (shown for anchor blocks)

surface layer of the clay till of importance to sliding, and hence this is the key parameter. It is, however, very difficult to quantify the degree of disturbance in the clay till interface, and hence an initially completely remoulded state provides a lower boundary in the design process.

Thus, the designers require the strength of the interface between foundation and clay till, τ_u, as a function of the applied normal stress σ. It is convenient to describe this strength by the sliding resistance ratio, H/V, where H and V are the loads parallel with and perpendicular to the sliding interface, respectively.

Experimental investigation of the strength of the interface between structure and clay till requires certain simplifications. Basically an element of the interface has been considered as shown schematically for an anchor block in Fig. 2. In practice the interface between the structure and the underlying clay till will be rough, but investigations of a 'smooth' surface comparable to precast concrete were also carried out.

Storebælt clay till

The clay till used for the experimental investigations is Storebælt Clay Till, a clay till which is so macroscopically striking that it can be recognized wherever it occurs in borehole profiles. It is a very dark grey, fine sandy clayey till which contains millimetre-sized 'augens' of calcareous material. The clay till originates from the Weichselian Ice Age. Due to the very extensive investigations in the Storebælt area, the clay till is differentiated in a number of distinct types. There are clay tills of low plasticity exhibiting varying degrees of overconsolidation. The grain size distributions vary slightly, but on average the clay content is 15% (silt content 32%, sand content 45%, gravel and boulders constitute the remainder). The clay minerals ($\leq 10\%$) are dominated by expandable clay minerals and illites. The classification parameters corresponding to anchor block conditions are summarized in Table 1.

Experimental investigation

The problem of sliding resistance for clay till was specifically addressed by:

(i) 28 field sliding tests using 1.2 m² concrete blocks
(ii) more than 70 Large Sliding Box tests on 400 mm cylindrical clay till specimens in the laboratory
(iii) 20 conventional direct shear box tests on 100 mm² and 30 mm high clay till specimens in the laboratory.

Due to the cost and complexity of field testing most of the parameter studies were performed in the laboratory. An element of the interface between a concrete structure, with or without a stone layer, and

Table 1. *Classification parameters for Storebælt Clay Till*

Soil parameter	Symbol	Average value
Natural water content	w	11%
Liquid limit	w_L	16%
Plastic limit	w_P	10%
Plasticity index	I_P	6%
Unit weight of solids	γ_s	26.9 kN/m^3
Lime content	$CaCO_3$	21%

(a)

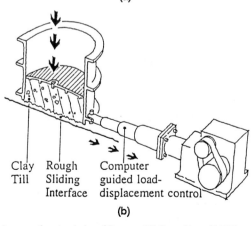

(b)

Fig. 3. *Overall view and principle of Large Sliding Box (LSB)*

underlying clay till was simulated in a purpose-built Large Sliding Box, the LSB (cf. Bak and Steenfelt, 1992). Quite literally, the interface problem was mirrored, as shown in Fig. 3(b). A cylindrical sample of clay till with a diameter of 400 mm and height 100 mm is forced to slide against a horizontal surface of steel or gravel-coated steel.

Two types of clay till interfaces were used in the model set-up corresponding to intermediate and maximum disturbance:

(i) *Disturbed*, where the intact clay till at natural water content is broken down to gravel size (d ≈ 30 mm).

(ii) *Remoulded*, where the clay till is completely remoulded (at $w \geq w_L$).

Thus, the clay till conditions closely resemble the non-intact conditions for the field sliding tests (cf. Hansen et al. (1991)). The specimen is loaded vertically by a load-controlled hydraulic jack. The horizontal sliding can be achieved either by a spindle with electronic displacement control or by a pressure or displacement controlled hydraulic jack, or by both systems simultaneously.

The displacements of the clay till specimen in the vertical and horizontal directions together with the vertical and horizontal loads are recorded by an automatic data acquisition system. The set-up allows displacement rates from 0.1 to 2000 mm/h and horizontal shear stress

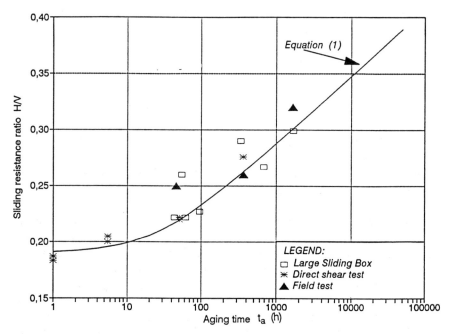

Fig. 4. *Sliding resistance ratio H/V as a function of ageing time t_a*

rates from 1 to 1000 kPa/h. However, the upper limit of the shear stress rate is intimately linked to the incurred displacement rate.

Evaluation of displacement controlled tests
General
The basic evaluation of parameters influencing the sliding resistance is based on remoulded clay till in a normally consolidated state. It is the most adverse condition and is covered by the majority of tests.

At an early stage in the experimental programme it was found that the magnitude of the vertical consolidation stress did not influence the sliding resistance ratio for OCR = 1. Thus most of the tests on remoulded clay till were carried out with a vertical consolidation stress of $\sigma'_c \approx 400$ kPa, as this is representative of the stress level for the caisson anchor blocks.

The very high degree of consistency in the trends from the conducted multi-factor analysis test programme and the few pairs of nearly identical tests suggest that the experimental uncertainty of the sliding resistance ratio H/V is lower than ± 0.02.

Effect of ageing
The test results from all three types of experiments (field and laboratory) confirm that the sliding resistance ratio increases with the logarithm of ageing time t_a, i.e. the time from initiation of the final vertical consolidation load step to start of sliding.

Figure 4 shows the test results for remoulded clay till (displacement rates 60–120 mm/h). As a pessimistic approximation eqn. (1) describes the increase in sliding resistance with time:

$$H/V = 0.19 + 0.06 \log_{10}\left(1 + \frac{t_a}{24 \text{ hours}}\right) \quad (1)$$

Note that the three field sliding tests, carried out with a displacement rate of $\dot{\delta} \approx 60$ mm/h rather than 120 mm/h as the rest of the tests in the graph, fit into the pattern. In evaluation of the influence of other parameters, the time dependent part of eqn. (1) may be used to adjust test results to the same ageing time t_a.

However, for tests with a low displacement rate the increase in sliding resistance is associated with relatively large displacements and a non-peak behaviour. For these tests the beneficial effects from ageing are 'lost' in the deformation process which ultimately leads to higher sliding resistances. Otherwise a conflict with the limiting drained sliding resistance, corresponding to $H/V \approx \tan \varphi'$ would result. The effective angle of friction, φ', is a constant at large strains. As a consequence eqn. (1), based on tests at $\dot{\delta} \approx 120$ mm/h, should not be used for $\dot{\delta} < 40$–60 mm/h.

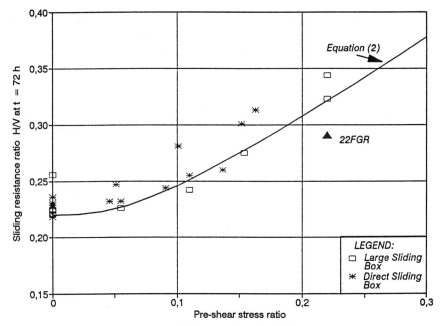

Fig. 5. Sliding resistance ratio H/V as a function of pre-shear stress ratio τ_0/σ'_c

Effect of pre-shearing

To simulate anchor block loading conditions, a series of tests was carried out in the LSB to quantify the effect on sliding resistance from a sub-failure shear stress applied during the consolidation process. The shear stress was applied after completion of primary vertical consolidation during the final vertical load step and the induced pore pressures were allowed to dissipate during an extended consolidation time.

Using eqn. (1) the sliding resistance ratios from the shear box and the LSB tests are adjusted to $t_a = 72$ h and shown versus applied pre-shear stress ratio in Fig. 5. The test results indicate a very consistent trend for increasing resistance with increasing applied pre-shear stress ratio, $\tau_0/\sigma'_c = H_0/V$, where the lower limit may be expressed by:

$$H/V = 0.22 + 0.75\, H_0/V - 0.05 \tan^{-1}(15 H_0/V) \tag{2}$$

Using eqn. (2) it is possible to compare test results obtained with different pre-shear stress ratios. It should be noted that the direct shearbox results indicate slightly higher effects of combined ageing and pre-shearing. The opposite was expected due to higher initial water contents in the direct shear box tests. However, the observed difference may be due to the smaller specimen height and to the fact that the sliding 'surface' here has inherent double-sided drainage where excess pore water pressure from the shearing process may dissipate faster.

In general, tests with combined ageing and pre-shearing give the highest resistances.

Effect from overconsolidation

All overconsolidated specimens, in the LSB, have initially been normally consolidated to $\sigma'_c = 600$ kPa and then allowed to swell back to σ'_{red}. Nominal overconsolidation ratios, OCR, from 1 to 6 and a displacement rate of $\dot{\delta} = 120$ mm/h were applied. The results are shown in Fig. 6, using eqn. (1) to give the peak values of sliding resistance ratio H/V for the average ageing time $t_a = 50$ h at the final load step (in the overconsolidated state).

Post peak values (residual values) are also indicated, and it is obvious that the effect from OCR remains constant for increasing displacement for a rough surface (56 or 77% gravel in sliding area). Hence, the relative gain in strength from overconsolidation is a permanent fixture of the clay till structure in undrained shearing. For steel surface (0% gravel) comparable to prefab concrete, however, the resistance drops considerably with increasing OCR.

For undrained failure the effect from OCR on the ratio of undrained

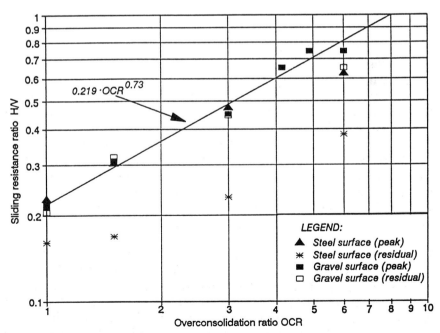

Fig. 6. Sliding resistance ratio as a function of OCR

shear strength, c_u, to effective vertical stress σ'_c, may be described by the SHANSEP approach (cf. Steenfelt and Foged (1992)):

$$(c_u/\sigma'_{red})_{oc} = (c_u/\sigma'_c)_{nc} OCR^\Lambda \qquad (3)$$

where $OCR = (\sigma'_c)_{nc}/(\sigma'_{red})_{oc}$ and Λ is the ratio of plastic to total volumetric strain in normal consolidation.

Assuming that the failure takes place undrained in the LSB at $\dot\delta \approx 120$ mm/h and that the measured sliding resistance and the undrained shear strength of the clay till are directly linked, eqn. (3) may be rewritten to:

$$(H/V)_{oc} = (H/V)_{nc} OCR^\Lambda \qquad (4)$$

According to eqn. (1), $t_a = 50$ h corresponds to $(H/V)_{nc} = 0.219$, and hence only Λ is unknown. A best estimate of $\Lambda = 0.73$ (± 0.05) is found from the test results with rough interface in Fig. 6. It appears that the same value of Λ applies for $OCR > 3$ for the tests on steel surface.

In three tests, all initially consolidated to $(\sigma'_c)_{nc} = 600$ kPa, a number of reloading/unloading cycles were performed after the primary failure phase. In each cycle the shear stress was reduced, corresponding to $H/V \approx 0$, and after a rapid change in total vertical stress σ_v, the shear stress was increased to failure again. All the failure points cluster closely around the line:

$$(H/V)_{load\ cycle} = (H/V)_{nc} [(\sigma'_c)_{nc}/(\sigma_v)_{load\ cycle}]^{0.78} \qquad (5)$$

i.e. corresponding to the upper limit indicated for eqn. (4). This means that the pore pressure developed in the sliding interface must be close to zero or slightly negative. Thus eqn. (4) with $OCR = V_c/V_{load\ cycle}$ can be used to predict the sliding resistance ratio even if $V_{load\ cycle} \neq V_{red}$ for rapid ('undrained') changes in the applied vertical load (V_c, V_{red} are the consolidation loads leading to the initial overconsolidation).

The hypothesis can be tested for one of the field tests where the vertical load was reduced from 500 kN (consolidation) to 280 kN immediately before the test with displacement rate 100 000 mm/h. Using the mean value $(H/V)_{nc} = 0.41$ for similar field tests with $OCR = 1$, eqn. (4) safely predicts $H/V = 0.41 \times 1.8^{0.73} = 0.63$ (or 0.65 using eqn. (5)) compared with the measured peak value of $H/V = 0.67$ (all values adjusted to $t_a = 50$ h using eqn. (1)).

Effect from displacement rate

The field sliding tests and the LSB tests clearly indicated the importance of the displacement rate. The combined picture for remoulded clay till as shown in Fig. 7, suggests a minimum resistance about $\dot\delta \approx 120$ mm/h.

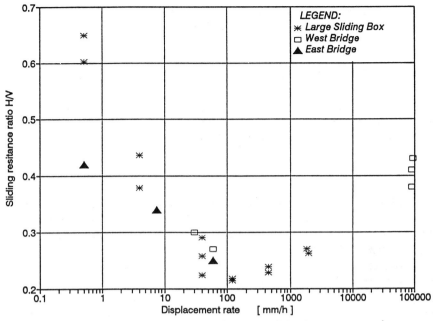

Fig. 7. Sliding resistance ratio H/V as a function of displacement rate $\dot{\delta}$

The sliding resistance ratios for $\dot{\delta} > 40$ mm/h have been corrected to a mean ageing time $t_a = 50$ h by means of eqn. (1). The minimum in H/V is rather flat and the resistance does not increase significantly before $\dot{\delta}$ is outside the range $60 \leq \dot{\delta} \leq 600$ mm/h. The minimum resistance ratio, H/V = 0.22, for $t_a = 50$ h and $\dot{\delta} = 120$ mm/h is borne out by the other tests.

At H/V = 0.22 the load–displacement curves exhibit a distinct change in curvature corresponding to transition from elastic to elastic–plastic behaviour. This corresponds to the point where soil elements in the sliding plane reach the yield surface of the soil.

For displacement rates $\dot{\delta} \geq 200$ mm/h both field and laboratory tests indicate a significant increase in resistance per log cycle of increase in displacement rate. Based on the LSB tests alone ($\dot{\delta} \approx 2000$ mm/h) the increase in H/V is 0.05 per log cycle of $\dot{\delta}$, whereas the combined field and laboratory tests suggest ΔH/V = 0.08 per log cycle for $\dot{\delta} > 450$ mm/h.

However, for displacement rates decreasing from $\dot{\delta} \approx 120$ mm/h, associated with very significant increases in sliding resistance, the pattern of behaviour in field and laboratory tests differs for $\dot{\delta} < 60$ mm/h.

For very low displacement rates a drained state of failure must exist in both test types. The LSB tests do approach the drained failure condition in the sliding plane for the lowest displacement rate:

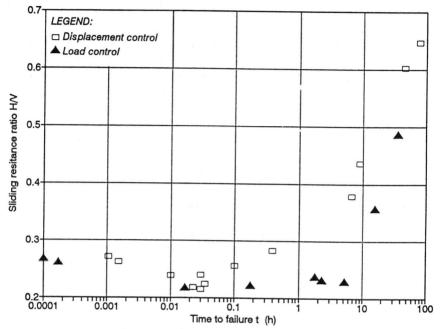

Fig. 8. *Sliding resistance ratios H/V for displacement and load controlled tests versus the time to failure*

$$H/V = \tau_h/\sigma'_v = c'/\sigma'_v + \tan \varphi' \approx 0.65\text{--}0.70 \qquad (6)$$

corresponding to $\varphi' \approx 33\text{--}34°$ and $c' = 0\text{--}20$ kPa, whereas the field sliding tests only give $H/V \approx 0.42$. The latter discrepancy may be explained by a closer inspection of the sliding conditions in the field which apart from serving as element tests also are full scale models of the bearing capacity behaviour of 1.2 m² footings.

Most adverse displacement rate

In conclusion, displacement controlled tests on remoulded clay till interface results in a minimum sliding resistance ratio of $H/V = 0.22$ for an ageing time of $t_a = 50$ h at the most adverse displacement rate of $\dot{\delta} = 100\text{--}200$ mm/h.

Since roughly the same pattern of behaviour with displacement rate is observed for both LSB and field tests, it is suggested that the inferred sliding resistance ratios can safely be applied for the full scale structure. At high displacement rates the development of the sliding resistance follows the trend for a sliding type failure.

At low displacement rate (≤ 0.6 mm/h) drained conditions are approached with near-drained sliding failure in the LSB element tests and possibly drained bearing capacity failure in the field tests.

Load controlled tests

In the LSB tests $\dot{\tau}$ was varied from 6 to ≈ 5000 kPa/h. For shear stress rates > 20–60 kPa/h the results indicate a constant sliding resistance ratio of H/V = 0.22, i.e. equivalent to the ratio for the most adverse displacement rate in the displacement controlled tests.

When the shear stress rate is reduced, the sliding resistance ratio increases. But although the time to failure for the lowest shear stress rate, $\dot{\tau} = 6$ kPa/h, is equivalent to the time to failure for the displacement controlled tests at the lowest displacement rate $\dot{\delta} = 0.5$ mm/h, the sliding resistance ratio is significantly lower and far from the value corresponding to drained sliding failure (cf. Fig. 8).

The correction formulas for ageing time and pre-shear stress eqns. (1) and (5) found in the displacement controlled tests for $\dot{\delta} = 120$ mm/h also work for the two load controlled tests with ageing or pre-shear at $\dot{\tau} = 60$ kPa/h.

In general the displacement rates induced by the load controlled tests are very low ($\dot{\delta} < 0.5$–1 mm/h) until the load approaches 70–80% of the maximum load. Then the displacement rate increases, through $\dot{\delta} \approx 20$ mm/h at failure, towards very high values.

To explain this behaviour let us consider tests with pre-shear. In these tests the horizontal shear stress is applied stepwise in order not to fail the specimen. However, for large stress increment the displacement rate increases significantly, but after a small amount of displacement it reduces to zero. This means that shear stress induced pore pressures in the sliding interface can dissipate for the applied constant shear stress. However, if the shear stress increment were large enough this would not be possible and sliding failure with large displacements would take place. Unless the shear stress increment was very large, the displacement rate would in all probability still reduce toward zero after considerable displacement.

Thus, without the restriction imposed in the sliding surface by a constant displacement rate, the load controlled tests exhibit a creep-type failure with ever increasing displacement rate due to the constant shear stress rate.

The sliding resistance ratios for displacement and load controlled tests on remoulded clay till in the LSB are compared in Fig. 8 using the time to failure as controlling parameter.

In general the same type of relationship between time to failure and resistance ratio is found in the two test types. However, for any given time to failure the load controlled tests offer the smallest resistance, although the most adverse value of sliding resistance is the same.

Judged from the test data a shear stress rate of the order 1–3 kPa/h would correspond to fully drained failure. This means that a transition from drained to most adverse sliding resistance ratio will take place for

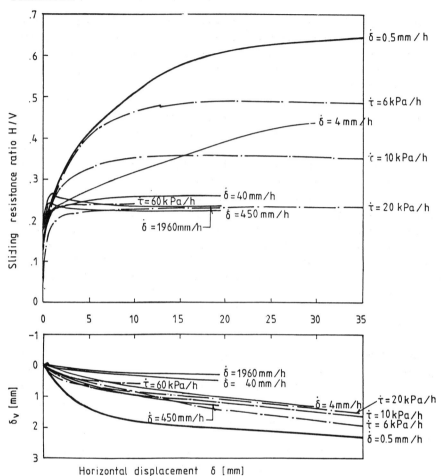

Fig. 9. Sliding resistance ratio H/V and vertical displacement δ_v for load and displacement controlled tests in the LSB versus horizontal displacement δ

one log cycle increase in shear stress rate. For comparison this transition requires three log cycles of increase in displacement rate for the displacement controlled tests.

The difference in behaviour can be elucidated by Fig. 9, where sliding resistance ratio H/V and vertical displacement δ_v (positive in compression) are shown versus the imposed or resulting horizontal displacement δ.

It would seem that a slow displacement controlled test (viz $\dot{\delta} = 0.5$ mm/h) can be used as a backbone test for the sliding tests in the LSB.

When the soil elements start yielding for H/V ≈ 0.22 the increase in H/V is associated with vertical compression, i.e. reduction in specific volume, and strength increase. As the critical state, with no further volume change, is approached, the maximum drained strength is exhausted for large displacements.

The increase in H/V with displacement is reduced as the sliding rate is increased. As the most adverse sliding rate is approached, no gains in sliding resistance are possible as the vertical compression before failure is arrested. At even higher displacement rates the specimen dilates slightly associated with peak strength development and reduction to the critical state at the original specific volume.

The tests with load control follow the backbone curve until the rate of pore pressure development in the sliding interface prevents any further gain in sliding resistance which is followed by an increasing displacement rate at constant sliding resistance ratio.

Essentially the effective stress state, which cannot be measured in the sliding interface, governs the sliding conditions. The time to failure, the displacement rate or the shear stress rate may all be used as substitutive parameters provided the basic phenomenological soil behaviour is kept in mind.

Information from other sources

The analysis of geotechnical problems, such as the foundation design for the East Bridge anchor blocks, requires the adoption of a soil behavioural model that should include all relevant soil properties. The measurement of soil properties can be enhanced and guided by empirical correlations established by the vast amount of already existing measurements. The calibre and applicability of the correlations vary almost as much as the available measurement methods. Some are global and reflect well-established, basic stress–strain–time behaviour of generalized soils, whereas others are local and often a regression analysis of available data points.

In order to enhance the findings from the test programme a review of information on soil properties, particularly relevant to the sliding resistance problem at hand, was carried out. The main objectives were:

(a) the unravelling of quantifiable effects on undrained soil strength from strain or displacement rates and possible physical explanations for applied rates

(b) the validation of observed soil resistance behaviour for sliding failure

(c) establishment of a physical interpretation and understanding of observations.

It should be borne in mind that the source, extent, and limitations of correlations are most often obscured in the presentation of the relationships. The simple lines presented as correlations may in reality be based on a veritable shotgun blast of data points. Furthermore, the rationale for specifications on test equipment and procedures warranted by specific site or problem considerations may subsequently have been lost or forgotten in the presentation of data in the form of correlations.

Despite these difficulties the review of existing information outside of Storebælt together with the specific information from Storebælt may be organized into a coherent framework.

Effect of test rates and rate changes

The quantification of rate effects on undrained (or drained) strength behaviour is inherently ambiguous. In standard triaxial or direct simple shear tests the rate is typically understood as vertical or shear strain rate $\dot{\varepsilon}$ %/h, $\dot{\gamma}$ %/h. In field tests and in direct shear tests a strain rate cannot be directly defined and hence a displacement rate $\dot{\delta}$ mm/h is usually indicated. In an NGI type direct simple shear apparatus with specimens of 80 mm diameter and 20 mm height a shear strain rate of $\dot{\gamma} = 3$ %/h translates into a displacement rate for the top of the specimen of $\dot{\delta} \approx 0.5$ mm/h.

Obviously, the applied rates in tests with incommensurable stress–strain conditions cannot be directly compared, and an agreed standard test rate does not exist. However, in laboratory testing a test rate of 1 %/h is probably an appropriate reference rate for evaluation of changes in undrained strength with test rate.

Two other very important factors, i.e. the degree of soil confinement and the type of test control, blur the influence from test rate.

In the laboratory tests with high control of test conditions the soil specimen is confined, usually by a latex membrane. Hence testing in an undrained state, i.e. with no volume change, is a meaningful concept. Changes in test rate will here mostly influence the magnitude of pore pressure development in the soil volume.

In contrast, a direct shear test and field tests have no such confinement, and hence *undrained* state is here a theoretical limiting condition. In practice the term *undrained* is conceived as a state where the test rate is sufficiently high to allow only insignificant pore pressure dissipation and change in volume in the part of the soil involved in the stress–strain changes. Conversely the term *drained* indicates that the test rate is sufficiently low not to allow any significant pore pressure build-up to take place.

Two different modes of test control are used in field and laboratory testing, i.e. strain (displacement) control or stress (load) control. Also a strict differentiation is not possible here as the two modes usually interact. The time to failure, however, might represent a viable means of comparison (cf. Fig. 8).

The review of the literature has confirmed that the described incommensurable test conditions allow only basic trends to be established for the rate dependence.

Field and laboratory tests
Standard field tests, vane, pile loading and CPT, are phenomenologically related to undrained sliding behaviour.

The field vane test is a hybrid test form as the torque of the vane rod is supposedly displacement controlled, but due to the large extensions of the dynamometer used for the force measurement, the test is rather load controlled. Hence the time to failure is used as the governing parameter. The literature suggests an increase of the order of 10% in measured vane strength per log cycle decrease in time to failure.

For the other two test types the review revealed no systematic investigations or even conflicting evidence on the effect from displacement rate.

Using a test rate of 1 %/h as reference, triaxial and direct simple shear tests in the literature suggest a 10% increase in undrained shear strength per log cycle of increase in test rate. The specific tests on Storebælt Clay Till relevant for sliding resistance of the anchor blocks indicate a 7% and 12% increase per log cycle of strain rate for triaxial and direct simple shear tests, respectively.

In conclusion, testing rates applied in different test types are somewhat fortuitous, albeit most often based on practical considerations of equipment and manpower. However, rate increases over a reference rate, deemed representative of undrained conditions, generally results in a 10% increase of undrained shear strength per log cycle of rate increase.

Effect from clay structure

The strength (and other engineering properties) of clays depends on the initial void ratio or water content and the stress history. More recently also the influence from structure has been recognized as an important aspect.

In general, natural clays in the intact state show superior strength characteristics compared with remoulded or reconstituted states. This is partly due to creep, which reduces the void ratio under constant effective stresses, and partly due to structure for instance caused by thixotropic hardening on ageing (bonding, cold welding at high pressures are other examples).

The change in structure of a remoulded clay due to thixotropic hardening is visualized in Fig. 10. The clay goes from a predominantly

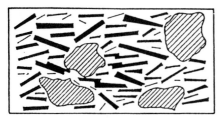 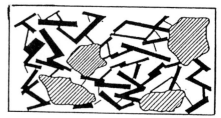

Fig. 10. Structure due to thixotropic hardening on ageing. (a) Immediately after remoulding; (b) final structure at end of thixotropic hardening (after Mitchell, 1960)

dispersed to a flocculated structure – in lay terms, a change in structure not dissimilar to coagulation.

The corresponding effects on the stress-strain-strength behaviour are illustrated in Fig. 11 for normal consolidation, where it is apparent that both creep and structure increase the shear strength. In terms of undrained strength, the soil acts as if it were overconsolidated. However, in contrast to the creep-induced strength, associated with decrease in void ratio, the structure-induced strength can be lost due to yield.

Shear-induced yield will in particular tend to bring the clay back to a more dispersed structure. A major part of the strength may be recovered by renewed ageing. Experience from tests in the Large Sliding Box suggests a high rate of strength recovery after sliding-induced failure, if the shear stress level is reduced below yield level. This, however, is associated with reduction in void ratio due to dissipation of the induced pore pressures during yielding.

Trends with specific relevance to sliding resistance

The literature review has not revealed any information from full or reduced scale experiments or measurements directly relevant to the sliding resistance problem for the East Bridge anchor blocks. However, some of the trends established from the sliding resistance tests for Storebælt may be indirectly checked by published results.

It seems that the direct simple shear box test (DSS), despite its inherent shortcomings, is the test type which most directly resembles the sliding resistance problem. There is overwhelming support, based on a large number of low to medium plasticity inorganic clays (e.g. Jamiolkowski et al., 1985; Ladd, 1991), for a shear stress ratio of:

$$(\tau_h/\sigma'_c)_{nc} = 0.23 \pm 0.04 \qquad (7)$$
$$(\tau_h/\sigma'_c)_{oc} = (\tau_h/\sigma'_c)_{nc} \, OCR^{0.8}$$

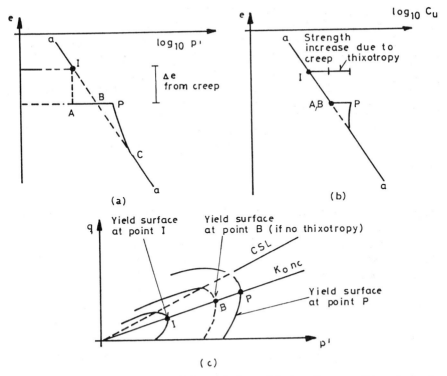

Fig. 11. *Normal consolidation of clay. (a) Consolidation diagram, (b) undrained strength development, (c) yield curve in p', q space (after Leroueil and Vaughan, 1990)*

This is very close to eqn. (2) established on the basis of Large Sliding Box tests. Note, however, that eqn. (7) is mainly based on 'intact' specimens, whereas eqn. (2) refers to remoulded specimens!

Ladd (1991) reports three DSS tests on Boston Blue Clay (I_P = 23.5%, I_L = 0.8) with pre-shear. Pre-shear stress ratios of 0, 0.1 and 0.2 produce increases in strength ratios of 0, 0.03 and 0.09, respectively. The increases suggested by eqn. (2) from the LSB are 0, 0.026 and 0.088, i.e. the same qualitative variation.

The ageing effect has not been reported specifically for DSS tests.

Information directly related to the trough-shaped form of the relationship between sliding resistance and logarithm of displacement rate (Fig. 7) was not found in the literature. However, based on the trends in literature a qualitative physical explanation of the phenomenon observed for *normally consolidated* specimens may be inferred.

All changes in soil behaviour are due to changes in effective stresses, irrespective of the type of stress path imposed, i.e. both for so-called undrained and drained conditions. At a given stress level the soil will exhibit creep with reduction in volume due to continuous dissipation of the strain-induced pore pressures. If, however, the time at the stress level is reduced the creep-induced pore pressures are reduced, and hence an increase in 'undrained' shear strength may result.

If the test rate in 'undrained' loading without confinement (i.e. the sliding experiment) is very high, the build-up of pore pressures is impeded by the high rate and high 'undrained' shear strengths at very limited strains may result. At a lower test rate the particles in the soil element are given more time to respond, and due to the pore pressure build-up the effective stresses may more readily reach the yield surface, accompanied by larger strains and reduced strength.

However, at a sufficiently low test rate the yielding of the soil is in itself impeded by the low rate, and due to the reduction in volume the soil strength increases. A drained failure with the associated larger strength corresponds to the lower limit for the test rate and the upper limit for the soil strength.

Recommendations and conclusions

Sliding resistance parameters to be used in design are intimately linked with:

(a) actual soil conditions in the excavated surface prior to positioning the structure
(b) the actual loading history for the structure
(c) the safety philosophy adopted in design
(d) engineering judgement in assessment of deviations from the basis for established correlations.

Combining eqns. (1), (2) and (4), the sliding resistance ratio for a rough interface to remoulded Clay Till 0–1 in normal or overconsolidated state can be calculated by eqn. (8):

$$(H/V)_{nc} = 0.19 + 0.06 \log_{10}\left(1 + \frac{t_a}{1 \text{ day}}\right)$$
$$+ 0.75 \, H_0/V_0 - 0.05 \tan^{-1}(15 \, H_0/V_0) \quad (8)$$
$$(H/V)_{oc} = (H/V)_{nc} \, OCR^{0.73}$$

In eqn. (8) it is assumed that the static load on a horizontal surface, with vertical component V_0 and horizontal component H_0, has been acting for t_a days prior to the increase in loading ratio which might take the structure to failure. Further, the pre-shear stress has been in the same direction as the shear stress increase and the pre-shear stress ratio has been H_0/V_0.

The effects of other loading histories may also be evaluated based on the data base. However, each case would require a specific analysis in order to assess the validity of application of the established trends.

Pre-sheared specimens with and without ageing conform to the general trend established for aged specimens without pre-shear. This suggests that a gradual application of pre-shear does not destroy structure developed by thixotropic hardening on ageing as long as the combination of ageing and pre-shearing expands the yield surface sufficiently to allow basically elastic behaviour of the soil.

A moderate increase in vertical load level will not invalidate the use of eqn. (8). If a permanent change takes place shortly (days, weeks) after the start of ageing, the ageing time will just have to be re-zeroed. If the change takes place after prolonged ageing, the soil will be strong enough to react elastically.

In general the established trends for sliding resistance are unequivocal. The fact that the sliding resistance development was checked at three different scales ranging from the 100 mm^2 specimens in the direct shear box to the 1 by 2 m^2 concrete blocks in the field tests lends credence to the possibility of extrapolation to full scale conditions. This optimism is partly based on theoretical considerations and partly on phenomenological observations.

Acknowledgements

The author gratefully acknowledges permission by the Client, Great Belt A.S., to publish the paper and appreciate the encouragement and comments by Mr Aage Hansen during execution of the project.

References

BAK, J.K. AND STEENFELT, J.S. (1992). Large sliding box test. Proc. XI Nordic Geotechnical Meeting, NGM-92, Aalborg, Vol. 1, pp. 113–118.

HANSEN, P.B., DENVER, H. AND MOLLERUP, E. (1991). Lateral sliding resistance – Large scale sliding tests. Proc. X ECSMFE, Florence, Vol. 1, pp. 433–436.

STEENFELT, J.S. AND FOGED, N. (1992). Clay till strength–SHANSEP and CSSM. Proc. XI Nordic Geotechnical Meeting, NGM-92, Aalborg, Vol. 1, pp. 81–86.

JAMIOLKOWSKI, M., LADD, C.C., GERMAINE, J.T. AND LANCELOTTA, R. (1985). New developments in field and laboratory testing of soils. Proc. 11th Int. Conference on Soil Mechanics and Foundation Engineering, San Francisco, Vol. 1, pp. 57–153.

LADD, C.C. (1991). Stability evaluation during staged construction. Journal of Geotechnical Engineering, Vol. 117, No. 4, pp. 540–615.

LEROUEIL, S. AND VAUGHAN, P.R. (1990). The general and congruent effects of structure in natural soils and weak rocks. Géotechnique, Vol. 40, No. 3, pp. 467–488.

MITCHELL, J.K. (1960). Fundamental aspects of thixotropy in soils. ASCE, Vol. 86, No. SM3.

Sampling disturbance — with particular reference to its effect on stiff clays

P.R. VAUGHAN, R.J. CHANDLER, J.P. APTED,
W.M. MAGUIRE and S.S. SANDRONI, Imperial College of
Science, Technology and Medicine, London

The behaviour of soft clays during tube-sampling is reviewed, followed by a more detailed discussion of various aspects of disturbance of stiff clays, particularly stiff plastic clays, during tube sampling. It is concluded that the main effect of tube sampling is to increase the effective stress in the sample, smaller diameter samples showing the greater increase. The reasons for this phenomenon are demonstrated, and its consequences, particularly for the measurement of undrained strength and the estimation of K_0, are discussed.

Introduction

In geotechnical engineering there will always be a need to compare the results of in situ tests with corresponding laboratory testing. High quality laboratory testing requires the highest possible quality of undisturbed sample, and it is well known that it is impossible to obtain a truly undisturbed soil sample for this purpose. Hand-cut block samples of clay soils probably remain the highest quality of sample presently available. However, it is often not practical to obtain block samples, and the necessary samples must be obtained by driving or jacking a tube into the base of a borehole.

It is the purpose of this paper to examine the causes of disturbance to tube samples of clay, referring particularly to the behaviour of stiff clay on which research at Imperial College has been concentrated over the years. As in many aspects of geotechnics, there is much to be gained by comparing soft and stiff clay behaviour. Consequently, in presenting our findings on the behaviour of stiff clays, we believe that it is also of considerable interest to review the corresponding behaviour of soft clays.

With high strength clays and clay shales tube sampling tends to initiate fracturing of the material; mechanical damage to the soil fabric of this nature is not considered.

The perfect sample

The stress changes that occur when a tube sample of clay is taken may be considered in two parts.

(a) The reduction in boundary total stress to zero which must occur at some stage in the transfer of the sample from the ground to the test apparatus. These stress changes must occur, even in the absence of other causes of mechanical disturbance.

(b) Additional stress changes which occur as a result of mechanical disturbance, both when taking the tube sample, and when extruding it in the laboratory.

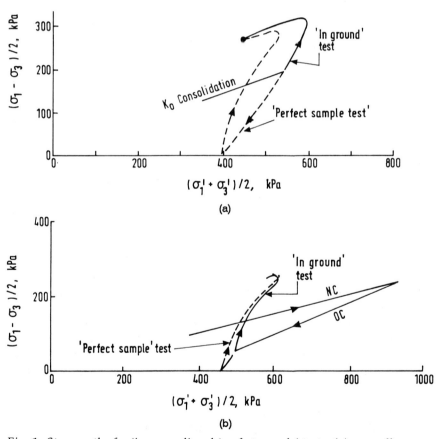

Fig. 1. Stress paths for 'in ground' and 'perfect sample' tests: (a) normally consolidated clay (Kawasaki clay, $I_p = 43\%$; reconsolidated Shelby tube samples; Ladd & Lambe, 1963), (b) lightly overconsolidated clay (Weald clay, $I_p = 24\%$; consolidated from $I_L = 0.48$; Skempton and Sowa, 1963)

'Perfect' sampling, as in (a) above, involves the release of total stress to zero, with uniform strains within the sample. The effect of these stress changes on soft clays was investigated by Skempton and Sowa (1963) and Ladd and Lambe (1963), using samples anisotropically consolidated in the laboratory. Pairs of matching samples were tested, one in undrained compression after consolidation (the 'perfect test', starting at the in situ stresses), the other being unloaded undrained to isotropic stress, before being loaded to failure in undrained compression (testing after 'perfect' sampling).

Skempton and Sowa used reconstituted Weald clay, with plasticity index = 24%. The samples were initially remoulded at the rather low liquidity index of 0.48, so that they would not have exhibited the degree of anisotropy expected of natural samples or of samples prepared at higher liquidity indices.

Ladd and Lambe used natural samples, which are likely to have retained much of the structure imposed by consolidation in situ. Figure 1 shows pairs of stress paths for a normally consolidated clay from Ladd and Lambe, and for a lightly overconsolidated clay from Skempton and Sowa.

The tests illustrated lead to the following conclusions for remoulded samples consolidated in the laboratory, and lacking geological structure.

(a) Peak strength in terms of effective stress is not changed significantly by perfect sampling.

(b) There is a considerable reduction in average effective stress when the total stresses acting in the 'ground' are removed from a normally consolidated sample. The stress paths only converge at large strains, as they arrive at the same critical state strength controlled by their common water content. Peak undrained strength is developed at an earlier stage of the test, and there is a significant difference between this strength for the two samples. The stiffnesses of the two samples (not shown) are similarly affected.

(c) For overconsolidated samples (OCR > 2), undrained unloading occurs at nearly constant mean effective stress; in any case, such samples in the ground are often in a stress state close to isotropic. Thus perfect sampling has a limited effect on average effective stresses, and there is little difference in the peak undrained strengths.

Cavitation in the sample

Such testing in the triaxial apparatus does not necessarily model the in situ sampling situation, as in these laboratory tests the samples retained their mean effective stress, and did not suffer from cavitation in the pore water when the total stress was reduced to zero. Had cavitation occurred, the conclusions given would not have applied.

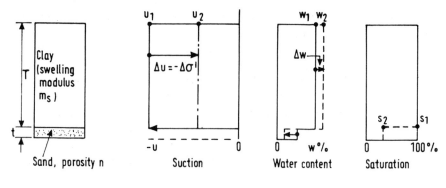

Fig. 2. *Equalisation of pore pressure within a clay sample containing a layer of sand, following sampling and cavitation in the sand*

Clay samples can sustain quite large suctions for long periods without cavitating and de-saturating. However, silt or sand layers within a clay will cavitate at quite low suctions. The result is illustrated in Fig. 2. After sampling, under zero total stress a uniform negative pore pressure (or suction; for which the symbol p_k is used) is established within the sample tube, with the silt or sand becoming partly saturated and water flowing to the clay layers. The final pore water suction in the clay will consequently be substantially reduced, and its water content increased. Some typical soil properties are used in Fig. 2 to illustrate the possible magnitude of this effect. Spuriously low undrained strengths will then be measured. Note that the same effect is obtained if the sample contains open fissures, root holes, or other imperfections which provide a source of free water during or after sampling.

In this event, correct water contents can only be obtained if samples are extruded immediately after sampling, and a water content determination made in the centre of a reasonably thick clay layer. If cavitation occurs in this manner, more representative test results can be obtained if samples are reconsolidated to estimated ground stresses before testing, particularly if a back pressure is used to encourage saturation of the silt or sand.

Similar problems to those resulting from cavitation will occur if water is present during sampling, when the clay at the bottom of the borehole may have been similarly exposed prior to sampling. The bottom of the sample will also be exposed when it is 'broken off' at withdrawal. If the cutting edge of the sampler is smaller than the inside diameter of the sample tube, then the sides of the sample may also be exposed to free water. These effects may be partly controlled if the affected material is removed from the sample ends before the sample is sealed.

Wetting up is avoided if a dry hole is used; this is preferable with stiff clays when the clay is strong enough to prevent bottom heave in the borehole. Air can be used for rotary drilling. With soft clays, the danger of bottom heave will usually require the use of a water to surcharge the base of the borehole.

Shear strains imposed on a tube sample

Shear strains are imposed on a sample as the tube is pushed or driven into the ground, a subject first studied by Ladd and Lambe (1963). The shear strains arise through two recognisable mechanisms.

(*a*) The soil to be sampled compresses vertically while the tube is above it, but expands vertically once it enters the sample tube. This mechanism affects the whole of the sample. Further strains of this nature are likely during sample extrusion.

(*b*) The side of the sample is subjected to large strains as the tube penetrates the ground, shearing a thin annulus of soil around the sample. Further shearing will occur as it is extruded from the tube. Deformation of this type can frequently be seen in samples of laminated soil.

The magnitudes of the strains caused by sampling cannot easily be measured, but they can be estimated using the Strain Path Method (SPM) (Baligh, 1985). The method assumes no volume change, as is appropriate for the undrained case. Figure 3 shows the vertical strains estimated by the SPM on the centre line of a sample as the sample tube is pushed into the ground. The results are expressed as a function of the ratio of sample diameter, B, to the tube wall thickness, t. Vertical compression occurs while the soil to be sampled remains below the sampler, followed by vertical extension as the soil enters the tube. Also shown are typical dimensions for standard and high quality samplers. With standard samplers ($B/t \approx 20$), centre-line strains are of the order of 2%. Even with high quality, thin-wall samplers, strains approach 1%.

The strain path method also predicts the strains along the sides of the

	B (mm)	t (mm)	B/t
U 100 (BS 5930)	115	4.6	25
Rodio-Nenzi-SGI	89	2	44.5
Fugro Piston	76	2	38
NGI 95mm Piston	101.6	2.65	38.3
Shelby Tube	50.8	1.25	40.8
University of Laval	218	5	43.6

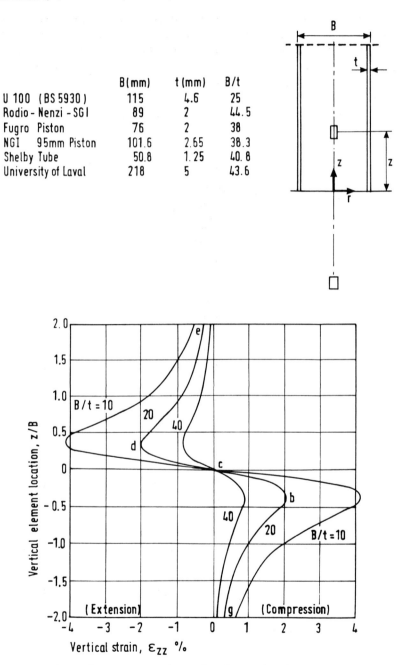

Fig. 3. Strain-path method predictions of vertical strain in a soil element during tube sampling (after Baligh (1985))

sample. Figure 4 (Baligh, 1985) shows these for a thick-walled sampler with B/t = 10. Shear strains are greater than 10% in a zone a little wider than the wall thickness of the sampler, and they are in excess of 100% in a thinner zone against the wall. Lesser, but still significant effects are predicted for thin-walled samples.

The probable effect of these strains on both the soil structure and the effective stresses within a sample are illustrated qualitatively by the

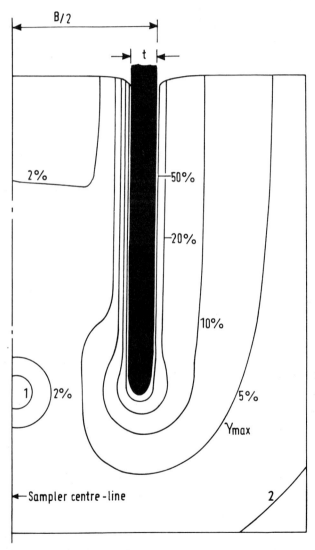

Fig. 4. Strain-path method predictions of maximum shear strains (γ_{max}) due to tube sampling for B/t = 10 (after Baligh 1985)

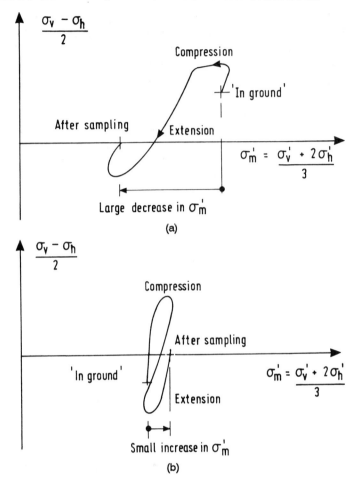

Fig. 5. Stress paths followed by centre-line soil elements during tube sampling: (a) normally consolidated clay, and (b) overconsolidated plastic clay

stress paths shown in Fig. 5. With normally consolidated clay (Fig. 5(a)), the compression strain may well cause the sample to reach peak deviatoric stress. Triaxial test data show that an axial strain of about 0.5% can be sufficient for this to occur, and reference to Fig. 3 shows that such strains are likely even with a high quality sampler. If failure occurs, then some of the structure of the soil will certainly be destroyed, and significant excess pore pressures will be generated. The effect of extension strain is similar, and a further increase in pore pressure occurs. It may be noted that these pore pressure changes will occur in all clays, but they are much greater in clays of low plasticity.

SAMPLING DISTURBANCE

In a stiff overconsolidated plastic clay (Fig. 5(b)) the effect of compression/extension strains on pore pressure is much less. Axial strains at failure are likely to be in the range 3–5%, and thus greater than those imposed by sampling with a thin-wall sampler. Consequently, the damage to the soil structure may well be limited. However, thin-wall samplers cannot normally be jacked or driven into very stiff clays without the tubes collapsing or the samples shattering in the tube.

The situation with stiff clays of low plasticity has not been examined in detail, and such clays are often stony and difficult to sample using tubes. However, cyclic triaxial tests show that compression/extension cycles may generate quite large undrained pore pressures, and consequently such samples may suffer a reduced effective stress during tube sampling.

The effect of shear on the sample periphery

Structure will be destroyed within the thin shear zone, but this is of limited area and will have only a comparatively small effect on the

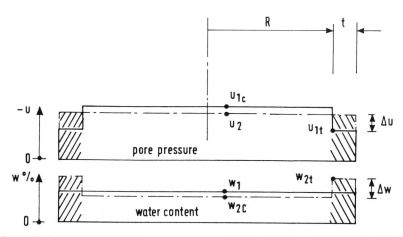

Consolidation: continuity of flow.

$(u_{1c} - u_2).\pi.R^2.m_c = (u_2 - u_{1t}).2.\pi.R.t.m_t$
m = coefficient of volume change

If $u_{1t} = u_{1c} - \Delta u$,
Then $u_{1c} - u_2 = \Delta u/[1 - (m_c.R/2m_t.t)]$

Soft clay; $m_t^1 > m_c^2$; stiff clay; $m_t^3 > m_c^4$
1. Consolidation of de-structured soil
2. Swelling of 'undisturbed' soil
3. Swelling of de-structured soil
4. Consolidation of 'undisturbed' soil

Change in water content of zone t by Δw.

By continuity of flow:

$w_1 - w_{2c} = \Delta w/[1 + (R/2t)]$.

Change in effective stress in centre:

$u_{1c} - u_2 = \Delta \sigma'_c$
$= \Delta w/\{1 + (R/2t)\}(1/m_c)\{1/(w_{1c} + (1/G)\}$.

If $R/t = 20$, $m_c = 0.05 m^2/MN$, $w_{1c} = 0.3$,
$G = 2.7$, then $\Delta \sigma'_c = 81 kPa$.

Fig. 6. *The effect of side-shear induced pore pressure changes on the effective stresses in a clay sample, subscripts (1) apply before equalisation, and (2) after*

measured properties. For instance, if structure were destroyed completely in an annulus 5 mm thick on the sides of a 100 mm diameter sample, then it would represent 20% of the area of the sample, and 20% of the effect of the sample's structure would not be measured.

The effect of side-shear on pore pressure is much greater. With a contractant soil, side-shearing will generate positive pore pressures in the peripheral zone of the sample, while the reverse will be true of a dilatant, stiff clay. The equilibration of pore pressure within the sample will occur quite quickly, often before it is extruded and tested. The pore pressure after equilibration will no longer be that of a perfect sample.

There are two ways of examining the potential change in pore pressure due to equilibration. The first is through consolidation theory;

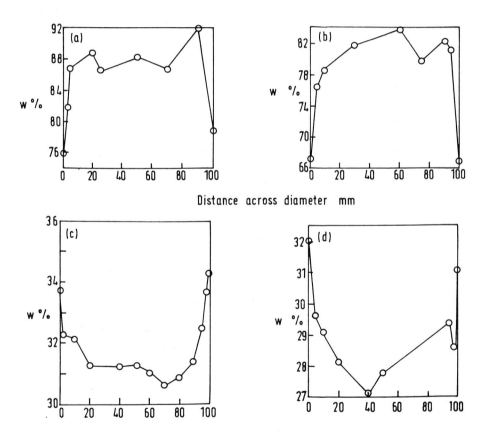

Fig. 7. The change of water content at the periphery of clay samples due to tube sampling: (a, b) reduction in water content at the periphery of a soft clay sample ($w_L = 92\%$, $w_P = 33\%$), and (c, d) increase at the periphery of a stiff clay sample ($w_L = 81\%$, $w_P = 30\%$).

however, it is difficult to determine the appropriate values of the coefficients of consolidation, c_v or c_s, for the thin zone of sheared and remoulded clay. A simpler method is to equate the difference in water content between the sheared zone and the sample centre (which can be measured after equilibration, at least approximately), with the change in effective stress in the undisturbed sample centre. The calculation is set out in Fig. 6. Because the sample centre is stiff, the change in effective stress can be large. In the example shown, with a shear zone thickness of 5% of sample radius and a change in water content of 3%, the increase in effective stress is 81 kPa.

Examples of such observations are given in Fig. 7 (Apted, 1977). Figure 7(a) shows the reduction in water content at the periphery of a sample of soft Thames alluvial clay. Here, the average water content of the periphery of the sample has been reduced by around 15%. Figure 7(b) shows an increase of water content of about 3% at the periphery of a London clay sample.

The change in water content in the outer skin of a sample of London clay cannot be explained from consolidation theory unless a large swelling modulus is assumed for the disturbed soil. It will be noted that the effect on the average effective stress in the sample reverses according to whether the soil dilates or contracts during shear. Thus effective stress is lost in a soft contractive clay, while it increases in a stiff, dilatant clay. These conclusions do not depend on the plasticity of the clay.

The effects which might be anticipated from these mechanisms are summarised in Table 1.

Table 1. *The effect of sample disturbance on the initial effective stress and undrained strength*

Material type	Axial strain in sample*	Side shear	Effect on s_u
Soft clay: low plasticity	Very large decrease	Large decrease	Very large decrease
high plasticity	Large increase	Large decrease	Large decrease
Stiff clay: low plasticity	Large decrease	Large increase	Negligible**
high plasticity	Slight	Large increase	Large increase

*As defined in Fig. 3.
**Because ultimate s_u is a function of water content only.

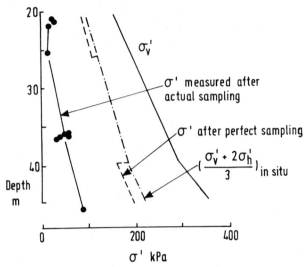

Fig. 8. Effective stresses measured in 70 mm Shelby piston samples, compared with effective stresses in situ and those expected after perfect sampling (after Ladd and Lambe (1963))

Effective stresses after sampling

The reduction of effective stress in tube samples of soft clays was recognised by Ladd and Lambe (1963), who give the data shown in Fig. 8 for normally consolidated Kawasaki clay. The effective stresses in samples of the clay obtained using a 76 mm diameter Shelby-tube piston sampler were measured in the laboratory by applying a confining stress to the samples in the triaxial apparatus. The effective stresses after

Table 2. Changes in effective stress, σ', on re-sampling into 38 mm tubes

Initial sample		Final 38 mm diameter sample	
Diameter, mm	Initial σ', kPa	σ' after sampling, kPa	
		Absolute	Change
150	174	266	92 ⎫
		266	92 ⎬ Av. = 95
		274	100 ⎭
100	143	238	95 ⎫
		295	152 ⎬ Av. = 124
70	411	450	39 ⎭

'perfect sampling' are estimated from laboratory data of the kind shown in Fig. 1. The comparison between the effective stresses predicted for the perfect sample and those measured shows that there has been a substantial loss of effective stress due to sampling. As a result of these observations, Ladd and Lambe recommended that samples should be reconsolidated to their in situ stresses before testing.

The effect of tube sampling on stiff plastic clays has been examined at Imperial College. Table 2 shows the changes in effective stresses observed when 38 mm tube samples were taken by jacking the tubes into larger samples in which the effective stresses had already been measured (Sandroni, 1977). Effective stresses were measured on saturated samples in the triaxial apparatus by applying a sufficiently large confining stress so as to record a positive pore pressure. It will be seen that there is a substantial increase in effective stress due to re-sampling. Table 3 shows the effective stresses measured in 38 mm diameter samples taken by jacking from a 265 mm diameter block of London clay. Sample tubes with different B/t ratios were used. The initial effective stress in the block were not known, but the results show an effective stress after sampling that increases with decreasing B/t.

In another similar experiment, effective stresses were measured in samples taken by jacking tubes of different sizes into Upper Lias clay in the floor of the borrow pit for the Empingham Dam in Leicestershire (Maguire, 1976). The results are shown as a series of histograms, Fig. 9. The borrow pit had been excavated rapidly, and the samples would have been undrained during excavation and sampling. Both the average effective stress and the scatter of results decrease as the samples become larger and the area ratio of the samplers improves. The suctions in the 38 mm samples which were hand trimmed from 100 mm diameter

Table 3. *Effect of 38 mm diameter tube sampler wall thickness on effective stress after sampling from a block*

Thickness, mm	B/t	σ' after sampling, kPa
0.7	54	74
0.8	47	92
1.2	32	163
1.3*	29	182
1.6	24	164
1.6	24	113
1.7	22	115

*No cutting edge.

samples show only a small change in suction. Hand trimming seems to have little effect on suction provided the removal of clay is more rapid than the rate of drying. The scatter of results has considerable practical significance, as is discussed later.

No alternative direct observations of in situ stresses were made at Empingham, but $K_0 = 1$ is a reasonable assumption as the clay is heavily brecciated and below a slope of 4°, both factors which would reduce K_0 (Vaughan, 1976). This value of K_0 also gave the best fit between field observations of deformation and the results of finite element analyses (Hamza, 1976).

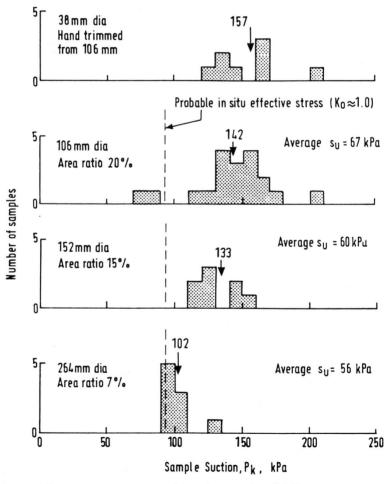

Fig. 9. *Effective stresses measured in tube samples of different diameter jacked into Upper Lias clay (after Maguire (1976))*

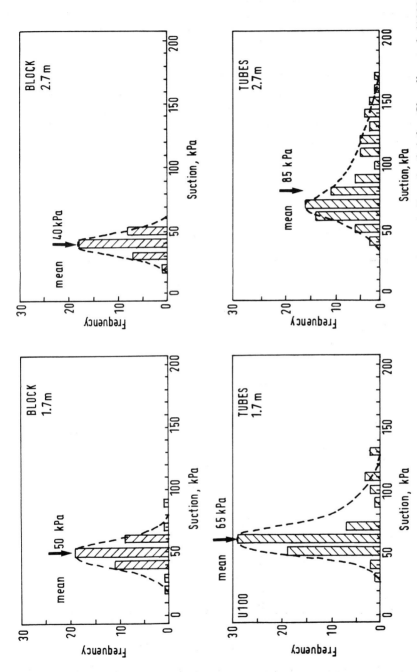

Fig. 10. Comparison of suctions measured in block and tube samples, London clay at shallow depth (after Chandler et al. (1992))

The equivalent suction after perfect sampling from this effective stress is indicated in Fig. 9. It is slightly lower than the suction recorded in the largest 264 mm diameter samples. It is reasonable to suppose that some effects of disturbance remain, even in these samples, which were taken with a tube of area ratio equivalent to that of a thin-walled 100 mm sampler.

More recent work (Chandler et al., 1992b) has used the filter paper technique (Chandler and Gutierrez, 1986; Chandler et al., 1992a) to compare suctions measured in block samples and in driven U100 tube samples of London clay taken at identical depths. The U100 samplers had an area ratio of 24% to 26%. Two sites were used, at one taking samples of fairly soft clay from shallow depths, at the other taking samples of very stiff clay from a depth of 22 m at the base of a shaft for an under-reamed pile.

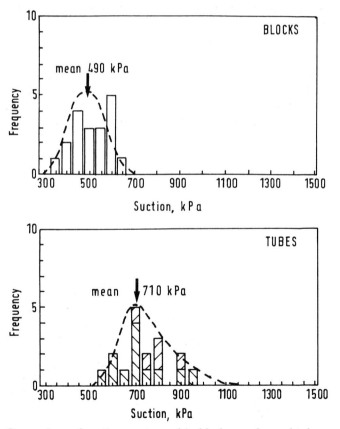

Fig. 11. *Comparison of suctions measured in block samples and tube samples, deep London clay (after Chandler et al. (1992))*

SAMPLING DISTURBANCE

The results of the filter paper tests are plotted as frequency diagrams of suction, Figs. 10 and 11. The consistency of the results will be noted, the scatter being probably due in almost equal parts to experimental variation and to real variations in the measured sample suctions.

The difference in suctions between block and tube samples shows clearly that the sample suction increases as a result of tube sampling. The slightly higher suctions at the depth of 1.7 m are presumed to be the consequence of desiccation, the samples being taken during a dry summer. Note, too, that the histograms of block sample suctions show a normal distribution, while the corresponding histograms for the U100 sample suctions are skewed towards the high suction end. This presumably represents the disturbance caused by driving the relatively thick-walled U100 samplers. As might perhaps be expected, the increase in suction is much greater with the stiffer clay.

The effect of changes in effective stress due to sampling on measured soil properties

The now well-established effect of a reduction in effective stress on the triaxial compression behaviour of normally consolidated clay is illustrated qualitatively in Fig. 12. This idealisation is based on work by Lacasse et al. (1985), and is confirmed by the recent Bothkennar Site Characterisation Study. The effect on undrained strength is very large, and it is generally accepted that reconsolidation to the field stress is desirable to minimise these effects. Even then the loss of structure may be significant, as discussed above.

The effect on dense dilatant clays depends on their plasticity. The likely behaviour of clays of low plasticity and low bonding, typical of Pennine glacial tills, is shown schematically in Fig. 13. Ultimate undrained strength is reached at a critical state point, and is dependent

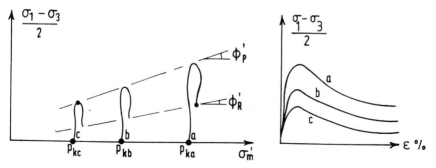

Fig. 12. Loss of undrained strength and stiffness due to loss of effective stress in normally consolidated clay at constant water content

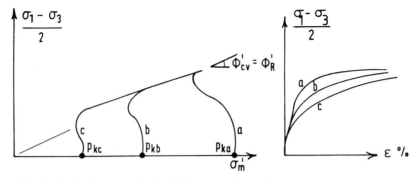

Fig. 13. Undrained strength of stiff low-plasticity clays at constant water content

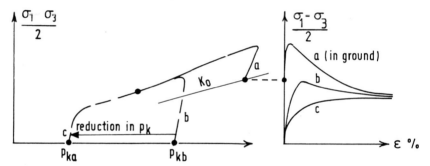

Fig. 14. Undrained strength of a stiff plastic clay; all samples at the same water content

only on water content. Stiffness will be affected by the degree of disturbance.

Provided that there is little change in water content due to sampling, which will often be the case even if the soil is exposed to water since the swelling modulus is low, sample disturbance will have little influence on undrained strength, or indeed on drained strength. It follows that a good approximation to both strengths may be measured in samples which have been fully remoulded, provided that the water content is correct. This has advantages in the testing of stony tills, in which structurally undisturbed tube sampling may be impossible, or where the large stones prevent the testing of samples directly (Vaughan et al., 1975). However, if stones are removed in remoulding, a correction to the water content must be used (Gens and Hight, 1979). Again, the effect of remoulding on the stiffness measured in laboratory tests will of course be considerable.

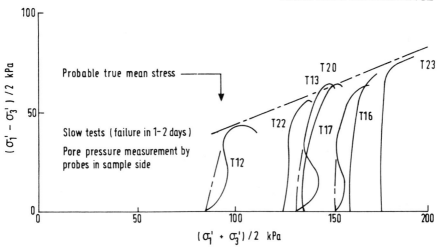

Fig. 15. Undrained strength of brecciated Upper Lias at constant water content, related to varying sample disturbance. 100 mm diameter diagonal samples, all at the same water content. The different initial effective stresses reflect different degrees of disturbance. Failure in 1–2 days; pore pressure measured with a mid-height probe (after Maguire (1976))

The undrained behaviour of dilatant plastic clays in triaxial compression, at least those with an aged structure, is shown schematically in Fig. 14. Such soils do not follow the critical state principle when 'on the dry side'. Rather than reaching the critical state point given by their water content, they form a failure surface when the maximum stress ratio is reached; that is, at peak, strength in terms of effective stress. Orientation of the dominant clay particles then starts to occur, and the strength in terms of effective stress drops back to the low residual value which develops when the clay is fully oriented. Thus, the peak undrained strength in triaxial compression becomes a direct function of initial suction in the sample, and hence is a function of sample disturbance. Moreover, sample disturbance increases the measured strength. Stiffness is also controlled by sample disturbance. Typical results are shown in Fig. 15; these tests are some of those for which initial suctions are shown in Fig. 9.

Sampling disturbance and the effect of sample size on undrained strength

Since the undrained strength of samples tested unconsolidated and undrained (UU) in the triaxial apparatus are a direct function of sample disturbance, it follows that, if disturbance varies from sample to sample,

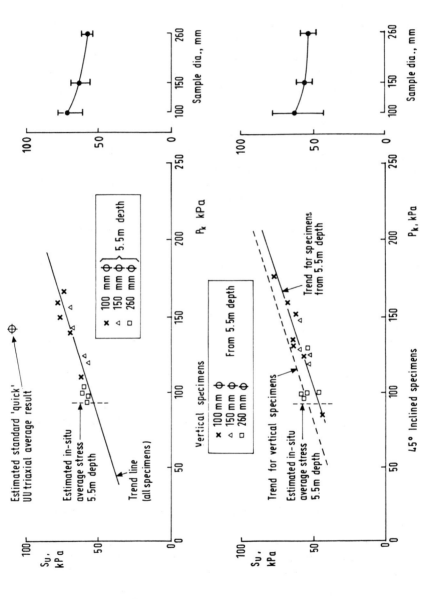

Fig. 16. *The spurious sample-size effect in Upper Lias clay due to sample disturbance; above vertical specimens, below inclined specimens (after Maguire (1976))*.

the strength measured will also vary from sample to sample. This variation in strength will reflect sample disturbance rather than any innate variability in the ground.

The data of Fig. 9 and Fig. 15 show this variability for Upper Lias clay. These samples were all taken from the same depth in the same clay, and there is no reason to suppose that the mean effective stress in the ground varied significantly from place to place. Neither was there any visible indication when this clay was inspected that the undrained strength varied to the extent shown in Fig. 15. In terms of effective stress, the strength shows little variation between samples, nor does it show much variation between samples of different size. As previously noted, the scatter of initial effective stress and undrained strength decreased with increasing sample size. The scatter of undrained strengths measured in the samples of standard size (100 mm) is therefore a function of varying sample disturbance, and not representative of actual variation of strength in the ground. Other effects, such as varying wetting of samples in a wet borehole, and will produce a similar spurious scatter of individual sample behaviour.

The decreasing effect of sample disturbance as sample size increases and with better sample tube proportions, can produce a spurious sample-size effect. Both the change in average initial stress and the change in measured undrained strength are shown plotted against sample diameter in Fig. 16. A significant decrease in strength with increasing sample size is shown, which is almost entirely a function of initial effective stress, and hence of sample disturbance.

These conclusions apply to the Upper Lias at Empingham, which had a brecciated fabric, giving the clay a closely spaced discontinuity pattern, even on the scale of the 38 mm diameter samples. Other clays, with more widely spaced discontinuities, will probably exhibit a sample-size effect that combines the effects of sample disturbance noted for the brecciated Upper Lias with that traditionally associated with the spacing of the discontinuities related to sample diameter.

Sample disturbance and its influence on the estimation of in situ stress

The in situ stress in stiff clays is one of the most difficult parameters to measure in soil mechanics, yet it is one of the most important. It controls side-shear in piles, deformations due to excavation, and, as discussed previously, it also determines undrained strength. It may be examined using in situ tests, including various pushed-in and drilled-in pressure cells, and different types of pressuremeter. The suction method, in which the mean effective stress in a sample of saturated soil is measured or deduced, and from which the stress prior to sampling is calculated, is an independent method of determining K_0 which is potentially valuable as a check on other methods. Measurements of this type are now commonly made using samples taken by jacking thin-walled tubes, and K_0 values thus obtained are usually comparable to those obtained by other in situ test methods.

By definition, suction measurements from tube samples must overestimate K_0. Figure 17 shows various comparative measurements made

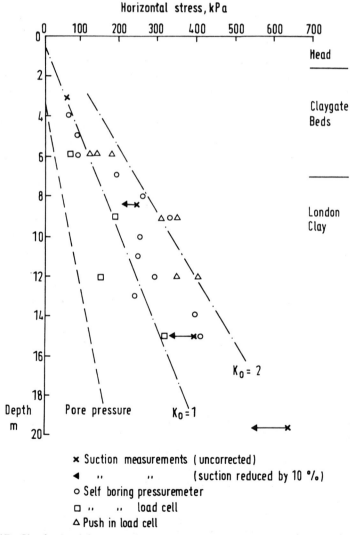

Fig. 17. K_0 obtained from suction measurements, compared with the results of other methods of measurement (modified, from Burland and Maswoswe (1982))

at the Bell Common site in London clay (Burland and Maswoswe, 1982). It can be seen that the uncorrected suction measurements give K_0 values on the high side of the range measured, although less than the push-in cells, which must also overestimate horizontal stress. Results are higher than those for self-boring load cells, which are likely to underestimate lateral stresses. A reduction in sample suction by 10% is a sensible correction which is compatible with the results of Fig. 9. The effect is small, as shown in Fig. 17. The results then suggest that the horizontal stresses estimated from self-boring pressuremeter tests slightly underestimate horizontal stress.

Conclusions

Sample disturbance will always affect the structure of a soil. It will reduce the influence of this structure on soil behaviour according to the strains imposed during sampling, and will reduce the sensitivity of the soil structure to these strains.

Cavitation of sand or silt layers adjacent to clay layers will allow the uptake of free water by the clay, reducing its effective stress. Similar results will be observed in wet boreholes, and where there are water-filled voids in the clay.

The effect of tube sampling, as compared with that in the 'perfect sample', are to increase the average effective stress in stiff plastic clays, to increase or reduce the effective stress in stiff clays of low plasticity (depending on whether the effects of cyclic compression/extension or of side shear predominate), and to reduce the effective stresses in tube samples of soft clays.

The decreasing effect of sample disturbance as sample size increases, and as the area ratio decreases, can produce a spurious sample-size effect.

The use of suction measurements from tube samples of saturated stiff clay will overestimate K_0 values, requiring a correction to allow for sample disturbance.

References

APTED, J.P. (1977). Effects of weathering on some geotechnical properties of London Clay. PhD Thesis, University of London.

BALIGH, M.M. (1985). The strain path method. Jour. Geotechnical Division ASCE, Vol. 3, pp. 1108–1136.

BURLAND, J.B. AND MASWOSWE, J. (1982). Discussion, Géotechnique, Vol. 32, pp. 285–286.

CHANDLER, R.J. AND GUTIERREZ, C.I. (1986). The filter-paper method of suction measurement. Géotechnique, Vol. 36, No. 2, pp. 265–268.

CHANDLER, R.J., CRILLY, M.S. AND MONTGOMERY-SMITH, G. (1992a). The filter paper method of soil suction determination applied to swelling clays. Proc. Instn. Civ. Engrs., Civ. Engng., Vol. 92, pp. 82–89.

CHANDLER, R.J., HARWOOD, A.H. AND SKINNER, P.J. (1992b). A study of sample disturbance in London clay. Géotechnique, Vol. 42, pp. 577–585.

GENS, A. AND HIGHT, D.W. (1979). The laboratory measurement of design parameters for a glacial till. Proc. 7th European Conf. S.M.& F.E., Brighton, Vol. 2, pp. 57–66.

HAMZA, M.M.A.F. (1976). The analysis of embankment dams by non-linear finite element analysis. PhD Thesis, University of London.

LACASSE, S., BERRE, T. AND LEFEBVRE, G. (1985). Block sampling of sensitive clays. Proc. 11th Int. Conf. S.M.& F.E., San Francisco, Vol. 2, pp. 887–892.

LADD, C.C. AND LAMBE, T.W. (1963). The strength of 'undisturbed' clay determined from undrained tests. Proc. Symp. Shear Testing of Soils, Ottawa, A.S.T.M. Spec. Tech. Pub. 361, pp. 342–371.

MAGUIRE, W.M. (1976). The undained strength and stress-strain behaviour of brecciated Upper Lias clay. PhD Thesis, University of London.

SANDRONI, S.S. (1977). The strength of London Clay in total and effective stress terms. PhD Thesis, University of London.

SKEMPTON, A.W. AND SOWA, V.A. (1963). Behaviour of saturated clays during sampling and testing. Géotechnique, Vol. 13, pp. 269–290.

VAUGHAN, P.R. (1976). In Horswill, P. and Horton, A. The valley of the River Gwash with special reference to cambering and valley bulging. Phil. Trans. Roy. Soc. A., Vol. 283, pp. 451–461.

VAUGHAN, P.R., LOVENBURY, H.T. AND HORSWILL, P. (1975). The design, construction and performance of Cow Green embankment dam. Géotechnique, Vol. 25, pp. 555–586.

Development and application of a critical state model for unsaturated soil

S.J. WHEELER and V. SIVAKUMAR, University of Sheffield

A critical state framework is proposed for unsaturated soil. Five state variables are included: mean net stress, deviator stress, suction, specific volume and water content. Isotropic normal consolidation states are defined by 3 equations linking the 5 state variables and critical states are defined by a further 3 equations. A section of state boundary linking isotropic normal consolidation states and critical states is defined by a final pair of equations relating the 5 state variables. Supporting evidence for the existence of the various relationships is provided by experimental data from controlled suction triaxial stress path tests on samples of unsaturated compacted kaolin. Possible applications of a critical state model for unsaturated soil include: providing a qualitative framework to aid understanding of unsaturated soil behaviour; guiding the choice of strength parameters to be used in conventional stability calculations; and providing a formalized mathematical stress–strain model for incorporation in finite element programs.

Introduction

Unsaturated soils occur throughout the world in the form of natural soils (in which the unsaturated zone exists as a surface layer above the water table) and compacted fills (which are inevitably in an unsaturated condition after placement). An understanding of the mechanics of these unsaturated materials is therefore vital for the effective design and analysis of many foundations, slopes, embankments and retaining structures.

It is now generally accepted that total stress σ, pore air pressure u_a and pore water pressure u_w must be combined in two independent stress parameters in order to describe the mechanical behaviour of unsaturated soils (Fredlund and Morgenstern, 1977). The two stress parameters normally selected are $\sigma - u_a$ (referred to hereafter as 'net stress') and $u_a - u_w$ (referred to as 'suction'). Many previous authors have related the volume change behaviour (Matyas and Radhakrishna, 1968) or the shear strength (Fredlund et al., 1978) of unsaturated soil to these two stress parameters.

Until very recently volume change and shear strength were treated independently for unsaturated soil. In the last few years, however, the first tentative steps have been made towards the development of a generalized constitutive model for unsaturated soil based on a critical state framework (Alonso et al., 1990). The intention is that such a model should be capable of predicting both volumetric and shear strains for any stress path and any drainage condition.

In this paper a possible critical state framework for unsaturated soil is described and then several features of the proposed framework are compared with experimental data from a series of triaxial tests on samples of unsaturated compacted kaolin. The paper concludes with a discussion of possible applications of an unsaturated critical state model within geotechnical analysis and design.

Development of an unsaturated critical state model
Critical state framework

Saturated critical state models are defined in terms of 3 state variables: the mean effective stress p′, deviator stress q and specific volume v (Schofield and Wroth, 1968). A critical state model for unsaturated soil must take account of the need to include net stress $\sigma - u_a$ and suction $u_a - u_w$ as independent stress parameters rather than dealing solely in terms of effective stresses. A second volumetric parameter (in addition to specific volume v), such as water content w or degree of saturation S_r, is also required to fully define the volumetric state of unsaturated soil. This means that an unsaturated critical state model is likely to involve 5 state variables: mean net stress p′, deviator stress q, suction s, specific volume v and water content w, where q, v and w are defined in the normal way and p′ and s are given by

$$p' = (\sigma_1 + \sigma_2 + \sigma_3)/3 - u_a \quad (1)$$

$$s = u_a - u_w \quad (2)$$

An unsaturated critical state model should therefore be defined in p′ : q : s : v : w space, and within this 5-dimensional space there should be equivalents of the isotropic normal consolidation line, the critical state line and the state boundary surface. Although the unsaturated model involves 2 more state variables than the saturated model, there should be only one more degree of freedom, because only one additional phase (the pore air) is involved. Critical states and normal consolidation states should therefore each be defined by 3 independent equations (rather than the pairs of equations used in the saturated model), and the state boundary should be defined by 2 equations (rather than a single relationship).

In the light of the above comments, isotropic normal consolidation relationships for unsaturated soil can be postulated. These would take the form

$$q = 0 \tag{3}$$

$$v = f(p', s) \tag{4}$$

$$w = f(p', s) \tag{5}$$

Equation (3) is true by definition for isotropic stress states and relationships of the form shown in eqns. (4) and (5) are already well-proven by previous research on the volume change behaviour of unsaturated soil (Matyas and Radhakrishna, 1968; Lloret and Alonso, 1985). The easiest way to visualise a relationship such as eqn. (4) is as a number of separate normal consolidation lines of v plotted against p', with each line corresponding to a different value of s.

By extension from saturated critical state soil mechanics, critical state relationships for unsaturated soil can be postulated, and these would take the form

$$q = f(p', s) \tag{6}$$

$$v = f(p', s) \tag{7}$$

$$w = f(p', s) \tag{8}$$

A relationship of the form shown in eqn. (6) is already well supported by existing shear strength data (Fredlund et al., 1978; Escario and Saez, 1986). In contrast, prior to completion of the series of experimental tests described below, little or no data existed to validate eqns. (7) and (8).

A section of state boundary can be postulated, linking the critical state line and the normal consolidation line. This state boundary would be defined by 2 equations

$$q = f(p', s, v) \tag{9}$$

$$w = f(p', s, v) \tag{10}$$

An elasto–plastic critical state model would suggest purely elastic behaviour when the soil state was inside the state boundary defined by eqns. (9) and (10) and the onset of plastic behaviour when the soil state reached the state boundary.

Alonso, Gens and Josa (1990) proposed a specific elasto–plastic critical state model for unsaturated soil that incorporates many of the features described above and tends to the modified Cam clay model for saturated soil (Roscoe and Burland, 1968) at zero suction. The model of Alonso et al. includes normal consolidation relationships of the form shown in eqns. (3) and (4), critical state relationships corresponding to eqns. (6) and (7) and a state boundary relationship corresponding to eqn. (9).

Water content w is not included as a state variable, and the relationships of eqns. (5), (8) and (10) are therefore missing from the model (the consequences of this are described later). The model assumes elastic behaviour inside the state boundary and plastic behaviour when the state boundary is reached. Due to the shape of the proposed state boundary, the model is able to predict many of the features of unsaturated soil behaviour, such as the complex pattern of swelling or collapse which occurs on wetting. If the soil state is inside the state boundary, wetting (reduction of s) leads to elastic swelling whereas if the soil state is on the boundary wetting leads to a much larger plastic compression (collapse).

Experimental apparatus and procedure
A programme of experimental research was devised to examine some of the unsaturated critical state concepts proposed above; namely the existence of the isotropic normal consolidation relationships of eqns. (3), (4) and (5), the critical state relationships of eqns. (6), (7) and (8) and the state boundary relationships of eqns. (9) and (10).

50 mm diameter triaxial samples of unsaturated compacted speswhite kaolin were prepared by compaction in a mould at a water content of 25% (4% less than optimum). All samples were compacted in 9 layers, with each layer statically compacted in a compression frame at a fixed displacement rate of 1.5 mm/min to a pressure of 400 kPa. This procedure corresponded to a compactive effort considerably less than the various standard dynamic compaction tests and resulted in a dry density of 1.20 Mg/m^3 and a degree of saturation of 54%. The intention was to produce samples with a low value of apparent pre-consolidation pressure, so that it was relatively easy subsequently to consolidate them to a virgin state.

The tests were conducted in two double-walled triaxial cells designed for testing unsaturated soil samples (Wheeler, 1988). Changes of sample volume were measured by monitoring the flow of water into the inner cell. A rolling diaphragm seal on the loading ram prevented leakage of water from the cell, and the system was calibrated for apparent volume changes caused by loading ram displacement, compression of the water within the cell, absorption of water by the acrylic cell wall and expansion of the cell with pressure (this last effect was minimized by the double-walled construction).

Pore water pressure u_w was applied or measured at the base of the sample via a porous filter with an air entry value of 500 kPa. Values of pore water pressure were maintained above atmospheric, using the 'axis translation' principle first proposed by Hilf (1956). Pore air pressure u_a was applied at the top of the sample via a filter with a low air entry value. The cell pressure σ_3, pore water pressure u_w and pore air

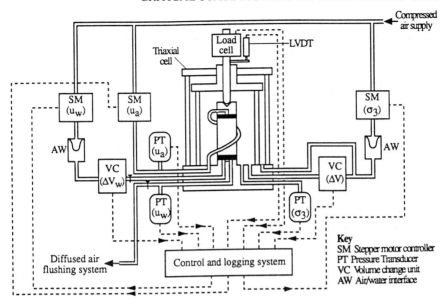

Fig. 1. *Experimental apparatus*

pressure u_a were each controlled independently by stepper motors operating regulators on a compressed air supply, as shown schematically in Fig. 1. The stepper motors were operated by a computerized control and logging system which enabled any required stress path to be followed, while simultaneously logging not only cell pressure, pore water pressure and pore air pressure, but also deviator load, axial displacement, sample volume change and flow of water from the sample.

After setting up in the triaxial cell each sample was brought to equilibrium at a mean net stress p' of 50 kPa and a suction of zero, 100, 200 or 300 kPa. The samples prepared at zero suction became saturated during this stage, but still retained the heterogeneous soil structure produced by the initial compaction process. Each sample was then isotropically consolidated to a pre-selected value of p', while maintaining suction constant, by ramping the cell pressure at a constant rate of 0.6 kPa/h. The data recorded during this stage indicated that each sample was at a virgin state after ramped consolidation, having been taken to a stress level that exceeded the apparent pre-consolidation pressure produced by the initial compaction process.

Following consolidation each sample was sheared in one of four different ways:

(a) Fully drained/constant s shearing. In these tests u_a and u_w were both held constant (maintaining constant suction) while the deviator stress was increased.

(b) Constant v/constant s shearing. In these tests the computerized control system was used to increase u_a and u_w by equal amounts (maintaining constant suction) in such a way as to keep the sample volume constant during shearing.

(c) Constant p'/constant s shearing. In these tests the control system was used to increase u_a and u_w by equal amounts (maintaining constant suction) in such a way as to keep the mean net stress constant during shearing.

(d) Constant w shearing. In these tests u_a was held constant but no drainage of the water phase was allowed, leading to variation in u_w. The suction s therefore varied during shearing, in contrast to test types (a), (b) and (c), which were all constant suction shear tests.

In all four types of shear stage the cell pressure σ_3 was held constant. Shearing was displacement controlled at a strain rate that gave a time to failure of approximately 15 days, resulting in almost complete equalization of pore water pressure throughout the sample at failure.

The primary objective of the test programme was to investigate whether all test paths followed a unique state boundary (defined by eqns. (9) and (10)) to a unique critical state line (defined by eqns. (6), (7) and (8)), independent of the type of shearing.

Experimental results

All samples tended towards a critical state, with mean net stress p', deviator stress q, suction s, specific volume v and water content w all either constant or approaching a final steady state value. In test types (a), (b) and (c) the water content w was still varying at a slow rate at the end of each test, whereas in test type (d) the suction s continued to vary at a slow rate.

Figure 2 shows the deviator stress q plotted against mean net stress p' for 8 constant suction shear tests (types (a), (b) and (c)) conducted at a suction of 200 kPa. The final critical states are represented by the square data points, and the figure shows that these critical states for a suction of 200 kPa appear to lie on a unique critical state line that is independent of the type of shear test.

Figure 3 shows specific volume v plotted against p' (with p' on a logarithmic scale) for the same 8 tests. The open data points represent the soil states at the start of shearing and these define the isotropic normal consolidation line for a suction of 200 kPa. The solid data points represent the critical states for a suction of 200 kPa, and the data support the suggestion of a unique critical state line that is independent of the test path applied.

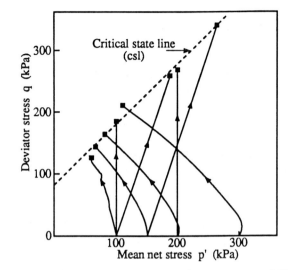

Fig. 2. *Deviator stress versus mean net stress for tests at* $s = 200\ kPa$

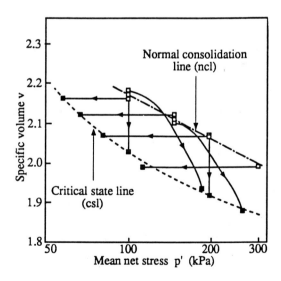

Fig. 3. *Specific volume versus mean net stress for tests at* $s = 200\ kPa$

Values of water content w are shown plotted against p' (with p' on a logarithmic scale) in Fig. 4 for the 8 tests conducted at a suction of 200 kPa. Despite the apparent scatter in the data points (which is mainly a consequence of the large scale selected for the w axis), Fig. 4 suggests a unique normal consolidation line and a unique critical state line for a

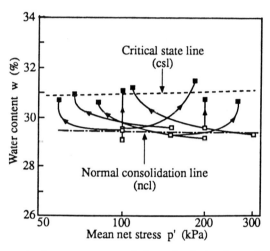

Fig. 4. *Water content versus mean net stress for tests at* s = 200 kPa

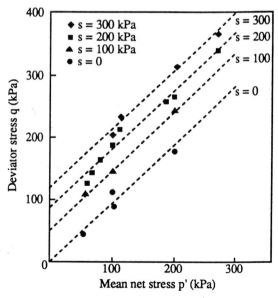

Fig. 5. *Critical state values of q versus p' for constant s tests*

suction of 200 kPa. Both lines of w plotted against p' appear almost horizontal for this particular value of suction.

Figure 5 shows critical state values of q plotted against p' for a total of 19 constant suction shear tests conducted at 4 different values of suction (zero, 100, 200 and 300 kPa). The critical state lines for the 4 different

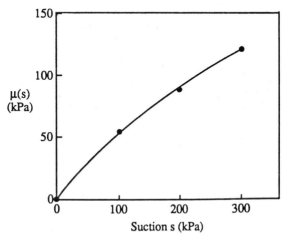

Fig. 6. Variation of $\mu(s)$ with suction

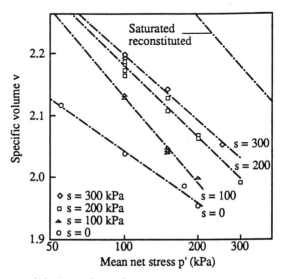

Fig. 7. Normal consolidation values of v versus p' for constant s tests

values of suction appear to be straight and parallel, and they can be fitted by an equation of the form

$$q = Mp' + \mu(s) \qquad (11)$$

where the slope M has a value of about 0.93, almost independent of suction, and the intercept $\mu(s)$ varies with suction in a non-linear fashion as shown in Fig. 6. This is consistent with the shear strength results reported by Escario and Saez (1986) and Gan et al. (1988).

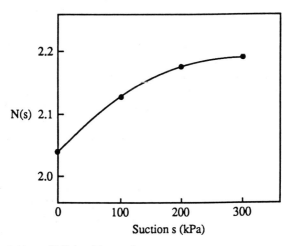

Fig. 8. Variation of N(s) with suction

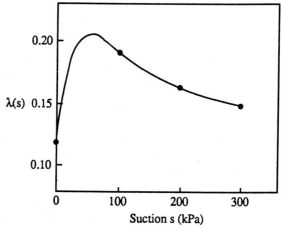

Fig. 9. Variation of λ(s) with suction

Figure 7 shows, for all tests, the values of v at the start of shearing plotted against the corresponding values of p' (with p' on a logarithmic scale). The normal consolidation lines for the 4 different values of suction can all be represented by straight lines of the form

$$v = N(s) - \lambda(s) \log_e \left(\frac{p'}{p_a}\right) \qquad (12)$$

Atmospheric pressure p_a (taken as 100 kPa) has been introduced into eqn. (12) to make the expression dimensionally consistent and to minimize any errors in the evaluation of the intercept N(s) (an intercept taken at p' = 1 kPa would be very susceptible to small errors in the slope $\lambda(s)$, because p' = 1 kPa falls well outside the range of experimental data). Both the intercept N(s) and the slope $\lambda(s)$ of the normal consolidation line are functions of suction, as shown in Figs. 8 and 9. The value of N(s) increases with increasing suction whereas $\lambda(s)$ decreases with increasing suction for values of s greater than 100 kPa. Both these features are consistent with the unsaturated critical state model proposed by Alonso et al. (1990). Somewhat unexpectedly, however, the value of $\lambda(s)$ appears to fall sharply as s is reduced to zero.

Figure 10 shows critical state values of v plotted against the corresponding values of p' (with p' on a logarithmic scale) for the 19 constant suction shear tests. The data provide clear evidence of a unique critical state line for each value of suction, independent of the test path to the critical state. There is some suggestion of curvature of the critical state lines for values of s greater than zero, and the most suitable choice of mathematical expression to relate critical state values of v to p' and s is still rather uncertain. It appears that the critical state lines for different values of suction could converge to a single unique line at values of p' greater than about 300 kPa, but additional tests at higher values of p' would be required to validate or disprove this suggestion.

The critical state line for zero suction shown in Fig. 10 and the normal consolidation line for zero suction shown in Fig. 7 both fall considerably below the corresponding lines for saturated reconstituted kaolin. The slopes of the lines for zero suction shown in Figs. 7 and 10 are also considerably less than the saturated reconstituted value. In contrast, the critical state plot of q versus p' for zero suction (shown in Fig. 5) is very similar to the corresponding plot for saturated reconstituted speswhite kaolin (for which M = 0·96). This pattern of results suggests that the heterogeneous soil structure produced by the compaction process is retained to some degree even after shearing to a critical state, and the heterogeneous structure has a major effect on the specific volume at critical states but a much smaller influence on the deviator stress.

Figure 11 shows normal consolidation values (open data points) and

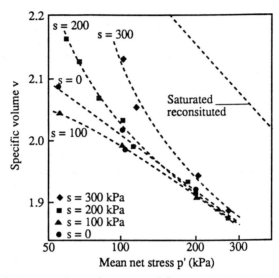

Fig. 10. Critical state values of v versus p' for constant s tests

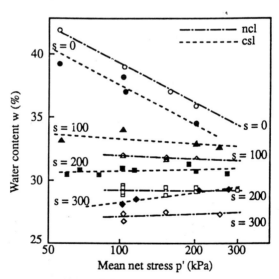

Fig. 11. Water content versus p' for constant s tests

CRITICAL STATE MODEL FOR UNSATURATED SOIL

critical state values (closed data points) of w plotted against p' (with p' on a logarithmic scale). It is interesting to note that, for suctions of 100, 200 and 300 kPa, each critical state line lies above the corresponding normal consolidation line whereas, of course, for saturated soil (s = 0) the critical state line lies below the normal consolidation line. Normal consolidation data for each value of suction can be represented by a straight line relationship

$$w = A(s) - \alpha(s) \log_e \left(\frac{p'}{p_a}\right) \qquad (13)$$

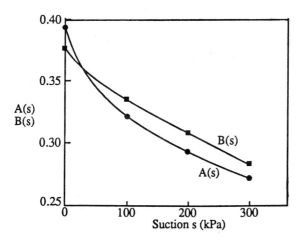

Fig. 12. Variation of A(s) and B(s) with suction

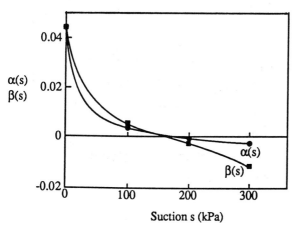

Fig. 13. Variation of $\alpha(s)$ and $\beta(s)$ with suction

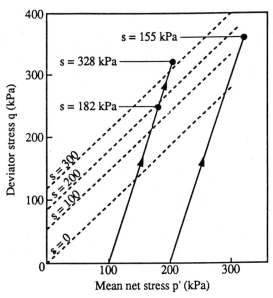

Fig. 14. Deviator stress versus p' for constant w tests

Critical state data can also be represented by straight lines

$$w = B(s) - \beta(s)\log_e\left(\frac{p'}{P_a}\right) \qquad (14)$$

The intercepts A(s) and B(s) of both the normal consolidation line and the critical state line decrease steadily with increasing suction, as shown in Fig. 12. The slopes $\alpha(s)$ and $\beta(s)$ decrease sharply as s is increased from zero (see Fig. 13), with approximately horizontal normal consolidation and critical state lines for suctions of 100, 200 and 300 kPa. There is some evidence of small negative slopes at suctions of 200 to 300 kPa.

Critical state values of q are shown in Fig. 14 for 3 constant water content shear tests (type (d)). In these tests the suction varied during shearing and the critical state values of suction are shown against the 3 data points. Also shown in the figure are the critical state lines deduced from the various constant suction shear tests, and the locations relative to these lines of the data points from the constant water content tests are seen to be consistent with the recorded values of suction. The critical state values of v and w from the 3 constant water content tests were also consistent with the critical state lines for v and w deduced from the various constant suction shear tests. This consistency of the data from the constant water content tests provides additional evidence that the critical state relationships are independent of test path.

Further analysis of the data from the series of triaxial tests indicated consistency with the suggestion that all test paths traversed a unique state boundary (defined by expressions of the general form shown in eqns. (9) and (10)) linking the normal consolidation and critical state lines. The complete analysis is presented by Sivakumar (1993) and, in the interests of brevity, is not given here.

Further experimental research is required to validate fully the proposal of a unique state boundary and to suggest the precise form of mathematical equations that would be most suitable for describing the shape of this state boundary. Other questions requiring investigation include: whether behaviour inside the state boundary can be treated as elastic and, if so, the form of this elastic behaviour; the shape of the yield surfaces formed by intersection of elastic constraints with the state boundary; and the flow rule defining plastic straining on the state boundary. An additional topic requiring further research is the role of the initial soil structure, produced by compaction or de-saturation, in determining the positions of the various lines and surfaces within the critical state model.

Application of an unsaturated critical state model

A critical state model for unsaturated soil would have applications at three different levels:

(1) in providing a qualitative framework that would enhance fundamental understanding of the mechanical behaviour of unsaturated soil
(2) in guiding the choice of drained and undrained strength parameters to be used in conventional limit equilibrium or plasticity calculations
(3) in providing a formalized elasto–plastic constitutive model that could be incorporated into numerical formulations, such as the finite element method.

The fields of possible application would include design and analysis of geotechnical constructions involving compacted fill (such as embankments, earth dams and retaining structures) and constructions involving unsaturated natural soils (particularly foundations and slopes in arid regions of the world, where the unsaturated zone extends to considerable depth).

Qualitative framework
A critical state framework could be used to provide qualitative predictions of soil behaviour under any loading or drainage conditions. The types of prediction that could be made would include: changes of soil volume resulting from wetting, drying or variation in external loading

(either drained or undrained); changes in suction caused by loading or unloading under undrained conditions; and the influence of suction, stress path and stress history on shear strength.

A critical state framework would be particularly useful for providing qualitative assessments of volume change. For example, a reduction of suction (wetting) would produce elastic swelling if the soil state were inside the state boundary but plastic collapse if the soil state were on the state boundary. Failure to appreciate this aspect of behaviour could therefore result in not only a failure to correctly predict the magnitude of ground deformations but also an error in the sign of the predicted deformations.

Limit equilibrium and plasticity calculations

Limit equilibrium and plasticity calculations are widely used in geotechnical engineering to analyse problems of collapse, such as bearing capacity of foundations, stability of slopes and active and passive earth pressures on retaining structures. In these methods the soil is idealized as a rigid-perfectly plastic material with cohesion c and friction angle ϕ. A critical state model for unsaturated soil would provide invaluable assistance in the selection of suitable drained and undrained values for c and ϕ. Many well-established analytical results, such as Terzaghi's bearing capacity formula or Rankine's expressions for active and passive earth pressures, could thus be used for unsaturated soils with an improved level of confidence.

For loading under drained conditions of the water phase (when the values of pore water pressure u_w, and hence suction s, are known) the shear strength τ can be calculated from the critical state relationship for q given in general form in eqn. (6) or more specifically in eqn. (11). Converting eqn. (11) to a conventional Mohr–Coulomb format (see Schofield and Wroth, 1968)

$$\tau = c' + \sigma' \tan \phi' \qquad (15)$$

where σ' is the net stress (total stress minus pore air pressure) perpendicular to the plane of shearing and ϕ' and c' are given by

$$\phi' = \sin^{-1}\left(\frac{3M}{6+M}\right) \qquad (16)$$

$$c' = \frac{(3 - \sin \phi')\mu(s)}{6 \cos \phi'} \qquad (17)$$

Equations (15), (16) and (17) indicate that, under drained conditions, unsaturated soil failing at the critical state can be treated as a Mohr–Coulomb material with a friction angle ϕ' that is equal to the saturated value (because the experimental results described above suggest that M is almost independent of suction) and an apparent cohesion intercept c' that varies with suction (via the term $\mu(s)$). A similar proposal was made by Fredlund (1979).

In many practical situations the suction would vary with depth, leading to variation of the apparent cohesion c'. It might therefore be necessary to treat the soil as a layered deposit, with a different value of c' for each layer. This would present no difficulties in slope stability calculations using a method of slices or in calculations of pressures on retaining structures using Rankine's earth pressure coefficients. Slightly greater difficulty would arise in attempting to apply Terzaghi's bearing capacity formula to a layered soil deposit.

The proposed methodology can be compared with standard current practice when analysing drained loading of unsaturated soil, which is to ignore the component of strength arising from suction and treat the soil as if it were saturated. Clearly this may sometimes lead to overly conservative design. It is also vital to ensure that any strength parameters used in a saturated analysis are acquired from tests on saturated samples. Back-calculation of saturated parameters from in situ or laboratory tests on unsaturated soil would lead to an overestimate of the saturated strength parameters if no allowance were made for the unsaturated condition. This could lead to an unsafe design (if the value of suction in the test was greater than in the field situation under analysis).

For rapid loading, under undrained conditions of the water phase, the suction s is initially unknown, but the shear strength τ can be calculated by combining the critical state relationships for q and w given in general form in eqns. (6) and (8) or more specifically in eqns. (11) and (14). Eliminating s from eqns. (6) and (8)

$$q = f(p', w) \qquad (18)$$

Equation (18) indicates that the undrained shear strength of unsaturated soil is not simply a function of water content, as is the case for saturated soil, but it also depends upon the level of mean net stress, i.e. ϕ_u is not zero.

Figure 15 shows the variation of deviator stress at critical states with mean total stress p (assuming $u_a = 0$) for 4 different values of w for the compacted speswhite kaolin tested in the series of triaxial tests. The curves were calculated by combining eqns. (11) and (14) (or the data in Figs. 5 and 11). The dashed lines in Fig. 15 are the critical state lines for constant values of suction, shown for reference purposes. For each value of w the soil reaches a saturated state (s = 0) at a particular value of p' and above this saturation stress level the deviator stress at critical states is independent of mean total stress p. For water contents of 30% and 32% the saturation values of p' were greater than 300 kPa (the upper

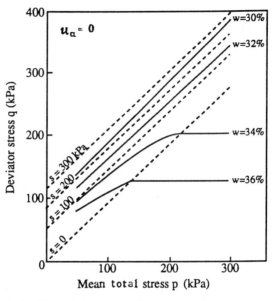

Fig. 15. *Predicted critical state values of q versus p for constant w*

limit of the experimental data). The saturation values of p' and q for each value of w are particularly easy to evaluate, as they can be calculated directly from the equations of the critical state line for zero suction.

The variation of critical state deviator stress with mean total stress p for each value of w, shown in Fig. 15, could be approximated by a bi-linear relationship; an inclined line for values of p below the saturation value and a horizontal line for higher values of p. This would correspond to a bi-linear failure envelope in the conventional Mohr–Coulomb plot. With a knowledge of the likely stress range for a particular practical problem, it should therefore be possible to select appropriate values of c_u and ϕ_u for use in limit equilibrium or plasticity calculations.

Numerical methods
The unsaturated elasto–plastic critical state model of Alonso et al. (1990) is currently being implemented within the finite element program CRISP. The program will then be capable of solving a wide range of boundary value problems involving unsaturated soil. As already mentioned, the constitutive model of Alonso et al. does not include expressions for the water content w, and therefore it can be used only for 'drained' conditions (where the variation of suction is externally specified). In the future it may be possible to incorporate additional

expressions for the water content into the finite element formulation, so that the program can also be used for analysing undrained problems. A long-term objective would be to incorporate the equations governing flow of water and air through unsaturated soil, in order to provide a fully coupled analysis of time-dependent processes such as consolidation or inundation.

Conclusions

A critical state framework, involving 5 state variables p', q, s, v and w, has been proposed for unsaturated soil. The framework includes an isotropic normal consolidation line (defined by 3 equations relating the 5 state variables), a critical state line (defined by 3 equations) and a state boundary (defined by 2 equations). Evidence in support of the proposed framework is provided by data from a series of triaxial shear tests on samples of unsaturated compacted kaolin. The critical state line for any constant value of suction appears to be straight in plots of q versus p' and w versus the logarithm of p' but is curved in plots of v versus the logarithm of p' (except for the critical state line corresponding to zero suction).

Further research is required in order to complete the development and validation of an elasto–plastic critical state model for unsaturated soil. A particular topic that will have to be tackled at a later stage in the model development will be the need to devise a relatively simple sequence of laboratory tests that can be used to ascertain the values of the relevant soil parameters within the model.

An unsaturated critical state framework could be used to provide qualitative predictions of soil behaviour under any loading or drainage conditions. It could also be used to provide guidance in the selection of suitable drained and undrained strength parameters for use in limit equilibrium and plasticity calculations. Under drained conditions the soil strength can be represented by a constant friction angle ϕ' and an apparent cohesion c' that varies with suction. Under undrained conditions the soil strength can be represented by a bi-linear failure envelope, with a ϕ_u value of zero above a saturation stress level and an approximately constant positive value of ϕ_u for lower stress levels. A final method of application of an unsaturated critical state model would occur via incorporation in finite element programs.

Acknowledgements

Financial support for the research described in this article was provided by the UK Science and Engineering Research Council.

References

ALONSO, E.E., GENS, A. AND JOSA, A. (1990). A constitutive model for partially saturated soils. Geotechnique, Vol. 40, No. 3, pp. 405–430.

ESCARIO, V. AND SAEZ, J. (1986). The strength of partly saturated soils. Geotechnique, Vol. 36, No. 3, pp. 453–456.

FREDLUND, D.G. (1979). Appropriate concepts and technology for unsaturated soils. Canadian Geotechnical Journal, Vol. 16, pp. 121–139.

FREDLUND, D.G. AND MORGENSTERN, N.R. (1977). Stress state variables for unsaturated soils. Proc. ASCE, Vol. 103, No. GT5, pp. 447–466.

FREDLUND, D.G., MORGENSTERN, N.R. AND WIDGER, R.A. (1978). The shear strength of unsaturated soils. Canadian Geotechnical Journal, Vol. 15, pp. 313–321.

GAN, J.K.M., FREDLUND, D.G. AND RAHARDJO, H. (1988). Determination of the shear strength parameters of an unsaturated soil using the direct shear test. Canadian Geotechnical Journal, Vol. 25, pp. 500–510.

HILF, J.W. (1956). An investigation of pore water pressure in compacted cohesive soils. US Bureau of Reclamation, Techn. Memo. 654, Denver.

LLORET, A. AND ALONSO, E.E. (1985). State surfaces for partially saturated soils. Proc. 11th International Conference on Soil Mechanics and Foundation Engineering, San Francisco, Vol. 2, pp. 557–562.

MATYAS, E.L. AND RADHAKRISHNA, H.S. (1968). Volume change characteristics of partially saturated soils. Geotechnique, Vol. 18, No. 4, pp. 432–448.

ROSCOE, K.H. AND BURLAND, J.B. (1968). On the generalised stress–strain behaviour of 'wet' clay. In Engineering Plasticity (J. Heyman and F.A. Leckie, eds.), Cambridge University Press, Cambridge, pp. 535–609.

SCHOFIELD, A.N. AND WROTH, C.P. (1968). Critical state soil mechanics. McGraw-Hill, London.

SIVAKUMAR, V. (1993). A critical state framework for unsaturated soil. PhD thesis, University of Sheffield.

WHEELER, S.J. (1988). The undrained shear strength of soils containing large gas bubbles. Geotechnique, Vol. 38, No. 3, pp. 399–413.

Predicting earthquake-caused permanent deformations of earth structures

R.V. WHITMAN, Massachusetts Institute of Technology

A classification of methods for calculating permanent displacements is introduced, and used as a framework for discussing the many different approaches that have been developed. Particular emphasis is placed upon validation of various methods using model tests and field observations.

Introduction

Prediction of permanent deformations and displacements has, during the past half-decade, become increasingly important to the field of earthquake engineering. This surge of interest has resulted in part from the evolution of computer-based technology enabling analysis of complex problems, and in part from increased recognition of the importance of the problem — especially with regard to pipelines, highways and port facilities. It is now clear that siting and design criteria based upon the possible or probable occurrence of 'liquefaction' (a rather loosely defined phenomenon that does not necessarily imply a catastrophic failure) can be excessively conservative and costly. There are renewed calls for design approaches and evaluative techniques based upon permanent displacements.

Geotechnical earthquake engineers are not well equipped to respond to these imperatives. Methods for making calculations are not lacking, but involve assumptions and have limitations that cast doubts upon the reliability of predictions. Validation of methods by comparison with full-scale 'experiments' has obvious problems. As a result, there has been interest in the use of model tests, especially with centrifuges. Often experimental results have been 'predicted' after the fact.

Background

While fault displacement can be very damaging to any engineered structure that sits astride a fault, the most damaging feature of earthquakes is the ground shaking that spreads outward from the rupture zone. This paper deals with the permanent deformations that may occur when an earthen mass — a dam, dike, embankment, slope, retained backfill, and just level ground — experiences such shaking. In extreme cases, the movements can be so large that there is a complete failure. Permanent horizontal or vertical movements of only a few feet or a few inches can be the cause of costly or disruptive damage to structures resting upon or passing through the earth. On the other hand, in many situations a movement of a few inches or even a few feet may cause little damage. While settlements are in some cases significant problems, this paper is concerned mainly with permanent movements related to shearing action.

The following sections delve into the strengths and weaknesses of a number of these methods. The aim is not to provide an exhaustive list of all proposed methods, but to develop an understanding of the various types of methods and to pinpoint the advantages and disadvantages associated with each.

Classification of analyses

To place into perspective the many methods proposed for evaluating permanent displacements, a classification system is convenient. While no simple system can capture completely the many variations in approach and methodology, Table 1 provides two useful types of classifications.

Classification by phenomenon causing permanent deformation
At the risk of oversimplification, there are two explanations for earthquake-induced permanent deformations:

(a) over-stressing by the inertial forces induced during shaking
(b) loss of strength or stiffness as a result of the shaking.

Table 1. Classification of analyses for permanent deformation

By Phenomenon Causing Permanent Deformation
1. Yielding but no loss of strength 2. Loss of strength 3. Both yielding and loss of strength
By Coupling of Dynamic Response and Permanent Straining
Fully coupled Partially coupled Uncoupled

These two aspects are interrelated, and both generally are present. However, many methods consider only one aspect.

Category 1 includes methods that account primarily for permanent displacements associated with yielding, whether this yielding occurs along a well-defined slip surface or throughout some zone. As an earthen mass is shaken, inertial forces are induced and cause additional stresses within the mass. If the static plus dynamic stresses reach the strength of the earth, yielding occurs. On the other hand, because of inertia the resulting permanent distortions generally are limited in magnitude. Unlike static problems, yielding during a dynamic loading — even if it occurs all along a 'failure surface' — does not necessarily cause 'failure'.

This principle is illustrated well by the 'block-on-a-plane problem' (Newmark, 1965). It was clearly expressed by Taylor in 1953, in a report that refers to an earlier study conducted in connection with a proposed sea-level Panama Canal.

Category 2 includes approaches that account for the changes of geometry necessary to reestablish equilibrium when there is a decrease in the strength of the earth such that the post-earthquake static safety factor for the original geometry is less than unity. Loss of strength is the major cause of significant earthquake-caused permanent displacements. Quoting from Taylor (1953), ". . . the threat of damage from earthquake action lies not in an increase in activating force but in a progressive decrease in shearing resistance as a result of many cycles of application of the activating force".

However, a post-earthquake safety factor less than unity is not necessarily 'fatal', provided that movements of tolerable magnitude are sufficient to restore equilibrium. Such displacements can be estimated by analysis of static equilibrium for deformed shapes of the earthen mass, without dynamic considerations. These are, in effect, 'gravity-turn-on' analyses.

Category 3 covers those methods (principally dynamic effective stress analyses) that incorporate both inertia-related yielding and loss of strength. These are potentially the most accurate methods, inasmuch as they can encapsulate many of the main aspects of the response of an earth mass to shaking. They are also the most complicated methods, generally requiring evaluation of uncommon soil parameters.

Classification by coupling of dynamic response and inelastic straining
The dynamic response of earth mass is affected by any inelastic straining that occurs during shaking. However, a variety of approximations have been introduced in the interest of simplification. Table 1 subdivides analyses according to whether or not dynamic response (accelerations,

stresses, etc.) and inelastic strains are calculated simultaneously within a single analysis.

In *fully coupled analyses*, the dynamic response of the earthen mass and the associated permanent strains are evaluated together at each time step in a single analysis. The constitutive models for these analyses are based upon plasticity theory, incorporating isotropic and sometimes kinematic hardening rules. Flow rules associated with yield surfaces control inelastic volume changes. Inside yield surfaces, linear or non-linear stress–strain relationships are used. These are the most complex and costly but potentially (subject to the correctness of the constitutive model) the most accurate analyses.

With *partially coupled analyses*, non-linear dynamic response is computed at each time step, using some stress–strain formulation in shear. Thus plastic shear straining is taken into account as the calculation proceeds. Increments of inelastic volumetric strain are injected at chosen times, based upon empirical expressions relating strains to the amplitude of cyclic strain.

In *uncoupled analyses*, one analysis is used to evaluate dynamic response. In a separate calculation, both inelastic shear strain and inelastic volume change are introduced at the end of a cycle, using experimentally measured relationships between these strains and the cyclic stress. At each such step, inelastic strains are first evaluated for each element of the grid, and then strains and stresses are redistributed through the grid so as to satisfy equilibrium and compatibility. The change of average (time-wise) stresses with time is evaluated along with the evolution of permanent displacements.

The major difference among fully coupled, partially coupled and uncoupled analyses appears with saturated sands. Excess pore pressures predicted for saturated soils are sensitive to inelastic volume changes. With fully coupled analyses, variation of pore pressure can be evaluated throughout each cycle. With uncoupled and partially coupled analyses, the change with time of average pore pressure is evaluated, but not the variation of pore pressure during a cycle. These issues are relatively unimportant when dry sands are analysed.

Sliding block analyses

For the idealized problem of a block-on-a-plane, with constant shear resistance, computation of permanent slip via a simple computer code is straightforward. Results from many analyses have been generalized into useful charts and equations for estimating slip (e.g. Whitman and Liao (1984); Makdisi and Seed (1977)). Computed slips are sensitive to the details of a ground motion recording, including duration.

There is little question concerning the essential correctness of the theory for a block slipping on a plane. Goodman and Seed (1966) placed a layer of sand on a slope and subjected the slope to a train of sinusoidal pulses. The magnitude of the computed slip agreed with that observed, although it was necessary to take into account the decrease in friction angle with increasing amount of accumulated slip.

Castro (1987) analysed the lateral spreading of ground having a very gentle slope in saturated sand, and backfigured from observed movement a reasonable value for residual strength of the sand after liquefaction.

Application to finite slopes

Sliding will in general not take place along a single plane. Rather, the surface of sliding will be curved — perhaps having roughly a circular shape — or may be composed of several planes. Taylor, Newmark, Makdisi and Seed and others have suggested use of equations for the block-on-a-plane to provide approximate estimates for the permanent movements that such slopes might experience during an earthquake. Generally, these computations ignore questions concerning compatibility of motions with assumed shape of failure surface, although in some cases (circular failure arcs) compatibility may exist.

An analysis begins with selection of an assumed location and shape for the failure surface. The assumed location may subsequently be varied to ascertain the 'most critical location'. The first step is a static analysis for the assumed free body, to find the acceleration that, when multiplied by the mass of the free body and shown as a horizontal inertia force through the center of gravity, gives a computed safety factor of one. With this maximum transmittable acceleration, permanent slip caused by the input motion is calculated. The input ground motion is the acceleration that the free body would experience were there no yielding. In many analyses, the motion is simply taken as some prescribed ground-level acceleration. More sophisticated analyses for dams take account of the change of acceleration over the height of the dam, by performing an elastic or non-linear dynamic analysis of the dam.

Application has been made to La Vallita Dam in Mexico (Elgamal et al., 1990). This dam was well instrumented at the time of construction, and experienced several major earthquake shakings causing permanent movement of the crest. During the largest earthquakes, acceleration recorded at the crest showed the non-symmetrical form typical of block-on-plane sliding. Values of yield acceleration were deduced from these accelerographs, and were used (together with the recorded crest accelerations) to predict permanent displacements. Predicted values ranged from about one-half to about equal with the measured values. Earlier applications to La Vallita and El Infernillo Dams were less successful (see Flores (1979)).

Application to retaining structures

Richards and Elms (1969) applied the sliding block concept to predicting permanent sliding of gravity walls. The wall plus wedge of soil is treated as a rigid block, with the maximum transmittable acceleration determined by frictional resistance at the base of the wall plus shear resistance along a failure plane though the soil. Various refinements have been suggested (see Whitman and Liao (1984)). Insights from analyses with this approach have helped to develop new design rules for gravity walls (Whitman, 1990).

Small-scale shaking table tests in New Zealand demonstrated the essential correctness of this approach, although predicted and recorded permanent slip differed by a factor of about 2. Such discrepancies are attributed mainly to uncertainty in appropriate choice of a friction angle for the soil (see Elms (1990)). Steedman (1984) performed similar tests on a geotechnical centrifuge. In these tests the backfill was quite dense, and decrease in post-peak friction angle with increased sliding had a great influence upon the amount of sliding.

Finite element and difference analyses

Analyses in which yielding is possible throughout some portion of the earthen mass (instead of being concentrated along sliding surfaces) require a continuum model together with a constitutive relation incorporating yielding, and a numerical calculation.

Fully coupled analyses

The best-known code in the United States today is DYNAFLOW, developed by Prevost (1978)[†]. The constitutive model is based upon the concept of multi-yield surface plasticity, with a linearized hyperbolic backbone curve used to generate stress–strain relations for loading, unloading and reloading. Segments of the backbone curve are fitted to stress–strain curves measured in monotonic loading laboratory tests for the particular soil. The shape of yield surfaces comes from fundamental studies upon soils. The flow rule associated with the yield surface provides the inelastic volume strain, and hence pore pressure changes. Validation has been achieved through comparisons between calculations and results of centrifuge model tests, although this information is not generally available in the literature.

Other fully coupled analyses use total stresses. Elgamal et al. (1985) describe such an analysis based upon Prevost's constitutive model, and apply it to La Vallita Dam. The results suggest that dispersed plastic deformation is less important than concentrated slip. The code FLEX

[†]The capabilities of this code have been incorporated into the code LINOS marketed by Earth Mechanics, Inc. of Fountain Valley, California.

(Vaughan and Richardson, 1989) has been used by Al-Homoud (1991) to predict successfully the response of tilting walls shaken on a centrifuge. This code uses a yield surface and a cap, plus linear behavior within the cap; dilative volume changes are overpredicted, but with dry sands this is not a serious difficulty.

Partially coupled analyses
Codes developed by Finn (1987), the latest versions of which are known as TARA-3 and TARA-3FL, use partially coupled analyses. Non-linear dynamic response is computed at each time step, using a hyperbolic expression for the backbone curve to generate stress-strain relations in shear. Thus plastic shear straining is automatically taken into account as the calculation proceeds. Increments of inelastic volumetric strain are injected at chosen times (e.g., at the end of a quarter-cycle), based upon empirically based expressions relating volume strains to cyclic strain. This approach thus avoids the use of yield surfaces and flow rules to determine inelastic volume change and hence excess pore pressures, and ties this aspect of behavior closely to data measured in conventional cyclic loading tests. The codes allow for dissipation and redistribution of excess pore pressures during shaking. TARA-3FL incorporates a feature whereby the shearing resistance after liquefaction can be switched to the residual strength, plus the capability to deal with very large strains. The figure shows the original and predicted deformed mesh for Sardis Dam.

Codes developed during evolution to the current versions of these codes have been checked against the results of centrifuged model tests. The calculations duplicated many of the phenomena observed during shaking of embankments and other earth structures with saturated sands. Siddharthan and Maragakis (1989) used this code for 'predictions' of the residual forces and deformations observed in model tests of

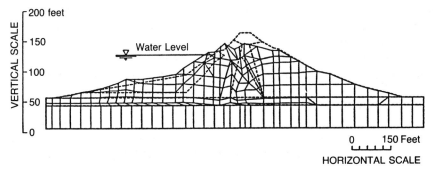

Fig. 1. Predicted deformation of Sardis Dam after liquefaction (U.S. Army Corps of Engineers Waterways Experiment Station)

cantilever walls retaining dry sand, carried out by Steedman (1984). Good agreement was found.

The code DSAGE (Roth et al., 1991) employs a finite difference scheme to compute dynamic response. Soil is taken to be elasto-plastic, with a Mohr–Coulomb yield surface. 'Predictions' were compared with centrifuged model tests of dry sand dams, with qualitatively correct results. Excess pore pressures are introduced via empirical relations, as a function of the ratio: cycles of strain to cycles to liquefaction.

Still other partially coupled models have been developed. Iai and Kameoka (1991) describe a method using hyperbolic stress-strain relations for shear plus an empirical relation between excess pore pressure and computed shear work. Computations for an anchored bulkhead are compared to observed movements at Akita Port. Alampalli (1990) uses a plasticity-based formulation for shear, and a pore-pressure versus cyclic strain relation similar to Finn's. It is used to compute response of gravity walls, employing Winkler springs to connnect soil to wall. No comparisons to experiments are reported.

Uncoupled analyses

The Residual Strain Method (Bouckovalas et al., 1991) is an uncoupled analysis. Cyclic stresses are evaluated — in an approximate way — from linear dynamic analysis that ignores the effect of residual strains upon dynamic response. In many cases it suffices to estimate the cyclic stresses from static analysis in which body forces are applied to the earthen mass. Required soil properties are:

(*a*) Relations between cyclic strain and permanent shear and volumetric strain, per cycle. A series of cyclic simple shear or triaxial tests is necessary.
(*b*) A stress–strain relation between change of average (time-wise) stress and average (time-wise) strain, for the part of the analysis restoring compatibility among elements of the grid. A hyperbolic relation is evaluated from results of monotonic triaxial tests.

This approach has been used for 'predictions' of permanent tilting of model gravity walls shaken aboard a centrifuge (Stamatopoulos and Whitman, 1989). The backfill in these tests was dry. No cyclic triaxial test results were available for the sand of these tests, and hence parameters evaluated for a different (and finer) sand were used. Despite this crudeness, predictions were generally within 30% of the observed results. Bouckovalas et al. (1991) discuss predictions for settlement of a structure resting upon saturated sand and shaken during centrifuging. These tests revealed complex behavior which were properly modeled by calculations, and the amount of settlement was predicted well.

Mixed analyses
Several methods combining the sliding block concept into finite element codes have been developed. Nadim and Whitman (1983), while studying the response of gravity retaining walls, used a finite element grid to incorporate a deformable backfill together with a Coulomb wedge. Succarieh and Elgamal (1990) developed a code for interaction among a number of irregularly shaped elements that can slide relative to each other as well as deforming. This code was applied to study of La Vallita Dam, using predetermined curved failure surfaces. It was found that inelastic deformations within elements were inconsequential compared to slip between elements along failure surfaces.

Methods considering only loss of strength

Earthquake-caused loss of shearing resistance is most often associated with liquefaction. The common perception is that liquefaction implies a total loss of strength so that soil behaves as a liquid. Actually, 'liquefied soils' have a residual or steady-state undrained strength that varies with the void ratio of the soil. Sensitive clays show a similar decrease in post-peak shearing resistance, and there have been major earthquake-caused slides in such soils. A similar phenomenon can occur with many soils and rocks.

As a failing slope moves, the driving moment reduces and because of upthrusting at the toe the resisting moment increases. The movement necessary to give a static safety factor of unity using residual strength can be calculated. In one such simple analysis using post-liquefaction residual strengths for a stratum of sand beneath a dike, Verruijt (1990) demonstrated that a slumping of about 1 m at the crest of the dike — less than the minimum freeboard in the particular case — would be adequate to restore equilibrium.

The motion to restore equilibrium is sensitive to the post-earthquake strength assigned to this zone. It is difficult to make reliable estimates for the residual strength of liquefiable sands (Marcuson et al., 1990), although less so in the case of clays with moderate sensitivities. For the analysis mentioned in the previous paragraph, a conservative value for the post-earthquake residual undrained strength was used.

An analysis applicable to situations where strength may be lost within a portion of an earthen mass can be analysed using a finite element procedure. Moriwaki et al. (1988) has shown that the displacements at the Upper San Fernando Dam during the 1971 earthquake are quite well 'predicted' by this approach. Two analyses are made: one with properties for preearthquake conditions, and a second with properties applicable immediately following an earthquake. If the zone losing strength is confined within the overall earthen mass, the predicted movement may not be sensitive to the residual strength assigned to this zone, but a reasonable choice for the stiffness of the non-liquefied material is important. In an application to a low dike, one difficult problem was to account for tension cracks. This observation serves as a warning concerning any analysis of possible permanent deformations for such a situation.

Analyses by Finn using TARA-3FL indicate that, where loss of strength occurs, permanent displacements associated with yielding during shaking are small compared to those required for readjustment to decreased strength — thus adding validity to the use of gravity-turn-on analyses. Such analyses may provide the basis for interpretation and generalization of empirical data from field studies (O'Rourke and Hamada, 1989).

Verification of theoretical predictions

As the foregoing discussions indicate, researchers have attempted to validate predictive theories by reference to field experiences and model tests. In general, 'predictions' have been qualitatively correct. With dry sands and cohesive soils, there often are errors by factors to 2 to 3 — depending primarily upon the care used in evaluation of assumed strengths. With saturated cohesionless soils, the magnitude of potential errors is not yet well established.

Such efforts must continue. There are few good case studies from past earthquakes, because reliable data were not recorded, because of inadequate knowledge concerning actual ground shaking, and because 'predictions' of movements were made only when the right answer was already known. It is exceedingly difficult to conduct full-scale blind prediction exercises. Such an effort would involve constructing an earthen structure whose sole purpose is to deform or displace during a major earthquake at some uncertain time in the future, and the details of the ground shaking during that future event cannot be known in advance. Model tests have and can partially fill this gap. While tests at normal gravity can be useful, recently attention has turned to model tests performed on centrifuges using simulated base shaking. The VELACS program (VErification of Liquefaction Analyses by Centrifuge Studies), sponsored by the National Science Foundation, is just now providing an opportunity for blind predictions of response for a variety of earthen masses shaken aboard centrifuges. The experimental effort, involving a number of laboratories in the United States plus the University of Cambridge, is expensive but is just the type of effort necessary to gain the needed confidence of methods for predicting permanent displacements. A symposium during which predictions will be compared to experimental results will take place in the fall of 1993.

Final comments

Viewing the growth of theoretical methods for predicting earthquake-caused permanent displacement, I am reminded of the observation by Wroth and Houlsby (1985):

"... the goal of developing comprehensive constitutive models for soils is overly ambitious .... a better approach is to tailor the complexity of the model to the accuracy of the solution required for a particular problem."

I agree strongly with regard to the choice of a method of analysis to apply to a specific problem — whether it involve an actual engineering undertaking or a research effort.

Despite the obvious crudeness of the sliding block model and difficulties in the choice of a suitable strength and input motion, it has played a very significant role in geotechnical earthquake engineering. It encapsulates a key feature of actual behavior: That yielding may lead only to small and often acceptable permanent displacements, unless the strength of soil decreases as a result of the earthquake.

If sophisticated finite element analyses are to be of value, they must be used. One potentially important use in an academic setting is to study a range of problems and develop generalizations and guidelines for use by engineers. However, academicians developing codes often lack exposure to actual engineering undertakings. The most important lessons likely will come when codes are used in connection with specific engineering efforts, which initially will emphasize special projects such as rehabilitation of existing earth dams. Efforts along these lines by the U.S. Army, and for major projects of the U.S. Navy and the Port of Los Angeles, hold considerable promise for eventual fall-out to the entire geotechnical earthquake engineering profession. In the meantime, for the engineering or projects where very expensive analyses cannot be justified, I urge use of simple sliding-block analyses for evaluation of slips associated with yielding and of gravity-turn-on models for analysis of effects of loss of strength.

References

ALAMPALLI, S. (1990). Earthquake response of retaining walls; full scale testing and computational modeling, PhD thesis, Dept. Civil Eng'g, Rensselaer Polytechnic Institute.

AL-HOMOUD, A. (1990). Evaluating tilt of gravity retaining wall during earthquakes, ScD thesis, Dept. Civil Eng'g, Mass. Institute of Technology.

BOUCKOVALAS, G., STAMATOPOULOS, C.A. AND WHITMAN, R.V. (1991). Analysis of seismic settlements and pore pressures in centrifuge tests, ASCE, Vol. 117 (GT10), pp. 1471–1508.

CASTRO, G. (1987). On the behavior of soils during earthquakes — liquefaction, Developments in Geotechnical Eng'g, Elsevier, Vol. 42, pp. 169–204.
ELGAMAL, A.-W.M., ABDEL-GHAFFAR, A.M. AND PREVOST, J.H. (1985). Earthquake-induced plastic deformation of earth dams, 2nd Int. Conf. Soil Dyn. and Earthquake Eng'g.
ELGAMAL, A.-W.M., SCOTT, R.F., SUCCARIEH, M.F. AND YAN, L. (1990). La Vallita Dam response during five earthquakes including permanent deformation, ASCE, Vol. 116 (GT10), pp. 1443–1462.
ELMS (1990). Seismic design of retaining walls, Design and Performance of Earth Retaining Structures, ASCE Geotechnical SP25, pp. 854–871.
IAI, S. AND KAMEOKA, T. (1991). Effective stress analysis of a sheet pile bulkhead, 2nd Int. Conf. Recent Advances in Geotechnical Earthquake Eng'g and Soil Dynamics, U. Missouri, I, pp. 649–656.
FINN, W.D.L. (1985). Dynamic effective stress analysis of soil structures; theory and centrifuge model studies, 5th Int. Conf. Num. Methods in Geomechanics, Vol. 1, pp. 35–46.
FLORES, B.R. (1979). Some comments on the dynamic behaviour of soils and its application to Civil Engineering projects, 6th Panamerican Conf. SMFE, III, pp. 165–185.
GOODMAN, R.E. AND SEED, H.B. (1966). Earthquake-induced displacements in sand embankments, ASCE, Vol. 92 (SM2), pp. 125–146.
MAKDISI, F.I. AND SEED, H.B. (1977). Simplified procedure for estimating dam and embankment earthquake-induced deformations, ASCE, Vol. 104 (GT7), pp. 849–867.
MARCUSON, W.F. III, HYNES, M.E. AND FRANKLIN, A.G. (1990). Evaluation and use of residual strength in seismic safety analysis of embankments, Earthquake Spectra, Vol. 6, No. 3, pp. 529–572.
MORIWAKI, Y., BEIKAE, M. AND IDRISS, I.M. (1988). Nonlinear seismic analysis of the Upper San Fernando Dam under the 1971 San Fernando Earthquake, 9th World Conf. Earthquake Eng'g, VIII, pp. 237–241.
NADIM, F. AND WHITMAN, R.V. (1983). Seismically induced movement of retaining walls, ASCE, Vol. 109 (GT7), pp. 915–931.
NEWMARK, N.M. (1965). Effects of earthquakes on dams and embankments, Geotechnique, Vol. 15(2), pp. 139–160.
O'ROURKE, T.D. AND HAMADA, M. (1989). Proc. 2nd U.S.-Japan Workshop on Liquefaction, Large Ground Deformation and Their Effects on Lifelines, NCEER, 89-0032, Nat. Center Earthquake Eng'g Research, Buffalo, New York.
PREVOST, J.H. (1983). Description of computer code DYNAFLOW, Dept. Civil Eng'g and Operations Research, Princeton University.
RICHARDS, R.J. AND ELMS, D.G. (1969). Seismic behavior of gravity retaining walls, ASCE, Vol. 105 (GT4), pp. 449–464.

ROTH, W.H., BUREAU, G. AND BRODT, G. (1991). Pleasant Valley Dam: An approach to quantifying the effect of foundation liquefaction. 16th Cong. Large Dams, pp. 1199–1123.

SIDDHARTHAN, R. AND MARAGAKIS, E.M. (1989). Performance of flexible retaining walls supporting dry cohesionless soils under cyclic loads, Int. J. Numerical and Analytical Methods in Geomechanics, Vol. 13(3), pp. 309–326.

STAMATOPOULOS, C. AND WHITMAN, R.V. (1990). Prediction of permanent tilt of gravity retaining walls by the residual strain method. 4th U.S. Nat. Conf. Earthquake Eng'g, III, p. 363.

STEEDMAN, R.S. (1984). Modelling the behavior of retaining walls in earthquakes, PhD thesis, Engieering Dept., Cambridge University.

SUCCARIEH, M.F. AND ELGAMAL, A.-W.M. (1990). Seismically induced stick-slip large displacements in earth structures, 3rd Japan-U.S. Workshop on Earthquake Resistant Design of Lifeline Facilities.

TAYLOR, D.W. (1953). Memorandum on stability analysis under dynamic conditions, South Pacific Division, Corps of Engineers, U.S. Army.

VAUGHAN, D.K. AND RICHARDSON, E. (1989). Flex user's manual, Weidlinger Assoc., Los Gatos, CA.

VERRUIJT, A. (1990) Personal communication.

WHITMAN, R.V. (1990). Seismic design and behavior of gravity retaining walls, Design and Performance of Earth Retaining Structures, ASCE Geotechnical SP25, pp. 817–842.

WHITMAN, R.V. AND LIAO, S. (1984). Seismic design of gravity retaining walls, 8th World Conf. Earthquake Eng'g, III, pp. 509–515.

WROTH, C.P. AND HOULSBY, G.T. (1985). Soil mechanics – property characterization and analysis procedures, XIth ICSMFE, Vol. 1, pp. 1–55.

The effects of installation disturbance on interpretation of in situ tests in clay

A.J. WHITTLE and C.P. AUBENY, Massachusetts Institute of Technology

In his 1984 Rankine lecture, Professor Wroth emphasized the need for a theoretical basis in order to develop meaningful correlations between engineering properties of soils and in situ test measurements. This paper describes an analytical framework for predicting in situ test measurements using: the Strain Path Method, to simulate soil disturbances caused by device installation; and a generalized effective stress soil model (MIT-E3), which is capable of modelling the anisotropic stress–strain response of soft clays. Predictions of piezocone penetration are evaluated through direct comparison with field measurements in Boston Blue clay. The analyses show that undrained shear strength can be estimated reliably from the measured cone resistance or tip pore pressures. Strain path analyses for flat plate penetrometers show the similarity between contact pressures measured by the dilatometer and lateral stresses around the shaft of an axisymmetric penetrometer. Installation disturbance also affects the interpretation of undrained shear strength from pressuremeter expansion tests. For displacement pressuremeters, the installation conditions lead to an underestimation of the derived shear strength; while simplified analyses of self-boring explain, in part, the overprediction of undrained shear strength frequently reported from self-boring pressuremeter tests.

Introduction

The mechanical installation of in situ test devices in the ground inevitably causes disturbance of the surrounding soil. For devices such as the piezocone, measurements of tip resistance and excess pore pressures during penetration are manifestations of stress changes induced in the soil by the installation process. Qualitatively similar disturbances can be expected from other 'displacement' penetrometers including earth pressure cells, field vanes, the Marchetti dilatometer, Iowa stepped blade, push-in pressuremeter, etc. In contrast, the design of the self-boring pressuremeter (SBPM; e.g. Wroth and Hughes (1972)) attempts to minimize disturbance by removing soil in order to accommodate the volume of the device.

This paper describes numerical analyses which quantify installation

INSULATION DISTURBANCE IN CLAYS

stresses and pore pressures for in situ tests in clays, and summarizes results that illustrate how soil disturbance affects the subsequent interpretation of engineering properties from these measurements. The study uses the Strain Path Method (SPM; Baligh (1985, 1986a, b)) to simulate the mechanics of quasi-static, undrained, deep penetration in clays. Changes in effective stresses and soil properties throughout the tests are computed using the Modified Cam Clay (MCC; Roscoe and Burland (1968)) and MIT-E3 (Whittle, 1990, 1992) effective stress soil models.

The presentation focuses on site-specific predictions of piezocone, dilatometer and pressuremeter measurements in Boston Blue Clay (BBC). Extensive laboratory test data for this material enable the predictive capabilities of the soil models to be evaluated thoroughly at the element level. These comparisons highlight the importance of anisotropic stress–strain properties for K_0-consolidated clays.

The main aim of the analyses is to predict the in situ measurements based on a known set of soil properties. Direct comparisons with field data can then be used to evaluate the reliability of the analytical methods. The analyses also guide the interpretation of in situ tests by establishing the relative merits of correlations between measurements and engineering properties of the soil. Results presented in this paper compare correlations of undrained shear strength with

(a) tip resistance and pore pressure measurements from piezocone tests
(b) dilatometer contact pressures
(c) expansion and contraction curves for displacement and self-boring pressuremeter tests.

Strain path analyses of in situ tests

The analysis of deep penetration in soils represents a highly complex problem due to:

(a) high gradients of the field variables around the penetrometer
(b) large deformations and strains which develop in the soil
(c) the complexity of the constitutive behaviour of soils, including non-linear, inelastic, anisotropic, and frictional response
(d) non-linear penetrometer–soil interface characteristics.

The Strain Path Method (SPM; Baligh (1985)) assumes that, due to the severe kinematic constraints in deep penetration, deformations and strains are essentially independent of the shearing resistance of the soil and can be estimated with reasonable accuracy based only on kinematic considerations and boundary conditions. For piezocone tests performed in low permeability clays, existing strain path analyses assume that there is no migration of pore water during penetration and hence, the soil is sheared in an undrained mode. The strain paths of individual soil elements are estimated using approximate velocity fields obtained from potential theory (i.e. treating the soil as an incompressible, inviscid and irrotational fluid).

The SPM analyses can simulate two or three-dimensional deformations of the soil elements and hence, provide a more realistic framework for describing the mechanics of deep penetration than one-dimensional (cylindrical or spherical) cavity expansion methods. On the other hand, the assumptions of strain-controlled behaviour greatly simplify the problem of deep penetration and avoid the computational complexity of comprehensive, non-linear finite element analyses.

Figure 1(a) shows contours of the octahedral shear strain, E (the second invariant of the deviatoric strain tensor) around an axisymmetric 'simple pile' penetrometer (Baligh, 1985, 1986). For this geometry, the soil deformations are proportional to the radius of the penetrometer, R, and strains can be expressed analytically at all locations. Baligh (1986a) shows that soil elements experience complex strain histories involving reversals of individual shear strain components which cannot be duplicated using existing laboratory equipment. For isotropic soils (modelled as linearly elastic–perfectly plastic materials), the zone of disturbance around the penetrometer can be equated with the contour, $E = E_y$, where E_y is the yield strain of the soil. Baligh (1986a) reports $E_y \approx 0.4\%$ for normally consolidated Boston Blue clay and hence, soil disturbance extends to a radial distance $r/R \approx 10$.

Whittle et al. (1991) have shown that strain path solutions for the simple pile geometry provide a close approximation for the stresses and pore pressures predicted around a standard 60° cone penetrometer. Thus, the simple pile geometry is used throughout this study to estimate installation disturbance for piezocone and 'cone–pressuremeter' devices.

Figure 1(b) presents similar strain paths solutions for a 'simple plate' penetrometer (Whittle et al., 1991), characterized by a length-to-width aspect ratio, $B/w = 20$. The simple plate geometry simulates the mechanics of penetration for flat plate-shaped penetrometers such as the Marchetti dilatometer ($B/w = 6.8$), earth pressure cells ($B/w = 15-25$), and the field vane ($B/w = 32.5$ for the standard Geonor field vane). The equivalent radius R_{eq} ($\approx \sqrt{4Bw/\pi}$) controls the lateral extent of disturbance caused by plate installation (i.e. at locations far from the penetrometer, soil strains and displacements depend only on the volume of soil displaced); while the aspect ratio B/w, affects the strain history of soil elements close to the penetrometer.

Strain path solutions have also been presented for open-ended piles and thin-walled sampling tubes (Baligh et al., 1987) which penetrate the soil in an 'unplugged mode'. Figure 1(c) illustrates the octahedral shear

INSULATION DISTURBANCE IN CLAYS

strains for a simple tube penetrometer with a diameter-to-wall thickness ratio, $B/t = 12$, which simulates approximately the disturbance caused by a push-in pressuremeter device (PIPM). The analyses indicate that the region of high shear strains ($E>10\%$) occurs within a thin annulus around the tube (with dimensions similar to the wall thickness), while the disturbance of the soil is controlled by the volume of soil displaced (for a thin walled tube, $R_{eq} \approx \sqrt{Bt}$).

In contrast to displacement penetrometers, the installation of the self-boring pressuremeter (SBPM) extracts soil from the system in order to accommodate the volume of the device. In practice, this is accomplished by grinding or jetting the soil into a slurry as it enters the cutting shoe, and then flushing it to the surface. A comprehensive analysis of this process is currently not conceivable. Figure 1(d) illustrates an approximate analysis which models the influence of soil extraction on

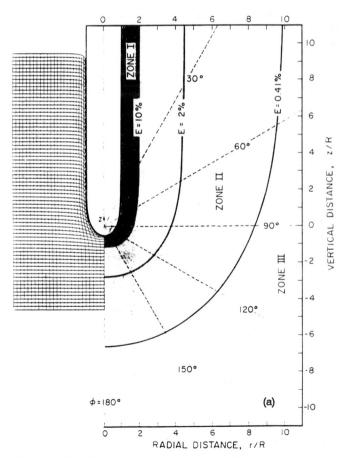

Fig. 1. *Shear strains for in situ tests: (a) simple pile*

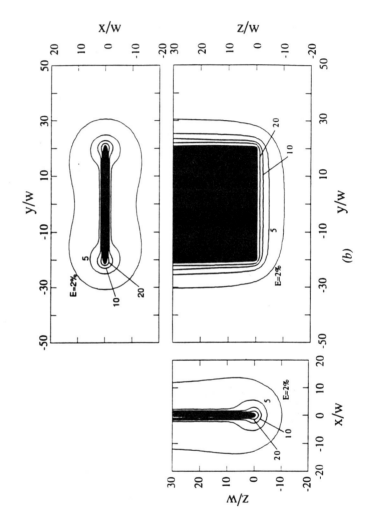

Fig. 1 (below and facing page). Shear strains for in situ tests: (b) simple plate, B/w = 20, (c) push-in pressuremeter, (d) ideal self-boring pressuremeter

INSULATION DISTURBANCE IN CLAYS

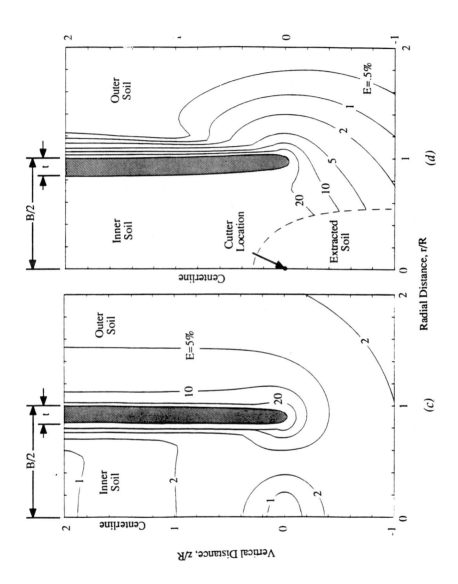

747

strains in the outer soil. The analysis simulates soil extraction using a point sink centrally located at the tip of the cutting shoe. For steady penetration, the rate of soil extraction can be conveniently expressed by the ratio, $f = V^-/V^+$, where V^+ and V^- are the strengths of the source (which generates the tube geometry) and sink, respectively. Ideal self-boring penetration occurs for $f = 1$, when the rate of soil extraction exactly balances the displacement due to the pressuremeter tube. The analyses presented in this paper consider a range of f values, $0 \le f \le 1$ which correspond to the transition between the PIPM and ideal SBPM tests. Figure 1(d) shows that, although self-boring greatly reduces the lateral extent of disturbances in the soil, significant shear strains do develop ahead of the cutting shoe.

One important issue not addressed directly in the strain path analyses is the partial drainage of pore pressures which can occur during penetration. Elghaib (1989) has developed a simplified analysis of partial drainage for piezocone tests in clays and silts. For a piezocone with radius, $R = 1.78$ cm, penetrating at a steady rate, $U = 2$ cm/s, these linear analyses show that partial drainage is not significant when the consolidation coefficient, $c \le 0.1$ cm^2/s. According to these calculations, piezocone penetration in Boston Blue clay ($0.3 \times 10^{-3} \le c_v \le 3.0 \times 10^{-3}$ cm^2/s) is effectively undrained. The equivalent radius is the primary factor affecting partial drainage for other displacement devices which are installed at the same nominal penetration rate. For devices with R_{eq} similar to the piezocone, the penetration process is effectively undrained. Rates of penetration by self-boring in Boston Blue Clay range from $0.03 \le U \le 0.2$ cm/s. In this case, partial drainage may affect the subsequent measurements during the pressuremeter test.

In the current application of the Strain Path Method, effective stresses, σ'_{ij}, are computed directly from the strain paths of individual soil elements using an effective stress–strain soil model (see next section). This approach can be contrasted with previous total stress analyses (Levadoux and Baligh, 1980; Baligh, 1986a; Teh and Houlsby, 1991) which compute shear stresses through a deviatoric stress–strain model, and introduce a separate model for shear induced pore pressures. The main advantage of the effective stress analysis is that the same soil model can be used to study stress changes after installation and throughout subsequent test procedures.

Equilibrium conditions control the penetration excess pore pressures in the Strain Path Method. However, due to the approximations used in the analysis, the equilibrium equations are not satisfied uniquely. Throughout this work, excess pore pressures are obtained by computing the divergence of the equilibrium equations and solving the resulting Poisson type equation using finite element methods (Aubeny, 1992).

Evaluation of soil models for Boston Blue Clay

Although simple models of soil behaviour provide useful physical insights into the underlying mechanics of undrained deep penetration in clays (e.g. Baligh, 1986a; Teh and Houlsby, 1991), more comprehensive constitutive equations are necessary in order to achieve reliable predictions of stress and pore pressure fields for real soils. The analyses presented in this paper use two particular effective stress models to describe clay behaviour:

(1) Modified Cam Clay (MCC; Roscoe and Burland (1968)) is the most widely used effective stress model in geotechnical analysis. The model formulation uses the incremental theory of rate independent elasto-plasticity and is characterized by an isotropic yield function, associated plastic flow and density hardening. The version of the model used in this study uses a von Mises generalization of the yield surface.

(2) MIT-E3 (Whittle, 1990, 1992) is a significantly more complex elasto-plastic model which describes many aspects of the rate-independent behaviour of K_0-consolidated clays, which exhibit normalized behaviour, including:

  (a) small-strain non-linearity
  (b) anisotropic stress–strain–strength
  (c) hysteretic and inelastic behaviour due to cyclic loading.

The model uses 15 input parameters which can be evaluated from standard types of laboratory tests comprising:

(a) 1–D compression tests with load reversals and lateral stress measurements

(b) resonant column (or similar tests) to estimate the small strain elastic shear modulus

(c) undrained triaxial shear tests on K_0-consolidated clay in compression (at OCR's = 1, 2) and extension (at OCR = 1) modes of shearing.

Whittle (1990) describes the selection of input parameters for several types of soil including Boston Blue Clay.

Figures 2 and 3 illustrate the predictive capabilities and limitations of the two models through comparisons with undrained plane strain shear data for K_0-normally consolidated Boston Blue clay measured in the Directional Shear Cell (DSC; Arthur et al., 1977). The DSC device can apply both shear and normal stresses to four faces of a cubical sample and hence control principal stress directions in the plane of loading. Seah (1990) performed a series of tests on K_0-normally consolidated BBC which apply incremental principal stresses oriented at an angle δ_{inc} to the principal consolidation stress direction (σ'_{yc}; Fig. 2). The data for $\delta_{inc} = 0°$ and $90°$ correspond to plane strain active and passive modes of shearing and match closely data reported previously by Ladd et al.

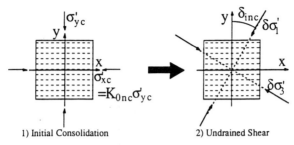

Fig. 2. *Test procedure for DSC tests at OCR = 1 (Seah, 1990): (1) initial consolidation, (2) undrained shear*

(1971). Tests at intermediate values of δ_{inc} produce continuous rotations of principal stress directions and hence, provide an initial basis for evaluating model predictions of in situ tests with principal stress rotations.

Figures 3(a) and (b) compare the predicted and measured maximum shear stress–strain response for the MIT-E3 and MCC models (Whittle et al., 1992). The MIT-E3 model describes accurately the undrained shear strengths at $\delta = 0$ and $90°$ ($s_{uPSA}/\sigma'_{vc} = 0.34$, $s_{uPSP}/\sigma'_{vc} = 0.18$; Fig. 3(a) as well as shear strains to peak resistance ($\gamma_p \approx 0.6\%$ and $\gamma_p \geq 6\%$, respectively), and post-peak strain softening in the active shear mode

Table 1. *Undrained shear strength ratios for Boston Blue clay at OCR = 1 and 4*

OCR	1.0		5.0	
K_0	0.48–0.53		0.75–1.0	
Undrained Strength Ratio	MIT-E3	Measured BBC	MIT-E3	Measured BBC
s_{uTC}/σ'_{vc}	0.33	0.33	1.16	1.04
s_{uTE}/σ'_{vc}	0.15	0.14	0.40	0.52–0.60
$s_{uPSA}/\sigma'^{(1)}_{vc}$	0.34	0.34	1.18	0.84–1.04
$s_{uPSP}/\sigma'^{(1)}_{vc}$	0.18	0.16–0.19	0.44	0.52–0.60
$s_{uDSS}/\sigma'^{(2)}_{vc}$	0.21	0.20	0.72	0.56–0.64
$s_{uPM}/\sigma'^{(3)}_{vc}$	0.21	0.21	0.68	0.64–0.76

(1) See Figs. 2, 3.
(2) Direct simple shear, $s_{uDSS}/\sigma'_{vc} = \tau_{hMAX}/\sigma'_{vc}$.
(3) Pressuremeter mode, see Fig. 7.

INSULATION DISTURBANCE IN CLAYS

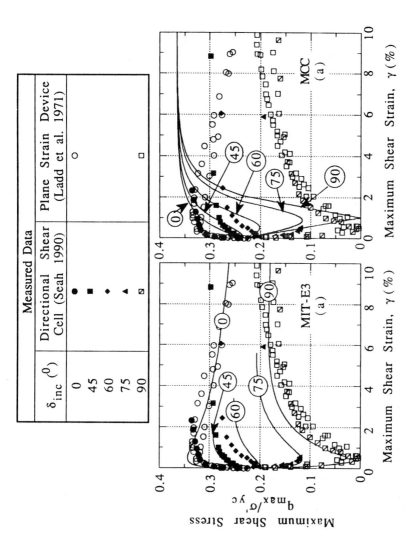

Fig. 3. Evaluation of model predictions for undrained plane strain DSC tests with principal stress rotation

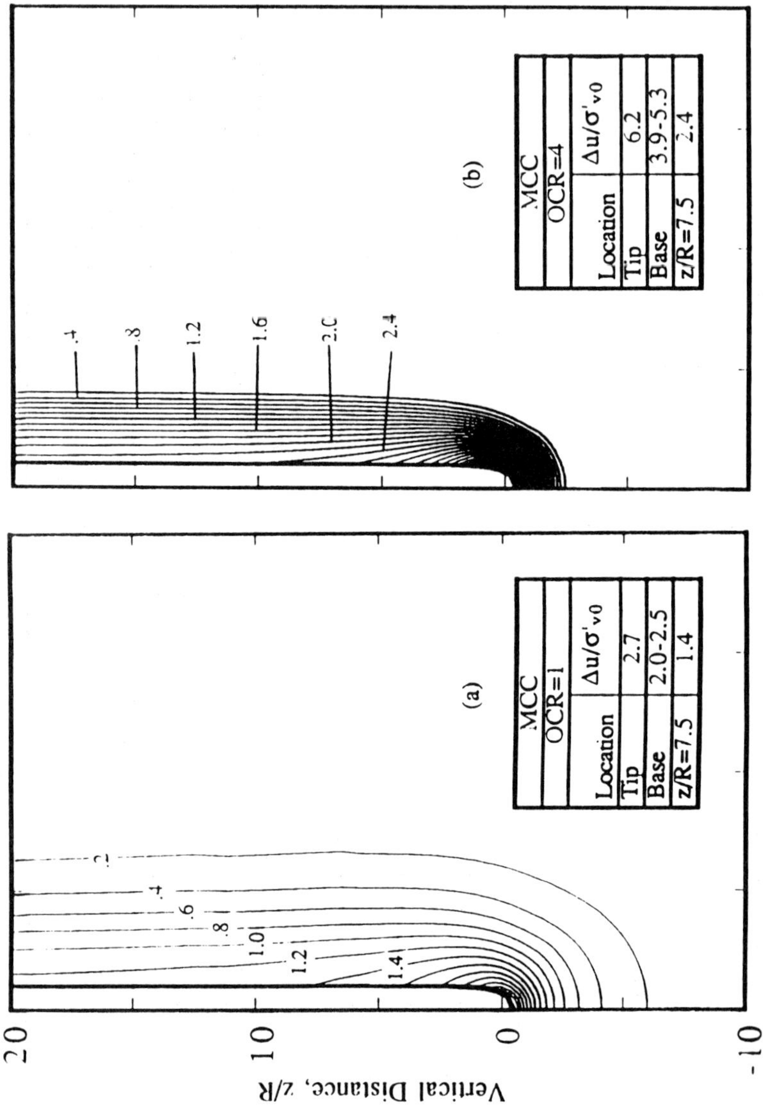

Fig. 4 (below and facing page). Excess pore pressures around the simple pile

INSULATION DISTURBANCE IN CLAYS

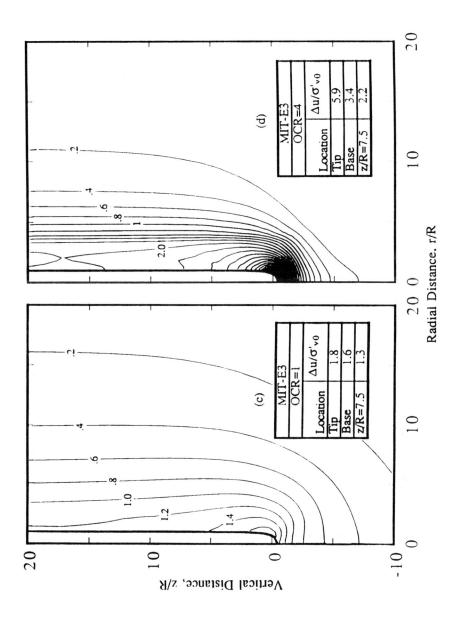

(Fig. 3(a))). The model is in good agreement with tests performed at intermediate load directions, δ_{inc} = 45, 60 and 75°.

The MCC model predicts a unique undrained shear strength which is mobilized at large strain, 'critical state' conditions and thus, overestimates significantly the measured shear strength for plane strain shear modes with $\delta > 0°$. The effective stress paths for δ_{inc} = 60, 75 and 90° (Fig. 3(b)) show clearly the elastic region of the model, which is not observed in the measured data. Overall, the comparisons show important limitations of simple, isotropic models for describing the anisotropic stress–strain behaviour which is typically measured for K_0-consolidated clays.

Table 1 summarizes MIT-E3 predictions of undrained shear strength ratios measured for Boston Blue clay in a variety of undrained shear modes for tests at OCR = 1.0 and 4.0. This table confirms the model capabilities for describing reliably the 'inelegant plethora' (Wroth, 1984) of undrained shear strengths measured in different modes of shearing.

Piezocone penetration

The simultaneous measurement of cone resistance and pore pressures during steady penetration give the piezocone unique capabilities for estimating soil stratification (e.g. Baligh et al. (1981)). However, the interpretation of engineering properties from these measurements relies primarily on empirical correlations. There are two main factors which limit the quantitative interpretation of piezocone data:

(a) There is no simple theory which is capable of describing the complex stress changes induced in the soil during the penetration process. Strain path analyses describe complex interactions of different shear modes, together with reversals in direction, which are not well approximated by simple cavity expansion methods. Similarly, the laboratory element tests demonstrate clearly the differences in the measured stress–strain response in different modes of shearing. This behaviour cannot be explained using simple, isotropic constitutive models.

(b) There is a lack of standardization in the design of piezocone equipment and test procedures. Numerous factors can affect significantly the quality of the measured data (e.g. Jamiolkowski et al. (1985)) including

    (1) load cell resolution
    (2) poor de-airing of filter elements
    (3) imprecise calibration of instrumentation.

From an interpretation perspective, there are two design features which are of particular importance:

    (a) a vertical equilibrium correction factor is usually applied to the measured tip resistance in order to account for pore pressures acting behind the cone

(b) there is no standard location for the porous filter element(s), although measurements show that pore pressures vary significantly from the tip of the cone to positions along the shaft (Levadoux and Baligh, 1986).

Figure 4 shows strain path predictions of excess pore pressure ($\Delta u/\sigma'_{v0}$) contours for a simple pile penetrating in Boston Blue clay using the Modified Cam Clay and MIT-E3 models at OCR's = 1 and 4. The results show the following:

(a) Undrained piezocone penetration develops large excess pore pressures in normally and lightly overconsolidated BBC. There are large gradients of excess pore pressure, especially at locations around the penetrometer tip and extending vertically to a distance, $z/R \approx 5$. Maximum excess pore pressures occur at or very close to the tip of the penetrometer.

(b) MIT-E3 (Figs. 4(c), (d)) predicts a much larger zone of excess pore pressures around the penetrometer than the MCC model (extending laterally to radial distances, $r/R \approx 30$ and 7, respectively). These distributions of penetration pore pressures are of particular importance for the interpretation of subsequent dissipation measurements.

(c) The two soil models predict very similar excess pore pressures acting at the pile shaft ($z/R \geq 10$). This result is surprising in view of the large differences in stress–strain response described in the previous section, and cannot be attributed to any single property of the clay. Large differences in the computed excess pore pressures at the tip of the penetrometer for normally consolidated BBC (MCC predictions are up to 40% higher than MIT-E3) have been explained, in part, by post-peak strain softening modelled by MIT-E3 (Whittle et al., 1991).

(d) The magnitudes of excess pore pressures, predicted at all locations around the penetrometer, increases very significantly with OCR (values are shown in Fig. 4). The changes in excess pore pressure are most pronounced at locations close to the tip, while more modest changes occur at locations around the pile shaft. These results indicate that filter locations on the tip or face of the piezocone are more sensitive indicators of variations in undrained shear strength or stress history than those elements located on the shaft.

The strain path predictions for Boston Blue clay can be evaluated through direct comparisons with piezocone measurements. For example, Fig. 5 compares predictions and measurements of net cone resistance, $(q_t - \sigma_{v0})/\sigma'_{v0}$, and excess pore pressures (measured at the

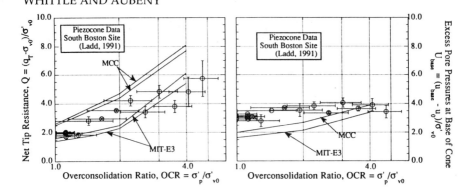

Fig. 5. *Comparison of predictions with piezocone measurements in BBC*

base of the cone), $(u_2 - u_0)/\sigma'_{v0}$, for a 60 m deep clay layer at a site in South Boston. The stress history (OCR) at this site is well defined from an extensive program of laboratory (incremental oedometer and CRS) consolidation tests (Ladd, 1991). The reported piezocone data represent average values recorded over 1.5 m penetration intervals, while the error bars indicate the extreme measured values and uncertainties in the stress history over this same interval.

The results in Fig. 5 show that MIT-E3 predictions are in good agreement with the measured net tip resistance (Fig. 5(a)) and with the base pore pressures measured at OCR ≈ 4 (Fig. 5(b)). However, this model underestimates significantly the pore pressures for OCR<4. In comparison, the MCC model overpredicts the measured tip resistance (typically by 30–50%), and gives slightly higher base pore pressures than MIT-E3. The quantitative agreement between tip resistance measurements and MIT-E3 predictions in Fig. 5(a) is encouraging. However, the underprediction of penetration pore pressures is consistent with all previous strain path analyses (e.g., Levadoux and Baligh (1980)) and further research is necessary in order to identify factors which contribute to this behaviour.

The numerical predictions provide a basis for evaluating empirical correlations between piezocone penetration measurements and engineering properties of Boston Blue clay. For example, Fig. 6 summarizes the predicted dimensionless measurement ratios $Q = (q_T - \sigma_{v0})/\sigma'_{v0}$, $U_{tip} = (u_1 - u_0)/\sigma'_{v0}$, $U_{base} = (u_2 - u_0)/\sigma'_{v0}$ and $U_{shaft} = (u_3 - u_0)/\sigma'_{v0}$ as functions of the reference undrained shear strength ratio, s_{uTC}/σ'_{v0}, for the MCC and MIT-E3 models at OCRs ≤ 4.

The correlation between net tip resistance, Q and undrained shear strength is almost linear with gradient, N_{KT}, referred to as the 'cone factor'. Existing empirical correlations (e.g. Rad and Lunne (1988)) report

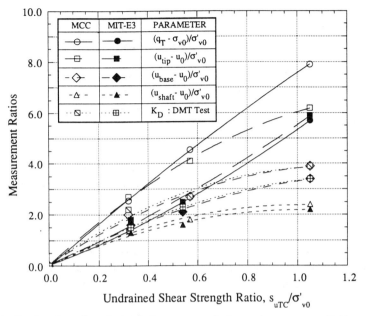

Fig. 6. *Evaluation of undrained shear strength from piezocone predictions for Boston Blue clay at OCR ≤ 4*

N_{KT} = 5–20 and thus, the predicted cone factors ($N_{KT} \approx 7.5$ for MCC; $N_{KT} \approx 5.4$ for MIT-E3) are at the low end of the measured range of behaviour. The predictions of tip resistance are controlled primarily by soil properties in the triaxial compression shear mode including

(a) the strain to peak shear resistance (yield strain)
(b) small strain non-linearity
(c) post-peak strain softening.

Although N_{KT} is not a universal constant for all clays, the results in Fig. 6 suggest that cone resistance is the most reliable measurement for estimating changes in undrained strength within a given soil deposit.

Excess pore pressures measured at the tip (or on the face) of the piezocone are also sensitive indicators of undrained shear strength. For the MIT-E3 model, the linear pore pressure factor $N^1_{\Delta u}$ ($=(u_1-u_0)/s_{uTC}) \approx 5.2$, is similar in magnitude to the cone factor. However, the MCC predictions show a non-linear relation between U_{tip} and undrained shear strength. The base and shaft pore pressures are less sensitive to changes in s_{uTC}/σ'_{v0} and hence, are less reliable measurements from which to estimate the undrained shear strength.

Plate penetration

The disturbance caused by undrained penetration affects the interpretation of in situ measurements using plate shaped penetrometers such as the dilatometer, earth pressure cells and the field vane. Whittle et al. (1991) describe comprehensive strain path analyses of these devices using the 'simple plate' geometry (Fig. 1(b)). Figure 7 summarizes contours of excess pore pressures, predicted in a horizontal plane far above the penetrating tip, for plates with aspect ratios, B/w = 1, 6.8, 20 and 32.5. The results show the following:

(a) The aspect ratio has little influence on the magnitude of excess pore pressure acting at the centre of the plate. However, pressures at

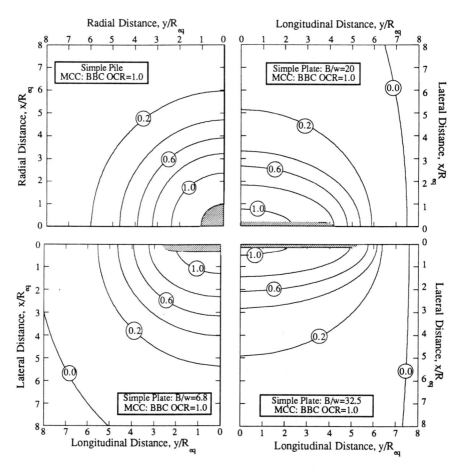

Fig. 7. *Excess pore pressures for flat plate penetrometers using the MCC model for normally consolidated BBC*

the edge of a slender plate (B/w = 20, 32.5) are typically 20% less than those predicted at the centre. These results are not affected significantly by the soil model or stress history (Whittle et al., 1991).

(b) The lateral extent of excess pore pressures is controlled by the equivalent radius, R_{eq}, of the plate. The MCC predictions (Fig. 7) show excess pore pressures extending to lateral distances, $x/R_{eq} \approx 7\text{--}8$ for all four plate geometries, while the MIT-E3 model predicts a much larger zone of disturbance (R_{eq} = 20–30, cf. fig. 4).

The lateral effective stress acting on the shaft of an axisymmetric or plate penetrometer during installation differ significantly from the in situ effective stress ratio, K_0. Figure 8 compares K_0 values for Boston Blue clay with strain path predictions of $K_i = \sigma'_{xx}/\sigma'_{v0}$ for different penetrometer geometries, soil models and initial OCRs. For normally and lightly overconsolidated BBC (OCR ≤ 4) these solutions show that penetrometer installation reduces significantly the lateral effective stress in the soil. The practical implications of these results are:

(a) excess pore pressures are similar in magnitude to the total lateral stress measurements during plate installation

(b) it is difficult to make reliable interpretation of K_0 stresses based on installation measurements.

Plate installation stresses and pore pressures have direct application in evaluating dilatometer measurements. The standard dilatometer (Mar-

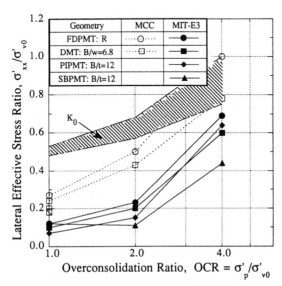

Fig. 8. *Lateral effective stresses at penetrometer shaft for Boston Blue clay*

chetti, 1980) test procedure measures 'contact pressures', p_0, acting over a centrally located, circular steel diaphragm ($A \approx 28\,\text{cm}^2$) immediately after installation. Empirical correlations then relate the dimensionless, horizontal stress index, K_D ($=(p_0 - u_0)/\sigma'_{v0}$) to K_0 and the stress history (OCR) of the soil. Figure 6 shows that strain path predictions of K_D for BBC are practically identical in magnitude to the excess pore pressures measured at the base of the cone. Further studies (Whittle et al., 1991) have found no simple correlations between contact pressures (using dimensionless groups, $(p_0 - \sigma_{h0}/\sigma'_{v0}$ etc.)) and either K_0, stress history or undrained shear strength (assuming normalized soil behaviour).

Undrained shear strength from pressuremeter tests

The interpretation of undrained shear strength from self-boring pressuremeter measurements can be obtained by the theoretical framework of one-dimensional, cylindrical cavity expansion analyses. This section describes simplified strain path analyses which evaluate the effects of installation disturbance on the subsequent interpretation of undrained shear strength for both displacement and self-boring pressuremeters.

The ideal, undrained pressuremeter shear mode can be simulated in a laboratory element test using the True Triaxial device or the Directional Shear Cell. The soil element is initially consolidated under one-dimensional (K_0) conditions (corresponding to the initial stress history in

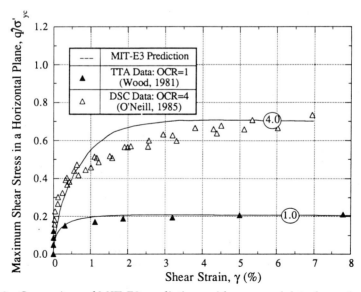

Fig. 9. *Comparison of MIT-E3 predictions with measured data for undrained pressuremeter mode of shearing for Boston Blue clay*

the ground), and is then sheared, in a plane strain mode at constant volume, in the plane normal to the direction of consolidation. Figure 9 shows the measured shear stress–strain response for pressuremeter element tests on K_0-consolidated Boston Blue Clay at OCR's = 1, 4. The measured data show that the undrained shear strength is mobilized at shear strains of $\gamma \approx 5\%$, and give no indication of post-peak strain softening. Table 1 shows that the undrained shear strength ratios, s_{uPM}/σ'_{v0}, are similar to those measured in the direct simple shear mode. Predictions of the MIT-E3 model (Fig. 9) match closely the measured behaviour (especially the undrained shear strengths) at both OCRs, but tend to overestimate the material stiffness.

In the field situation, pressuremeter installation induces spatial variations in the effective stresses and soil properties prior to membrane expansion. The strain path analyses initially focus on radial disturbance effects, which represent conditions far above the tip (or cutting shoe) of the penetrometer. The calculations compare disturbance effects for the 'full displacement' (FDPMT); push-in (PIPMT; B/t = 10–12) and self-boring pressuremeter (SBPMT) devices. Figure 10 reports predictions of the net applied pressure, $P = (p - u_0)/\sigma'_{v0}$, as a function of the current volumetric strain, $\Delta V/V$ (on a logarithmic scale) for K_0-normally consolidated Boston Blue Clay. In the ideal pressuremeter test, there is no installation disturbance and the initial net pressure is equal to the in situ earth pressure coefficient, K_0 (=0.48). The undrained shear strength

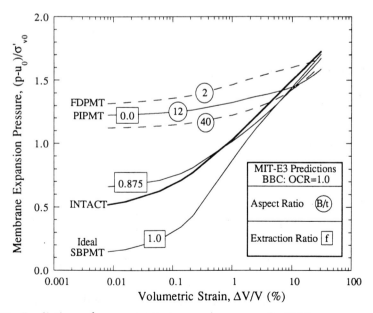

Fig. 10. Predictions of pressuremeter expansion curves for BBC

ratio is computed from the maximum slope of the expansion curve (i.e. $s_u/\sigma'_{v0} = \max\{dP/d \ln \Delta V/V\}$), according to generalized cavity expansion analyses (e.g. Palmer (1972)). For the example shown in Fig. 10, the slope of the undisturbed (intact) expansion curve is approximately linear for $\Delta V/V > 1\%$, and the computed undrained shear strength matches the elemental behaviour described previously.

Installation disturbance caused by full displacement (FDPMT) and push-in pressuremeters (PIPMT) generate large excess pore pressures in the soil (cf. Fig. 4) while reducing the lateral effective stress adjacent to the membrane (Fig. 7). Hence, the net pressure at 'lift-off' (i.e. $\Delta V/V \to 0$) is much larger than the initial K_0 in the soil as shown in Fig. 10. The expansion curves from these displacement pressuremeters are notably more non-linear than the reference undisturbed response. The undrained shear strength is mobilized at large expansion strains and is therefore very difficult to estimate reliably.

The strain path analyses represent steady, self-boring penetration in terms of the volume balance between soil displaced by the pressuremeter and that extracted by the cutting process. Figure 10 shows the pressuremeter expansion curve for an ideal self-boring pressuremeter test (with extraction ratio, $f = 1$). In this case, excess pore pressures are negligible (lateral effective stresses are shown in Fig. 8) and hence, the

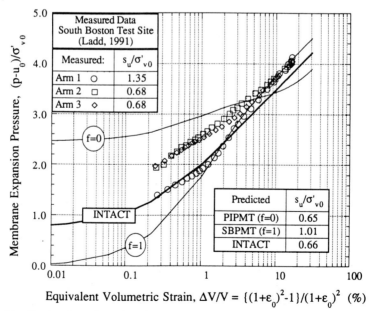

Fig. 11. *Evaluation of predicted and measured pressuremeter expansion curves for Boston Blue clay at OCR = 4*

lift-off pressure is much smaller than the in situ K_0 condition. Visual inspection of the expansion curve shows that the peak shear resistance occurs at $\Delta V/V \approx 1\%$, while the subsequent curvature implies an apparent post-peak softening.

It is difficult to equate the idealized simulation of self-boring with the parameters which control the cutting process in a real test. As a result, it is not possible to make a true 'prediction' of the field measurements. Figure 11 compares the measured data from a self-boring pressuremeter test in Boston Blue clay at OCR = 4 with the predicted expansion behaviour for three cases:

(1) no disturbance (intact)
(2) push-in penetration (f = 0, B/t = 12)
(3) an ideal SBPMT (f = 1).

The equivalent volumetric strain is computed from the circumferential strains (ε_θ) measured by three feeler arms. Although there are significant initial differences in the measured expansion curves, the data coalesce for volumetric strains $\Delta V/V \geq 5\%$, and coincide with the predicted expansion curve for the ideal SBPM test. The apparent shear resistance computed from these curves at large strains (i.e. for all three arms and f = 1) are in good agreement with the reference, undisturbed pressuremeter strength. The measurements for arm 1 coincide with idealized SBPM curve for $\Delta V/V > 1.5\%$ with an interpreted peak shear strength approximately 50–100% higher than that of the intact clay. Differences in the simulated and measured expansion curves at small strain levels may indicate the importance of factors not considered in the analysis such as internal pressures inside the cutting shoe, or soil consolidation prior to membrane expansion (the test in Fig. 11 was performed after a 30 min 'relaxation period').

Figure 12 summarizes the predictions of undrained shear strength for normally consolidated Boston Blue clay from the pressuremeter expansion curves as functions of:

(a) the aspect ratio, B/t, for displacement penetration
(b) the extraction ratio, f, for self-boring penetration.

The error bars correspond to tests in which the peak shear resistance develops at large volumetric strains. Undrained shear strengths from displacement pressuremeter (FDPM, PIPM) expansion curves underestimate significantly the true pressuremeter shear strength ($s_{uPM}/\sigma'_{vc} = 0.21$). In contrast, the predictions for self-boring tests with extraction ratios, $0.85 \leq f \leq 1$, estimate peak shear strengths up to 50% higher than the true behaviour. The interpreted shear strengths agree qualitatively with measurements from different types of pressuremeter reported by Lacasse et al. (1990). Further studies are now required to

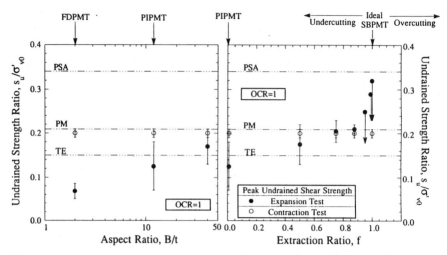

Fig. 12. Summary of interpreted undrained shear strengths from pressuremeter predictions for Boston Blue clay

establish the effects of installation disturbance for the overcutting mode of penetration (f > 1), and to refine the analyses through the inclusion of finite membrane length, proximity to the cutting shoe, etc.

Figure 12 also shows undrained shear strengths interpreted from predictions of the pressuremeter contraction response, as originally proposed by Houlsby and Withers (1988) for the full displacement pressuremeter (FDPMT). The predictions show that:

(a) there is much less ambiguity in the interpretation of undrained shear strength from the contraction curve

(b) the interpreted undrained shear strength ratio is in excellent agreement with s_{uPM}/σ'_{vc} (measured in laboratory tests), and is not affected by the installation disturbance (for the range of conditions considered in this paper).

These results suggest that undrained shear strength can be reliably estimated from the contraction curve. Further evaluations of this method are currently in progress.

Conclusions

This paper has presented a systematic analysis of the effects of installation disturbance on in situ measurements in clay. The analyses apply the strain path method, together with generalized effective stress soil models (MCC and MIT-E3), in order to predict in situ measurements for tests performed in Boston Blue clay (BBC). Detailed evaluations of the constitutive models demonstrate that MIT-E3 provides reliable predictions of the anisotropic, stress-strain behaviour measured in laboratory tests on K_0-consolidated BBC with $1 \leq OCR \leq 4$.

The strain path predictions are in good agreement with measurements of cone resistance obtained from piezocone tests at a site in South Boston, but generally underpredict the penetration pore pressures at the base of the cone. The numerical predictions also show that the net tip resistance ($q_T - \sigma_{v0}$) responds almost linearly to cones in the undrained shear strength for a given soil model (i.e. this implies N_{KT} is a constant for a given soil type). Excess pore pressures measured at the tip (or on the face of) the piezocone are also well correlated through the factor $N_{\Delta u}$; while measurements at the base of the cone, or at locations along the shaft, are significantly less responsive to changes in the shear strength.

The strain path analyses show that there are strong similarities in the stresses and pore pressures predicted around flat plate and axisymmetric penetrometers. The plate aspect ratio, B/w, controls the distribution of stresses around the penetrometer, while the equivalent radius, R_{eq}, controls the lateral extent of the disturbance. Predictions of contact pressures, measured immediately after installation by the Marchetti dilatometer, are similar in magnitude to the excess pore pressures measured at the base of a piezocone.

The analyses show that installation procedures affect the interpretation of undrained shear strength from pressuremeter expansion tests. For displacement pressuremeter tests in K_0-normally consolidated BBC, the interpreted undrained shear strength ratio, $s_u/\sigma'_{vc} \approx 0.07$–$0.12$, is significantly lower than the true material response ($s_{uPM}/\sigma'_{vc} = 0.21$). In contrast, simplified simulations of self-boring show interpreted strength ratios up to $s_u/\sigma'_{vc} = 0.32$ for 'idealized' SBPM tests. Further analyses show that interpretation methods using pressuremeter contraction measurements give $s_u/\sigma'_{vc} = 0.20$ and are insensitive to the effects of installation disturbance.

Acknowledgements

This work was supported by the Air Force Office of Scientific Research through grant AFOSR-89-0060. The authors are grateful to Professor Charles Ladd for his assistance in the interpretation of the field data from the South Boston test site. These data were obtained by Haley and Aldrich, Inc. as part of the Central Artery/Third Harbour Tunnel project in Boston.

References

ARTHUR, J.R.F., CHUA, K.S. AND DUNSTAN, T. (1977). Induced anisotropy in a sand. Géotechnique, Vol. 27, No. 1, pp. 13–36.

AUBENY, C.P. (1992). Rational interpretation of in-situ tests in cohesive soils. PhD Thesis. MIT, Cambridge, MA.

BALIGH, M.M. (1985). Strain Path Method. ASCE Journal of Geotechnical Engineering, Vol. 111, No. GT9, pp. 1108–1136.

BALIGH, M.M. (1986a). Undrained deep penetration: I. Shear stresses. Géotechnique, Vol. 36, No. 4, pp. 471–485.

BALIGH, M.M. (1986b). Undrained deep penetration: II. Pore pressures. Géotechnique, Vol. 36, No. 4, pp. 487–501.

BALIGH, M.M., AZZOUZ, A.S., WISSA, A.E.Z., MARTIN, R.T. AND MORRISON, M.J. (1981). The piezocone penetrometer. Proceedings ASCE Symposium on Cone Penetration Testing and Experience, St Louis, MO., pp. 247–263.

BALIGH, M.M., AZZOUZ, A.S. AND CHIN, C.T. (1987). Disturbances due to 'ideal' tube sampling. ASCE Journal of Geotechnical Engineering, Vol. 113, No. GT7, pp. 739–757.

ELGHAIB, M.K. (1989). Prediction and interpretation of piezocone tests in clays, sands and silts. PhD Thesis, MIT, Cambridge, MA.

HOULSBY, G.T. AND WITHERS, N.J. (1988). Analysis of cone pressuremeter test in clay. Géotechnique, Vol. 38, No. 4, pp. 575–589.

JAMIOLKOWSKI, M., LADD, C.C., GERMAINE, J.T. AND LANCELLOTTA, R. (1985). New developments in field and laboratory testing of soils. Proceedings 11th Intl. Conf. on Soil Mechs. and Foundation Engrg., San Francisco, CA. Vol. 1, pp. 57–153.

LACASSE, S.L., D'ORAZIO, T.D. AND BANDIS, T.B. (1990). Interpretation of self-boring and push-in pressuremeter tests. Pressuremeters, ICE, London, pp. 273–285.

LADD, C.C., BOVEE, R., EDGERS, L. AND RIXNER, J.J. (1971). Consolidated undrained plane strain tests on Boston Blue Clay. MIT Research Report R71-13.

LADD, C.C. (1991). Personal communication.

LEVADOUX, J.-N. AND BALIGH, M.M. (1980). Pore pressures in clays due to cone penetration. MIT Research Report R80-15.

LEVADOUX, J.-N. AND BALIGH, M.M. (1986). Consolidation after undrained piezocone penetration. I: Prediction. ASCE Journal of Geotechnical Engineering, Vol. 112, No. GT7, pp. 707–726.

MARCHETTI, S. (1980). In-situ tests by flat dilatometer, ASCE Journal of Geotechnical Engineering, Vol. 106, No. GT3, pp. 299–321.

O'NEILL, D.A. (1985). Undrained strength anisotropy of an overconsolidated, thixotropic clay. SM Thesis. MIT, Cambridge, MA.

PALMER, A.C. (1972). Undrained plane strain expansion of a cylindrical cavity in clay: a simple interpretation of the pressuremeter test. Géotechnique, Vol. 22, No. 3, pp. 451–457.

RAD, N.S. AND LUNNE, T. (1988). Direct correlations between piezocone test results and undrained shear strength of clay. Proceedings ISOPT 1, Orlando, FL, Vol. 2, pp. 911–918.

Roscoe, K.H. and Burland, J.B. (1968). On the generalized stress–strain behaviour of 'Wet' clay. Engineering Plasticity, Eds. J. Heyman and F.A. Leckie, Cambridge University Press, pp. 535–609.

Seah, T.H. (1990). Anisotropy of resedimented Boston Blue Clay. PhD Thesis, MIT, Cambridge, MA.

Teh, C.-I. and Houlsby, G.T. (1991). An analytical study of the cone penetration test in clay. Géotechnique, 41, No. 1, pp. 1–17.

Whittle, A.J. (1990). A constitutive model for overconsolidated clays. MIT Sea Grant Report, MITSG 90–15.

Whittle, A.J. (1992). Evaluation of a constitutive model for overconsolidated clays. Géotechnique, 1993, Vol. 43.

Whittle, A.J., Aubeny, C.P., Rafalovich, A., Ladd, C.C. and Baligh, M.M. (1991). Prediction and interpretation of in-situ penetration tests in cohesive soils. MIT Research Report, R91–01.

Whittle, A.J., Degroot, D.J., Seah, T.H. and Ladd, C.C. (1992). Evaluation of model predictions of the anistropic behaviour of Boston Blue Clay. Submitted to ASCE Journal of Geotechnical Engineering.

Wood, D.M. (1981). True triaxial tests on Boston Blue Clay. Proceedings 10th Intl. Conf. on Soil Mechs. and Foundation Engrg., Stockholm, pp. 825–830.

Wroth, C.P. (1984). The interpretation of in-situ soil tests, Géotechnique, 34, No. 4, pp. 449–489.

Wroth, C.P. and Hughes, J.M.O. (1973). An instrument for the in-situ measurement of the properties of soft clays. Proceedings 8th Intl. Conf. on Soil Mechs. and Foundation Engrg., Moscow, Vol. 1.2, pp. 487–494.

Shear modulus and strain excursion in the pressuremeter test

R.W. WHITTLE, J.C.P. DALTON and P.G. HAWKINS, Cambridge Insitu

Shear modulus, G, is the single most important parameter that the pressuremeter test yields and its variation with strain is of great interest to the designer. It is usually measured by taking the gradients of lines drawn centrally through unload–reload loops in the pressure/strain graph. Specifications for the pressuremeter test usually ask for a record of the strain range over which these reload loops are taken. It is rare that any statement is made about the size of such loops beyond the remark first made by Wroth (1982) that the pressure drop should not exceed $2c_u$. This paper demonstrates that, provided reload loops do not exceed the Wroth limit above, practically any reload loop can be made to yield a complete mapping of the variation of observed G against strain increment. All that is necessary is proper selections of the origins for strain and for pressure in the reload loop. Since strain changes down to the resolution limit of the pressuremeter, well below 0.01% strain, can be examined it becomes possible seriously to narrow the gap between shear modulus as measured by the pressuremeter and as measured by geophysical methods.

Introduction

The results of a series of tests in a borehole near the centre of London are presented in Table 1. These tests are noteworthy because rebound loops were taken as close to the origin of the expansion as possible. This resulted in modulus parameters being obtained whose variation was often 4 to 1 within a single test. Although the results indicate a clear pattern of stiffness declining with increase of total strain there is not the uniformity that is frequently offered as a feature of pressuremeter modulus derived in this manner.

Although it is common knowledge that stiffness will vary as a function of shear strain amplitude, pressuremeter modulus has been considered a large strain stiffness because of the appreciable size of the total strain changes within the usual rebound loop. At small cavity strain and with high resolution equipment it is possible to take very small rebound loops and these result in higher values of stiffness being derived. Nor is it the absolute level of cavity strain which is significant. At small cavity strain there are fewer problems with material creep

SHEAR MODULUS AND STRAIN EXCURSION

affecting the unload/reload process, hence permitting good loops of small strain amplitude.

If a plot is made of the variation of modulus against the coordinates of the centre of the 'loop' it shows that the variation of observed modulus with strain reduces markedly for loops taken above 1% total cavity strain. The large variations below 1% strain will now be examined though before doing so we shall have to consider the mechanics of the pressuremeter that Wroth and Hughes designed.

The classic analysis for the pressuremeter test is that due to Gibson and Anderson (1961) which, if only by frequent endorsement, is effectively a pressuremeter test standard solution. This analysis depends on there being a single value for shear stiffness applying throughout the pressuremeter expansion.

This is not the same as stating that shear modulus is a unique soil parameter, only that a single value applies to a particular deformation condition. In a test the condition is the application of stresses sufficiently high to ensure that the cavity continues to expand. Not only is the Gibson and Anderson shear modulus a single value, but it is a limiting

Table 1. Parameters derived from conventional analyses

Test No.	Depth (mbg)	σ_{h0} kPa	c_u^1 kPa	c_u^2 kPa	P_l kPA	G_i MPa	G_{ur}^1 MPa	G_{ur}^2 MPa	G_{ur}^3 MPa	G_{ur}^4 MPa	G_{ru} MPa
1	4.2	260	56	65	635	19	67	12	10*		
2	5.7	140	63	71	541	26	34	15	12*		
3	7.2	127	138	140	843	20	48	24	19*		
4	9.1	338	60	76	744	26	33	20	15*		
5	10.5	346	68	80	839	52	41	17	15*		
6	12.5	301	136	146	1077	41	52	36	16*		16
7	14.5	435	91	158	1140	42	50	44	25*		26
8	16.5	490	86	166	1215	46	106	66	44	35	26
9	18.5	625	145	241	1631	59	–	77	49		39
10	20.5	730	269	318	2432	103	133	98	66	43*	
11	22.5	690	213	213	1911	46	95	60	46	29*	

(1) Values for insitu lateral stress have been obtained from either 'lift-off' or 'modified' Marsland and Randolph analyses.
(2) c_u^1 is the 'ultimate strength' or residual value of undrained shear strength derived from applying the method suggested by Gibson and Anderson (1961).
(3) c_u^2 is the peak value of undrained shear strength obtained from applying the method suggested by Palmer (1972).
(4) G_i is the shear modulus obtained from the slope of the initial part of the loading curve.
(5) G_{ur}^n is the shear modulus derived from inspection of the slope of unload/reload loops—'n' denotes the order of the loop. If the value has been followed by * then it is the value derived from the slope of the final unloading, using a streess decrease of $2c_u$.
(6) G_{ru} is the shear modulus derived from the slope of a reload/unload loop.

case, being the stiffness applicable at maximum mobilised shear strength. It may be usefully described as G_{min}.

With this in mind the Gibson and Anderson model of soil behaviour as simple elastic/perfectly plastic continues to be useful in describing much of the pressuremeter testing in London clay.

It is not within the scope of this paper to consider the work done by Jardine et al. (1986) or the more recent insights of Muir Wood (1990, 1991). They have considered and utilised some aspects of the variation of stiffness seen in a pressuremeter test in stiff clay, and Muir Wood has clarified the connection between pressuremeter secant modulus and triaxial tangential stiffness. However the curve of stiffness against strain that is offered here will fit without difficulty into the theoretical framework that has been established.

All the plots and figures used within this paper are taken from SBP tests in London clay and, with the exception of Figs. 3 and 8, from the same borehole.

Reducing the data

There are three displacement measuring arms in the Cambridge Self Boring Pressuremeter. They are at the same level but spaced 120° apart. At the start of a test the centre of the instrument is the centre of the borehole and in principle a separate loading curve can be plotted for each strain arm. The parameters so obtained ought to be compatible but often are not, the problem being that all arms use the body of the instrument as a reference to measure the current radius of the cavity. Expecting the individual loading curves to be valid means assuming that the reference does not move at all. As even a low pressure test in stiff clay results in aggregate loads of several tonnes being applied to the soil this is not a reasonable assumption.

A typical rebound cycle sees a total strain change of 0.1% which for the SBP is a change in radius only 40 μm. This change is plotted in steps of about 5 μm. Unless the instrument axis is perfectly stationary the average measurement will always be better than considering the arms separately. It is the average loading curve from which the data in this paper have been derived.

The mechanics of the displacement measuring system

Because much of what follows is concerned with resolving small displacements it is worth examining the mechanics of the strain arm in more detail. Figure 1 gives a view of one arm in the current model of SBP.

The arm itself is brass and pivoted on a hardened and ground steel roller running in a hard steel bush. This combination is chosen as the

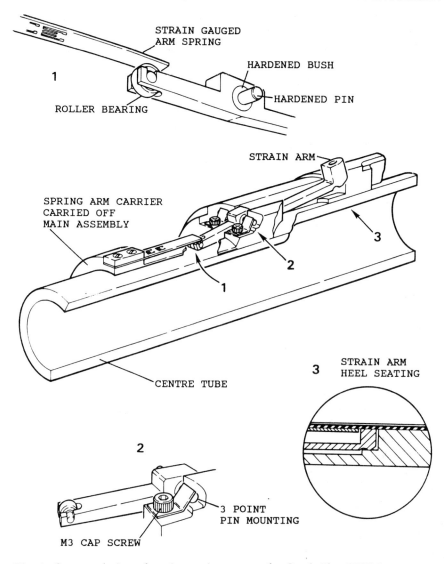

Fig. 1. Current design of strain sensing system for Cambridge SBPM

clearances, and hence the backlash, can be made very small. The short end of the arm has a leaf spring pressing on it, forcing the long end to follow any movements of the membrane. The spring presses on the arm via a small ball race to reduce axial friction (Fahey and Jewell, 1990).

The spring is strain gauged with a full bridge circuit so converting arm movements to an electrical output. The signals from all the internal transducers are amplified and converted to a serial digital data train within the instrument. The system has proved capable of resolving movements of less than 0.5 μm.

Typical performance figures for this arrangement show that it is linear to 1% over the range 0 to 8 mm with a maximum hysteresis of less than 0.3%. However because even with the ball race present there is still some friction in the system the consequences of hysteresis are not evenly distributed — the constant frictional offset appears immediately the direction of measurement changes. This is important because such a reversal happens twice in every rebound cycle.

Pressuremeter shear modulus

Initial modulus
For an SBP test in clay the slope of the pressure/strain graph after 'lift-off' but before the onset of plasticity can give a reasonable value for elastic stiffness. An appropriate choice for this slope is indicated in Fig. 2. At zero cavity strain the initial shear modulus is given by

$$G = \frac{\text{change in shear stress}}{\text{change in shear strain}} = \frac{\Delta P}{2\delta\varepsilon}$$

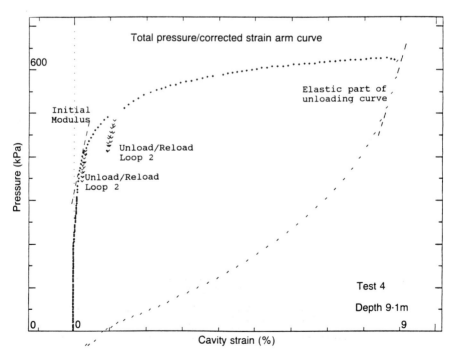

Fig. 2. Typical SBPM test in London clay with initial modulus G_i shown

where ΔP is the change in applied cavity pressure and $\delta\varepsilon$ is the associated change in cavity strain, being approximately one half the change in shear strain. For modulus derived at larger cavity strains the value should be increased by a factor R/R_0, but the correction is negligible for small strains.

A good SBP test in London clay reveals that the early part of the test before the onset of plasticity is markedly non-linear. As a consequence there tends to be no single choice of slope for the initial modulus. Figure 3 is an example of non-linearity from the early part of a test in London clay (but a site different from that used to compile Table 1). The plot shows a mistake in that an unload/reload loop was attempted close to the origin but was allowed to go too far, seriously exceeding the shear stress then mobilised. However the zone of plastic material around the pressuremeter was of such small radius that, as the reloading path shows, the entire process resulted in largely recoverable but non-linear

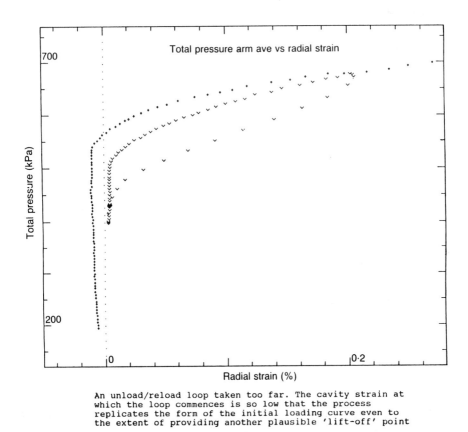

An unload/reload loop taken too far. The cavity strain at which the loop commences is so low that the process replicates the form of the initial loading curve even to the extent of providing another plausible 'lift-off' point

Fig. 3. Load disturbance test showing non-linear elasticity

strains. This error gives an interesting demonstration of non-linear elasticity.

As the plot shows, there is no difficulty in deriving a value for the in situ lateral stress (from the point where the cavity begins to expand) but the onset of plasticity is not so easily identified. The reduction in secant stiffness as the strain increases towards the elastic limit conceals it.

The conventional method of deriving usable and repeatable stiffness parameters from pressuremeter tests is to take the slope of the chord of small rebound cycles carried out after the onset of plasticity. Any unload of the expanding cavity will for an appreciable change in stress result in the whole of the soil surrounding the pressuremeter responding elastically, a state that ceases when the change of stress is sufficient to cause reverse plasticity, that is when $\sigma_r - \sigma_\theta \geq -2c_u$.

If however the unloading is never allowed to go so far then the strains will be largely elastic and recoverable, permitting the elastic properties of the soil to be derived from any point within the plastic phase of the test.

Although on the scale of the plot in Fig. 2 the slopes of all the reload cycles appear linear an expanded view shows that this is not so. Figure 4, which is a plot of the second cycle from test 6, reveals that the cycle is non-linear and shows hysteresis. The origin of the cycle is also in doubt.

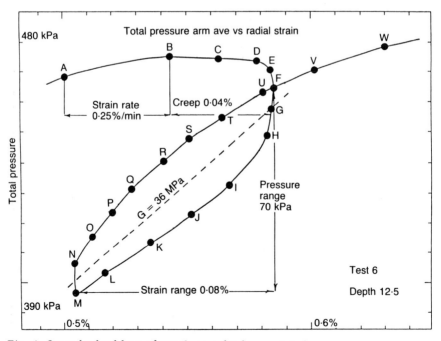

Fig. 4. *Second reload loop of test 6, completely annotated*

Although the direction of loading is reversed the direction of straining remains for some time unaltered.

An explanation frequently put forward to account for this is that some of the excess pore water pressure generated by the rapid expansion of the cavity is dissipating leading to time dependent deformations.

This may be true in some materials but the evidence presented here suggests that this is not necessarily the case in London clay. For example, Fig. 5 shows test 8 plotted for both total and effective pressure. It is apparent that, to quite a good first approximation, the entire test is conducted at a constant effective stress. Although not obvious on the scale of Fig. 5, it happens that the measured pore water pressures exactly follow the movements of the total pressures. Drainage cannot be the explanation.

We suggest that a more likely reason for this creep behaviour is that it is due to unavoidable rate effects. Even if the expansion test is conducted at a constant rate of strain this is only true for one radius. Within the whole annulus of soil being loaded the rate of strain will vary inversely as the square of the radius. It is probable (Wroth, 1975) that the stress–strain curve being followed is not unique. Thus whenever the

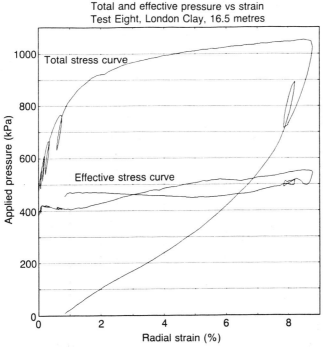

Fig. 5. *A typical quick undrained test showing that it is carried out essentially at constant effective stress*

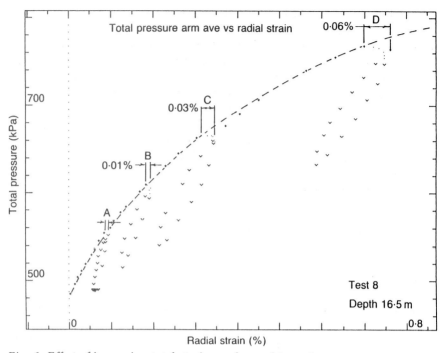

Fig. 6. *Effect of increasing total strain on observed 'creep'*

direction of loading changes the soil transfers from one set of stress/strain curves to another and this results in some additional deformation.

This implies that rebound cycles taken at small levels of cavity strain will be much less affected than those taken at large strain. Some evidence for this can be seen in Fig. 6, which is an expanded view of the initial part of the test shown in Fig. 5. The start of the unload for each of the loops A to D shows a pattern of increasing 'creep' even though the change in cavity strain during this interval is comparatively modest.

Coping with this unwanted deformation takes two forms. The first is to ignore it. In normal circumstances at some point during the unloading the strain will change direction. As long as the pressure change required to reach this state does not exceed the elastic range then a loop will be obtained.

The second is to halt the expansion and wait for the creep to slow to insignificance before reversing the deformation process. This invites, but does not necessarily incur, drainage. The waiting time often need be no more than a single minute.

Figure 4 shows a typical loop from these tests. Between the reversal of loading to the point where the loading curve picks up its original path there are a number of data points taken, not all of which are used to derive the modulus value. If following the usual practice, most observers would consider that points F to U in Fig. 4 enclose the cycle, and would quote a stiffness derived from the slope of the chord FM through the apices of the cycle. However, the point F is an arbitrary choice for origin, set by the magnitude of any creep and the speed of the unloading process and also subject to operator judgement. Taking this as an origin implies that the loop is modelled as linear and symmetrical which in this material it clearly is not. If a single value for the loop is to be quoted then the logical choice would be derived from the slope of the chord of maximum symmetry, that which distributes the unload and reload points in the most regular manner. There is, however, a much better strategy available.

Varying values of elastic shear modulus

Figure 4 is a typical unload/reload cycle for this material and is the second rebound cycle from test 6. The loading was stopped at point B when the cavity was expanding at a rate of 0.25% per minute. Between points B and F the cavity expanded a further 0.04% for no increase in pressure. As the strain amplitude of the cycle is only 0.08% this is a significant movement. In theory somewhere between B and C should be the origin for the cycle; in practice due to material creep point F seems to make a plausible origin.

If F is treated as the origin of the cycle, then had the cycle been terminated at any of the points H to L higher values of stiffness would have resulted. It follows that stiffness derived in this manner is not a function of the cavity strain at which the cycle is initiated but is related to the strain amplitude of the cycle. Furthermore, were we to be certain of the validity of our origin, this cycle could be used to obtain a range of different stiffnesses merely by taking the slope of the chord F to G, F to H and so on.

This leads to the idea that there is a much better choice of start point available if the origin for strain and the origin for pressure can be considered separately.

Any point at the beginning of the loop is going to be affected by creep, by consolidation and by friction. The bottom of the loop, on the other hand, is operating in the elastic region and will only be affected by friction of which the effect will be merely to displace the pressure co-ordinate upwards by the fixed amount necessary to overcome the frictional force.

In the loop shown we can take the origin for strain as the point N and the origin for pressure as the point M without serious error. Strictly one might define the strain origin as the bottom left corner of the loop and the pressure origin as the point where the loop departs from the ordinate through the strain origin and there are data points conveniently

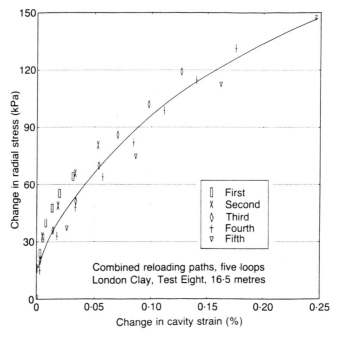

Fig. 7. Moduli computed from all possible strain intervals in a test with five reload loops

close to these places. Most computer data handling systems now provide a large number of readings thus ensuring a convenient selection is available in most loops.

Having established a reliable origin for the cycle, all strains and stresses recorded on the reloading path can be recalculated from this point. If this is done for every cycle and for every data point and if all these data are plotted on common axes it will be found that every cycle within a test gives consistent answers for the same change in strain.

This is illustrated in Fig. 7. The data from the reloading of four cycles carried out on the expansion phase of test 8, and the unloading of a single cycle carried out on the contraction phase of the same test are shown. Given that the first cycle was taken at a cavity strain of 0.07%, whereas the last cycle was taken at a cavity strain of 8.04%, the general agreement is very encouraging.

The procedure has been repeated in Fig. 8. This is a recent test, also in London clay but some miles from the test shown in Fig. 7. It happened that the rebound cycles shown in Fig. 8 had an abundance of data points and are particularly convenient for confirming the observations made in Fig. 7.

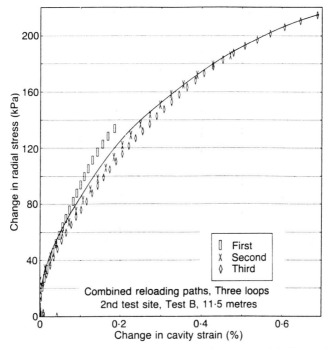

Fig. 8. *Moduli from all possible strain intervals in a test with three reload loops but with many more data points*

A plot similar to those in Figs. 7 and 8 has been published by Muir Wood (1991), who applied a curve matching procedure directly to his data in order to derive the tangential shear modulus/shear strain relationship.

The tests here suggest that if observed stiffness is correlated against the shear strain excursion over which the stiffness value was obtained then the greater part of any variability can be accounted for. The relationship is sufficiently close to be able to state that wherever in the test the rebound cycle is taken the same change in shear strain will produce the same stiffness. As the strain change can only be generated by employing some of the available shear stress the effect is that stiffness becomes an inverse function of the fraction of the utilised shear strength.

Conclusions

Designs frequently require shear moduli measured at small absolute strains and over small strain changes. Reload loops interpreted in the conventional way give a single shear modulus value corresponding to a large strain.

Table 2. Details of the rebound loops

Test No.	Loop No.	Type	Value (MPa)	Loop Coordinate Strain	Loop Coordinate Pressure	Loop Amplitude Strain	Loop Amplitude Pressure
1	1	U	67	0.036	263	0.04	47
	2	U	12	1.477	385	0.33	78
	3	U*	10	8.285	453	0.67	120
2	1	U	34	0.106	168	0.07	45
	2	U	15	0.601	228	0.18	52
	3	U*	12	8.297	359	0.58	126
3	1	U	48	0.260	173	0.05	50
	2	U	24	1.089	287	0.12	60
	3	U*	19	8.299	473	0.58	205
4	1	U	33	0.165	378	0.11	70
	2	U	20	0.913	446	0.18	71
	3	U*	15	8.558	548	0.48	135
5	1	U	41	0.126	422	0.11	93
	2	U	17	0.913	496	0.36	122
	3	U*	15	8.457	614	0.58	162
6	1	U	52	0.251	375	0.06	58
	2	U	36	0.543	429	0.08	64
	3	U*	16	8.466	719	0.59	175
	4	R	16	7.904	642	0.49	146
7	1	U	50	0.323	556	0.07	69
	2	U	44	0.889	657	0.10	89
	3	U*	25	8.540	831	0.45	210
	4	R	26	7.747	666	0.43	207
8	1	U	106	0.065	518	0.03	66
	2	U	66	0.150	562	0.07	91
	3	U	44	0.270	599	0.14	124
	4	U	35	0.660	695	0.17	120
	5	R	26	8.037	813	0.37	178
9	1	U	221	−0.014	551	0.02	91
	2	U	77	0.263	673	0.07	114
	3	U	49	0.922	877	0.15	145
	4	R	39	7.406	952	0.26	188
10	1	U	133	0.427	1051	0.05	145
	2	U	98	0.746	1163	0.11	216
	3	U	66	2.477	1388	0.21	277
	4	U*	43	6.962	1554	0.70	569
11	1	U	95	0.133	739	0.08	145
	2	U	60	0.405	885	0.16	189
	3	U	46	1.020	981	0.22	198
	4	U	29	7.652	1214	0.76	407

(1) 'Type' indicates whether the loop was an U(nload/reload) loop or a R(eload/unload) loop. U* indicates that the 'loop' is actually the slope of the initial part of the final unloading path. The stress change for such a 'loop' will be about $2c_u$.
(2) 'Loop Coordinate' is the radial strain and total stress value of the mid-point of the loop.
(3) 'Loop Amplitude' is the size of the loop in terms of the change in radial strain and total stress

However, by examining reload loops data point by data point and by being certain about the correct origin for strain and for pressure it is possible to explore the variation of stiffness with increments of shear strain over a wide range. Although reload cycles at small levels of cavity strain remain desirable they are not necessary; a single loop taken at large cavity strain is able to provide data for strains smaller than 0.01%.

This observation allows pressuremeter moduli and moduli measured by very small strain triaxial testing and geophysical methods to be much more closely related than before.

Notation

c_u	undrained shear strength
P_l	limit pressure (using the Gibson and Anderson convention of the pressure needed to double the initial volume)
G	horizontal shear modulus
G_i	initial elastic horizontal shear modulus
G_{ur}	horizontal elastic shear modulus derived from the slope of an unload-reload loop.
G_{ru}	horizontal elastic shear modulus derived from the slope of a reload/unload loop
P	the pressure being applied to the soil by the pressuremeter
P_{mx}	maximum pressure reached during a test
R	current radius of the pressuremeter
R_0	at-rest radius of the pressuremeter
R_{mx}	maximum radius reached during a test
ε	strain $[(R - R_0)/R_0]$
σ_{h0}	horizontal in situ stress

Acknowledgements

All parameters and plots are taken from real tests conducted under the usual commercial pressures and restrictions. For this and associated reasons no site can be identified and the sources have preferred to remain anonymous. We are grateful for permission to quote freely from the data.

References

GIBSON, R.E. AND ANDERSON, W.F. (1961). Insitu Measurement of Soil Properties with the Pressuremeter. Civil Engineering and Public Works Review, Vol. 56, No. 658, 615–618.

HOULSBY, G.T., CLARKE, B.G. AND WROTH, C.P. (1986). Analysis of the Unloading of a Pressuremeter in Sand. Proc. 2nd Int. Symposium, The Pressuremeter and its Marine Applications. ASTM STP 950, pp. 245–262.

HUGHES, J.M.O. (1982). Interpretation of Pressuremeter Tests for the Determination of the Elastic Shear Modulus. Proc. Conference on Updating Subsurface Sampling of Soils and Rocks and Their Insitu-Testing, Santa Barbara, California.

JARDINE, R.J., POTTS, D.M., FOURIE, A.B. AND BURLAND, J.B. (1986). Studies of the Influence of Non-Linear Stress–Strain Characteristics in Soil-Structure Interaction. Géotechnique, Vol. 36, No. 3, pp. 377–396.

JARDINE, R.J. (1991). Discussion of Strain-Dependent Moduli and Pressuremeter Tests. Géotechnique, Vol. 41, No. 4, pp. 621–626.

MUIR WOOD, D. (1990). Strain–Dependent Moduli and Pressuremeter Tests. Géotechnique, Vol. 40, No. 3, pp. 509–512.

MUIR WOOD, D. (1991). Discussion of Strain-Dependent Moduli and Pressuremeter Tests. Géotechnique, Vol. 41, No. 4, pp. 621–626.

WINDLE, D. AND WROTH, C.P. (1977). Insitu Measurement of the Properties of Stiff Clays with Self-Boring Instruments. Proc. 9th International Conference on Soil Mechanics and Foundation Engineering, 1977, Tokyo, Vol. 1, pp. 347–352.

WROTH, C.P. (1975). Insitu Measurement of Initial Stresses and Deformation Characteristics. Proc. ASCE Specialty Conf. on Insitu Measurement of Soil Properties, Raleigh, N.C., 1–4 June, 1975, pp. 181–230.

WROTH, C.P. (1982). British Experience with the Self-Boring Pressuremeter. Proc. Symp. on the Pressuremeter and its Marine Applications, Institut Français du Pétrole, Laboratoires des Ponts et Chaussées, Paris.

WROTH, C.P. (1984). The Interpretation of In-Situ Soil Tests. 24th Rankine Lecture. Géotechnique, Vol. 39, No. 4, pp. 449–489.

Analysis of the dilatometer test in undrained clay

H.S. YU, The University of Newcastle, J.P. CARTER and
J.R. BOOKER, The University of Sydney

The Marchetti dilatometer is being used increasingly in geotechnical practice to obtain design parameters for a variety of soils. Various authors have claimed success in obtaining strength and deformation parameters from this test, as well as knowledge of the stress state in the ground. All of these parameters are important for the prediction of soil behaviour in practical application. To date, the interpretation of the test has been performed almost exclusively using empirical methods. Curiously, there seems to have been little attempt so far to develop rigorous theoretical methods of the dilatometer. In this paper, a theoretical treatment of the dilatometer test in undrained clays is presented. Results of a series of numerical analyses have been used to develop an interpretation of the field test, that is not based on empiricism but on the rigorous application of an elastoplastic soil model to the behaviour of undrained clay soils.

Introduction

Site investigation and assessment of the characteristics of soils are essential parts of the geotechnical design process. The principal parameters of interest to designers are strength, deformation moduli, in situ horizontal stress and permeability. In situ testing to complement laboratory tests in obtaining these fundamental soil properties is becoming increasingly important in practice. Demand for in situ testing has developed with a growing appreciation of the inadequacy of conventional laboratory testing. Inevitable sample disturbance affects laboratory test results and raises questions as to the validity of the soil strength and deformation properties measured. Difficulties associated with the sampling of undisturbed soil specimens have led to the development of indirect methods in which the strength and deformation characteristics of the soil under field conditions are related to parameters derived from in situ penetration tests. Because of the complex behaviour of the soil when subject to the sophisticated loading conditions imposed by the in situ tests, the interpretation of test data is beset with difficulties (see, for example, Wroth (1984)). The quality of the interpretation, however, directly affects the accuracy of the soil properties obtained from the in situ tests. It is therefore vital to develop rigorous methods for the interpretation of in situ penetration tests so that the accuracy of the derived soil properties can be assured.

A relatively new in situ penetration test device, the flat dilatometer, is being used increasingly in geotechnical practice to obtain design parameters for a variety of soils because

(*a*) it is simple to operate and maintain
(*b* it does not rely on minimizing disturbance during insertion
(*c*) it provides a repeatable and continuous profile of the measured parameters.

The dilatometer is a flat-bladed penetrometer 14 mm thick, 95 mm wide, 220 mm long, which has a flexible stainless steel membrane 60 mm in diameter located on one face of the plate (Marchetti, 1980). Usually the dilatometer is inserted into the ground in a quasi-static manner and then its circular membrane is pressurized by way of a control console. Pressure expansion of the membrane occurs at a given rate and pressures are recorded at two instants: at the first to initiate movement of the membrane off the plane of the blade into the soil, and at the second to cause a membrane deformation of 1 mm against the soil.

In the past, the interpretation of the dilatometer test has been performed almost exclusively using empirical correlations, but this is not considered to be ideal. It is now considered preferable to interpret in situ tests using a rigorous analysis, based on a well-defined soil model and taking proper account of the appropriate boundary conditions. This has not always been possible due to the lack of suitable analytical or numerical solutions.

This paper presents a preliminary study aimed at the development of rigorous theoretical models of the dilatometer test in clay. In particular, the behaviour of undrained clay has been modelled as an elastic–perfectly plastic material obeying the Tresca yield criterion, and the dilatometer test has been analysed as a boundary value problem using the finite element technique.

Methods of analysis

As is the case with other mechanical in situ tests, the dilatometer does not measure any particular properties of the soil directly; rather, the load response to an imposed deformation is measured.

As mentioned previously, the conventional interpretation of the dilatometer test data is based predominantly on empirical correlations, using measurements of the lift-off pressure of the membrane P_0 and the pressure required to cause a further 1 mm expansion of the membrane P_1. These two pressures are related to the undrained shear strength (S_u),

DILATOMETER TEST IN UNDRAINED CLAY

the initial total horizontal stress (σ_{h0}), the soil rigidity index (G/S_u, where G is the elastic shear modulus of the soil) and the stress history. Before using any data from the dilatometer tests it is important to account for the fact that the dilatometer is a soil displacing probe. The initial state of stress in the soil is altered by the installation of the dilatometer. The magnitude of these changes depends on many factors such as stress history and soil stiffness. Thus, a sensible interpretation of the dilatometer test data requires a more fundamental understanding of the soil behaviour during penetration as well as during membrane expansion.

The penetration of the dilatometer can be regarded as a complex loading test on the soil. A possible way of analysing the penetration process is to model it as the expansion of a flat cavity, tractable as the enforcement into the soil of an opposing pair of rigid vertical strips. The penetration of the dilatometer causes a horizontal displacement of the soil elements originally on the vertical axis of 7 mm (the half thickness of the dilatometer). During the penetration of the dilatometer there is a

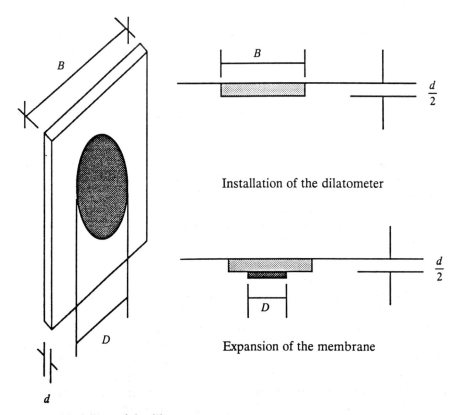

Fig. 1. Modelling of the dilatometer test process

concentration of shear strain near the edges of the blade, so that the volume of soil facing the membrane undergoes a smaller shear strain than other areas. The subsequent expansion of the circular membrane to 1 mm outward displacement generates smaller increments of strain in the soil and this expansion may be satisfactorily modelled as a continued loading of a circular area after the penetration stage. This approach is shown graphically in Fig. 1.

Analysis of the dilatometer test

To gain a better understanding of the dilatometer test process, numerical simulations have been carried out for the test in undrained clay. The soil mass has been idealized as an elastic–perfectly plastic medium which obeys the Tresca yield criterion and deforms under constant volume conditions. The numerical work was carried out using a general finite element program, AFENA, developed at the University of Sydney (Carter and Balaam, 1989). Figure 2 shows the geometry and boundary conditions used for modelling the installation of the dilatometer. A mesh that consists of 400 eight-noded plane strain elements and 1281 nodes was used. The outer boundaries of the mesh were placed at a distance of approximately 25 times the width of the dilatometer from its centre so that the behaviour of an infinite medium could be appropriately modelled. Since only a quarter of the dilatometer is analysed and due to the symmetric nature of the problem, a rigid vertical boundary condition is used for the upper boundary of the finite element mesh. Care was taken in designing the mesh so that regions likely to have high stress gradients had a higher density of elements. A reduced integration scheme (2×2) was used to approximate the stiffness matrix of the soil mass so that the incompressibility condition could be adequately approximated (Sloan and Randolph, 1982; Yu, 1991). Furthermore, elastic incompressibility was approximated by assigning a value of 0.49 to Poisson's ratio.

As mentioned above, the analysis can be divided into two different stages. First of all, the installation of the dilatometer was modelled as a pair of opposing rigid strips. The expansion of the dilatometer membrane could also be modelled as a continuous loading of a smaller circular area. Obviously, the modelling of the second stage involving the expansion of the dilatometer membrane is more complex than the first stage. This is mainly due to the fact that accurate modelling of the expansion of the membrane would require a three dimensional analysis. As this paper represents the first stage of an on-going research programme on the analysis of the dilatometer, only the installation process has been considered here. The analysis of the expansion of the dilatometer membrane is currently in progress and will be treated in a later paper.

DILATOMETER TEST IN UNDRAINED CLAY

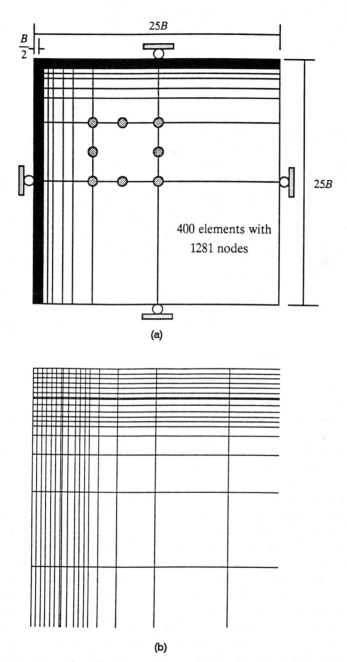

Fig. 2. (a) The overall mesh and boundary conditions for the dilatometer analysis, and (b) the detailed mesh for the region around the dilatometer

The dilatometer installation is modelled as a plane strain problem with no strain in the vertical direction. For simplicity in this preliminary study, it is also assumed that the initial state of stress in the ground is isotropic.

The dilatometer analysis simulates the full-displacement penetration of a flat spade into the ground, and in some respects this is similar to the installation of a full-displacement pressuremeter and a cone. Recent research suggests that the installation of the full-displacement pressuremeter can be satisfactorily modelled as the expansion of a cylindrical cavity within the soil (Houlsby and Withers, 1988; Yu, 1990; Houlsby and Yu, 1990). It has been pointed out (Baligh, 1986) that this modelling of installation provides only an approximate solution, but nevertheless, this simple approach has been largely supported by other theoretical studies on the cone penetration test (e.g. Teh (1987)). Teh showed that the stress distribution far behind the cone-tip predicted from Baligh's strain path method is similar to the distribution created by the expansion of a cylindrical cavity from zero initial radius. In detail, the stresses seem to correspond more closely to a cavity expansion followed by a small contraction, but the magnitude of the appropriate contraction is difficult to assess. According to the cavity expansion theory, the lift-off pressure of the cone-pressuremeter test in undrained Tresca materials, P_i, can be expressed in a normalized manner as follows:

$$N_{P_i} = \frac{P_i - \sigma_{h0}}{S_u} = 1 + \ln \frac{G}{S_u} \qquad (1)$$

Equation (1) indicates that the stress changes due to the installation are mainly controlled by the rigidity index and undrained shear strength. Because of the similarity between the installation of the dilatometer and the full-displacement pressuremeter (or cone-pressuremeter), it would be expected that the lift-off pressure of the dilatometer should also be primarily controlled by the undrained shear strength and the rigidity index of the soil. It is being increasingly recognized that the rigidity index is one of the most important parameters in the understanding of many in situ tests in undrained clays (Houlsby and Wroth, 1989; Yu, 1992; Yeung and Carter, 1990). It is for this reason that the effects of the rigidity index of the soil will be studied in detail in this paper.

Numerical results

Figure 3 shows the typical result of a numerical simulation of the dilatometer test in undrained clay. The pressure–expansion curve from each numerical simulation is interpreted as if it were derived from a real test. The average value of the dilatometer–soil contact pressure at end of

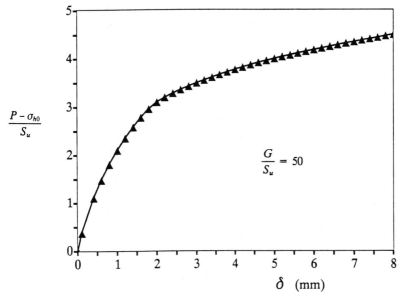

Fig. 3. *Typical result of a numerical simulation of the dilatometer test*

installation, i.e. when the outward horizontal displacement of each rigid strip (δ) reaches 7 mm, is denoted by P_0. The values of 25, 50, 100, 200, 300 and 500 have been selected for the rigidity index in the calculations so that its effects could be fully investigated.

As discussed earlier, the lift-off pressure P_0 against the side of a penetrating dilatometer is generally a function of initial total horizontal stress σ_{h0}, the undrained shear strength and the rigidity index of the soil. The results of the analyses can be represented by a non-dimensional parameter defined by $N_{P_0} = (P_0 - \sigma_{h0})/S_u$.

Figure 4 shows the predicted variation of the normalized lift-off pressure N_{P_0} with the rigidity index. It is found that N_{P_0} increases significantly with the rigidity index. The value of N_{P_0} ranges from 3.64 for very soft clays to 8.3 for stiff soils. The numerical data presented in Fig. 4 may be approximately represented by the following expression

$$N_{P_0} = \frac{P_0 - \sigma_{h0}}{S_u} = -1.75 + 1.57 \ln \frac{G}{S_u} \qquad (2)$$

To highlight the significance of the numerical correlation (2), the results of field dilatometer tests in clays reported by Marchetti (1980) have been reanalysed. A total of 16 tests which have independent measurements of K_D, K_0, OCR and S_u/σ'_{v0} are particularly useful for this purpose. The quantity K_D denotes the lateral stress index of the

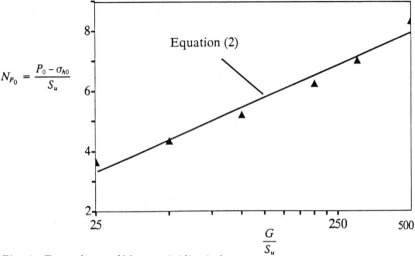

Fig. 4. Dependence of N_{P_0} on rigidity index

dilatometer test, defined as $(P_0 - u)/\sigma'_{v0}$, where σ'_{v0} and u are the vertical effective stress and the pore water pressure respectively prior to the dilatometer insertion. K_0 represents the initial lateral stress ratio given by $(\sigma_{h0} - u)/\sigma'_{v0}$ and OCR stands for the overconsolidation ratio of the soil. The normalized lift-off pressure defined previously in this paper for

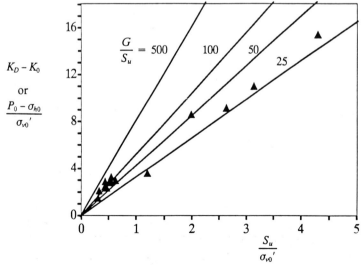

Fig. 5. Correlation between $K_D - K_0$ and S_u/σ'_{v0}

the dilatometer can be calculated using the known values of K_D, K_0 and S_u/σ'_{v0} according to $N_{P_0} = (P_0 - \sigma_{h0})/S_u = (K_D - K_0)/(S_u/\sigma'_{v0})$. Plotted in Fig. 5 is the variation of $K_D - K_0$ with S_u/σ'_{v0}. The triangles represent the actual test data reported by Marchetti (1980), while the solid lines drawn on this figure represent a form of the theoretical relationship expressed in eqn. (2), for selected values of rigidity index. It is easily shown that the slope of a line passing through the origin and each individual data point represents the value of N_{P_0} for that test. Using the correlation (2), the rigidity index for the soil in each test can be estimated from the measured value of N_{P_0}.

It has been well established by both experimental studies (Ladd et al., 1971; Ladd and Edgers, 1972) and theoretical research using the concepts of Critical State Soil Mechanics (Wroth and Houlsby, 1985) that the rigidity index of the soil increases with OCR until OCR is about 2.0, after which it starts to decrease with OCR. With the theoretical result (2), it is possible to estimate the variation of the rigidity index with overconsolidation ratio OCR from the field test results.

Figure 6 presents the measured variation of N_{P_0} with OCR, obtained from the field tests. It is found that the experimental value of normalized lift-off pressure range from 3.03 to 6.72. As discussed before, an estimate can also be made of the rigidity index from the value of N_{P_0}, by using eqn. (2) for each test. The results obtained using this approach are shown in Fig. 7. Although there is a scatter in the data, it can be clearly seen that the variation of rigidity index with OCR deduced from the

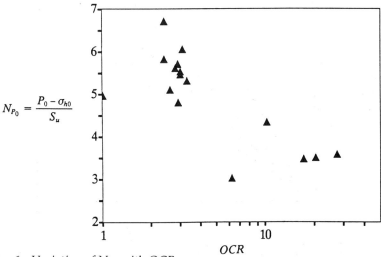

Fig. 6. Variation of N_{P_0} with OCR

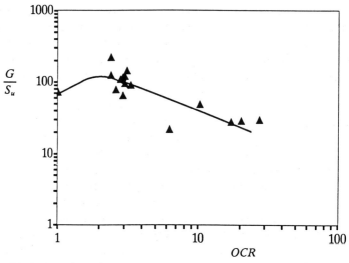

Fig. 7. *Variation of rigidity index with OCR*

dilatometer test results follows a pattern that is consistent with independent estimates, specifically the experimental data cited previously and the theoretical prediction using the concepts of Critical State Soil Mechanics.

It is interesting to compare the lift-off pressures predicted for the flat blade dilatometer and the full displacement pressuremeter (or cone-

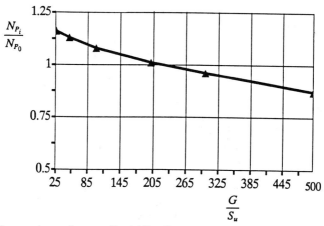

Fig. 8. *Comparison of normalized lift-off pressures for the cone–pressuremeter and the dilatometer*

pressuremeter). As argued before, the lift-off pressure P_i for the full-displacement pressuremeter may be estimated by eqn. (1), and this estimate has been obtained from the theory for a cylindrical cavity expansion. The numerical results presented in Fig. 4 (or eqn. (2)) may be used to provide an approximation for the lift-off pressure for the dilatometer P_0. Figure 8 shows the predicted variation of the ratio of the normalized lift-off pressures for the cone-pressuremeter and the dilatometer with the rigidity index. It can be seen that the predicted lift-off pressure for the cone pressuremeter is slightly higher than that of the dilatometer for soft clays, but the reverse is true for stiff clays. In general, however, the lift-off pressures for both the cone–pressuremeter and the dilatometer are very close. This predicted trend is consistent with the field data reported by Lutenegger and Blanchard (1990), who conducted many field with tests both the full-displacement pressuremeter and the flat dilatometer at a number of different clay sites. The soils tested ranged in stiffness from very soft sensitive marine clays to very stiff glacial clay tills. The comparison of measured lift-off pressures for the cone-pressuremeter and the dilatometer is shown in Fig. 9.

Conclusion

This paper has described a numerical study of the flat dilatometer test in undrained clay. It represents the first stage of an on-going research

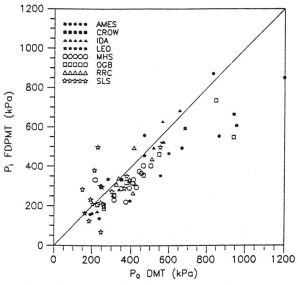

Fig. 9. Comparison of measured lift-off pressures for the cone–pressuremeter and the dilatometer (after Lutenegger and Blanchard (1990))

programme on the analysis of the dilatometer test, and so attention has been focused on the modelling of the dilatomer installation. The penetration of the dilatometer into the ground was modelled as the expansion of a flat cavity. A plane strain condition was assumed, with no strain permitted in the vertical direction.

The results of a series of numerical analyses indicate that the lift-off pressure of the flat dilatometer is a function of initial horizontal stress, undrained shear strength and rigidity index of the soil. It was found that the normalized lift-off pressure of the dilatometer, as defined by $N_{P_0} = (P_0 - \sigma_{h0})/S_u$, is not a constant but increases very significantly with the rigidity index of the soil. With the numerical correlation between the normalized lift-off pressure and the rigidity index, the shear modulus can be calculated provided K_D, K_0 and S_u can be estimated. The application of the numerical predictions to field test results leads to a similar pattern for the variation of the rigidity index with overconsolidation ratio OCR as observed in laboratory tests. The theoretical correlation has also been used to predict the relative magnitudes of the lift-off pressures for both the full-displacement pressuremeter and the flat dilatometer. In particular, it suggests that the lift-off pressure of the full-displacement pressuremeter is generally close to that of the dilatometer, with the lift-off pressure of the pressuremeter being slightly lower for stiff clays. This prediction is validated by the results of many field tests in clays carried out with both the full-displacement pressuremeter and the dilatometer.

The numerical study presented in this paper is of limited scope. A comprehensive method for the interpretation of the dilatometer test will not be achieved until the expansion of the circular membrane has also been analysed. This task will form a second stage of the theoretical study of the dilatometer test in undrained clay. Because of the complex boundary conditions corresponding to the expansion of the circular membrane, a three dimensional analysis may be necessary. In addition, it is also proposed to extend the analysis of the dilatometer test to include a treatment of purely frictional materials, so that a rigorous method for the interpretation of the test in sand may be available. This work will form the subject of future papers.

References

Baligh, M.M. (1986). Undrained deep penetration, I: Shear stresses. Geotechnique, Vol. 36, No. 4.

Carter, J.P. and Balaam, N.P. (1989). AFENA – Users' Manual. Centre for Geotechnical Research, University of Sydney.

Houlsby, G.T. and Withers, N.J. (1988). Analysis of the cone pressuremeter in clay. Geotechnique, Vol. 38, No. 4, pp. 575–587.

Houlsby, G.T. and Wroth, C.P. (1989). The influence of soil stiffness and lateral stress on the results of in-situ tests. Proc. XII Int. Conf. on Soil Mechanics and Foundation Engineering, Vol. 1, pp. 227–232, Rio de Janeiro.

HOULSBY, G.T. AND YU, H.S. (1990). Finite element analysis of the cone pressuremeter test. Proc. 3rd Int. Symp. on Pressuremeters, Oxford, pp. 221–230.

LADD, C.C., BOVEE, R.B. AND RIXNER, J.J. (1971). Consolidated undrained plane strain shear tests on Boston blue clay. MIT research report, R71-13.

LADD, C.C. AND EDGERS, L. (1972). Consolidated undrained direct simple shear tests on saturated clays. MIT research report, R72–82.

LUTENEGGER, A.J. AND BLANCHARD, J.D. (1990). A comparison between full displacement pressuremeter tests and dilatometer tests in clay. Proc. 3rd Int. Symp. on Pressuremeters, Oxford, pp. 309–320.

MARCHETTI, S. (1980). In-situ tests by flat dilatometer. J. Geotech. Engg, ASCE, GT3, pp. 299–321.

SLOAN, S.W. AND RANDOLPH, M.F. (1982). Numerical prediction of collapse loads using finite element methods. Int. J. Num. Analy. Geomech., Vol. 6, pp. 47–76.

TEH, C.I. (1987). An analytical study of the cone penetration test. DPhil Thesis, University of Oxford.

WROTH, C.P. (1984). The interpretation of in-situ soil tests. 24th Rankine Lecture, Geotechnique, Vol. 34, No. 4, pp. 449–489.

WROTH, C.P. AND HOULSBY, G.T. (1985). Soil mechanics — property characterization and analysis procedures. Theme lecture No. 1, Proc. XI Int. Conf. on Soil Mechanics and Foundation Engineering, pp. 1–55, San Francisco.

YEUNG, S.K. AND CARTER, J.P. (1990). Interpretation of the pressuremeter test in clay allowing for membrane end effects and material non-homogeneity. Proc. 3rd Int. Symp. on Pressuremeters, Oxford, pp. 199–208.

YU, H.S. (1990). Cavity expansion theory and its application to the analysis of pressuremeters. DPhil Thesis, University of Oxford.

YU, H.S. (1991). A rational displacement interpolation function for axisymmetric finite element analysis of nearly incompressible materials. Finite Elements in Analysis and Design, Vol. 10, pp. 205–219.

YU, H.S. (1992). Finite element analysis of pressuremeter tests in soil. Proc. 6th ANZ Conference on Geomechanics, Christchurch, New Zealand.

Predictions associated with the pile downdrag study at the SERC soft clay site at Bothkennar, Scotland

J.A. LITTLE and K. IBRAHIM, University of Paisley, Scotland

The Symposium has provided an ideal opportunity to invite predictions associated with the pile downdrag study at the SERC soft clay test site at Bothkennar, Scotland. The effects of negative skin friction on both single piles and pile groups are investigated in this study. Piles are instrumented for the measurement of axial shaft load, pore pressure and total lateral pressure at the pile/soil interface. This instrumentation was positioned at four strategic levels, coinciding with the levels for pore pressure and settlement measurements in the adjacent ground. After pile driving, 2.5 m of surcharge was laid and compacted to induce settlement in the clay and to induce downdrag on the piles. While the primary aim of the full-scale experiment is to improve understanding of pile downdrag, the study has also provided opportunities to investigate other aspects of geotechnical engineering. These have included the effects on soil pore pressure and settlement of loading soft clay by the application of a surcharge. During the Symposium, various predictions were compared with the field observations, made public for the first time. A detailed description of the predictions and observations are presented in the paper.

Introduction

It was Professor Wroth who originally conceived the idea of a national soft clay site in the United Kingdom. With the purchase of such a site in 1987 by the Science and Engineering Research Council, a facility existed whereby experiments in geotechnical engineering at large and full scale could be carried out.

Located in the Parish of Bothkennar near Grangemouth on the Forth estuary in Scotland, the 11 ha site has been the subject of much activity since 1987, the geotechnical outcome of which has been largely summarized in Volume 42 (1992) of *Géotechnique*.

The first full-scale geotechnical engineering research project to be undertaken on the site commenced in November 1988. At that time, with SERC, industry and BRE funding, an experiment was started on the site with the aim of providing basic data from two pile groups subjected to downdrag. Two years of intensive preparation were required to bring this work to the stage when, in July 1991, a fill surcharge (comprising a 2.5 m high embankment) could be laid and compacted around the 22 piles. Since that time, the arrays of ground and pile instrumentation previously installed have been monitored at monthly intervals.

In view of the interest generated by this study, it was suggested to the organizers of the Wroth Memorial Symposium that a 'Bothkennar prediction event' might be included as a part of the programme. This seemed particularly apposite in view of Peter Wroth's connection with the site, and the theme of the symposium ('predictive soil mechanics').

After discussions with the organizers, it was planned to set aside a part of the programme when some of the experimental results from the downdrag study would be made public for the first time. The results would then be compared with predictions submitted by individuals and groups of individuals beforehand, who had attempted to predict performance associated with certain events related to the study, but without the benefit of data.

Description of the study

A detailed description of the downdrag study at Bothkennar in terms of its aims, pile types, equipment and instrumentation, experimental layout and installation techniques has been reported by Little et al. (1991); a shorter description is provided below.

Figure 1 shows a plan and cross-section of the experimental area. The installation of 62 no. ground piezometers at four levels (and at different locations), together with two inclinometer boreholes down to the gravel layer at approximately 20 m depth, took place in 1989. The H-shaped area of ground in Fig. 1 was carpeted with a geotextile (Lotrak) and overlain with 500 mm uniform depth of compacted burnt oil shale (known locally as 'blaes') in February 1990. The purpose of this was to provide a working surface for piling plant subsequently. Monitoring of the effects of laying the 500 mm of blaes continued for approximately 12 months. During July 1991, two groups of nine piles (8 Westpile shell and 1 steel tubular, nominal outside diameter 406 mm) were driven (sleeved) through the 500 mm layer, at the positions indicated in Fig. 1. One group (friction, region Y) was driven to set approximately 1 m above the gravel layer, the depth of which is 21.5 m below original ground level in this part of the site. The length of the friction piles is 20.4 m. The other group of nine piles (end-bearing, region Z) was driven to set onto this gravel layer, here at a depth of 20.8 m below original ground level. The spacing of the piles in the groups was 1.624 m (4 pile diameters) centre to centre. Single control piles (one slip-coated and one uncoated per group) were also driven at this time, away from the area of influence of each group (approximately 20 pile diameters).

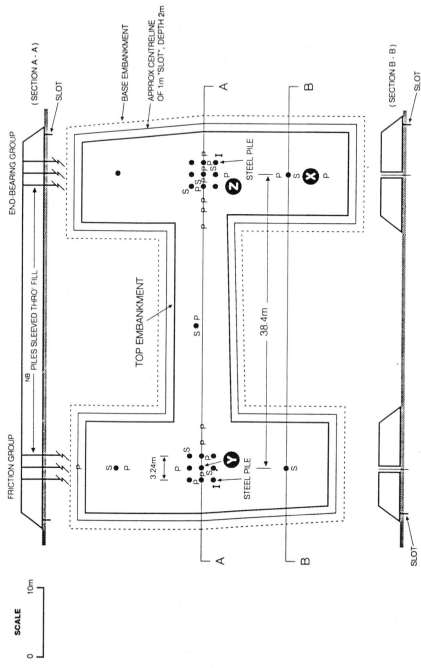

Fig. 1. Bothkennar pile downdrag experiment layout. ● = pile, P = piezometer, S = settlement gauge, I = inclinometer

Eight of the Westpile shell piles were instrumented for the measurement of axial shaft load as well as pore pressure and total lateral pressure at the pile/soil wall. This instrumentation was positioned at four strategic levels, coinciding with the levels for pore pressure and settlement measurements in the adjacent ground. In each group of piles, an edge, a corner and the centre pile was thus instrumented. The single 'control' pile belonging to each group was similarly instrumented.

Monitoring of all instrumentation proceeded from the beginning of pile driving, and continued at one-monthly intervals until approximately 12 months later.

During a five days' period at the end of June and the beginning of July 1991, the 0.50 m layer of blaes was raised by an additional 2.50 m, thus producing an embankment of total uniform height 3.0 m. The perimeter of this embankment had been 'slotted' with a trench of 2 m depth, 1 m width, through the desiccated crust, to decouple the surcharged area from the surrounding ground prior to the placing of the 500 mm layer of blaes. This trench was subsequently infilled with the excavated destructured material. Its position, in plan, is mid-way between the solid line marking the perimeter of the top of the embankment and the dotted line marking the perimeter of the base of the embankment in Fig. 1.

The 2.5 m of blaes laid in 1991 was placed in 500 mm layers and compacted using a heavy-duty Bomag vibrator; in between the piles a hand-held vibrator was used to compact the fill. Measurements of compacted density by both nuclear densimeter and the sand replacement methods indicated average compacted densities of 19 kN/m^3 and 15 kN/m^3 outside and within the groups of piles respectively.

Immediately after the compacted blaes was laid, monitoring of all instrumentation resumed, complete sets of readings being taken at monthly intervals since that time. This has involved manually reading

(a) ground piezometers
(b) ground magnet extensometers
(c) pile Glotzl cells
(d) pile piezometers
(e) pile load cells
(f) precise levelling on the tops of the piles.

Additional information

Participants in the exercise were directed to the paper by Hawkins et al. (1989) for more detailed geological and geotechnical information for the site. For the convenience of those participating, a number of figures in this paper were reproduced and included in the mailed prediction package. These included profiles of natural water content and Atterberg limits, profiles of cone resistance and friction ratio, and peak/remoulded

Table 1. Summary of predictions

		Δu_i (kPa) 1991	ρ_i (mm)	Δu (kPa) 1992	ρ_{consol} (mm) 1992	D_f, centre friction group (kN)	D_f, centre end-bearing group	D_f max for 5, 6	Comments
Wong and Teh	X	47.5	60	33	350	324	600		
Rojas and Houlsby	X	37.0	63	22	101	158	315	269	Also single piles
	Y	40.4	58	28.2	119			359	
Muir Wood	Z	45	23.5	31	101	365	≃600	≃700	No group effect
Pender	X	45.3	11	8.5	555	834	1122	288	Single pile
	Z	47							
Lehane and Jardine	X	47.5	18	12.5	101	192	273	410	
Observed	X	19	28	13	83	162			
	Y	26	32	14	101				
	Z	24	35	12	95				

PILE DOWNDRAG AT BOTHKENNAR

undrained vane shear strengths. Additional one-dimensional oedometer consolidation data, as well as K_0 consolidated undrained triaxial compression tests on intact specimens of Bothkennar clay (various depths), were also provided.

Prediction requirements
Predictions were invited for any of the following:

(a) The *initial* pore pressure response of the clay from original ground level to the top of the Buried Gravel Layer (approximately 20 m depth) due to the additional 2500 mm blaes surcharge laid in June 1991 (all excess pore pressure due to initial 500 mm fill dissipated).
(b) The *initial* settlement of the ground surface (i.e. under the blaes fill) due to the additional 2500 mm blaes surcharge.
(c) The excess pore pressure distribution in the clay (i.e. with depth) in July 1992.
(d) The consolidation settlement of the clay surface due to the additional 2500 mm of blaes, as of July 1992.
(e) The profile of force with depth in the centre pile in the friction group in July 1992.
(f) The profile of force with depth in the centre pile in the end-bearing group in July 1992.
(g) The anticipated maximum downdrag forces in these piles (centre friction, centre end-bearing) after full consolidation.

Participants were also invited to make estimates as in (e)-(g), for the single friction and end-bearing piles.

Predictions
Five sets of predictions were submitted:

(a) K.S. Wong and C.I. Teh (Nanyang Technological University)
(b) E. Rojas and G.T. Houlsby (University of Oxford)
(c) D. Muir Wood (University of Glasgow)
(d) M.J. Pender (University of Auckland)
(e) B. Lehane and R. Jardine (Imperial College of Science, Technology and Medicine).

For completeness, the basis of each prediction, as provided by each predictor, is contained in the Appendix to this paper.

A summary of the predicted quantities is shown in Table 1.

Commentary
In order to provide a framework within which the predicted quantities may be viewed, a brief commentary of the observed results is presented

here. A more detailed presentation and analysis of the results of the Bothkennar downdrag study is currently being prepared for publication at a later date.

Initial pore pressure response

Figures 2, 4 and 6 describe the predicted and observed excess pore water pressure, in the clay, as a result of applying the additional 2.5 m of fill in

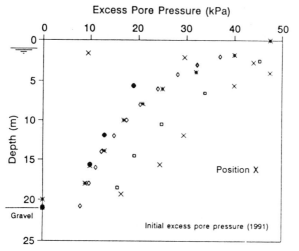

Fig. 2

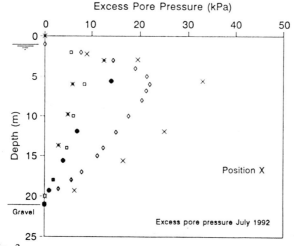

Fig. 3

June 1991. It is difficult to be precise about surcharge pressure anywhere directly underneath the loaded area, due to its irregular shape. Most predictors assumed a contact pressure of approximately 48 kPa (2.5 m of fill having a unit volume weight of 19 kN/m^3). In a simplistic way this would indicate a maximum excess pore pressure of approximately 50 kPa near to the surface. Figure 2 describes the excess pore pressure observed at location X (adjacent to the single end-bearing pile); Figs. 4

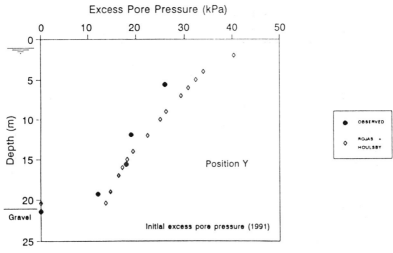

Fig. 4

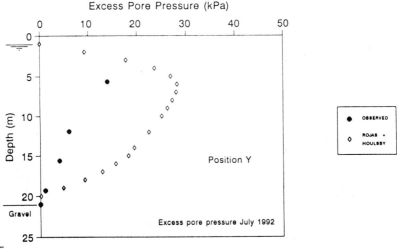

Fig. 5

and 6 describe the excess pore pressure observed within the friction and end-bearing pile groups respectively.

There were marked differences between the excess pore pressures observed at position X, and those observed at positions Y and Z. The maximum recorded excess pore water pressure was 26 kPa at position Y, within the friction pile group. This was recorded by the piezometer installed at the highest level, positioned 5.6 m below original ground

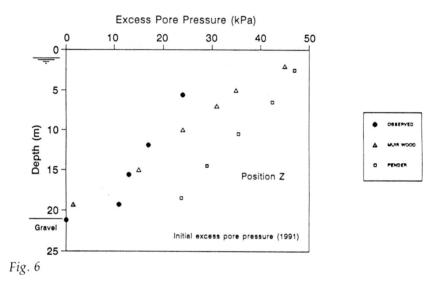

Fig. 6

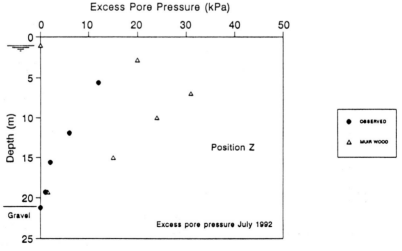

Fig. 7

level. The corresponding pore pressure within the end-bearing group (position Z) was 24 kPa. The maximum excess pore pressure observed adjacent to the single end-bearing pile (position X) was 19 kPa.

Generally, the predictions over-estimated the maximum excess pore pressures between 5.6 m and 20 m. The predictions for excess pore pressure above 5.6 m depth cannot be corroborated, due to their being no instrumentation above that depth. In certain cases (for example Fig.

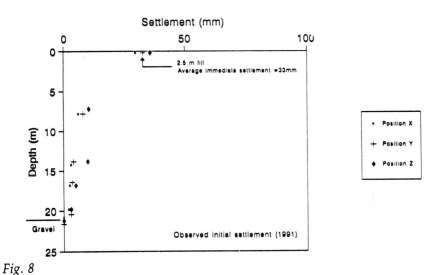

Fig. 8

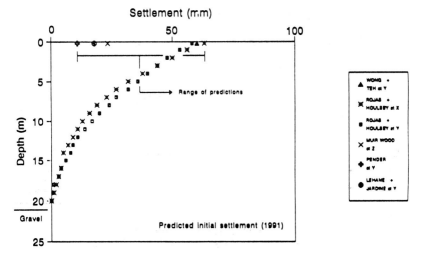

Fig. 9

4) the distribution of excess pore pressure was modelled reasonably satisfactorily, particularly below 15 m.

Bothkennar clay has been shown to have a permeability-enhancing fabric (Little et al., 1992) a fact which might explain why the values for the maximum observed excess pore pressures were somewhat less than those predicted, particularly during and immediately post-surcharging. Similar differences between observed and predicted pore pressures were

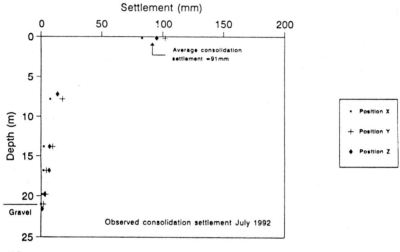

Fig. 10

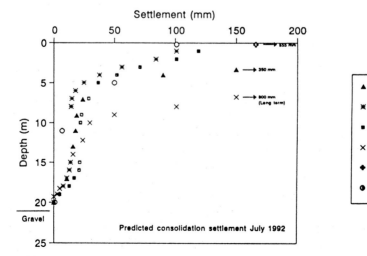

Fig. 11

reported by Leroueil et al. (1978) for embankment B at Saint-Alban. They suggested partial drainage during surcharging as a possible cause.

Initial settlement

Figure 8 describes the observed initial, or immediate, settlements of the clay from ground level to the gravel layer at positions X, Y, Z. The average immediate settlement at ground level, due to the 2.5 m surcharge, was 33 mm (note the relatively narrow range of values). At 7.5 m depth the settlement was approximately 6 mm, reducing to zero at the clay/gravel interface.

Figure 9 indicates the much larger range of predicted initial settlements (10–65 mm) at ground level. As with excess pore pressure, the closest correspondence between predicted and observed initial settlement was for depths below 12 m.

Excess pore pressure July 1992

Figures 3, 5 and 7 describe the predictions for pore pressure in July 1992, approximately one year after surcharging.

Whilst the observations indicated a dissipation of almost 50% of the maximum excess pore pressures recorded in 1991, a number of predictions showed excess pore pressures that had hardly dissipated during this twelve months. This is a little surprising in view of the fact that predictions for consolidation settlement over this same period (see below) had been modelled with more success.

Consolidation settlement July 1992

Figures 10, 11 describe the observed and predicted consolidation settlements down to the gravel up to July 1992. The average measured consolidation settlement was 91 mm (again, a relatively narrow range was evident between X, Y, Z). As with initial settlement, there had been a considerable reduction in the measured consolidation settlement at 7 m (approximately 12 mm at this depth). The consolidation settlement of the clay at the top of the gravel was negligible. A number of predictions grossly over-estimated the consolidation settlement at this time — presumably for the same reasons that the initial and subsequent excess pore pressures were also over-estimated. The distribution of settlement below 3 m had, however, been modelled satisfactorily by two of the five predictors.

Force versus depth, centre pile friction group, July 1992

Figure 12 shows the observed distribution of total axial force for the centre pile in the friction group. Six distributions are shown, corresponding to 1 m, 2 m, 2.5 m of fill, and to 22, 90, 300 days after surcharging.

The distributions conform with the expectation of increasing downdrag with both increasing surcharge and time. Thus, immediately after surcharging the maximum axial force ($\simeq 75$ kN) was recorded at approximately 12 m depth in the pile. After 300 days this had increased to more than 150 kN. The measured force at the pile toe was negligible.

Figure 13 describes the maximum predicted distributions of axial force with depth. (The majority of the predictors preferred not to make

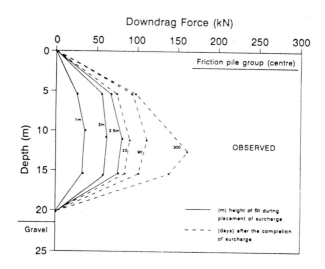

Fig. 12

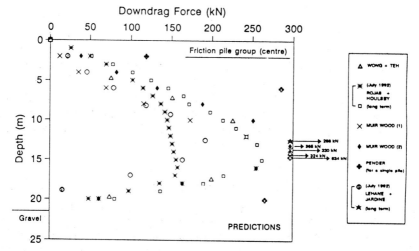

Fig. 13

estimates for 1992, nor to estimate for the effects of pile grouping; Fig. 13 therefore represents a distillation of disparate predictions varying with both time and pile position.)

Generally, predictions for downdrag over-estimated that measured in July 1992. In some cases the over-prediction was 500–600%. The size and distribution of the downdrag prediction submitted by Rojas and Houlsby (using the method of Zeevaert) was considered to be very close to the actual distribution.

Force versus depth, centre pile end-bearing group, July 1992
During the presentation of the results of the predictions in Oxford 1992 the authors described the distribution of force with pile depth for the end-bearing piles. Whilst the observed distribution of downdrag above 16 m was in accord with expectations, the performance of the base load cells raised uncertainties regarding the measured distribution below that depth. The authors are currently investigating the reasons for this and, in view of these uncertainties, did not make comparisons of the predictions for the end-bearing piles in the same way.

Acknowledgements
The work described in this paper is funded by the Science and Engineering Research Council. The collaboration of Westpile Ltd and the Building Research Establishment is also gratefully acknowledged.

References
HAWKINS, A.B., LARNACH, W.J., LLOYD, I.M. AND NASH, D.F.T. (1989). Selecting the location, and the initial investigation of the SERC soft clay test bed site. Quart. J. Eng. Geol., Vol. 22, No. 4, pp. 281–316.

LEROUEIL, S., TAVENAS, F., TRAK, B., LA ROCHELLE, P. AND ROY, M. (1978). Construction pore pressures in clay foundations under embankments, Part 1: The Saint-Alban test fills. Canadian Geotechnical J., Vol. 15, No. 1, pp. 54–65.

LITTLE, J.A., MUIR WOOD, D., PAUL, M.A. AND BOUAZZA, A. (1992). Some laboratory measurements of permeability of Bothkennar clay in relation to soil fabric. Geotechnique, Vol. 42, No. 2, pp. 355–361.

Appendix: Predictions
K.S. Wong and C.I. Teh, Nanyang Technological University
The prediction of consolidation settlements and pore pressures is done with the aid of the computer program CONSOL (Wong and Duncan, 1984). The program is based on a small strain formulation of Terzaghi's

one-dimensional consolidation theory using an implicit finite difference scheme. It can model the construction of a large area fill or a strip fill of finite width. The presence of intermediate drainage layers and the reduction in stress due to submergence with time can also be taken into consideration.

The following assumptions were adopted for this prediction exercise:

(a) The soil is homogeneous with drainage at the top and bottom of the clay deposit.
(b) The ground water is at 1 m below the top of the original ground surface.
(c) The clay has an overconsolidation ratio of 1.5.
(d) The compression and recompression indices are 1.6 and 0.1 respectively. The coefficients of consolidation for virgin loading and recompression are 1.34 and 50 m²/y respectively.
(e) The surcharge fill is a continuous strip with a width of 14 m and is placed instantaneously.

The immediate settlement of the clay layer is computed based on the elastic solution by Christian and Carrier (1978). The shape of fill was assumed to be a continuous strip.

Prediction of downdrag forces in piles

The problem of a pile group resting in a consolidating soil layer is shown diagrammatically in Fig. 14. The piles are assumed to be linear elastic with Young's modulus, E_p. For the purpose of analysis, the problem can

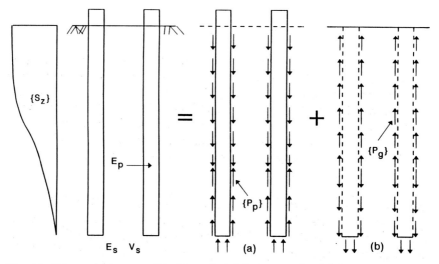

Fig. 14. Pile group in consolidating soil: (a) forces on pile, (b) forces on soil

be decomposed into two sub-systems, namely,

(a) the piles subjected to pile–soil interface forces (P_p),
(b) the soil subjected to an opposite set of forces (P_s).

The piles in the decomposed system are modelled by one-dimensional bar elements. The load deformation relationship of the piles may be written as

$$[K_p]\{W_p\} = \{P_p\}$$

where $[K_p]$ = assembled stiffness matrix of the pile elements

$\{W_p\}$ = vertical deformation of the pile nodes.

Considering the soil system alone, the relative soil movement due to the system of discretized loading $\{P_s\}$ may be written as

$$\{W_p\} - \{S_z\} = [F_s]\{P_s\}$$

where $[F_s]$ is the flexibility matrix and $\{S_z\}$ is the vector of soil displacement if the piles are not present. The elements of the flexibility matrix can be computed based on Mindlin's solution for an elastic half-space. For a non-homogeneous soil layer, the averaging procedure of the soil modulus as proposed by Poulos and Davies (1975) is adopted. Combining the equilibrium and compatibility conditions for the two sub-systems, the governing equation for the problem can be shown to be (Chow et al., 1990),

$$([K_p] + [K_s])\{W_p\} = [K_s]\{S_z\}$$

where $[K_s] = [F_s]^{-1}$. The magnitude of the downdrag forces acting on the pile is limited by the limiting shaft friction. In the current method of analysis, the governing equation is solved assuming that no slip occurs between the pile and the soil nodes. The magnitude of the nodal forces is then computed. If this exceeds the limiting shaft friction values, then the nodal forces are set to the limiting values and the problem is reanalysed iteratively. Typically, three or four iterations are necessary to get a converged solution.

Predictions of settlements, pore pressures and downdrag at Bothkennar
Pore pressure response in June 1991. The predicted total pore pressure response at location X in June 1991 immediately after the placement of the additional 2.5 m blaes surcharge is shown in Fig. 2. The computed pore pressure distribution is likely to be the upper bound.

Initial settlement due to 2.5 m of blaes surcharge. The predicted initial settlement of the ground surface is 60 mm, Fig. 9. This is based on the method by Christian and Carrier (1978) using an average undrained shear strength of 20 kPa and an E_u/C_u ratio between 200 and 300.

Excess pore pressure distribution in July 1992. The computed pore pressure distribution in July 1992 is shown in Fig. 3. This is likely to be the upper bound.

Consolidation settlement as of July 1992. The consolidation settlement due to the additional 2.5 m blaes surcharge from June 1991 to July 1992 is 350 mm, Fig. 11. The profile in Fig. 11 reflects the settlements which have occurred since the installation of the piles up until July 1992. The final settlement is calculated to be 600 mm.

Downdrag forces in piles. The following input parameters are used in the computation:

(a) The Young's modulus for the pile is assumed to be 36 GPa.

(b) The limiting friction on the pile shaft is assumed to be the same as the undrained shear strength of the soil, varying from 10 kPa at ground surface to 50 kPa at 20 m depth.

(c) The shear modulus is determined by assuming a rigidity index of 150 for the soil.

A pile length of 20 m is assumed for all the analyses. Since no information was given on the sectional properties of the centre pile, it is assumed to be of standard Westpile section. The results of the analyses are shown below and, for the centre pile in the friction group, in Fig. 13.

Single pile:	Friction pile:	365 kN
	End-bearing pile:	733 kN
Centre pile in group:	Friction pile:	324 kN
	End-bearing pile:	600 kN

(These values represent the downdrag force in July 1992.)

References

Chow, Y.K., Chin, Y.T. and Lee, S.L. (1990). Negative skin friction on pile groups. Int. Journal of Numerical and Analytical Methods in Geomechanics, Vol. 14, pp. 75–91.

Christian, J.L. and Carrier, W.D. (1978). Janbu, Bjerrum and Kjaernsli's chart reinterpreted. Canadian Geotechnical Journal, Vol. 15, pp. 124–128.

Poulos, H.G. and Davies, E.H. (1975). Prediction of downdrag forces in end-bearing piles. Journal of Geotechnical Engineering Division, ASCE, Vol. 101 (GT2), pp. 189–204.

Wong, K.S. and Duncan, J.M. (1984). CONSOL: A Computer Program for 1-D Consolidation Analysis of Layered Soil Masses: Microcomputer Version. Geotechnical Engineering Research Report No. UCB/GT/84-06, Dept of Civil Engineering, University of California, Berkeley.

E. Rojas and G.T. Houlsby, Oxford University

These predictions were made according to common engineering procedures:

(a) The initial pore pressure response of the clay was obtained from Henkel's expression $\Delta u = \alpha \Delta p' + \beta \Delta q$ where the parameters α and β were established from the results of the K_0 consolidated undrained triaxial compression tests on intact specimens of the clay ($\alpha = 1.0$, $\beta = 0.2$). The stress increments $\Delta p'$ and Δq were obtained from the elastic solution for a uniformly loaded rectangular area and employing the superposition method for the entire H-shape embankment.

(b) Initial settlements were also obtained according to elastic theory. The value of the shear modulus at different depths was obtained from the expression $G = 360 \, \sigma'_v$ which was derived from the seismic cone test results reported by Nash, Powell and Lloyd (1992). This expression corresponds to the sole available data of the shear modulus at small shear strains and was employed as no information on the variation of the shear modulus versus the shear strain was provided. A value of 0.5 was assumed for Poisson's ratio.

(c) The excess pore pressure in the clay was computed considering both vertical and radial drainage and employing the finite difference method. The coefficient of consolidation was deduced from the consolidation test results, reported by Nash et al., by taking into account the stress increments induced in the clay by the embankment ($c_v = 3.2 \, m^2/y$). The value of this parameter was considered to be equal in the vertical and horizontal directions as the overconsolidation ratio of the soil was reported to be small (1.5). Although this parameter is slightly higher to the one proposed in the invitation package for pressures ranging between 20 and 50 kPa ($c_v = 1.7 \, m^2/y$) the former was preferred as it agreed better (although not completely) with the fact that the initial pore pressure produced by the first embankment layer of 500 mm was completely dissipated after 6 months.

(d) The consolidation settlement of the surface was obtained from the classic theory of consolidation. The coefficient of volume compressibility (m_v) at each level was obtained by associating the plots of the consolidation tests for three different depths with the stress increment produced by the embankment at that depth. The values of the coefficient ranged from $m_v = 0.0012–0.0022 \, m^2/kN$.

(e) The profile of force with depth for friction and end-bearing concrete piles for both the isolated pile and the central pile in the group were obtained from the method proposed by Zeevaert (1973) on the basis of the initial and long-term final effective stresses computed for each case. For the short-term behaviour of the pile, a correction had to be applied as it was considered that the displacements of the soil produced by the consolidation settlements near the neutral point of the pile were too small to mobilize the total shear stress between soil and pile. This correction was made according to the theoretical stress–displacement curve of the pile as proposed by Poulos and Davis (1980).

Due to the fact that Zeevaert's analysis is based on the effective stress distribution around a single pile, this analysis may become rather complex for a group of piles as the distribution of stresses depends on the position and contribution of each one of the piles in the group. However an approximate solution can be obtained if the analysis is constrained to the tributary area of the analysed pile in the group which in this case is $4D \times 4D$.

For the steel piles it is considered that the forces on the shaft will be the same as those given for the concrete piles reduced by a factor of 2/3.

References
1. NASH, D.F.T., POWELL, J.J.M. AND LLOYD, I.M. (1992). Initial investigations of the soft clay test bed site at Bothkennar. Geotechnique, Vol. 42, No. 2.
2. ZEEVAERT, L. (1973). Foundation engineering for difficult subsoil conditions. Van Nostrand Reinhold Company, New York.
3. POULOS, H.G. AND DAVIS, E.H. (1980). Pile foundations analysis and design. John Wiley & Sons.

D. Muir Wood, University of Glasgow

This is intended as a 'back of the envelope' estimate of the various items of the prediction exercise.

Vertical effective stresses were calculated from given water contents. Preconsolidation pressures were estimated from oedometer data provided: after inspection a constant ratio of preconsolidation pressure to in situ vertical effective stress was assumed throughout.

The applied loading is approximated to an H of three rectangles. Elastic stress distributions are superimposed to obtain an estimate of vertical stress variation with depth at position Z as a result of the initial 0.5 m and subsequent 2.5 m of blaes.

Settlements in the long term are calculated by dividing the vertical stress increase into part below and part above the preconsolidation pressure, with appropriate compressibilities taken from the oedometer data.

The findings of Leroueil et al. suggest that while the new vertical stress is below the preconsolidation pressure the loading will be partially drained, but once it is beyond the preconsolidation pressure the loading will be undrained: from their work an estimate is obtained of the degree of consolidation with depth during the partially drained phase, and the total pore pressure distribution is then found to be roughly triangular with depth (Fig. 6).

The consolidation data provided do not lead to very plausible estimates of the rate of consolidation. A guess is made to produce the pore pressure distribution after 1 year, Fig. 7. The assumed effective stress increase during the partially drained stages of loading can be used to estimate initial settlements and settlement after 1 year. In calculating downdrag forces, no attempt will be made to allow for pile group effects. It is assumed that the clay movements are sufficient to mobilise full shaft friction at all depths. It is assumed that $\tau/\sigma'_{vc} = 0.2$. A parabolic distribution of force in the pile with depth emerges. For the floating pile a neutral point is found from equilibrium. With time, the preconsolidation pressure increases as the clay consolidates and the shaft friction $\tau = 0.2\sigma'_{vc}$ also increases. Corresponding new distributions of force with depth in the pile can be calculated for both end-bearing and floating piles.

M.J. Pender, University of Auckland

Initial pore water pressure
The predictions of the initial pore water pressure are as follows:

Depth (m)	2.5	6.5	10.5	14.5	18.5
Δu at Z (kPa)	47.0	42.5	35.5	29.0	23.7
Δu at X (kPa)	45.3	33.9	24.8	19.2	15.7

These were obtained by estimating the increase in vertical stress due to the surface loading. My rationale for the use of $\Delta\sigma_v$ is that this is greater than Δp and thus includes some contribution to cover the shear generated positive pore water pressure one would expect in a soft clay.

Initial (undrained) settlement of the ground surface
The prediction for the settlement at positions X, Y and Z is 11 mm. This was calculated as an undrained elastic settlement using the Steinbrenner approach as set out on p. 81 of Harr and p. 162 of Poulos and Davis. Initially the undrained Young's modulus was set to $4500s_u$ and the settlement for a soil profile calculated, consisting of four layers each 5 m thick. As part of the settlement calculation the vertical strains in the soil layers were obtained. The stiffness was then modified to account for the strains using the Seed and Sun G-γ relation for a soil with the appropriated plasticity index range.

Since the calculations were for undrained conditions, $\nu = 0.5$, the settlement was controlled by the fill above and there was negligible contribution form the other arms of the 'H'. Thus the settlement for points X, Y and Z are the same.

Excess pore water pressure distribution with depth at July 1992

The prediction of the excess pore water pressure with depth in July 1992 at position X is:

Depth (m)	2	6	10	14	18
Δu (kPa)	5.7	8.5	6.2	4.8	1.9

It is assumed that the dominant mechanism for consolidation is dissipation in the horizontal direction, due to thin permeable layers. The information supplied suggested that the excess pore water pressures were all zero 18 months after the initial 500 mm of blaes fill was placed. Thus $U = 75\%$ was assumed at July 1992.

Consolidation settlement at position X at July 1992

At point X (away from the pile groups) the estimate for the final consolidation settlement is 740 mm. With $U = 75\%$ this gives a consolidation settlement at July 1992 of 555 mm.

This was determined by taking a layered soil profile, and associating the e-log p curves with the appropriate layer and determining m_v for each layer.

Force profile with depth for the isolated friction pile

The prediction of the force carried by the isolated friction pile (without slip coating) at July 1992 is:

z (m)	2	6	10	14	14.6	18	Base
Pile force (kN)	119	285	508	786	834	602	266

The method of calculation is the same as that below. The maximum force occurs at 14.6 m.

Force profile with depth for the isolated end bearing pile

The prediction of the force carried by the isolated end bearing pile (without slip coating) at July 1992 is:

z (m)	2	6	10	14	18
Pile force (kN)	119	285	508	786	1122

To get this estimate it was assumed that the full slip condition was relevant to the full length of the pile shaft. Effective stress calculations were performed with $K\tan\phi = 0.4$.

B. Lehane and R. J. Jardine, Imperial College of Science, Technology and Medicine

The following are notes only, and merely give an indication of the methods adopted in the calculations.

Initial pore pressure

p' = constant, assume $\Delta u = \Delta p$. Δp calculated from Boussinesq equations with $\nu = 0.5$. (OCR estimated for site, little, if any, undrained yielding.)

Initial settlement

p' = constant, $\nu = 0.5$, $\Delta q = 3GE_z$. G/p'_0 variation with E_z backfigured from the results of other footing experiments at Bothkennar ($G/p'_0 = 520$ at $E_z = 0.01\%$, and 230 at $E_z = 9.1\%$). Max E_z evaluated = 0.12% at $Z \approx 5$ m.

Excess pore pressure after 1 year

Calculate $\Delta \sigma_v$ due to addition of surcharge (Boussinesq):

(a) for $\sigma'_v = \sigma'_{v0} + \Delta \sigma_v \leq \sigma'_{vy}$ (yield stress) assumed triangular distribution of Δu; hence degree of dissipation from standard solution using $c_v = 30$ m²/y (solution indicated $\bar{U} = 70\%$)
(b) for $\sigma'_v > \sigma'_{vy}$, as $c_v \to 0$, σ'_v limited to 1.25 σ'_{vy} (see Smith, 1992)
$\therefore \Delta u = \sigma'_{v0} + \Delta \sigma_v - 1.25 \sigma'_{vy}$

Consolidation settlement

(Crude) estimate based on Smith (1992), together with calculated vertical stresses.

for depth $Z \leq 5$ m, σ'_v was limited to 1.25 σ'_{vy}, with an assumed maximum vertical strain $(E_z) \approx 1.0\%$.
for $5 < Z \leq 12$ m, $\sigma'_{vy} < \sigma'_v < 1.25 \sigma'_{vy}$; E_z assumed $\approx 0.3\%$.
for $20 < Z < 20$ m, $E_z \approx 0.05\%$ assumed.

(No additional 'elastic' settlement has been included; assumption also that $E_v = E_z$.)

Profile of force with depth (friction pile)

Limiting local shear stress profile derived using the method of Lehane (1992). First estimate assumed a rigid pile, and that 14 mm relative displacement was required to mobilise τ (this value was extrapolated from the Imperial College research on pushed steel tubular piles at Bothkennar). A value for pile head movement was selected, as was the point where $\tau = 0$.

Below this depth the τ versus depth relationship was extrapolated from the Imperial College pushed pile tests. This, together with an estimate of pile end-bearing, led to an estimated pile settlement $\simeq 8$ mm in addition to 2 mm of elastic compression. After 1 year the corresponding value was calculated to be 10 mm.

Profile of force with depth (end-bearing pile)
It was assumed that the pile was rigid, and that the elastic compression of the soil at the pile base was $\simeq 3$ mm. Thus the total settlement was $\simeq 5$ mm. After 1 year the corresponding value is 5 mm.

Long-term behaviour
Secondary consolidation will move the point of zero shear stress downwards. It is estimated that this may result in an increase of the axial load in the friction pile by $\simeq 50\%$. The calculation for consolidation settlement after 1 year indicates $\rho_{consol} \simeq 101$ mm at a point away from the influence of the pile (i.e. at point X), and an immediate settlement $\rho_i \simeq 18$ mm at that same position. The total settlement after 1 year is therefore calculated to be $\simeq 120$ mm.

Recollections from the Wroth Memorial Symposium: Predictive Soil Mechanics

S.E. STALLEBRASS, City University, S.M. SPRINGMAN, Cambridge University, and J.P. LOVE, Geotechnical Consulting Group

These recollections were culled during the three days of the Wroth Memorial Symposium, and delivered in the final session of that meeting by the Assistant Reporters, with assistance and advice from many others, and in particular, the General Reporters. That so many people had travelled from around the world, and that the symposium was so open and discursive, was a tribute to a man who had made contributions to soil mechanics in so many areas, and who was always at the forefront of important new developments in our field. 'Predictive Soil Mechanics' proved to be a highly appropriate title for the Wroth Memorial Symposium.

In technical terms, it was a consolidation of the many Wroth ideas from constitutive relations, laboratory and in situ testing, with some exciting developments in the field of numerical modelling and some excellent predictive case studies. This paper is not an exhaustive record, and will merely focus on some interesting points that arose during presentation of the papers to this conference and the subsequent discussions.

The Assistant Reporters have contributed as follows: Dr Stallebrass focused on constitutive modelling and laboratory testing, Dr Springman discussed in situ testing and Dr Love introduced some thoughts on prediction and design.

Constitutive modelling

The constitutive modelling discussed at the symposium fell into two main areas: the use of a constitutive model to provide a framework for the interpretation of soil behaviour, and the formulation of new models to include specific features of the soil response. All the models discussed or presented at the symposium were based on either specific testing programmes or used recent high-quality data obtained by others through modern laboratory testing techniques. The importance of the link between good experimental data and the derivation of new or improved constitutive models was clearly made.

Within the first category, critical state soil mechanics was recognised as the best general framework to characterise the behaviour of soils including sands and unsaturated soils. This proved particularly valuable in drawing together apparently disparate aspects of soil behaviour into a common framework, which can then be used to predict the general response of new soil deposits. Using test data from three very different deposits, the behaviour of granular soils was found to be dominated by the mechanism of particle crushing, yet standard normal compression lines, critical state lines and the shearing behaviour were clearly and uniquely defined. For both the sands and unsaturated soils it was also possible to identify those features of the soil behaviour which do not correspond to the basic framework. A full description of the state of an unsaturated soil requires the use of two extra state parameters and is thus described in a five-dimensional hyperspace rather than the more conventional three-dimensional space. Additional new relationships were required to link these parameters and the standard state parameters, but the basic theories of critical state soil mechanics still held.

The specific features which new constitutive models had been developed to predict were anisotropy and the variation of soil stiffness (particularly at low values of strain). Throughout the symposium there was considerable emphasis and discussion on the prediction and implications of the non-linearity of soil stiffness at overconsolidated stress states. The deformation of overconsolidated soils is clearly elasto-plastic and this is influenced markedly by recent stress history. These characteristics of the behaviour were taken into account by some of the new models proposed, either using kinematic surfaces inside the state boundary surface or by controlling plastic strains using a physical analogy, as in the 'brick' model. This latter, rather novel, model is unusual in that it is defined in strain space rather than stress space. However, many simpler non-linear elastic models have been proposed and it was suggested that even simple models can provide a reasonable answer to certain boundary value problems when implemented in finite element analysis. Methods were also proposed for optimising the choice of soil parameters for use with simpler models, so that the analyses reproduced the observed data more closely.

There is clearly a need for comparison between the different models implemented in finite element programs or other design methods, and used to predict the behaviour of standard boundary value problems. Comparisons of this kind would make it possible to choose an appropriate model for the type of analysis to be carried out. The need to suit the complexity of model to the design situation was stressed frequently during the symposium.

Laboratory testing

Over the past decade, there has been particular emphasis on the improvement of laboratory testing techniques in order to make accurate measurements of the deformation characteristics of stiff soils at small strains. This area of research dominated the recent European Conference on Soil Mechanics held in Florence, although at this symposium, more interest was shown in the problems of measuring small strain stiffness using in situ tests. Nevertheless, stiffness data obtained from high-quality stress path triaxial tests and the technical advances associated with the measurement of these data were also discussed. In addition, the results of several laboratory testing programmes were presented and, more generally, the problems of obtaining truly undisturbed samples of soil were reviewed.

The considerable amount of work that has been carried out to measure the effect of disturbance on the stress state of the soil was summarised. The effect of disturbance has been quantified by various methods including comparing tube and block samples and also by using finite element analyses. Different sampling techniques cause variations in the type of disturbance undergone by the sample and so correction factors derived from this work are specific to particular sampling methods.

Once again the importance of measuring soil properties at appropriate stress states was emphasised. Additionally, it was noted that both the stress state and recent stress history of the soil need to be replicated in tests to measure the site-specific stiffness moduli of overconsolidated soils.

The wide range of devices which have been used to measure the deformation of soil samples inside stress path triaxial cells were described. There were interesting comparisons between internally measured strains, both on the sample and between the end platens, and external measurements. These comparisons provided some insight into the effectiveness of internal strain-measuring devices and sources of error in both internal and external measurements. However, the work on the measurement of the small strain stiffness of soils generally concentrated on the relationship between shear modulus and shear strain, ignoring the relevance of the link between shear and volumetric strains.

Significant programmes of tests had been undertaken in order to improve the understanding and characterisation of different soil types, in particular sands and unsaturated soils. There were two features common to these testing programmes. Firstly, the use of remoulded samples prepared in a standard way in the laboratory with well-controlled water contents and densities. Secondly, the development of laboratory test equipment which was particularly appropriate to the soils under investigation, in the same way that local strain-measuring devices were developed to measure a particular characteristic of soil behaviour. The tests on sand used high-pressure stress path cells and the unsaturated soil tests required carefully instrumented double-walled triaxial cells. In both cases it would not have been possible to identify the unifying features of the behaviour, discussed in the previous section on constitutive modelling, without this new equipment.

In conclusion, laboratory testing and constitutive modelling are inextricably linked and the results of the laboratory testing presented during the conference inevitably led to improved interpretation of data within new and old theoretical frameworks. In general the discussion on laboratory testing produced two major ideas, the importance of carrying out laboratory tests which were appropriate for the properties or soil under investigation and the interdependence of laboratory testing and in situ testing in the characterisation of soil behaviour.

In situ testing
Pressuremeter
There appeared to be more consensus than before that significant disturbance arose from insertion of the self-boring pressuremeter. Previous investigations by Hughes (1973), Wroth & Hughes (1973) using telltale lead threads subsequently exposed to X-radiation, to detect the extent of the influence of installation, had indicated that this apparatus was capable of giving a good estimate of both horizontal effective stress σ'_h and coefficient of earth pressure at rest, K_0. However, Whittle and Aubeny's numerical analysis using the strain path method to model the pushing in of a Menard-type pressuremeter (PMT) and the action of the self-boring pressuremeter (SBPT) by extracting soil from a 'sink' showed significant strain adjacent to the outer shell, even for the SBPT, where the effects of over- or under-cutting should be carefully considered. Further comments from both Clarke and Fahey (during the discussion) confirmed the impression that the SBPT cannot be considered an instrument for giving an accurate determination of σ'_h or K_0.

A consequence of this installation disturbance is that the horizontal effective stresses derived directly from interpretations of the pressuremeter lift-off pressure can give a lower bound to values measured by other in situ techniques (from Fig. 2 of Benoit and Lutenegger), although in London clay, the opposite effect may be observed. With respect to the earth pressure coefficient, these results seemed to fall towards and even above the postulated passive limit in the upper regions of the soil deposit and as low as the active value at depth. However, the difficulty in selecting the appropriate lift-off pressure was stated by Hight et al., who showed data in which the σ'_h deduced from SBPT tests at various depths was mainly higher than values obtained from suction measurements, dilatometer tests and other analyses.

So, it is instructive to consider what effect the installation disturbance has on stiffness measurement. Hight et al. found that the 1st unload loop from SBPT tests gave stiffer small strain values than those obtained from extension or compression laboratory tests. The 2nd unload loop gave values towards the stiffer region within this envelope of laboratory test results. The implication appears to be that the laboratory samples were even more disturbed or that the loading path was incorrectly modelled.

Another aspect to be questioned during the symposium was the arbitrary nature of the unload–reload loop amplitude. Two methods were mentioned: either unloading until radial stress equalled circumferential stress, or reducing the shear stress by up to, but not more than, twice the undrained shear strength. However, this requires up-to-date information from the test feedback system in order to be able to estimate when these states are being approached.

The repeatability of the stiffness–shear strain plots for unloading loops at radial strains from <0.01% up to yield, as reported by Whittle, Dalton and Hawkins, is encouraging. It will be interesting to see how these results compare with the equivalent laboratory tests, and whether there is room for improvement in the interpretation of the origins of the shear stress and strain increments.

Following General Reporter Graham's opening questions to the symposium about the ability of PMT/SBPT to determine strength, it was felt that it was still extremely difficult to obtain meaningful values. The combination of end and rate effects needs to be considered carefully, and in consequence, there is a large body of past data that is probably inaccurate. In some cases, the analysis of the behaviour of stiff clays can follow the drained analysis more closely than that of the undrained. This is probably due to drainage via fissures. In soft clays, it is vital that the loading rate is fast enough to ensure that partial drainage does not occur.

Future developments in combination instruments may be greater use of a seismic geophone in the SBPT and perhaps even inclusion of one in the cone pressuremeter (CPMT). It would be most interesting to see these devices developed at 1/100th scale for use in the centrifuge, given the interesting problem of size combined with modelling time in strain rate and diffusion terms!

Cone devices

Other questions posed by Graham for consideration at the Symposium concerned the cone penetrometer (CPT) and later derivations, the cone pressuremeter (CPMT). Modelling the penetration behaviour has been done in many ways. Was this achieving a measure of consensus? There were several methods of analysis which included critical state soil mechanics (CSSM) as the backbone, with cavity expansion theory as the key. Together with Whittle and Aubeny's analysis using the strain path method, it would seem that there is reasonable understanding of the event when the instrument is driven into a clay. However, the effect of dilatancy causes more doubt about the analysis of the installation of this device in sand. Empirical methods remain dominant, and we should look towards confirmation of these findings by fundamental theory — perhaps via the Coop CSSM framework for sands with an appropriately (non-)associated flow rule?

There appeared to be more support for using cone factors corrected for the pore pressure measured behind the shoulder of the cone. Comparing the range of uncorrected values of cone factor, N_K (as a function of undrained shear strength determined by field vane, $s_{u\ FV}$) lying between 8–20, with those for the corrected values of cone factor, N_{KT} as 10–14 (from Schnaid et al.) shows more consistency when this source of error is accounted for.

However, the ability to derive overconsolidation ratio (OCR) from the results shows encouraging comparisons (Mayne), although the trend of over- or underpredicting is not clear. Perhaps the influence of plasticity index, OCR and rate effects on the measured value of s_u (by SBPT, PMT, CPT and CPMT) need to be considered, much in the way a reliability envelope was deduced for the field vane (Bjerrum, 1973). By contrast, there was agreement that there has been significant progress since the Wroth Rankine lecture on the understanding of how the different values of s_u may be related when obtained from a variety of laboratory tests. Values ranging from +30% to −65% were reported by Kulhawy in his Fig. 6, with respect to triaxial CIUC tests.

Geophysical tests
Wroth also commented that 'more attention needs to be paid to various geophysical tests . . .', a development which has followed rather more slowly than small-strain advocates would prefer. Reports from X ECSMFE Florence (1991) showed that the Italians are heavily committed to the use of seismic testing to obtain shear wave velocities and estimates of small strain stiffness, G_{max}, but specification of these non-destructive tests by site investigators is less prevalent elsewhere. There has been substantive agreement that small strain stiffnesses measured by laboratory tests seem to overlap those obtained from in situ tests. However, an example which illustrates the importance of small strain in situ testing was observed from Crova et al. in their Fig. 13, where the destructuring and disturbance from sampling caused re-moulded resonant column test samples to show stiffnesses that were

RECOLLECTIONS FROM THE WROTH MEMORIAL SYMPOSIUM

Table 1

People/dates	Event	Analogy
Casagrande, Rendulic, Hvorslev (all 1930s), Taylor (1948)	Assembly of four components of soil constitutive modelling, obtained from laboratory tests	Steady scoring, solid opening partnerships.
Roscoe, Schofield, Wroth (1953–68)	Yielding of soil => initiation critical state soil mechanics	Acceleration of run scoring, elegant centuries!!
Wroth (1984)	Rankine lecture, rise of in situ testing. Laboratory sample disturbance, appropriate nature of stress paths and boundary conditions questioned.	In situ team take new ball, middle-order batting collapse.
Debate: Wroth Memorial Symposium. July 1992. Motion: Laboratory testing is now redundant! Proposers: Peter Robertson, Martin Fahey. Opposers: Peter Vaughan, Matthew Coop.	Measurement of small strain stiffness in the laboratory, development of stress path cells. Disturbance in in situ tests, appropriate nature of in situ stress paths and boundary conditions questioned. Centrifuge modellers claimed as lab testers! Redefinition of motion with claims that 90% of the world relies solely on in situ testing for design. Savage attack on laboratory sample treatment.	Tailenders engage in a hearty battle, a few balls slogged to the boundary. Peter Vaughan snatches some leg byes . . . . . . and a quick single (to be run by Matthew Coop!) Peter Robertson inspires a stumping . . . . . . and a run-out. Martin Fahey delivers final wickets.

20–60% of those measured by in situ seismic tests. Values deduced from SPT and CPT results fell between these two limits.

Application
Wroth was an advocate for 'appropriate' models combined with the optimisation of parameter selection. The numerical modellers were particularly apposite in their choice of Gunn's criteria: the simplest model to give a reasonable answer. Based on extensive CPT testing, Schnaid used a three-parameter linear elastic perfectly plastic model to predict and then back-analyse local movements under a reinforced road, but for St John et al., and Hight et al., accurate predictions of far-field displacements were crucial in the design and construction of excavations of deep basements in London. In both cases, small strain stiffness was modelled using Jardine et al.'s (1986) non-linear elastic perfectly plastic model with about 10 parameters. These were determined by a complementary series of laboratory and in situ tests, confirming the feeling of the symposium when the motion that 'Laboratory testing was now redundant' was defeated in a staged debate.

To mark this occasion, and in deference to Rachel Wroth's request for the inclusion of one of Peter's favourite sports, a cricketing analogy was proposed . . . with the laboratory supporters' team in to bat (see Table 1). . . . leaving the match finely balanced after the 1st innings, with all to play for, and an appreciation that both sides need to be strong and effective for a truly complementary contribution towards a fine game of cricket (i.e. the further development and understanding of the behaviour of soil).

The words of the final paragraph of the Wroth Rankine Lecture (1984) are still pertinent today:

> 'It is concluded that there is an important need for further research and development in conduct and interpretation of both laboratory and in situ testing of soils; this must go hand in hand with practice so that field experience of full-scale structures can be used to test new theories, new equipment and better interpretation.'

Prediction and design

This conference has been rather like a deep-sea trawler that has collected an impressive load of very high-level research, comprising much excellent analytical, laboratory and field work. But practising design engineers may have a problem in being unable to use this information yet, because it is either unprocessed or is too specific. The cartoon in Fig. 1 depicts design engineers as a flock of rather desperate seagulls following behind the trawler, trying to snatch up some practical 'snippets' which can be digested and used immediately. It is therefore

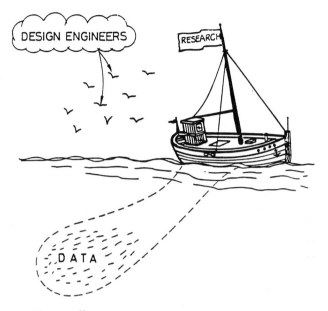

Fig. 1. Trawler with seagulls

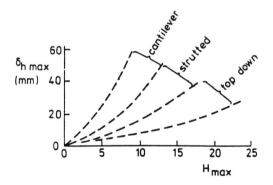

Type of wall	$\delta h/H$ (%)
Cantilever	0.40
Strutted	0.25
Top down	0.15

Fig. 2. Deep excavations in London Clay (St John et al., Fig. 16)

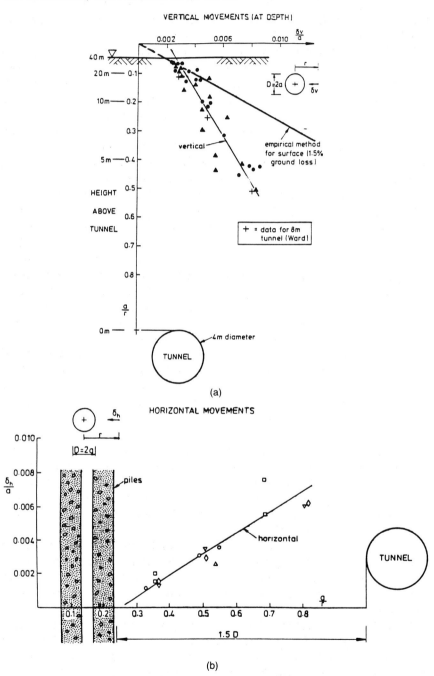

Fig. 3. (a) Vertical movements (at depth) for ground above tunnels (Mair and Taylor, Fig. 5); (b) horizontal movements adjacent to tunnels

for the 'seagulls' that these notes have been collected. It is of course an inexhaustive list, and apologies are due for this.

St John et al.'s simple chart (Fig. 2) summarising significant empirical data is just such a useful 'snippet'. It shows maximum horizontal wall movements, δ'_{hmax} for different types of excavation carried out in London clay to a depth of H_{max}, and while it has obvious limitations (e.g. it does not tell you exactly where in the wall δ'_{hmax} occurs, or what the maximum bending moments are), it gives an instant feel to the problem and allows some simple rule-of-thumb statements to be made about wall movements associated with different types of propping constraints. This can be used either in preliminary design, or as a backcloth against which to check more sophisticated calculations.

Mair and Taylor's Fig. 5 (shown here as Figs 3(a) and (b)) has been presented several times during the conference; it represents measurements of subsurface ground movements as a result of tunnelling. The tunnels were all 4 m diameter in London Clay, between 20 m and 40 m in depth. A practical application of this data is given when rotating their Fig. 5 to imply vertical movements (Fig. 3(a)). Note that the depth scale is not linear. Current practice is to estimate *subsurface* settlements by assuming that the ground surface (for which established empirical solutions already exist) lies at the point of interest. As pointed out by Dr Mair, this would overpredict settlements by approximately a factor of two at distances of 5–10 m above the tunnel (see line drawn on Fig. 3(a)).

Valuable confirmation of Mair and Taylor's data was provided by Dr Ward in discussion, when he showed vertical movements observed for an 8 m diameter tunnel at a depth of 40 m in London Clay; plotted here on Fig. 3(a) with normalised axes.

The data also include measurements of horizontal ground movements at the level of the tunnel axis. By overlaying this second worked example (Fig. 3(b); horizontal movements), it is of great interest to see that horizontal ground movements appear to be negligible beyond approximately 1.5 D (where D = tunnel diameter) measured from the outside wall of the tunnel.

These findings are likely to be of immediate practical importance to two major tunnelling projects in London (the Jubilee Line Extension and Crossrail) where there is a particular need to estimate subsurface ground movements beneath existing overlying tunnels and near existing piled foundations.

Tunnel stability rather than deformation is shown by Sloan and Assadi in Fig. 4. Design engineers still do hand calculations and appreciate a simple model or a convenient set of design charts. These charts appear to be useful, by extending existing solutions to the case of strength increasing with depth. A simple worked example is given in Fig. 4 for a 4 m diameter tunnel constructed in firm clay with strength

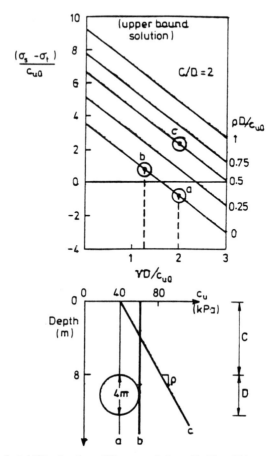

Fig. 4. Tunnel stability in clays (Sloan and Assadi, Fig. 10)

increasing with depth (where c_{u0} is defined in Fig. 4 as the undrained shear strength at ground level). The chart demonstrates that if the designers were to take an average value of the strength profile (e.g. either surface value 40 kPa, or 60 kPa, shown by the vertical lines a and b) they would significantly underestimate the stability of the tunnel $\{(\sigma_s - \sigma_t)/c_{u0}\}$ compared to the true strength profile (shown by the inclined line c). It should be noted that the solutions are for a plane strain circular tunnel. In reality, it is often the three-dimensional tunnel heading which is of relevance, in which case the plane strain solutions are conservative.

This summary of Vaughan et al.'s findings (see Fig. 5) demonstrates the typical amount by which the undrained shear strength of stiff plastic clays is overestimated by tests on samples from standard U100 sample

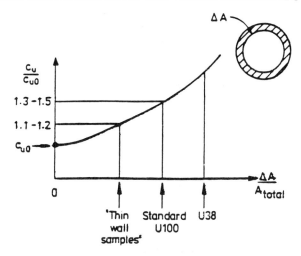

Fig. 5. Sampling disturbance in stiff plastic clays (after Vaughan et al.)

tubes (i.e. 30–50%). This is due to the increase in effective stress caused by sampling (due to the thickness of the sample tube wall). Even tests performed on the state-of-the-art 'thin' wall push samples are thought to overpredict undrained shear strength, c_u by 10–20%. (For soft clays the reverse effect applies and undrained strength is generally underestimated, but the effect of different sample tube wall thicknesses tends to be much less marked.)

Whether corrections should be made to the value of c_u in design depends on the nature of the design method. If the design method is empirically based (e.g. $\propto c_u$ approach for friction piles), then the uncorrected values should be used, but if the design uses fundamental theory (e.g. bearing capacity or slope stability analysis), then corrections should be made. (The above does not take into account the softening effect of free water, if present, at the base of the borehole during sampling.)

Figure 7 of the paper by Lehane and Jardine shows radial effective stress, σ'_r, mobilised on two identical piles in soft clay (Bothkennar). This demonstrates that σ'_r is not merely a function of depth below ground level (as is often assumed in current design methods), but also a function of height above the pile tip. For example at 3 m depth, the pile founded at 3.15 m mobilises about 45 kPa, whereas the other pile which is founded at 6 m mobilises something less than 30 kPa at the same depth (3 m).

In the case of the design of offshore jackup rigs, the end fixity condition provided by the spud-can legs has been a grey area for a while, with uncertainty over the magnitude of the restraining moment at

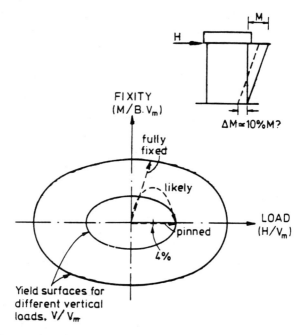

Fig. 6. Offshore jackup design: moment fixity in sand (Dean et al., Fig. 4)

the base of the leg ranging between fully fixed and fully pinned. There are obvious implications to the savings in design bending moment at the top of the leg; whatever moment can be provided at the bottom can be subtracted from the top (Fig. 6; inset). Values in the region of 10% (for both sands and clays) seem to be emerging, however this value should be used with caution. The two elliptical yield surfaces shown on Fig. 6 (Dean et al.) represent allowable combinations of horizontal and moment loading on sand for two different magnitudes of vertical loading. For a fully fixed leg the load path would follow a steep line to a larger yield surface, but for a fully pinned leg, it would move horizontally to meet a smaller yield surface. The likely actual load path would typically follow an intermediate path, as shown by James. This typical load path is seen to reach a maximum value of 'moment fixity' at a horizontal loading, H, on the structure of approximately 4% of the maximum vertical load capacity, V_m, but it then decreases again to zero under further horizontal loading.

In order to justify the design saving offered by moment fixity in the foundation, the designer would therefore have to ensure that H/V_m will not exceed approximately 4%. (The post-peak behaviour of the moment fixity curve may not be so evident in the case of clays, due to the greater embedment of the spud-cans and hence the ability to generate tension, preventing a significant reduction in vertical load on the lifting leg.)

It is stressed that the work to date (on spud cans founded in sands) is based on the assumption of no significant movement of the legs and no generation of pore pressures.

Finally, these are some additional scraps, gleaned from notes taken during the recent discussion periods (for the seagulls, of course):

(a) Insertion of the self-boring pressuremeter is *not* undisturbed, and as a result we should use a 'soft focus' lens when trying to use apparent lift-off pressures to derive K_0.

(b) The cone pressuremeter can be used to derive K_0 in sands, but will it be possible also to use it to derive K_0 (reliably) in stiff clays for which there is a desperate need?

(c) As a result of back analysis of a field trial, the cone factor, N_k (against corrected shear vane c_u) was narrowed down to a value of 16 for the soft clays at Shellhaven (Schnaid et al.).

(d) A reinforced sand fill was deduced to be mobilising an angle of friction as high as 52° (Jewell et al.).

(e) A case history was described in which the existence of a high stiffness at low strain levels was relevant; shear wave testing had detected light cementation in gravels, whereas conventional penetration testing had failed to do so (Crova et al.).

(f) A simple inequality ($a/g \leq c_u/\gamma D$), where a is horizontal acceleration and D is depth of clay layer, may be used to determine whether a linear elastic analysis is justifiable for a given magnitude of seismic acceleration (Pender).

In conclusion, it should be said that extracting 'snippets' from a range of papers can of course be dangerous as well as useful, and that when used out of context or inappropriately, an over-simplified rule-of-thumb can backfire. So the Assistant Reporters hope that they have not *saved* anyone from ever reading the papers again, but rather that they have *encouraged* one to re-read the papers.

Future developments

Several themes emerged throughout the Symposium. The importance of non-linearity was reiterated, with the focus on soil strain: small strain stiffness, strain-controlled 'brick' elements (Simpson), and Bolton's strain deformation mechanisms. Anisotropy of stress and soil fabric, the three-dimensional influence in modelling and prediction—which was noticed especially with respect to tunnelling at the Waterloo International Terminal when combined with the effects of construction sequence, disturbance and ageing of field samples were discussed by many contributors, with a general conclusion that there was more work to be done in these areas. Sloan's upper/lower-bound methods of analysis using finite elements, which gave reasonable bounds to results obtained from centrifuge modelling tests on tunnels, are to be developed to deal with many of these concerns, in particular three dimensions, axisymmetry, anisotropy, structural aspects. This may prove a powerful tool in the hands of the predictor.

References

BJERRUM, L. (1973). Problems of soil mechanics and construction on soft clays and structurally unstable soils. Proc. 8th ICSMFE. Moscow, Vol. 3, pp. 111–159.

JARDINE, R.J., POTTS, D.M., FOURIE, A.B. AND BURLAND, J.B. (1986). Studies of the influence of non-linear stress–strain characteristics in soil-structure interaction. Geotechnique Vol. 36, pp. 377–396.

HUGHES, J.M.O. (1973). An instrument for the in-situ measurement of the properties of soft clays. PhD thesis. Cambridge University.

KULHAWY, F. AND MAYNE, P. (1990) Manual on estimation of soil properties. Report EL-6800. Electric Power Res. Inst. Palo Alto. 306 pp.

WROTH, C.P. AND HUGHES, J.M.O. (1973). Undrained plane strain expansion of a cylindrical cavity in clay: a simple interpretation of the pressuremeter test. Geotechnique Vol. 23, pp. 284–7.

WROTH, C.P. (1984). The interpretation of in situ soil tests. Geotechnique Vol. 34, No. 4, pp. 449–489.

Author index

Allman, M. A., 436
Almeida, M. S. S., 73
Apted, J. P., 685
Arulanandan, K., 94
Assadi, A., 644
Atkinson, J. H., 111
Aubeny, C. P., 742
Been, K., 121
Benoît, J., 135
Bolton, M. D., 50
Booker, J. R., 783
Burd, H. J., 38, 378
Calabresi, G., 544
Carter, J. P., 783
Chan, A. H. C., 496
Chandler, R. J., 685
Clarke, B. G., 156
Clayton, C. R. I., 173
Coop, M. R., 186
Crova, R., 199
Dalton, J. C. P., 768
Davies, M. C. R., 219
Dean, E. T. R., 230
de Josselin de Jong, G., 254
De Moor, E. K., 317
Fahey, M., 261
Ferreira, C. A. M., 73
Ferreira, R. S., 562
Fraser, R. A., 279
Gibson, R. E., 293
Graham, J., 1
Gunn, M. J., 173, 304
Hagiwara, T., 404
Hawkins, P. G., 768
Higgins, K. G., 317, 581
Hight, D. W., 317
Hope, V. S., 173
Houlsby, G. T., 38, 339, 359, 813
Ibrahim, K., 796

Iizuka, A., 513
James, R. G., 230
Jamiolkowski, M., 199
Jardine, R. J., 317, 421, 581, 817
Jefferies, M. G., 121
Jewell, R. A., 378
Jewell, R. J., 261
Jubb, P., 609
Khorshid, M. S., 261
Kulhawy, F. H., 394
Kusakabe, O., 404
Lancellotta, R., 199
Lee, I. K., 186
Lehane, B., 421, 817
Lewin, P. I., 436
Little, J. A., 796
Lo Presti, D. C. F., 199
Love, J. P., 819
Lutenegger, A. J., 135
MacKenzie, N. L., 496
Maeda, Y., 404
Maguire, W. M., 685
Mair, R. J., 449
Mandolini, A., 464
Martin, C. M., 339
Mayne, P. W., 483
Milligan, G. W. E., 378
Morita, Y., 513
Muir Wood, D., 496, 814
Newson, T. A., 219
Nishihara, A., 513
Nutt, N. R. F., 359
Nyirenda, Z. M., 317
Ohta, H., 513
Ohuchi, M., 404
Pender, M. J., 529, 815
Pickles, A. R., 317
Potts, D. M., 317, 581
Rampello, S., 544

835

Randolph, M. F., 261
Robertson, P. K., 562
Rojas, E., 813
St John, H. D., 581
Sagaseta, C., 19
Sandroni, S. S., 685
Schnaid, F., 609
Schofield, A. N., 230
Simpson, B., 628
Sivakumar, V., 709
Sloan, S. W., 644
Smith, A. K. C., 609
Stallebrass, S. E., 819
Springman, S. M., 819
Steenfelt, J. S., 664

Sybico, J., 94
Tamagnini, C., 544
Tan, F. S. C., 230
Taylor, R. N., 449
Teh, C. I., 809
Tsukamoto, Y., 230
Vaughan, P. R., 685
Viggiani, C., 464
Wheeler, S. J., 709
Whitman, R. V., 729
Whittle, A. J., 742
Whittle, R. W., 768
Wong, K. S., 809
Wood, W. R., 609
Yu, H. S., 783

Subject index

All references are given to the first page of the relevant paper

Accretion, 293
Ageing effects, 199, 664
Analysis, 254, 293, 742, 783
Anchors, 664
Anisotropy, 219
Bearing capacity, 230, 339, 609
Bothkennar test site, 421, 436, 796
Braced excavation, 279, 544, 581
Calcareous soil, 261
Cam clay models, 111, 219, 483, 709
Carbonate soil, 261
Case history, 279, 317, 464, 544, 581, 609, 796
Centrifuge tests, 94, 230, 404, 729
Clays, 219, 317, 339, 359, 483, 496, 544, 581, 664, 685, 709, 742, 783, 796, *see also* Compacted clay, Overconsolidated clay, Soft clay, Stiff clay
Compacted clay, 709
Compressibility, 496, 709

Cone penetration test, 121, 359, *see also* Piezocone
Cone pressuremeter test, 359
Constitutive models, 219, 496, 513, 628, 709, 729, 742, 819
Consolidation, 73, 293, 513
Coupled consolidation analysis, 513
Critical state, 219, 709
Deformation, 317, 449, 729
Design methods, 50, 544, 819
Dilatometer test, 135, 742, 783
Direct shear, 664
Downdrag, 796
Drilled-and-grouted piles, 261
Dynamic analysis, 729
Earth pressure, 279
Earthquake, 529, 729
Earth structures, 729
Elasticity, 254, 562
Field tests: *see* In situ tests

SUBJECT INDEX

Finite element analysis, 279, 304, 404, 544, 581, 609, 628, 783
Flow rule, 219
Footings, 230, 339, 404
Gravels, 199, 404
Ground movements, 304, 449, 529, 544, *see also* Settlements
High pressure tests, 186
Horizontal stress, 135, 156, 359, 685
Inhomogeneous soil, 644
In situ stress: *see* Horizontal stress
In situ tests, 1, 135, 359, 483, 742, 783, 819, *see also* Dilatometer test, Cone penetration test, Cone pressure-meter test, Large penetration test, Piezocone, Pressuremeter test, Self-boring pressuremeter test, Standard penetration test, Vane test
Instrumentation, 279, 317, 421, 581, 609
Jack-up units, 230, 339
Laboratory tests, 1, 394, 436, 819, *see also* Simple shear test, Triaxial test
Large penetration test, 199
Limit equilibrium, 378
Liquefaction, 94, 729
Loading tests, 404, 609
Messina Strait crossing, 199
Model tests, 230, 339
Non-linear elasticity, 19, 111, 449, 529, 768, *see also* Stiffness
Numerical models, 111, 317, *see also* Finite element analysis
Offshore foundations, 230, 261, 339
Optimisation, 496, 562
Overconsolidated clay, 483
Performance measurements, 38, 544, 581, 609
Permeability, 94
Piezocone, 73, 121, 483
Piles, 261, 421, 464, 796
Plane strain, 436
Plasticity, 230, 254, 339, 449, 496
Pore pressures, 73, 449
Preconsolidation, 483
Predictions, 38, 394, 529, 581, 796, 819
Pressuremeter test, 156, 359, 562, 628, 742, 768, *see also* Cone pressure-meter test, Self-boring pressuremeter test

Pyroclastic soils, 464
Rate effects, 664
Reinforcement, 50, 378
Retaining walls, 38, 50, 378
Sampling, 685
Sands, 186, 230, 359, 404
Sedimentation, 293
Seismic response, 529, 729
Seismic tests, 173, 199, 562
Self-boring pressuremeter test, 135, 156, 768
Settlements, 73, 94, 449, 464, 544, 609
Shear modulus, 111, 768, *see also* Stiffness
Shear strength, 394, 664
Sidewall friction, 378
Simple shear test, 254
Site investigation, 173, 261
Small strain stiffness, 304, 768, *see also* Non-linear elasticity
Soft clay, 73, 135, 421, 544, 609
Soil classification, 121
Soil fabric, 94
Soil properties, 1, 19, 394, 742
Soil-structure interaction, 317, 464, 544
Spudcan foundations, 230, 339
Stability analysis, 513, 644
Standard penetration test, 199
State parameter, 121
Stiff clay, 685
Stiffness, 1, 19, 156, 173, 304, 317, 359, 496, 768, *see also* Non-linear elasticity, Shear modulus
Strains, 436
Strength measurements, 156, 359
Three-dimensional test, 436
Tomography, 173
Triaxial test, 186, 436, 709
Tunnels, 38, 50, 304, 449, 644, 819
Uncertainty, 394
Unconfined compression test, 513
Undrained shear strength, 394, 513
Unloading, 768
Unsaturated soils, 709
Vane test, 513
Vertical drains, 73
Wave equation, 293

Symposium delegates

SYMPOSIUM DELEGATES

Back row: Professor P. R. Vaughan[4], Dr D. W. Smith, Mr G. Viggiani, Professor C. Viggiani, Dr P. I. Lewin, Dr V. Fioravante, Mr C. Tamagnini, Professor R. Lancellotta[2], Dr D. C. F. Lo Presti, Mr L. Callisto, Dr S. Rampello, Mr A. Mandolini, Mr S. Miliziano, Mr C. Golightly, Mr D. Cook, Dr C. C. Hird, Dr I. C. Pyrah, Mr Y. Tsukamoto, Mr K. Kawasaki, Dr P. Avgherinos, Dr A. K. C. Smith, Mr I. C. Martorano

Row 5: Professor C. R. I. Clayton[1], Mr C. S. Eccles, Dr B. G. Clarke[2], Mr C. M. Martin, Dr J. A. Little[3], Dr T. Lunne, Dr S. M. Springman[6], Dr S. E. Stallebrass[6], Dr W. F. Anderson, Mr W. R. Wood, Mr T. Palmer, Mr N. R. F. Nutt, Dr J. A. M. Teunissen, Mr R. W. Whittle, Mr M. G. Jefferies, Dr R. J. Fannin[2], Dr W. R. Ward, Mr D. G. Boden, Dr R. G. James

Row 4: Mr F. E. Toolan[1], Mr P. G. Hawkins, Mr G. Y. Yang, Mr K. Ibrahim, Dr B. Simpson, Professor G. Wiseman, Dr A. Iizuka, Mr J. J. M. Powell, Professor S. Frydman, Dr R. J. Jardine, Mr T. Hagiwara, Professor A. J. Whittle, Dr T. Orr, Dr S. J. Wheeler, Dr E. Farrell, Professor P. K. Robertson[4], Mr L. J. Kennedy, Professor J. L. Davidson, Mr F. M. Jardine, Dr M. A. Allman, Mr T. H. By, Dr J. P. Love[6], Mr T. A. Newson, Mr F. Bransby, Dr M. C. R. Davies, Mr J. Bohac, Professor J. P. Carter[2], Professor S. F. Brown[1]

Row 3: Professor S. Marchetti, Dr P. A. Vermeer[2], Dr M. Lojander, Dr A. Swain, Professor M. J. Pender[1], Mr R. Fernie, Dr P. Loudon, Dr S. F. Cinicioglu, Dr B. Lehane, Dr M. R. Coop[4], Professor F. H. Kulhawy, Dr P. W. Mayne, Professor R. V. Whitman, Professor S. Sture[1], Professor A. Verruijt[2], Professor D. Muir Wood[3], Dr F. Schnaid, Dr H. S. Yu, Dr H. Van Langen[3], Mr E. J. den Haan, Miss E. J. Ward, Dr S. W. Sloan, Professor K. T. Chang, Dr E. Rojas, Dr S. G. Chung, Dr S. D. Thomas

Row 2: Mr A. R. Chaudhry, Mr M. Ohuchi, Professor M. S. S. Almeida, Mr L. Ortuno, Mr D. Costanzo, Dr T. Shiomi, Dr M. D. Bolton[5], Professor K. Arulanandan, Professor J. S. Steenfelt[2], Professor J. Graham[5], Dr I. Towhata, Dr K. Been, Professor M. Jamiolkowski[1], Mr T. Edstam, Ms K. Rankka, Mr M. G. Smith, Mr F. Yuan

Front row: Miss N. S. Andrews, Professor O. Kusakabe, Professor H. Ohta[2], Professor J. H. Atkinson, Professor C. Sagaseta[5], Mr J. C. P. Dalton[2], Dr R. J. Mair[3], Professor J. B. Burland[1], Professor G. T. Houlsby[5], Dr R. H. G. Parry[1], Mrs R. A. Wroth, Professor G. de Josslin de Jong, Professor A. N. Schofield[1], Dr H. J. Burd[3,5], Dr G. W. E. Milligan[3], Dr G. C. Sills[3], Dr R. N. Taylor, Mr M. J. Gunn

Symposium delegates not in the photograph: Dr G. Amaniampong, Mr R. H. Coe, Mr R. Driscoll[3], Dr M. Fahey[4], Dr R. A. Fraser, Professor M. Fuchsberger, Dr D. A. Greenwood[1], Professor E. C. Hambly[1], Mr P. Jubb, Mr J. W. Li, Professor M. Maugeri, Mr R. Newman, Mr A. Pickles, Professor N. E. Simons, Dr H. D. St John, Dr R. S. Steedman[3], Mr M. Toriihara, Dr H. J. Walbancke

[1]Session chairman. [2]Session co-chairman. [3]Discussion organiser. [4]Debater. [5]Reporter. [6]Co-reporter.